Structural Stability Theory and Practice

Structural Stability Theory and Practice

Buckling of Columns, Beams, Plates, and Shells

Sukhvarsh Jerath
Professor Emeritus of Civil Engineering
University of North Dakota
Grand Forks, ND, USA

Registered Office(s)
John Wiley & Sons, Inc., 111 River Street, Hoboken, NJ 07030, USA

Editorial Office
111 River Street, Hoboken, NJ 07030, USA

For details of our global editorial offices, customer services, and more information about Wiley products visit us at www .wiley.com.

Wiley also publishes its books in a variety of electronic formats and by print-on-demand. Some content that appears in standard print versions of this book may not be available in other formats.

Library of Congress Cataloging-in-Publication Data

Names: Jerath, Sukhvarsh, author.
Title: Structural stability theory and practice : buckling of columns,
  beams, plates, and shells / Sukhvarsh Jerath.
Description: Hoboken : Wiley, 2021. | Includes index.
Identifiers: LCCN 2020020408 (print) | LCCN 2020020409 (ebook) | ISBN
  9781119694526 (paperback) | ISBN 9781119694502 (adobe pdf) | ISBN
  9781119694496 (epub)
Subjects: LCSH: Buckling (Mechanics)
Classification: LCC TA656.2 .J47 2021 (print) | LCC TA656.2 (ebook) | DDC
  624.1/76—dc23
LC record available at https://lccn.loc.gov/2020020408
LC ebook record available at https://lccn.loc.gov/2020020409

Cover Design: Wiley
Cover Images: Weisse Bruecke 02 © Viaframe/Getty Images, Study of patterns and Lines © Roland Shainidze
Photogaphy/Getty Images

Set in 9.5/12.5pt STIXTwoText by SPi Global, Chennai, India
SKY10022523_111620

Dedicated to my wife Saroj and children Ashley, Rahul, Roger, and Purabi for patiently enduring the years of preparation

Contents

Foreword

Structural design is most efficient and usually most economical if the full strength of the material of the member or frame can be achieved under the factored design load. For example, a steel tension strut should have a strength equal to its area times the yield stress, and a beam should be able to achieve the plastic moment at the critical cross-section. This desirable state cannot always be attained: the material may fracture, or the member, the frame, the plate, or the shell may buckle, that is, the structure may become unstable. Most buckling phenomena are sudden, they may not announce their imminent occurrence, and they can result in catastrophic failure. Design codes have extensive clauses that the structural designer must consider in order to safeguard against this undesirable situation. These design specifications provide many formulas and rules that are then routinely applied during the design of a structure. The formulas often are based on approximations and they frequently include shortcuts and numerical coefficients devised by the advisory committees of the code authority. Furthermore, the design codes are not able to account for all possible design situations, especially those occurring during fabrication and erection. It is thus vitally important that the structural engineer be educated in the theory of structural stability and be able to apply this knowledge when using the code formulas or when dealing with stability issues outside the purview of the codes.

The subject of this book is the analysis of the stability of structures and structural elements. Its contents are directed to students and to the practicing structural engineering community. After an introductory chapter that introduces the basic theoretical phenomena and methods of analysis of the structural mechanics of stability design, the following seven chapters lead the student to an understanding of the major practical subdivisions of the structural elements that comprise a designed built system: Columns, beams, frames, plates, and shells. The treatment of each topic follows classical mathematics, emphasizing rigorous analysis in ample detail so the student will have no problem following the derivation and can understand the practical significance of the method and the outcome. Following the derived problems, there is a discussion of advanced methods of analysis, such as finite element analysis or other numerical procedures, and a presentation of the code rules for the topical element of the given chapter. Design rules from the USA, Canada, Europe, and Australia are introduced, and the chapters conclude with several detailed examples of design problems. Several chapters in this book are suitable for a text in an advanced senior course in structural engineering, and it is recommended in its entirety for a course in the first year of a graduate course.

T.V. Galambos

February, 2020

Preface

The material in this book covers the failure of structures due to buckling and structural instability. The book represents an effort to comprehensively describe the principles and theory of structural stability of different types of structures. The failures due to structural instability depend on the structural geometry, size, and the stiffness. It is important to understand structural instability failures because using higher strength materials will not prevent these failures. More structures are failing due to instability because of the increased use of high strength materials and larger structures, such as long bridges, tall buildings, vast sports and public arenas, large storage tanks, bigger airplanes, larger engines, turbines, motors, etc. The enlarged size increases the slenderness of the members of a structure, such that these members reach their stability limit before their material strength. If one looks at the design procedures of different codes, it becomes clear that in many situations the maximum forces members can support is governed by structural instability.

The author felt there is a need for a book on the subject that should be comprehensive, covering theory and practice, and at the same time should be relatively easy to read. To this end, theoretical derivations are given in the most detailed form possible and practical problems are solved to illustrate the use of theory. An attempt is made to tie the theory and the codes of practice.

The author has taught the subject for about 20 years to senior level undergraduate and graduate students, and advised research projects of graduate students in this area. The book should be a valuable text for upper-level undergraduate and graduate students who wish to study the stability of structures courses in aerospace, civil engineering, mechanics, and mechanical engineering departments. It can also be used by practicing engineers as a reference for their designs. It will be a very useful book for engineering libraries. The book uses both FPS and SI units to solve the problems so that the book can be used by students and engineering professionals worldwide. In many cases the nondimensional quantities are used in the formulation of equations.

Chapter 1 presents general concepts of structural stability problem. This chapter introduces the subject by covering the stability of rigid bars using both the equilibrium and energy methods. In this, subject equilibrium of the displaced structure is considered, whereas in the ordinary structural analysis equilibrium equations are formed for the original undeformed geometry. The energy method uses the principle of stationary potential energy. The governing equations for rigid bars are algebraic equations because these bars have infinite stiffness and are easier to solve and formulate. After gaining some fundamental understanding of the subject, Chapter 2 deals with the buckling of columns. Columns with different boundary conditions and supports are covered and critical loads are found by using the classical column theory and energy method. The governing differential equations for different cases and general differential equations applicable to all boundary conditions are derived and solved. Eccentrically loaded columns and imperfect columns are dealt with, in order to know the practical range of failure loads. Large deflection theory is used to know the post-buckling behavior of columns. When the stresses at buckling failure are greater than the elastic limit, the failure is called inelastic buckling. Inelastic behavior is studied in Chapter 3 for metal columns made of aluminum and steel. Different codes of practice used around the world are discussed to solve practical problems in column design.

When structural members are acted on by transverse loads in addition to axial forces, these are called beam columns. In Chapter 4, the concept for solving the instability problems of beam columns is discussed. Basic differential equations of beam columns for different support conditions acted on by different loads are derived and solved. The beam column instability problems are also solved by the slope deflection method of structural analysis except that the equations are derived for the displaced shape of the structure. The continuous and inelastic beam columns are also studied. The design procedures given in different codes are discussed. The critical loads in single-story, multistory, and multibay frames are found in Chapter 5. Frames with and without sidesway and different types of loads and support conditions are studied. Critical loads in frames are found by the equilibrium, slope deflection, matrix and finite element methods. Inelastic buckling of frames is also considered. Design problems for two-story and two-bay frames are solved under the action of gravity and wind loads.

The torsional and lateral buckling of beams is discussed in Chapter 6. The chapter starts with pure and non-uniform torsion in thin-walled open cross-sections. After dealing with general thin-walled open sections, the commonly used I and channel sections are studied. The concepts of St. Venant's and warping torsion are elaborated. Torsional flexural buckling loads of different cross-sections with different boundary conditions are found. The problems are solved by the equilibrium and energy methods. In the second half of Chapter 6, lateral buckling of beams is studied. The lateral buckling phenomenon is studied for beams that are supported differently. The design equations for torsional and lateral buckling are given in order to know the capacity of members under such forces. The concepts of lateral torsional buckling and lateral buckling of beams derived in the book are shown to be related to the equations used by codes.

The buckling of thin rectangular plates with and without stiffeners under different kinds of loading and support conditions is discussed in Chapter 7. Post-buckling behavior and inelastic buckling of plates are also studied. The buckling of circular plates is introduced with solved problems. Linear and nonlinear theories of cylindrical shells are derived, and some problems are solved using linear stability equations. The failure and the post-buckling behavior of cylindrical shells are discussed. Nonlinear and linear equations of stability are derived using curvilinear coordinates for general shells. In Chapter 8, linear equations of general shells are converted to solve problems of the shells of revolution, shallow spherical caps, conical shells, and toroidal shells.

The book was written by keeping in mind those who wish to learn the concepts of buckling and instability of structures. Theoretical derivations are given in detail and are made as simple as possible. The book is also written for those who are in practice and are designing structures so they do not fail due to instability. The author has covered the subject in detail as well as keeping it simple and practical. It is hoped the readers will like and enjoy the book, and find it useful.

Sukhvarsh Jerath
Grand Forks, ND
USA

About the Companion Website

Don't forget to visit the companion website for this book: www.wiley.com/go/jerath

There you will find valuable material designed to enhance your learning, including:

- Learning Outcomes for all chapters
- Exercises for all chapters

1

Structural Stability

1.1 Introduction

Structures fail mainly either due to material failure or because of buckling or structural instability. Material failures are governed by the material strength that may be the ultimate strength or the yield point strength of the material. The failure due to structural instability depends on the structural geometry, size, and its stiffness. It does not depend on the strength of the material. It is important to understand the failure due to structural instability, because using a higher strength material will not prevent this type of failure. More and more structures are failing because of stability problems because of the present trend to use high strength materials and large structures. The increase in size increases the slenderness ratio of the members of a structure, and these members reach their stability limit before their material strength. A look at different design codes makes it clear that in many situations the maximum force a system can support is governed by structural instability than by material strength.

An interesting question to ask is, if the material strength is not exceeded, then why does the member fail?. The answer may be that all systems take the path of least resistance when they deform, a basic law of nature. For slender members, it is easier to bend than to shorten under a compressive force resulting in the buckling of the member before it fails by exceeding its material strength. For short members it is easier to shorten than to bend under a compressive force. In practice, there is always a tendency of a slender member to bend sideways even if the intended force is an axial compression. This tendency is due to small accidental eccentricity, unintended lateral disturbing force, imperfections, or other irregularities in the member. For small compressive forces the internal resistance of a member to bending exceeds external action forcing it to bend. As the external forces increase, a limiting load is reached where their overturning effect to bend exceeds the internal resistance to bending of the member. As a result, more and more bending of the system called buckling occurs. The maximum compressive force at which the member can remain in equilibrium in the straight configuration without bending is called the buckling load. A system is called stable if small disturbances cause small deformations of the system configuration. Displaced shape equilibrium and the energy methods are the two most commonly used procedures to solve the buckling loads problem and to study the stability of equilibrium.

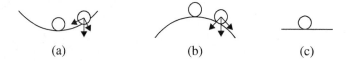

Figure 1.1 Types of equilibrium: (a) stable; (b) unstable; (c) neutral.

1.2 General Concepts

Concepts of stability can be explained by considering the equilibrium of a ball resting on three different surfaces [1] shown in Figure 1.1. The ball on the concave surface in Figure 1.1a is in stable equilibrium because any small displacement will increase the potential energy of the ball. The component of the self-weight parallel to the sliding surface will bring the ball back to its original equilibrium position. In Figure 1.1b, the ball rests on a convex surface, a small displacement from its equilibrium position will decrease the potential energy of the ball. The parallel component of the self-weight will slide the ball further from its initial configuration, and the equilibrium is unstable. If the ball is displaced on the flat surface, the potential energy of the ball remains the same, and the ball assumes a new equilibrium position. Thus, potential energy, Π, is a minimum for stable equilibrium, whereas it is a maximum for the unstable equilibrium position, and the potential energy remains the same for the position of neutral equilibrium. Energy methods are based on these concepts for solving the structural stability problems. If $\Delta\Pi > 0$, the displaced configuration is stable, whereas for $\Delta\Pi < 0$, the displaced shape is in unstable equilibrium, the transition $\Delta\Pi = 0$, which is the position of neutral equilibrium gives critical load at which the system becomes unstable by energy method.

Also, since we are studying the state of equilibrium in the slightly displaced position of the body, the equilibrium equations are written based on the displaced shape of the body in the displaced shape equilibrium method. Both methods can be used to formulate the equilibrium equations and calculate the critical loads. However, the displaced equilibrium approach does not give the nature of equilibrium when the critical load is reached. To answer that question, the second variation of potential energy $\delta^2\Pi$ is to be considered. The potential energy may be expanded into a Taylor series about the equilibrium state and written as

$$\Delta\Pi = \delta\Pi + \delta^2\Pi + \delta^3\Pi + \text{-----} \tag{1.1a}$$

where

$$\delta\Pi = \sum_{i=1}^{n} \frac{\partial \Pi}{\partial q_i} \delta q_i \tag{1.1b}$$

$$\delta^2\Pi = \frac{1}{2!} \sum_{i=1}^{n} \sum_{j=1}^{n} \frac{\partial^2 \Pi}{\partial q_i \partial q_j} \delta q_i \delta q_j \tag{1.1c}$$

$$\delta^3\Pi = \frac{1}{3!} \sum_{i=1}^{n} \sum_{j=1}^{n} \sum_{k=1}^{n} \frac{\partial^3 \Pi}{\partial q_i \partial q_j \partial q_k} \delta q_i \delta q_j \delta q_k \tag{1.1d}$$

$\delta\Pi$, $\delta^2\Pi$, and $\delta^3\Pi$ are called the first, second and third derivatives respectively of the potential energy Π. The critical load P_{cr} is obtained from the conditions of equilibrium given by $\delta\Pi = 0$

for any δq_i, or $\dfrac{\partial \Pi}{\partial q_i} = 0$ for each i [2]. The equilibrium state is stable if $\Delta \Pi > 0$. Therefore, the equilibrium state is stable for $\delta^2 \Pi > 0$, and is unstable for $\delta^2 \Pi < 0$.

Because energy is quadratic, it can also be written as

$$2\Pi = \sum_{i=1}^{n} \sum_{j=1}^{n} K_{ij} q_i q_j = \boldsymbol{q}^T \boldsymbol{K} \boldsymbol{q} \tag{1.1e}$$

where

$\boldsymbol{q}$ = column vector of the generalized displacements

$\boldsymbol{q}^T$ = transpose of the column vector

$\boldsymbol{K}$ = square matrix (n × n) with elements K_{ij}

For elastic structures, matrix $\boldsymbol{K}$ represents the stiffness matrix of the structure with regard to its generalized displacements, and Π is the potential energy. The stiffness elements are given by

$$K_{ij} = \frac{\partial^2 \Pi}{\partial q_i \partial q_j} = \frac{\partial^2 \Pi}{\partial q_j \partial q_i} = K_{ji} \tag{1.1f}$$

That shows the stiffness matrix is symmetric. The second variation of the potential energy from Eq. (1.1c) is

$$2\delta^2 \Pi = \sum_{i=1}^{n} \sum_{j=1}^{n} K_{ij} \delta q_i \delta q_j \tag{1.1g}$$

For $\delta^2 \Pi > 0$, the matrix with elements K_{ij} will be positive definite. A real symmetric matrix is positive definite if and only if all its principal minors are positive, that is,

$$D_1 = K_{11} > 0, \quad D_2 = \begin{bmatrix} K_{11} & K_{12} \\ K_{21} & K_{22} \end{bmatrix} > 0, \text{-----------} D_n = \begin{bmatrix} K_{11} & - & - & K_{1n} \\ - & - & - & - \\ - & - & - & - \\ K_{n1} & - & - & K_{nn} \end{bmatrix} > 0 \tag{1.1h}$$

or

$$D_1 = K_{11} > 0, \quad |D_2| = \begin{vmatrix} K_{11} & K_{12} \\ K_{21} & K_{22} \end{vmatrix} > 0, \text{----------} D_n = \begin{vmatrix} K_{11} & - & - & K_{1n} \\ - & - & - & - \\ - & - & - & - \\ K_{n1} & - & - & K_{nn} \end{vmatrix} > 0 \tag{1.1i}$$

When systems are subjected to compressive forces three types of instabilities can occur: (i) bifurcation of equilibrium; (ii) maximum or limit load instabilities; and (iii) Finite disturbance instability.

1.2.1 Bifurcation of Equilibrium

Equilibrium paths are shown as load displacement plots in Figure 1.2. The equilibrium path starting from the unloaded configuration is called the fundamental or primary path. At a certain load the equilibrium path can continue to be the fundamental path or it could change to an alternate configuration if there is a small lateral perturbation. This alternate path is called

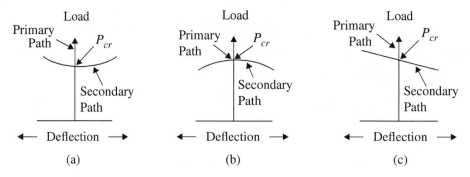

Figure 1.2 Bifurcation equilibrium paths: (a) Symmetric stable bifurcation; (b) Symmetric unstable bifurcation; (c) Asymmetric bifurcation.

the secondary or post-buckling path [3]. The point of intersection between the primary and secondary paths is called the point of bifurcation, and the load corresponding to this point is called the critical load. In Figures 1.2a and 1.2b, the secondary paths are symmetrical. In the symmetric bifurcation the post-buckling load deflection behavior remains the same irrespective of the direction in which the structure bends. It is a stable bifurcation in Figure 1.2a because the load increases with deflection after buckling, axially loaded columns and thin plates subjected to in-plane forces exhibit this behavior. The load decreases below the critical as the deflection increases in the post-buckling stage in Figure 1.2b, and the structure has an unstable bifurcation at the critical load. Guyed towers exhibit this behavior because some of the cables come under compression and are unable to sustain the external forces. If the post-buckling load deflection diagram is affected by the direction of buckling, then the bifurcation is asymmetric as shown in Figure 1.2c. Some framed structures show this kind of behavior.

1.2.2 Limit Load Instability

This type of instability is also called snap-through buckling. In this type of buckling, the primary path is nonlinear and once the load reaches a maximum, the point P in Figure 1.3a jumps to Q on another branch of the curve. The load at point P is the critical load in this type of instability. The structure snaps through to a nonadjacent equilibrium position represented by point Q. Spherical caps and shallow arches exhibit this behavior.

1.2.3 Finite Disturbance Instability

This type of instability occurs in cylindrical shells under the action of axial forces shown in Figure 1.3b. The load capacity of the structure drops suddenly at the critical load in Figure 1.3c. The structure takes a non-cylindrical shape after the critical load. The structure continues to take more axial compression in Figure 1.3c after taking another equilibrium configuration. In this type of instability, a finite disturbance of the cylinder or imperfection in the cylinder will lower the critical load considerably and the structure will change equilibrium configuration upon reaching the ideal critical load.

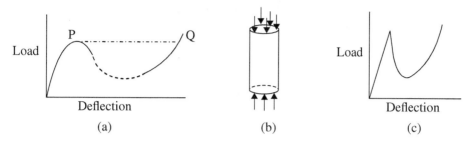

Figure 1.3 Post-buckling equilibrium paths: (a) Limit load instability; (b) Cylindrical shell under axial compression; (c) Finite difference instability.

1.3 Rigid Bar Columns

Columns consisting of rigid bars supported by springs and acted on by axial compression are studied by the displaced shape equilibrium, or by energy methods. At first, the small deflection analysis is considered. The study of rigid bar columns provides a good background on the nature of stability problems and the different methods used to solve them because these systems have limited degrees of freedom.

1.3.1 Rigid Bar Supported by a Translational Spring

1.3.1.1 The Displaced Shape Equilibrium Method
Consider a perfect rigid vertical column supported by a hinge at the bottom and a linear spring of stiffness "k" at the top. The bar is acted on by an axial load shown in Figure 1.4. If there is an accidental lateral disturbance, the spring force, $kL \sin \theta$, will bring it back to the vertical position for small axial loads. In this case the restoring moment due to spring force is larger than the overturning moment due to the force P as shown in Eq. (1.2a):

$$k L^2 \sin \theta \, \cos \theta > PL \, \sin \theta \tag{1.2a}$$

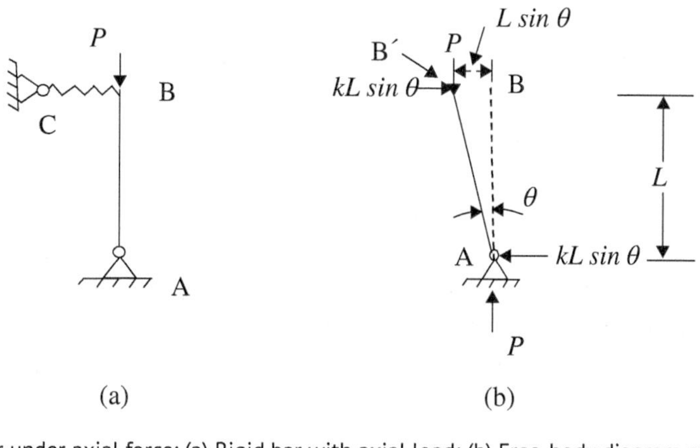

Figure 1.4 Rigid bar under axial force: (a) Rigid bar with axial load; (b) Free-body diagram of displaced shape.

and the vertical position of the bar is stable. The spring force will not be able to bring back the rigid bar to its vertical position for large axial force, because the overturning moment will be larger than the restoring moment shown below

$$k\,L^2 \sin\theta \, \cos\theta < PL \, \sin\theta \qquad (1.2b)$$

and the vertical position of the bar is unstable. The minimum axial force at which the bar becomes unstable is called the critical load. It is the force at which the equilibrium changes from stable to unstable, and

$$k\,L^2 \sin\theta \, \cos\theta = PL \, \sin\theta$$

$$\text{or } P = kL \, \cos\theta \qquad (1.2c)$$

The critical load, P_{cr}, can be found by considering the equilibrium of the slightly displaced position of the bar by taking moments of all forces about A in Figure 1.4b as follows:

$$\sum M_A = 0$$

$$PL \, \sin\theta - kL^2 \sin\theta \, \cos\theta = 0$$

$$\text{or } P = kL \, \cos\theta \qquad (1.2d)$$

The same result is obtained from Eqs. (1.2c and 1.2d), hence the critical load can be found by considering the equilibrium of the slightly displaced shape. For small deflections, $\cos\theta \approx 1$, therefore,

$$P_{cr} = kL \qquad (1.2e)$$

1.3.1.2 The Energy Method

The first law of thermodynamics can be used to derive equations used in the energy method. This law, which is a statement of the law of conservation of energy, can be stated as "The work that is performed on a mechanical system by external forces plus the heat that flows into the system from the outside equals the increase of kinetic energy plus the increase of internal energy."

$$W_e + Q = \Delta T + \Delta U \qquad (1.3a)$$

Here, W_e, is the work performed on the system by the external forces, Q is the heat that flows into the system, ΔT is the increase of kinetic energy, and ΔU is the increase of internal energy [4]. For an adiabatic change, $Q = 0$, and for a body in equilibrium, $\Delta T = 0$. This reduces Eq. (1.3a) to

$$W_e = \Delta U \qquad (1.3b)$$

The change in internal energy of an elastic body is determined by the strains, and is called the strain energy. If the system is subjected to conservative forces, W_e is independent of the path the system takes from the configuration X_0 to another configuration X. In this case, the W_e depends only on the two terminal configurations, and is denoted by $-V(X_0, X)$. The function $V(X_0, X)$

is called the potential energy of the external forces, and it is always measured as the change in the potential energy, ΔV, from one configuration to another configuration of the system.

$$W_e = -\Delta V \tag{1.3c}$$

Equations (1.3b and 1.3c) can be combined to write

$$\Delta V + \Delta U = 0 \tag{1.3d}$$

or

$$\Delta(V + U) = 0 \tag{1.3e}$$

V is the potential energy due to external forces, and U is considered the potential energy of the internal forces. Total potential energy of the system is

$$\Pi = V + U \tag{1.3f}$$

The total potential energy of a system is a minimum in the position of stable equilibrium, whereas it is a maximum for unstable equilibrium. The critical load can be obtained by equating the first derivative of the total potential energy equal to zero. In Figure 1.4

$$U = \tfrac{1}{2}k(L\,\sin\theta)^2 \tag{1.3g}$$
$$V = -PL(1 - \cos\theta) \tag{1.3h}$$
$$\Pi = -PL(1 - \cos\theta) + \tfrac{1}{2}k(L\,\sin\theta)^2 \tag{1.3i}$$
$$\frac{d\Pi}{d\theta} = -PL\,\sin\theta + kL^2\sin\theta\,\cos\theta \tag{1.3j}$$

Substituting $\dfrac{d\Pi}{d\theta} = 0$, we get

$$P = kL\cos\theta \tag{1.3k}$$

$\cos\theta \approx 1$ for small values of θ

or

$$P_{cr} = kL \tag{1.3l}$$

giving the same critical load as in Eq. (1.2e). From Eq. (1.3j)

$$\frac{d^2\Pi}{d\theta^2} = -PL\cos\theta + kL^2\cos 2\theta \tag{1.3m}$$

For the initial position,

$$\theta = 0, \quad \frac{d^2\Pi}{d\theta^2} = -PL + kL^2 \tag{1.3n}$$

For $P < P_{cr}$, $\dfrac{d^2\Pi}{d\theta^2} > 0$, and for $P > P_{cr}$, $\dfrac{d^2\Pi}{d\theta^2} < 0$ in Eq. (1.3n). So the system is in stable equilibrium if $P < P_{cr}$, and is in unstable equilibrium for $P > P_{cr}$, in the initial position.

1.3.2 Two Rigid Bars Connected by Rotational Springs

1.3.2.1 The Displaced Shape Equilibrium Method

Consider two rigid bars as shown in Figure 1.5. The lower bar is connected to a pin support and a linear rotational spring of stiffness c_1 at the bottom. At the top the lower bar is connected to another bar by a linear rotational spring of stiffness c_2. The upper bar is free at the top, and the bars are subjected to an axial force of P.

Taking the equilibrium of the lower bar in Figure 1.5c, the sum of the moments of all forces about A is equal to zero,

$$c_1\theta_1 - c_2(\theta_2 - \theta_1) - PL_1 \sin\theta_1 = 0 \tag{1.4a}$$

From Figure 1.5d, sum the moments of all the forces about B and equate it to zero,

$$c_2(\theta_2 - \theta_1) - PL_2 \sin\theta_2 = 0 \tag{1.4b}$$

$\sin\theta \approx \theta$ in radians for small values of θ, and Eqs. (1.4a and 1.4b) can be written in the matrix form as

$$\begin{bmatrix} c_1 + c_2 - PL & -c_2 \\ -c_2 & c_2 - PL_2 \end{bmatrix} \begin{Bmatrix} \theta_1 \\ \theta_2 \end{Bmatrix} = \begin{Bmatrix} 0 \\ 0 \end{Bmatrix} \tag{1.4c}$$

$$\text{or} \quad \left\{ \begin{bmatrix} c_1 + c_2 & -c_2 \\ -c_2 & c_2 \end{bmatrix} - P \begin{bmatrix} L_1 & 0 \\ 0 & L_2 \end{bmatrix} \right\} \begin{Bmatrix} \theta_1 \\ \theta_2 \end{Bmatrix} = \begin{Bmatrix} 0 \\ 0 \end{Bmatrix} \tag{1.4d}$$

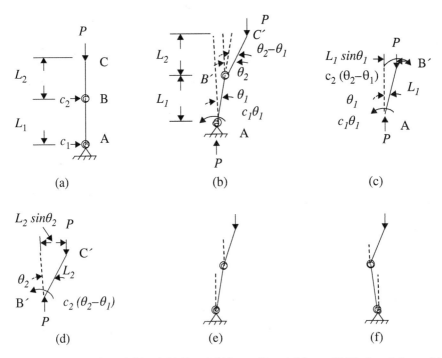

Figure 1.5 Two rigid bars under axial load: (a) Two rigid bars with axial force; (b) Displaced shape; (c) Free body diagram of lower bar; (d) Free body diagram of upper bar; (e) First buckling mode; (f) Second buckling mode.

Equation (1.4d) is an eigenvalue problem. The critical loads P are the eigenvalues, and the angular displacements, θ_1 and θ_2 are given as eigenvectors. For a nontrivial solution the determinant of the coefficient matrix is zero [5],

$$\begin{vmatrix} c_1 + c_2 - PL_1 & -c_2 \\ c_2 & c_2 - PL_2 \end{vmatrix} = 0 \tag{1.4e}$$

or

$$(c_1 + c_2 - PL_1)(c_2 - PL_2) - (c_2)^2 = 0$$

or

$$P^2 - P\left(\frac{c_1}{L_1} + \frac{c_2}{L_2} + \frac{c_2}{L_1}\right) + \frac{c_1 c_2}{L_1 L_2} = 0 \tag{1.4f}$$

The solution of the quadratic Eq. (1.4f) is given by

$$P = \frac{\frac{c_1}{L_1} + \frac{c_2}{L_2} + \frac{c_2}{L_1} \pm \sqrt{\left(\frac{c_1}{L_1} + \frac{c_2}{L_2} + \frac{c_2}{L_1}\right)^2 - 4\left(\frac{c_1 c_2}{L_1 L_2}\right)}}{2} \tag{1.4g}$$

If $c_1 = c_2 = c$, and $L_1 = L_2 = L$

$$P = \frac{\frac{3c}{L} \pm \sqrt{5}\frac{c}{L}}{2}$$

$$P = 0.382\frac{c}{L}, \quad \text{or } 2.618\frac{c}{L} \tag{1.4h}$$

The corresponding eigenvectors are:

$$\left\{\begin{matrix} \theta_1 \\ \theta_2 \end{matrix}\right\} = \left\{\begin{matrix} 1 \\ 1.618 \end{matrix}\right\} \quad \text{and}$$

$$\left\{\begin{matrix} \theta_1 \\ \theta_2 \end{matrix}\right\} = \left\{\begin{matrix} 1 \\ -0.618 \end{matrix}\right\} \tag{1.4i}$$

1.3.2.2 The Energy Method

The critical load for the two rigid bars shown in Figure 1.5a subjected to an axial force P can be found by using the principle of stationery potential energy. The strain energy of the system in the displaced shape is given by

$$U = \frac{1}{2}c_1(\theta_1)^2 + \frac{1}{2}c_2(\theta_2 - \theta_1)^2 \tag{1.5a}$$

The potential energy of the external force P is

$$V = -P\left[L_1(1 - \cos\theta_1) + L_2(1 - \cos\theta_2)\right] \tag{1.5b}$$

Total potential energy of the system is

$$\Pi = \frac{1}{2}c_1(\theta_1)^2 + \frac{1}{2}c_2(\theta_2 - \theta_1)^2 - P\left[L_1(1 - \cos\theta_1) + L_2(1 - \cos\theta_2)\right] \tag{1.5c}$$

The potential energy of the system must be stationary for equilibrium. The first derivatives of the potential energy function, Π, with respect to θ_1 and θ_2 are:

$$\frac{\partial\Pi}{\partial\theta_1} = c_1\theta_1 + c_2(\theta_2 - \theta_1)(-1) - PL_1\sin\theta_1 = 0 \tag{1.5d}$$

$$\frac{\partial \Pi}{\partial \theta_2} = c_2(\theta_2 - \theta_1) - PL_2 \sin \theta 2 = 0 \tag{1.5e}$$

For small values of θ, $\sin \theta \approx \theta$ in radians, and Eqs. (1.5d and 1.5e) can be written in matrix form as:

$$\begin{bmatrix} c_1 + c_2 - PL_1 & -c_2 \\ -c_2 & c_2 - PL_2 \end{bmatrix} \begin{Bmatrix} \theta_1 \\ \theta_2 \end{Bmatrix} = \begin{Bmatrix} 0 \\ 0 \end{Bmatrix} \tag{1.5f}$$

Equations (1.4c and 1.5f) are the same, giving the same solution for the critical load, P_{cr}, by the energy method as given before by the displaced shape equilibrium method. For $c_1 = c_2 = c$, and $L_1 = L_2 = L$, from Eqs. (1.5d and 1.5e)

$$\frac{\partial^2 \Pi}{\partial \theta_1{}^2} = c + c + -PL \cos \theta_1 \tag{1.5g}$$

$$\frac{\partial^2 \Pi}{\partial \theta_2{}^2} = c - PL \cos \theta_2 \tag{1.5h}$$

$$\frac{\partial^2 \Pi}{\partial \theta_1 \partial \theta_2} = -c \tag{1.5i}$$

For the two degrees of freedom systems from Eq. (1.1g),

$$2\delta^2 \Pi = K_{11} \delta q_1{}^2 + 2K_{12} \delta q_1 \delta q_2 + K_{22} \delta q_2{}^2 \tag{1.5j}$$

For the initial position, $\theta_1 = \theta_2 = 0$, from Eqs. (1.1f, 1.5g, 1.5h, and 1.5i),

$$K_{11} = \frac{\partial^2 \Pi}{\partial \theta_1{}^2} = 2c - PL, K_{22} = \frac{\partial^2 \Pi}{\partial \theta_2{}^2} = c - PL, K_{12} = K_{21} = \frac{\partial^2 \Pi}{\partial \theta_1 \partial \theta_2} = -c \tag{1.5k}$$

The two degrees of freedom system in Figure 1.5 is in stable equilibrium if $\delta^2 \Pi > 0$, or from Eq.(1.1i) we have

$$D_1 = K_{11} > 0, \text{ or } 2c-PL > 0, \quad \text{ or } P < \frac{2c}{L} \tag{1.5l}$$

$$\text{and } |D_2| = \begin{vmatrix} K_{11} & K_{12} \\ K_{21} & K_{22} \end{vmatrix} = \begin{vmatrix} 2c - PL & -c \\ -c & c - PL \end{vmatrix} > 0$$

$$\text{or } P^2 - \frac{3Pc}{L} + \frac{c^2}{L^2} > 0$$

$$\text{and } \left(P - \frac{0.382c}{L} \right) \left(P - \frac{2.618c}{L} \right) > 0 \tag{1.5m}$$

Therefore, the two degrees of freedom system is in stable equilibrium if $P < P_{cr} = \frac{0.382c}{L}$, because the inequalities (1.5l and 1.5m) are satisfied in the initial position. It is unstable if

$$P > P_{cr} = \frac{0.382c}{L}.$$

1.3.3 Three-Member Truss

1.3.3.1 The Energy Method

Consider a three-member truss where the bars AB and AC are rigid. These bars are pin-connected at A, and the truss is simply supported at B and C as shown in Figure 1.6a. Points B and C are connected by a linear spring of stiffness k. The bars make an initial angle of θ_0 with the horizontal initially. When a vertical force of P is applied at A, the truss deforms, and the bars make an angle of θ with the horizontal in Figure 1.6b.

The strain energy of the system after deformation is given by

$$U = \tfrac{1}{2}k[\,2L(\cos\theta - \cos\theta_0)]^2 = 2kL^2(\cos\theta - \cos\theta_0)^2 \tag{1.6a}$$

Potential energy of the external force P is

$$V = -PL(\sin\theta_0 - \sin\theta) \tag{1.6b}$$

Total potential energy of the system is

$$\Pi = -PL(\sin\theta - \sin\theta_0) + 2kL^2(\cos\theta - \cos\theta_0)^2$$

$$\frac{d\Pi}{d\theta} = PL\cos\theta - 4kL^2(\cos\theta - \cos\theta_0)\sin\theta \tag{1.6c}$$

By making $d\Pi/d\theta = 0$, the equilibrium equation is

$$\frac{P}{4kL} = \sin\theta - \cos\theta_0 \tan\theta \tag{1.6d}$$

$$\frac{d^2\Pi}{d\theta^2} = -PL\sin\theta - 4kL^2[\cos 2\theta - \cos\theta_0 \cos\theta] \tag{1.6e}$$

Substitute (1.6d) into (1.6e)

$$\frac{d^2\Pi}{d\theta^2} = -4kL[\sin\theta - \cos\theta_0 \tan\theta]L\sin\theta - 4kL^2[\cos 2\theta - \cos\theta_0 \cos\theta]$$

Simplifying the above expression gives

$$\frac{d^2\Pi}{d\theta^2} = \frac{4kL^2}{\cos\theta}(\cos\theta_0 - \cos^3\theta) \tag{1.6f}$$

For stable equilibrium, $\dfrac{d^2\Pi}{d\theta^2} > 0$, or $\cos\theta_0 > \cos^3\theta$ is the desired condition.

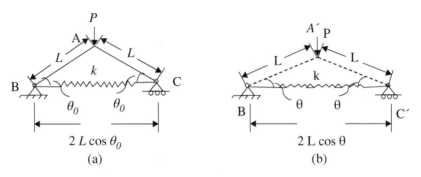

Figure 1.6 Three-member truss with rigid bars: (a) Three member truss; (b) Displaced shape.

Therefore, at $\theta_c < \theta < -\theta_c$, the truss is in stable equilibrium. Where $\cos\theta_0 = \cos^3\theta_c$ at the critical equilibrium, and θ_c is the angle the rigid bars make at the critical equilibrium with the horizontal.

For unstable equilibrium, $\dfrac{d^2\Pi}{d\theta^2} < 0$, or $\cos\theta_0 < \cos^3\theta$ is the desired condition. Therefore, when $-\theta_c < \theta < \theta_c$, the truss is in unstable equilibrium. We can also look at the stability of the truss with respect to the load P. At the critical equilibrium, $\theta = \theta_c$, now substituting $\dot{n}$ in Eq. (1.6d) gives

$$\frac{P}{4kL} = \sin\theta_c - \cos^3\theta_c \tan\theta_c = \sin^3\theta_c$$

Therefore, at $\dfrac{P}{kL} < 4\sin^3\theta_c$, the equilibrium is stable and at $\dfrac{P}{kL} > 4\sin^3\theta_c$, the equilibrium is unstable. From Eq. (1.6d)

$$\frac{P}{4kL} = \sin\theta - \cos\theta_0 \tan\theta$$

or $\quad \dfrac{P}{kL} = 4\sin\theta\left(1 - \dfrac{\cos\theta_0}{\cos\theta}\right)$

or $\quad \dfrac{P}{kL} = 0$ for $\theta = 0$, and $\theta = \pm\theta_0$. $\hspace{2cm}$ (1.6g)

Assume the initial inclination of the truss members is $\theta_c = 20^0$, then for critical equilibrium

$$\cos 20^0 = \cos^3\theta_c, \quad \text{or } \theta_c = 11.62^0.$$

For stable equilibrium,

$$\cos\theta_0 > \cos^3\theta, \quad \text{or } \cos\theta < (\cos\theta_0)^{\frac{1}{3}}$$

or

$$\theta > \left\{\cos^{-1}\left[\cos(\theta_0)^{\frac{1}{3}}\right]\right\}$$

hence,

$$\theta > \theta_c = 11.62^0 \text{ and } \theta < \theta_c = -11.62^0$$

For unstable equilibrium, $\cos\theta_0 < \cos^3\theta$, or $\cos\theta > (\cos\theta_0)^{\frac{1}{3}}$

or

$$\theta < \left\{\cos^{-1}\left[\cos(\theta_0)^{\frac{1}{3}}\right]\right\}$$

hence,

$$\theta < \theta_c = 11.62^0 \text{ and } \theta > \theta_c = -11.62^0$$

A plot of Eq. (1.6g) is shown for $\theta_0 = 20^0$ in Figure 1.7.

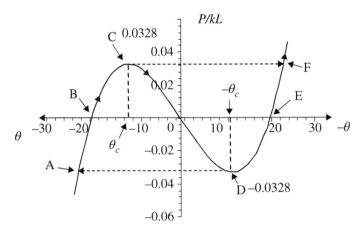

Figure 1.7 Displacement path of three-member truss.

In Figure 1.7 the stable equilibrium paths lie on the lines ABC and DEF, while the unstable equilibrium path lie on the segment CD. As the load P increases from zero, first the stable equilibrium path BC is followed until the critical load at point C is reached. At point C the structure snaps through from point C to F as shown by the dashed line in Figure 1.7. This occurs because as the load is increased infinitesimally from the peak point C, the stable equilibrium available at that load is corresponding to point F. Therefore, there is a large deformation for a small change in the load until the state corresponding to point F is reached. The structure is in stable equilibrium beyond F in new configuration. The change of state from point C to F does not occur through equilibrium paths but occurs dynamically and the structure is unstable during this change. This type of instability is called snap through or limit point instability. If the load P is decreased, the structure follows the path FED, and at point D snaps through to point A. The load deflection curve in Figure 1.7 also shows that this problem is nonlinear even at small deformations. We cannot obtain meaningful results if linearization simplification is used for angles θ and θ_0, even if these angles are small.

1.3.4 Three Rigid Bars with Two Linear Springs

1.3.4.1 The Displaced Shape Equilibrium Method

Three rigid bars are shown in Figure 1.8a. the system is supported by a hinge at A and a roller support at B. The bars are joined by pins C and D, the supports at C and D consist of two linear springs each of stiffness k. The system is subjected to an axial force P as shown. As the force increases, the system deflects as shown in Figure 1.8b, the vertical deflections at C and D are δ_1 and δ_2 respectively. It is a two degrees of freedom system because these two deflections are needed to define the displaced shape. The deflections are assumed to be small.

In Figure 1.8b,

$$\Sigma M_B = 0$$

$$V_A(3L) - k\,\delta_1(2L) - k\,\delta_2(L) = 0$$

$$V_A = \frac{2}{3}k\delta_1 + \frac{1}{3}k\delta_2, \text{ and}$$

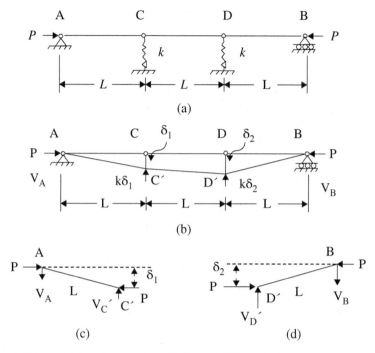

Figure 1.8 Three rigid bars with two linear springs: (a) Three rigid bars; (b) Displaced shape; (c) Free body diagram of AC'; (d) Free body diagram of BD'.

$$\Sigma F_{vertical} = 0$$

$$\frac{2}{3}k\,\delta_1 + \frac{1}{3}k\,\delta_2 - k\delta_1 - k\,\delta_2 + V_B = 0$$

$$V_B = \frac{1}{3}k\delta_1 + \frac{2}{3}k\,\delta_2$$

From Figures 1.8c and 1.8d, and small deformations

$$\Sigma M_{C'} = 0$$

$$\left(\frac{2}{3}k\delta_1 + \frac{1}{3}k\delta_2\right)L - P\delta_1 = 0 \qquad\qquad (1.7a)$$

$$\Sigma M_{D'} = 0$$

$$\left(\frac{1}{3}k\delta_1 + \frac{2}{3}k\delta_2\right)L - P\delta_2 = 0 \qquad\qquad (1.7b)$$

Equations (1.7a and 1.7b) can be written in the matrix form as

$$\begin{bmatrix} \frac{2}{3}kL - P & \frac{1}{3}kL \\ \frac{1}{3}kL & \frac{2}{3}kL - P \end{bmatrix} \begin{Bmatrix} \delta_1 \\ \delta_2 \end{Bmatrix} = \begin{Bmatrix} 0 \\ 0 \end{Bmatrix} \qquad\qquad (1.7c)$$

For a nontrivial solution the determinant of the coefficient matrix is zero,

$$\begin{vmatrix} \frac{2}{3}kL - P & \frac{1}{3}kL \\ \frac{1}{3}kL & \frac{2}{3}kL - P \end{vmatrix} = 0 \tag{1.7d}$$

The characteristic equation is

$$P^2 - \frac{4}{3}kLP + \frac{1}{3}k^2L^2 = 0 \tag{1.7e}$$

The two roots of Eq. (1.7e) are $P_1 = \frac{1}{3}kL$, and $P_2 = kL$.

The first eigenvector is

$$\begin{Bmatrix} \delta_1 \\ \delta_2 \end{Bmatrix} = \begin{Bmatrix} 1 \\ -1 \end{Bmatrix}$$

for $P_1 = P_{cr} = \frac{1}{3}kL$, and the deflected shape is the buckling mode as given in Figure 1.9a.

The second eigenvector is

$$\begin{Bmatrix} \delta_1 \\ \delta_2 \end{Bmatrix} = \begin{Bmatrix} 1 \\ 1 \end{Bmatrix}$$

for $P_2 = kL$, and the deflected shape is symmetric as shown in Fig. 1.9b.

1.3.4.2 The Energy Method
The strain energy of the system in Figure 1.8b is given by

$$U = \frac{1}{2}k\delta_1^2 + \frac{1}{2}k\delta_2^2 \tag{1.8a}$$

The potential energy of the external force is

$$V = -P\left[\left(L - \left(1 - \cos\frac{\delta_1}{L}\right) + L\left(1 - \cos\frac{\delta_2}{L}\right) + L\left[1 - \cos\frac{\delta_2 - \delta_1}{L}\right]\right] \tag{1.8b}$$

Total potential energy of the system is

$$\Pi = U + V \tag{1.8c}$$

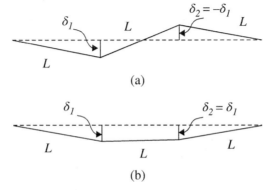

Figure 1.9 Mode shapes of the three rigid bars with linear springs: (a) Asymmetrical deflected shape; (b) Symmetrical deflected shape.

or

$$\Pi = \frac{1}{2}k\delta_1{}^2 + \frac{1}{2}k\delta_2^2 - PL\left[3 - \cos\frac{\delta_1}{L} - \cos\frac{\delta_2}{L} - \cos\frac{\delta_2 - \delta_1}{L}\right] \tag{1.8d}$$

The first derivatives of the potential energy function, Π, with respect to δ_1 and δ_2 must be zero for potential energy to be stationary. Therefore,

$$\frac{\partial\Pi}{\partial\delta_1} = k\delta_1 - PL\left[\frac{1}{L}\sin\frac{\delta_1}{L} - \frac{1}{L}\sin\frac{\delta_2 - \delta_1}{L}\right] \tag{1.8e}$$

$$\frac{\partial\Pi}{\partial\delta_1} = k\delta_2 - PL\left[\frac{1}{L}\sin\frac{\delta_2}{L} + \frac{1}{L}\sin\frac{\delta_2 - \delta_1}{L}\right] \tag{1.8f}$$

For small angle approximation, $\sin\frac{\delta_1}{L} \approx \frac{\delta_1}{L}$, $\sin\frac{\delta_2}{L} \approx \frac{\delta_2}{L}$, and $\sin\frac{\delta_2 - \delta_1}{L} \approx \frac{\delta_2 - \delta_1}{L}$, therefore,

$$\frac{\partial\Pi}{\partial\delta_1} = k\delta_1 - P\left(\frac{\delta_1}{L} - \frac{\delta_2 - \delta_1}{L}\right) = 0 \tag{1.8g}$$

$$\frac{\partial\Pi}{\partial\delta_2} = k\delta_2 - P\left(\frac{\delta_2}{L} + \frac{\delta_2 - \delta_1}{L}\right) = 0 \tag{1.8h}$$

Equations (1.8g and 1.8h) can be written in the matrix form as:

$$\begin{bmatrix} k - P\frac{2}{L} & \frac{P}{L} \\ \frac{P}{L} & k - P\frac{2}{L} \end{bmatrix}\begin{Bmatrix} \delta_1 \\ \delta_2 \end{Bmatrix} = \begin{Bmatrix} 0 \\ 0 \end{Bmatrix} \tag{1.8i}$$

For a nontrivial solution the determinant of the coefficient matrix is zero,

$$\begin{bmatrix} k - P\frac{2}{L} & \frac{P}{L} \\ \frac{P}{L} & k - P\frac{2}{L} \end{bmatrix} = 0 \tag{1.8j}$$

or

$$P^2 - \frac{4}{3}PLk + \frac{L^2}{3}k^2 = 0 \tag{1.8k}$$

Equation (1.8k) is the same characteristic equation as Eq. (1.7e), giving the same two roots of $P_1 = P_{cr} = \frac{kL}{3}$, and $P_2 = kL$ as before by the displaced shape equilibrium method.

$$\frac{\partial^2\Pi}{\partial\delta_1^2} = K_{11} = k - 2\frac{P}{L} \tag{1.8l}$$

$$\frac{\partial^2\Pi}{\partial\delta_2^2} = K_{22} = k - 2\frac{P}{L} \tag{1.8m}$$

$$\frac{\partial^2\Pi}{\partial\delta_1\partial\delta_2} = K_{12} = K_{21} = \frac{P}{L} \tag{1.8n}$$

The three rigid bar system in Figure 1.8 is in stable equilibrium if $\delta^2\Pi > 0$, therefore from Eq. (1.1i)

$$D_1 = K_{11} = k - 2\frac{P}{L} > 0, \text{ or } P < \frac{kL}{2} \tag{1.8o}$$

and $\quad |D_2| = \begin{vmatrix} K_{11} & K_{12} \\ K_{21} & K_{22} \end{vmatrix} = \begin{vmatrix} k - \frac{2P}{L} & \frac{P}{L} \\ \frac{P}{L} & k - \frac{2P}{L} \end{vmatrix} > 0$

or $\quad P^2 - \frac{4}{3}PkL + \frac{k^2L^2}{3} > 0$

and $\quad \left(P - \frac{kL}{3} \right)(P - kL) > 0$ $\hfill (1.8\text{p})$

Therefore, the three bar rigid system is in stable equilibrium if $P < P_{cr} = \frac{kL}{3}$, because the inequalities in Eqs. (1.8o and 1.8p) are satisfied. It is unstable if $P > P_{cr} = \frac{kL}{3}$.

1.4 Large Displacement Analysis

So far, the analysis has been limited to the linear, small deflection theory that applies to infinitely small deformations from the initial stressed state of the structure. The small deflection theory gives information about the critical load and it is also possible to determine the state of equilibrium in the initial position by studying the second derivatives of the total potential energy by this theory. This is sufficient for most structural engineering problems. However, nonlinear finite displacement theory is needed to gain a full understanding of the post-buckling behavior of a system. We can plot the post buckling equilibrium path using this large displacement theory. It also gives us an indication of the stability of bifurcation.

1.4.1 Rigid Bar Supported by a Translational Spring

1.4.1.1 The Displaced Shape Equilibrium Method
The rigid bar given in Figure 1.4 will be considered here without making the assumption of small deformation. The equilibrium equation is

$\quad P = kL \cos\theta$ $\hfill (1.2\text{d})$

and the critical load is given by $P_{cr} = k\,L$. The equilibrium diagram, $\frac{P}{P_{cr}}$ versus θ, giving the post-buckling path is plotted in Figure 1.10 using Eqs. (1.2d and 1.2e). The initial inclination of the column to the right or left causes a decrease in the load capacity of the column and values of $\frac{P}{P_{cr}}$ continually decrease with increasing θ. The post-buckling displacement path is also symmetric about the initial position of the column, therefore, the bifurcation is called symmetric unstable bifurcation.

1.4.1.2 The Energy Method
This method can also be used to find the critical load and the load deflection graph as shown in Figure 1.10. In addition, it can give the nature of equilibrium in the system initially when the applied load reaches the critical load value as well as during post-buckling. Equations (1.3i–1.3m) can be rewritten from Figure 1.4 as

$\quad \Pi = -P\,L(1 - \cos\theta) + {}^1\!/_2 k(L \sin\theta)^2$ $\hfill (1.3\text{i})$

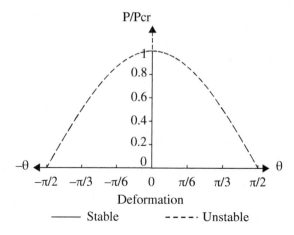

Figure 1.10 Equilibrium path of rigid bar in Figure 1.4.

$$\frac{d\Pi}{d\theta} = -PL \sin\theta + kL^2 \sin\theta \cos\theta \tag{1.3j}$$

$\frac{d\Pi}{d\theta} = 0$, and we get the equilibrium equation as

$$P = kL \cos\theta \tag{1.3k}$$

$\cos\theta \approx 1$ for small values of θ
or

$$P_{cr} = kL \tag{1.3l}$$

$$\frac{d^2\Pi}{d\theta^2} = -PL\cos\theta + kL^2\cos 2\theta \tag{1.3m}$$

From Eq. (1.3k) and Eq. (1.3m)

$$\frac{d^2\Pi}{d\theta^2} = -kL^2\cos\theta + kL^2\cos 2\theta \tag{1.9a}$$

Substituting $P = P_{cr} = kL$, and $\theta = 0$ in Eq. (1.9a), $\frac{d^2\Pi}{d\theta^2} = 0$. This does not give us an idea of the nature of equilibrium at the bifurcation. Therefore, to determine the initial post-critical behavior near bifurcation we may write total potential energy, Π, as a Taylor series as follows:

$$\Pi = \Pi\Big|_{\theta=0} + \frac{d\Pi}{d\theta}\Big|_{\theta=0}\theta + \frac{1}{2!}\frac{d^2\Pi}{d\theta^2}\Big|_{\theta=0}\theta^2 + \frac{1}{3!}\frac{d^3\Pi}{d\theta^3}\Big|_{\theta=0}\theta^3 + \frac{1}{4!}\frac{d^4\Pi}{d\theta^4}\Big|_{\theta=0}\theta^4 + ---- \tag{1.9b}$$

$$\frac{d^3\Pi}{d\theta^3} = PL\sin\theta - 2kL^2\sin 2\theta \tag{1.9c}$$

At

$$P = P_{cr} = kL, \text{ and } \theta = 0, \frac{d^3\Pi}{d\theta^3} = 0$$

$$\frac{d^4\Pi}{d\theta^4} = PL\cos\theta - 4kL^2\cos 2\theta \tag{1.9d}$$

At $P = P_{cr} = kL$ and $\theta = 0$, $\dfrac{d^4 \Pi}{d\theta^4} = -3kL^2$ (1.9e)

Therefore,

$$\Pi = \frac{-3kL^2}{24}\theta^4 = -\frac{1}{8}kL^2\theta^4$$ (1.9f)

This indicates that the total potential energy, Π, is negative or it decreases with increasing θ, at the initial position, $\theta = 0$ and $P = P_{cr}$. The bifurcation is symmetric and unstable from Eqs. (1.3k and 1.9f), as shown in Figure 1.10.

During the post-buckling path when $\theta \neq 0$, Eqs. (1.3k and 1.3m) give

$$\frac{d^2 \Pi}{d\theta^2} = -kL^2\cos^2\theta + kL^2 \cos 2\theta$$

or $\qquad \dfrac{d^2 \Pi}{d\theta^2} = -kL^2\sin^2\theta$ (1.9g)

Therefore, $\dfrac{d^2 \Pi}{d\theta^2} < 0$ for different values of θ, and the post-buckling path is unstable.

1.4.2 Rigid Bar Supported by Translational and Rotational Springs

1.4.2.1 The Displaced Shape Equilibrium Method

A rigid bar connected to a translational spring at the top and a rotational spring at the bottom is acted on by an axial force P as shown in Figure 1.11a. The free body diagram of the deflected system is shown in Figure 1.11b. Taking the moment of all the forces acting on the system in Figure 1.11b about A and equating to zero, we have

$$PL \sin\theta - kL \sin\theta (L\cos\theta) - c\theta = 0$$ (1.10a)

or

$$P = kL \cos\theta + \frac{c}{L}\frac{\theta}{\sin\theta}$$ (1.10b)

For small values of θ, $\cos\theta \approx 1$, $\sin\theta \approx \theta$, hence, the critical load is

$$P_{cr} = kL + \frac{c}{L}.$$ (1.10c)

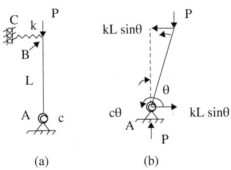

(a)   (b)

Figure 1.11 Rigid bar connected to translational and rotational springs: (a) Rigid bar with two springs; (b) Free body diagram of displaced shape.

1.4.2.2 The Energy Method

The strain energy of the system in Figure 1.11 is

$$U = \frac{1}{2}k(L \sin \theta)^2 + \frac{1}{2}c\theta^2 \tag{1.11a}$$

and the potential energy of the external forces is

$$V = -P(L - L \cos \theta) \tag{1.11b}$$

Total potential energy is given by

$$\Pi = \frac{1}{2}k(L \sin \theta)^2 + \frac{1}{2}c\theta^2 - P(L - L \cos \theta) \tag{1.11c}$$

Taking the first derivative of the total potential energy with respect to θ and equating it to zero gives the equilibrium equation

$$\frac{d\Pi}{d\theta} = kL^2 \sin \theta \cos \theta + c\theta - PL \sin \theta = 0 \tag{1.11d}$$

giving the same relation between the force P and θ as in Eq. (1.10b) and the same critical load P_{cr} as before.

$$\text{or} \frac{d\Pi}{d\theta} = \frac{kL^2}{2} \sin 2\theta + c\theta - PL \sin \theta$$

$$\frac{d^2\Pi}{d\theta^2} = kL^2 \cos 2\theta + c - PL \cos \theta \tag{1.11e}$$

Substituting $P = P_{cr} = kL + \frac{c}{L}$, and $\theta = 0$ in Eq. (1.11e) gives

$$\frac{d^2\Pi}{d\theta^2} = kL^2 \cos(0) + c - \left(kL + \frac{c}{L}\right)L \cos(0) = 0$$

Therefore, use higher terms in the Taylor series in Eq. (1.9b) to know whether the total potential energy is relative maximum or minimum at the bifurcation.

$$\frac{d^3\Pi}{d\theta^3} = -2kL^2 \sin 2\theta + PL \sin \theta \tag{1.11f}$$

At the bifurcation,

$$P = P_{cr} = kL + \frac{c}{L}, \theta = 0,$$

$$\frac{d^3\Pi}{d\theta^3} = 0$$

and

$$\frac{d^4\Pi}{d\theta^4} = -4kL^2 \cos 2\theta + PL \cos \theta$$

$$= -4kL^2 + \left(kL + \frac{c}{L}\right)L = -3kL^2 + c \tag{1.11g}$$

Therefore, from the Taylor series of Eq. (1.9b) we have

$$\Pi = \frac{1}{24}(-3kL^2 + c)\theta^4 = \left(-\frac{kL^2}{8} + \frac{c}{24}\right)\theta^4 \tag{1.11h}$$

Π is positive, if $\dfrac{c}{24} > \dfrac{kL^2}{8}$, or $\dfrac{kL^2}{c} < \dfrac{1}{3}$ for stable equilibrium at the bifurcation. On the other hand, Π is negative if $\dfrac{kL^2}{c} > \dfrac{1}{3}$, and the equilibrium at the bifurcation is unstable.

During the post-buckling path when $\theta \neq 0$, from Eqs. (1.10a and 1.11e)

$$\frac{d^2\Pi}{d\theta^2} = kL^2\cos^2\theta - kL^2\sin^2\theta + c - \left(kL\cos\theta + \frac{c\theta}{L\sin\theta}\right)L\cos\theta$$

or $\dfrac{d^2\Pi}{d\theta^2} = -kL^2\sin^2\theta + c - c\theta\cot\theta$

For $\dfrac{d^2\Pi}{d\theta^2} > 0$, the post-buckling path is stable, and at $\dfrac{d^2\Pi}{d\theta^2} < 0$, it is unstable. Therefore,

if $\dfrac{kL^2}{c} < \dfrac{\sin\theta - \theta\cos\theta}{\sin^3\theta}$ \hfill (1.11i)

the post-buckling path is stable, and for

$$\frac{kL^2}{c} > \frac{\sin\theta - \theta\cos\theta}{\sin^3\theta} \tag{1.11j}$$

the post-buckling path is unstable.

At $\theta = 0$, $\dfrac{\sin\theta - \theta\cos\theta}{\sin^3\theta} = \dfrac{0}{0}$. Therefore, differentiate numerator and denominator with respect to θ, and applying Le Hospital's rule, we get $\dfrac{\sin\theta - \theta\cos\theta}{\sin^3\theta} = \dfrac{1}{3}$ for $\theta = 0$. Hence, if $\dfrac{kL^2}{c} < \dfrac{1}{3}$, it is stable bifurcation, and for $\dfrac{kL^2}{c} > \dfrac{1}{3}$, it is unstable at $\theta = 0$ as shown before.

Let $\dfrac{kL^2}{c} = 0.35$, or $kL = 0.35\dfrac{c}{L}$, and from Eq. (1.10c)

$$P_{cr} = 0.35\frac{c}{L} + \frac{c}{L} = \frac{1.35c}{L}$$

From Eq. (1.10b)

$$\frac{PL}{c} = \frac{kL^2}{c}\cos\theta + \frac{\theta}{\sin\theta}, \text{ dividing both sides of the equation by } 1.35$$

$$\frac{PL}{1.35c} = \frac{P}{P_{cr}} = \frac{0.35\cos\theta + \frac{\theta}{\sin\theta}}{1.35} \tag{1.11k}$$

$\dfrac{P}{P_{cr}}$ versus θ graph is plotted in Figure 1.12, and it shows that post-buckling path is unstable at the bifurcation because $\dfrac{kL^2}{c} = 0.35 > \dfrac{1}{3}$ and it continues to be unstable until $\dfrac{kL^2}{c} = 0.35 < \dfrac{\sin\theta - \theta\cos\theta}{\sin^3\theta}$, when it becomes stable.

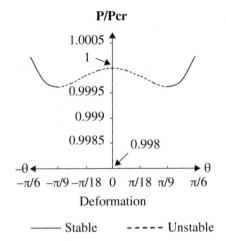

Figure 1.12 Displacement path of rigid bar supported by translational and rotational springs.

1.4.3 Two Rigid Bars Connected by Rotational Springs

1.4.3.1 The Energy Method

The two rigid bars of equal length L and connected by rotational springs of equal spring stiffness c shown in Figure 1.5 are analyzed here by large displacement analysis. The first and second derivatives of the total potential energy function, Π, from Eqs. (1.5d, 1.5e, 1.5g, 1.5h, and 1.5i) are as follows:

$$\frac{\partial \Pi}{\partial \theta_1} = 2c\,\theta_1 - c\,\theta_2 - PL\,\sin\theta_1 = 0 \tag{1.12a}$$

$$\frac{\partial \Pi}{\partial \theta_1} = -c\theta_1 + c\theta_2 - PL\,\sin\theta_2 = 0 \tag{1.12b}$$

$$\frac{\partial^2 \Pi}{\partial \theta_1{}^2} = 2c - PL\cos\theta_1 \tag{1.12c}$$

$$\frac{\partial^2 \Pi}{\partial \theta_2{}^2} = c - PL\cos\theta_2 \tag{1.12d}$$

$$\frac{\partial^2 \Pi}{\partial \theta_1 \partial \theta_2} = -c \tag{1.12e}$$

Equations (1.12a and 1.12b) are equilibrium equations of the system. These are solved by eliminating θ_2. From Eq. (1.12a)

$$\theta_2 = \frac{2c\theta_1 - PL\sin\theta_1}{c} \tag{1.12f}$$

Substituting Eq. (1.12f) in Eq. (1.12b), we get

$$-c\theta_1 + 2c\theta_1 - PL\sin\theta_1 - PL\sin\frac{2c\theta_1 - PL\sin\theta_1}{c} = 0$$

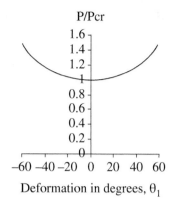

P/Pcr

Deformation in degrees, θ_1

Figure 1.13 Displacement path of two rigid bars connected by rotational springs.

or

$$\frac{P}{\frac{c}{L}} = \frac{\theta_1}{\sin \theta_1 + \sin \left(2\theta_1 - \frac{PL}{c} \sin \theta_1 \right)} \tag{1.12g}$$

$P_{cr} = 0.382\frac{c}{L}$, from Eq. (1.5i).
Therefore,

$$\frac{P}{P_{cr}} = \frac{1}{0.382} \left[\frac{\theta_1}{\sin \theta_1 + \sin \left(2\theta_1 - \frac{PL}{c} \sin \theta_1 \right)} \right] \tag{1.12h}$$

$\frac{P}{P_{cr}}$ vs. θ_1 graph is plotted in Figure 1.13, and it shows that the post-buckling path is stable.

1.5 Imperfections

So far, it has been assumed that the rigid bars considered were geometrically perfect. In general, the columns may be imperfect, having a certain amount of deformation present in the initial state when the springs are unrestrained at the load P = 0.

1.5.1 Rigid Bar Supported by a Rotational Spring at the Base

1.5.1.1 The Displaced Shape Equilibrium Method
Consider a rigid bar of length L supported by a rotational spring of stiffness c. The column is initially imperfect and inclined by an angle α as shown in Figure 1.14. From the equilibrium of the column in the displaced position making an angle of θ with the vertical, we have

$$PL \sin \theta - c (\theta - \alpha) = 0 \tag{1.13a}$$

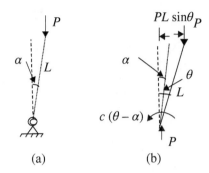

Figure 1.14 Imperfect rigid-bar column with rotational spring at the base: (a) Imperfect rigid bar; (b) Free body diagram of displaced shape.

or

$$P = \frac{c(\theta - \alpha)}{L \sin \theta} \tag{1.13b}$$

1.5.1.2 The Energy Method

Strain energy of the imperfect column in Figure 1.14 is

$$U = \frac{1}{2}c(\theta - \alpha)^2 \tag{1.14a}$$

and the potential energy of the external forces is

$$V = -PL(\cos \alpha - \cos \theta) \tag{1.14b}$$

The total potential energy is given by

$$\Pi = \frac{1}{2}c(\theta - \alpha)^2 - PL(\cos \alpha - \cos \theta) \tag{1.14c}$$

$$\frac{d\Pi}{d\theta} = c(\theta - \alpha) - PL \sin \theta \tag{1.14d}$$

Setting $\dfrac{d\Pi}{d\theta} = 0$, we have the equilibrium condition, and $P = \dfrac{c(\theta - \alpha)}{L \sin \theta}$ as before. For a perfect column and small θ, $P_{cr} = \dfrac{c}{L}$.

The equilibrium diagrams, $\dfrac{P}{P_{cr}}$ versus θ, are plotted in Figure 1.15 for initial imperfections of $\alpha = -10, -5, 5,$ and $10°$ of inclination with the vertical. The points where columns change from stable to unstable state lie on the critical curve defined by $\dfrac{d^2\Pi}{d^2\theta} = 0$.

$$\frac{d^2\Pi}{d\theta^2} = c - PL \cos \theta \tag{1.14e}$$

If $\dfrac{d^2\Pi}{d^2\theta} = 0$, $P = \dfrac{c}{L \cos \theta}$, or $\dfrac{P}{P_{cr}} = \dfrac{1}{\cos \theta}$, and $\dfrac{P}{P_{cr}}$ versus θ critical curve is plotted in Figure 1.15.

The column is stable if $\dfrac{d^2\Pi}{d^2\theta} > 0$, and it is unstable if $\dfrac{d^2\Pi}{d^2\theta} < 0$. Substituting the value of $P = P_{cr}$ from Eq. (1.13b) into Eq. (1.14e), we have

$$\frac{d^2\Pi}{d\theta^2} = c\left[1 - \frac{(\theta - \alpha)}{\tan\theta}\right] \tag{1.14f}$$

So the equilibrium path is stable if $\tan\theta > \theta - \alpha$ if $\alpha < \theta < \pi/2$, and $\tan\theta < \theta - \alpha$ if $-\pi/2 < \theta < 0$. This can also be seen from the slopes of the equilibrium curves in Figure 1.15.

$$P = \frac{c}{L}\left(\frac{\theta - \alpha}{\sin\theta}\right)$$

$$\frac{dP}{d\theta} = \frac{c}{L\sin\theta}\left[1 - \frac{\theta - \alpha}{\tan\theta}\right] \tag{1.14g}$$

For $\alpha > 0$ and $\theta > \alpha$, $\tan\theta > \theta - \alpha$; and from Eq. (1.14g), $\dfrac{dP}{d\theta} > 0$. Similarly, for $\alpha > 0$ and $\theta < 0$, $\tan\theta < \theta - \alpha$; and from Eq. (1.14g), $\dfrac{dP}{d\theta} < 0$. Therefore, equilibrium curves are stable when their slope is positive in the bottom right and negative in the top left in Figure 1.15. The same way it can be proved that for $\alpha < 0$, the equilibrium curves are stable when their slope is negative for $\theta < 0$ in the bottom left; and the slope is positive for $\theta > 0$ in the top right in Figure 1.15. The results are symmetrical. For the critical state, $\dfrac{d^2\Pi}{d^2\theta} = 0$, therefore, from Eqs. (1.14f and 1.14g), $\dfrac{dP}{d\theta} = 0$ and

$$\tan\theta = \theta - \alpha \tag{1.14h}$$

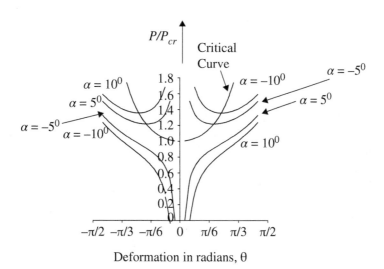

Figure 1.15 Equilibrium path of the rigid bar imperfect column with rotational spring at the base.

The zero slope point on each equilibrium curve gives the critical state. Figure 1.15 also shows that the imperfect column can be stable at loads higher than that of the perfect column.

1.5.2 Two Rigid Bars Connected by Rotational Springs

1.5.2.1 The Displaced Shape Equilibrium Method

Consider the column shown in Figure 1.16a that has two degrees of freedom. The deflected shape of the column is defined by the angles θ_1 and θ_2. Initially the column is imperfect shown by the angles of inclination α_1 and α_2 of the two bars with the vertical.

Taking the sum of the moments about A in Figure 1.16b equal to zero,

$$PL \sin \theta_1 + c[(\theta_2-\alpha_2)-(\theta_1-\alpha_1)]-c(\theta_1-\alpha_1) = 0$$

or

$$c\,\theta_2 = 2c\theta_1-2c\alpha_1 + c\alpha_2-PL \, \sin \theta_1 \tag{1.15a}$$

Similarly taking the sum of the moments about B in Figure 1.16c equal to zero,

$$PL \, \sin \theta_2-c[(\theta_2-\alpha_2)-(\theta_1-\alpha_1)] = 0 \tag{1.15b}$$

Eliminating θ_2 from Eqs. (1.15a and 1.15b), we have

$$\frac{P}{\frac{c}{L}} = \frac{\theta_1-\alpha_1}{\sin \theta_1 + \sin \left(2\theta_1-2\alpha_1 + \alpha_2-\frac{PL}{c} \sin \theta_1 \right)} \tag{1.15c}$$

1.5.2.2 The Energy Method

The strain energy of the column in Figure 1.16 is given by

$$U = \frac{1}{2}c(\theta_1-\alpha_1)^2 + \frac{1}{2}c[\theta_2-\alpha_2-(\theta_1-\alpha_1)]^2 \tag{1.16a}$$

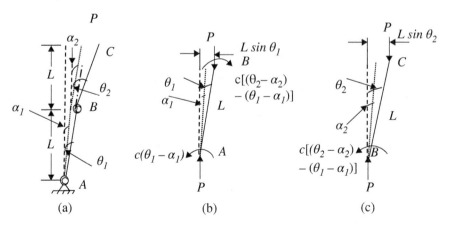

(a) (b) (c)

Figure 1.16 Imperfect column with two rigid bars and two rotational springs: (a) Displaced shape of the column; (b) Free body diagram of the lower bar; (c) Free body diagram of the upper bar.

$$V = -P[L \cos \alpha_2 - L \cos \theta_2 + L \cos \alpha_1 - L \cos \theta_1] \tag{1.16b}$$

The total potential energy is $\Pi = U + V$, or

$$\Pi = \frac{1}{2}c(\theta_1 - \alpha_1)^2 + \frac{1}{2}c[(\theta_2 - \alpha_2) - (\theta_1 - \alpha_1)]^2 - P[L \cos \alpha_2 - L \cos \theta_2 + L \cos \alpha_1 - L \cos \theta_1]$$

By differentiating, we obtain the equilibrium conditions:

$$\frac{\partial \Pi}{\partial \theta_1} = c(\theta_1 - \alpha_1) - c(\theta_2 - \alpha_2 - \theta_1 + \alpha_1) - PL \sin \theta_1 = 0 \tag{1.16c}$$

or $c\,\theta_2 = 2c\theta_1 - 2c\alpha_1 + c\alpha_2 - PL \sin \theta_1$, same as Eq. (1.15a).

$$\frac{\partial \Pi}{\partial \theta_2} = PL \sin \theta_2 - c[\theta_2 - \alpha_2 - (\theta_1 - \alpha_1)] = 0 \tag{1.16d}$$

Equation (1.16d) is the same as Eq. (1.15b), therefore, eliminating θ_2 from above equations will lead to the same P versus θ_1 relation as in Eq. (1.15c). For the column in Figure 1.16 if it is perfect, i.e. $\alpha_1 = \alpha_2 = 0$, and if the displacements, θ_1 and θ_2 are small, the critical load $P_{cr} = 0.382\frac{c}{L}$. From Eq. (1.15c) we get

$$\frac{P}{P_{cr}} = \frac{\theta_1 - \alpha_1}{0.382\left[\sin \theta_1 + \sin\left(2\theta_1 - 2\alpha_1 + \alpha_2 - \frac{PL}{c}\sin \theta_1\right)\right]} \tag{1.16e}$$

Assume

$$\alpha_1 = \alpha_2 = \alpha.$$

The equilibrium path given by Eq. (1.16e) is plotted in Figure 1.17.

Discrete systems with one or two degrees of freedom have been analyzed in this chapter. In the analysis for stability of discrete systems, algebraic equations were developed and solved. Differential equations are formed when the analysis of continuous systems such as beams and

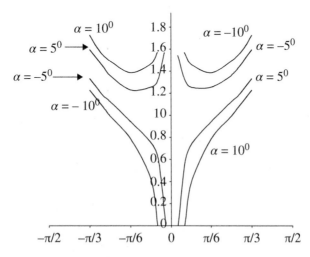

Figure 1.17 Displacement path of two imperfect rigid bars column connected by two rotational springs.

columns is performed. The solution of these differential equations is much more difficult than the algebraic equations. These differential equations can be converted to algebraic equations by discretizing a structure or assuming a Fourier series expansion for its displacements. So the analysis of discrete systems is also valuable to analyze the continuous systems. The methods for solving stability problems learned here will be useful in later chapters.

Problems

1.1 Find the critical load P_{cr} for the rigid bar column in Figure P1.1 by using the equilibrium method. The column is restricted by a rotational spring of stiffness c at the support

Figure P1.1

1.2 Solve Problem 1.1 by the energy method.

1.3 Determine the critical load P_{cr} for the rigid bar column in Figure P1.3 a, b by using the equilibrium method.

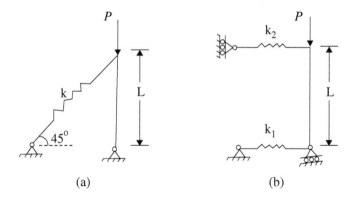

(a)   (b)

Figure P1.3

1.4 Solve Problem 1.3 by the energy method.

1.5 Analyze the stability behavior of the rigid bar system in Figure P1.5.

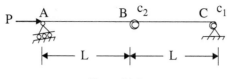

Figure P1.5

1.6 Analyze the stability behavior of the rigid bar and spring system in Figure P1.6. The column is initially imperfect and is inclined by an angle α.

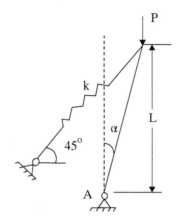

Figure P1.6

References

1 Timoshenko, S.P. and Gere, J.M. (1961). *Theory of Elastic Stability*, 2e. New York: McGraw-Hill.

2 Bazant, Z.P. and Cedolin, L. (1991). *Stability of Structures: Elastic, Inelastic, Fracture, and Damage Theories*. New York: Oxford University Press, Inc.

3 Chen, W.F. and Lui, E.M. (1987). *Structural Stability: Theory and Implementation*. Upper Saddle River, NJ: Prentice Hall.

4 Langhaar, H.L. (1962). *Energy Methods in Applied Mechanics*. New York: Wiley.

5 Kreyszig, E. (1972). *Advanced Engineering Mathematics*, 3e. New York: Wiley.

2

Columns

2.1 General

Vertical straight members whose lengths are considerably greater than their lateral dimensions are usually called columns. These structural members are generally thought of as carrying axial compressive force, and are important load-carrying elements in a structure. An understanding of their behavior is therefore important for the overall safety of a structure. Columns can usually be classified as short, intermediate, and long, depending on the ratio of their length to the lateral dimensions and are defined by structural codes. A column can fail either due to material failure or because of lateral deflection called instability or buckling. Short columns fail due to material failure at a load that causes a yield point in a steel column, or crushing in a concrete column. In this failure, capacity of the column material to bear the load is exceeded. On the other hand, long columns reach a load that causes lateral deflection called buckling before the material capacity is reached. In long columns, buckling may occur when the material is still elastic in all fibers of the cross-section. When elastic stress of the material occurs before buckling takes place in the case of intermediate columns, this is called inelastic buckling.

The maximum axial load a column can support when it is on the verge of buckling is called the critical or buckling load. Though these two terms are used interchangeably, there is a difference between the two. The critical load is the load obtained theoretically for an ideal structure, whereas the buckling load is the actual load obtained for a real structure. A straight column subjected to a small axial force P remains straight and undergoes only axial displacement. This straight form of elastic equilibrium is stable because a lateral deflection produced by a small lateral force disappears when the lateral force is removed. If the axial force P is gradually increased, a condition will be reached when a small lateral force will produce a displacement that does not disappear when the lateral force is removed. This happens because the column in its straight equilibrium position becomes unstable. The smallest axial force at which this happens is called the critical load. This axial force value where a column can remain in equilibrium in the straight position as well as in a slightly deflected position is called the point of bifurcation.

2.2 The Critical Load According to Classical Column Theory

We first study the behavior of an idealized perfect column subjected to axial force. The axially loaded member is assumed to be made of homogeneous material and has a constant cross-section throughout its length. We will use the displaced shape or neutral equilibrium method to write equilibrium equations of the slightly bent column. This approach is called the Euler method or an eigenvalue analysis. The critical load is given by the eigenvalues and the deflected shape is obtained from the eigenvectors. The solution gives only the deflected shape called the mode shape and not the magnitude of deflections. The solution was originally given by Swiss mathematician Leonhard Euler [1] in 1744. Though Euler analyzed an axially loaded column fixed at the base and free at the other end [2], nowadays a member that is simply supported at both ends is usually called a Euler column. Certain assumptions are made in the analysis:

i) The column is perfectly straight.
ii) The load is applied along the centroidal axis of the column.
iii) The material of the column is homogeneous and obeys Hooke's law.
iv) The assumption of the theory of bending, i.e. plane sections before deformation remain plane after deformation applies.
v) The deformation of the column is small so that the curvature can be assumed to be $1/\rho = d^2y/dx^2$, because the term $(dy/dx)^2$ is negligible in comparison to 1 in the curvature expression, $\dfrac{1}{\rho} = \dfrac{\dfrac{d^2y}{dx^2}}{\left[1+\left(\dfrac{dy}{dx}\right)^2\right]^{\frac{3}{2}}}$ [3]. Where ρ is the radius of curvature, and y is the lateral deflection of the column.

2.2.1 Pinned-Pinned Column

Considering the equilibrium of the deflected shape of the column in Figure 2.1a under the action of axial load P, we can write the equilibrium equation of the free body diagram shown in Figure 2.1b.

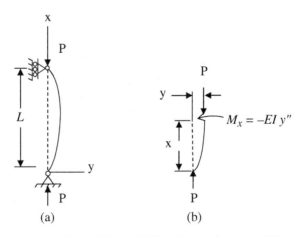

Figure 2.1 Pinned-pinned column under axial load: (a) Pinned-pinned column; (b) Free body diagram.

From bending theory [4],

$$\frac{M}{I} = \frac{E}{\rho} \tag{2.1a}$$

where M is the internal mending moment at a section, I is the moment of inertia of the section about the axis of bending, E is the modulus of elasticity of the material, and ρ is the radius of curvature of the bent shape at the section. The radius of curvature as shown before is given by

$$\frac{1}{\rho} = y'' \tag{2.1b}$$

where $y'' = \dfrac{d^2y}{dx^2}$, a prime shows the derivative with respect to x. In Figure 2.1b, $y' = \dfrac{dy}{dx}$, is a positive quantity for the coordinate axes chosen. The slope, y', is decreasing with the increase of coordinate x, therefore, the rate of change of slope, y'', is negative. The internal moment M_x at a distance x, in Figure 2.1b is positive, so the expression relating the moment M_x and the curvature is given by $M_x = -EIy''$. The moment equilibrium equation in Figure 2.1b is given by

$$P y - M_x = 0$$

or

$$EI\, y'' + P y = 0 \tag{2.1c}$$

and

$$y'' + k^2 y = 0 \tag{2.1d}$$

where $k^2 = P/EI$. Equation (2.1d) is a linear, second-order, homogeneous differential equation with constant coefficients. The general solution for the equation is

$$y = A \sin kx + B \cos kx \tag{2.1e}$$

Constants A and B are calculated from the boundary conditions

$$y = 0 \text{ at } x = 0 \tag{2.1f}$$

and

$$y = 0 \text{ at } x = L \tag{2.1g}$$

Substituting Eq. (2.1f) into Eq. (2.1e) gives

$$B = 0$$

That gives

$$y = A \sin kx \tag{2.1h}$$

Substituting Eq. (2.1g) into Eq. (2.1h) we get

$$A \sin kL = 0 \tag{2.1i}$$

Equation (2.1i) is satisfied if either $A = 0$ or $\sin kL = 0$.

If $A = 0$, it states that $y = 0$ always, or the straight line position is the equilibrium position. It is called the trivial solution. Hence, A cannot be zero, therefore, $\sin kL = 0$. This gives the equilibrium position of the column in a slightly deformed position, called the nontrivial solution. Therefore,

$$kL = n\,\pi, \quad \text{where} \ \ n = 1, 2, 3, \text{-------} \tag{2.1j}$$

or

$$k^2 = \frac{n^2\pi^2}{L^2}$$

and

$$P = \frac{n^2\pi^2 EI}{L^2} \tag{2.1k}$$

The deformed shape is given by

$$y = A\sin\frac{n\pi x}{L} \tag{2.1l}$$

The smallest value of load P is when $n = 1$, and the associated axial load on the column is called the critical load P_{cr} or the Euler load. It is the smallest load at which the column ceases to be in stable equilibrium. Thus,

$$P_{cr} = \frac{\pi^2 EI}{L^2} \tag{2.1m}$$

and the corresponding deflected shape is

$$y = A\sin\frac{\pi x}{L} \tag{2.1n}$$

Equation (2.1n) gives the deflected shape and not the exact value of the deflection because constant A is unknown. This is because the governing equilibrium condition given by the homogeneous differential Eq. (2.1c) became linear because of the linearization of curvature expression in assumption (v). The solution of Eq. (2.1c) is called an eigenvalue problem that leads to eigenvalues and eigenvectors given by Eqs. (2.1k) and (2.1l) respectively. The load versus the displacement of the column is shown in Figure 2.2. The column remains straight up to the Euler critical load, P_{cr}. After this load, the column can remain straight or displace, signifying the state of neutral equilibrium exists at the Euler critical load, therefore, it marks the transition from stable to unstable equilibrium.

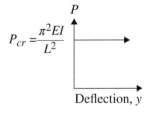

Figure 2.2 Load displacement path.

Critical stress σ_{cr} is given by $\sigma_{cr} = \frac{P_{cr}}{A}$, where
$A = $ cross-sectional area of the column

or

$$\sigma_{cr} = \frac{\pi^2 EI}{L^2 A}$$

and

$$\sigma_{cr} = \frac{\pi^2 E}{L/r^2} \tag{2.1o}$$

where $I = A r^2$, r is the radius of gyration of the column cross-section about its axis of bending. For values of n larger than 1, higher forces can be obtained from Eq. (2.1k) and the corresponding displaced shapes called mode shapes can be obtained from Eq. (2.1l), shown in Figure 2.3.

2.2.2 Fixed-Fixed Column

2.2.2.1 Symmetric Mode

If a column is fixed at both ends, it will bend as shown in Figure 2.4a when it is displaced slightly in a symmetric mode. Fixed end bending moments, M_F, are generated at the ends of the column. The column will have zero displacements and rotations at both the ends. For the deformed shape to be in equilibrium, the resisting internal moment should be equal to the external moment in the free body diagram in Figure 2.4b at a distance x from the origin. Taking the equilibrium of moments in the free body diagram we get

$$P y - M_x - M_F = 0 \tag{2.2a}$$

Since $M_x = -EI y''$, Eq. (2.2a) becomes

$$y'' + k^2 y = \frac{M_F}{EI} \tag{2.2b}$$

where $k^2 = P/EI$, as before. Equation (2.2b) is a non-homogeneous differential equation. The complete solution consists of a complementary part satisfying the homogeneous equation and a particular part of the total solution satisfying the entire equation. The complementary part is

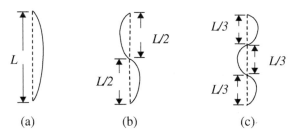

(a)   (b)   (c)

Figure 2.3 Mode shapes of buckling: (a) First mode shape; (b) Second mode shape; (c) Third mode shape.

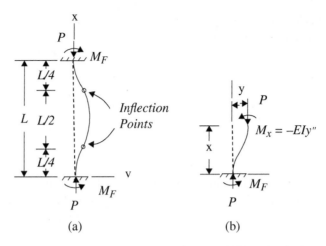

Figure 2.4 Fixed-fixed column under axial load and symmetric mode: (a) Fixed fixed column; (b) Free body diagram.

given by Eq. (2.1e). The right-hand side of Eq. (2.2b) is constant, so we can assume the particular solution also a constant y_p, that can be obtained by observation and substituting it in Eq. (2.2b).

$$y_p = \frac{M_F}{k^2 EI} \qquad (2.2c)$$

Therefore, the complete solution of Eq. (2.2b) is given by

$$y = A \sin kx + B \cos kx + \frac{M_F}{k^2 EI}$$

Now, $k^2 = \frac{P}{EI}$, thus,

$$y = A \sin kx + B \cos kx + \frac{M_F}{P} \qquad (2.2d)$$

Boundary conditions at the bottom of the column in Figure 2.4a are

$$y = 0, \text{ and } y' = 0 \text{ at } x = 0$$

Substituting these boundary conditions in Eq. (2.2d), we have

$$B = -\frac{M_F}{P}, \text{ and } A = 0$$

or

$$y = \frac{M_F}{P}(1 - \cos kx) \qquad (2.2e)$$

Substitute the boundary condition, $y = 0$ at $x = L$ in Eq. (2.2e). Since M_F and P are not zero,

$$1 - \cos kL = 0, \text{ or } \cos kL = 1$$

or

$$kL = 2n\pi \quad n = 1, 2, 3, \text{-----}$$

and the load P is given by $P = \dfrac{4n^2\pi^2 EI}{L^2}$

The lowest load is obtained when $n = 1$, giving

$$P_{cr} = \frac{4\pi^2 EI}{L^2} \tag{2.2f}$$

and the corresponding deformed shape from Eq. (2.2e) is

$$y = \frac{M_F}{P}\left(1 - \cos\frac{2\pi x}{L}\right) \tag{2.2g}$$

$$y'' = \frac{M_F}{P}\left(\frac{4\pi^2}{L^2}\sin\frac{2\pi x}{L}\right) \tag{2.2h}$$

when $x = L/2$ and $3L/4$, $y'' = 0$. Therefore, at these sections the bending moment is zero, and these sections are the points of inflection on the column bent shape. The column of length $L/2$ between the inflection points bends in the shape of a pinned-pinned column, and is thus equivalent to a pinned-pinned column of length $L/2$. The critical load can be obtained by using this observation as

$$P_{cr} = \frac{\pi^2 EI}{\left(\frac{L}{2}\right)^2} \tag{2.2i}$$

It can be seen that the critical load for a fixed-fixed column is that of an equivalent pinned-pinned column of length $L/2$, from Eq. (2.1m). It is called the effective length, KL, for the particular column, thus the effective length of a fixed-fixed column is half its actual length.

2.2.2.2 Anti-Symmetric Mode

The fixed-fixed column will bend as shown in Figure 2.5a in an anti-symmetric displaced shape. The fixed end bending moment, M_F, and the end shears are also shown in the Figure 2.5a by taking the equilibrium of the entire column as a free body diagram in the deflected position. The column will have zero displacements and rotations at the fixed supports.

From the moment equilibrium of the free body diagram in Figure 2.5b

$$P_y + \frac{2M_F}{L}x - (-EIy'') - M_F = 0$$

or

$$y'' + k^2 y = \frac{M_F}{EI} - \frac{2M_F x}{EIL} \tag{2.3a}$$

where $k^2 = P/EI$. The total solution of the differential Eq. (2.3a) is given by

$$y = A\sin kx + B\cos kx + \frac{M_F}{P} - \frac{2M_F x}{PL} \tag{2.3b}$$

Boundary conditions at the bottom of the column in the Figure 2.5 are $y = 0$ and $y' = 0$ at $x = 0$.

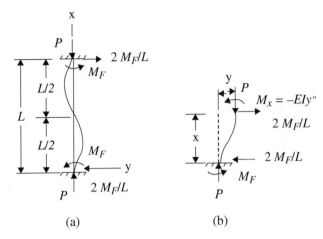

Figure 2.5 Fixed-fixed column under axial load and anti-symmetric mode:(a) Fixed-fixed column; (b) Free body diagram.

Substituting these boundary conditions in the Eq. (2.3b), we get

$$B = -\frac{M_F}{P} \text{ and } A = \frac{2M_F}{kPL}$$

or

$$y = \frac{2M_F}{kPL} \sin kx - \frac{M_F}{P} \cos kx + \frac{M_F}{P} - \frac{2M_F x}{PL} \tag{2.3c}$$

At the other end of the column, the boundary condition is $y = 0$, at $x = L$. Substituting it into Eq. (2.3c) we have

$$\frac{M_F}{P} \left(\frac{2 \sin kL}{kL} - \cos kL - 1 \right) = 0 \tag{2.3d}$$

Since $\frac{M_F}{P}$ is not zero,

$$\frac{2 \sin kL}{kL} - \cos kL - 1 = 0 \tag{2.3e}$$

Use the trigonometric expressions: $\sin kL = 2 \sin \frac{kL}{2} \cos \frac{kL}{2}$ and $\cos kL = 2\cos^2 kL - 1$ in Eq. (2.3e) to obtain

$$\tan \frac{kL}{2} = \frac{kL}{2} \tag{2.3f}$$

The smallest root of the transcendental Eq. (2.3f) is $kL/2 = 4.4934$, or $k^2 L^2 = 80.76$, and

$$P_{cr} = \frac{80.76 EI}{L^2} \tag{2.3g}$$

The critical load for the anti-symmetric buckling given by Eq. (2.3g) is much larger than that given by Eq. (2.2f) for symmetrical buckling. Hence, unless the column is restrained in the middle of its length, the column will buckle in the symmetrical mode at a much lesser critical load.

For practical purposes the critical load for the fixed-fixed column is given by $P_{cr} = 4\pi^2 EI/L^2$ in Eq. (2.2f) and the corresponding deflected shape is given by Eq. (2.2g).

2.2.3 Cantilever Column

A cantilever column will bend as shown in Figure 2.6a when it is displaced laterally by a small amount. The lateral deflection at the free end is δ, and a moment, $P\delta$, is developed at the fixed end at the base. The slope, y', is increasing with the increase of coordinate x, therefore, the rate of change of slope, y'', is positive. The internal moment M_x at a distance x, in Figure 2.6b is negative, so the expression relating the moment M_x and the curvature is $M_x = -EI\,y''$. The moment equilibrium in Figure 2.6b is given by

$$P\ y - (-EI\,y'') - P\delta = 0 \tag{2.4a}$$

or

$$y'' + k^2 y = k^2 \delta \tag{2.4b}$$

where $k^2 = P/EI$. The general solution of Eq. (2.4b) is

$$y = A\sin kx + B\cos kx + \delta \tag{2.4c}$$

Applying the boundary conditions at the base of the column, at $x = 0$: $y(0) = y'(0) = 0$, we get $A = 0$, and $B = -\delta$. Therefore,

$$y = \delta(1 - \cos kx) \tag{2.4d}$$

Now, applying the boundary condition at the upper end of the member, at $x = L$: $y(L) = \delta$, we have

$$\cos kL = 0 \tag{2.4e}$$

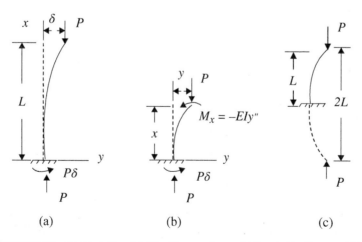

(a)   (b)   (c)

Figure 2.6 Cantilever column under axial load: (a) Cantilever column; (b) Free body diagram; (c) Equivalent pinned-pinned column.

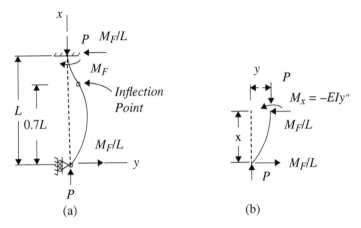

Figure 2.7 Fixed-pinned column under axial load: (a) Fixed-pinned column; (b) Free body diagram.

or

$$kL = \frac{n\pi}{2}, n = 1, 3, 5, ----- \tag{2.4f}$$

and

$$y = \delta \left(1 - \cos \frac{n\pi x}{2L}\right) \tag{2.4g}$$

The critical load is obtained when $n = 1$, or from Eq. (2.4f), $kL = \frac{\pi}{2}$, and

$$P_{cr} = \frac{\pi^2 EI}{4L^2} \tag{2.4h}$$

The corresponding mode shape is given by

$$y = \delta \left(1 - \cos \frac{\pi x}{2L}\right) \tag{2.4i}$$

The displacement curve of the cantilever column can be thought of as half the displacement curve of a pinned-pinned column as shown in Figure 2.6c. The effective length of a cantilever column is thus $KL = 2L$, and the critical load is expressed as

$$P_{cr} = \frac{\pi^2 EI}{(2L)^2} \tag{2.4j}$$

2.2.4 Fixed-Pinned Column

A column that is pinned at the base and fixed at the top is shown in Figure 2.7 when it is displaced laterally by a small amount. There will be moment M_F at the fixed support and shear forces M_F/L at each support that can be found by considering the equilibrium of the entire column in Figure 2.7a. From the moment equilibrium of the free body diagram in Figure 2.7b, we have

$$EIy'' + Py = \frac{M_F x}{L} \tag{2.5a}$$

or

$$y'' + k^2 y = \frac{M_F x}{EIL} \tag{2.5b}$$

where $k^2 = P/EI$. The complete solution of Eq. (2.5b) is

$$y = A \sin kx + B \cos kx + \frac{M_F x}{PL} \tag{2.5c}$$

Boundary conditions are

$$y = 0 \text{ at } x = 0$$
$$y = 0 \text{ at } x = L$$
$$y' = 0 \text{ at } x = L$$

Applying the first boundary condition gives $B = 0$, and using the second boundary condition leads to

$$A = -\frac{M_F}{P \sin kL}$$

Hence, from Eq. (2.5c) the deflection is given by

$$y = \frac{M_F}{P} \left(\frac{x}{L} - \frac{\sin kx}{\sin kL} \right) \tag{2.5d}$$

or

$$y' = \frac{M_F}{P} \left(\frac{1}{L} - \frac{k \cos kx}{\sin kL} \right) \tag{2.5e}$$

Substituting the boundary condition $y' = 0$ at $x = L$, we have

$$0 = \frac{M_F}{P} \left(\frac{1}{L} - \frac{k \cos kL}{\sin kL} \right) \tag{2.5f}$$

Since M_F/P is not zero, Eq. (2.5f) leads to

$$\tan kL = kL \tag{2.5g}$$

The lowest root of this transcendental equation is $kL = 4.4934$, therefore

$$P_{cr} = \frac{20.19 EI}{L^2} \tag{2.5h}$$

Substituting $kL = 4.4934$ into Eq. (2.5d), the deflected shape of the column at the critical load is

$$y = \frac{M_F}{P}\left[\frac{x}{L} + 1.0245 \sin\left(4.4934\frac{x}{L}\right)\right] \tag{2.5i}$$

$$y' = \frac{M_F}{P}\left[\frac{1}{L} + 1.0245\left(\frac{4.4934}{L}\right)\cos\left(\frac{4.4934x}{L}\right)\right] \tag{2.5j}$$

$$y'' = \frac{M_F}{P}\left[-1.0245\left(\frac{4.4934}{L}\right)^2 \sin\left(\frac{4.4934x}{L}\right)\right] \tag{2.5k}$$

From Eq. (2.5k), bending moment M_x at a section is zero, if

$$\sin\left(\frac{4.4934x}{L}\right) = 0$$

or

$$\frac{4.4934x}{L} = n\pi, \quad n = 0, 1, 2, 3 - - - - - -$$

The smallest nonzero value is $\frac{4.4934x}{L} = \pi$, or $x = 0.699\,L \approx 0.7\,L$. The point of inflection is at a distance of $0.7L$ from the origin on the bent shape of the column in Figure 2.7a. The column of length $0.7L$ between the inflection points bends in the shape of a pinned-pinned column, and is thus equivalent to a pinned-pinned column of length $0.7L$. The effective length of a fixed-pinned column is thus $KL = 0.7L$. and the critical load can be obtained from

$$P_{cr} = \frac{\pi^2 EI}{(0.7L)^2} \tag{2.5l}$$

2.3 Effective Length of a Column

It is seen in Eqs. (2.1m), (2.2i), (2.4j), and (2.5l) that the critical load for an elastic straight perfect column loaded with an axial force depends on the boundary conditions and can be written as

$$P_{cr} = \frac{\pi^2 EI}{(KL)^2} \tag{2.6}$$

where KL is called the effective length of the column. Theoretical and design values for the effective lengths are given in Table 2.1 for various boundary conditions. Theoretical values have been derived for the ideal boundary conditions. Recommended design values are given by the American Institute of Steel Construction (AISC) [5] in their *Specification for Structural Steel Buildings* for use when these boundary conditions are not fully realized in practice. The design values are higher than the theoretical values because joint fixity is generally not completely realized.

Table 2.1 Effective length factor K for columns.

Buckled shape of the column is shown by solid line.	L	L $L/2$	L $2L$	L $0.7L$
Theoretical K value	1.0	0.5	2.0	0.7
Recommended AISC design value	1.0	0.65	2.1	0.8

2.4 Special Cases

Four standard cases of column buckling have been discussed above. Critical loads were obtained as eigenvalues and the eigenvectors gave the buckled shapes of those columns. Usually the lowest eigenvalue is of interest because it gives the lowest force at which the column will buckle and hence gives the critical load. The corresponding eigenvector is the deflected shape of the column when it bends. Now we will examine some more cases where the axial force and the column cross-section are not constant along the length of the column.

2.4.1 Pinned-Pinned Column with Intermediate Compressive Force

The case where axial forces are applied at the ends and at an intermediate cross-section of a column that is hinged at both ends is discussed here. If the axial forces slightly exceed their critical value, the column will buckle as shown in Figure 2.8a. I_1 and I_2 are the moments of inertia of the cross-sections for the portions BC and AB portions of the column. Let δ be the

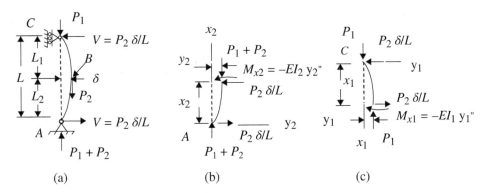

(a)   (b)   (c)

Figure 2.8 Pinned-pinned column under axial forces at the ends and at the intermediate cross-section: (a) Pinned-pinned column; (b) Free body diagram within length L_2; (c) Free body diagram within length L_1.

deflection at the cross-section at point B where the intermediate force P_2 is applied. Taking the entire column in its displaced shape as a free body diagram and taking the moment equilibrium of the column, the lateral reactions at the supports are obtained from $P_2 \delta - VL = 0$, or $V = \dfrac{P_2 \delta}{L}$ as shown in Figure 2.8a.

In Figure 2.8c, $y_1' = \dfrac{dy_1}{dx_1}$, is a positive quantity for the coordinate axes chosen. The slope, y_1', is decreasing with the increase of coordinate x_1, therefore, the rate of change of slope, y_1'', is negative. The internal moment M_{x1} at a distance x_1, in Figure 2.8c is positive, so the expression relating the moment M_{x1} and the curvature is given by $M_{x1} = -EI\,y_1''$. From the moment equilibrium of the free body diagram in Figure 2.8c of length B to C, we have

$$EI_1 y_1'' + P_1 y_1 = -\frac{P_2 \delta}{L} x_1$$

or

$$y_1'' + k_1^2 y_1 = -k_4^2 \frac{\delta}{L} x_1 \tag{2.7a}$$

where $k_1^2 = P_1/EI_1$, and $k_4^2 = P_2/EI_1$.

For the length A to B, the moment equilibrium of the free body diagram in Figure 2.8b gives

$$EI_2 y_2'' + (P_1 + P_2) y_2 = \frac{P_2 \delta}{L} x_2$$

or

$$y_2'' + k_3^2 y_2 = k_2^2 \frac{\delta}{L} x_2 \tag{2.7b}$$

where $k_2^2 = P_2/EI_2$ and $k_3^2 = (P_1 + P_2)/EI_2$. Eqs. (2.7a) and (2.7b) have the following general solutions

$$y_1 = A_1 \sin k_1 x_1 + B_1 \cos k_1 x_1 - \frac{k_4^2}{k_1^2} \frac{\delta}{L} x_1 \quad 0 < x_1 < L_1 \tag{2.7c}$$

$$y_2 = A_2 \sin k_3 x_2 + B_2 \cos k_3 x_2 + \frac{k_2^2}{k_3^2} \frac{\delta}{L} x_2 \quad 0 < x_2 < L_2 \tag{2.7d}$$

The constants of integration are obtained from the following end conditions of the two portions of the buckled bar: BC and AB.

Portion BC

$$y_1 = 0 \text{ at } x_1 = 0, \text{ and } y_1 = \delta \text{ at } x_1 = L_1$$

From these conditions we find

$$A_1 = \frac{\delta(k_1^2 L + k_4{}^2 L_1)}{k_1^2 L \sin k_1 L_1}, \quad B_1 = 0$$

Portion AB

$$y_2 = 0 \text{ at } x_2 = 0, \text{ and } y_2 = \delta \text{ at } x_2 = L_2$$

From these conditions we find

$$A_2 = \frac{\delta(k_3^2 L - k_2^2 L_2)}{k_3^2 L \sin k_3 L_2}, \quad B_2 = 0$$

The continuity condition at B is $\left(\frac{dy_1}{dx_1}\right)_{x_1=L_1} = -\left(\frac{dy_2}{dx_2}\right)_{x_2=L_2}$. Differentiate Eqs. (2.7c) and (2.7d) to get

$$\frac{dy_1}{dx_1} = y_1' = \frac{\delta(k_1^2 L + k_4^2 L_1)}{k_1^2 L \sin k_1 L_1} k_1 \cos k_1 x_1 - \frac{k_4^2}{k_1^2} \frac{\delta}{L}$$

$$\frac{dy_2}{dx_2} = y_2' = \frac{\delta(k_3^2 L - k_2^2 L_2)}{k_3^2 L \sin k_3 L_2} k_3 \cos k_3 x_2 + \frac{k_2^2}{k_3^2} \frac{\delta}{L}$$

Substituting these values in the continuity condition at B, we get

$$\frac{k_4^2}{k_1^2} - \frac{k_1^2 L + k_4^2 L_1}{k_1 \tan k_1 L_1} = \frac{k_2^2}{k_3^2} + \frac{k_3^2 L - k_2^2 L_2}{k_3 \tan k_3 L_2} \tag{2.7e}$$

If $P_2 = 0$, $I_1 = I_2 = I$, and $L_1 = L_2 = L/2$, then

$$k_1^2 = P_1/EI_1 = P_1/EI, \quad k_2^2 = P_2/EI_2 = 0,$$
$$k_3^2 = (P_1 + P_2)/EI_2 = P_1/EI, \quad k_4^2 = P_2/EI_1 = 0$$

and

$$k_1^2 = k_3^2 = k^2 = P_1/EI.$$

Substituting the above values of k_1, k_2, k_3, and k_4 in Eq. (2.7e) we obtain

$$-\frac{k}{\tan \frac{kL}{2}} = \frac{k}{\tan \frac{kL}{2}}$$

or

$$\tan \frac{kL}{2} = \infty, \quad \frac{kL}{2} = \frac{n\pi}{2}, \quad n = 1, 3, 5, ---$$

The lowest value of the compressive force is given by

$$kL = \pi, \quad \text{or} \quad P_{cr} = \frac{\pi^2 EI}{L^2}$$

Thus, we get critical compressive force that is the same as for the pinned-pinned column with a single axial force in Figure 2.1. For the two-force system, assume

$$\frac{P_1 + P_2}{P_1} = m, \quad \frac{I_2}{I_1} = n, \quad \text{and} \quad \frac{L_2}{L_1} = p \tag{2.7f}$$

The critical or the smallest values of the axial compressive forces $P_1 + P_2$ that satisfy Eq. (2.7e) were obtained using trial and error by Gere and Timoshenko [4]. This critical force is represented by the formula

$$(P_1 + P_2)_{cr} = \frac{\pi^2 EI_2}{(KL)^2} \qquad (2.7g)$$

in which KL is the effective length of the bar. For the case where $L_1 = L_2$, or $p = 1$, the values of the effective length factor K are given from the above reference [4] in Table 2.2.

2.4.2 Cantilever Column with Intermediate Compressive Force

Consider a cantilever column subjected to axial compressive forces at the ends and at an intermediate cross-section. Let a force P_1 act at the intermediate cross-section and a force P_2 act at the free end of the column. The moments of inertia of the column vary along the length of the column. I_1 is the moment of inertia in the lower portion AB and I_2 is the moment of inertia in the upper portion BC of the column as shown in Figure 2.9. The deflected shape of the entire column is shown in Figure 2.9a. Consider the equilibrium of the entire column in its displaced shape to find the external reactions at the bottom of the column. The deflection at the cross-section of the application of load P_1 is δ_1 and the deflection at the free end of the column in its displaced position is δ_2. The reactions at the fixed end at the base of the column are found to be $(P_1 + P_2)$ in the vertical direction and a moment of $(P_1\delta_1 + P_2\delta_2)$ by considering the equilibrium of the displaced shape of the entire column.

Considering the moment equilibrium of the portion AB, $0 \le x \le L_1$ in Figure 2.9b

$$(P_1 + P_2)y_1 - (-EI_1 y_1'') - M_A = 0, \quad \text{where } M_A = P_1\delta_1 + P_2\delta_2$$

$$y_1'' + k_1^2 y_1 = \frac{M_A}{EI_1} \qquad (2.8a)$$

where

$$\frac{P_1 + P_2}{EI_1} = k_1^2$$

Table 2.2 Effective length factor K for column in Figure 2.8, with $L_1 = L_2 = L/2$.

m/n	1.00	1.25	1.50	1.75	2.00
1.00	1.00	0.95	0.91	0.89	0.87
1.25	1.06	1.005	0.97	0.94	0.915
1.50	1.12	1.06	1.02	0.99	0.96
1.75	1.18	1.11	1.07	1.04	1.005
2.00	1.24	1.16	1.12	1.08	1.05

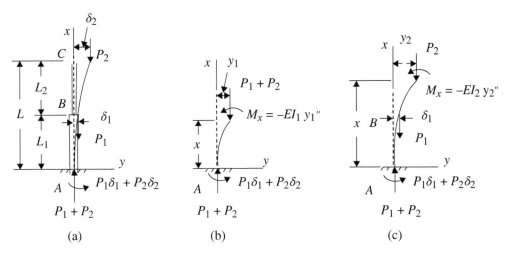

Figure 2.9 Cantilever column under axial forces at the free end and at the intermediate cross-section: (a) Cantilever column; (b) Free body diagram within portion AB; (c) Free body diagram within portion BC.

The general solution of Eq. (2.8a) is

$$y_1 = A_1 \sin k_1 x + B_1 \cos k_1 x + \frac{M_A}{EI_1 k_1^2} \tag{2.8b}$$

$$y_1' = k_1 A_1 \cos k_1 x - k_1 B_1 \sin k_1 x \tag{2.8c}$$

Boundary conditions in the portion AB are: $y_1 = 0$ at $x = 0$, and $y_1' = 0$ at $x = 0$. Substituting these conditions in Eq. (2.8b) we get

$$A_1 = 0, \quad B_1 = -\frac{M_A}{EI_1 k_1^2}$$

Therefore,

$$y_1 = \frac{M_A}{EI_1 k_1^2}(1 - \cos k_1 x) \tag{2.8d}$$

$$y_1' = \frac{M_A}{EI_1 k_1} \sin k_1 x \tag{2.8e}$$

Considering the moment equilibrium of the portion BC, $L_1 \le x \le L$ in Figure 2.9(c)

$$P_2 y_2 + EI_2 y_2'' + P_1 \delta_1 - (P_1 \delta_1 + P_2 \delta_2) = 0 \tag{2.8f}$$

or

$$y_2'' + k_2^2 y_2 = k_2^2 \delta_2 \tag{2.8g}$$

where

Where : $\dfrac{P_2}{EI_2} = k_2^2$

and the general solution of Eq. (2.8g) is

$$y_2 = A_2 \sin k_2 x + B_2 \cos k_2 x + \delta_2 \tag{2.8h}$$

Boundary conditions in the portion BC are: $y_2 = \delta_1$, at $x = L_1$, and $y_2 = \delta_2$ at $x = L$. Substituting these conditions in Eq. (2.8h) we get

$$A_2 = \frac{\delta_1 - \delta_2}{\sin k_2 L_1 - \tan k_2 L \cos k_2 L_1}$$

$$B_2 = -\frac{\delta_1 - \delta_2}{\sin k_2 L_1 - \tan k_2 L \cos k_2 L_1} \tan k_2 L$$

$$y_2 = \frac{\delta_1 - \delta_2}{\sin k_2 L_1 - \tan k_2 L \cos k_2 L_1}[\sin k_2 x - \tan k_2 L \cos k_2 x] + \delta_2 \tag{2.8i}$$

$$y_2 = \frac{\delta_1 - \delta_2}{\sin k_2 L_1 - \tan k_2 L \cos k_2 L_1}[k_2 \cos k_2 x + k_2 \tan k_2 L \sin k_2 x] \tag{2.8j}$$

Continuity conditions between the portions AB and BC are:

$$y_1(L_1) = y_2(L_1)$$

or

$$\frac{M_A}{EI_1 k_1^2}(1 - \cos k_1 L_1) = \frac{\delta_1 - \delta_2}{\sin k_2 L_1 - \tan k_2 L \cos k_2 L_1}(\sin k_2 L_1 - \tan k_2 L \cos k_2 L_1) + \delta_2$$

$$\text{or} \quad \frac{M_A}{EI_1 k_1^2}(1 - \cos k_1 L_1) = \delta_1 \tag{2.8k}$$

and $y_1'(L_1) = y_2'(L_1)$, from Eqs. (2.8e) and (2.8j)

$$\frac{M_A}{P_1 + P_2} k_1 \sin k_1 L_1 = \frac{\delta_1 - \delta_2}{\sin k_2 L_1 - \tan k_2 L \cos k_2 L_1}(k_2 \cos k_2 L_1 + k_2 \tan k_2 L \sin k_2 L_1) \tag{2.8l}$$

Substituting $M_A = P_1 \delta_1 + P_2 \delta_2$, and reorganizing the terms in Eqs. (2.8k) and (2.8l), we have

$$\left[\frac{P_1}{P_1 + P_2}(1 - \cos k_1 L_1) - 1\right]\delta_1 + \frac{P_2}{P_1 + P_2}(1 - \cos k_1 L_1)\delta_2 = 0$$

$$\left[\frac{P_1}{P_1 + P_2}k_1 \sin k_1 L_1 - \frac{k_2 \cos k_2 L_1 + k_2 \tan k_2 L \sin k_2 L_1}{\sin k_2 L_1 - \tan k_2 L \cos k_2 L_1}\right]\delta_1 +$$

$$\left[\frac{P_2}{P_1 + P_2}k_1 \sin k_1 L_1 + \frac{k_2 \cos k_2 L_1 + k_2 \tan k_2 L \sin k_2 L_1}{\sin k_2 L_1 - \tan k_2 L \cos k_2 L_1}\right]\delta_2 = 0$$

or

$$\begin{bmatrix} -\dfrac{P_2}{P_1 + P_2} - \dfrac{P_1}{P_1 + P_2}\cos k_1 L_1 & \dfrac{P_2}{P_1 + P_2}(1 - \cos k_1 L_1) \\[2ex] \dfrac{P_1}{P_1 + P_2}k_1 \sin k_1 L_1 - & \dfrac{P_2}{P_1 + P_2}k_1 \sin k_1 L_1 + \\[1ex] \dfrac{k_2 \cos k_2 L_1 + k_2 \tan k_2 L \sin k_2 L_1}{\sin k_2 L_1 - \tan k_2 L \cos k_2 L_1} & \dfrac{k_2 \cos k_2 L_1 + k_2 \tan k_2 L \sin k_2 L_1}{\sin k_2 L_1 - \tan k_2 L \cos k_2 L_1} \end{bmatrix}\begin{Bmatrix} \delta_1 \\ \delta_2 \end{Bmatrix} = \begin{Bmatrix} 0 \\ 0 \end{Bmatrix}$$

$$\tag{2.8m}$$

2.4.2.1 Case 1

Suppose $P_1 = P_2 = P$, and $L_1 = L_2 = \frac{L}{2}$, Eq. (2.8m) will become

$$
\begin{bmatrix}
-1 - \cos k_1 \frac{L}{2} & 1 - \cos k_1 \frac{L}{2} \\
\frac{1}{2} k_1 \sin k_1 \frac{L}{2} \sin k_2 \frac{L}{2} + k_2 \cos k_2 \frac{L}{2} & k_1 \sin k_1 \frac{L}{2} \sin k_2 \frac{L}{2} - k_2 \cos k_2 \frac{L}{2}
\end{bmatrix}
\begin{Bmatrix} \delta_1 \\ \delta_2 \end{Bmatrix}
=
\begin{Bmatrix} 0 \\ 0 \end{Bmatrix}
\tag{2.9a}
$$

If $I_1 = I_2 = I$, $k_1 = \sqrt{2} k_2$, substituting in Eq. (2.9a) we have

$$
\begin{bmatrix}
-1 - \cos \frac{k_2 L}{\sqrt{2}} & 1 - \cos \frac{k_2 L}{\sqrt{2}} \\
\frac{1}{\sqrt{2}} \sin \frac{k_2 L}{\sqrt{2}} \sin \frac{k_2 L}{2} + \cos \frac{k_2 L}{2} & \frac{1}{\sqrt{2}} \sin \frac{k_2 L}{\sqrt{2}} \sin \frac{k_2 L}{2} - \cos \frac{k_2 L}{2}
\end{bmatrix}
\begin{Bmatrix} \delta_1 \\ \delta_2 \end{Bmatrix}
=
\begin{Bmatrix} 0 \\ 0 \end{Bmatrix}
\tag{2.9b}
$$

For a nontrivial solution by Cramer's rule

$$
\begin{vmatrix}
-1 - \cos \frac{k_2 L}{\sqrt{2}} & 1 - \cos \frac{k_2 L}{\sqrt{2}} \\
\frac{1}{\sqrt{2}} \sin \frac{k_2 L}{\sqrt{2}} \sin \frac{k_2 L}{2} + \cos \frac{k_2 L}{2} & \frac{1}{\sqrt{2}} \sin \frac{k_2 L}{\sqrt{2}} \sin \frac{k_2 L}{2} - \cos \frac{k_2 L}{2}
\end{vmatrix} = 0
\tag{2.9c}
$$

or

$$
\cos \frac{k_2 L}{\sqrt{2}} \cos \frac{k_2 L}{2} - \frac{1}{\sqrt{2}} \sin \frac{k_2 L}{\sqrt{2}} \sin \frac{k_2 L}{2} = 0
\tag{2.9d}
$$

By trial and error the lowest value of $k_2 L$ satisfying Eq. (2.9d) is $k_2 L = 1.4378$.

$$
k_2^2 = \frac{P}{EI} = \left(\frac{1.4378}{L} \right)^2
$$

$$
P_{cr} = \frac{\pi^2 EI}{4.775 L^2}
\tag{2.9e}
$$

2.4.2.2 Case 2

Suppose $P_1 = 0$, and $P_2 = P$, then $k_1^2 = \frac{P}{EI_1}$, and $k_2^2 = \frac{P}{EI_2}$

From Eq. (2.8m) we obtain

$$
\begin{bmatrix}
-1 & (1 - \cos k_1 L_1) \\
-\dfrac{k_2 \cos k_2 L_1 + k_2 \tan k_2 L \sin k_2 L_1}{\sin k_2 L_1 - \tan k_2 L \cos k_2 L_1} & k_1 \sin k_1 L_1 + \dfrac{k_2 \cos k_2 L_1 + k_2 \tan k_2 L \sin k_2 L_1}{\sin k_2 L_1 - \tan k_2 L \cos k_2 L_1}
\end{bmatrix}
$$

$$
\times \begin{Bmatrix} \delta_1 \\ \delta_2 \end{Bmatrix} = \begin{Bmatrix} 0 \\ 0 \end{Bmatrix}
$$

For a nontrivial solution by Cramer's rule, the determinant of the coefficient matrix is equal to zero.

$$\begin{vmatrix} -1 & (1 - \cos k_1 L_1) \\ -\dfrac{k_2 L_1 + k_2 \tan k_2 L \sin k_2 L_1}{\sin k_2 L_1 - \tan k_2 L \cos k_2 L_1} & \begin{matrix} k_1 \sin k_1 L_1 + \\ \dfrac{k_2 \cos k_2 L_1 + k_2 \tan k_2 L \sin k_2 L_1}{\sin k_2 L_1 - \tan k_2 L \cos k_2 L_1} \end{matrix} \end{vmatrix} = 0$$

or $-k_1 \sin k_1 L_1 \sin k_2 L_1 + k_1 \sin k_1 L_1 \dfrac{\sin k_2 L}{\cos k_2 L} \cos k_2 L_1 - k_2 \cos k_1 L_1 \cos k_2 L_1 -$

$$k_2 \cos k_1 L_1 \dfrac{\sin k_2 L}{\cos k_2 L} \sin k_2 L_1 = 0$$

Thus,

$$k_1 \sin k_1 L_1 \sin(k_2 L - k_2 L_1) - k_2 \cos k_1 L_1 \cos(k_2 L - k_2 L_1) = 0$$

Therefore, $k_1 \sin k_1 L_1 \sin k_2 L_2 - k_2 \cos k_1 L_1 \cos k_2 L_2 = 0$

Finally, we get:

$$\tan k_1 L_1 \tan k_2 L_2 = \frac{k_2}{k_1} \tag{2.10a}$$

If $L_1 = L_2 = \dfrac{L}{2}$, $I_1 = I_2 = I$, $k_1 = k_2 = k = \sqrt{\dfrac{P}{EI}}$

Eq. (2.10a) becomes: $\tan \dfrac{kL}{2} \tan \dfrac{kL}{2} = 1$

or

$$\tan \frac{kL}{2} = 1 \tag{2.10b}$$

Lowest value of $\dfrac{kL}{2}$ satisfying Eq. (2.10b) is

$$\frac{kL}{2} = \frac{\pi}{4}$$

or $P_{cr} = \dfrac{\pi^2 EI}{4L^2}$, same as Eq. (2.4h)

2.5 Higher-Order Governing Differential Equation

The second-order differential equations were derived for particular loads and boundary conditions for various columns in the previous sections. In this section, a general differential equation that is applicable to all boundary conditions in a column is derived by taking an element of length dx in Figure 2.10a.

From the equilibrium of horizontal forces in Figure 2.10b we have

$$V + \left(\frac{dV}{dx}\right) dx - V = 0$$

or

$$\frac{dV}{dx} = 0 \tag{2.11a}$$

Similarly, from the equilibrium of the moments in Figure 2.10b we get

$$P\,dy + V\,dx + M - \left(M + \frac{dM}{dx} dx\right) = 0$$

$$V = \frac{dM}{dx} - P\frac{dy}{dx} \tag{2.11b}$$

$$\frac{dV}{dx} = \frac{d^2M}{dx^2} - P\frac{d^2y}{dx^2} \tag{2.11c}$$

Substituting Eq. (2.11a) into Eq. (2.11c) we have

$$\frac{d^2M}{dx^2} - P\frac{d^2y}{dx^2} = 0$$

$$M = -EI\frac{d^2y}{dx^2} \tag{2.11d}$$

For a uniform homogeneous column EI is constant, therefore:

$$EI\frac{d^4y}{dx^4} + P\frac{d^2y}{dx^2} = 0$$

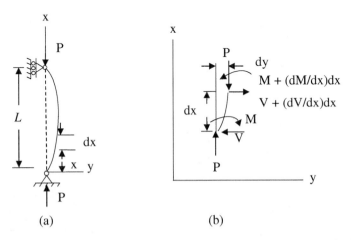

(a)   (b)

Figure 2.10 Column under axial force shown with internal forces: (a) Column under axial force; (b) Free body diagram of an element of length dx.

or

$$y^{IV} + k^2 y'' = 0 \qquad (2.11e)$$

where

$$k^2 = \frac{P}{EI}$$

The general solution of the fourth- order differential Eq. (2.11e) is

$$y = A \sin kx + B \cos kx + Cx + D \qquad (2.11f)$$

Boundary conditions are used to find the constants and the critical load from Eq. (2.11f) for various columns.

2.5.1 Boundary Conditions for Different Supports

2.5.1.1 Pinned Support
At this support the lateral deflection and bending moment are zero, and hence we have the conditions:

$$y = 0 \text{ and } y'' = 0 \qquad (2.12a)$$

2.5.1.2 Fixed Support
In this case, the lateral deflection and the slope of the deflected shape of the column are zero, and hence the boundary conditions are:

$$y = 0 \text{ and } y' = 0 \qquad (2.12b)$$

2.5.1.3 Free End
At the free end the bending moment and shear force must be zero. Referring to Eqs. (2.11b) and (2.11d), we get the following conditions:

$$y'' = 0 \text{ and } y''' + k^2 y' = 0 \qquad (2.12c)$$

2.5.1.4 Guided Support
The slope of the deflected shape and shear force are zero at the guided support. Therefore, using Eq. (2.12c), we have

$$y' = 0 \text{ and } y''' = 0 \qquad (2.12d)$$

2.5.2 Pinned-Pinned Column

Consider a pinned-pinned column in Figure 2.11. The general solution is given by

$$y = A \sin kx + B \cos kx + Cx + D \qquad (2.13a)$$

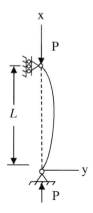

Figure 2.11 Pinned-pinned column under axial load.

where

$$k^2 = \frac{P}{EI}$$

Taking the derivatives of Eq. (2.13a), we have

$$y' = kA \cos kx - kB \sin kx + C \tag{2.13b}$$

$$y'' = -k^2 A \sin kx - k^2 B \cos kx \tag{2.13c}$$

Substituting $y = 0$ and $y'' = 0$ from Eq. (2.12a) into Eqs. (2.13a) and (2.13c) at $x = 0$, we get

$$B = 0, \text{ and } D = 0$$

Equations (2.13a) and (2.13c) reduce to

$$y = A \sin kx + Cx \tag{2.13d}$$

and

$$y'' = -k^2 A \sin kx \tag{2.13e}$$

Now, substitute $y = 0$ and $y'' = 0$ in Eqs. (2.13d) and (2.13e) at $x = L$

$$A \sin kL + CL = 0$$

and

$$-k^2 \sin kL = 0$$

or

$$\begin{bmatrix} \sin kL & L \\ -k^2 \sin kL & 0 \end{bmatrix} \begin{Bmatrix} A \\ C \end{Bmatrix} = \begin{Bmatrix} 0 \\ 0 \end{Bmatrix} \tag{2.13f}$$

For a nontrivial solution, the determinant of the coefficient matrix in Eq. (2.13f) must vanish, therefore,

$$\begin{vmatrix} \sin kL & L \\ -k^2 \sin kL & 0 \end{vmatrix} = 0$$
(2.13g)

$$k^2 L \sin kL = 0, \quad \text{or} \quad \sin kL = 0$$

Therefore,

$$kL = n\,\pi, \quad n = 1, 2, 3, 4, \text{-------}$$

The critical load is the least valued obtained by setting $n = 1$,

$$P_{cr} = \frac{\pi^2 EI}{L^2}$$

which is same as Eq. (2.1m), and the corresponding deflection from Eq. (2.13d), since $C = 0$, is

$$y = A \sin \frac{\pi x}{L}$$

That is the same as in Eq. (2.1n).

2.5.3 Cantilever Column

Consider a cantilever column in Figure 2.12. Substituting the boundary conditions, $y = 0$ and $y' = 0$, at the fixed end $x = 0$, in Eqs. (2.13a) and (2.13b), we have

$$B + D = 0$$
(2.14a)

$$A k + C = 0$$
(2.14b)

The third derivative of Eq. (2.13a) is

$$y''' = -k^3 A \cos kx + B k^3 \sin kx$$
(2.14c)

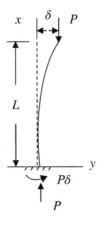

Figure 2.12 Cantilever column under axial load.

At the free end, $x = L$, the boundary conditions are: $y'' = 0$, and $y''' + k^2 y' = 0$.
By substituting the free end boundary conditions in Eqs. (2.13c) and (2.14c) we have:

$$A \, \sin \, kL + B \cos kL = 0 \qquad (2.14d)$$

and $C = 0$

Therefore, from Eq. (2.14b) $A = 0$

Now, Eqs. (2.14a) and (2.14d) can be written as

$$\begin{bmatrix} 1 & 1 \\ \cos kL & 0 \end{bmatrix} \begin{Bmatrix} B \\ D \end{Bmatrix} = \begin{Bmatrix} 0 \\ 0 \end{Bmatrix} \qquad (2.14e)$$

For nontrivial solution, the determinant of the coefficient matrix in Eq. (2.14e) is zero, or

$$\begin{vmatrix} 1 & 1 \\ \cos kL & 0 \end{vmatrix} = 0$$

and $\cos kL = 0$, this is satisfied if

$$kL = \frac{n\pi}{2}, \quad n = 1, 3, 5, - - -$$

When $n = 1$, we get the critical load

$$P_{cr} = \frac{\pi^2 EI}{4L^2}$$

It is the same as Eq. (2.4h). At the free end, $y = \delta$, at $x = L$, so from Eqs. (2.13a) and (2.14a) we have

$$y = \delta \left(1 - \cos \frac{\pi x}{2L} \right)$$

which is the deflection of the deformed shape at the critical load, and is the same as in Eq. (2.4i).

2.5.4 Pinned-Guided Column

Consider a pinned-guided column in Figure 2.13. The general solution and its derivatives are given by Eqs. (2.13a), (2.13b), (2.13c), and (2.14c) as before.

$$y = A \sin kx + B \cos kx + Cx + D \qquad (2.13a)$$

$$y' = kA \cos kx - kB \sin kx + C \qquad (2.13b)$$

$$y'' = -k^2 A \sin kx - k^2 B \cos kx \qquad (2.13c)$$

$$y''' = -k^3 A \cos kx + B k^3 \sin kx \qquad (2.14c)$$

where

$$k^2 = \frac{P}{EI}$$

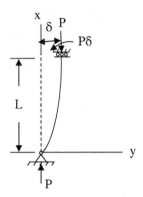

Figure 2.13 Pinned-guided column under axial load.

Boundary conditions: $y = 0$ and $y'' = 0$, at the pinned end at $x = 0$, substituting these values in Eqs. (2.13a) and (2.13c) we have

$$B = 0, \text{ and } D = 0$$

Therefore, Eq. (2.13a) becomes

$$y = A \ \sin \ kx + Cx \tag{2.15a}$$

At the guided end, substituting $y''' = 0$, at $x = L$, into Eq. (2.14c) we have

$$A \, k^3 \cos kL = 0$$

or $\cos kL = 0$, this is satisfied if

$$kL = \frac{n\pi}{2}, \quad n = 1, 3, 5, - - -$$

The smallest value of $kL = 1$ gives the critical load as

$$P_{cr} = \frac{\pi^2 EI}{4L^2} \tag{2.15b}$$

Another boundary condition at the guided end, $x = L$, is $y' = 0$. Therefore, Eq. (2.13b) gives

$$A \, k \cos kL + C = 0, \ \text{ or } C = 0$$

The deflected shape at buckling is

$$y = \delta \sin \frac{\pi x}{2L} \tag{2.15c}$$

2.6 Continuous Columns

Let us consider a continuous column supported on three supports and compressed by forces P applied at the ends as shown in Figure 2.14. It is intended to find the critical load for this column

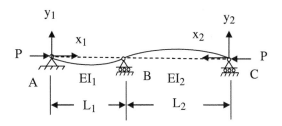

Figure 2.14 Continuous column under axial load.

at which buckling takes place. Divide the column into two spans, AB and BC with the axes for each span shown in Figure 2.14. For the span AB, the fourth-order differential equation is

$$y_1^{IV} + k_1 y_1'' = 0 \tag{2.16a}$$

where $k_1^2 = \frac{P}{EI_1}$, and the general solution of Eq. (2.16a) is

$$y_1 = A_1 \sin k_1 x_1 + B_1 \cos k_1 x_1 + C_1 x_1 + D_1 \tag{2.16b}$$

The first and second derivatives of Eq. (2.16b) are

$$y_1' = A_1 k_1 \cos k_1 x_1 - B_1 k_1 \sin k_1 x_1 + C_1 \tag{2.16c}$$

$$y_1'' = -A_1 k_1^2 \sin k_1 x_1 - B_1 k_1^2 \cos k_1 x_1 \tag{2.16d}$$

Using the boundary conditions, $y_1(0) = y_1''(0) = y_1(L_1) = 0$, for the span AB, we get

$$B_1 = D_1 = 0, \text{ and}$$

$$A_1 \sin k_1 L_1 + C_1 L_1 = 0 \tag{2.16e}$$

For the span BC the fourth-order differential equation is

$$y_2^{IV} + k_2 y_2'' = 0 \tag{2.16f}$$

where $k_2^2 = \frac{P}{EI_2}$, and the general solution of Eq. (2.16f) is

$$y_2 = A_2 \sin k_2 x_2 + B_2 \cos k_2 x_2 + C_2 x_2 + D_2 \tag{2.16g}$$

The first and second derivatives of Eq. (2.16g) are

$$y_2' = A_2 k_2 \cos k_2 x_2 - B_2 k_2 \sin k_2 x_2 + C_2 \tag{2.16h}$$

$$y_2'' = -A_2 k_2^2 \sin k_2 x_2 - B_2 k_2^2 \cos k_2 x_2 \tag{2.16i}$$

Using the boundary conditions, $y_2(0) = y_2''(0) = y_2(L_2) = 0$, for the span CB, we get

$$B_2 = D_2 = 0,$$

and

$$A_2 \sin k_2 L_2 + C_2 L_2 = 0 \tag{2.16j}$$

From the continuity condition at B, $y_1'(L_1) = -y_2'(L_2)$, we have

$$A_1 k_1 \cos k_1 L_1 + A_2 k_2 \cos k_2 L_2 + C_1 + C_2 = 0 \tag{2.16k}$$

From the equilibrium of moments at B, we have

$$M_{BA} + M_{BC} = 0$$

or

$$E I_1 y_1'' - E I_2 y_2'' = 0$$

or

$$-A_1 I_1 k_1^2 \sin k_1 L_1 + A_2 I_2 k_2^2 \sin k_2 L_2 = 0 \tag{2.16l}$$

Equations (2.16e), (2.16j), (2.16k), and (2.16l) can be expressed in the matrix form as

$$\begin{bmatrix} \sin k_1 L_1 & L_1 & 0 & 0 \\ 0 & 0 & \sin k_2 L_2 & L_2 \\ k_1 \cos k_1 L_1 & 1 & k_2 \cos k_2 L_2 & 1 \\ -I_1 k_1^2 \sin k_1 L_1 & 0 & I_2 k_2^2 \sin k_2 L_2 & 0 \end{bmatrix} \begin{Bmatrix} A_1 \\ C_1 \\ A_2 \\ C_2 \end{Bmatrix} = \begin{Bmatrix} 0 \\ 0 \\ 0 \\ 0 \end{Bmatrix} \tag{2.16m}$$

By Cramer's rule for a nontrivial solution of Eq. (2.16m), the determinant of the coefficient matrix is zero. Therefore,

$$\begin{vmatrix} \sin k_1 L_1 & L_1 & 0 & 0 \\ 0 & 0 & \sin k_2 L_2 & L_2 \\ k_1 \cos k_1 L_1 & 1 & k_2 \cos k_2 L_2 & 1 \\ -I_1 k_1^2 \sin k_1 L_1 & 0 & I_2 k_2^2 \sin k_2 L_2 & 0 \end{vmatrix} = 0 \tag{2.16n}$$

or

$$\sin k_1 L_1 \sin k_2 L_2 (I_1 k_1^2 L_1 + I_2 k_2^2 L_2)$$
$$- L_1 L_2 k_1 k_2 (I_1 k_1 \sin k_1 L_1 \cos k_2 L_2 + I_2 k_2 \sin k_2 L_2 \cos k_1 L_1) = 0 \tag{2.16o}$$

If $I_1 = I_2 = I$, then $k_1 = k_2 = k$, and Eq. (2.16o) becomes

$$(L_1 + L_2) \sin k L_1 \sin k L_2 - k L_1 L_2 \sin k(L_1 + L_2) = 0 \tag{2.16p}$$

Let us take $L_1 = 1.5\,L$ and $L_2 = L$, then from Eq. (2.16p) we have

$$5 \sin \frac{3}{2} kL \, \sin kL - 3kL \, \sin \frac{5}{2} kL = 0 \tag{2.16q}$$

The lowest value of kL that satisfies Eq. (2.16q) is obtained by trial and error as 2.427.

$$kL = 2.427$$

or

$$P_{cr} = \frac{5.89 EI}{L^2} \tag{2.16r}$$

If $L_2 = 2L_1$, then from Eq. (2.16p)

$$3 \sin kL_1 \sin 2kL_1 - 2kL_1 \sin 3kL_1 = 0 \tag{2.16s}$$

or

$$kL_1 = 1.93, \quad P_{cr} = \frac{3.725EI}{L_1^2}$$

In terms of L_2, Eq. (2.16p) can be written as

$$3 \sin \frac{kL_2}{2} \sin kL_2 - kL_2 \sin \frac{3kL_2}{2} = 0 \tag{2.16t}$$

or

$$kL_2 = 3.86, \quad P_{cr} = \frac{14.9EI}{L_2^2}$$

which is the same as in Timoshenko and Gere [4]. If the two bars were separate spans with hinged ends, the critical loads for spans AB and BC will be given by $\frac{\pi^2 EI}{L_1^2}$ and $\frac{\pi^2 EI}{L_2^2}$ respectively. The stability of the shorter span is reduced due to the action of the longer span, while the stability of the longer span is increased due to the action of the shorter span.

2.7 Columns on Elastic Supports

2.7.1 Column Pinned at One End and Elastic Support at the Other End

The end conditions considered for columns were ideal boundary conditions, but in reality the columns are usually connected to other members, which are elastic in nature. This type of support is referred to as elastically restrained support. The support fixity depends on the relative rigidities of the members at the support. The members connected at the elastically restrained supports are considered to be joined by a rigid joint, the angle between the members at the support remains the same before and after deformation.

Consider a column AB of length L_1 and stiffness rigidity EI_1 that is hinged at the base A, and elastically restrained at the end B by a beam BC. The beam is fixed at the end C, it has a length of L_2 and stiffness rigidity of EI_2 as shown in Figure 2.15. It is assumed that the joint B is a rigid joint. It is assumed that there is no bending in the beam prior to column buckling, and the beam bends after the column buckles. The beam resists bending due to its stiffness and in turn exerts a restraining moment M_B on the column as shown in Figure 2.15. It is assumed that deformations are small and hence the shear forces and the axial forces acting on the beam are small, and can be neglected. The axial force acting on the column can still be assumed to be equal to P. The governing equation is given by

$$y^{IV} + k^2 y'' = 0 \tag{2.17a}$$

where

$$k^2 = \frac{P}{EI_1}$$

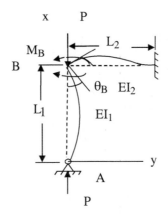

Figure 2.15 Elastically restrained column: hinged base.

The general solution of the fourth-order differential Eq. (2.17a) is

$$y = A \sin kx + B \cos kx + Cx + D \tag{2.17b}$$

The first and second derivatives of the general solution are

$$y' = Ak \cos kx - Bk \sin kx + C \tag{2.17c}$$

$$y'' = -Ak^2 \sin kx - Bk^2 \cos kx \tag{2.17d}$$

Four boundary conditions are needed to solve for the constants in Eq. (2.17b), two of them are at the base A of the column and another two are obtained at the upper end B. At the base, at $x = 0$, the deflection, y, and the moment, M, are both equal to zero, so

$$y(0) = y''(0) = 0$$

and substituting these boundary conditions in Eqs. (2.17b) and (2.17d) we get

$$B = D = 0 \tag{2.17e}$$

At the upper end of the column, at $x = L_1$, $y(L_1) = 0$, giving

$$A = -\frac{CL_1}{\sin kL_1}$$

hence,

$$y = -\frac{CL_1}{\sin kL_1} \sin kx + Cx \tag{2.17f}$$

$$\frac{dy}{dx} = CL_1 \left(\frac{1}{L_1} - \frac{k \cos kx}{\sin kL_1} \right)$$

$$\frac{d^2y}{dx^2} = \frac{CL_1 k^2 \sin kx}{\sin kL_1} \tag{2.17g}$$

$$\left.\frac{dy}{dx}\right|_{x=L_1} = CkL_1 \left(\frac{1}{kL_1} - \frac{1}{\tan kL_1}\right) \tag{2.17h}$$

The slope deflection equation at the end B of the beam BC is

$$M_B = \frac{EI_2}{L_2}(4\theta_B + 2\theta_C) + M_{FB} \tag{2.17i}$$

where θ_C, the rotation at end C of the beam, is zero. Also the fixed end moment at the end C, $M_{FC} = 0$, because there is no transverse force on the beam BC.

$$\theta_B = \frac{M_B L_2}{4EI_2} \tag{2.17j}$$

At the rigid joint B, $-\left.\frac{dy}{dx}\right|_{x=L_1} = \theta_B$, the negative sign is used because the slope at the upper end of the column given by Eq. (2.17h) is negative, whereas θ_B is positive. Using Eqs. (2.17h) and (2.17j) we have

$$\frac{M_B L_2}{4EI_2} = -CkL_1 \left(\frac{1}{kL_1} - \frac{1}{\tan kL_1}\right) \tag{2.17k}$$

At the upper end of the column AB, $M_B = EI_1 y''$, the moment M_B and the y'' are both of the same sign, i.e. negative. Also, at the upper end of the column, $y'' = CL_1 k^2$, from Eq. (2.17g). Substituting in Eq. (2.17k), we have

$$\frac{EI_1 CL_1 k^2 L_2}{4EI_2} = -CkL_1 \left(\frac{1}{kL1} - \frac{1}{\tan kL_1}\right) \tag{2.17l}$$

$$\frac{k^2 I_1 L_2}{4I_2} = -\frac{1}{L_1} + \frac{k}{\tan kL_1}$$

$$\text{or } \frac{I_1 L_2}{I_2 L_1}(kL_1)^2 \tan kL_1 - 4kL_1 + 4\tan kL_1 = 0 \tag{2.17m}$$

Equation (2.17m) is a transcendental equation that can be solved for the smallest root of kL_1 for a given $\left(\frac{I_1}{L_1} / \frac{I_2}{L_2}\right)$, and the corresponding critical load is found as

$$P_{cr} = \frac{(kL_1)^2 EI_1}{L_1^2} \tag{2.17n}$$

If $L_1 = L_2 = L$, and $I_1 = I_2 = I$, Eq. (2.17m) reduces to

$$\tan kL = \frac{4kL}{(kL)^2 + 4} \tag{2.17o}$$

The smallest root satisfying Eq. (2.17o) is $kL = 3.83$, and the corresponding critical load is

$$P_{cr} = \frac{14.7EI}{L^2} \tag{2.17p}$$

The critical load in Eq. (2.17p) falls between P_{cr} for a hinged-hinged column, $\frac{\pi^2 EI}{L^2}$, and that of a hinged-fixed column, $\frac{20.19EI}{L^2}$. The upper elastically restrained end falls between a hinge and a fixed end.

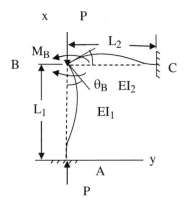

Figure 2.16 Elastically restrained column: fixed base.

2.7.2 Column Fixed at One End and Elastic Support at the Other End

Consider a column AB similar to the one shown in Figure 2.15, except that it is fixed at the bottom. At the upper end B, it is elastically restrained by a beam as shown in Figure 2.16. The governing equation and its derivatives are given by Eqs. (2.17a), (2.17b), (2.17c), and (2.17d) as before

$$y^{IV} + k^2 y'' = 0 \tag{2.17a}$$

where

$$k^2 = \frac{P}{EI_1}$$

The general solution of the fourth-order differential Eq. (2.17a) is

$$y = A \sin kx + B \cos kx + Cx + D \tag{2.17b}$$

The first and second derivatives of the general solution are

$$y' = Ak \cos kx - Bk \sin kx + C \tag{2.17c}$$

$$y'' = -Ak^2 \sin kx - Bk^2 \cos kx \tag{2.17d}$$

The boundary conditions at the base of the column are

$$y(0) = y'(0) = 0$$

and substituting these boundary conditions in Eqs. (2.17b) and (2.17c) we get

$$B + D = 0 \tag{2.18a}$$

$$Ak + C = 0 \tag{2.18b}$$

At the upper end of the column, $y(L_1) = 0$, giving

$$A \sin kL_1 + B \cos kL_1 + CL_1 + D = 0 \tag{2.18c}$$

From Eq. (2.17c) at the upper end of the column

$$\frac{dy}{dx}\bigg|_{x=L_1} = Ak\cos kL_1 - Bk\sin kL_1 + C \tag{2.18d}$$

The slope θ_B at the end of the beam from the slope deflection equation is given by Eq. (2.17j) as before by $\theta_B = \dfrac{M_B L_2}{4EI_2}$. At the rigid joint B, $-\dfrac{dy}{dx}\bigg|_{x=L_1} = \theta_B$, the negative sign is used because the slope at the upper end of the column given by Eq. (2.18d) is negative, whereas θ_B is positive. From Eqs. (2.17j) and (2.18d) we have

$$\frac{M_B L_2}{4EI_2} = -Ak\cos kL_1 + Bk\sin kL_1 - C \tag{2.18e}$$

At the upper end of the column AB, $M_B = EI_1 y''$, because the moment M_B and the y'' are both of the same sign, i.e. negative. Thus, the moment M_B at the upper end of the column from Eq. (2.17d) is

$$M_B = EI_1(-Ak^2\sin kL_1 - Bk^2\cos kL_1) \tag{2.18f}$$

Substituting M_B from Eq. (2.18f) into Eq. (2.18e) we have

$$A\left(\frac{k^2 I_1 L_2}{4I_2}\sin kL_1 - k\cos kL_1\right) + B\left(\frac{k^2 I_1 L_2}{4I_2}\cos kL_1 + k\sin kL_1\right) - C = 0 \tag{2.18g}$$

Equations (2.18a), (2.18b), (2.18c), and (2.18g) can written in matrix form as

$$\begin{bmatrix} 0 & 1 & 0 & 1 \\ k & 0 & 1 & 0 \\ \sin kL_1 & \cos kL_1 & L_1 & 1 \\ \frac{k^2 I_1 L_2}{4I_2}\sin kL_1 - -k\cos kL_1 & \frac{k^2 I_1 L_2}{4I_2}\cos kL_1 + k\sin kL_1 & -1 & 0 \end{bmatrix} \begin{Bmatrix} A \\ B \\ C \\ D \end{Bmatrix} = \begin{Bmatrix} 0 \\ 0 \\ 0 \\ 0 \end{Bmatrix} \tag{2.18h}$$

For a nontrivial solution of Eq. (2.18h), the determinant of the coefficient matrix should be zero.
Therefore,

$$\begin{vmatrix} 0 & 1 & 0 & 1 \\ k & 0 & 1 & 0 \\ \sin kL_1 & \cos kL_1 & L_1 & 1 \\ \frac{k^2 I_1 L_2}{4I_2}\sin kL_1 - k\cos kL_1 & \frac{k^2 I_1 L_2}{4I_2}\cos kL_1 + k\sin kL_1 & -1 & 0 \end{vmatrix} = 0 \tag{2.18i}$$

Solving this determinant gives

$$8I_2 - k\sin kL_1(4I_2 L_1 - I_1 L_2) - k^2 I_1 L_1 L_2\cos kL_1 - 8I_2\cos kL_1 = 0$$

which can be written as

$$\left(\frac{I_1}{L_1}\bigg/\frac{I_2}{L_2}\right)(kL_1\sin kL_1 - k^2 L_1^2\cos kL_1) + 8(1 - \cos kL_1) - 4kL_1\sin kL_1 = 0 \tag{2.18j}$$

If $L_1 = L_2 = L$, and $I_1 = I_2 = I$, Eq. (2.18j) reduces to

$$3kL \sin kL + \cos kL(8 + k^2L^2) - 8 = 0 \qquad (2.18k)$$

The smallest root satisfying the transcendental Eq. (2.18k) is $kL = 5.33$, and the corresponding critical load is given by

$$P_{cr} = \frac{28.41EI}{L^2} \qquad (2.18l)$$

The critical load in Eq. (2.18l) falls between P_{cr} for a fixed-hinged column, $\frac{20.19EI}{L^2}$, and that of a fixed-fixed column, $\frac{4\pi^2 EI}{L^2}$, because the upper elastically restrained end falls between a hinge and a fixed end. The values of the critical load, P_{cr}, are shown in Table 2.3 for different ratios of $\frac{I_1}{L_1} / \frac{I_2}{L_2}$ for two columns shown in Figures 2.15 and 2.16.

Table 2.3 shows that as the ratio $\frac{I_1}{L_1} / \frac{I_2}{L_2}$ decreases, the column at the upper end approaches a fixed end, whereas with the increase in this ratio, the column at the upper end approaches a hinged end as expected. The critical loads also tend to approach the values for the corresponding columns with ideal boundary conditions.

2.8 Eccentrically Loaded Columns

In practice, most of the time columns are not perfectly loaded by an axial load, there is always some accidental eccentricity present. Critical load given by Euler's theory does not match that obtained from experiments due to the presence of this eccentricity. The values given by Euler's load are on the non-conservative side, they are higher than the experimental force. Therefore, in

Table 2.3 Critical loads for elastically restrained columns.

One end elastically restrained, the other hinged (Figure 2.15)			One end elastically restrained, the other fixed (Figure 2.16)		
$\frac{I_1}{L_1} / \frac{I_2}{L_2}$	kL_1	$P_{cr}\left(\frac{L_1^2}{EI_1}\right)$	$\frac{I_1}{L_1} / \frac{I_2}{L_2}$	kL_1	$P_{cr}\left(\frac{L_1^2}{EI_1}\right)$
0.5	4.065	16.52	0.5	5.660	32.03
1.0	3.830	14.67	1.0	5330	28.41
1.5	3.685	13.58	1.5	5.140	26.42
2.0	3.590	12.89	2.0	5.019	25.19
10.0	3.259	10.62	10.0	4.624	21.38

design of even axially loaded columns AISC [5] and American Concrete Institute [6] specifications take into account the lowering of critical load due to this accidental eccentricity. A simply supported column acted on by an axial force with an eccentricity of e is shown in Figure 2.17a.

It is assumed that the member is initially straight, elastic, and small deflection theory is applied. Considering the moment equilibrium of the free body diagram in Figure 2.17b, we have

$$P(e + y) - (-EIy'') = 0$$
$$EIy'' + Py = -Pe$$

or

$$y'' + k^2y = -k^2e \qquad (2.19a)$$

where $k^2 = P/EI$. The general solution of the Eq. (2.19a) is

$$y = A \sin kx + B \cos kx - e \qquad (2.19b)$$

Boundary conditions are:

$$\text{At } x = 0, y = 0$$
$$x = L, y = 0$$

Substituting these boundary conditions in Eq. (2.19b) we get

$$B = e, \text{ and } A = \frac{e(1 - \cos kL)}{\sin kL}$$

Therefore, the general solution of differential Eq. (2.19a) is

$$y = e\left(\frac{1 - \cos kL}{\sin kL} \sin kx + \cos kx - 1\right) \qquad (2.19c)$$

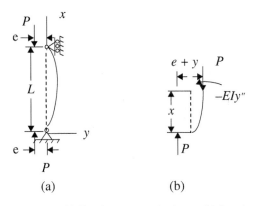

Figure 2.17 Eccentrically loaded column: (a) Simply supported column; (b) Free body diagram.

Maximum deflection occurs at the mid-height, $x = L/2$, given by

$$y_{max} = e\left(\frac{1 - \cos kL}{\sin kL}\sin\frac{kL}{2} + \cos\frac{kL}{2} - 1\right) \tag{2.19d}$$

Using trigonometric relations

$$\cos kL = 1 - 2\sin^2\frac{kL}{2}, \text{ and } \sin kL = 2\sin\frac{kL}{2}\cos\frac{kL}{2}$$

and substituting in Eq. (2.19d), we obtain

$$y_{max} = e\left(\sec\frac{kL}{2} - 1\right) \tag{2.19e}$$

$$\frac{kL}{2} = \sqrt{\frac{P}{EI}}\frac{L}{2} = \frac{\pi}{2}\sqrt{\frac{PL^2}{\pi^2 EI}} = \frac{\pi}{2}\sqrt{\frac{P}{P_e}}$$

where Euler's buckling load for a simply supported column is $P_e = \frac{\pi^2 EI}{L^2}$.
Therefore,

$$y_{max} = e\left(\sec\frac{\pi}{2}\sqrt{\frac{P}{P_e}} - 1\right) \tag{2.19f}$$

From Eq. (2.19c),

$$y' = e\left(\frac{1 - \cos kL}{\sin kL}k\cos kx - k\sin kx\right) \tag{2.19g}$$

$$y'' = e\left[\frac{1 - \cos kL}{\sin kL}(-k^2\sin kx) - k^2\cos kx\right] \tag{2.19h}$$

Maximum bending moment also occurs at the mid-span, $M_{max} = -EIy''|_{x=\frac{L}{2}}$

$$M_{max} = EIek^2\sec\frac{kL}{2} \tag{2.19i}$$

or

$$M_{max} = Pe\sec\frac{\pi}{2}\sqrt{\frac{P}{P_e}} \tag{2.19j}$$

The end moment P_e, is amplified by an amplification factor $A_f = \sec\frac{\pi}{2}\sqrt{\frac{P}{P_e}}$ to give the maximum moment in the column. As P approaches the Euler buckling load P_e, the maximum moment tends to become very large. A plot of load versus total maximum deflection (y_{total}) is shown in Figure 2.18, where

$$y_{total} = y_{max} + y = e\sec\frac{\pi}{2}\sqrt{\frac{P}{P_e}} \tag{2.19k}$$

As the load P acting eccentrically on the column approaches the Euler critical load P_e, the deflection increases very rapidly and the load-deflection curve tends to become almost horizontal. It indicates the maximum load an eccentrically loaded column can take is below the Euler critical load, the greater the eccentricity, the less force the column can take before failing. This observation is very important from a practical point of view because the failure load in practice is lower than the theoretical Euler load. In this case, the theory gives non-conservative values.

2.8.1 The Secant Formula

The maximum stress in elastic eccentrically loaded columns can be obtained by combining the axial stress and the bending stress as follows:

$$\sigma_{max} = \frac{P}{A} + \frac{M_{max}c}{I} \tag{2.20a}$$

where

> P = Axial force on the column
>
> A = Area of cross-section of the column
>
> I = Moment of inertia of the column cross-section about the bending axis
>
> c = Distance from the neutral axis to the outer fiber of the column

Substituting the maximum bending moment from Eq. (2.19j) in Eq. (2.20a) we have

$$\sigma_{max} = \frac{P}{A}\left(1 + \frac{ec}{r^2}\sec\frac{\pi}{2}\sqrt{\frac{P}{P_e}}\right) \tag{2.20b}$$

where $I = Ar^2$, r is the radius of gyration of the cross-section of the column about the bending axis. For a given value of σ_{max}, plots of Eq. (2.20b) can be plotted as L/r versus P/A for various values of eccentricity ratio ec/r^2 as shown in [3]. The curves show that the eccentricity values matter more for short columns than long columns. Therefore, it is important to know more accurately the eccentricity for the design of short columns.

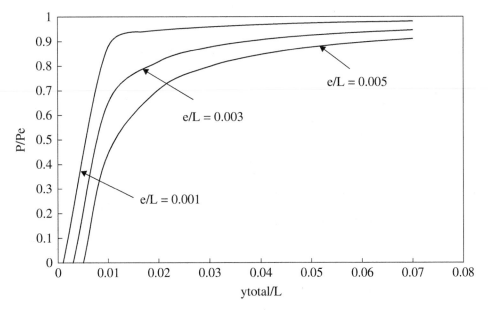

Figure 2.18 Load versus maximum total deflection for eccentrically loaded column.

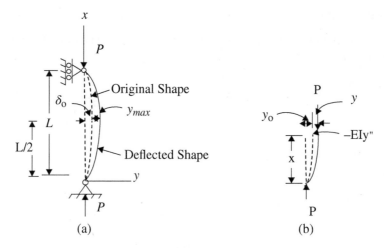

Figure 2.19 Geometrically imperfect column: (a) Initially bent column; (b) Free body diagram.

2.9 Geometrically Imperfect Columns

It is assumed that a column is perfectly straight and the loading axis passes through the center of gravity of every cross-section in the derivation of Euler's formula. In real-life columns, perfectly straight columns are hard to find, and certain amount of geometric imperfections exist in all columns. It is assumed here that the column is initially bent, the material remains elastic, and the deformations are small.

Let the initial bent shape be given by

$$y_0 = \sum_{n=1}^{\infty} a_n \sin \frac{n\pi x}{L} \tag{2.21a}$$

Considering the moment equilibrium of the free body diagram in Figure 2.19b, we have

$$P(y + y_0) - (-EIy'') = 0$$
$$EIy'' + Py = -Py_0$$
$$y'' + k^2 y = -k^2 y_0$$

or

$$y'' + k^2 y = -k^2 \sum_{n=1}^{\infty} a_n \sin \frac{n\pi x}{L} \tag{2.21b}$$

The general solution of the differential Eq. (2.21b) is

$$y = y_c + y_p$$

The complementary solution is

$$y_c = A \sin kx + B \cos kx$$

And the particular solution is given by

$$y_p = \sum_{n=1}^{\infty} b_n \sin \frac{n\pi x}{L}$$

or

$$y_p' = \sum_{n=1}^{\infty} b_n \frac{n\pi}{L} \cos \frac{n\pi x}{L}$$

and

$$y_p'' = \sum_{n=1}^{\infty} -b_n \left(\frac{n\pi}{L}\right)^2 \sin \frac{n\pi x}{L}$$

Substituting the particular solution and its second derivative in Eq. (2.21b) we have

$$\sum_{n=1}^{\infty} -\left(\frac{n\pi}{L}\right)^2 b_n \sin \frac{n\pi x}{L} + k^2 \sum_{n=1}^{\infty} b_n \sin \frac{n\pi x}{L} = -k^2 \sum_{n=1}^{\infty} a_n \sin \frac{n\pi x}{L}$$

or

$$\sum_{n=1}^{\infty} b_n \left(k^2 - \left(\frac{n\pi}{L}\right)^2\right) = -k^2 \sum_{n=1}^{\infty} a_n$$

or

$$b_n = -\frac{k^2 a_n}{k^2 - \frac{n^2 \pi^2}{L^2}} = \frac{a_n}{\frac{n^2 \pi^2}{k^2 L^2} - 1} = \frac{a_n}{n^2 \frac{P_e}{P} - 1}$$

where $P_e = \frac{\pi^2 EI}{L^2}$. Let $\alpha = \frac{P}{P_e}$, then

$b_n = \frac{a_n \alpha}{n^2 - \alpha}$, and the particular solution of the Eq. (2.21b) is

$$y_p = \sum_{n=1}^{\infty} \frac{a_n \alpha}{n^2 - \alpha} \sin \frac{n\pi x}{L}$$

Hence the general solution of Eq. (2.21b) is given by

$$y = A \sin kx + B \cos kx + \sum_{n=1}^{\infty} \frac{a_n \alpha}{n^2 - \alpha} \sin \frac{n\pi x}{L} \tag{2.21c}$$

Boundary conditions, at $x = 0$, $y = 0$; and at $x = L$, $y = 0$ are applied to get the values of the constants A and B. Substituting $x = 0$, $y = 0$, gives $B = 0$; and $x = L$, $y = 0$, we get $0 = A \sin kL + \sum_{n=1}^{\infty} \frac{a_n \alpha}{n^2 - \alpha} \sin n\pi$, or $A \sin kL = 0$, because $\sin n\pi = 0$. If $\sin kL = 0$, is assumed, it gives $P = P_e$, which is not the desired solution. Therefore, $A = 0$. So, the general solution given by Eq. (2.21c) becomes

$$y = \sum_{n=1}^{\infty} \frac{\alpha}{n^2 - \alpha} a_n \sin \frac{n\pi x}{L} \tag{2.21d}$$

If the initial column is bent, the problem is not an eigenvalue problem because for every load there is a definite deflection given by Eq. (2.21d). It gives the deflection, y, from the initial bent position, whereas the total defection from the vertical x axis is given by

$$y_{total} = y_o + y$$

Combining Eqs. (2.21a) and (2.21d), we get

$$y_{total} = \sum_{n=1}^{\infty} \frac{1}{1 - \frac{\alpha}{n^2}} a_n \sin \frac{n\pi x}{L} \tag{2.21e}$$

If

$n = 1$,

$$y_{total} = \frac{1}{1 - \alpha} \delta_0 \sin \frac{\pi x}{L} \tag{2.21f}$$

The mid-height total deflection, δ_{total}, as shown in Figure 2.19a is obtained from Eq. (2.21f) by substituting, $x = L/2$:

$$\delta_{total} = \delta_0 + y_{max} = \frac{\delta_0}{1 - \alpha} \tag{2.21g}$$

$$\delta_{total} = \frac{1}{1 - \frac{P}{P_e}} \delta_0$$

$$\delta_{total} = A_f \delta_0 \tag{2.21h}$$

where A_f is the amplification factor given by

$$A_f = \frac{1}{1 - \frac{P}{P_e}} \tag{2.21i}$$

if n is more than 1, then

$$y_{total} = \frac{a_1}{1 - \alpha} \sin \frac{\pi x}{L} + \frac{a_2}{1 - \frac{\alpha}{2^2}} \sin \frac{2\pi x}{L} + \frac{a_3}{1 - \frac{\alpha}{3^2}} \sin \frac{3\pi x}{L} + - - - - - - \tag{2.21j}$$

Moment in the column at any distance x is given by

$$M_x = -EIy''$$

From Eq. (2.21d)

$$y' = \sum_{n+1}^{\infty} \frac{\alpha}{n^2 - \alpha} \left(\frac{n\pi}{L} \right) a_n \sin \frac{n\pi x}{L}$$

or

$$y'' = \sum_{n=1}^{\infty} \frac{\alpha}{n^2 - \alpha} \left(\frac{n^2 \pi^2}{L^2} \right) \left(-a_n \sin \frac{n\pi x}{L} \right)$$

Therefore,

$$M_x = EI \sum_{n=1}^{\infty} \frac{\alpha}{n^2 - \alpha} \left(\frac{n^2 \pi^2}{L^2} \right) \left(a_n \sin \frac{n\pi x}{L} \right)$$

For
$n = 1,$

$$M_x = EI \frac{\alpha}{1 - \alpha} \left(\frac{\pi^2}{L^2} \right) \delta_0 \sin \frac{\pi x}{L} = P_e \frac{\frac{P}{P_e}}{1 - \frac{P}{P_e}} \delta_0 \sin \frac{\pi x}{L}$$

or

$$M_x = \frac{1}{1 - \frac{P}{P_e}} P \delta_0 \sin \frac{\pi x}{L}$$

The same result can be obtained alternatively as

$$M_x = P(y_0 + y) = P y_{total}$$

Substituting from Eq. (2.21f)

$$M_x = \frac{1}{1 - \frac{P}{P_e}} P \delta_0 \sin \frac{\pi x}{L} \tag{2.21k}$$

If M_I, is the first-order moment in the column caused by the load P based on the initial geometry of the geometrically imperfect member, then M_x in Eq. (2.21k) is the second-order moment, M_{II}, based on the deformed geometry of the column at any distance x. Thus,

$$M_{II} = A_f M_I \tag{2.21l}$$

The maximum second-order moment at the mid-height is given by substituting $x = L/2$, and is

$$M_{II\,max} = A_f P \delta_o \tag{2.21m}$$

The mid-height deflection δ_{total}/L versus load ratio P/P_e is plotted in Figure 2.20 for three values of imperfection ratios. Initially, the imperfect column starts to bend as soon as the load is applied. As the load approaches the Euler's buckling load, the deflection increases very rapidly. The larger the initial deflection, the larger is the deflection corresponding to any load. The failure load for any initially imperfect column is smaller than the Euler load. Columns with small initial imperfections fail at a slightly smaller than the Euler load, whereas if the initial imperfection is large, the failure load will be much smaller than the Euler load.

2.9.1 The Southwell Plot

From Eq. (2.21d)

$$y = \frac{\alpha}{1 - \alpha} a_1 \sin \frac{\pi x}{L} + \frac{\alpha}{2^2 - \alpha} a_2 \sin \frac{2\pi x}{L} + \frac{\alpha}{3^2 - \alpha} a_3 \sin \frac{3\pi x}{L} + - - - \tag{2.22a}$$

If in the imperfection, a_1 existed, then near the critical load P_e, the ratio $\alpha = P/P_e \to 1$. In the series given by Eq. (2.22a), the first term predominates, and we can assume the mid-length deflection, δ, as

$$\delta = \frac{\alpha a_1}{1 - \alpha} \tag{2.22b}$$

$$\delta = \frac{\frac{P}{P_e}}{1 - \frac{P}{P_e}} a_1 = \delta \frac{P}{P_e} + a_1 \frac{P}{P_e}$$

or

$$\frac{\delta}{P} = \frac{\delta}{P_e} + \frac{a_1}{P_e} \tag{2.22c}$$

δ and P can be measured from an experiment conducted on an initially imperfect column. δ/P versus δ is plotted as a straight line given by Eq. (2.22c). This is known as the Southwell plot [7] as shown in Figure 2.21. The intercepts in the plot give a_1/P_e and a_1, from which the critical load P_e can be found.

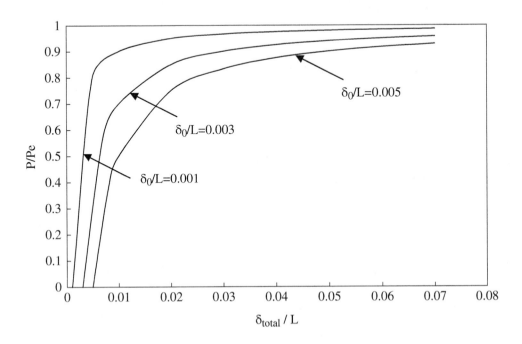

Figure 2.20 Load versus maximum total deflection for imperfect column.

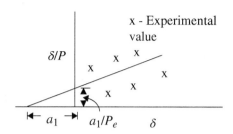

Figure 2.21 The Southwell plot.

2.10 Large Deflection Theory of Columns

2.10.1 Pinned-Pinned Column

The curvature, $1/\rho$, is given by

$$\frac{1}{\rho} = \frac{\frac{d^2y}{dx^2}}{\left[1 + \left(\frac{dy}{dx}\right)^2\right]^{\frac{3}{2}}} \tag{2.23a}$$

So far, we have assumed that the deflections are small, so the curvature was approximated by $\frac{1}{\rho} = y''$ in Eq. (2.1b). Furthermore, the magnitude of the deflection at critical load remained indeterminate because of this assumption. In the large deflection analysis of columns, the exact expression for the curvature is used. This enables us to determine at various axial forces on the column the magnitude of the corresponding deflections. Consider the pinned-pinned column shown in Figure 2.22a subjected to an axial force. Other assumptions made to derive the Euler buckling load, e.g. the column is straight, the material is homogeneous and elastic, etc. are still valid.

Considering the moment equilibrium of the free body diagram in Figure 2.22b we have

$$\frac{EI}{\rho} + Py = 0 \tag{2.23b}$$

For a curve, the relationship between the subtended angle, $d\theta$, the arc length, ds, and the radius of curvature, ρ, is given by

$$\frac{ds}{\rho} = d\theta \tag{2.23c}$$

The equilibrium equation for the column free body diagram can be found from Eqs. (2.23b) and (2.23c) as

$$EI\frac{d\theta}{ds} + Py = 0 \tag{2.23d}$$

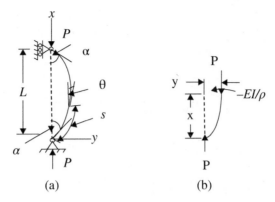

Figure 2.22 Large deflection under axial load of a pinned-pinned column: (a) Large deflection of pined-pinned column; (b) Free body diagram.

The solution of Eq. (2.23d) is given by Wang [8]. Differentiating Eq. (2.23d) with respect to s, and using the relation, $\frac{dy}{ds} = \sin\theta$, and substituting $P/EI = k^2$, we have

$$\frac{d^2\theta}{ds^2} + k^2 \sin\theta = 0 \tag{2.23e}$$

Multiply both sides by $d\theta$ and integrating gives

$$\int \frac{d^2\theta}{ds^2}\frac{d\theta}{ds}ds + \int k^2 \sin\theta\, d\theta = 0 \tag{2.23f}$$

Equation (2.23f) can be expressed in the form

$$\frac{1}{2}\int \frac{d}{ds}\left(\frac{d\theta}{ds}\right)^2 ds + k^2 \int \sin\theta\, d\theta = 0 \tag{2.23g}$$

Integrating, we obtain

$$\frac{1}{2}\left(\frac{d\theta}{ds}\right)^2 - k^2 \cos\theta = C \tag{2.23h}$$

where C is the constant of integration. At the lower end of the column in Figure 2.22a, $\theta = \alpha$, and $d\theta/ds = 0$, since the bending moment is zero. Applying these conditions to Eq. (2.23h) gives

$$C = -k^2 \cos\alpha$$

And, therefore,

$$\left(\frac{d\theta}{ds}\right)^2 = 2k^2(\cos\theta - \cos\alpha) \tag{2.23i}$$

or

$$\frac{d\theta}{ds} = \pm k\sqrt{2(\cos\theta - \cos\alpha)} \tag{2.23j}$$

$d\theta/ds$ is negative from Figure (2.22a), because θ decreases as s increases, therefore

$$\frac{d\theta}{ds} = -k\sqrt{2(\cos\theta - \cos\alpha)} \tag{2.23k}$$

$$ds = -\frac{d\theta}{k\sqrt{2(\cos\theta - \cos\alpha)}}$$

The total length of the column is obtained by integrating ds from 0 to L. So

$$L = \int_0^L ds = -\int_\alpha^{-\alpha} \frac{d\theta}{k\sqrt{2(\cos\theta - \cos\alpha)}}$$

Reversing the limits of integration we obtain

$$L = \int_{-\alpha}^{\alpha} \frac{d\theta}{k\sqrt{2(\cos\theta - \cos\alpha)}}$$

Using trigonometric relations: $\cos\theta = 1 - 2\sin^2\frac{\theta}{2}$, and $\cos\alpha = 1 - 2\sin^2\frac{\alpha}{2}$, we obtain

$$L = \frac{1}{2k}\int_{-\alpha}^{\alpha} \frac{d\theta}{\sqrt{\sin^2\frac{\alpha}{2} - \sin^2\frac{\theta}{2}}} \tag{2.23l}$$

Substitute, $p = \sin\frac{\alpha}{2}$, and introduce a new variable ϕ, such that

$$\sin\frac{\theta}{2} = p\sin\phi = \sin\frac{\alpha}{2}\sin\phi \tag{2.23m}$$

When $\theta = -\alpha$, $\sin\phi = -1$, then $\phi = -\frac{\pi}{2}$; and when $\theta = \alpha$, $\sin\phi = 1$, then $\phi = \frac{\pi}{2}$.
Taking differential of terms in Eq. (2.23m) gives

$$\frac{1}{2}\cos\frac{\theta}{2}d\theta = p\cos\phi\,d\phi$$

or

$$d\theta = \frac{2p\cos\phi\,d\phi}{\cos\frac{\theta}{2}} = \frac{2p\cos\phi\,d\phi}{\sqrt{1-\sin^2\frac{\theta}{2}}}$$

Therefore,

$$d\theta = \frac{2p\cos\phi\,d\phi}{\sqrt{1 - p^2\sin^2\phi}} \tag{2.23n}$$

Now,

$$\sqrt{\sin^2\frac{\alpha}{2} - \sin^2\frac{\theta}{2}} = \sqrt{\sin^2\frac{\alpha}{2} - \sin^2\frac{\alpha}{2}\sin^2\phi} = p\cos\phi \tag{2.23o}$$

Substituting Eqs. (2.23n) and (2.23o) into Eq. (2.23l) we obtain

$$L = \frac{1}{k}\int_{-\frac{\pi}{2}}^{\frac{\pi}{2}} \frac{d\phi}{\sqrt{1 - p^2\sin^2\phi}}$$

$$L = \frac{2}{k} \int_0^{\frac{\pi}{2}} \frac{d\phi}{\sqrt{1 - p^2 \sin^2 \phi}} \tag{2.23p}$$

or

$$L = \frac{2K(p)}{k} \tag{2.23q}$$

where

$$K(p) = \int_0^{\frac{\pi}{2}} \frac{d\phi}{\sqrt{1 - p^2 \sin^2 \phi}} \tag{2.23r}$$

The integral in Eq. (2.23r) is known as the complete elliptic integral of the first kind and is designated by $K(p)$. Consider now the case of small end rotation α, then $p = \sin(\alpha/2)$ is $\ll 1$. The integral in Eq. (2.23r) can be written by expanding the integrand by binomial theorem. Binomial theorem can be written as

$$(1 - x)^{-n} = 1 + nx + \frac{n(n+1)}{2!} x^2 + \frac{n(n+1)(n+2)}{3!} x^3 + - - - - (x^2 < 1)$$

or

$$(1 - p^2 \sin^2 \phi)^{-\frac{1}{2}} = 1 + \frac{1}{2} p^2 \sin^2 \phi + \frac{3}{8} p^4 \sin^4 \phi + \frac{15}{48} p^6 \sin^6 \phi + ---- (p^2 < 1)$$

Hence,

$$K(p) = \int_0^{\frac{\pi}{2}} \left(1 + \frac{1}{2} p^2 \sin^2 \phi + \frac{3}{8} p^4 \sin^4 \phi + \frac{15}{48} p^6 \sin^6 \phi + - - - - \right) d\phi$$

or

$$K(p) = \frac{\pi}{2} \left[1 + \left(\frac{1}{2}\right)^2 p^2 + \left(\frac{3}{8}\right)^2 p^4 + \left(\frac{15}{48}\right)^2 p^6 + - - - - - - - \right] \tag{2.23s}$$

The value of $K(p)$ can be obtained from integral tables or from Eq. (2.23s) for different values of p.

We can write Eq. (2.23q) as

$$L = \frac{2K(p)}{\sqrt{\frac{P}{EI}}}$$

or

$$\frac{P}{P_e} = \frac{4K^2(p)}{\pi^2} \tag{2.23t}$$

where $P_e = \frac{\pi^2 EI}{L^2}$, is the Euler buckling load for the column. When the deflection of the column is very small, α and p will also be very small and the higher terms containing p in Eqs. (2.23r) and (2.23s) can be neglected in comparison with unity. Then value of $K(p)$ tends to $\pi/2$, and $P = P_e$ from Eq. (2.23t). Large deflection nonlinear theory gives the same critical load as the small deflection linear theory, since both theories apply for small deflections.

Knowing $dy = \sin\theta\, ds$, and making use of Eq. (2.23k) we get

$$dy = -\frac{\sin\theta\, d\theta}{k\sqrt{2(\cos\theta - \cos\alpha)}}$$

Mid-height deflection is given by

$$\delta = \int_0^\delta dy = \int_\alpha^0 -\frac{\sin\theta\, d\theta}{k\sqrt{2(\cos\theta - \cos\alpha)}}$$

As θ varies from α to 0, y varies from 0 to δ. Using trigonometry and switching the limits of integration, we obtain

$$\delta = \frac{1}{2k}\int_0^\alpha \frac{\sin\theta\, d\theta}{\sqrt{\left(\sin^2\frac{\alpha}{2} - \sin^2\frac{\theta}{2}\right)}} \tag{2.23u}$$

Using Eq. (2.23m),

$$\sin\theta = 2\sin\frac{\theta}{2}\cos\frac{\theta}{2} = 2p\sin\phi\sqrt{1 - p^2\sin^2\phi} \tag{2.23v}$$

Equations (2.23n), (2.23o), (2.23u), and (2.23v) are used to give

$$\delta = \frac{2p}{k}\int_0^{\frac{\pi}{2}}\sin\phi\, d\phi = \frac{2p}{k} \tag{2.23w}$$

$$\delta = \frac{2p}{\sqrt{\frac{P}{EI}}} = \frac{2pL}{\pi\sqrt{\frac{P}{P_e}}}$$

or

$$\frac{\delta}{L} = \frac{2p}{\pi\sqrt{\frac{P}{P_e}}} \tag{2.23x}$$

Now assume α, and find $K(p)$ from Eq. (2.23s). Then P/P_e can be found from Eq. (2.23t) and the corresponding δ/L can be obtained from Eq. (2.23x) for various values of α. This procedure is illustrated in Table 2.4 and the results are plotted in Figure 2.23.

It can be seen from Table 2.4 and Figure 2.23 there is considerable increase in deflection when there is a slight increase in P above the Euler buckling load of P_e. At $\alpha = 30°$, $\delta/L = 0.162$, a considerable increase in the deflection in Table 2.4, but the axial force is only 3.5% above P_e. The curve in Figure 2.23 is tangent to the horizontal line at $P = P_e$, where the deflection is zero. Thus, the increase in load P, corresponding to a small increment of deflection, is a small quantity of second order. That is why the deflection was found to be indefinite in magnitude when the approximation for curvature was used as stated by Timoshenko and Gere [9]. Large deflection analysis gives the amplitude of deflection corresponding to a particular load, whereas the small deflection theory gives only the deflected shape. Post-buckling behavior of the column is stable because the column can support additional load beyond the Euler load P_e as the deflection increases. The graph in Figure 2.23 is valid only as long as the material remains linearly elastic.

Table 2.4 Load and mid-length deflection from the large deflection analysis.

α(deg)	$p = \sin \alpha/2$	$K(p)$	P/P_e	δ/L
0	0.00	1.571	1.000	0
30	0.259	1.598	1.035	0.162
60	0.500	1.686	1.151	0.297
90	0.707	1.852	1.390	0.382
120	0.866	2.114	1.811	0.410
150	0.966	2.410	2.354	0.401

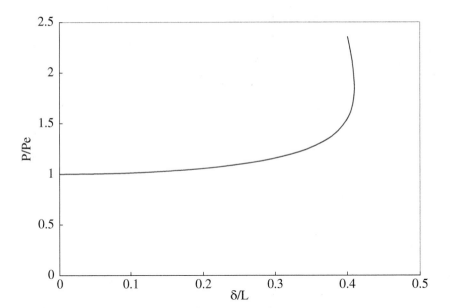

Figure 2.23 Load deflection graph for pinned-pinned column from large deflection analysis.

Consider now the case of small end rotation α, then $p = \sin(\alpha/2)$ is $\ll 1$. Since p is small, taking the first two terms only in Eq. (2.23s) we have

$$K(p) = \frac{\pi}{2}\left(1 + \frac{1}{4}p^2\right)$$

$$p = \sin\left(\frac{\alpha}{2}\right) \approx \frac{\alpha}{2},$$

because α is small, therefore

$$K(p) = \frac{\pi}{2}\left(1 + \frac{\alpha^2}{16}\right) \tag{2.23y}$$

Substituting $K(p)$ from Eq. (2.23y) into Eq. (2.23q) we get

$$\sqrt{\frac{P}{EI}}\,L = \pi\left(1 + \frac{\alpha^2}{16}\right)$$

Squaring both sides and neglecting small terms of higher order of α gives

$$\frac{P}{EI}L^2 = \pi^2\left(1 + \frac{\alpha^2}{8}\right)$$

Therefore, for a pinned-pinned column

$$\frac{P}{P_e} = 1 + \frac{\alpha^2}{8} \tag{2.23z}$$

2.10.2 Cantilever Column

Consider a column fixed at the base and free at the upper end. The coordinate axes are taken as shown in Figure 2.24a, and measure the distance s along the axis of the column from the origin O.

The exact expression for the curvature of the bar is $1/\rho = d\theta/ds$. Considering the equilibrium of the free body of the column in the deformed position in Figure 2.24b, we have the same expression as for a pinned-pinned column

$$EI\frac{d\theta}{ds} + Py = 0 \tag{2.23d}$$

Following the same steps as for the pinned-pinned column and solving for ds gives

$$ds = -\frac{d\theta}{k\sqrt{2(\cos\theta - \cos\alpha)}} \tag{2.24a}$$

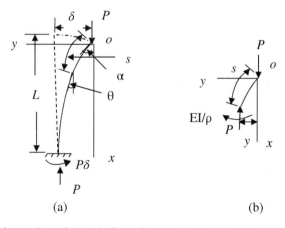

(a)   (b)

Figure 2.24 Large deflection under axial load of cantilever column: (a) Large deflection of cantilever column; (b) Free body diagram.

The total length of the column after reversing the limits of integration is

$$L = \int_0^\alpha \frac{d\theta}{k\sqrt{2(\cos\theta - \cos\alpha)}} \qquad (2.24b)$$

Substituting $p = \sin\frac{\alpha}{2}$ and introducing a new variable ϕ, such that

$$\sin\frac{\theta}{2} = p\sin\phi = \sin\frac{\alpha}{2}\sin\phi$$

It can be concluded from these relations that when θ varies from 0 to α, the quantity $\sin\phi$ varies from 0 to 1. Hence ϕ varies from 0 to $\frac{\pi}{2}$. Again following the same steps as before for the pinned-pinned column we obtain total length of the column as

$$L = \frac{1}{k}\int_0^{\frac{\pi}{2}} \frac{d\phi}{\sqrt{1 - p^2\sin^2\phi}} \qquad (2.24c)$$

or

$$L = \frac{1}{k}K(p) \qquad (2.24d)$$

and

$$L = \frac{K(p)}{\sqrt{\frac{P}{EI}}}$$

or

$$\frac{P}{P_e} = \frac{4K^2(p)}{\pi^2} \qquad (2.24e)$$

where $P_e = \frac{\pi^2 EI}{4L^2}$ is the Euler buckling load for the cantilever column. The complete integral of the first kind, $K(p)$ can be obtained from Eq. (2.23s). When the deflection of the column is very small, $K(p)$ tends to $\pi/2$, and

$$P = P_e = \frac{\pi^2 EI}{4L^2} \qquad (2.24f)$$

Maximum deflection at the top of this column is also given by Eq. (2.23u), that can be simplified and written as Eq. (2.23w). Therefore,

$$\delta = \frac{2p}{k} = \frac{2p}{\sqrt{\frac{P}{EI}}} = \frac{2p}{\sqrt{\frac{4PL^2}{\pi^2 EI}}} \cdot \frac{2L}{\pi} = \frac{4pL}{\pi\sqrt{\frac{P}{P_e}}}$$

or

$$\frac{\delta}{L} = \frac{4p}{\pi\sqrt{\frac{P}{P_e}}} \qquad (2.24g)$$

Assume α, and find $K(p)$ from Eq. (2.23s). Then P/P_e can be found from Eq. (2.24e) and the corresponding δ/L can be obtained from Eq. (2.24g) for various values of α. This procedure is illustrated in Table 2.5 and Figure 2.25.

Table 2.5 Load and deflection at the top of the cantilever column by the large deflection analysis.

α(deg)	$p = \sin(\alpha/2)$	$K(p)$	P/P_e	δ/L
0	0	1.571	1.000	0
30	0.259	1.598	1.035	0.324
60	0.500	1.686	1.151	0.593
90	0.707	1.852	1.390	0.764
120	0.866	2.114	1.811	0.819
150	0.966	2.410	2.354	0.801

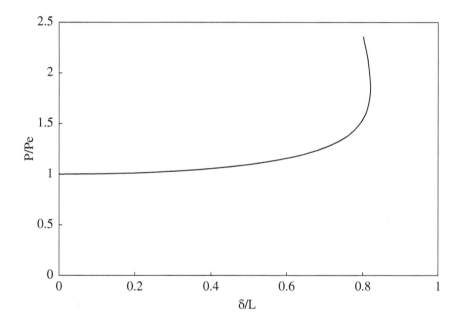

Figure 2.25 Load deflection graph for a cantilever column from large deflection analysis.

Substituting $K(p)$ from Eq. (2.23y) into Eq. (2.24d) we get

$$\sqrt{\frac{P}{EI}}\,L = \frac{\pi}{2}\left(1 + \frac{\alpha^2}{16}\right)$$

Squaring both sides gives

$$\frac{P}{EI}L^2 = \frac{\pi^2}{4}\left(1 + \frac{\alpha^2}{8}\right)$$

neglecting small terms of higher order of α

Therefore, for a cantilever column

$$\frac{P}{P_e} = 1 + \frac{\alpha^2}{8} \tag{2.24h}$$

which is valid when α is not too large. For $\alpha \to 0$ we have $P = P_e$. With increasing end rotation α, the load P increases, therefore the post-buckling behavior of the column is stable. $\alpha = 15°$, $P/P_e = 1.00857$ from Eq. (2.24h), or the error in P from small deflection linear theory is 0.857%. Also, the substantial increase in load above the Euler load can be achieved at a very large end rotation or lateral deflection, when the material of the column becomes inelastic. Consequently, within the range of elastic behavior, results obtained from the small deflection linear theory can be used to obtain the critical load of the column.

2.10.3 Effective Length Approach

We have shown solutions for two end conditions. The solutions for other types of supports can be obtained directly from the solution of the pinned-pinned column by using effective length approach as in Section 2.3. Let KL be the effective length of the column. Eq. (2.23q) can be written as

$$KL = \frac{2K(p)}{k} \tag{2.25a}$$

And Eq. (2.23w) can be written as

$$\delta = \frac{2p}{k} = \frac{2p}{\sqrt{\dfrac{P}{EI}\dfrac{K^2 L^2}{\pi^2}}} \frac{KL}{\pi}$$

or

$$\frac{\delta}{KL} = \frac{2p}{\pi\sqrt{\dfrac{P}{P_e}}} \tag{2.25b}$$

2.10.3.1 Pinned-Pinned Column
The effective length is equal to L, because $K = 1$. Therefore, from Eqs. (2.25a) and (2.25b) we obtain

$$L = \frac{2K(p)}{k}$$

same as before in Eq. (2.23q), and

$$\frac{\delta}{L} = \frac{2p}{\pi\sqrt{\dfrac{P}{P_e}}}$$

same as before in Eq. (2.23x)

2.10.3.2 Cantilever Column

The effective length is equal to $2L$, because $K = 2$. Therefore, from Eqs. (2.25a) and (2.25b) we get

$$L = \frac{K(p)}{k}$$

same as Eq. (2.24d), and

$$\frac{\delta}{L} = \frac{4p}{\pi\sqrt{\frac{P}{P_e}}}$$

same as before in Eq. (2.24g).

2.10.3.3 Fixed-Fixed Column

The effective length is equal to $0.5L$, because $K = 0.5$. Therefore, from Eqs. (2.25a) and (2.25b) we obtain

$$L = \frac{4K(p)}{k} \tag{2.26a}$$

And

$$\frac{\delta}{L} = \frac{p}{\pi\sqrt{\frac{P}{P_e}}} \tag{2.26b}$$

The deflection δ in Eq. (2.26b) is the maximum mid height deflection in the fixed-fixed column.

2.10.3.4 Fixed-Pinned Column

The effective length is equal to $\frac{L}{\sqrt{2}}$, because $K = \frac{1}{\sqrt{2}}$. Therefore, from Eqs. (2.25a) and (2.25b) we get

$$L = \frac{2\sqrt{2}K(p)}{k} \tag{2.27a}$$

$$\frac{\delta}{L} = \frac{\sqrt{2}p}{\pi\sqrt{\frac{P}{P_e}}} \tag{2.27b}$$

The deflection δ in Eq. (2.27b) is the maximum horizontal deflection in the fixed-pinned column.

2.11 Energy Methods

2.11.1 Calculus of Variations

The calculus of variation is a generalization of the minimum and maximum problem of the ordinary calculus. It seeks to determine a function $y = f(x)$ that minimizes/maximizes a definite integral called functional (function of functions) and whose integrant contains y and its derivatives and the independent variable x given by Eq. (2.28a).

$$I = \int_{x_1}^{x_2} F(x, y, y', y'', \ldots \ldots \ldots, y^n)\, dx \qquad (2.28a)$$

In ordinary calculus, one obtains the actual value of a variable at which a given function has a stationary value. In the calculus of variations, one does not get a function that extremizes a given integral. Here, one only gets the differential equation that the function must satisfy so that the function has a stationary value. Thus the calculus of variations is used to obtain the governing differential equation of a stationary value problem. It is not a computational tool to solve the problem [10].

In structural mechanics the method is used to find the deformed shape of a system at which the system has a stationary potential energy or, in other words, finding the deformation corresponding to the equilibrium state of the system. To illustrate the calculus of variation, consider a pinned-pinned column in Figure 2.26a and find the conditions under which it will be in equilibrium under a deformed shape. The strain energy of bending for the column free body diagram

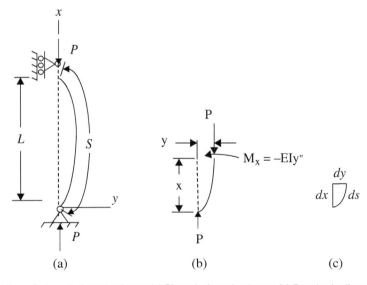

Figure 2.26 Buckling of pinned-pinned column: (a) Pinned-pinned column; (b) Free body diagram; (c) Differential element.

in Figure 2.26b is

$$U = \int_0^L \frac{M_x^2 dx}{2EI} = \frac{1}{2} \int_0^L EI \left(\frac{d^2y}{dx^2} \right)^2 dx \tag{2.28b}$$

The potential energy V of the external forces is given by

$$V = -P\Delta \tag{2.28c}$$

From Figure 2.26c

$$ds^2 = dx^2 + dy^2$$

or

$$ds^2 = \left[1 + \left(\frac{dy}{dx} \right)^2 \right] dx^2$$

or

$$ds = \left[1 + \left(\frac{dy}{dx} \right)^2 \right]^{\frac{1}{2}} dx$$

or

$$ds = \left[1 + \frac{1}{2} \left(\frac{dy}{dx} \right)^2 \right] dx$$

Hence,

$$\int_0^S ds = \int_0^L \left[1 + \frac{1}{2} \left(\frac{dy}{dx} \right)^2 \right] dx \tag{2.28d}$$

$$\Delta = S - L = \frac{1}{2} \int_0^L \left(\frac{dy}{dx} \right)^2 dx \tag{2.28e}$$

$$V = -\frac{P}{2} \int_0^L \left(\frac{dy}{dx} \right)^2 dx \tag{2.28f}$$

$$\Pi = \frac{EI}{2} \int_0^L \left(\frac{d^2y}{dx^2} \right)^2 dx - \frac{P}{2} \int_0^L \left(\frac{dy}{dx} \right)^2 dx \tag{2.28g}$$

It is intended to find $y(x)$ which will make the total potential energy of the system stationary, that is

$$\delta(U + V) = 0 \tag{2.28h}$$

$y(x)$ must be continuous and it must satisfy the boundary conditions $y(0) = y(L) = 0$. Assume

$$\bar{y}(x) = y(x) + \varepsilon\eta(x) \tag{2.28i}$$

which satisfies only the geometric boundary conditions. $\eta(x)$ is an arbitrary function satisfying boundary conditions and is twice differentiable, and ε is a small parameter.

$$\eta(0) = \eta(L) = 0 \tag{2.28j}$$

The function $\bar{y}(x)$ is drawn graphically in Figure 2.27.

The total potential energy in terms of the displacement $\bar{y}\ (x)$ is

$$\Pi = U + V = \int_0^L \left[\frac{EI}{2}(y'' + +\varepsilon\eta'')^2 - \frac{P}{2}(y' + \varepsilon\eta')^2 \right] dx \qquad (2.28k)$$

Π is a function of ε for a given $\eta\ (x)$. If $\varepsilon = 0$, then $\bar{y}\ (x) = y(x)$, which is the curve that provides a stationary value to Π. Hence

$$\left. \frac{d(U + V)}{d\varepsilon} \right|_{\varepsilon=0} = 0 \qquad (2.28l)$$

$$\frac{d(U + V)}{d\varepsilon} = \int_0^L [EI(y'' + \varepsilon\eta'')\eta'' - P(y' + \varepsilon\eta')\eta']\,dx \qquad (2.28m)$$

Eq. (2.28m) is zero at $\varepsilon = 0$, hence

$$\int_0^L [EIy''\eta'' - Py'\eta']\,dx = 0 \qquad (2.28n)$$

Integrate Eq. (2.28n) by parts using $\int u\,dv = uv - \int v\,du$

$$\int_0^L \eta'y'\,dx = y'\eta|_0^L - \int_0^L \eta y''\,dx$$

Use Eq. (2.28j) to get

$$\int_0^L \eta'y'\,dx = -\int_0^L \eta y''\,dx$$

$$\int_0^L y''\eta''\,dx = y''\eta'|_0^L - \int_0^L \eta'y'''\,dx = y''\eta'|_0^L - y'''\eta|_0^L + \int_0^L \eta y^{IV}\,dx$$

Thus, Eq. (2.28n) becomes

$$\int_0^L (EIy^{IV} + Py'')\,\eta\,dx + (EIy''\eta')_0^L = 0 \qquad (2.28o)$$

Each of the two parts of Eq. (2.28o) is separately equal to zero because η is arbitrary. Hence

$$\int_0^L (EIy^{IV} + Py'')\,\eta\,dx = 0 \qquad (2.28p)$$

$$(EIy''\eta')_0^L = 0 \qquad (2.28q)$$

Since $\eta'\,(0),\ \eta'\,(L)$ are not zero, $\eta(x)$ is arbitrary, also $\eta'\,(0) \neq \eta'\,(L)$, therefore, $y\,(x)$ must satisfy

$$EIy^{IV} + Py'' = 0 \qquad (2.28r)$$

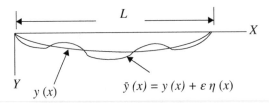

Figure 2.27 Deflected shape.

$$EIy''|_{x=0} = 0 \tag{2.28s}$$

$$EIy''|_{x=L} = 0 \tag{2.28t}$$

Equation (2.28r) is the Eulerian differential equation of an axially loaded column as was found in Eq. 2.11e by considering the moment equilibrium of the deformed column. Equations (2.28s) and (2.28t) give the natural boundary conditions and indicate that the bending moments at the ends of a simply supported column are zero. For simple systems, such as simply supported columns, the governing differential equation can be obtained by considering the equilibrium of the deformed shape. For complex systems such as plate and shell buckling, the stationary potential energy method is simpler to obtain the governing differential equation. The geometric or kinematic boundary conditions involve displacements (deflection and slope), where natural boundary conditions give force conditions (bending moment or shear force) at the boundary.

2.11.2 The Rayleigh-Ritz Method

The principle of stationary potential energy is used here to find the critical load for a hinged-hinged column. The total potential energy of the column is given by

$$\Pi = U + V \tag{2.29a}$$

where U is the strain energy in the column due to bending, and V is the potential energy due to external force P.

$$U = \int_0^L \frac{M_x^2 dx}{2EI} = \frac{1}{2} \int_0^L EI \left(\frac{d^2y}{dx^2} \right)^2 \tag{2.29b}$$

$$V = -\frac{P}{2} \int_0^L \left(\frac{dy}{dx} \right)^2 dx \tag{2.29c}$$

$$\Pi = \frac{EI}{2} \int_0^L \left(\frac{d^2y}{dx^2} \right)^2 dx - \frac{P}{2} \int_0^L \left(\frac{dy}{dx} \right)^2 dx \tag{2.29d}$$

Assume

$$y = A \sin \frac{\pi x}{L} \tag{2.29e}$$

The function in Eq. (2.29e) satisfies the boundary conditions at the ends of the hinged-hinged column given as

$$y = 0 \text{ at } x = 0, \text{ and } y = 0 \text{ at } x = L \tag{2.29f}$$

Substitute Eq. (2.29e) in Eq. (2.29d) to obtain

$$\Pi = \frac{EIA^2\pi^4}{2L^4} \int_0^L \sin^2 \frac{\pi x}{L} dx - \frac{PA^2\pi^2}{2L^2} \int_0^L \cos^2 \frac{\pi x}{L} dx \tag{2.29g}$$

The definite integrals

$$\int_0^L \sin^2 \frac{\pi x}{L} dx = \int_0^L \cos^2 \frac{\pi x}{L} dx = \frac{L}{2} \tag{2.29h}$$

Hence,

$$\Pi = \frac{EIA^2\pi^4}{4L^3} - \frac{PA^2\pi^2}{4L} \tag{2.29i}$$

At the critical load, the neutral equilibrium is possible, therefore, $\frac{d\Pi}{dA} = 0$. Therefore,

$$\frac{d\Pi}{dA} = \frac{AEI\pi^4}{2L^3} - \frac{AP\pi^2}{2L} = 0$$

or

$$P_{cr} = \frac{\pi^2 EI}{L^2} \tag{2.29j}$$

In this case, the critical force P_{cr} is the same as obtained in Eq. (2.1m). In this case, the exact critical load, P_{cr}, is obtained because we could assume the correct deflected shape given by Eq. (2.29e). In other cases of column boundary conditions where the exact deflected shape is not available, the critical load obtained will be an approximate value. The error will depend on the deflected shape assumed.

2.11.3 The Galerkin Method

The Galerkin method is another approximate method of finding the buckling load of a structure. The difference between the Rayleigh-Ritz method and the Galerkin method is that the former requires the integration of energy expression, whereas the latter requires the integration of the governing differential equation. We can approximate the deflection of a column by assuming a series consisting of n independent functions $g_i(x)$ each multiplied by undetermined coefficients a_i. Hence,

$$y \approx a_1 g_1(x) + a_2 g_2(x) + - - - - - + a_n g_n(x) \approx \sum_{i=1}^{n} a_i g_i(x) \tag{2.30a}$$

For the potential energy of a hinged-hinged column to be stationary, Eq. (2.28o) can be written as follows:

$$\int_0^L (EIy^{IV} + Py'')\, \delta y dx + (EIy''\delta y')_0^L = 0 \tag{2.30b}$$

where δy is a virtual displacement. If each $g_i(x)$ function satisfied the geometric and the natural boundary conditions, the second term in Eq. (2.30b) containing the natural boundary conditions given by Eqs. (2.28s) and (2.28t) vanishes when y is replaced by y_{approx}. To make the first term in Eq. (2.30b) equal to zero, the coefficients a_i are chosen such that y_{approx} satisfies the differential equation given by Eq. (2.28r). The column differential equation can be written as

$$Q(\phi) = 0 \tag{2.30c}$$

where

$$Q = EI\frac{d^4}{dx^4} + P\frac{d^2}{dx^2} \tag{2.30d}$$

and

$$\phi = \sum_{i=1}^{n} a_i g_i(x) \tag{2.30e}$$

The first term in Eq. (2.30b) vanishes for y_{approx} if

$$\int_0^L Q(\phi)\delta\phi dx = 0 \tag{2.30f}$$

ϕ is a function of n parameters a_i, hence

$$\delta\phi = \sum_{i=1}^n \frac{\partial\phi}{\partial a_i}\delta a_i = \sum_{i=1}^n g_i(x)\delta a_i \tag{2.30g}$$

or

$$\int_0^L Q(\phi)\sum_{i=1}^n g_i(x)\delta a_i dx = 0 \tag{2.30h}$$

Since $g_i(x)$ are independent of each other, the only way the Eq. (2.30h) is satisfied (identically zero) if each of the n terms in the Eq. (2.30h) vanish individually, that is

$$\int_0^L Q(\phi)g_i(x)\delta a_i dx = 0 \text{ for } i = 1, 2, \text{---}n \tag{2.30i}$$

Since a_i is arbitrary

$$\int_0^L Q(\phi)g_i(x)dx = 0 \text{ for } i = 1, 2, \text{---}n \tag{2.30j}$$

The equations given by Eq. (2.30j) are called Galerkin equations.

A hinged-hinged column is used to illustrate the Galerkin method to compute the critical load. In this case, we know the deflected shape can be assumed as

$$y = A\sin\frac{\pi x}{L} \tag{2.30k}$$

$$Q(\phi) = \left(EI\frac{d^4}{dx^4} + P\frac{d^2}{dx^2}\right)A\sin\frac{\pi x}{L} \tag{2.30l}$$

or

$$Q(\phi) = A\frac{\pi^2}{L^2}\left(EI\frac{\pi^2}{L^2} - P\right)\sin\frac{\pi x}{L} \tag{2.30m}$$

$$g(x) = \sin\frac{\pi x}{L} \tag{2.30n}$$

$$\int_0^L Q(\phi)g(x)dx = 0 \tag{2.30o}$$

$$\frac{A\pi^2}{L^2}\int_0^L \left(EI\frac{\pi^2}{L^2} - P\right)\sin\frac{\pi x}{L}\sin\frac{\pi x}{L}dx = 0 \tag{2.30p}$$

or

$$A\frac{\pi^2 L^2}{L^2}\left(EI\frac{\pi^2}{L^2} - P\right)\frac{L}{2} = 0$$

or

$$P_{cr} = \frac{\pi^2 EI}{L^2} \tag{2.30q}$$

Again, the critical load given by Eq. (2.30q) is the same as given by Eq. (2.1m). In this case, the exact critical load, P_{cr}, is obtained because we could assume the deflected shape given in Eq. (2.30n). In other cases of column boundary conditions where the exact deflected shape is not available, the critical load obtain will be an approximate value. The error will depend on the deflected shape assumed.

Problems

2.1 Find the critical load of the fixed-guided column in Figure P2.1. The column is completely fixed at the lower end and is free to translate laterally but prevented from rotating at the upper end.

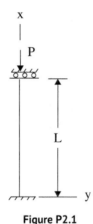

Figure P2.1

2.2 Determine the critical load of the structure in Figure P2.2.

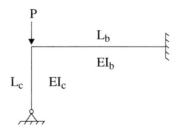

Figure P2.2

2.3 A steel bar of 10 ft. (3048 mm) length and cross-section of 2 in. × 3 in. (50.8 mm × 76.2 mm) is hinged at both ends with respect to bending about the strong axis, and is clamped at both ends with respect to bending about the weak axis. Determine the critical load for the elastic buckling if $E = 10.6 \times 10^3$ ksi (73 GPa).

2.4 Use the Rayleigh-Ritz method to obtain the critical elastic buckling load of the cantilever stepped column in Figure P2.4.

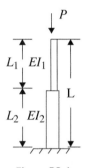

Figure P2.4

2.5 Consider the stepped column in Figure P2.4. Find the critical load by using the Galerkin method.

2.6 Consider a continuous column supported on three supports shown in Figure P2.6 and is compressed by forces P applied at the ends. Determine the critical load for the column.

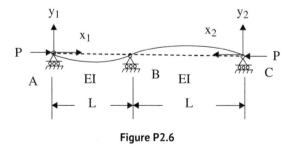

Figure P2.6

2.7 Find the critical load for the column in Figure P2.7. Modulus of elasticity $= E$.

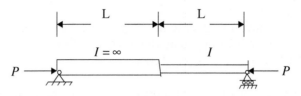

Figure P2.7

2.8 An imperfect column is hinged at both ends. If the imperfection along the length of the column is given by

$$y = \delta \sin \frac{\pi x}{L}$$

Show that the maximum stress due to axial load on a circular column of diameter d is given by

$$\sigma_{\max} = \frac{4P}{\pi d^2}\left[1 + \left(\frac{1}{1 - \frac{P}{P_e}}\right)\frac{8\delta_0}{d}\right]$$

2.9 Find the critical load for a pinned-pinned column where moment of inertia varies from I_o to $2I_o$ shown in Figure P2.9 by the energy method. Assume the deflection is given by $y = a \sin \frac{\pi x}{L}$.

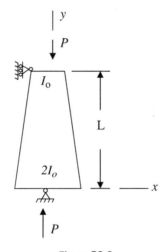

Figure P2.9

References

1 Euler, L. (1744). *De Curvis Elasticis*. Lausanne and Geneva.
2 Euler, L. (1759). Sur le force de colonnes. *Mémoires de Ĭ Académie Royale des Sciences et Belles Lettres* 13.
3 Crandall, S.H., Dahl, N.C., and Lardner, T.J. (1972). *An Introduction to the Mechanics of Solids*, 2e. New York: McGraw-Hill.
4 Gere, J.M. and Timoshenko, S.P. (1997). *Mechanics of Materials*, 4e. Boston, MA: PWS Publishing.
5 AISC (2017). *Steel Construction Manua*, 15e. Chicago, IL: AISC.
6 ACI (2019). *Building Code Requirements for Structural Concrete*. Farmington Hills, MI: American Concrete Institute.
7 Southwell, R.V. (1932). *Proceedings of the Royal Society* series A (135): 601.
8 Wang, C.T. (1953). *Applied Elasticity*. New York: McGraw-Hill.
9 Timoshenko, S.P. and Gere, J.M. (1961). *Theory of Elastic Stability*, 2e. New York: McGraw-Hill.
10 Chajes, A. (1974). *Principles of Structural Stability Theory*. Englewood Cliffs, NJ: Prentice Hall, Inc.

3

Inelastic and Metal Columns

3.1 Introduction

Structures can fail either due to material failure or to instability. But they can also fail due to a combination of both. The material failure is usually preceded by inelastic phenomena, which generally has a destabilizing influence on structures, and must therefore be taken into consideration. Even for structures that are elastic under service loads, achievement of a uniform safety margin requires the consideration of overloads, and overloads inevitably involve inelastic deformations. The strength of a perfectly straight prismatic column with concentric loading is the Euler buckling load, as long as the material is still elastic when the buckling occurs. Many practical columns are found in a range of slenderness where, at buckling, portions of the column are no longer elastic, and thus one of the key assumptions underlying the Euler column theory is violated. Essentially the stiffness of the column is reduced by yielding.

Before considering the theory of inelastic column behavior, let us briefly review the historical development of column buckling theories. The Euler formula was derived by Leonhard Euler in 1744. Later on, it was found that the formula gave higher values of critical load than found from experiments on short columns, in other words the formula was unconservative for these columns. Engesser and Considère in 1889 and 1891 respectively [1, 2] found that Euler's formula was valid only for slender columns. They also realized that the Euler formula can be used to find the critical load for short columns if the modulus of elasticity which is a material constant is replaced by an effective modulus whose value depends on the magnitude of stress at buckling. Engesser proposed using tangent modulus as the effective modulus for inelastic column buckling. Considère did not give any specific value for the effective modulus. Instead he suggested that as a column begins to bend at the critical load, stresses on the concave side, which is in compression, increase in accordance with the tangent modulus given by the slope of the tangent CC', and the stresses on the convex side, which is in tension, decrease in accordance with the modulus of elasticity given by the slope of the line CC'' in Figure 3.1. Point C corresponds to the critical condition and the curve OBC represents the compression stress strain diagram for the column material. Thus, the thought of combining the tangent modulus and the elastic modulus to get the reduced modulus that forms the basis for the double modulus theory was born. Engesser [3] was the first to derive the value of reduced modulus on the basis of double modulus theory. However, the double modulus theory did not gain wide acceptance

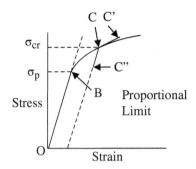

Figure 3.1 Stress strain diagram for a column under compression.

until von Kármán [4] also derived the double modulus theory independently in 1910. He later conducted experiments on short rectangular columns of mild steel and supported the double modulus theory. The double modulus theory or reduced modulus theory was accepted for some time as the correct theory for inelastic buckling. Shanley [5, 6] experimented with columns of different shapes of an aluminum alloy in 1947 and found that the results found from the double modulus or reduced modulus theory were different than those obtained from the experiments. The experimental values in his case were closer to the theoretical values obtained from the tangent modulus theory. In the double modulus theory, it is assumed that when the column bends at the critical load, strains decrease on the convex side while strains increase on the concave side of the column. This strain reversal will take place only if the load remains constant when the column bends. If the load continues increasing, the strain reversal may not take place at any point in the column cross-section. In that case, tangent modulus theory governs the behavior of the column at buckling. Tangent modulus theory gives a lower buckling load than the double modulus theory and gives closer values to the experimental results, so it has been accepted for inelastic buckling. It is lot easier to find the tangent modulus than the double modulus, because the double modulus depends on the shape of the cross-section in addition to the properties of the material.

3.2 Double Modulus Theory

The double modulus theory, also called the reduced modulus theory, was put forward by Engesser [3] in 1895 using the concepts enunciated by Considère [2]. It is assumed in this theory that

1. A column is initially perfectly straight and is concentrically loaded.
2. Both ends of the column are hinged.
3. Bending deformations are small so that the curvature can be assumed as y''.
4. Plane sections before bending remain plane after bending. Therefore, strains at a point in the cross-section of a column can be assumed to vary linearly with the distance from the neutral axis.

5. Up to the critical condition the column remains straight, and the critical load P_{cr} is calculated as the force at which a column can also remain in equilibrium in a slightly deflected shape from the straight position.
6. Central compressive force is applied first and then maintained at this constant value while a small lateral deflection was given to the column.

The critical load is found by considering the equilibrium of the inelastic column in its deflected shape in Figure 3.2a. As the axial force is assumed to remain constant during bending, there will be a small increase in the strain and stress on the concave side, and a decrease in the strain and stress on the convex side. It is assumed that the critical stress $\sigma_{cr} = P_{cr}/A$ is above the elastic limit of the material.

Consider a column cross-section where the centroidal and neutral axes are as shown in Figure 3.3. Deformation is linear with the distance from the neutral axis because of the fourth assumption as shown in Figure 3.4a. The stresses also vary linearly but with different slopes on two sides of the neutral axis as shown in Figure 3.4b.

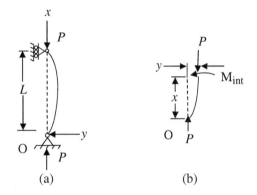

(a)   (b)

Figure 3.2 Inelastic pinned-pinned column under axial force: (a) Deflected shape of column; (b) Free body diagram of deflected shape.

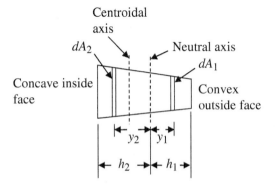

Figure 3.3 Cross-section of the column.

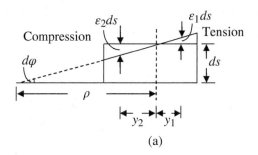

(a)

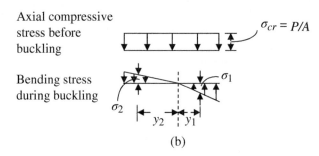

(b)

Figure 3.4 Stresses and strains by double modulus theory: (a) Deformation due to bending; (b) Axial and bending stresses.

ρ denotes the radius of curvature of the deflection curve in Figure 3.4a. It is assumed that the axial force remains constant during bending, therefore, the first equation of equilibrium is that the resultant of the compressive and tensile forces due to bending should be zero in Figure 3.4b.

$$\int_0^{h_1} \sigma_1 dA_1 - \int_0^{h_2} \sigma_2 dA_2 = 0 \tag{3.1a}$$

Stresses σ_1 and σ_2 are the tensile and compressive bending stresses at the distances of y_1 and y_2 respectively from the neutral axis in Figure 3.4b. The stress-strain relation on the tensile (convex) side is governed by the slope of line CC'', i.e. modulus of elasticity E of the material because the fibers unload, whereas the stress-strain relation on the compression (concave) side is governed by the slope of the line CC', which is the tangent modulus E_t at C, in Figure 3.1. Therefore,

$$\sigma_1 = E\varepsilon_1 \tag{3.1b}$$

$$\sigma_2 = E_t\varepsilon_2 \tag{3.1c}$$

Similar triangles in Figure 3.4a give

$$\frac{\varepsilon_1 ds}{y_1} = \frac{ds}{\rho}$$

or

$$\varepsilon_1 = \frac{y_1}{\rho}$$

Since the curvature is given by

$$\frac{1}{\rho} = \frac{d\phi}{ds} = -y''$$

$$\varepsilon_1 = -y_1 y'' \qquad\qquad (3.1d)$$

Similarly,

$$\varepsilon_2 = -y_2 y'' \qquad\qquad (3.1e)$$

Now substitute the quantities from Eqs. (3.1b), (3.1c), (3.1d), and (3.1e) in Eq. (3.1a) and we obtain

$$E y'' \int_0^{h_1} y_1 dA_1 = E_t y'' \int_0^{h_2} y_2 dA_2 \qquad\qquad (3.1f)$$

or

$$EQ_1 = E_t Q_2 \qquad\qquad (3.1g)$$

Q_1 and Q_2 are the moments of the areas on either side of the neutral axis about the neutral axis in Figure 3.3. The position of the neutral axis is obtained from Eq. (3.1g). Since the modulus of elasticity, E, and the tangent modulus, E_t, are not equal, the moments of areas Q_1 and Q_2 are not equal. So, the neutral axis does not pass through the centroid of the cross-section for double modulus theory for inelastic buckling.

The second equation of equilibrium is obtained by taking the moment of the forces about the centroidal axis of the column at O in Figure 3.2b, and is written as

$$P\, y - M_{int} = 0 \qquad\qquad (3.1h)$$

$$M_{int} = \int_0^{h_1} \sigma_1 y_1 dA_1 + \int_0^{h_2} \sigma_2 y_2 dA_2$$

or

$$M_{int} = -E y'' \int_0^{h_1} y_1^2 dA_1 - E_t y'' \int_0^{h_2} y_2^2 dA_2 \qquad\qquad (3.1i)$$

or

$$M_{int} = -y''(EI_1 + E_t I_2) \qquad\qquad (3.1j)$$

where

$$I_1 = \int_0^{h_1} y_1^2 dA_1 \qquad \text{and} \qquad I_2 = \int_0^{h_2} y_2^2 dA_2$$

are the moments of inertias of the areas on either side of the neutral axis about the neutral axis in Figure 3.3. Equations (3.1h) and (3.1j) are combined to give

$$(EI_1 + E_t I_2)y'' + Py = 0$$

or

$$\frac{(EI_1 + E_tI_2)}{I}y'' + \frac{Py}{I} = 0$$

and

$$E_ry'' + \frac{Py}{I} = 0 \tag{3.1k}$$

where

$$E_r = \frac{EI_1 + E_tI_2}{I} \tag{3.1l}$$

I is the moment of inertia of the column cross-section about the axis of bending passing through the centroid. E_r is called the reduced modulus. Its value depends on the stress–strain relation of the material of column and on the shape of the cross-section. E_r is always smaller than the modulus of elasticity of the material E. E_r does not vary along the length of the column, i.e. it is not a function of x, and is constant along a given column. Equation (3.1k) can be written as

$$y'' + k^2y = 0 \tag{3.1m}$$

where $k^2 = P/E_rI$. Equation. (3.1m) is of the same form as Eq. (2.1d) for elastic buckling except that E_r takes the place of E. The critical load for a pinned-pinned column in which the stress exceeds the elastic limit prior to buckling is therefore given by integrating Eq. (3.1m) as

$$(P_r)_{cr} = \frac{\pi^2 E_r I}{L^2} \tag{3.1n}$$

We see that the Euler formula used previously for elastic columns can also be used for inelastic buckling if reduced modulus is used instead of the modulus of elasticity. $(P_r)_{cr}$ is referred to as the reduced modulus load. Since $E_r < E$, the reduced modulus load is always smaller than the Euler load in elastic buckling. The corresponding reduced modulus critical stress is obtained from

$$(\sigma_r)_{cr} = \frac{\pi^2 E_r}{(L/r^2)} \tag{3.1o}$$

3.2.1 Rectangular Section

Consider now a perfect pin-ended column of a rectangular section of width b and depth h as shown in Figure 3.5. Assume that the column buckles at constant axial force P. At the start of buckling, the concave face of the column, i.e. the one toward the center of curvature undergoes further shortening, that is loading, and the convex face undergoes extension, unloading. At the neutral axis within the cross-section the axial strain does not change. The distances of the neutral axis from the convex and concave faces of the column are denoted as h_1 and h_2 respectively in Figure 3.5.

For a rectangular cross-section, Eq. (3.1g) gives

$$Eh_1^2 = E_th_2^2 \tag{3.2a}$$

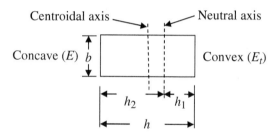

Figure 3.5 Rectangular cross-section.

or

$$\frac{h_1}{h_2} = \frac{\sqrt{E_t}}{\sqrt{E}} \tag{3.2b}$$

Total depth of the section h can be written as

$$h = h_1 + h_2 \tag{3.2c}$$

Substitute Eq. (3.2c) in Eq. (3.2b) and we obtain

$$h_1 = \frac{h\sqrt{E_t}}{\sqrt{E} + \sqrt{E_t}}, \qquad h_2 = \frac{h\sqrt{E}}{\sqrt{E} + \sqrt{E_t}}$$

$$I_1 = \frac{1}{3}bh_1^3 = \frac{1}{3}b\left(\frac{h\sqrt{E_t}}{\sqrt{E} + \sqrt{E_t}}\right)^3 \tag{3.2d}$$

$$I_2 = \frac{1}{3}bh_2^3 = \frac{1}{3}b\left(\frac{h\sqrt{E}}{\sqrt{E} + \sqrt{E_t}}\right)^3 \tag{3.2e}$$

and the moment of inertia of the rectangular section about the centroidal bending axis is

$$I = \frac{1}{12}bh^3 \tag{3.2f}$$

For a rectangular section substituting moment of inertia values in Eq. (3.1l) gives

$$E_r = \frac{4EE_t}{(\sqrt{E} + \sqrt{E_t})^2} \tag{3.2g}$$

For a rectangular section $(P_r)_{cr}$ and $(\sigma_r)_{cr}$ can be determined from Eqs. (3.1n) and (3.1o) substituting E_r from Eq. (3.2g).

3.3 Tangent Modulus Theory

Although the early test results agreed with the reduced modulus theory well, some experimental studies conducted after the publication of von Kármán's work in 1910 [4] revealed that columns can fail by buckling at loads that are significantly lower than the reduced modulus critical load. The reason for this discrepancy was explained by Shanley in 1947 [5]. He showed that in a normal practical situation the column does not buckle at a constant load, as assumed in the reduced modulus theory. The first five assumptions made in the double modulus theory are valid in the tangent modulus theory to calculate the critical load. However, the sixth assumption where the axial force applied to the column remains constant during buckling is no longer retained.

During the testing of an actual column, the axial force increases simultaneously with lateral deflection as shown in Figure 3.6a. In that case, it is possible that the tensile strain increments caused by the deflection may be compensated for by the axial shortening increment due to the increase of the axial load, so that there is no strain reversal in the cross-section anywhere. Thus, the actual deformation may proceed without any release of stress in the fibers on the convex side. The deformations beyond the critical load are small, so the increase in stress during bending, $\Delta\sigma$, is very small in comparison to the critical stress σ_{cr}. Therefore, E_t corresponding to the stress σ_{cr} is taken as the governing modulus for the entire cross-section. The only difference between this case and the bending during elastic buckling is that stresses are related to strain by the tangent modulus E_t, rather than by the modulus of elasticity of the column material, E. The column remains in the straight position until the axial force reaches the critical load. Then the axial force is increased by ΔP and the column moves from the straight position to a bent shape. The incremental force ΔP is negligible in comparison to the critical load P, so the moment equilibrium of the free body diagram in Figure 3.6b is taken by neglecting ΔP, and it leads to

$$E_t I y'' + P y = 0 \tag{3.3a}$$

or

$$y'' + k^2 y = 0 \tag{3.3b}$$

where $k^2 = P/E_t I$. Equation (3.3b) is of the same form as Eq. (2.1d) for elastic buckling except that E_t takes the place of E. The critical load for a pinned-pinned column in which the stress

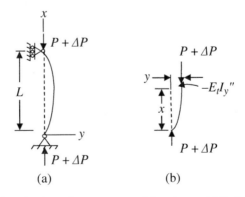

Figure 3.6 Inelastic pinned-pinned column by tangent modulus theory: (a) Deflected shape of column; (b) Free body diagram of the deflected shape.

exceeds the elastic limit prior to buckling is therefore given by tangent modulus theory as

$$(P_t)_{cr} = \frac{\pi^2 E_t I}{L^2} \qquad (3.3c)$$

$(P_r)_{cr}$ is referred to as the tangent modulus load. Since $E_t < E_r$ as shown in Eq. (3.11), the tangent modulus load is less than the reduced modulus load and is independent of the shape of the cross-section. The corresponding tangent modulus critical stress is obtained from

$$(\sigma_t)_{cr} = \frac{\pi^2 E_t}{(L/r^2)} \qquad (3.3d)$$

3.4 Shanley's Theory for Inelastic Columns

In Shanley's [5] model, it is assumed that the load is increasing while buckling is taking place. If deflections are small as assumed in both the tangent and reduced modulus theories, the tangent modulus can be assumed to be constant over the part of the cross-section where the compressive stress is increasing. However, for the finite deformations necessary to be considered when studying post-buckling behavior, the tangent modulus varies along the various fibers of a cross-section as well as along the length of the member. Since it is not possible to find an analytical solution to study the post-buckling behavior, the problem can be solved by numerical methods only.

In 1947, Shanley [5] used a simple column model consisting of an idealized pin-ended column to study the post-buckling behavior of an inelastic column. An approximate analytical solution can be obtained for this model. The model consists of two rigid bars connected at the mid-span by a very short elastic–plastic link of length $h < < L$ that deforms, as shown in Figure 3.7a. This short deformable link has the cross-section of an ideal I-beam whose height is also h. The I-beam consists of two flange elements, each of area $A/2$. All the deformable material is concentrated in the elements, which permit us to avoid variation of the tangent modulus along the length and across the section of the column.

The moment of inertia of the deformable link is $I = Ah^2/4$, where A is the cross-sectional area of the link. The model is supposed to remain straight until the critical load is reached. The initial equilibrium of the straight column at load P_0 is disturbed by applying a small disturbing lateral displacement q while the axial load is raised by a small increment ΔP to $P = P_0 + \Delta P$. Taking the equilibrium of moments in Figure 3.7b, we have

$$P(q) - M_i = 0 \qquad (3.4a)$$

The incremental strains in the deformable link due to bending and any change in axial load during bending in the two elements of the deformable link are ε_1 and ε_2, as shown in Figure 3.7c. These strains do not include axial strains before bending. We can write from Figures 3.7b and c

$$q = \theta L/2 \qquad (3.4b)$$

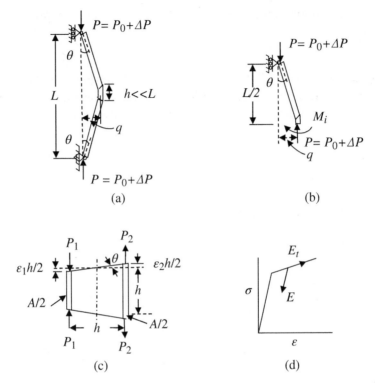

Figure 3.7 Shanley's rigid bar column with the elastic-plastic link: (a) Shanley's rigid bar column; (b) Free body diagram of the deflected shape; (c) Strain increments and internal forces in the link; (d) Loading and unloading moduli.

and

$$\theta = \frac{(\varepsilon_1 + \varepsilon_2)h}{2h} = \frac{\varepsilon_1 + \varepsilon_2}{2} \tag{3.4c}$$

or

$$q = \frac{(\varepsilon_1 + \varepsilon_2)L}{4} \tag{3.4d}$$

In Figure 3.7c, the internal forces in the two links of the deformable link are

$$P_1 = \varepsilon_1 E_1 \frac{A}{2}, \quad \text{and} \quad P_2 = \varepsilon_2 E_2 \frac{A}{2} \tag{3.4e}$$

where the modulus E_1 is applicable on the concave side, and the modulus E_2 is applicable on the convex side. The internal moment in the deformable link, M_i, in Figure 3.7c using Eq. (3.4e) is

$$M_i = (P_1 + P_2)\frac{h}{2} = \frac{Ah}{4}(E_1\varepsilon_1 + E_2\varepsilon_2) \tag{3.4f}$$

Equations (3.4a), (3.4d), and (3.4f) give

$$\frac{PL}{4}(\varepsilon_1 + \varepsilon_2) = \frac{Ah}{4}(E_1\varepsilon_1 + E_2\varepsilon_2) \tag{3.4g}$$

or

$$P = \frac{Ah\,(E_1\varepsilon_1 + E_2\varepsilon_2)}{L(\varepsilon_1 + \varepsilon_2)} \tag{3.4h}$$

If it is assumed that there is no strain reversal while bending takes place, then $E_1 = E_2 = E_t$, and the tangent modulus load from Eq. (3.4h) is given by

$$P_t = \frac{AhE_t}{L} \tag{3.4i}$$

Now, assume that the applied load on the column is increased during the bending of the column after the critical load has been reached. Then

$$E_1 = E_t \tag{3.4j}$$

And let

$$E_2 = \xi E_t \tag{3.4k}$$

The value of ξ depends on whether strain reversal takes place on the convex side or not. If strain reversal takes place, then $\xi = E/E_t > 1$, and $E_2 = E$; if strain reversal does not take place, then $\xi = 1$, and $E_2 = E_t$.

Substitute Eqs. (3.4j) and (3.4k) into Eq. (3.4h), and we have

$$P = \frac{AhE_t(\varepsilon_1 + \xi\varepsilon_2)}{L(\varepsilon_1 + \varepsilon_2)} \tag{3.4l}$$

From Eq. (3.4d) we have

$$\varepsilon_1 = \frac{4q}{L} - \varepsilon_2 \tag{3.4m}$$

Substitute Eqs. (3.4d) and (3.4m) into Eq. (3.4l), and we get

$$P = \frac{AhE_t}{L}\left[1 + \frac{L}{4q}(\xi - 1)\varepsilon_2\right]$$

Or using Eq. (3.4i) we obtain

$$P = P_t\left[1 + \frac{L}{4q}(\xi - 1)\varepsilon_2\right] \tag{3.4n}$$

If bending starts at the critical load given by the tangent modulus load P_t, then

$$P = P_0 + \Delta P = P_t + \Delta P$$

The increase of load during bending is given from the equilibrium of vertical forces as

$$\Delta P = P_1 - P_2 \tag{3.4o}$$

Combine Eqs. (3.4e) and (3.4o) and we get

$$\Delta P = \frac{\varepsilon_1 E_1 A}{2} - \frac{\varepsilon_2 E_2 A}{2} \tag{3.4p}$$

Use Eqs. (3.4j) and (3.4k) to give

$$\Delta P = \frac{\varepsilon_1 E_t A}{2} - \frac{\varepsilon_2 \xi E_t A}{2} \tag{3.4q}$$

Combine Eqs. (3.4m) and (3.4q) to get

$$\Delta P = \frac{AE_t}{2}\left[\frac{4q}{L} - (\xi + 1)\varepsilon_2\right] \tag{3.4r}$$

$$P = P_t + \Delta P = P_t + \frac{AE_t}{2}\left[\frac{4q}{L} - (\xi + 1)\varepsilon_2\right]$$

Rearranging the terms, we have

$$P = P_t\left[1 + \frac{2q}{h} - \frac{L}{2h}(\xi + 1)\varepsilon_2\right] \tag{3.4s}$$

Both Eqs. (3.4n) and (3.4s) give value of P, so by equating these values we obtain

$$P_t\left[1 + \frac{L}{4q}(\xi - 1)\varepsilon_2\right] = P_t\left[1 + \frac{2q}{h} - \frac{L}{2h}(\xi + 1)\varepsilon_2\right]$$

or

$$\frac{L}{4q}(\xi - 1)\varepsilon_2 = \frac{2q}{h} - \frac{L}{2h}(1 + \xi)\varepsilon_2$$

or

$$\varepsilon_2 = \frac{4q}{L(\xi - 1)}\frac{1}{\frac{h}{2q} + \left(\frac{\xi+1}{\xi-1}\right)} \tag{3.4t}$$

Substitute ε_2 from Eq. (3.4t) into Eq. (3.4n) to get

$$P = P_t\left[1 + \frac{1}{\frac{h}{2q} + \left(\frac{\xi + 1}{\xi - 1}\right)}\right] \tag{3.4u}$$

Equation (3.4u) gives the relationship between load P and the lateral deflection q after the critical load P_t was reached, giving the post-buckling behavior of the idealized column. It is assumed in the derivation of Eq. (3.4u) that the bending begins at the tangent modulus load P_t, therefore, when $q = 0$, $P = P_t$. As q increases, P increases, and the P - q relationship is guided by the value $\xi = E_2/E_t$. If there is no strain reversal on the convex side, $\xi = 1$, and $P = P_t$ from Eq. (3.4u). That means the load during bending is constant, which should lead to strain reversal, a contradiction, so this is discarded. That means strain reversal takes place on the convex side, and $\xi = E/E_t$, so P increases as q increases. As q becomes very large in comparison to h, P from Eq. (3.4u) tends to the following value:

$$P = P_t\left[1 + \frac{\xi - 1}{\xi + 1}\right] \tag{3.4v}$$

The reduced modulus value, P_r, for the model can be derived, assuming the bending takes place at the constant load as follows:

For constant load, $P_1 - P_2 = \Delta P = 0$, thus, $\varepsilon_1 E_1 = \varepsilon_2 E_2$ from Eq. (3.4p). Hence

$$\varepsilon_2 = \frac{\varepsilon_1 E_1}{E_2} = \frac{\varepsilon_1 E_t}{E} = \frac{\varepsilon_1}{\xi} \tag{3.4w}$$

Substitute Eq. (3.4w) in Eq. (3.4d), and it gives

$$\varepsilon_2 = \frac{4q}{L(1 + \xi)}$$

and substituting the value of ε_2 in Eq. (3.4n) we get

$$P = P_r = P_t \left(1 + \frac{\xi - 1}{\xi + 1} \right)$$

The same result as in Eq. (3.4v). This means the load on the idealized column reaches the reduced modulus load P_r, as the lateral deflection q becomes very large.

A plot of Eq. (3.4u) where load P vs lateral deflection q at the mid-height of the idealized column is shown in Figure 3.8. The solid curve in the plot shows post-buckling behavior in the model column after the critical load is reached at P_t. The tangent modulus E_t is assumed to be constant in the model column, whereas in the actual column, the tangent modulus reduces with the increase in the compressive strain and varies both across the cross-section and along the length of the column. The dashed line gives the behavior of the actual column. The maximum load for an actual column lies between the tangent modulus and reduced modulus loads. Hence the tangent modulus load represents the lower bound and the reduced modulus load represents an upper bound for a real column that is straight, inelastic, and is loaded concentrically.

The most significant finding of the analysis of the idealized column is that the maximum load reached for a real column lies somewhere between the tangent and reduced modulus loads. Experiments on real columns show that the maximum load they can reach is closer to the tangent modulus load than to the reduced modulus load [7] because of imperfections and accidental eccentricity. Some experimental studies suggest that $P_t/P_{max} \approx 1.02$ to 1.10 [8]. In practice, tangent modulus theory can be used to find the buckling load of inelastic straight columns loaded concentrically, because it is easier to find P_t in comparison to P_r, and it gives conservative results of buckling load. Von Kármán, in response to Shanley's work, stated that it

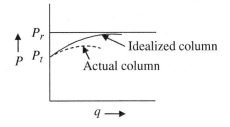

Figure 3.8 Post-buckling behavior of an idealized column.

determines "What is the smallest value of the axial load at which bifurcation of the equilibrium position can occur, regardless of whether or not the transition to the bent position requires an increase of the axial load." Thus, the tangent modulus load gives a safe lower bound of the critical load of an elasto-plastic structure. This is the principal merit of Shanley's work [9].

3.5 Columns with Other End Conditions

In Section 3.4, the critical load for an axially loaded elastoplastic column was found with hinged ends. A similar approach can be used to find the critical load for axially loaded inelastic columns with other end conditions. Elastic columns with other end conditions were discussed in Section 2.2 of Chapter 2. The differential equation of the slightly bent column is of the same form when compressed beyond the elastic limit as that of the elastic column. The only difference is that the modulus of elasticity, E, is replaced by the tangent modulus, E_t. The mathematical expressions for the end conditions also remain the same. Hence the formulas for the critical loads beyond the proportional limit remain unchanged except for replacing E by E_t. Thus, the effective lengths derived previously for elastic conditions can be used in the inelastic columns to find the critical loads for various end conditions from the critical load for the hinged-hinged column, as given in Section 2.3.

3.6 Eccentrically Loaded Inelastic Columns

In practice, there are situations where the columns are acted on by forces that are not passing through the centroids of the cross-sections of the column throughout its length. Hence, it is necessary to consider eccentrically loaded columns where stresses exceed the elastic limit. It is not possible to get a closed form solution because of the complexity of the variation of stress from point to point in the column. Von Kármán (see Bleich) [4, 10] suggested a numerical solution of the problem that is laborious. Instead an approximate solution given by Chajes [8] is given here. The approximate solution is based on the following assumptions:

1. The deflected shape of the column is assumed as a half sine wave.
2. The stress varies linearly across the section of the column. The actual stress-strain diagram for the material, which is a curve, is assumed to obtain the stresses at the extreme fibers, and then a linear variation between these stresses is assumed to exist. For an idealized I-section consisting of two flanges connected by a web of negligible area, the linear variation is an exact stress distribution.
3. Plane sections before bending remain plane after bending and normal to the center line of the column.
4. The deformations are small, so that the curvature can be approximated by the second derivative of the deflection of the column.

An eccentrically loaded column is shown in Figure 3.9a. The line joining the extreme stresses σ_1 and σ_2 is taken as the chord modulus, E_{CH}, in the stress–strain diagram shown in Figure 3.9b.

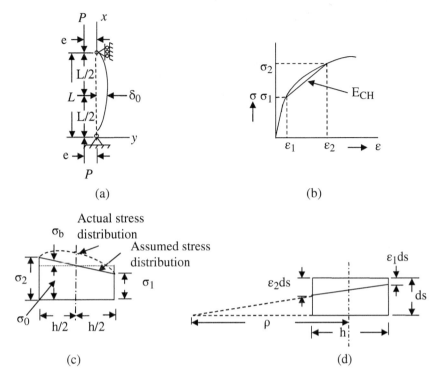

Figure 3.9 Inelastic column under eccentric load: (a) Eccentrically loaded column; (b) Stress strain diagram for the material; (c) Stress variation across the column section; (d) Deformation across the column section.

The stresses at any section consist of an axial stress $\sigma_0 = P/A$, and a bending stress σ_b. Extreme fiber stresses, are shown as σ_1 and σ_2 in Figure 3.9c, also the actual and assumed stress variations are shown by dashed and solid lines respectively. Deformations across the column depth h are shown in Figure 3.9d.

From the mechanics of materials, $\varepsilon = y/\rho$, where ε is the strain in the fiber of a member in bending that lies at a distance of y from the neutral axis, and ρ is the radius of curvature of the bent shape of the column. Therefore,

$$\varepsilon_1 = \frac{\dfrac{h}{2}}{\rho} \tag{3.5a}$$

$$\varepsilon_2 = \frac{\dfrac{-h}{2}}{\rho} \tag{3.5b}$$

In Eq. (3.5b) a negative sign is used because the distance $h/2$ is in the negative direction of y. ε_1 and ε_2 are the strains corresponding to the stresses σ_1 and σ_2 in the outer fibers of the column in Figure 3.9c. Equations (3.5a) and (3.5b) give

$$\frac{\varepsilon_1 - \varepsilon_2}{h} = \frac{1}{\rho} \tag{3.5c}$$

Because of assumption (4) above, the curvature, $1/\rho$, can be written as

$$y'' = \frac{1}{\rho} \tag{3.5d}$$

Combining Eqs. (3.5c) and (3.5d) we have

$$\frac{\varepsilon_1 - \varepsilon_2}{h} = y'' \tag{3.5e}$$

The chord modulus in Figure 3.9b is given by

$$E_{CH} = \frac{\sigma_2 - \sigma_1}{\varepsilon_2 - \varepsilon_1} \tag{3.5f}$$

Substituting Eq. (3.5e) into Eq. (3.5f) we obtain

$$y'' = \frac{\sigma_1 - \sigma_2}{h E_{CH}} \tag{3.5g}$$

In Figure 3.9c $\sigma_1 = \frac{P}{A} - \frac{Mh}{2I}$, and $\sigma_2 = \frac{P}{A} + \frac{Mh}{2I}$
or

$$\sigma_1 - \sigma_2 = -\frac{Mh}{I}$$

or

$$y'' = -\frac{M}{E_{CH}I} \tag{3.5h}$$

Assume the deformed shape of the member as per assumption no. 1 as

$$y = \delta_0 \sin \frac{\pi x}{L}$$

and

$$y'' = -\delta_0 \frac{\pi^2}{L^2} \sin \frac{\pi x}{L} \tag{3.5i}$$

where δ_0 is the deflection of the column at the mid-height. Substituting Eq. (3.5i) into Eq. (3.5h), we have

$$\delta_0 \frac{\pi^2}{L^2} \sin \frac{\pi x}{L} = \frac{M}{E_{CH}I} \tag{3.5j}$$

Substitute $x = L/2$ and $M = P(e + \delta_0)$, into Eq. (3.5j)

$$\delta_0 = \frac{L^2}{\pi^2} \frac{P(e + \delta_0)}{E_{CH}I}$$

Since $I = Ar^2$ and $P = \sigma_0 A$, we can write

$$\delta_0 \left(1 - \frac{L^2}{\pi^2} \frac{\sigma_0}{E_{CH}r^2}\right) = \frac{L^2}{\pi^2} \frac{\sigma_0 e}{E_{CH}r^2}$$

or

$$\delta_0 = \frac{e}{\dfrac{\pi^2 E_{CH}}{(L/r)^2 \sigma_0} - 1} \tag{3.5k}$$

The maximum stress in the column cross-section at the mid-height is given by

$$\sigma_2 = \frac{P}{A} + \frac{P(e + \delta_0)h}{2Ar^2}$$

or

$$\sigma_2 = \sigma_0 \left[1 + \frac{eh}{2r^2} \left(1 + \frac{\delta_0}{e} \right) \right] \tag{3.5l}$$

Substituting δ_0/e from Eq. (3.5k) into Eq. (3.5l) we have

$$\sigma_2 = \sigma_0 \left[1 + \frac{eh}{2r^2} \frac{1}{1 - \dfrac{\sigma_0(L/r)^2}{\pi^2 E_{CH}}} \right] \tag{3.5m}$$

Load P versus deflection δ_0 can be plotted using Eq. (3.5k) if E_{CH} is known. Equation (3.5m) cannot be used directly to determine E_{CH} corresponding to a value $\sigma_0 = P/A$, because σ_2 is also unknown. Therefore, Eq. (3.5m) is solved by iteration. Assume a value of E_{CH}, and solve for σ_2 from Eq. (3.5m) for a certain value of P ($\sigma_0 = P/A$). Now, find new value of E_{CH} using Eq. (3.5f), where the strain values are obtained from the stress–strain graph of the material of the column. σ_1 is obtained assuming a doubly symmetric section using Figure 3.9c as follows:

$$\sigma_0 = \frac{\sigma_1 + \sigma_2}{2}, \quad \text{or} \quad \sigma_1 = 2\sigma_0 - \sigma_2 \tag{3.5n}$$

The solution of the eccentrically loaded inelastic column is demonstrated by solving an example.

Example 3.1 Consider an eccentrically loaded pinned-pinned column as shown in Figure 3.10a. It is made of an idealized symmetric I section of an aluminum alloy consisting of two flanges and a web whose area is negligible as shown in Figure 3.10b. Each flange area is 12, 900 mm^2, and the flanges are 254 mm apart. The stress-strain diagram of the material of the column is shown in Figure 3.10c. The slenderness ratio of the column (L/r) is equal to 30.4. The average compression stress $\sigma_0 = 262$ MPa, and the external force P is applied at an eccentricity of 25.4 mm. Find the deflection of the column loaded in inelastic range.

Moment of inertia, $I = 12,900 \ (127)^2(2) = 416,128, 200 \text{ mm}^4$

$$r^2 = \frac{I}{A} = \frac{416,128, 200}{12,900 \text{x} 2} = 16,129 \text{ mm}^2$$

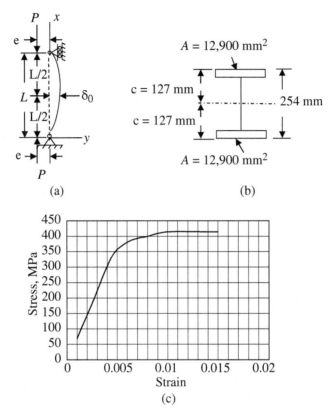

Figure 3.10 Inelastic eccentrically loaded column: (a) Eccentrically loaded column; (b) Cross-section of the column; (c) Stress strain diagram of the column material.

Substituting the given values in Eq. (3.5m) we have

$$\frac{eh}{r^2} = \frac{ec}{r^2} = \frac{25.4 \times 127}{16,129} = 0.2$$

$$\sigma_2 = 262 \left[1 + 0.2 \frac{1}{1 - \frac{262(30.4)^2}{\pi^2 E_{CH}}} \right]$$

or

$$\sigma_2 = 262 + \frac{52.4}{1 - \frac{24532.9}{E_{CH}}} \quad \text{MPa} \tag{3.5o}$$

First trial:
Assume

$\sigma_2 = 332\,\text{MPa}$

$\sigma_1 = 2\,\sigma_0 - \sigma_2 = 2(262) - 332 = 192\,\text{MPa}$

From Figure 3.10c,

$\varepsilon_2 = 0.00460, \quad \varepsilon_1 = 0.00262$

Using Eq. (3.5f) we have

$$E_{CH} = \frac{\sigma_2 - \sigma_1}{\varepsilon_2 - \varepsilon_1} = \frac{332 - 192}{0.00460 - 0.00262} = 70707\,\text{MPa}$$

From Eq. (3.5o) we get

$$\sigma_2 = 262 + \frac{52.4}{1 - \dfrac{24532.9}{70707}} = 342\,\text{MPa}$$

Second trial:
Assume

$\sigma_2 = 342\,\text{MPa}$

$\sigma_1 = 2\sigma_0 - \sigma_2 = 2(262) - 342 = 182\,\text{MPa}$

From Figure 3.10c,

$\varepsilon_2 = 0.00475, \quad \varepsilon_1 = 0.00245$

Using Eq. (3.5f) we have

$$E_{CH} = \frac{\sigma_2 - \sigma_1}{\varepsilon_2 - \varepsilon_1} = \frac{342 - 182}{0.00475 - 0.00245} = 69565\,\text{MPa}$$

From Eq. (3.5o) we get

$$\sigma_2 = 262 + \frac{52.4}{1 - \dfrac{24532.9}{69565}} = 343\,\text{MPa}$$

Third trial:
Assume

$\sigma_2 = 343\,\text{MPa}$

$\sigma_1 = 2\sigma_0 - \sigma_2 = 2(262) - 343 = 181\,\text{MPa}$

From Figure 3.10c,

$\varepsilon_2 = 0.00480, \quad \varepsilon_1 = 0.00244$

Using Eq. (3.5f) we have

$$E_{CH} = \frac{\sigma_2 - \sigma_1}{\varepsilon_2 - \varepsilon_1} = \frac{343 - 181}{0.00480 - 0.00244} = 68644 \text{ MPa}$$

From Eq. (3.5o) we get

$$\sigma_2 = 262 + \frac{52.4}{1 - \dfrac{24532.9}{68644}} = 343.54 \text{ MPa} \approx 343 \text{ MPa}$$

Therefore, $\sigma_2 = 343$ MPa, and $E_{CH} = 68644$ MPa corresponding to $\sigma_0 = 262$ MPa. The mid-height deflection of the column from Eq. (3.5k) is

$$\delta_0 = \frac{e}{\dfrac{\pi^2 E_{CH}}{(L/r)^2 \sigma_0} - 1} = \frac{25.4}{\dfrac{68644}{24532.9} - 1} = 14.12 \text{ mm}$$

Two solutions of mid-height deflection exist for each value of σ_0. By trial and error as before, assume $\sigma_2 = 412$ MPa. Thus, from Eq. (3.5n) we get

$$\sigma_1 = 2(262) - 412 = 112 \text{ MPa}$$

From Figure 3.10c corresponding to stress values of 412 MPa and 112 MPa we obtain

$$\varepsilon_2 = 0.0094, \text{ and } \varepsilon_1 = 0.0016$$

and Eq. (3.5f) gives

$$E_{CH} = \frac{412 - 112}{0.0094 - 0.0016} = 38462 \text{ MPa}$$

Substituting in Eq. (3.5o) we get

$$\sigma_2 = 262 + \frac{52.4}{1 - \dfrac{24532.9}{38462}} = 407 \text{ MPa}$$

Repeating the above process

$$\sigma_1 = 2(262) - 407 = 117 \text{ MPa}.$$

From Figure 3.10c corresponding to stress values of 407 and 117 MPa we obtain

$$\varepsilon_2 = 0.0094 \text{ and } \varepsilon_1 = 0.0016$$

and Eq. (3.5f) gives

$$E_{CH} = \frac{412 - 112}{0.0094 - 0.0016} = 38462 \text{ MPa}$$

Substituting in Eq. (3.5o) we get

$$\sigma_2 = 262 + \frac{52.4}{1 - \dfrac{24532.9}{38462}} = 407 \text{ MPa}$$

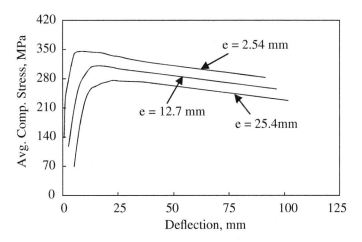

Figure 3.11 Average compressive stress σ_0 versus mid-height deflection δ_0.

Using $\sigma_2 = 407$ MPa, and $E_{CH} = 38462$ MPa in Eq. (3.5k) we have

$$\delta_0 = \frac{e}{\dfrac{\pi^2 E_{CH}}{(L/r)^2 \sigma_0} - 1} = \frac{25.4}{\dfrac{38462}{24532.9} - 1} = 44.74 \text{ mm}$$

The procedure used here can be used to obtain and plot mid-height deflections versus average compressive stress for different values of eccentricity of the external loading, as shown in Figure 3.11 for three different eccentricities of 2.54, 12.70, and 25.4 mm. It is seen from these curves that in the beginning the load increases as deflection increases until a maximum value of the external force is reached. The load decreases with the increase of deflection after the maximum load is reached. So, there are two branches of the load deflection curves, ascending and descending. The ascending branch shows the region when the column exhibits stable buckling, whereas the descending branch of the curve depicts the unstable buckling behavior of the column.

When the load corresponding to the maximum load occurs in a column as shown in Figure 3.11, it collapses because the load decreases with the increase in deflection. The maximum load drops substantially as the eccentricity of the load increases in the column, thus in the case of short columns where buckling occurs at stresses above the elastic limit, the buckling and collapse of a column occur simultaneously, as seen in Figure 3.11. In the case of long columns that buckle elastically, there is a delay between the onset of buckling and the collapse of a column. It is also found from the quantitative curves by Chajes [8] that the maximum load is approaching the tangent modulus load as the eccentricity in the column approaches zero. In addition, as seen before, the tangent modulus load is conservative and easier to calculate in comparison to the reduced modulus buckling load. Therefore, the tangent modulus load is preferred to the reduced modulus load to measure the inelastic buckling of columns.

3.7 Aluminum Columns

Tangent modulus theory is used in the design applications for aluminum-alloy columns in the inelastic range. The tangent modulus theory can be used only if the mechanical properties of the column are constant throughout the cross-section and the length of the member. Research by Mazzolani and Frey in 1977 [11] has shown that the residual stresses are insignificant in the aluminum columns. Residual stresses in the extruded aluminum members are small because of the method of production and the straightening of the finished member by stretching. There is also good agreement between column test results and the calculated values using the tangent modulus theory in the case of initially straight aluminum alloy 6061-T6 columns shown by Batterman and Johnston in 1967 [12]. So, the design and analysis of aluminum-alloy columns have generally been based on the tangent modulus theory. The maximum strength of initially curved aluminum-alloy columns reduces as the imperfectness in the form of initial curvature increases [12]. Tests by Hariri in 1967 [13] have also indicated that the effect of end eccentricity was more harmful than the effect of the same magnitude of initial curvature.

The buckling load, P_{cr}, using the tangent modulus theory is given by

$$P_{cr} = P_t = \frac{\pi^2 E_t I}{(KL)^2}$$

or

$$\sigma_{cr} = \frac{P_{cr}}{A} = \frac{\pi^2 E_t}{(KL/r)^2}$$

$$I = Ar^2$$

and

$$\frac{KL}{r} = \pi \sqrt{\frac{E_t}{\sigma_{cr}}} \tag{3.6a}$$

where A = area of cross-section, r = radius of gyration of the cross-section, I = moment of inertia of the cross-section about the bending axis, KL = effective length of the column, and E_t is the tangent modulus at the critical stress σ_{cr}. The relation between KL/r versus σ_{cr} is called the column strength curve. To plot this curve, the tangent modulus, E_t, should be known which is a function of σ_{cr}. That means the stress-strain graph of the column material should be known. The Ramberg-Osgood formula [14] can be used to fit the experimental results of stress-strain diagrams of aluminum alloys in the form given below:

$$\varepsilon = \frac{\sigma}{E} + 0.002 \left(\frac{\sigma}{\sigma_{0.2}} \right)^n \tag{3.6b}$$

where E = elastic Young's modulus, $\sigma_{0.2}$ = 0.2 % offset yield stress, n = hardening parameter. The tangent modulus E_t is the slope of the stress-strain curve, therefore, differentiating

Eq. (3.6b) with respect to strain ε, we get

$$\frac{d\sigma}{d\varepsilon} = E_t = \frac{E}{1 + \frac{0.002nE}{\sigma_{0.2}}\left(\frac{\sigma}{\sigma_{0.2}}\right)^{n-1}} \tag{3.6c}$$

For a particular aluminum alloy E, $\sigma_{0.2}$, and n are known from the stress-strain diagram for that material. The coefficient, n, the hardening parameter (it reflects the shape of the stress-strain diagram) can be taken as the value of the yield strength in kN/cm^2 as suggested by Steinhart and Galambos [15, 16]. For a particular column, assume $\sigma = \sigma_{cr}$, E_t can be obtained from Eq. (3.6c), then use Eq. (3.6a) to calculate the slenderness ratio, KL/r, of the column. This way we choose different values of $\sigma = \sigma_{cr}$, find tangent modulus E_t at those values, and then obtain corresponding slenderness ratios, KL/r.

The stress-strain graph of the aluminum alloy 6061-T6 is given by Batterman and Johnston [12] from experimental values. The non-dimensional stress-strain diagram showing $\sigma/\sigma_{0.2}$ versus strain ε is shown in Figure 3.12. The graph is used to calculate $E = 10, 100$ ksi (69, 690 MPa), $\sigma_{0.2} = 40.25$ ksi (277.72 MPa), then, $n = 27.77$ (277.72/10 = 27.77 kN/cm^2) [15] for the aluminum alloy 6061-T6. Now we use Eqs. (3.6c) and (3.6a) to obtain E_t and KL/r [a] corresponding to $\sigma = \sigma_{cr}$, and prepare Table 3.1. The column strength curve for the aluminum alloy 6061-T6 is plotted in Figure 3.13.

3.7.1 North American and Australian Design Practice

The Aluminum Association design manual-AA 1994 [17], CSA 1980 [18], and SAA 1979 [19] specifications provide formulas based on the tangent modulus formula, that is simplified in the

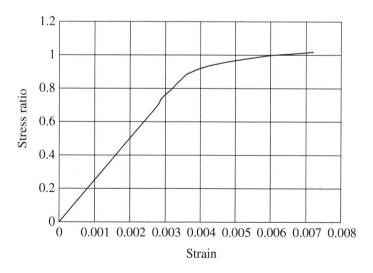

Figure 3.12 Stress ratio ($\sigma/\sigma_{0.2}$) versus strain diagram of aluminum alloy 6061-T6.

Table 3.1 Calculated quantities for column strength curve.

$\sigma = \sigma_{cr}$ ksi(MPa)	$\frac{\sigma_{cr}}{\sigma_{0.2}}$	$\frac{E_t}{E}$ a)	$\frac{KL}{r}$ a)	KL/r[b)
24(165.6)	0.596	1.000	64.45	71.18
28(193.2)	0.696	0.999	59.64	57.83
30(207.0)	0.745	0.995	57.49	51.16
32(220.8)	0.795	0.971	54.99	44.49
34(234.6)	0.845	0.868	50.44	37.81
36(248.4)	0.894	0.587	40.33	31.14
38(262.2)	0.944	0.251	25.65	24.47
40(276.0)	0.994	0.078	13.96	17.79
40.25(277.72)	1.000	0.067	12.88	16.96
40.5(279.45)	1.006	0.057	11.88	16.12
41(282.9)	1.019	0.042	10.10	14.45

a) Tangent modulus theory.
b) [17].

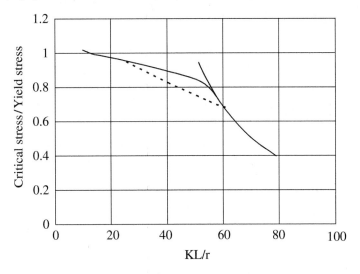

Figure 3.13 Column strength curve for aluminum alloy 6061-T6.

inelastic range to a straight line as follows [16]:

$$\frac{\sigma_c}{\sigma_y} = \alpha^2(1 - k\alpha\lambda) \qquad \text{for } \lambda < C \tag{3.7a}$$

in which $\lambda = (KL/r)/\pi g$ and $g = (E/\sigma_y)^{1/2}$. For fully heat-treated alloys,

$$\alpha = (1 + 2/g)^{1/2}, k = 0.31, \text{ and } C = 1.3/\alpha$$

and for non-heat-treated aluminum alloys,

$$\alpha = (1 + 3/g)^{1/2}, k = 0.38, \text{ and } C = 1.75/\alpha$$

where σ_c = critical stress, σ_y = yield strength, g = distance from the shear center of girder to the point of application of transverse load, K = effective length factor, L = length of the column, r = radius of gyration of the cross-section of the column, E = modulus of elasticity of the material of the column. The above formulas are based on the ratios $\sigma_{0.2}/\sigma_{0.1}$ of 1.04 for fully heat-treated alloys and 1.06 for non-heat-treated alloys, where $\sigma_{0.2}$ and $\sigma_{0.1}$ are the yield strengths of the aluminum alloy at 0.2 and 0.1% offset strains in the stress-strain diagram of the material under axial compression. For $\lambda > C$, the Euler formula is used.

If the above-mentioned substitutions are made in Eq. (3.7a) we get the following:

For fully heat-treated aluminum alloys

$$\frac{\sigma_c}{\sigma_y} = \left(1 + 2\sqrt{\frac{\sigma_y}{E}}\right) - 0.31\left(1 + 2\sqrt{\frac{\sigma_y}{E}}\right)^{3/2} \frac{KL}{\pi r}\sqrt{\frac{\sigma_y}{E}} \tag{3.7b}$$

for

$$\lambda = \frac{KL}{\pi r}\sqrt{\frac{\sigma_y}{E}} < C = \frac{1.3}{\left(1 + 2\sqrt{\frac{\sigma_y}{E}}\right)^{1/2}}$$

And for non-heat-treated aluminum alloys

$$\frac{\sigma_c}{\sigma_y} = \left(1 + 3\sqrt{\frac{\sigma_y}{E}}\right) - 0.38\left(1 + 3\sqrt{\frac{\sigma_y}{E}}\right)^{3/2} \frac{KL}{\pi r}\sqrt{\frac{\sigma_y}{E}} \tag{3.7c}$$

for

$$\lambda = \frac{KL}{\pi r}\sqrt{\frac{\sigma_y}{E}} < C = \frac{1.75}{\left(1 + 3\sqrt{\frac{\sigma_y}{E}}\right)^{1/2}}$$

For both fully heat-treated and non-heat-treated aluminum alloys in the elastic range

$$\frac{\sigma_c}{\sigma_y} = \frac{\pi^2 E}{\left(\frac{KL}{r}\right)^2 \sigma_y} \quad \text{for} \quad \lambda > C \tag{3.7d}$$

The above-mentioned formulas to calculate the critical buckling stress in the inelastic range are given in the Specifications for Aluminum Structures, the design manual of the AA (2005) [20] and the Standards Association of Australia SSA 1997 [21] in a different form as

$$\sigma_c = \left(B_c - D_c\frac{kL}{r}\right) \quad \text{for} \quad \frac{KL}{r} < C_c \tag{3.7e}$$

Formulas for buckling constants for aluminum alloy products whose temper designation begins with T5, T6, T7, T8, or T9 are given in AA 2005 [20] by

$$B_c = \sigma_y \left[1 + \left(\frac{\sigma_y}{2250} \right)^{1/2} \right], \text{where } B_c \text{ is in kips/in.}^2, \tag{3.7f}$$

$$B_c = \sigma_y \left[1 + \left(\frac{\sigma_y}{15510} \right)^{1/2} \right], \text{where } B_c \text{ is in MPa}, \tag{3.7g}$$

$$D_c = \frac{B_c}{10} \left(\frac{B_c}{E} \right)^{1/2}, \text{where } D_c \text{ is in kips/in.}^2 \text{ or MPa}, \tag{3.7h}$$

$$\text{and } C_c = 0.41 \frac{B_c}{D_c}, \text{where both } B_c \text{ and } D_c \text{ are in kips/in.}^2 \text{ or MPa}. \tag{3.7i}$$

Substituting the values of B_c and D_c from Eqs. (3.7f), (3.7h), and (3.7i) into Eq. (3.7e) we have

$$\frac{\sigma_c}{\sigma_y} = \left[1 + \left(\frac{\sigma_y}{2250} \right)^{1/2} \right] - \frac{1}{10} \left[1 + \left(\frac{\sigma_y}{2250} \right)^{1/2} \right]^{3/2} \frac{KL}{r} \sqrt{\frac{\sigma_y}{E}} \tag{3.7j}$$

for

$$\frac{KL}{\pi r} \sqrt{\frac{\sigma_y}{E}} < \frac{1.3}{\left[1 + \left(\frac{\sigma_y}{2250} \right)^{1/2} \right]^{1/2}}$$

where σ_y and E are in kips/in.^{2}.

Now, substituting the values of B_c and D_c from Eqs. (3.7g), (3.7h), and (3.7i) into Eq. (3.7e) we have

$$\frac{\sigma_c}{\sigma_y} = \left[1 + \left(\frac{\sigma_y}{15510} \right)^{1/2} \right] - \frac{1}{10} \left[1 + \left(\frac{\sigma_y}{15510} \right)^{1/2} \right]^{3/2} \frac{KL}{r} \sqrt{\frac{\sigma_y}{E}} \tag{3.7k}$$

for

$$\frac{KL}{\pi r} \sqrt{\frac{\sigma_y}{E}} < \frac{1.3}{\left[1 + \left(\frac{\sigma_y}{15510} \right)^{1/2} \right]^{1/2}}$$

where σ_y and E are in MPa. If the modulus of elasticity E values of an aluminum alloy are substituted in Eq. (3.7b) in kips/in.^{2}, Eq. (3.7j) will be obtained. On the other hand, if the E values are substituted in Eq. (3.7b) in MPa, Eq. (3.7k) will be obtained.

Formulas for buckling constants for aluminum alloy products whose temper designation begins with O, H, T1, T1, T3, or T4 are given by (AA 2005 [20]):

$$B_c = \sigma_y \left[1 + \left(\frac{\sigma_y}{1000} \right)^{1/2} \right], \text{where } B_c \text{ is in kips/in.}^2 \tag{3.7l}$$

$$B_c = \sigma_y \left[1 + \left(\frac{\sigma_y}{6900} \right)^{1/2} \right], \text{where } B_c \text{ is in MPa} \tag{3.7m}$$

$$D_c = \frac{B_c}{20} \left(\frac{6B_c}{E} \right)^{1/2}, \text{where } D_c \text{ is in kips/in.}^2 \text{ or MPa} \tag{3.7n}$$

and $C_c = \dfrac{2B_c}{3D_c}$, where both B_c and D_c are in kips/in.^{2} or MPa. \hfill (3.7o)

Substituting the values of B_c and D_c from Eqs. (3.7l), (3.7n), and (3.7o) in Eq. (3.7e) we have

$$\frac{\sigma_c}{\sigma_y} = \left[1 + \left(\frac{\sigma_y}{1000} \right)^{1/2} \right] - \frac{\sqrt{6}}{20} \left[1 + \left(\frac{\sigma_y}{1000} \right)^{1/2} \right]^{3/2} \frac{KL}{r} \sqrt{\frac{\sigma_y}{E}} \tag{3.7p}$$

for

$$\frac{KL}{\pi r} \sqrt{\frac{\sigma_y}{E}} < \frac{1.75}{\left[1 + \left(\frac{\sigma_y}{1000} \right)^{1/2} \right]^{1/2}}$$

where σ_y and E are in kips/in.^{2}.

Now, substituting the values of B_c and D_c from Eqs. (3.7m), (3.7n), and (3.7o) in Eq. (3.7e) we have

$$\frac{\sigma_c}{\sigma_y} = \left[1 + \left(\frac{\sigma_y}{6900} \right)^{1/2} \right] - \frac{\sqrt{6}}{20} \left[1 + \left(\frac{\sigma_y}{6900} \right)^{1/2} \right]^{3/2} \frac{KL}{r} \sqrt{\frac{\sigma_y}{E}} \tag{3.7q}$$

for

$$\frac{KL}{\pi r} \sqrt{\frac{\sigma_y}{E}} < \frac{1.75}{\left[1 + \left(\frac{\sigma_y}{6900} \right)^{1/2} \right]^{1/2}}$$

where σ_y and E are in MPa. If the modulus of elasticity E values of an aluminum alloy are substituted in Eq. (3.7c) in kips/in.^{2}, Eq. (3.7p) will be obtained. On the other hand if the E values are substituted in Eq. (3.7c) in MPa, Eq. (3.7q) will be obtained.

For all columns in compression in the elastic range, buckling stress is given by the Euler elastic buckling formula [20, 21]

$$\frac{\sigma_c}{\sigma_y} = \frac{\pi^2 E}{\left(\frac{KL}{r} \right)^2 \sigma_y} \quad \text{for} \quad \frac{KL}{r} > C_c \tag{3.7r}$$

which is the same equation as Eq. (3.7d). Thus, the *Guide to Stability Design Criteria for Metal Structures* [16, 22] and the aluminum design manual [17, 20] give the same equations to find the buckling load for aluminum alloy columns. Equations (3.7b) and (3.7j) are used to calculate

KL/r values in Table 3.1 and to plot the straight line as per AA 1994 [17] and AA 2005 [20], in the column strength curve given by Figure 3.13.

For most alloys used in structural applications, welding reduces the strength of the metal in a narrow zone around the weld, thereby diminishing the capacity of columns of low and intermediate slenderness ratios. Welding can also introduce residual stresses and crookedness in the column. For columns with longitudinal welds or with transverse welds that affect only part of the cross-section, the test results can be predicted by the equation [16, 22]

$$\sigma_{pw} = \sigma_n - \frac{A_w}{A}(\sigma_n - \sigma_w) \tag{3.7s}$$

where

σ_{pw} = Critical stress for columns with part of the cross-section affected by welding

σ_n = Critical stress for the same column if there were no welds

σ_w = Critical stress for the same column if the entire cross-section was affected by welding

A_w = Cross-sectional area of affected zone

A = Total area of cross-section

Details of the effects of welds on the buckling strength of aluminum alloy columns are given by Brungraber and Clark [23]. Mazzolani [24], Hong [25], and Lai and Nethercot [26].

3.8 Steel Columns

In the case of steel columns for buckling in the inelastic range even if the elastoplastic behavior and the imperfections are taken into account, there still is a discrepancy between the theoretical and experimental values. For aluminum columns, the nonlinear stress-strain behavior is primarily due to material nonlinearity and all fibers exhibit the same stress-strain relationship. This is not true for steel columns. In steel columns there is an additional effect due to residual stresses that are produced during the manufacture of the structural steel shapes. Therefore, direct application of the tangent modulus theory is not possible in order to calculate the critical load. Several research studies by Osgood [27], Yang, Beedle, and Johnston (1952) [28], Beedle and Tall [29], have shown the residual stresses in steel columns have a major effect on the load capacity of short hot-rolled steel columns.

Residual stresses arise from non-uniform cooling of structural steel shapes during their manufacture. The tips of the flanges of wide flange shapes cool faster than the flange web intersections. When the central parts of the flanges cool and try to shrink, the contraction is resisted by the already hardened flange tips. This causes tension in the central part of flanges and compression in the flange tips to maintain the equilibrium of forces as shown in Figure 3.14a. The central part of the web also cools faster than the flange-web junction, creating compressive residual stresses there, as shown in Figure 3.14a. Residual stresses in the flanges are more important than in the web because it is closer to the neutral axis. The stress-strain curve of mild steel is elastic-perfectly plastic, which can be observed from a coupon test, whereas the

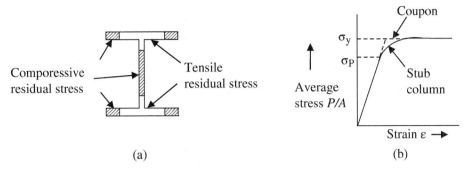

Figure 3.14 Hot rolled wide flange steel sections: (a) Residual stress in a W-section; (b) Strain curve for a steel column.

stress-strain graph for a stub column under compression has a gradually decreasing slope, as shown in Figure 3.14b.

The apparent yield strength σ_p is much lower in a stub column than the actual yield strength σ_y, i.e. $\sigma_p = \sigma_y - \sigma_r$, where σ_r is the maximum residual stress. Parts of the cross-section that have large residual compressive stress yield first before the applied stress reaches the yield strength of the material. As the load increases, the yielding occurs in other fibers with less or no initial compressive residual stresses. Finally, fibers with initial tensile residual stress yield. Because of this, there is slowly increasing curve between stresses σ_p and σ_y in Figure 3.14b. A triangular distribution of residual stresses is assumed in the flanges from a maximum compressive stress of σ_r at the flange tip to a maximum tensile stress of σ_r at the flange web junction, as shown in Figure 3.15a. It is not possible to calculate the critical load in the case of a steel column directly from the tangent modulus theory due to the presence of residual stresses. An analytical approach is presented here to find the critical load corresponding to a particular value of slenderness ratio in the inelastic range. The column is assumed to be perfect, and there is no deflection until the tangent modulus load is reached. Axial load and bending moment increase simultaneously after the critical load has reached, and there is no strain reversal as buckling takes place. For calculation purposes an ideal I section is chosen in which the area of the web is neglected in Figure 3.15.

Initially residual stresses are present in the cross-section before any external axial compressive force is applied. The flange tips are in compression whereas the central portion of the flange near the flange web junction is in tension, as shown in Figure 3.15b. As the external axial force, P, increases, the stresses consist of the algebraic sum of the residual and the stresses due to applied load. Total stresses at various stages of loading are shown in Figures 3.15c–f, where A is the area of the entire cross-section. A positive sign for tensile and a negative sign for compressive stress are used. The yielding begins at the flange tips and progresses inward until the entire cross-section has yielded, as shown in Figure 3.15f.

3.8.1 Buckling of Idealized Steel I-Section

In a steel column, if we assume that there is no stress reversal as the buckling proceeds, it means that the increased load is taken by the elastic part of the cross-section because the yielded portion cannot carry any further load because the material is elastic-perfectly plastic as observed

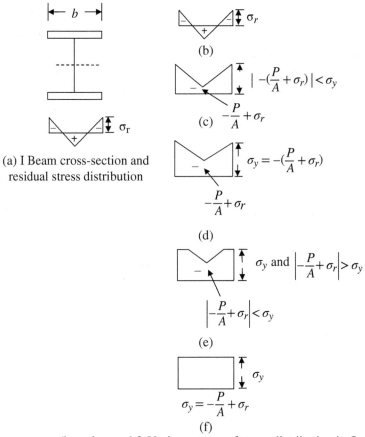

(a) I Beam cross-section and residual stress distribution

(b)

(c) $-\dfrac{P}{A}+\sigma_r$

$\left|-(\dfrac{P}{A}+\sigma_r)\right|<\sigma_y$

(d) $-\dfrac{P}{A}+\sigma_r$

$\sigma_y=-(\dfrac{P}{A}+\sigma_r)$

(e) $\left|-\dfrac{P}{A}+\sigma_r\right|<\sigma_y$

σ_y and $\left|-\dfrac{P}{A}+\sigma_r\right|>\sigma_y$

(f) $\sigma_y=-\dfrac{P}{A}+\sigma_r$

(b, c, d, e, and f) Various stages of stress distribution in flanges

Figure 3.15 Idealized steel I section and effect of residual stresses.

by Yang, Beedle, and Johnston [28]. In this case, the tangent modulus, $E_t = E$, is the elastic modulus outside the yielded zone. Therefore, the critical load of the column is given by

$$P_{cr} = \frac{\pi^2 E I_e}{(KL)^2} \tag{3.8a}$$

or

$$P_{cr} = P_e \frac{I_e}{I} \tag{3.8b}$$

Where KL = effective length of the column, P_e = Euler's buckling load, I_e moment of inertia of the cross-section that is elastic, I = moment of inertia of the whole cross-section. The tangent modulus, E_t, at a point on the curved portion of the stress-strain relation in Figure 3.14b is given by

$$E_t = \frac{d\sigma}{d\varepsilon} = \frac{\dfrac{dP}{A}}{\dfrac{dP}{A_e E}} = \frac{EA_e}{A} \tag{3.8c}$$

or

$$\frac{E_t}{E} = \frac{A_e}{A} = \varsigma \tag{3.8d}$$

The A and A_e are the areas of cross-section of the entire section and of elastic portion of the section, respectively. If the stress-strain relation of a stub column is known experimentally, the ratio $\varsigma = A_e/A$ can be obtained from there using Eq. (3.8d). Otherwise considering the linear residual stress distribution shown in Figure 3.15a, for the idealized I section, the quantity ς can be calculated analytically. The value of ς depends on the shape of the cross-section, the residual stress distribution, and the bending axis.

3.8.1.1 Strong Axis Bending

Neglecting the resistance of the web, and ignoring the moment of inertia of the flanges about the axes parallel to the strong axis and passing through their centroids, the ratio I_e/I can be written as

$$\frac{I_e}{I} = \frac{2(b_e t)\dfrac{h^2}{4}}{2(bt)\dfrac{h^2}{4}} = \frac{A_e}{A} = \varsigma \tag{3.9a}$$

Where $A = 2tb$, $A_e = 2tb_e$; b, b_e are the widths of the entire flange and the elastic portion of the flange respectively; and t is the flange thickness shown in Figure 3.16.

The critical load for strong axis bending can be obtained from Eqs. (3.8a) and (3.9a) as

$$P_{cr} = \varsigma \frac{\pi^2 EI}{(KL)^2} \tag{3.9b}$$

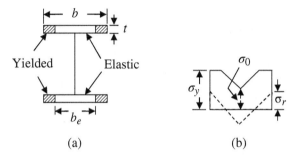

(a)   (b)

Figure 3.16 Elastic–plastic idealized I section.

3.8.1.2 Weak Axis Bending

Similarly, the ratio I_e/I for weak axis bending can be derived as follows:

$$\frac{I_e}{I} = \frac{2\dfrac{tb_e^3}{12}}{2\dfrac{tb^3}{12}} = \left(\frac{b_e}{b}\right)^3 = \left(\frac{A_e}{A}\right)^3 = \varsigma^3 \tag{3.10a}$$

and the critical load for weak axis is given by Eqs. (3.8a) and (3.10a) as

$$P_{cr} = \varsigma^3 \frac{\pi^2 EI}{(KL)^2} \tag{3.10b}$$

3.8.2 Column Strength Curves for Steel Columns

It is seen from Eqs. (3.9b) and (3.10b) that the effect of residual stresses is more pronounced for bending about the weak axis than for the strong axis. The P_{cr} and ς are not known in Eqs. (3.9b) and (3.10b). One extra Eq. (3.11a) can be formed by considering the equilibrium of the column in the vertical direction.

$$P = 2\left[\sigma_y(b - b_e)t + \frac{\sigma_y + \sigma_0}{2}(b_e t)\right] \tag{3.11a}$$

From parallel lines in Figure 3.16b, we have

$$\frac{\sigma_y - \sigma_0}{\dfrac{b_e}{2}} = \frac{2\sigma_r}{\dfrac{b}{2}}$$

or

$$\sigma_0 = \sigma_y - 2\sigma_r\frac{b_e}{b}$$

Substituting for σ_0 in Eq. (3.11a)

$$P = 2t\left(\sigma_y b - \sigma_r\frac{b_e^2}{b}\right)$$

or

$$\sigma_{avg} = \frac{P}{A} = \sigma_y - \sigma_r\left(\frac{A_e}{A}\right)^2,$$

or

$$\varsigma = \frac{A_e}{A} = \left(\frac{\sigma_y - \sigma_{avg}}{\sigma_r}\right)^{\frac{1}{2}} \tag{3.11b}$$

Strong axis bending: Eqs. (3.9b) and (3.11b) give by substituting $\sigma_{cr} = \sigma_{avg}$

$$\frac{P_{cr}}{P_y} = \left(\frac{\sigma_y - \sigma_{cr}}{\sigma_r}\right)^{\frac{1}{2}}\frac{\pi^2 EI}{(KL)^2}\frac{1}{P_y}$$

or

$$\frac{P_{cr}}{P_y} = \frac{\left(\dfrac{\sigma_y - \sigma_{cr}}{\sigma_r}\right)^{\frac{1}{2}}}{\lambda^2} \tag{3.11c}$$

where

$$\lambda = \left(\frac{KL}{r}\right)\left(\frac{1}{\pi}\right)\sqrt{\frac{\sigma_y}{E}} \tag{3.11d}$$

Weak axis bending: Eqs. (3.10b) and (3.11b) give by substituting $\sigma_{cr} = \sigma_{avg}$

$$\frac{P_{cr}}{P_y} = \left(\frac{\sigma_y - \sigma_{cr}}{\sigma_r}\right)^{\frac{3}{2}} \frac{\pi^2 EI}{(KL)^2} \frac{1}{P_y}$$

or

$$\frac{P_{cr}}{P_y} = \frac{\left(\dfrac{\sigma_y - \sigma_{cr}}{\sigma_r}\right)^{\frac{3}{2}}}{\lambda^2} \tag{3.11e}$$

Now, taking $\sigma_r = 0.3\,\sigma_y$ for hot rolled steel sections, and substituting $P_{cr} = \sigma_{cr}\,A$, $P_y = \sigma_y$ A, λ can be found for a specific value of σ_{cr} from Eqs. (3.11c) and (3.11e). Thus, separate column strength curves can be plotted for I-sections with respect to strong and weak axis bending. The analytical derivation above is based on the linear distribution of the residual stresses. If the residual stress variation is assumed to be different, say, a parabolic, then different column strength curves will be obtained [7]. Column research at Lehigh University [29–32, 34] has shown the residual stress distribution lies between linear and parabolic variations.

The Euler's curve and the column curves for the strong and weak axes bending meet at $\sigma_{cr} = (1 - 0.3)\,\sigma_y = 0.7\sigma_y$ as shown in Figure 3.17. From Eq. (3.11c), we have

$$\frac{P_{cr}}{P_y} = \left(\frac{1 - \dfrac{\sigma_{cr}}{\sigma_y}}{\dfrac{\sigma_r}{\sigma_y}}\right)^{\frac{1}{2}} \frac{1}{\lambda^2} = \left(\frac{1 - \dfrac{\sigma_{cr}}{\sigma y}}{0.3}\right)^{\frac{1}{2}} \frac{1}{\lambda^2}$$

or

$$\lambda^2 = 1.8257 \frac{\left(1 - \dfrac{\sigma_{cr}}{\sigma_y}\right)^{\frac{1}{2}}}{\left(\dfrac{\sigma_{cr}}{\sigma_y}\right)}, \quad \text{or} \quad \lambda = 1.2$$

Similarly, it can be shown, $\lambda = 1.2$, where the weak axis bending curve for inelastic buckling meets the elastic buckling given by Euler's curve, as shown in Figure 3.17.

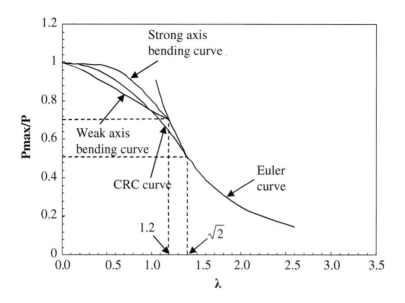

Figure 3.17 Column strength curves for wide flange steel sections.

3.8.3 Column Research Council Curve

Based on analysis and experiments, the Column Research Council (CRC) gave the following column formula [33] in the inelastic range to take into account both the strong and weak axis bending.

$$\sigma_{cr} = A - B\left(\frac{KL}{r}\right)^2 \tag{3.12a}$$

whereas, in the elastic range, the column strength is represented by the Euler formula. The constants A and B are calculated from the conditions,

$$\sigma_{cr} = \sigma_y \quad \text{at} \quad \frac{KL}{r} = 0$$

This condition gives $A = \sigma_y$. Another condition is that the parabola of Eq. (3.12a) intersects the Euler's curve at the proportional limit, $\sigma_P = \sigma_y - \sigma_r$, of a stub column. Thus, from Eq. (3.12a) and the Euler formula we have at the intersection

$$\sigma_y - \sigma_r = \sigma_y - B\left(\frac{KL}{r}\right)^2 \quad \text{and}$$

$$\sigma_y - \sigma_r = \frac{\pi^2 E}{\left(\frac{KL}{r}\right)^2} \quad \text{(Euler's formula)}$$

or

$$B = \frac{\sigma_r(\sigma_y - \sigma_r)}{\pi^2 E}$$

It is assumed that the maximum residual stress is $\sigma_r = 0.5\,\sigma_y$, instead of $0.3\,\sigma_y$, to accommodate both strong and weak axes bending and to be more conservative. Now the CRC formula can be written as

$$\sigma_{cr} = \sigma_y - \frac{\sigma_y^2}{4\pi^2 E}\left(\frac{KL}{r}\right)^2 \tag{3.12b}$$

The KL/r value that is at the junction of the Euler's curve and the CRC curve is called the critical slenderness ratio, C_c. This slenderness ratio corresponds to $\sigma_{cr} = 0.5\,\sigma_y$, and is given by

$$C_c = \sqrt{\frac{2\pi^2 E}{\sigma_y}} \tag{3.12c}$$

In terms of the slenderness parameter, $\lambda = \left(\frac{KL}{\pi r}\right)\sqrt{\frac{\sigma_y}{E}}$, Eq. (3.12b) can be written as

$$\frac{\sigma_{cr}}{\sigma_y} = 1 - \frac{\lambda^2}{4} \tag{3.12d}$$

Substitute $\sigma_{cr} = 0.5\,\sigma_y$ into Eq. (3.12d), and we get $\lambda = \sqrt{2}$. In the elastic range, the Euler formula gives

$$\sigma_{cr} = \frac{\pi^2 E}{\left(\frac{KL}{r}\right)^2}\frac{\sigma_y}{\sigma_y}, \quad \text{or} \quad \frac{\sigma_{cr}}{\sigma_y} = \frac{1}{\lambda^2}$$

So, the CRC formula in terms of the slenderness parameter is given by

$$\frac{\sigma_{cr}}{\sigma_y} = \begin{cases} 1 - \dfrac{\lambda^2}{4} & \text{for } \lambda \leq \sqrt{2} \\[2ex] \dfrac{1}{\lambda^2} & \text{for } \lambda > \sqrt{2} \end{cases} \tag{3.12e}$$

For idealized wide flange sections with residual stresses, the CRC column strength curve meets the Euler curve at $\lambda = \sqrt{2}$ as shown in Figure 3.17.

3.8.4 Structural Stability Research Council Curves

The CRC was later renamed the Structural Stability Research Council (SSRC). It adopted curves 1, 2, and 3, introducing the concept of multiple column curves. The critical load for steel columns depends on the length, cross-sectional and material properties, end restraints, magnitude and distribution of residual stresses, and the shape and magnitude of the initial out-of-straightness. On the basis of numerical analysis and column tests of pin-ended members, where the actual magnitude and distribution of residual stress were known, a set of 112 column curves were generated. For these curves with the initial out-of-straightness of sinusoidal shape of 1/1000, the column length was assumed at the mid-height of the columns. The column tests included rolled and welded shapes, light and heavy cross-sections, and different steels. The concept of multiple column curves was developed from the investigation to take into account multiple variables for different steel grades, shapes, and manufacturing methods by Bjorhovde and Tall [34] and Bjorhovde [35]. These curves were divided into three subgroups, identified by an average curve for the subgroup. These three column curves were adopted by the SSRC, earlier known as the CRC, in the third edition of its *Guide to Stability Design Criteria for Metal Structures* [36], indicating the column types to which these curves are applicable. The equations for the three curves were obtained by curve fitting and are as follows:

$$
\textbf{SSRC Curve 1}: \frac{\sigma_u}{\sigma_{cr}} =
\begin{cases}
1 \ \text{(Material yields)} & \text{for } 0 \le \lambda \le 0.15 \\
0.990 + 0.122\lambda - 0.376\lambda^2 & \text{for } 0.15 \le \lambda \le 1.2 \\
0.051 + 0.801\lambda^{-2} & \text{for } 1.2 \le \lambda \le 1.8 \\
0.008 + 0.942\lambda^{-2} & \text{for } 1.8 \le \lambda \le 2.8 \\
\lambda^{-2} \ \text{(Euler buckling curve)} & \text{for } \lambda \ge 2.8
\end{cases}
\tag{3.13a}
$$

$$
\textbf{SSRC Curve 2}: \frac{\sigma_u}{\sigma_{cr}} =
\begin{cases}
1 \ \text{(Material yields)} & \text{for } 0 \le \lambda \le 0.15 \\
1.035 - 0.202\lambda - 0.222\lambda^2 & \text{for } 0.15 \le \lambda \le 1.0 \\
-0.111 + 0.636\lambda^{-1} + 0.087\lambda^{-2} & \text{for } 1.0 \le \lambda \le 2.0 \\
0.009 + 0.877\lambda^{-2} & \text{for } 2.0 \le \lambda \le 3.6 \\
\lambda^{-2} \ \text{(Euler buckling curve)} & \text{for } \lambda \ge 3.6
\end{cases}
\tag{3.13b}
$$

$$
\textbf{SSRC Curve 3}: \frac{\sigma_u}{\sigma_r} =
\begin{cases}
1 \ \text{(Material yields)} & \text{for } 0 \le \lambda \le 0.15 \\
1.093 - 0.622\lambda & \text{for } 0.15 \le \lambda \le 0.8 \\
-0.128 + 0.707\lambda^{-1} - 0.102\lambda^{-2} & \text{for } 0.8 \le \lambda \le 2.2 \\
0.008 + 0.792\lambda^{-2} & \text{for } 2.2 \le \lambda \le 5.0 \\
\lambda^{-2} \ \text{(Euler buckling curve)} & \text{for } \lambda \ge 5.0
\end{cases}
\tag{3.13c}
$$

The SSRC curves can be approximated by one equation presented by Rondal and Maquoi [37], and Lui and Chen [38] and given in the fifth edition of the *Guide to Stability Design Criteria for Metal Structures* [16]:

$$
\sigma_u = \frac{\sigma_y}{2\lambda^2}(Q - \sqrt{Q^2 - 4\lambda^2}) \le \sigma_y
\tag{3.13d}
$$

where

$$
Q = 1 + \alpha(\lambda - 0.150 + \lambda^2), \text{ and}
$$

$$
\alpha =
\begin{cases}
0.103 & \text{for SSRC curve 1} \\
0.293 & \text{for SSRC curve 2} \\
0.622 & \text{for SSRC curve 3}
\end{cases}
$$

In 1974, the Canadian Standards Association (CSA) adopted SSRC curve 2 to calculate the basic column strength. The CSA also adopted SSRC curve 1 for heat-treated tubes in 1984, and later incorporated SSRC curve 1 for the design of columns made of welded wide flange sections in 1994 [16]. The Euler, CRC, and three different curves of SSRC are given in Figure 3.18.

3.8.5 European Multiple Column Curves

The concept of multiple column curves was also developed in Europe, beginning in the late 1950s. German standard DIN 4114 used one curve for tubes and another for different other shapes. Research was conducted under the sponsorship of European Convention for Constructional Steelwork (ECCS) by Beer and Schultz [39], Jacquet [40], and Sfintesco [41]. The ECCS curves are now part of the column design procedure described in the Eurocode 3 [42]. The ECCS also used the initial out-of-straightness of 1/1000 of the length in the development of their curves. The European multiple column strength curves show the strength of columns made of different steels and shapes of cross-section by different curves. These curves vary for the shapes of the same category, e.g. rolled or welded W and H sections, but different dimensions. There are five curves, a^0, a, b, c, and d [16, 22] shown in Figure 3.19.

Curve a^0 is applicable to welded W and H annealed sections made of high strengths steels. Curve a is used for rolled W, hot formed tubes, welded box, W and H annealed sections. Curve b is suitable for rolled W, welded box and H, and rolled W annealed sections. Curve c gives the strength of columns made of rolled W, welded H, Tee, Channel, circular and rectangular tubes, and welded heavy H sections. Curve d is used for rolled heavy W, and welded heavy H sections made of steel.

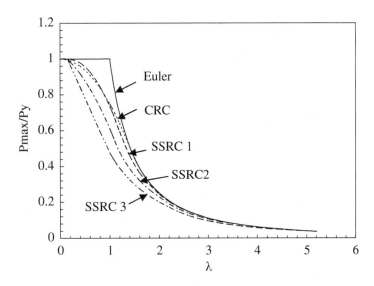

Figure 3.18 Euler, CRC, and SSRC column strength curves.

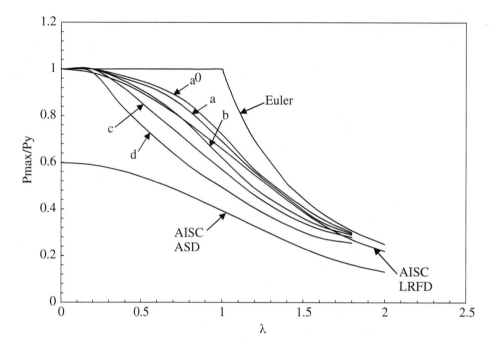

Figure 3.19 ECCS (a^0, a, b, c, and d), AISC (LRFD and ASD), and Euler column strength curves.

3.8.6 AISC Design Criteria for Steel Columns

The use of multiple curves seems to be better but it is unclear whether there is any significant gain in accuracy from the results obtained for real columns because some effects are not precisely known in the formulation of these curves. The uncertainties of end restraint, the inelastic behavior, the calculation of effective length, initial imperfectness, residual stress distribution, etc. all affect the column strength. The American Institute of Steel Construction (AISC) has adopted a single curve for the design of steel columns. In the ninth edition of the AISC (1989) *Manual of Steel Construction: Allowable Stress Design (ASD)* [43], the allowable stress was obtained by dividing the CRC formula given by Eq. (3.12d) by a variable factor of safety for slenderness parameter, $\lambda \leq \sqrt{2}$, as shown below:

$$\text{Safety factor} = \frac{5}{3} + \frac{3}{8}\left(\frac{\lambda}{\sqrt{2}}\right) - \frac{1}{8}\left(\frac{\lambda}{\sqrt{2}}\right)^3$$

The safety factor depended on the slenderness ratio, KL/r, of a column. The higher the slenderness ratio, the higher was the safety factor. This is supported by the fact that the effect of imperfection is larger for more slender columns. For columns in which the slenderness parameter is $\lambda > \sqrt{2}$, allowable stress was obtained from the Euler's formula divided by a constant factor of safety of 23/12. The two formulas of AISC ASD [43] design were:

$$\frac{\sigma_a}{\sigma_y} = \frac{1 - \dfrac{\lambda^2}{4}}{\dfrac{5}{3} + \dfrac{3}{8}\left(\dfrac{\lambda}{\sqrt{2}}\right) - \dfrac{1}{8}\left(\dfrac{\lambda}{\sqrt{2}}\right)^3} \quad \text{for} \quad \lambda \leq \sqrt{2} \tag{3.14a}$$

$$\frac{\sigma_a}{\sigma_y} = \frac{12}{23\lambda^2} \quad \text{for} \quad \lambda > \sqrt{2} \tag{3.14b}$$

where σ_a is the allowable stress in the column, and σ_y is the yield strength of column steel. The allowable load that a concentrically loaded column can carry is given by $F_a = \sigma_a A_g$, where A_g is the gross area of the cross-section of the column.

In the third edition of the AISC (2001), *Manual of Steel Construction: Load Resistance Factor Design (LRFD)* [44], the critical stress was given by

$$\frac{\sigma_{cr}}{\sigma_y} = 0.658^{\lambda^2} \quad \text{for} \quad \lambda \leq 1.5 \tag{3.14c}$$

$$\frac{\sigma_{cr}}{\sigma_y} = \frac{0.877}{\lambda^2} \quad \text{for} \quad \lambda > 1.5 \tag{3.14d}$$

The nominal load that a column can sustain was $P_n = \sigma_{cr} A_g$. The maximum design load for a concentrically loaded column was given by $P_u = \phi_c P_n$, where ϕ_c is the resistance factor for direct compression, taken as 0.85.

In 2005, the thirteenth edition of the AISC manual combined the allowable stress design and the load resistance factor design methods into one manual and the same design practice is continued in the fifteenth edition [45].

The allowable compressive strength (ASD) is given by

$$P_a = P_n / \Omega_c \tag{3.14e}$$

and the design compressive strength (LRFD) is given by

$$P_u = \phi_c P_n \tag{3.14f}$$

where

$$P_n = \sigma_{cr} A_g \tag{3.14g}$$

and the critical stress σ_{cr} is given by

$$\frac{\sigma_{cr}}{\sigma_y} = 0.658^{\frac{\sigma_y}{\sigma_e}} \quad \text{for} \quad \frac{KL}{r} \leq 4.71\sqrt{\frac{E}{\sigma_y}} \quad \text{or} \quad (\sigma_e \geq 0.44\,\sigma_y) \tag{3.14h}$$

$$\frac{\sigma_{cr}}{\sigma_y} = 0.877\frac{\sigma_e}{\sigma_y} \quad \text{for} \quad \frac{KL}{r} > 4.71\sqrt{\frac{E}{\sigma_y}} \quad \text{or} \quad (\sigma_e < 0.44\,\sigma_y) \tag{3.14i}$$

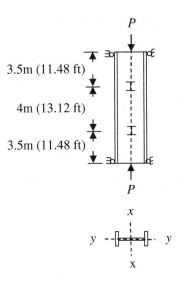

Figure 3.20 Wide flange steel column.

where $\sigma_e = \dfrac{\pi^2 E}{\left(\dfrac{KL}{r}\right)^2} = \dfrac{\sigma_y}{\lambda^2}$ is the Euler critical stress. Eqs. (3.14h) and (3.14i) are the same as

Eqs. (3.14c) and (3.14d) respectively in terms of the slenderness parameter λ. So, the equations to find the critical stress in the fifteenth edition of the AISC manual are the same as in the third edition of the AISC LRFD design manual:

$\Omega_c = 1.67$, is the factor of safety for the AISC allowable stress design (ASD)

and $\phi_c = 0.90$ is the resistance factor for the AISC load resistance factor design (LRFD).

The European design curves and the AISC (2017) ASD and LRFD design curves along with the Euler curve for concentrically loaded columns are given in Figure 3.19. The AISC ASD curve shows the column allowable load whereas the others show the maximum critical load.

Example 3.2 Determine the allowable design strength and the design strength using the AISC (2005) ASD and LRFD methods for the pin-ended axially loaded steel column that is supported in the weak direction as shown in Figure 3.20. Its cross-section is a W12 × 50 (in. × lb/ft) or W310 × 74 (mm × kg/m) wide flange shape. The yield strength of steel is $\sigma_y = 50$ ksi (345 MPa). The area of the cross-section is $A = 14.6$ in.^{2} (9419.3 mm^2). The radii of gyration about the x and y axes are: $r_x = 5.18$ in. (131.57 mm), and $r_y = 1.96$ in. (49.78 mm). The effective length factors in the x and y directions are: $K_x = 1.0$, and $K_y = 1.0$.

$K_x L_x = 36.08$ ft (11 m), and $K_y L_y = 13.12$ ft (4 m)

The slenderness ratios in the x and y directions are given by

$$\frac{K_x L_x}{r_x} = \frac{36.08x12 \text{ in.}}{5.18 \text{ in.}} \text{ or } \left(\frac{11x1000 \text{ mm}}{131.57 \text{ mm}}\right) = 83.61$$

$$\frac{K_y L_y}{r_y} = \frac{13.12x12 \text{ in.}}{1.96 \text{ in.}} \text{ or } \left(\frac{4x1000 \text{ mm}}{49.78 \text{ mm}}\right) = 80.35$$

The slenderness ratio about the x axis is larger, so it governs the buckling strength of the column.

$$\lambda = \frac{KL}{\pi r}\sqrt{\frac{\sigma_y}{E}} = \frac{83.61}{\pi}\sqrt{\frac{50,000 \text{ lb/in.}^2}{29x10^6 \text{ lb/in.}^2}} \text{ or } \left(\frac{83.61}{\pi}\sqrt{\frac{345 \text{ MPa}}{200,000 \text{ MPa}}}\right) = 1.105$$

$\lambda < 1.5$, so it is a case of inelastic buckling.

$$\frac{\sigma_{cr}}{\sigma_y} = 0.658^{\lambda^2} = 0.658^{(1.105)^2} = 0.6$$

$$\sigma_{cr} = 0.6 \times 50 \text{ ksi or } (0.6 \times 345 \text{ MPa}) = 30 \text{ ksi or } (207 \text{ MPa})$$

$$P_n = \sigma_{cr} A = 30 \text{ ksi} \times 14.6 \text{ in.}^2 \text{ or } (207 \text{ MPa} \times 9419.3 \text{ mm}^2/1000)$$

or

$$P_n = 438 \text{ kips or } (1949.80 \text{ kN})$$

Maximum design strength from LRFD

$$P_u = \phi_c P_n = 0.9 \times 438 = 394.20 \text{ kips or } (0.9 \times 1949.80 = 1754.82 \text{ kN})$$

Allowable strength from ASD

$$P_a = \frac{P_n}{\Omega_c} = \frac{438 \text{ kips}}{1.67} \text{ or } \left(\frac{1949.80 \text{ kN}}{1.67}\right) = 262.28 \text{ kips or } (1167.54 \text{ kN})$$

Problems

3.1 Construct a column design curve, i.e. tangent modulus stress σ_{cr} versus slenderness ratio, KL/r.

It is an aluminum alloy column whose stress-strain values are given.

σ ksi (MPa)	ε in./in. (mm/mm)	σ ksi (MPa)	ε in./in. (mm/mm)
10 (69)	0.001	55 (380)	0.006
21 (145)	0.002	57 (393)	0.007
30 (207)	0.0028	58 (400)	0.008
41 (280)	0.0037	60 (414)	0.01
50 (345)	0.0047	60 (414)	0.015

3.2 Plot the reduced modulus column design curve for the aluminum column in Problem 3.1. The cross-section is an idealized cross-section in Figure 3.10b.

3.3 A steel column with a wide flange cross-section W14 × 53 (in. × lb/ft) [W310 × 79 (mm × kg/m)] and length of 30 ft. (9.14 m) has a yield strength of 36 ksi (250 MPa). The cross-sectional dimensions are: $A = 15.6$ in.^{2} (10100 mm^2), the moment of inertia about the x and y axes, $I_x = 425$ in.^{4} (177 × 10^6 mm^4) and $I_y = 95.8$ in.^{4} (39.9 × 10^6 mm^4), respectively. Find the allowable strength and design strength using the AISC ASD and LRFD methods, respectively. The column is pinned-pinned with respect to the strong axis and clamped at both ends with respect to the weak axis bending. Modulus of elasticity of the steel, $E = 29 × 10^6$ psi (200, 000 MPa).

References

1 Engesser, F. (1889). Über die Knickfestigkeit gerader Stäbe. *Zeitschrift für Architektur und Ingenieurwesen* 35: 455.

2 Considère, A. (1891) Résistance des pièces comprimées, Congrès International des Procédés de Construction, Paris, 3, 371.

3 Engesser, F. (1895). Knickfragen. *Schweitzerische Bauzeitung* 26.

4 von Kármán, T. (1910) Untersuchungen über Knickfestigkeit, Mitteilungen über Forschungsarbeiten auf dem Gebiete des Ingenieurwesens, Berlin, 81.

5 Shanley, F.R. (1947). Inelastic column theory. *Journal of Aeronautical Sciences* 14: 261–264.

6 Iyengar, N.G.R. (2007). *Elastic Stability of Structural Elements*. Delhi, India: Macmillan India, Ltd.

7 Chen, W.F. and Lui, E.M. (1987). *Structural Stability: Theory and Implementation*. Upper Saddle River, NJ: Prentice Hall.

8 Chajes, A. (1974). *Principles of Structural Stability Theory*. Englewood Cliffs, NJ: Prentice Hall, Inc.

9 Bazant, Z.P. and Cedolin, L. (1991). *Stability of Structures: Elastic, Inelastic, Fracture and Damage Theories*. New York: Oxford University Press, Inc.

10 Bleich, F. and Bleich, H.H. (1952). *Buckling Strength of Metal Structures*. New York: McGraw-Hill.

11 Mazzolani, F.M. and Frey, F. (1977) Buckling behavior of aluminum alloy extruded members. Paper presented at 2nd International Colloquium on Stability in Steel Structures, Liege, Belgium, pp. 85–94.

12 Batterman, R.H. and Johnston, B.G. (1967). Behavior and maximum strength of metal columns. *Journal of the Structural Division (ASCE)* 93: 205–230.

13 Hariri, R. (1967) Post-buckling behavior of tee-shaped aluminum columns, Ph.D. Dissertation, University of Michigan, Ann Arbor, MI.

14 Ramberg, W. and Osgood, W.R. (1943) Description of stress-strain curves by three parameters, National Advisory Committee on Aeronautics (NACA) Technical Note, 902.

15 Steinhart, G. (1971) Aluminum in engineered construction, Aluminum, 47 (in German).

16 Galambos, T.V. (ed.) (1998). *Guide to Stability Design Criteria for Metal Structures*, 5e. Hoboken, NJ: Wiley.

17 AA (1994). *Specifications for Aluminum Structures*. Washington, DC: Aluminum Association.

18 CSA (1980). *Specifications for the Design of Highway Bridges, CAN3-S6-M78*. Rexdale, Ontario, Canada: Canadian Standards Association.

19 SAA (1979). *Rules for the Use of Aluminum in Structures*. North Sydney, New South Wales, Australia: AS1664, Standards Association of Australia.

20 AA (2005). *Specifications for Aluminum Structures*. Washington, D.C.: Aluminum Association.

21 SAA (1997). *Aluminum Structures-Limit State: Design*. North Sydney, New South Wales, Australia: AS1664.1, Standards Association of Australia.

22 Ziemian, R.D. (ed.) (2010). *Guide to Stability Design Criteria for Metal Structures*, 6e. Hoboken, NJ: Wiley.

23 Brungraber, R.J. and Clark, J.W. (1962). Strength of welded aluminum columns. *Transactions of the American Society of Civil Engineers* 127: 202–226.

24 Mazzolani, F.M. (1985). *Aluminum Alloy Structures*. Marshfield, MA: Pitman.

25 Hong, G.M. (1991). Effects of non-central transverse welds on aluminum columns. In: *Aluminum Structures: Recent Research and Development* (ed. S.I. Lee). London: Elsevier Applied Science.

26 Lai, Y.F.W. and Nethercot, D.A. (1992). Strength of aluminum members containing local transverse welds. *Engineering Structures* 14: 241–254.

27 Osgood, W.R. (1951) The effect of residual stress on column strength, Proceedings of the 1st U.S. National Congress of Appied Mechanics, 415.

28 Yang, C.H., Beedle, L.S., and Johnston, B.G. (1952). Residual stress and the yield strength of steel beams. *Welding Journal (Supplement)* 31 (4): 202–229.

29 Beedle, L.S. and Tall, L. (1960). Basic column strength. *Journal of Structural Engineering, ASCE* 86 (7): 139–173.

30 Huber, A.W. (1956) Residual stresses in wide flange beams and columns, Fritz Engineering laboratory, Report No. 220A.25, Lehigh University, Bethlehem, PA.

31 Ketter, R.L. (1958) The influence of residual stresses on the strength of structural members, Welding Research Council Bulletin No. 44.

32 Tabedge, N. and Tall, L. (1973) Residual stresses in structural steel shapes – a summary of measured values, Fritz Engineering Laboratory Report No. 337.34, Lehigh University, Bethlehem, PA.

33 Johnston, B.G. (1960). *Guide to Design Criteria for Metal Compression Members*, 2e. New York: Wiley.

34 Bjorhovde, R. and Tall, L. (1971) Maximum column strength and the multiple column curve concept, Fritz Engineering Laboratory Report No. 337.29, Lehigh University, Bethlehem, PA.

35 Bjorhovde, R. (1972) Deterministic and probabilistic approaches to the strength of steel columns, Ph.D. dissertation, Lehigh University, Bethlehem, PA.

36 Johnston, B.G. (ed.) (1976). *Guide to Stability Design Criteria for Metal Structures: Column Research Council (CRC)*, 3e. New York: Wiley.

37 Rondal, J. and Maquoi, R. (1979). Single equation for SSRC column strength curves. *Journal of the Structural Division, ASCE* 105: 247–250.

38 Lui, E.M. and Chen, W.F. (1984). Simplified approach to the analysis and design of columns with imperfections. *Engineering Journal* 21: 99–117.

39 Beer, H. and Schultz, G. (1970). Theoretical basis for the European column curves. *Construction Métalique* 3: 58.

40 Jacquet, J. (1970). Column tests and analysis of their results. *Construction Métalique.* 3: 13–36.

41 Sfintesco, D. (ed.) (1976). *ECCS Manual on the Stability of Steel Structures*, 2e. Brussels, Belgium: European Convention for Constructional Steelwork.

42 CEN (2005), Eurocode 3: Design of Steel Structures, Part 1.1: General Rules and Rules for Buildings, EN 1993-1-1, Comité Européen de Normalisation (CEN), European Committee for Standardization, Brussels, Belgium.

43 AISC (1989) Manual of Steel Construction: Allowable Stress Design (ASD), Ninth Edition, American Institute of Steel Construction (AISC), Chicago, IL.

44 AISC (2001) Manual of Steel Construction: Load and Resistance Factor Design (LRFD), Third Edition, AISC, Chicago, IL.

45 AISC (2017) Steel Construction Manual, Fifteenth Edition, AISC, Chicago, IL

4

Beam columns

4.1 Introduction

In the previous chapters we studied the axial loading on columns and its effect on the stability of columns. The presence of bending moment either due to small accidental eccentricities or imperfections was studied. In these cases, the predominant effect was the axial load that was accompanied by small bending effects. In this chapter we study the effect of combined axial load and bending moment on the bending and buckling of a member. Members are supposed to carry simultaneously significant values of both moments and axial loads. Members that are subjected to axial loads and bending moments simultaneously are called beam columns. In this chapter, beam columns containing different supports and loadings will be analyzed for buckling. In the case of beams subjected to lateral loads and moments, it is assumed for small deflection analysis that the deformed shape of the beam does not significantly affect the analysis. Thus, it is possible to make calculations for shear forces, bending moments, stresses, and deformations on the basis of the initial shape of the member. However, the transverse forces and moments cause bending moments and lateral deflections in beam columns that are called primary effects. Furthermore, the axial force acts along with this lateral deflection causing additional lateral deflections and moments in the member that are called secondary effects. Therefore, the problem becomes nonlinear and the principle of superposition cannot be applied.

4.2 Basic Differential Equations of Beam Columns

Consider a beam column of length L simply supported on two end supports, as shown in Figure 4.1a. It carries an axial compressive force of P and a distributed lateral load of $w(x)$ that varies along the length of the beam column. For stability analysis by the equilibrium method, the equilibrium of the beam column is considered in its deflected shape. The assumed lateral force is considered positive in the positive direction of y axis, which is downward here. The axial force is taken along the horizontal axis x, the undeformed axis of the beam column, whereas the shear force is shown along the y-axis perpendicular to the x-axis.

The basic equations for the stability analysis of the beam column can be derived by considering the equilibrium of an element of projected length dx shown in Figure 4.1b. Shearing force

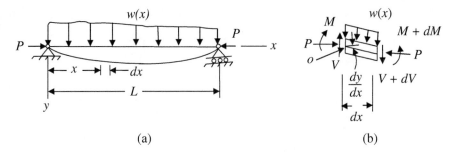

Figure 4.1 Beam column with a distributed lateral load and axial force: (a) Beam column; (b) Element of length *dx*.

and bending moments are shown in the positive directions for the differential element. Over the length *dx* measured along the *x*-axis, the shear force varies by *dV*, the internal moment changes by *dM*, and the axial force *P* remains constant. Considering the equilibrium of the forces acting on the differential element in the *y* direction we obtain

$$V - w(dx) - (V + dV) = 0$$

or

$$\frac{dV}{dx} = -w \tag{4.1a}$$

Considering the equilibrium of moments about the point o of the forces acting on the differential
element we get

$$M + P\frac{dy}{dx}dx + (V + dV)dx - (M + dM) + wdx\frac{dx}{2} = 0$$

Neglecting second-order terms, we have

$$\frac{dM}{dx} = V + P\frac{dy}{dx} \tag{4.1b}$$

For small deformations, $M = -EI\,d^2y/dx^2$ as before in Chapter 1. Substituting this expression into Eq. (4.1b) we get

$$EI\frac{d^3y}{dx^3} + P\frac{dy}{dx} = -V \tag{4.1c}$$

Differentiating Eq. (4.1c) with respect to *x*, and substituting from Eq. (4.1a), the following is obtained

$$EI\frac{d^4y}{dx^4} + P\frac{d^2y}{dx^2} = w \tag{4.1d}$$

Now substitute $k^2 = P/EI$, in Eqs. (4.1c) and (4.1d), and we have

$$\frac{d^3y}{dx^3} + k^2\frac{dy}{dx} = -\frac{V}{EI} \tag{4.1e}$$

and

$$\frac{d^4y}{dx^4} + k^2 \frac{d^2y}{dx^2} = \frac{w}{EI} \tag{4.1f}$$

Equations (4.1e) and (4.1f) are the basic differential equations for analyzing stability for beam columns. If the axial force P is zero, we get same equations as for bending of beams by lateral forces only.

4.3 Beam Column with a Lateral Concentrated Load

Consider a beam column simply supported at the ends and subjected to an axial force P and a lateral concentrated load of Q at a distance of a from the left-hand support in Figure 4.2a. Free body diagrams are shown in Figures 4.2b and c for the left and right portions of the beam column with respect to a section at a distance x from the left support.

Considering the moment equilibrium of the free body diagram in Figure 4.2b, we have

$$Py + \frac{Q(L-a)}{L}x - M = 0 \tag{4.2a}$$

Substitute $M = -EId^2y/dx^2$, and $P/EI = k^2$ in Eq. (4.2a) and we get

$$y'' + k^2y = -\frac{Q(L-a)}{EIL}x \quad \text{for} \quad 0 \le x \le a \tag{4.2b}$$

Now, considering the moment equilibrium of the free body diagram in Figure 4.2c, we have

$$Py + \frac{Qa}{L}(L-x) - M = 0 \tag{4.2c}$$

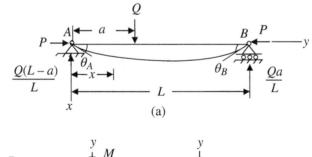

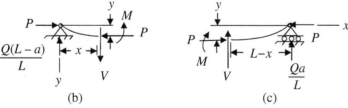

Figure 4.2 Beam column with a concentrated lateral load and axial force: (a) Beam column; (b) Free body diagram left portion of the beam; (c) Free body diagram right portion of the beam.

Substitute $M = -EId^2y/dx^2$, and $P/EI = k^2$ in Eq. (4.2c), and we get

$$y'' + k^2y = -\frac{Qa}{EIL}(L - x) \quad \text{for} \quad a \leq x \leq L \tag{4.2d}$$

Equations (4.2b) and (4.2d) are second-order differential equations because these have been derived for particular end conditions, in this case of simply supported ends, as opposed to the general case for developing Eqs. (4.1e) and (4.1f).

The general solution of Eq. (4.2b) is given by

$$y = A \sin kx + B \cos kx - \frac{Q(L - a)}{EILk^2}x \quad \text{for} \quad 0 \leq x \leq a \tag{4.2e}$$

or

$$y' = Ak \cos kx - Bk \sin kx - \frac{Q(L - a)}{EILK^2} \quad \text{for} \quad 0 \leq x \leq a \tag{4.2f}$$

The general solution of Eq. (4.2d) is given by

$$y = C \sin kx + D \cos kx - \frac{Qa}{EILk^2}(L - x) \quad \text{for} \quad a \leq x \leq L \tag{4.2g}$$

or

$$y' = Ck \cos kx - Dk \sin kx + \frac{Qa}{EILK^2} \quad \text{for} \quad a \leq x \leq L \tag{4.2h}$$

The constants of integration A, B, C, and D are found from the boundary conditions at the ends of the beam column and the conditions of continuity at the point of application of the lateral load Q. The boundary conditions at the ends of the member are

$$y = 0 \text{ at } x = 0, \quad \text{and} \quad y = 0 \text{ at } x = L \tag{4.2i}$$

Substituting $y = 0$ at $x = 0$, in Eq. (4.2e) we obtain

$$B = 0 \tag{4.2j}$$

And substituting $y = 0$ at $x = L$ into Eq. (4.2g) we have

$$D = -C \tan kL \tag{4.2k}$$

The continuity conditions are as follows:
At

$$x = a$$

$$y \text{ in Eq.}(4.2e) = y \text{ in Eq.}(4.2g) \tag{4.2l}$$

and

$$y' \text{ in Eq.}(4.2f) = y' \text{ in Eq.}(4.2h) \tag{4.2m}$$

The continuity condition from Eq. (4.2l) combined with Eqs. (4.2j) and (4.2k) gives

$$A \sin ka = C \sin ka - C \tan kL \cos ka$$

or

$$A = C - C\frac{\tan kL}{\tan ka} \tag{4.2n}$$

The continuity condition from Eq. (4.2m) combined with Eqs. (4.2j) and (4.2k) gives

$$Ak \cos ka - \frac{Q}{EIk^2} = Ck \cos ka + Ck \tan kL \sin ka \tag{4.2o}$$

Combining Eqs. (4.2n) and (4.2o) we have

$$-Ck \cos ka \frac{\tan kL}{\tan ka} - \frac{Q}{EIk^2} = Ck \tan kL \sin ka$$

$$C = -\frac{Q \sin ka}{EIk^3 \tan kL} \tag{4.2p}$$

The constants A and D can be determined from Eqs. (4.2k), (4.2n) and Eq. (4.2p), and are as follows:

$$A = \frac{Q}{EIk^3 \sin kL} \sin k(L - a) \tag{4.2q}$$

$$D = \frac{Q \sin ka}{EIk^3} \tag{4.2r}$$

The constants A, B, C, and D are now substituted in Eqs. (4.2e), (4.2f), (4.2g), and (4.2h) to give the following deflection and slope relations along the length of the beam column:

$$y = \frac{Q \sin k(L - a)}{EIk^3 \sin kL} \sin kx - \frac{Q(L - a)}{EILk^2}x \quad \text{for} \ \ 0 \le x \le a \tag{4.2s}$$

$$y = -\frac{Q \sin ka}{EIk^3 \tan kL} \sin kx + \frac{Q \sin ka}{EIk^3} \cos kx - \frac{Qa(L - x)}{EILk^2} \quad \text{for} \ \ a \le x \le L$$

$$y = \frac{Q \sin ka}{EIk^3 \sin kL} \sin k(L - x) - \frac{Qa(L - x)}{EILk^2} \quad \text{for} \ \ a \le x \le L \tag{4.2t}$$

or

$$y' = \frac{Q \sin k(L - a)}{EIk^2 \sin kL} \cos kx - \frac{Q(L - a)}{EILk^2} \quad \text{for} \ \ 0 \le x \le a \tag{4.2u}$$

$$y' = -\frac{Q \sin ka}{EIk^2 \sin kL} \cos k(L - x) + \frac{Qa}{EILk^2} \quad \text{for} \ \ a \le x \le L \tag{4.2v}$$

The rotations at the ends A and B of the beam column are given by

$$\theta_A = y'|_{x=0} = \frac{Q \sin k(L - a)}{EIk^2 \sin kL} - \frac{Q(L - a)}{EILk^2} \tag{4.2w}$$

$$\theta_B = -y'|_{x=L} = \frac{Q \sin ka}{EIk^2 \sin kL} - \frac{Qa}{EILk^2} \tag{4.2x}$$

Now differentiate Eqs. (4.2u) and (4.2v) with respect to x, and the following relations are obtained:

$$y'' = -\frac{Q \sin k(L - a)}{EIk \sin kL} \sin kx \quad \text{for} \ \ 0 \le x \le a \tag{4.2y}$$

$$y'' = -\frac{Q \sin ka}{EIk \sin kl} \sin k(L - x) \quad \text{for} \ \ a \le x \le L \tag{4.2z}$$

The absolute maximum deflection and bending moment in a beam column subjected to a concentrated lateral load occur at the center of the span when the concentrated load acts at the mid-span.

4.3.1 Concentrated Lateral Load at the Mid-span

If the concentrated lateral load acts at the mid-span, $a = L/2$, then the maximum deflection and bending moment occur at $x = L/2$. Substituting these values of a, and x into Eqs. (4.2s) or (4.2t), Eqs. (4.2w) or (4.2x), and Eqs. (4.2y) or (4.2z), we have

$$y_{max} = \frac{Q}{2EIk^3} \tan \frac{kL}{2} - \frac{QL}{4EIk^2} \tag{4.3a}$$

and

$$\theta_A = \theta_B = \frac{Q \sin \frac{kL}{2}}{EIk^2 \sin kL} - \frac{QL}{2EIk^2 L}$$

or

$$\theta_A = \theta_B = \frac{QL^2}{16EI} \left[\frac{2(1 - \cos u)}{u^2 \cos u} \right]$$

where

$$u = \frac{kL}{2}, \quad \text{or } k = \frac{2u}{L}$$

or

$$\theta_A = \theta_B = \frac{QL^2}{16EI} \lambda(u) \tag{4.3b}$$

where

$$\lambda(u) = \frac{2(1 - \cos u)}{u^2 \cos u} \tag{4.3c}$$

and

$$M_{max} = -EIy'' = \frac{Q}{2k} \tan \frac{kL}{2} \tag{4.3d}$$

Therefore, from Eq. (4.3a)

$$y_{max} = \frac{QL^3}{16EI} \left[\frac{\tan u}{u^3} - \frac{1}{u^2} \right]$$

or

$$y_{max} = y_0 \left[\frac{3(\tan u - u)}{u^3} \right] \tag{4.3e}$$

where $y_0 = \frac{QL^3}{48EI}$ is the deflection in a simply supported beam at the center of span, when a concentrated load Q acts at the mid-span without the axial load P acting on it

or

$$y_{max} = \frac{QL^3}{48EI} \chi(u) \tag{4.3f}$$

where

$$\chi(u) = \left[\frac{3(\tan u - u)}{u^3} \right] \tag{4.3g}$$

The $\tan u$ can be written in series form as [1]

$$\tan u = u + \frac{u^3}{3} + \frac{2}{15}u^5 + \frac{17}{315}u^7 + \frac{62}{2835}u^9 + - - - - \tag{4.3h}$$

Therefore, Eq. (4.3e) can be written as

$$y_{max} = \frac{3y_0}{u^3} \left[u + \frac{u^3}{3} + \frac{2}{15}u^5 + \frac{17}{315}u^7 + \frac{62}{2835}u^9 + - - - - - -u \right]$$

or

$$y_{max} = y_0 \left[1 + \frac{2}{5}u^2 + \frac{17}{105}u^4 + \frac{62}{945}u^6 + - - - -- \right] \tag{4.3i}$$

Now

$$u^2 = \frac{k^2 L^2}{4} = \frac{P}{EI} \frac{L^2}{4} \frac{\pi^2}{\pi^2} = \frac{\pi^2}{4} \frac{P}{P_e} = 2.467 \frac{P}{P_e}$$

where P_e is the Euler's critical load for a pinned-pinned column with axial compressive force. Substituting for u^2 in Eq. (4.3i) we have

$$y_{max} = y_0 \left[1 + 0.987 \frac{P}{P_e} + 0.986 \left(\frac{P}{P_e} \right)^2 + 0.986 \left(\frac{P}{P_e} \right)^3 + - - - - \right]$$

or

$$y_{max} \approx y_0 \left[1 + \frac{P}{P_e} + \left(\frac{P}{P_e} \right)^2 + \left(\frac{P}{P_e} \right)^3 + - - - - - - \right] \tag{4.3j}$$

Binomial theorem gives [1]

$$(1 - x)^{-1} = 1 + x + x^2 + x^3 + x^4 + x^5 + \text{-------} \tag{4.3k}$$

Using binomial theorem and Eq. (4.3j) we get

$$y_{max} \approx \frac{y_0}{1 - \frac{P}{P_e}} \tag{4.3l}$$

where $\left[\frac{1}{1-(P/P_e)} \right]$ is the amplification factor for the deflection due to the presence of an axial force in the beam column. Similarly, an expression giving the magnification factor for the maximum bending moment can be derived as follows:

Substitute $u = kL/2$ in Eq. (4.3d)

$$M_{max} = \frac{QL}{4u} \tan u$$

or

$$M_{max} = M_0 \left[\frac{\tan u}{u} \right] \tag{4.3m}$$

where $M_0 = \frac{QL}{4}$ is the bending moment in a simply supported beam at the center of span, when a concentrated load Q acts at the mid-span without the axial load P acting on it. Substitute $\tan u$ from Eq. (4.3h) in Eq. (4.3m)

$$M_{max} = M_0 \left[1 + \frac{u^2}{3} + \frac{2}{15}u^4 + \frac{17}{315}u^6 + \frac{62}{2835}u^8 + - - - \right]$$

Now, $u^2 = 2.467 \, P/P_e$, therefore

$$M_{max} \approx M_0 \left[1 + 0.82\frac{P}{P_e} \left\{ 1 + \frac{P}{P_e} + \left(\frac{P}{P_e}\right)^2 + \left(\frac{P}{P_3}\right)^3 + \left(\frac{P}{P_4}\right)^4 + - - - \right\} \right]$$

or

$$M_{max} \approx M_0 \left[1 + 0.82\frac{P}{P_e} \left(\frac{1}{1 - (P/P_e)} \right) \right]$$

or

$$M_{max} \approx M_0 \left[\frac{1 - 0.18(P/P_e)}{1 - (P/P_e)} \right] \tag{4.3n}$$

where $\left[\frac{1-0.18(P/P_e)}{1-(P/P_e)} \right]$ is the amplification factor for the bending moment due to the presence of an axial force in the beam column.

4.3.2 Beam Columns with Several Concentrated Loads

Results for one concentrated load in Section 4.3 show that the deflections and their derivatives of a beam column are linearly proportional to the transverse load Q acting on them. Whereas the relationship between the axial load and the deflections and their derivatives is nonlinear because $k^2 = P/EI$ is contained in the trigonometric functions. This fact can be used to obtain deflections, slopes, and bending moments in a beam column by the method of superposition when several transverse concentrated loads are acting on it along with an axial load.

Assume there are n lateral forces, $Q_1, Q_2, - - -, Q_m, Q_{m+1}, - - - - -, Q_n$, that are acting on the beam column at $a_1, a_2, - - - -, a_m, a_{m+1}, - - - -, a_n$ distances respectively from the left support, where $a_1 < a_2 < - - - < a_m < a_{m+1} - - - - < a_n$ as shown in Figure 4.3. Then using Eqs. (4.2s) and (4.2t) for a single lateral load and the method of superposition, the deflection between the loads

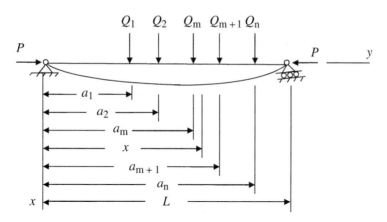

Figure 4.3 Beam column with several concentrated loads.

Q_m and Q_{m+1} in the beam column subjected to several lateral loads and an axial load P is given by

$$y = \frac{\sin k(L-x)}{EIk^3 \sin kL} \sum_{i=1}^{m} Q_i \sin ka_i - \frac{(L-x)}{EILk^2} \sum_{i=1}^{m} Q_i a_i$$

$$+ \frac{\sin kx}{EIk^3 \sin kL} \sum_{i=m+1}^{n} Q_i \sin k(L-a_i) - \frac{x}{EIk^2 L} \sum_{i=m+1}^{n} Q_i (L-a_i) \tag{4.4a}$$

Differentiating Eq. (4.4a) with respect to x, the slope between the loads Q_m and Q_{m+1} in the beam column subjected to several lateral loads and an axial load P is given by

$$y' = -\frac{\cos k(L-x)}{EIk^2 \sin kL} \sum_{i=1}^{m} Q_i \sin ka_i + \frac{1}{EIk^2 L} \sum_{i=1}^{m} Q_i a_i$$

$$+ \frac{\cos kx}{EIk^2 \sin kL} \sum_{i=m+1}^{n} Q_i \sin k(L-a_i) - \frac{1}{EIk^2 L} \sum_{i=m+1}^{n} Q_i(L-a_i) \tag{4.4b}$$

Differentiating Eq. (4.4b) with respect to x, following relation is obtained:

$$y'' = -\frac{\sin k(L-x)}{EIk \sin kL} \sum_{i=1}^{m} Q_i \sin ka_i - \frac{\sin kx}{EIk \sin kL} \sum_{i=m+1}^{n} Q_i \sin k(L-a_i) \tag{4.4c}$$

The bending moment between the loads Q_m and Q_{m+1} at the beam column subjected to several lateral loads and an axial load P is given by

$$M = -EIy'' = \frac{\sin k(L-x)}{k \sin kL} \sum_{i=1}^{m} Q_i \sin ka_i + \frac{\sin kx}{k \sin kL} \sum_{i=m+1}^{n} Q_i \sin k(L-a_i) \tag{4.4d}$$

4.3.3 Beam Column with Lateral Uniformly Distributed Load

Consider a simply supported beam column loaded with a uniformly distributed load w throughout its span, and axial force P as shown in Figure 4.4a. The free body diagram of the left portion at a distance x from the left support is shown in Figure 4.4b.

Considering the moment equilibrium of the free body diagram in Figure 4.4b we have

$$Py + \frac{wL}{2}x - \frac{wx^2}{2} - M = 0 \tag{4.5a}$$

Substitute $M = -EId^2y/dx^2$, and $P/EI = k^2$ in Eq. (4.5a) and we get

$$y'' + k^2y = \frac{wx^2}{2EI} - \frac{wLx}{2EI} \tag{4.5b}$$

The general solution of Eq. (4.5b) is given by

$$y = A \sin kx + B \cos kx + y_p \tag{4.5c}$$

where y_p is the particular solution. Assume y_p in the form of a polynomial

$$y_p = Cx^2 + Dx + E \tag{4.5d}$$

The degree of the polynomial is assumed to be the same as in the right term in Eq. (4.5b).

$$y_p' = 2Cx + D, \quad \text{and} \quad y_p'' = 2C$$

Substituting y_p and its derivatives into Eq. (4.5b) we have

$$2C + k^2(Cx^2 + Dx + E) = \frac{wx^2}{2EI} - \frac{wLx}{2EI}$$

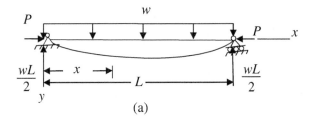

(a)

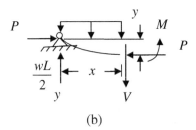

(b)

Figure 4.4 Beam column with a uniformly distributed load and axial force: (a) Beam Column; (b) Free body diagram of left portion of the beam.

or

$$C = \frac{w}{2EIk^2} \tag{4.5e}$$

$$D = -\frac{wL}{2EIk^2} \tag{4.5f}$$

and

$$E = -\frac{w}{k^4EI} \tag{4.5g}$$

Substituting the coefficients C, D, and E from Eqs. (4.5e), (4.5f), and (4.5g) in Eqs. (4.5d) and (4.5c) we have

$$y = A \sin kx + B \cos kx + \frac{w}{2EIk^2}x^2 - \frac{wL}{2EIk^2}x - \frac{w}{EIk^4} \tag{4.5h}$$

Now applying boundary conditions at the ends of the member:

$$y = 0 \text{ at } x = 0, \quad \text{and} \quad y = 0 \text{ at } x = L$$

Substituting $y = 0$ at $x = 0$, in Eq. (4.5h) we obtain

$$B = \frac{w}{EIk^4} \tag{4.5i}$$

And substituting $y = 0$ at $x = L$, in Eq. (4.5h) we have

$$A = \frac{w}{EIk^4} \tan \frac{kL}{2} \tag{4.5j}$$

Substitute A and B from Eqs. (4.5i) and (4.5j) into Eq. (4.5h), and we have

$$y = \frac{w}{EIk^4} \left(\tan \frac{kL}{2} \sin kx + \cos kx - 1 \right) - \frac{w}{2EIk^2}x(L - x) \tag{4.5k}$$

Let

$$u = \frac{kL}{2}, \quad k^2 = \frac{4u^2}{L^2}, \quad \text{and} \quad k^4 = \frac{16u^4}{L^4}$$

Therefore,

$$y = \frac{wL^4}{16EIu^4} \left(\tan u \sin \frac{2ux}{L} + \cos \frac{2ux}{L} - 1 \right) - \frac{wL^2}{8EIu^2}x(L - x) \tag{4.5l}$$

and

$$y' = \frac{wL^4}{16EIu^4} \left(\frac{2u}{L} \tan u \cos \frac{2ux}{L} - \frac{2u}{L} \sin \frac{2ux}{L} \right) - \frac{wL^2}{8EIu^2}(L - 2x) \tag{4.5m}$$

The rotations at the ends A and B are

$$\theta_A = \frac{dy}{dx}\bigg|_{x=0} = \frac{wL^4}{16EIu^4} \left[\frac{2u}{L} \tan u \right] - \frac{wL^3}{8EIu^2}$$

and

$$\theta_B = -\frac{dy}{dx}\bigg|_{x=L} = -\frac{wL^4}{16EIu^4}\left[\frac{2u}{L}\tan u \cos 2u - \frac{2u}{L}\sin 2u\right] - \frac{wL^3}{8EIu^2}$$

or

$$\theta_A = \theta_B = \frac{wL^3}{24EI}\left[\frac{3(\tan u - u)}{u^3}\right] = \frac{wL^3}{24EI}\chi(u) \tag{4.5n}$$

$$y'' = \frac{wL^4}{16EIu^4}\left(-\frac{4u^2}{L^2}\tan u \sin\frac{2ux}{L} - \frac{4u^2}{L^2}\cos\frac{2ux}{L}\right) + \frac{wL^2}{4EIu^2} \tag{4.5o}$$

$$M = -EIy''$$

or

$$M = \frac{wL^2}{4u^2}\left(\tan u \sin\frac{2ux}{L} + \cos\frac{2ux}{L} - 1\right) \tag{4.5p}$$

The maximum deflection occurs at $x = \frac{L}{2}$

$$y_{max} = \frac{wL^4}{16EIu^4}[\tan u \sin u + \cos u - 1] - \frac{wL^4}{32EIu^2}$$

or

$$y_{max} = \frac{5wL^4}{384EI}\left[\frac{12(2\sec u - u^2 - 2)}{5u^4}\right] \tag{4.5q}$$

or

$$y_{max} = \frac{5wL^4}{384EI}\eta(u) \tag{4.5r}$$

where

$$\eta(u) = \frac{12(2\sec u - u^2 - 2)}{5u^4} \tag{4.5s}$$

Expressing $\sec u$ [1] in an infinite series as

$$\sec u = 1 + \frac{1}{2}u^2 + \frac{5}{24}u^4 + \frac{61}{720}u^6 + \frac{277}{8064}u^8 + - - - - - \tag{4.5t}$$

and substituting this series in Eq. (4.5q), we get

$$y_{max} = y_0[1 + 0.4067u^2 + 0.1649u^4 + - - -] \tag{4.5u}$$

where $y_0 = \frac{5wL^4}{384EI}$ [2] is the maximum deflection that would occur in a beam that is loaded with a uniform transverse load without axial force.

$$u = \frac{kL}{2} = \frac{L}{2}\sqrt{\frac{P}{EI}} = \frac{\pi}{2}\sqrt{\frac{P}{P_e}}, \text{where } P_e = \frac{\pi^2 EI}{L^2}$$

or

$$y_{max} = y_0 \left[1 + 1.003 \frac{P}{P_e} + 1.004 \left(\frac{P}{P_e} \right)^2 + - - - - \right]$$

(4.5v)

or

$$y_{max} \approx y_0 \left[1 + \frac{P}{Pe} + \left(\frac{P}{P_e} \right)^2 + - - - - \right]$$

The binomial theorem [1] states

$$(1 - x)^{-1} = 1 + x + x^2 + x^3 + - - - - - \text{for } x^2 < 1$$

Therefore,

$$y_{max} = y_0 \left[\frac{1}{1 - \frac{P}{P_e}} \right]$$

(4.5w)

The term in the bracket is the amplification factor for the deflection due to the presence of axial load along with the transverse load on the beam column.

Maximum bending moment occurs at $x = \frac{L}{2}$, so from Eq. (4.5p)

$$M_{max} = \frac{wL^2}{4u^2} (\tan u \, \sin u + \cos u - 1)$$

or

$$M_{max} = \frac{wL^2}{8} \left[\frac{2(\sec u - 1)}{u^2} \right] = \frac{wL^2}{8} \lambda(u)$$

(4.5x)

Substituting $\sec u$ as an infinite series from Eq. (4.5t), we get

$$M_{max} = M_0[1 + 0.4167u^2 + 0.1694u^4 + 0.0687u^6 + - - -]$$

(4.5y)

where M_0 [2] is the maximum bending moment that would occur in a beam that is loaded with a uniform transverse load without axial force.

Substitute

$$u = \frac{\pi}{2} \sqrt{\frac{P}{P_e}}$$

or

$$M_{max} = M_0 \left[1 + 1.028 \frac{P}{P_e} + 1.029 \left(\frac{P}{P_e} \right)^2 + 1.032 \left(\frac{P}{P_e} \right)^3 - - - \right]$$

or

$$M_{max} = M_0 \left[1 + 1.028 \frac{P}{P_e} \left\{ 1 + \frac{P}{P_e} + \left(\frac{P}{P_e} \right)^2 + - - \right\} \right]$$

or

$$M_{max} = M_0\left[1 + 1.028\frac{P}{P_e}\left(\frac{1}{1 - \frac{P}{P_e}}\right)\right]$$

or

$$M_{max} = M_0\left[\frac{1 + 0.028\frac{P}{P_e}}{1 - \frac{P}{P_e}}\right] \tag{4.5z}$$

The term in the bracket is the amplification factor for the bending moment due to the presence of the axial load along with the transverse load on the beam column.

4.3.4 Beam Columns with Uniformly Distributed Load Over a Portion of Their Span

Consider the case of a simply supported beam column where the load is distributed along only a portion of the span shown in Figure 4.5.

The principle of superposition is applied in the case of partially loaded beam column to find deflection at a distance x in Figure 4.5. There can be three cases depending on the value of x as shown below:

$$0 \le x \le a$$

In this case, the section under consideration is to the left of the partial distributed load. We consider a small element of length dc at a distance c from the left support. The load acting on the element can be considered a concentrated load of $w\,dc$ acting at a distance c from the left support. The deflection dy at a distance x due to the elemental load is given by Eq. (4.2s) by substituting c for a and w dc for Q in the equation.

$$dy = \frac{w\,dc\sin k(L-c)}{EIk^3\sin kL}\sin kx - \frac{w\,dc(L-c)}{EILk^2}x$$

The deflection produced by the total load is then found by integrating between the limits $c = a$ and $c = b$, and is given by

$$y = \int_a^b \frac{w\sin k(L-c)}{EIk^3\sin kL}\sin kx\,dc - \int_a^b \frac{w(L-c)}{EILk^2}x\,dc \tag{4.6a}$$

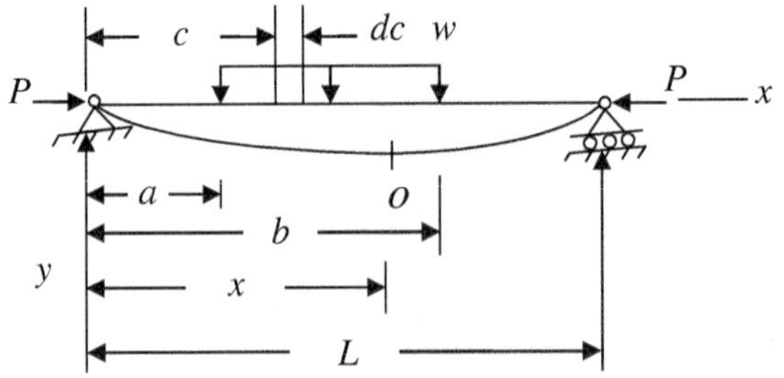

Figure 4.5 Beam column with partial distributed load.

$$a \leq x \leq b$$

If it is intended to find the deflection at a point o under the load, then use Eq. (4.2t) for the load to the left of x, and Eq. (4.2s) for the load to the right of o. Then the deflection at o produced by the total load is obtained from

$$y = \int_a^x \frac{\sin k(L-x)}{EIk^3 \sin kL} w \sin kc \ dc - \int_a^x \frac{(L-x)}{EILk^2} wc \ dc + \int_x^b \frac{\sin kx}{EIk^3 \sin kL} w \sin k(L-c) dc$$

$$- \int_x^b \frac{x}{EIk^2 L} w(L-c) \ dc \tag{4.6b}$$

If the beam is fully loaded with a uniformly distributed load as shown in Figure 4.4a, then Eq. (4.6b) can be used to find the deflection at a point o distance x from the left support by changing the integration limits and substituting $a = 0$ and $b = L$. The results obtained will be the same as in Section 4.3.3 given by Eqs. (4.5l) and (4.5p).

$$b \leq x \leq L$$

If the deflection is to be determined at a point distance x from the left support, the point is to the right of the partial uniform load. The deflection y at a distance x produced by the total load is given by Eq. (4.2t) by substituting c for a, and $w \, dc$ for Q in the equation. Then integrate the expression between $c = a$ and $c = b$, and we get

$$y = \int_a^b \frac{w \sin kc}{EIk^3 \sin kL} \sin k(L-x) \ dc - \int_a^b \frac{wc(L-x)}{EILk^2} dc \tag{4.6c}$$

4.3.5 Beam Columns with Uniformly Increasing Load Over a Portion of Their Span

If w is not uniformly distributed but is a certain function of c, we can obtain the deflections in the beam column for various values of x by substituting for w the given function of c from Eqs. (4.6a), (4.6b), and (4.6c). Consider the case of a uniformly increasing load covering part of the span of the beam column shown in Figure 4.6.

The deflections are obtained by substituting $w = \frac{w_0(c-a)}{b-a}$ into Eqs. (4.6a), (4.6b), and (4.6c) for various values of x. If the triangular load in Figure 4.6 spans the entire length of the beam column increasing from 0 at the left support to w_0 at the right support, the deflection at any point distance x from the left support is given by Eq. (4.6b) and the integration limits are taken as $a = 0$, and $b = L$. The method of superposition can be used to find the deflection and moment at any section of a beam column subjected to a given loading combination consisting of distributed and concentrated loads.

4.4 Beam Columns Subjected to Moments

So far the effects of transverse loads combined with axial load have been considered to find the deflections and bending moments in beam columns. Now let us consider the effects of different cases of moments combined with an axial load on the beam columns.

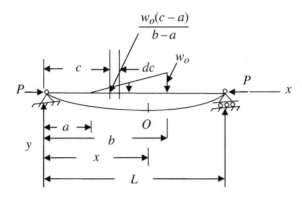

Figure 4.6 Beam column with triangular load.

4.4.1 Span Moment on Beam Column

A moment M_a is applied at a distance a from the left support in Figure 4.7a, and an equivalent couple consisting of two equal and opposite forces of magnitude Q is applied as shown in Figure 4.7b. The magnitude of the couple is $Q\,\Delta a = M_a$. Assume that the distance Δa approaches zero and at the same time Q increases, so that $Q\,\Delta a = M_a$ remains finite. This way, we obtain a moment M_a acting at a distance a, from the left support.

For $0 \le x \le a$, the deflection to the left of Q is given by

$$y = Q[-f(a + \Delta a) + f(a)] \tag{4.7a}$$

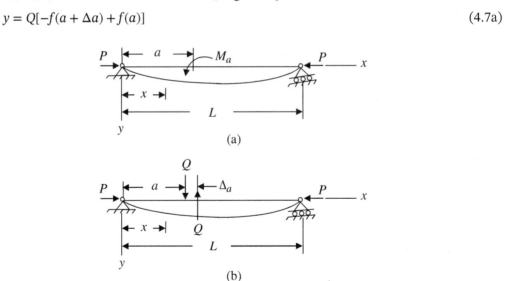

Figure 4.7 Beam column with span moment and axial force: (a) Span moment on beam column; (b) Beam column with equivalent couple.

The deflection given by the right-hand force Q acting upward in Figure 4.7b is upward, so it is negative in Eq. (4.7a). The $f(a)$ is obtained from Eq. (4.2s) as given below:

$$f(a) = \frac{\sin k(L-a)}{EIk^3 \sin kL} \sin kx - \frac{(L-a)}{EIlk^2} x$$

Dividing and multiplying the right-hand side of Eq. (4.7a) by Δa, we get

$$y = -Q\Delta a \frac{f(a+\Delta a) - f(a)}{\Delta a} \tag{4.7b}$$

Taking the limit as $\Delta a \to 0$, we have

$$y = -M_a \frac{df(a)}{da} \tag{4.7c}$$

or

$$y = \frac{M_a}{P}\left[\frac{\cos k(L-a)\sin kx}{\sin kL} - \frac{x}{L}\right] \tag{4.7d}$$

where
$P = EIk^2$.

$$y' = \frac{dy}{dx} = \frac{M_a}{P}\left[\frac{\cos k(L-a)k\cos kx}{\sin kL} - \frac{1}{L}\right] \tag{4.7e}$$

$$y'' = \frac{d^2y}{dx^2} = \frac{M_a}{P}\left[-\frac{k^2 \cos k(l-a)\sin kx}{\sin kL}\right] \tag{4.7f}$$

The bending moment at any section at a distance x is given by

$$M = -EIy'' = M_a\left[\frac{\cos k(L-a)\sin kx}{\sin kL}\right] \tag{4.7g}$$

For $a \le x \le L$, the deflection is given by Eq. (4.7c), where $f(a)$ is obtained from Eq. (4.2t) as given below:

$$f(a) = \frac{\sin ka}{EIk^3 \sin kL}\sin k(L-x) - \frac{a(L-x)}{EILk^2}$$

Hence, from Eq. (4.7c) we have

$$y = \frac{M_a}{P}\left[-\frac{\cos ka}{\sin kL}\sin k(L-x) + \frac{(L-x)}{L}\right] \tag{4.7h}$$

$$y' = \frac{M_a}{P}\left[\frac{\cos ka\,(k)\cos k(L-x)}{\sin kL} - \frac{1}{L}\right] \tag{4.7i}$$

$$y'' = \frac{M_a}{P}\left[\frac{k^2 \cos ka \sin k(L-x)}{\sin kL}\right] \tag{4.7j}$$

The bending moment at any distance x is given by

$$M = -EIy'' = -M_a \left[\frac{\cos ka \sin k(L-x)}{\sin kL} \right] \tag{4.7k}$$

If the moment M_a is applied at the center of span of the beam column, i.e. $a = L/2$, then the deflected shape of the beam column is antisymmetric about the mid-span, and Eqs. (4.7d) and (4.7h) give $y\,(x=L/2)=0$. In this case, the slope of the deflected shape at the center of the span is obtained by substituting $a = \frac{L}{2}$ and $x = \frac{L}{2}$ in Eqs. (4.7e) or (4.7i). Hence

$$y'|_{x=L/2} = \frac{M_a k}{P} \left[\frac{1}{2 \tan \frac{kL}{2}} - \frac{1}{kL} \right] \tag{4.7l}$$

4.4.2 End Moment on a Beam Column

If a beam column is subjected to moment M_A at one end in addition to an axial force P in Figure 4.8a, the deflection can be obtained by the same procedure as shown for the span moment. The moment M_A can be represented by two equal and opposite transverse vertical forces, each of magnitude Q acting at the distance, a, apart as shown in Figure 4.8b.

Assume that the distance a approaches zero and at the same time Q increases so that the magnitude of the equivalent couple $Qa = M_A$ remains finite. The deflection at a distance x is given by Eq. (4.2t) as follows:

$$y = \frac{Q \sin ka}{EIk^3 \sin kL} \sin k(L-x) - \frac{Qa(L-x)}{EILk^2}$$

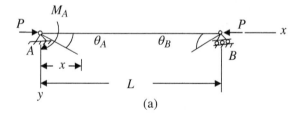

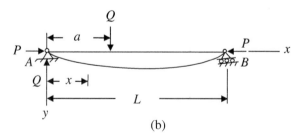

Figure 4.8 Beam column with end moment and axial force: (a) End moment on beam column; (b) Beam column with equivalent couple.

For $a \to 0$, substitute $\sin ka = ka$, and we get

$$y = \frac{Qka}{EIk^3 \sin kL} \sin k(L - x) - \frac{Qa(L - x)}{EILk^2}$$

Substituting $Qa = M_A$, and $P = EIk^2$, we have

$$y = \frac{M_A}{P} \left[\frac{\sin k(L - x)}{\sin kL} - \frac{(L - x)}{L} \right] \tag{4.8a}$$

Taking the derivative of y we get

$$\frac{dy}{dx} = \frac{M_A}{P} \left[\frac{-k \cos k(L - x)}{\sin kL} + \frac{1}{L} \right] \tag{4.8b}$$

$$\theta_A = \left(\frac{dy}{dx} \right)_{x=0} = -\frac{M_A}{P} \left[\frac{k \cos kL}{\sin kL} - \frac{1}{L} \right] \tag{4.8c}$$

$$\theta_B = -\left(\frac{dy}{dx} \right)_{x=L} = \frac{M_A}{P} \left[\frac{k}{\sin kL} - \frac{1}{L} \right] \tag{4.8d}$$

Now, substitute $u = \dfrac{kL}{2}$, or $k = \dfrac{2u}{L}$

Also $u = \dfrac{L}{2} \sqrt{\dfrac{P}{EI}}$, or $P = \dfrac{4u^2}{L^2} EI$

Therefore,

$$\theta_A = \frac{M_A L}{3EI} \frac{3}{2u} \left[\frac{1}{2u} - \frac{1}{\tan 2u} \right] \tag{4.8e}$$

and

$$\theta_B = \frac{M_A L}{6EI} \frac{3}{u} \left[\frac{1}{\sin 2u} - \frac{1}{2u} \right] \tag{4.8f}$$

The expressions $M_A L/3EI$ and $M_A L/6EI$ give the angles produced by the moment M_A alone. The influence of the axial force P on the angles of rotations of the beam column is given by the terms inside the square brackets of Eqs. (4.8e) and (4.8f). Using the following notations

$$\psi_u = \frac{3}{2u} \left[\frac{1}{2u} - \frac{1}{\tan 2u} \right] \tag{4.8g}$$

$$\phi_u = \frac{3}{u} \left[\frac{1}{\sin 2u} - \frac{1}{2u} \right] \tag{4.8h}$$

We can write

$$\theta_A = \frac{M_A L}{3EI} \psi_u \tag{4.8i}$$

and

$$\theta_B = \frac{M_A L}{6EI} \phi_u \tag{4.8j}$$

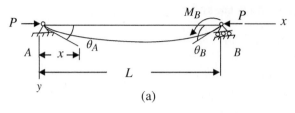

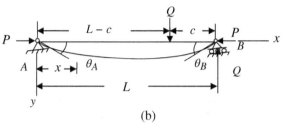

Figure 4.9 Beam column with the end moment on the right support and axial force: (a) End moment at the right support in beam column; (b) Beam column with equivalent couple.

The numerical values of the functions in Eqs. (4.3c), (4.3g), (4.5s), (4.8g), and (4.8h) are given by Timoshenko and Gere [3] for different u values.

If a moment M_B is acting on the right end as shown in Figure 4.9a, the deflection can be obtained by the same procedure by replacing the moment by two vertical forces, each of magnitude Q acting at the distance, c, apart forming a couple of magnitude M_B as illustrated in Figure 4.9b.

Assume as c approaches zero, the quantity $Qc = M_B$ remains finite. The deflection at a distance x is given by substituting $a = L - c$, or $c = L - a$ in Eq. (4.2s).

$$y = \frac{Q \sin kc}{EIk^3 \sin kL} \sin kx - \frac{Qc}{EILk^2} x \text{ for } 0 \leq x \leq L - c$$

For $c \to 0$, substitute $\sin kc = kc$, and we get

$$y = \frac{Qkc}{EIk^3 \sin kL} \sin kx - \frac{Qc}{EILk^2} x$$

Substituting $Qc = M_B$, we obtain

$$y = \frac{M_B}{P} \left[\frac{\sin kx}{\sin kL} - \frac{x}{L} \right] \tag{4.8k}$$

Taking the derivative of y we get

$$\frac{dy}{dx} = \frac{M_B}{P} \left[\frac{k \cos kx}{\sin kL} - \frac{1}{L} \right] \tag{4.8l}$$

$$\theta_A = \left(\frac{dy}{dx} \right)_{x=0} = \frac{M_B}{P} \left[\frac{k}{\sin kL} - \frac{1}{L} \right] \tag{4.8m}$$

$$\theta_B = -\left(\frac{dy}{dx} \right)_{x=L} = -\frac{M_B}{P} \left[\frac{k \cos kL}{\sin kL} - \frac{1}{L} \right] \tag{4.8n}$$

Now, substitute $u = \dfrac{kL}{2}$, or $k = \dfrac{2u}{L}$

Also $u = \dfrac{L}{2}\sqrt{\dfrac{P}{EI}}$, or $P = \dfrac{4u^2}{L^2}EI$

Therefore,

$$\theta_A = \frac{M_B L}{6EI}\frac{3}{u}\left[\frac{1}{\sin 2u} - \frac{1}{2u}\right] \tag{4.8o}$$

and

$$\theta_B = \frac{M_B L}{3EI}\frac{3}{2u}\left[\frac{1}{2u} - \frac{1}{\tan 2u}\right] \tag{4.8p}$$

or

$$\theta_A = \frac{M_B L}{6EI}\phi_u \tag{4.8q}$$

and

$$\theta_B = \frac{M_B L}{3EI}\psi_u \tag{4.8r}$$

4.4.3 Moments at Both Ends of Beam Column

If two moments are applied at the ends A and B of the beam column, the deflection at any point on the deflected shape can be obtained by superposing Eqs. (4.8a) and (4.8k), giving the deflections due to M_A and M_B respectively. Adding these results together, we get the deflection for the moments applied in Figure 4.10a. This type of load occurs when two eccentrically applied forces act as shown in Figure 4.10b.

$$y = \frac{M_A}{P}\left[\frac{\sin k(L-x)}{\sin kL} - \frac{(L-x)}{L}\right] + \frac{M_B}{P}\left[\frac{\sin kx}{\sin kL} - \frac{x}{L}\right] \tag{4.9a}$$

θ_A is obtained by adding the results from Eqs. (4.8c) and (4.8m), whereas θ_B is obtained by adding the results of Eqs. (4.8d) and (4.8n) as follows:

$$\theta_A = -\frac{M_A}{P}\left[\frac{k\cos kL}{\sin kL} - \frac{1}{L}\right] + \frac{M_B}{P}\left[\frac{k}{\sin kL} - \frac{1}{L}\right] \tag{4.9b}$$

$$\theta_B = \frac{M_A}{P}\left[\frac{k}{\sin kL} - \frac{1}{L}\right] - \frac{M_B}{P}\left[\frac{k\cos kL}{\sin kL} - \frac{1}{L}\right] \tag{4.9c}$$

or

$$\theta_A = \frac{M_A L}{3EI}\psi(u) + \frac{M_B L}{6EI}\phi_u \tag{4.9d}$$

and

$$\theta_B = \frac{M_B L}{3EI}\psi(u) + \frac{M_A L}{6EI}\phi(u) \tag{4.9e}$$

Similarly, the slope at any point on the deflected shape can be obtained by superposing Eqs. (4.8b) and (4.8l) giving the slopes due to moments M_A and M_B respectively. Adding these results together, we get the slope for the moments applied in Figure 4.10a.

$$y' = \frac{M_A}{P}\left[-k\frac{\cos k(L-x)}{\sin kL} + \frac{1}{L}\right] + \frac{M_B}{P}\left[\frac{k\cos kx}{\sin kL} - \frac{1}{L}\right] \tag{4.9f}$$

$$y'' = \frac{M_A}{P}\left[-k^2\frac{\sin k(L-x)}{\sin kL}\right] + \frac{M_B}{P}\left[-k^2\frac{\sin kx}{\sin kL}\right] \tag{4.9g}$$

$$y''' = \frac{M_A}{P}\left[k^3\frac{\cos k(L-x)}{\sin kL}\right] + \frac{M_B}{P}\left[-k^3\frac{\cos kx}{\sin kL}\right] \tag{4.9h}$$

The bending moment at a point in the span of the beam column is given by

$$M = -EIy'', \quad \text{and substituting} \quad P = EIk^2$$

$$M = M_A\left[\frac{\sin k(L-x)}{\sin kL}\right] + M_B\left[\frac{\sin kx}{\sin kL}\right] \tag{4.9i}$$

The maximum bending moment will occur where $\frac{dM}{dx} = 0$, or $y''' = 0$, therefore, from Eq. (4.9h), we have

$$\frac{M_A}{P}\left[k^3\frac{\cos k(L-x)}{\sin kL}\right] + \frac{M_B}{P}\left[-k^3\frac{\cos kx}{\sin kL}\right] = 0$$

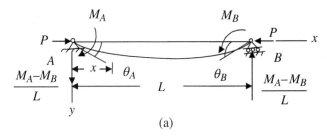

(a)

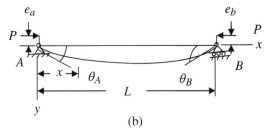

(b)

Figure 4.10 Beam column loaded with two end moments: (a) Beam column with two end moments; (b) Eccentrically loaded column.

or

$$\tan k\bar{x} = \frac{-M_A \cos kL + M_B}{M_A \sin kL} \tag{4.9j}$$

Eq. (4.9i) gives the position $\bar{x}$ of the maximum bending moment in the span of the beam column.

Therefore,

$$\sin k\bar{x} = \frac{-M_A \cos kL + M_B}{\sqrt{M_A^2 - 2M_A M_B \cos kL + M_B^2}} \tag{4.9k}$$

and

$$\cos k\bar{x} = \frac{M_A \sin kL}{\sqrt{M_A^2 - 2M_A M_B \cos kL + M_B^2}} \tag{4.9l}$$

Eq. (4.9i) can be written as

$$M = M_A \left[\frac{\sin kL \cos kx - \cos kL \sin kx}{\sin kL} \right] + M_B \frac{\sin kx}{\sin kL} \tag{4.9m}$$

Substitute the values of $\sin k\bar{x}$ and $\cos k\bar{x}$ from Eqs (4.k and 4.9l) in the Eq. (4.9m) and we have

$$M_{max} = M_B \left[\sqrt{ \frac{\left(\frac{M_A}{M_B} \right)^2 - 2 \left(\frac{M_A}{M_B} \right) \cos kL + 1}{\sin^2 kL} } \right] \tag{4.9n}$$

The calculated value of $\bar{x}$ may not lie within the span of the beam column, i.e. within the range $0 \leq x \leq L$. In that case, Eq. (4.9m) is not applicable and the maximum moment on the beam column is equal to the larger of the two end moments and acts at the end.

4.4.3.1 Two Equal Moments
If two equal moments act at the two ends given by

$$M_A = M_B = M_e$$

then the defection curve is obtained from Eq. (4.9a) as follows:

$$y = \frac{M_e}{P} \left[\frac{\sin k(L - x)}{\sin kL} - \frac{(L - x)}{L} + \frac{\sin kx}{\sin kL} - \frac{x}{L} \right] \tag{4.10a}$$

or

$$y = \frac{M_e}{P \cos \frac{kL}{2}} \left[\cos \left(\frac{kL}{2} - kx \right) - \cos \frac{kL}{2} \right] \tag{4.10b}$$

Substituting, $u = \frac{kL}{2}$, we get

$$y = \frac{M_e L^2}{8EI} \frac{2}{u^2 \cos u} \left[\cos \left(u - \frac{2ux}{L} \right) - \cos u \right] \qquad (4.10c)$$

In this case, the maximum deflection always occurs at the center of the span, it is obtained by substituting $x = L/2$, and the result is

$$y_{max} = (y)_{x=\frac{L}{2}} = \frac{M_e L^2}{8EI} \frac{2}{u^2 \cos u} (1 - \cos u) = \frac{M_e L^2}{8EI} \lambda(u) \qquad (4.10d)$$

Differentiating Eq. (4.10c) we have

$$y' = \frac{M_e L^2}{8EI} \frac{2}{u^2 \cos u} \left[\frac{2u}{L} \sin \left(u - \frac{2ux}{L} \right) \right] \qquad (4.10e)$$

The angles at the end can be obtained by taking

$$\theta_A = (y')_{x=0}, \text{ and } \theta_B = -(y')_{x=L}$$

or

$$\theta_A = \theta_B = \frac{M_e L}{2EI} \frac{\tan u}{u} \qquad (4.10f)$$

Taking the second derivative of Eq. (4.10c)

$$y'' = -\frac{M_e}{EI \cos u} \cos \left(u - \frac{2ux}{L} \right) \qquad (4.10g)$$

The bending moment is given by

$$M = -EIy'' = \frac{M_e}{\cos u} \cos \left(u - \frac{2ux}{L} \right) \qquad (4.10h)$$

The maximum bending moment occurs at the middle of the beam column at $x = L/2$, therefore

$$M_{max} = M_e \sec u \qquad (4.10i)$$

When the axial force P is small in comparison to its critical value, $P = \frac{\pi^2 EI}{L^2}$, the term u is small and $\sec u$ can be taken equal to unity, that means the bending moment is constant along the span of the beam column. As P approaches P_{cr}, the value of u increases to $\pi/2$, and $\sec u$ goes to infinity. At such a value of P, the bending moment at the mid-span is very large.

4.4.3.2 Moments at Both Ends of the Beam Column: Alternate Method

The beam column in Figure 4.10 can also be analyzed by considering the equilibrium of the free body diagram of the deflected shape as shown in Figure 4.11.

Taking moments of all the forces in Figure 4.11 about the right end of the free body diagram we have

$$-\frac{M_A - M_B}{L} x + Py + M_A - M_x = 0 \qquad (4.11a)$$

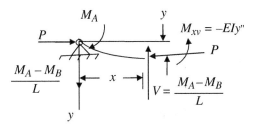

Figure 4.11 Free body diagram of beam column loaded with two end moments.

Substitute $M_x = -EIy''$, and $P = EIk^2$, then Eq. (4.11a) becomes

$$y'' + k^2 y = \frac{M_A - M_B}{LEI} x - \frac{M_A}{EI} \tag{4.11b}$$

The general solution of the differential Eq. (4.11b) is

$$y = A \sin kx + B \cos kx + \frac{M_A - M_B}{EIk^2 L} x - \frac{M_A}{EIk^2} \tag{4.11c}$$

The boundary conditions at the end of the member are

$$x = 0, \ y = 0; \ \text{and} \ x = L, \ y = 0$$

Substituting these boundary conditions, the constants A and B are obtained as

$$B = \frac{M_A}{EIk^2}, \ \text{and} \ A = -\frac{1}{EIk^2 \sin kL}(M_A \cos kL - M_B) \tag{4.11d}$$

The general solution of Eq. (4.11b) is given by

$$y = \frac{1}{EIk^2 \sin kL}(-M_A \cos kL + M_B) \sin kx + \frac{M_A}{EIk^2} \cos kx$$
$$+ \frac{M_A - M_B}{EIk^2 L} x - \frac{MA}{EIk^2} \tag{4.11e}$$

or

$$y = \frac{M_A}{P}\left[-\frac{\cos kL \sin kx}{\sin kL} + \cos kx + \frac{x}{L} - 1\right] + \frac{M_B}{P}\left[\frac{\sin kx}{\sin kL} - \frac{x}{L}\right] \tag{4.11f}$$

or

$$y = \frac{M_A}{P}\left[\frac{\sin k(L-x)}{\sin kL} - \frac{L-x}{L}\right] + \frac{M_B}{P}\left[\frac{\sin kx}{\sin kL} - \frac{x}{L}\right] \tag{4.11g}$$

Equation (4.11g) is the same as Eq. (4.9a) before being obtained by the method of superposition.

4.4.3.3 End Moments of the Same Sign Giving Double Curvature
Equation (4.11g) is derived for a beam column bent in a single curvature, but it can also be used when the member is bent in a double curvature by two clockwise end moments as shown in

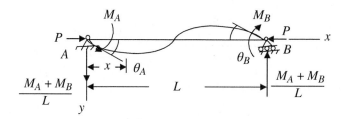

Figure 4.12 Beam column in a double curvature due to end moments.

Figure 4.12. Replacing M_B in the Eq. (4.11g) by $-M_B$, we have

$$y = \frac{M_A}{P}\left[\frac{\sin k(L-x)}{\sin kL} - \frac{L-x}{L}\right] - \frac{M_B}{P}\left[\frac{\sin kx}{\sin kL} - \frac{x}{L}\right] \tag{4.12a}$$

$$\frac{dy}{dx} = \frac{M_A}{P}\left[\frac{-k\cos k(L-x)}{\sin kL} + \frac{1}{L}\right] - \frac{M_B}{P}\left[\frac{k\cos kx}{\sin kL} - \frac{1}{L}\right] \tag{4.12b}$$

$$\theta_A = \frac{dy}{dx}\Big|_{x=0} = \frac{M_A}{P}\left[-\frac{k\cos kL}{\sin kL} + \frac{1}{L}\right] - \frac{M_B}{P}\left[\frac{k}{\sin kL} - \frac{1}{L}\right]$$

Now, substitute $u = \dfrac{kL}{2}$, or $k = \dfrac{2u}{L}$

Also $u = \dfrac{L}{2}\sqrt{\dfrac{P}{EI}}$, or $P = \dfrac{4u^2}{L^2}EI$

$$\theta_A = \frac{M_A L}{3EI}\psi(u) - \frac{M_B L}{6EI}\phi(u) \tag{4.12c}$$

$$\theta_B = \frac{dy}{dx}\Big|_{x=L} = \frac{M_A}{P}\left[-\frac{k}{\sin kL} + \frac{1}{L}\right] - \frac{M_B}{P}\left[\frac{k\cos kL}{\sin kL} - \frac{1}{L}\right]$$

or

$$\theta_B = -\frac{M_A L}{6EI}\phi(u) + \frac{M_B L}{3EI}\psi(u) \tag{4.12d}$$

The maximum bending moment due to end moments causing a double curvature is given by replacing M_B by $-M_B$ in Eq. (4.9n) as follows:

$$M_{max} = -M_B\left[\sqrt{\frac{\left(\frac{M_A}{M_B}\right)^2 + 2\left(\frac{M_A}{M_B}\right)\cos kL + 1}{\sin^2 kL}}\right] \tag{4.12e}$$

This maximum moment occurs at $\bar{x}$ obtained from Eq. (4.9j) by replacing M_B by $-M_B$

$$\tan k\bar{x} = \frac{-(M_A\cos kL + M_B)}{M_A\sin kL} \tag{4.12f}$$

If the calculated value of $\bar{x}$ lies outside the span, $0 \leq x \leq L$, the maximum moment lies at the member end given by the larger of the two end moments in Figure 4.12. The maximum absolute moments for single curvature and double curvature bent shapes of a beam column can be written together in the following form:

$$M_{\max} = |M_B| \left[\sqrt{\frac{\left(\frac{M_A}{M_B}\right)^2 + 2\left(\frac{M_A}{M_B}\right)\cos kL + 1}{\sin^2 kL}} \right] \qquad (4.12g)$$

The ratio $\frac{M_A}{M_B}$ is taken as positive for a double curvature bending and is taken as negative if the member is bent in a single curvature.

4.5 Columns with Elastic Restraints

Most of the columns in practice do not have idealized end conditions. These members are supported by other members such as beams or columns at the ends. Hence the end conditions are elastically restrained, the degree of restraint depends on the stiffness of the supporting members. A column that is supported against translation but is allowed to rotate partially by two rotational springs at the ends is considered here as shown in Figure 4.13. Since these rotational springs exert moments at the ends, the problem is similar to the beam columns subjected to end moments.

The end moments caused by bending of the member are shown in the positive direction in Figure 4.12. Those end moment that cause compression on the top of the beam column are considered positive. The angles are considered positive when the ends rotate in the direction of positive moments. Let β_A and β_B be the stiffness of the rotational springs at the ends A and B of the member. The rotational stiffness is defined as the reactive moment at the end when the angle of rotation is equal to unity. If θ_A and θ_B are the end rotations, there will be couples M_A and M_B at the ends of the member given by

$$M_A = -\beta_A \theta_A, \text{ and } M_B = -\beta_B \theta_B \qquad (4.13a)$$

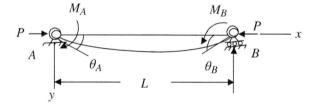

Figure 4.13 Elastically restrained column.

Using Eqs. (4.9d) and (4.9e) and Eq. (4.13a) yields

$$-\frac{M_A}{\beta_A} = \frac{M_A L}{3EI}\psi(u) + \frac{M_B L}{6EI}\phi u$$

and

$$-\frac{M_B}{\beta_B} = \frac{M_B L}{3EI}\psi(u) + \frac{M_A L}{6EI}\phi u \tag{4.13b}$$

Thus, we get the following homogeneous linear algebraic equations in M_A and M_B:

$$\begin{bmatrix} \frac{1}{\beta_A} + \frac{L}{3EI}\psi(u) & \frac{L}{6EI}\phi(u) \\ \frac{L}{6EI}\phi(u) & \frac{1}{\beta_B} + \frac{L}{3EI}\psi(u) \end{bmatrix} \begin{Bmatrix} M_A \\ M_B \end{Bmatrix} = \begin{Bmatrix} 0 \\ 0 \end{Bmatrix} \tag{4.13c}$$

For a nontrivial solution, the determinant of the coefficient matrix should be zero. Therefore

$$\begin{vmatrix} \frac{1}{\beta_A} + \frac{L}{3EI}\psi(u) & \frac{L}{6EI}\phi(u) \\ \frac{L}{6EI}\phi(u) & \frac{1}{\beta_B} + \frac{L}{3EI}\psi(u) \end{vmatrix} = 0 \tag{4.13d}$$

The characteristic equation is

$$\left[\frac{1}{\beta_A} + \frac{L}{3EI}\psi(u)\right]\left[\frac{1}{\beta_B} + \frac{L}{3EI}\psi(u)\right] - \left[\frac{L}{6EI}\phi(u)\right]^2 = 0 \tag{4.13e}$$

If $\beta_A = \beta_B = \beta$, then

$$\frac{1}{\beta} + \frac{L}{3EI}\psi(u) = \pm\frac{L}{6EI}\phi(u) \tag{4.13f}$$

From the first equation of Eq. (4.13c)

$$\left[\frac{1}{\beta_A} + \frac{L}{3EI}\psi(u)\right]M_A + \frac{L}{6EI}\phi(u)M_B = 0 \tag{4.13g}$$

From Eqs. (4.13f) and (4.13g)

$$M_B = \mp M_A \tag{4.13h}$$

Consider the first case, $M_B = -M_A$. In this case, the end moments are acting as shown in Figure 4.12 and the member bends in a double curvature. Equation (4.13g) yields

$$\frac{1}{\beta_A} + \frac{L}{3EI}\psi(u) = \frac{L}{6EI}\phi(u)$$

Substitute values of $\psi(u)$ and $\varphi(u)$ from Eqs. (4.8g) and (4.8h)

$$\frac{1}{\beta} + \frac{L}{3EI}\left(\frac{3}{2u}\right)\left[\frac{1}{2u} - \frac{1}{\tan 2u}\right] = \frac{L}{6EI}\left(\frac{3}{u}\right)\left[\frac{1}{\sin 2u} - \frac{1}{2u}\right] \tag{4.13i}$$

or

$$\tan u = \frac{u}{1 + \frac{2EIu^2}{\beta L}} \tag{4.13j}$$

The quantity u can be calculated for a particular value of β from Eq. (4.13j), which then gives the critical load. For example, as As $\beta \to 0$, the end conditions are pinned, $\tan u \to 0$, $u_{cr} = \pi$. Therefore, $P_{cr} = \frac{4\pi^2 EI}{L^2} = \frac{\pi^2 EI}{\left(\frac{L}{2}\right)^2}$, which is the same result as for a pinned-pinned column of length $L/2$. The member acts as two pinned-pinned columns, each of length equal to $L/2$.

As $\beta \to \infty$, the end conditions are fixed, $\tan u \to u$, $u_{cr} = 4.493$. Therefore, $P_{cr} = \frac{8.18\pi^2 EI}{L^2} = \frac{\pi^2 EI}{\left(0.7\frac{L}{2}\right)^2}$, which is the same result as for a fixed-pinned column of length $L/2$. The member acts as two fixed-pinned columns, each of length equal to $L/2$.

Consider the second case, $M_B = M_A$. In this case the end moments are acting as shown in Figure 4.10 and the member bends in a single curvature. Eq. (4.13g) yields

$$\frac{1}{\beta_A} + \frac{L}{3EI}\psi(u) = -\frac{L}{6EI}\phi(u)$$

Substitute values of ψ_u and φ_u from Eqs. (4.8g) and (4.8h)

$$\frac{1}{\beta} + \frac{L}{3EI}\left(\frac{3}{2u}\right)\left[\frac{1}{2u} - \frac{1}{\tan 2u}\right] = -\frac{L}{6EI}\left(\frac{3}{u}\right)\left[\frac{1}{\sin 2u} - \frac{1}{2u}\right] \qquad (4.13k)$$

or

$$\tan u = -\frac{2EI}{\beta L}u \qquad (4.13l)$$

The u can be calculated for a particular value of β, which then gives the critical load. For example as $\beta \to 0$, the end conditions are pinned, $\tan u \to \infty$, or $u_{cr} = \pi/2$.

Therefore, $P_{cr} = \frac{\pi^2 EI}{L^2}$, and we get the same result as for a pinned-pinned column. The member acts as a pinned-pinned column of length L.

As $\beta \to \infty$, the end conditions are fixed, $\tan u \to 0$, or $u_{cr} = \pi$.

Therefore, $P_{cr} = \frac{4\pi^2 EI}{L^2}$, and we get the same result as for a fixed-fixed column. The member acts as a fixed-fixed column of length L.

Out of the two cases presented by Eq. (4.13h), a critical load for the column whose ends are elastically restrained occurs when the member bends in a single curvature. Therefore, the minimum critical load is given by Eq. (4.13l) for the column that is elastically restrained at the ends.

4.6 Beam Columns with Different End Conditions and Loads

4.6.1 Pinned-fixed Beam Columns with a Concentrated Load

Consider the beam column that is simply supported at the left end and is fixed at the right support. A concentrated force Q is acting on it at a distance a from the left support as shown

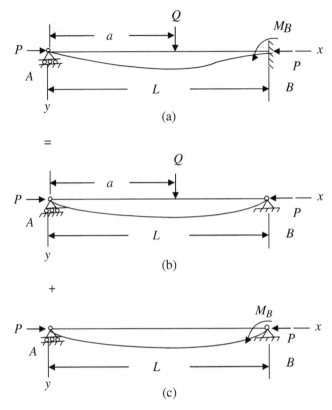

Figure 4.14 Pinned-fixed beam column with concentrated force: (a) Fixed-pinned beam column; (b) Pinned-pinned beam column with concentrated force; (c) Pinned-pinned beam column with end moment.

in Figure 4.14. The method of superposition is used here to find the deflection and bending moment at any section of the member.

The beam column in Figure 4.14a can be considered to be the sum of the simply supported beam acted upon by a concentrated load Q in Figure 4.14b, and the simply supported beam column subjected to an end moment M_B as shown in Figure 4.14c. The fixed end moment M_B is shown in the positive direction in Figure 4.14c. The fixed end moment M_B is found by adding the rotation produced by the concentrated force Q in Figure 4.14b (Eq. 4.2x) and the rotation produced by the moment M_B in Figure 4.14c (Eq. 4.8r), and equating the sum equal to zero.

$$\frac{Q\sin ka}{EIk^2 \sin kL} - \frac{Qa}{EILk^2} + \frac{M_B L}{3EI}\psi(u) = 0 \tag{4.14a}$$

If we consider the concentrated force Q acting at the mid-span, then $a = L/2$.

or

$$\frac{Q\sin \frac{kL}{2}}{EIk^2 \sin kL} - \frac{QL}{2EIk^2 L} + \frac{M_B L}{3EI}\psi(u) = 0$$

or

$$\frac{QL^2}{16EI}\frac{2(1-\cos u)}{u^2}+\frac{M_B L}{3EI}\psi(u)=0$$

or

$$M_B=-\frac{3QL}{16}\frac{\lambda(u)}{\psi(u)} \tag{4.14b}$$

The negative sign in Eq. (4.14b) shows that the moment M_B acts opposite to the assumed direction.

Having found M_B, the deflection at the mid-span, $x = L/2$, produced by the end moment M_B in Figure 4.14c can be found from Eq. (4.8k) as follows:

$$y=\frac{M_B}{2EIk^2}\left[\frac{1-\cos u}{\cos u}\right]=\frac{M_B L^2}{16EI}\left[\frac{2(1-\cos u)}{u^2\cos u}\right] \tag{4.14c}$$

The deflection at the mid-span in Figure 4.14a when $a = L/2$ is obtained by superposing deflections due to the concentrated force Q in Figure 4.14b is given by Eq. (4.3e) and the deflection produced by the end moment M_B given by Eq. (4.14c) is as follows:

$$y|_{x=L/2}=\frac{QL^3}{48EI}\left[\frac{3(\tan u - u)}{u^3}\right]+\frac{M_B L^2}{16EI}\left[\frac{2(1-\cos u)}{u^2\cos u}\right]$$

or

$$y|_{x=L/2}=\frac{QL^3}{48EI}\chi(u)+\frac{M_B L^2}{16EI}\lambda(u) \tag{4.14d}$$

If there was no axial load P, the deflection at the center of span produced by the concentrated load Q acting at the mid-span in Figure 4.14a is given by discarding the terms of u in Eqs. (4.14b) and (4.14d).

$$y|_{x=\frac{L}{2}}=\frac{QL^3}{48EI}+\frac{M_B L^2}{16EI}=\frac{QL^3}{48EI}-\frac{3QL}{16}\left[\frac{L^2}{16EI}\right]$$

or

$$y|_{x=\frac{L}{2}}=\frac{7QL^3}{768}$$

a result that can be verified from the structural analysis of pinned-fixed beam subjected to a concentrated load Q at the center of span.

4.6.2 Pinned-fixed Beam Columns Subjected to Uniformly Distributed Load

A uniformly distributed load w is applied on the column shown in Figure 4.15. The beam column is assumed to consist of the sum of the uniformly loaded, simply supported member in Figure 4.4a and the simply supported member subjected to an end moment M_B in Figure 4.14c to solve the problem. The method of superposition is used here.

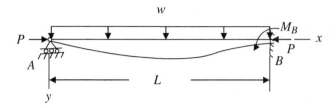

Figure 4.15 Pinned-fixed beam column with uniformly distributed load.

The fixed end moment M_B is obtained by adding the rotations at the end B produced by the uniformly distributed load in Figure 4.4a (Eq. 4.5n) and the moment M_B in Figure 4.14c (Eq. 4.8r) and equating the sum equal to zero

$$\frac{wL^3}{24EI}\chi(u) + \frac{M_BL}{3EI}\psi(u) = 0$$

or

$$M_B = -\frac{wL^2}{8}\frac{\chi(u)}{\psi(u)} \tag{4.15a}$$

The negative sign shows that the moment is in the opposite direction to that assumed. Knowing the end moment, the deflection curve is obtained by superposing the deflections produced by the uniform load (Eq. 4.5l) and the deflections produced by the end moment M_B (Eq. 4.8k). If $x = L/2$, then the deflection at the mid-span is found by superposing the deflection produced by the uniformly distributed load (Eq. 4.5r) and the deflection at the mid-span produced by the end moment M_B (Eq. 4.8k, $x = L/2$).

$$y|_{x=\frac{L}{2}} = \frac{5wL^4}{384EI}\eta(u) + \frac{M_BL^2}{16EI}\lambda(u) \tag{4.15b}$$

It can easily be shown that if the axial force P is zero, the deflection at the mid-span of the beam column is $\frac{wL^4}{192EI}$, which is obtained by discarding the terms of u in Eqs. (4.15a) and (4.15b)

Similarly, the bending moment at a section of the member can be found by superposing the bending moment produced by the uniform load (Eq. 4.5p) and the moment produced by the end moment M_B (second part of Eq. 4.9i).

$$M = \frac{wL^2}{4u^2}\left[\tan u \sin\frac{2ux}{L} + \cos\frac{2ux}{L} - 1\right] + M_B\left[\frac{\sin kx}{\sin kL}\right] \tag{4.15c}$$

If

$$x = \frac{L}{2}$$

$$M|_{x=\frac{L}{2}} = \frac{wL^2}{8}\left[\frac{2(1-\cos u)}{u^2\cos u}\right] + \frac{M_B}{2}\sec u$$

Substituting M_B from Eq. (4.15a) we have

or

$$M|_{x=\frac{L}{2}} = \frac{wL^2}{8}\lambda(u) - \frac{wL^2}{16}\frac{\chi(u)}{\psi(u)}\sec u \qquad (4.15d)$$

It can easily be shown that if the axial force $P = 0$, the bending moment at the center of the span is $\frac{wL^2}{8}$.

4.6.3 Fixed-fixed Beam Column with Concentrated Force

The method of superposition can be used to evaluate the deflection and bending moment in the fixed-fixed beam column acted on by a concentrated force shown in Figure 4.16. The beam column can be considered as a sum of a simply supported beam column acted on by a transverse concentrated force and the axial force P (Figure 4.2a) plus the simply supported beam column acted on by two end moments and the axial force P (Figure 4.10a). The moments at the ends, M_A and M_B, are found from the conditions that the slopes at the ends are zero. The equations for finding the end moments are

$$\theta_A = \theta_{A0} + \theta_{A(M_A)} + \theta_{A(M_B)} = 0 \qquad (4.16a)$$

$$\theta_B = \theta_{B0} + \theta_{B(M_A)} + \theta_{B(M_B)} = 0 \qquad (4.16b)$$

The rotations θ_{A0} and θ_{B0} (Eqs. 4.2w and 4.2x) are the angles of rotation at the ends of the simply supported beam column when the concentrated load acts along with the axial force. The angles $\theta_{A(M_A)}$ (Eq. 4.8i) and $\theta_{B(MA)}$ (Eq. 4.8j) are the rotations at the ends of the simply supported member due to the end moment M_A. Whereas the angles $\theta_{A(M_B)}$ (Eq. 4.8q) and $\theta_{B(M_B)}$ (Eq. 4.8r) are the rotations at the ends of the simply supported member due to the end moment M_B. The axial force P also acts along with the end moments on the member. Equations (4.16a) and (4.16b) can be written as

$$\theta_A = \theta_{A0} + \frac{M_A L}{3EI}\psi(u) + \frac{M_B L}{6EI}\phi(u) = 0 \qquad (4.16c)$$

$$\theta_B = \theta_{B0} + \frac{M_A L}{6EI}\phi(u) + \frac{M_B L}{3EI}\psi(u) = 0 \qquad (4.16d)$$

If the concentrated force Q acts at the center of the span of the member, $a = L/2$, then

$$M_A = M_B = M_F$$

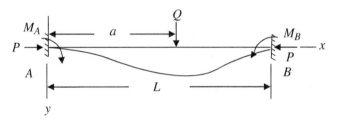

Figure 4.16 Fixed-fixed beam column with a concentrated force.

and

$$\theta_{A0} = \theta_{B0} = \frac{QL^2}{16EI} \lambda(u) \tag{4.3b}$$

Substituting in Eq. (4.16c) or (4.16d), we get

$$M_F = -\frac{QL}{8} \left[\frac{u}{\tan u} \lambda(u) \right]$$

or

$$M_F = M_{\max} = -\frac{QL}{8} \left[\frac{2(1 - \cos u)}{u \sin u} \right] \tag{4.16e}$$

The minus sign indicates that the moments are in the opposite directions to those assumed. The term, $\frac{QL}{8}$, is the absolute value of the fixed end moments when the axial force P is zero. The maximum deflection occurs at the mid-span and from Eqs. (4.3f) and (4.10d) is equal to

$$y_{\max} = y|_{x=\frac{L}{2}} = \frac{QL^3}{48EI} \chi(u) + \frac{M_F L^2}{8EI} \lambda(u) \tag{4.16f}$$

In the absence of the axial load, $P = 0$, the y_{max} becomes

$$y_{\max} = \frac{QL^3}{48EI} + \frac{M_F L^2}{8EI} = \frac{QL^3}{48EI} - \frac{QL}{8} \left[\frac{L^2}{8EI} \right] = \frac{QL^3}{192EI} \tag{4.16g}$$

The bending moment at the mid-span from Eqs. (4.3m) and (4.10i) is given by

$$M|_{x=\frac{L}{2}} = \frac{QL}{4} \left[\frac{\tan u}{u} \right] + M_F \sec u$$

When P is absent

$$M|_{x=\frac{L}{2}} = \frac{QL}{4} - \frac{QL}{8} = \frac{QL}{8} \tag{4.16h}$$

4.6.4 Fixed-fixed Beam Column with Uniformly Distributed Load

Consider the case of beam column that has both built-in ends and is loaded with a uniformly distributed load w and an axial force P in Figure 4.17. The member is considered to be the sum of a simply supported beam column subjected to uniform load and an axial force plus a simply supported member acted on by end moments and the axial force P.

The deflection curve in this case is symmetrical and the moments at the built in ends are equal, $M_A = M_B = M_F$. The end moments are obtained from the condition that the sum of the rotations produced by the uniformly distributed transverse load acting on a simply supported member are canceled by the rotations produced by the end moments on the simply supported member. In both cases, the beam column is also subjected to the axial force P in addition to the transverse load or the end moments. Use Eqs. (4.5n) and (4.10f) to get

$$\frac{wL^3}{24EI} \chi(u) + \frac{M_F L}{2EI} \left[\frac{\tan u}{u} \right] = 0 \tag{4.17a}$$

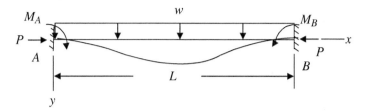

Figure 4.17 Fixed-fixed beam column with uniformly distributed load.

or

$$\frac{wL^3}{24EI}\left[\frac{3(\tan u - u)}{u^3}\right] + \frac{M_F L}{2EI}\left[\frac{\tan u}{u}\right] = 0$$

or

$$M_F = M_{max} = -\frac{wL^2}{12}\left[\frac{3(\tan u - u)}{u^2 \tan u}\right] \tag{4.17b}$$

The minus sign indicates the end moments are in the opposite direction to those assumed.

In the absence of axial force P, $M_F = -\frac{wL^2}{12}$.

The deflection at any cross-section is given by superposing the deflection produced by the uniform load (Eq. 4.5l) and the deflection produced by the two equal end moments (Eq. 4.10c). Similarly, the bending moment at any section is obtained by superposing the bending moment produced by the uniform load and the bending moment produced by the two equal end moments (Eqs. 4.5p and 4.10h). The maximum deflection occurs at the center of the span given by adding the deflections from Eqs. (4.5q) and (4.10d).

$$y_{max} = \frac{5wL^4}{384EI}\left[\frac{12(2\sec u - u^2 - 2)}{5u^4}\right] + \frac{M_F L^2}{8EI}\left[\frac{2(1 - \cos u)}{u^2 \cos u}\right]$$

or

$$y_{max} = \frac{wL^4}{384EI}\left[\frac{12(2 - 2\cos u - u\sin u)}{u^3 \sin u}\right] \tag{4.17c}$$

If the axial force is absent, $P = 0$, and, $y_{max} = \frac{wL^4}{384EI}$.

The bending moment at the mid-span is obtained by adding the bending moments from Eqs. (4.5x) and (4.10i).

$$M|_{x=\frac{L}{2}} = \frac{wL^2}{8}\left[\frac{2(1 - \cos u)}{u^2 \cos u}\right] + M_F \sec u$$

or

$$M|_{x=\frac{L}{2}} = \frac{wL^2}{8}\left[\frac{2(1 - \cos u)}{u^2 \cos u}\right] - \frac{wL^2}{12}\left[\frac{3(\tan u - u)}{u^2 \tan u}\right]\frac{1}{\cos u}$$

or

$$M|_{x=\frac{L}{2}} = \frac{wL^2}{24}\left[\frac{6(u - \sin u)}{u^2 \sin u}\right] \tag{4.17d}$$

If

$$P = 0, M|_{x=\frac{L}{2}} = \frac{wL^2}{24}.$$

4.7 Alternate Method Using Basic Differential Equations

4.7.1 Fixed-fixed Beam Column with Uniformly Distributed Load

The beam column with the clamped supports and a uniformly distributed load acting on it as shown in Figure 4.17 is analyzed here using the basic differential Eq. (4.1f).

$$\frac{d^4y}{dx^4} + k^2 \frac{d^2y}{dx^2} = \frac{w}{EI} \tag{4.1f}$$

The general solution is

$$y = A \sin kx + B \cos kx + Cx + D + y(p) \tag{4.18a}$$

Where $y(p)$ is the particular solution. Assume the particular solution as a polynomial

$$y(p) = Ex^4 + Fx^3 + Gx^2 + Hx + I \tag{4.18b}$$

Take the derivatives of Eq. (4.18b) with respect to x and substitute into Eq. (4.18a) to get the particular solution

$$y(p) = \frac{wx^2}{2EIk^2} \tag{4.18c}$$

Therefore, the general solution becomes

$$y = A \sin kx + B \cos kx + Cx + D + \frac{wx^2}{2EIk^2} \tag{4.18d}$$

The constants A, B, C, and D are obtained from the boundary conditions

At $x = 0$, $y = 0$, and $y' = 0$

and $x = L$, $y = 0$, and $y' = 0$

Substitute $y = 0$ at $x = 0$

$$B = -D \tag{4.18e}$$

$$y' = Ak \cos kx - Bk \sin kx + C + \frac{wx}{EIk^2} \tag{4.18f}$$

Substitute $y' = 0$ at $x = 0$

$$C = -Ak \tag{4.18g}$$

Now substitute at $x = L$, $y = 0$, and $y' = 0$, we obtain

$$A \sin kl + B \cos kL + CL + D + \frac{wL^2}{2EIk^2} = 0 \tag{4.18h}$$

and

$$Ak \cos kL - Bk \sin kL + C + \frac{wL}{EIk^2} = 0 \tag{4.18i}$$

Solving Eqs. (4.18e), (4.18g), (4.18h), and (4.18i), we get

$$A = \frac{wL}{2EIk^3} \tag{4.18j}$$

$$B = \frac{wL}{2EIk^3 \tan \frac{kL}{2}} \tag{4.18k}$$

$$C = -\frac{wL}{2EIk^2} \tag{4.18l}$$

$$D = -\frac{wL}{2EIk^3 \tan \frac{kL}{2}} \tag{4.18m}$$

Now, Eq. (4.18a) becomes

$$y = \frac{wL}{2EIk^3} \sin kx + \frac{wL}{2EIk^3 \tan \frac{kl}{2}} \cos kx - \frac{wLx}{2EIk^2} - \frac{wL}{2EIk^3 \tan \frac{kl}{2}} + \frac{wx^2}{2EIk^2}$$

or

$$y = \frac{wL}{2EIk^3} \left[\frac{\tan \frac{kL}{2} \sin kx + \cos kx - 1}{\tan \frac{kl}{2}} \right] - \frac{w}{2EIk^2} [x(L - x)] \tag{4.18n}$$

$$y' = \frac{wL}{2EIk^3} \left[k \cos kx - \frac{k \sin kx}{\tan \frac{kL}{2}} - k + \frac{2kx}{L} \right] \tag{4.18o}$$

$$y'' = \frac{wL}{2EIk^3} \left[-k^2 \sin kx - \frac{k^2 \cos kx}{\tan \frac{kL}{2}} + \frac{2k}{L} \right] \tag{4.18p}$$

The maximum deflection occurs at $x = L/2$

$$y_{max}|_{x=\frac{L}{2}} = \frac{wL^4}{384EI} \left[\frac{12(2 - 2\cos u - u\sin u)}{u^3 \sin u} \right] \tag{4.18q}$$

That is the same as Eq. (4.17c).

$$M = -EIy''$$

The bending moment at a section at a distance x from the left support is given by

$$M = \frac{wL}{2k^3} \left[k^2 \sin kx + \frac{k^2 \cos kx}{\tan \frac{kL}{2}} - \frac{2k}{L} \right] \tag{4.18r}$$

The maximum bending moment occurs at the fixed end, $x = 0$, and is equal to

$$M = -\frac{wL^2}{12} \left[\frac{3(\tan u - u)}{u^2 \tan u} \right]$$

That is the same as Eq. (4.17b).

The bending moment at the mid-span, $x = L/2$, is given by

$$M|_{x=\frac{L}{2}} = \frac{wl^2}{24u^2}\left[\frac{6(u - \sin u)}{\sin u}\right] \tag{4.18s}$$

That is the same as Eq. (4.17d).

4.7.2 Pinned-fixed Beam Column with Uniformly Distributed Load

Consider the beam column in Figure 4.15, that is pinned at the left-hand support and fixed at the right-hand support. The basic differential equation governing the equilibrium in the deflected shape is given by

$$\frac{d^4y}{dx^4} + k^2\frac{d^2y}{dx^2} = \frac{w}{EI} \tag{4.1f}$$

The general solution is

$$y = A \sin kx + B \cos kx + Cx + D + \frac{wx^2}{2EIk^2} \tag{4.18d}$$

$$y' = Ak \cos kx - Bk \sin kx + C + \frac{wx}{EIk^2} \tag{4.18f}$$

$$y'' = -k^2A \sin kx - Bk^2 \cos kx + \frac{w}{EIk^2} \tag{4.19a}$$

The constants A, B, C, and D are obtained from the boundary conditions

At $x = 0$, $y = 0$, and $y'' = 0$

and $x = L$, $y = 0$, and $y' = 0$

Substituting $y = 0$ at $x = 0$

$$B = -D \tag{4.19b}$$

Substitute $y'' = 0$ at $x = 0$

$$B = \frac{w}{EIk^4} \tag{4.19c}$$

Now substituting at $x = L$, $y = 0$, and $y' = 0$, we obtain

$$A \sin kl + B \cos kL + CL + D + \frac{wL^2}{2EIk^2} = 0 \tag{4.19d}$$

and

$$Ak \cos kL - Bk \sin kL + C + \frac{wL}{EIk^2} = 0 \tag{4.19e}$$

Solving Eqs. (4.19b), (4.19c), (4.19d), and (4.19e), we get

$$A = \frac{w}{2EIk^4}\left[\frac{k^2L^2 - 2\cos kL - 2kL \sin kL + 2}{\sin kL - kL \cos kL}\right] \tag{4.19f}$$

$$B = \frac{w}{EIk^4} \tag{4.19g}$$

$$C = \frac{w}{2EIk^3} \left[\frac{k^2L^2 \cos kL - 2\cos kL - 2kL \sin kL + 2}{\sin kL - kL \cos kL} \right] \tag{4.19h}$$

$$D = -\frac{w}{EIk^4} \tag{4.19i}$$

Substitute the constants A, B, C, and D from Eqs. (4.19f–4.19i) in Eq. (4.18d) to get the deflection curve for the beam column. It can be shown that the deflection at the mid-span, $x = L/2$, is the same as obtained in Eqs. (4.15a) and (4.15b). Similarly, the bending moment at any section is obtained from

$$M = -EIy''$$

or

$$M = \frac{w}{2k^2} \left[\frac{k^2L^2 - 2\cos kL - 2kL \sin kL + 2}{\sin kL - kL \cos kL} \right] \sin kx + \frac{w}{k^2}(\cos kx - 1) \tag{4.19j}$$

It can easily be seen that the moment is zero at the left support, $x = 0$. It can also be shown that the moment at the right support, $x = L$, and the bending moment at the mid-span, $x = L/2$, are the same obtained from Eq. (4.15a) and Eq. (4.15d) respectively.

4.8 Continuous Beam Columns

If the beam columns are supported continuously on more than two supports, then they are called continuous beam columns and these structures are statically indeterminate. Continuous beam columns on rigid supports with a lateral load and axial forces are considered here. In these cases the support bending moments are considered as redundant forces. Let 1, 2, 3, ------, m denote the consecutive supports; $M_1, M_2, M_3, - - - - - - -, M_m$ be the corresponding support moments; and $L_1, L_2, L_3, - - - - - - -, L_{m-1}$ are the span lengths between the supports. For each span the axial force is given by $P_1, P_2, P_3, - - - - - - - -, P_{m-1}$; and the flexural rigidity is given by $EI_1, EI_2, EI_3, - - - - - -, EI_{m-1}$. The compressive force and the flexural rigidity may vary from span to span but it is constant within the span.

Now, consider any two consecutive spans between the supports $n - 1$, n, and $n + 1$, as shown in Figure 4.18. The bending moments at the supports are shown in their positive directions. They are considered positive if they produce compression at the top of the member. The angles are considered positive when they occur in the same direction as the positive bending moments. At the intermediate support n, the tangent to the deflection curves in the two spans is a straight line. Hence,

$$\theta_n = -\theta_n' \tag{4.20a}$$

The rotations are found for each span considering it as a simply supported member subjected to lateral load and the end moments. The total angle of rotation is obtained by adding the rotations due to lateral load and the end moments from Eqs. (4.9d) and (4.9e) as follows:

$$\theta_n = \theta_{0n} + \frac{M_n L_{n-1}}{3EI_{n-1}} \psi(u_{n-1}) + \frac{M_{n-1} L_{n-1}}{6EI_{n-1}} \phi(u_{n-1}) \tag{4.20b}$$

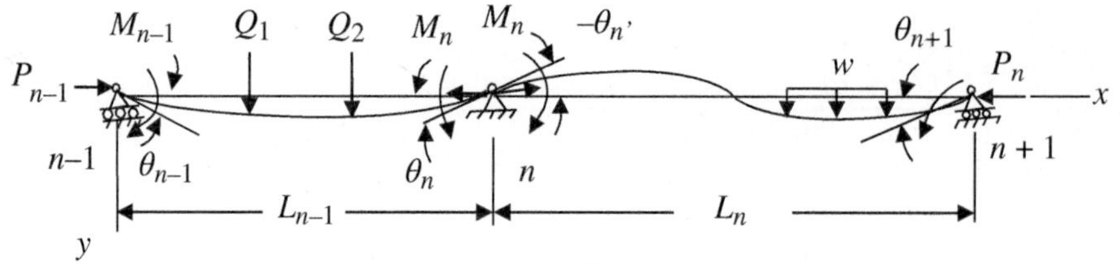

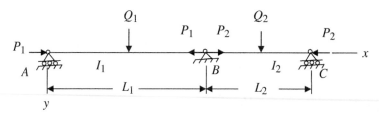

Figure 4.18 Continuous beam column.

and

$$\theta'_n = \theta'_{0n} + \frac{M_n L_n}{3EI_n}\psi(u_n) + \frac{M_{n+1}L_n}{6EI_n}\phi(u_n) \tag{4.20c}$$

where the angles θ_{0n} and θ'_{0n} are the angles produced by the lateral load on the two spans shown in Figure 4.18, if each span was a simply supported beam column. Thus, Eq. (4.20a) becomes

$$\theta_{0n} + \frac{M_n L_{n-1}}{3EI_{n-1}}\psi(u_{n-1}) + \frac{M_{n-1}L_{n-1}}{6EI_{n-1}}\phi(u_{n-1}) = -\theta'_{0n} - \frac{M_n L_n}{3EI_n}\psi(u_n) - \frac{M_{n+1}L_n}{6EI_n}\phi(u_n) \tag{4.20d}$$

or

$$M_{n-1}\phi(u_{n-1}) + 2M_n \left[\psi(u_{n-1}) + \frac{I_{n-1}L_n}{I_n L_{n-1}}\psi(u_n)\right] + M_{n+1}\frac{I_{n-1}L_n}{I_n L_{n-1}}\phi(u_n)$$

$$= -\frac{6EI_{n-1}}{L_{n-1}}(\theta_{0n} + \theta'_{0n}) \tag{4.20e}$$

If the Eq. (4.20e) is written for each intermediate support, and also using the boundary conditions at the two end supports, we obtain a sufficient number of equations to solve for all the unknown moments (see Examples 4.1 and 4.2).

Example 4.1   Consider a two-span continuous beam column shown in Figure 4.19 with two spans loaded by concentrated forces Q_1 and Q_2 at the middle of each span. The ends A and C are simply supported, and the end B is a continuous support. The modulus of elasticity of the entire member is constant, E. The span lengths are L_1 and L_2, whereas the moments of inertias are I_1 and I_2.

If the ends of the continuous beam column are simply supported, then the equations of the type given by Eq. (4.20e), when written for each intermediate support, will provide as many

Figure 4.19 Two-span continuous beam column.

equations as there are statically indeterminate moments. In this problem just one equation will be sufficient to find the bending moment at the intermediate support B. Thus, Eq. (4.20a) at the support B becomes

$$M_A\phi(u_1) + 2M_B\left[\psi(u_1) + \frac{I_1 L_2}{I_2 L_1}\psi(u_2)\right] + M_C\frac{I_1 L_2}{I_2 L_1}\phi(u_2) = -\frac{6EI_1}{L_1}(\theta_{02} + \theta'_{02}) \qquad (4.20f)$$

Now, $M_A = M_C = 0$, therefore,

$$M_B\left[\psi(u_1) + \frac{I_1 L_2}{I_2 L_1}\psi(u_2)\right] = -\frac{3EI_1}{L_1}(\theta_{02} + \theta'_{02}) \qquad (4.20g)$$

From Eq. (4.3b)

$$\theta_{02} = \frac{Q_1 L_1^2}{16 EI_1}\lambda(u_1) \qquad (4.20h)$$

$$\theta'_{02} = \frac{Q_2 L_2^2}{16 EI_2}\lambda(u_2) \qquad (4.20i)$$

where

$$u_1 = \frac{k_1 L_1}{2} = \sqrt{\frac{P_1}{EI_1}}\frac{L_1}{2}; \qquad u_2 = \frac{k_2 L_2}{2} = \sqrt{\frac{P_2}{EI_2}}\frac{L_2}{2}$$

Knowing u_1 and u_2; $\psi(u_1)$ and $\psi(u_2)$, $\lambda(u_1)$ and $\lambda(u_1)$, θ_{02}, and θ'_{02} can be calculated. Then the unknown moment M_B can be calculated as

$$M_B = \frac{\frac{-3EI_1}{L_1}(\theta_{02} + \theta'_{02})}{\left[\psi(u_1) + \frac{I_1 L_2}{I_2 L_1}\psi(u_2)\right]} \qquad (4.20j)$$

At the critical value of the axial compressive force P, the moment M_B becomes infinite, therefore

$$\psi(u_1) + \frac{I_1 L_2}{I_2 L_1}\psi(u_2) = 0 \qquad (4.20k)$$

Equation (4.20k) is the characteristic equation to determine the critical load for a two-span continuous column.

Let $L_1 = 1.5\,L$, $L_2 = L$, $I_1 = I_2 = I$, and $P_1 = P_2 = P$, then Eq. (4.20k) becomes

$$\frac{\psi(u_1)}{\psi(u_2)} = -\frac{L_2}{L_1}$$

or

$$\frac{(\tan 2u_1 - 2u_1)\tan 2u_2}{(\tan 2u_2 - 2u_2)\tan 2u_1} = -\frac{L_2}{L_1}\frac{u_1^2}{u_2^2}$$

$$k_1 = k_2 = k = \sqrt{\frac{P}{EI}}, \qquad u_1 = \frac{3kL}{4}, \qquad u_2 = \frac{kL}{2}, \qquad \text{and} \qquad \frac{u_1}{u_2} = \frac{3}{2}$$

or

$$2\tan 2u_2(\tan 2u_1 - 2u_1) + 3\tan 2u_1(\tan 2u_2 - 2u_2) = 0$$

or

$$5 \sin \frac{3kL}{2} \sin kL - 3kL \sin \frac{5kL}{2} = 0 \tag{4.20l}$$

Equation (4.20l) is the same as the Eq. (2.13g) derived before to find the critical axial compressive force for the two-span continuous column in Chapter 2.

Example 4.2 Consider a two-span continuous beam column shown in Figure 4.20 with two spans loaded by concentrated forces Q_1 and Q_2 at the middle of each span. The end A is simply supported, end C is built in, and end B is a continuous support. The modulus of elasticity of the entire member is constant, E. The span lengths are L_1 and L_2, whereas the moments of inertias are I_1 and I_2.

In this case, there are two redundant moments, M_B and M_C. To get the three moment equations at the supports B and C, add an imaginary span to the right of support C of length zero. This way we obtain

At B

$$0 + 2M_B \left[\psi(u_1) + \frac{I_1 L_2}{I_2 L_1} \psi(u_2) \right] + M_C \frac{I_1 L_2}{I_2 L_1} \phi(u_2) = -\frac{6EI_1}{L_1}(\theta_{0B} + \theta'_{0B}) \tag{4.20m}$$

At C

$$M_B \phi(u_2) + 2M_C [\psi(u_2) + 0] + 0 = -\frac{6EI_2}{L_2}(\theta_{0C} + 0) \tag{4.20n}$$

or

$$\begin{bmatrix} 2\left[\psi(u_1) + \frac{I_1 L_2}{I_2 L_1}\psi(u_2)\right] & \frac{I_1 L_2}{I_2 L_1}\phi(u_2) \\ \\ \phi(u_2) & 2\psi(u_2) \end{bmatrix} \begin{Bmatrix} M_B \\ \\ M_C \end{Bmatrix} = \begin{Bmatrix} -\frac{6EI_1}{L_1}(\theta_{0B} + \theta'_{0B}) \\ \\ -\frac{6EI_2}{L_2}\theta_{0C} \end{Bmatrix} \tag{4.20o}$$

Knowing the properties of the beam column, the axial force, and the lateral loads, the bending moments M_B and M_C can be determined from Eq. (4.20o). The critical values of the axial forces are those at which the bending moments become infinitely large. This requires that the

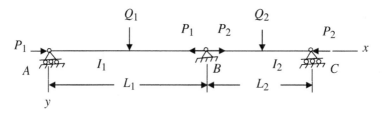

Figure 4.20 Two-span beam column with one end fixed.

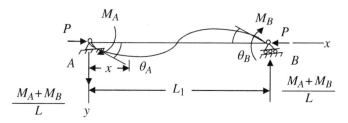

Figure 4.21 Beam column with ends rotated through θ_A and θ_B.

In such cases, we can use L'Hospital's rule, according to which we continue taking derivatives of the numerator and the denominator terms with respect to u, until substituting $u = 0$ gives a finite value for the function $\psi(u)$. In this case, we will see that as u goes to zero, $\psi(u)$ becomes 1. Similarly, when the axial force is zero

$$\phi(u) = \frac{3}{u}\left[\frac{1}{\sin 2u} - \frac{1}{2u}\right] = \frac{3}{2}\left[\frac{2u - \sin 2u}{u^2 \sin 2u}\right] = \frac{0}{0}$$

By using L'Hospital's rule, it can be shown that as u goes to zero, $\phi(u)$ becomes 1. Therefore, $f_{11} = f_{22} = 2$, and $f_{12} = f_{21} = -1$. Thus, when there is no axial force, the Eq. (4.21b) becomes

$$\left\{\begin{matrix} \theta_A \\ \theta_B \end{matrix}\right\} = \frac{L}{6EI}\begin{bmatrix} 2 & -1 \\ -1 & 2 \end{bmatrix}\left\{\begin{matrix} M_A \\ M_B \end{matrix}\right\} \tag{4.21c}$$

Eq. (4.21c) is the same relation between the end rotations and the end moments as given in structural analysis texts [4] for beams subjected to clockwise end moments with no axial force.

4.9.1 Matrix Inversion

Let the matrix be $[A] = \begin{bmatrix} a_{11} & a_{12} & a_{13} \\ a_{21} & a_{22} & a_{23} \\ a_{31} & a_{32} & a_{33} \end{bmatrix}$

$$[A]^{-1} = \frac{\text{Adjoint}[A]}{|A|}$$

The minor, M_{ij}, is the determinant of the $(n-1)\times(n-1)$ submatrix of an $n \times n$ matrix $[A]$ obtained by deleting the ith row and jth column. The cofactor, A_{ij}, associated with M_{ij}, is defined to be $A_{ij} = (-1)^{i+j}M_{ij}$. The adjoint $[A]$ is the transpose of the matrix obtained by replacing each element of $[A]$ by its cofactor. Then the adjoint $[A]$ is given by

$$\text{Adjoint }[A] = \begin{bmatrix} A_{11} & A_{21} & A_{31} \\ A_{12} & A_{22} & A_{32} \\ A_{13} & A_{23} & A_{33} \end{bmatrix}$$

$|A|$ is the determinant of matrix $[A]$. From Eq. (4.21a)

$$\left\{\begin{matrix} M_A \\ M_B \end{matrix}\right\} = \left\{\frac{L}{6EI}\begin{bmatrix} 2\psi(u) & -\phi(u) \\ -\phi(u) & 2\psi(u) \end{bmatrix}\right\}^{-1}\left\{\begin{matrix} \theta_A \\ \theta_B \end{matrix}\right\}$$

$$\begin{Bmatrix} M_A \\ M_B \end{Bmatrix} = \frac{6EI}{L} \frac{1}{4\psi^2(u) - \phi^2(u)} \begin{bmatrix} 2\psi(u) & \phi(u) \\ \phi(u) & 2\psi(u) \end{bmatrix} \begin{Bmatrix} \theta_A \\ \theta_B \end{Bmatrix} \tag{4.21d}$$

$$\begin{Bmatrix} M_A \\ M_B \end{Bmatrix} = \frac{EI}{L} \begin{bmatrix} k_{11} & k_{12} \\ k_{21} & k_{22} \end{bmatrix} \begin{Bmatrix} \theta_A \\ \theta_B \end{Bmatrix} \tag{4.21e}$$

where

$$k_{11} = k_{22} = \frac{12\psi(u)}{4\psi^2(u) - \phi^2(u)} \tag{4.21f}$$

$$k_{12} = k_{21} = \frac{6\phi(u)}{4\psi^2(u) - \phi^2(u)} \tag{4.21g}$$

Equation (4.21e) gives two slope deflection equations for a beam column subjected to axial force and end rotations. The slope deflection coefficients for the beam column buckling, k_{11}, k_{22}, k_{12}, and k_{21} are given in Appendix A for various values of kL.

The coefficients k_{ij}, $i = 1, 2$ and $j = 1, 2$; when multiplied by EI/L are the stiffness coefficients for a beam column subjected to axial and lateral forces in Eq. (4.21e). When the axial force is zero, $u = 0$, and $\psi(u) = \phi(u) = 1$ as shown before. Therefore $k_{11} = k_{22} = 4$, and $k_{12} = k_{21} = 2$. Thus, when there is no axial force on the beam column, Eq. (4.21e) becomes

$$\begin{Bmatrix} M_A \\ M_B \end{Bmatrix} = \frac{EI}{L} \begin{bmatrix} 4 & 2 \\ 2 & 4 \end{bmatrix} \begin{Bmatrix} \theta_A \\ \theta_B \end{Bmatrix} \tag{4.21h}$$

Equation (4.21h) is the same relation as given in structural analysis texts [4], giving induced moments at the ends of a beam whose ends have been rotated clockwise with no axial force.

4.9.2 Beam Columns Subjected to Rotations and Relative Displacement at the Ends

Let θ_A and θ_B be the total rotations including Δ/L rotation due to the relative displacement of the end A with respect to the end B as shown in Figure 4.22. therefore, the moments at the ends are produced by the rotations, $\theta'_A = \theta_A - (\Delta/L)$, and $\theta'_B = \theta_B - (\Delta/L)$.

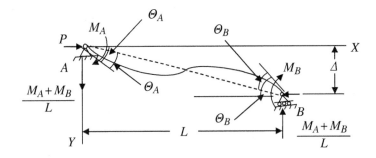

Figure 4.22 Beam column with end rotations and relative end displacement.

Thus, the slope deflection equations relating the end rotations, the relative displacement of the ends, and the end moments using Eq. (4.21e) are

$$M_A = \frac{EI}{L}\left[k_{11}\left(\theta_A - \frac{\Delta}{L}\right) + k_{12}\left(\theta_B - \frac{\Delta}{L}\right)\right]$$

$$M_A = \frac{EI}{L}\left[k_{11}\theta_A + k_{12}\theta_B - (k_{11} + k_{12})\frac{\Delta}{L}\right] \tag{4.22a}$$

and

$$M_B = \frac{EI}{L}\left[k_{21}\left(\theta_A - \frac{\Delta}{L}\right) + k_{22}\left(\theta_B - \frac{\Delta}{L}\right)\right]$$

$$M_B = \frac{EI}{L}\left[k_{21}\theta_A + k_{22}\theta_B - (k_{21} + k_{22})\frac{\Delta}{L}\right] \tag{4.22b}$$

Equations (4.22a) and (4.22b) are the slope deflection equations for a beam column that is subjected to end rotations and the relative joint translation of the ends. When the axial force is zero, Equations (4.22a) and (4.22b) become

$$M_A = \frac{EI}{L}\left(4\theta_A + 2\theta_B - 6\frac{\Delta}{L}\right) \tag{4.22c}$$

$$M_B = \frac{EI}{L}\left(2\theta_A + 4\theta_B - 6\frac{\Delta}{L}\right) \tag{4.22d}$$

Equations (4.22c) and (4.22d) are the same equations as given in structural analysis texts [4] for a beam subjected to end rotations and relative joint translation.

4.9.3 Beam Columns Having One End Hinged

In this case, there is a hinge at either end A or B.

If there is a hinge at the end A, $M_A = 0$, the beam column will bend in a single curvature and the two slope deflection equations given by Eq. (4.21e) can be written as

$$M_A = 0 = \frac{EI}{L}[k_{11}\theta_A + k_{12}\theta_B]$$

or

$$\theta_A = -\frac{k_{12}}{k_{11}}\theta_B$$

and

$$M_B = \frac{EI}{L}\left[k_{22} - \frac{k_{12}^2}{k_{11}}\right]\theta_B \tag{4.23a}$$

Similarly, if there is a hinge at the end B, $M_B = 0$, we get

$$M_A = \frac{EI}{L}\left[k_{11} - \frac{k_{12}^2}{k_{22}}\right]\theta_A \tag{4.23b}$$

4.9.4 Beam Columns with Transverse Loading

When a transverse load is acting on the beam column, say, a uniformly distributed load, and the member is given end rotations, as shown in Figure 4.23, the total rotations at the ends are given by

$$\theta_A = \theta'_A + \theta''_A$$
$$\theta_B = \theta'_B - \theta''_B$$

Where θ''_A and θ''_B are the rotations at the ends due to transverse load; θ_A and θ_B are the total rotations at the ends. The angles θ'_A and θ'_B induce moments M_A and M_B at the ends.

$$\theta'_A = \theta_A - \theta''_A$$
$$\theta'_B = \theta_B + \theta''_B$$

The slope deflection equation at the end A is

$$M_A = \frac{EI}{L}[k_{11}(\theta_A - \theta_A'') + k_{12}(\theta_B + \theta_B'')] \tag{4.24a}$$

$$M_A = \frac{EI}{L}[k_{11}\theta_A + k_{12}\theta_B] + \frac{EI}{L}[-k_{11}\theta_A'' + k_{12}\theta_B''] \tag{4.24b}$$

Substituting the values of k_{11}, k_{12}, and $\theta_A'' = \theta_B''$, from Eqs. (4.21f), (4.21g), and (4.5n), we have

$$\frac{EI}{L}[-k_{11}\theta_A'' + k_{12}\theta_B''] = \frac{EI}{L}\left[-\frac{12\psi(u)}{4\psi^2(u) - \phi^2(u)}\left(\frac{wL^3}{24EI}\chi(u)\right) + \frac{6\phi(u)}{4\psi^2(u) - \phi^2(u)}\left(\frac{wL^3}{24EI}\chi(u)\right)\right]$$

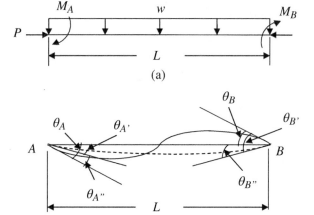

Figure 4.23 Beam column subjected to end moments and transverse load: (a) Transverse loading and end moments on beam column; (b) Deflected shape of beam column subjected to end moments and transverse load.

Substitute the values of the functions $\psi(u)$ and $\phi(u)$ from Eqs. (4.8g) and (4.8h), we have

$$\frac{EI}{L}[-k_{11}\theta_A'' + k_{12}\theta_B''] = -\frac{wL^2}{12}\chi(u)\frac{u}{\tan u}$$

Substituting the value of the function $\chi(u)$ from Eq. (4.3g), we have

$$\frac{EI}{L}[-k_{11}\theta_A'' + k_{12}\theta_B''] = -\frac{wL^2}{12}\left[\frac{3(\tan u - u)}{u^2 \tan u}\right] \tag{4.24c}$$

The right-hand side of Eq. (4.24c) is the fixed end moment at the support A as shown in Eq. (4.17b). The negative sign shows that the fixed end moment at the support A is counter-clockwise due to uniformly distributed transverse load.

$$M_{FA} = -\frac{wL^2}{12}\left[\frac{3(\tan u - u)}{u^2 \tan u}\right] \tag{4.24d}$$

Thus, Eq. (4.24b) becomes

$$M_A = \frac{EI}{L}[k_{11}\theta_A + k_{12}\theta_B] + M_{FA} \tag{4.24e}$$

The slope deflection equation at the end B is

$$M_B = \frac{EI}{L}[k_{21}(\theta_A - \theta_A'') + k_{22}(\theta_B + \theta_B'')] \tag{4.24f}$$

$$M_B = \frac{EI}{L}[k_{21}\theta_A + k_{22}\theta_B] + \frac{EI}{L}[-k_{21}\theta_A'' + k_{22}\theta_B''] \tag{4.24g}$$

It can be shown as before that

$$\frac{EI}{L}[-k_{21}\theta_A'' + k_{22}\theta_B''] = \frac{wL^2}{12}\left[\frac{3(\tan u - u)}{u^2 \tan u}\right] = M_{FB} \tag{4.24h}$$

The positive sign shows that the fixed end moment, M_{FB}, at the support B is clockwise due to uniformly distributed transverse load. Thus, Eq. (4.24g) becomes

$$M_B = \frac{EI}{L}[k_{21}\theta_A + k_{22}\theta_B] + M_{FB} \tag{4.24i}$$

Equations (4.24e and 4.24i) are the slope deflection equations for a beam column subjected to an axial force, the end rotations, and a uniformly distributed load of intensity, w, per unit length over the entire span. The extra terms are added in Eq. (4.24e) and Eq. (4.24i) if there is relative translation between the end joints, as shown in Eq. (4.22a) and Eq. (4.22b) respectively. The fixed end moments, M_{FA} and M_{FB} reduce to the values of $-wL^2/12$ and $wL^2/12$ respectively, if there is no axial force, that is $u = 0$.

4.9.5 Beam Columns in Single Curvature

If the end rotations are as shown in Figure 4.24, the beam column is bent in a single curvature. In this case Eqs. (4.21e) can be used by substituting $\theta_B = -\theta_B$.

Slope deflection equations are

$$M_A = \frac{EI}{L}[k_{11}\theta_A + k_{12}(-\theta_B)] \tag{4.25a}$$

$$M_B = \frac{EI}{L}[k_{21}\theta_A + k_{22}(-\theta_B)] \tag{4.25b}$$

4.10 Inelastic Beam Columns

So far we have considered the problems of beam columns where the material remained elastic. The elastic buckling analysis is possible if we are not concerned about the failure of the members. If the members are to be analyzed for failure, then inelasticity in the material is to be taken into account. The inelastic behavior makes the problem much more difficult because stress-strain relationship is not linear and the rigidity of a member changes with the level of stress. Under these conditions it is very cumbersome to find the curvature of the bending curve. Inelastic beam column problems are solved numerically or in some limited cases these can be solved by making simplified assumptions. The solution presented here follows the procedure given by Ježek [5, 6] and Bleich [7]. The solution is based on the following assumptions:

1) The cross-section of the beam column is rectangular, as shown in Figure 4.25b.
2) The stress-strain diagram of beam column material is elastic-perfectly plastic, as shown in Figure 4.25d. The modulus of elasticity, E, is assumed to be constant up to the yield point, σ_y, from there on, it is zero. The yield point is considered to be the same in tension and compression.
3) The deflection curve of the member is a half sine wave. This assumption makes it possible to use the method that is valid for eccentrically loaded and laterally loaded beam columns.
4) The problem is solved by assuming equilibrium conditions at the center of the member span.
5) Deformations are small so that the curvature can be approximated by the second derivative of the deflection curve.

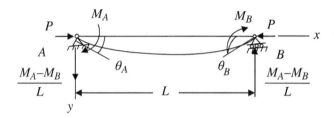

Figure 4.24 Beam column in a single curvature.

Consider a simply supported straight beam column of rectangular cross-section that is subjected to axial force and lateral load, that produces a bending moment varying over the span but is symmetrical about the center of the span.

At any cross-section distance x, the total bending moment is

$$M_{ext} = M_x + Py \tag{4.26a}$$

where M_x is the bending moment due to the eccentricity of the axial force and lateral load as shown in Figure 4.25c, and y is the deflection of the member at distance x in Figure 4.25a. The moment equilibrium condition gives

$$M_x + Py = M_{int} \tag{4.26b}$$

where M_{int} is the moment produced by the internal stress distribution. In the case of elastic members, the internal moment is given by

$$M_{int} = -EI\frac{d^2y}{dx^2} \tag{4.26c}$$

where E and I are the modulus of elasticity of the material and the moment of inertia of the member cross-section about the neutral axis, respectively.

$$\frac{1}{\rho} = -\frac{d^2y}{dx^2} \tag{4.26d}$$

where ρ is the curvature of the deflection curve, Eqs. (4.26b), (4.26c), and (4.26d) give the moment curvature relation for an elastic member. Equations (4.26a, 4.26b, and 4.26d) are valid

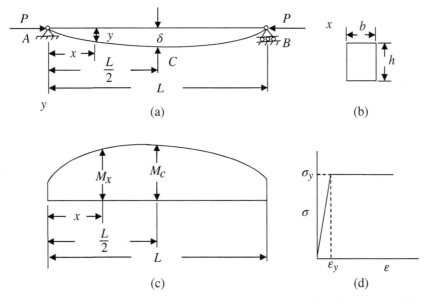

Figure 4.25 Elastic-perfectly plastic beam column subjected to axial force and bending moment: (a) Inelastic beam column; (b) Cross section; (c) Bending moment diagram; (d) Stress strain diagram.

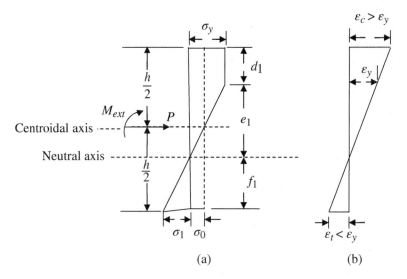

Figure 4.26 Stress distribution in case 1 of an inelastic beam column: (a) Stress diagram; (b) Strain diagram.

for both elastic and inelastic cases. However, Eq. (4.26c) giving the moment curvature relation within the elastic limit is not valid for the inelastic case. The inelastic moment curvature relation is derived by taking into consideration the stress distribution across the depth of the member. There are two different cases of stress distribution possible. In Case 1 shown in Figure 4.26, the yield point is reached on the concave, i.e. the compressive side of the bent column. The tensile stresses on the convex side still remain within the elastic limit. In Case 2, the yield strength is reached on both compression and tension sides of the member cross-section.

4.10.1 Case 1: Yielding on the Compression Side Only

In this case, the yield point is reached on the concave, the compressive side of a bent column. The tensile stresses on the convex side still remain within the elastic limit. The yielding in the compression zone is assumed to extend a distance d_1 into the cross-section shown in Figure 4.26a. Two equilibrium equations are used in the analysis. The first states that the sum of the internal normal forces acting on a cross-section is equal to the axial force on the beam column. The second equation states that the moment of the internal normal forces about the centroidal axis is equal to M_{ext}, the external moment about the centroidal axis.

The equilibrium conditions are

$$\int_{-h/2}^{h/2} \sigma dA = P \tag{4.27a}$$

$$\int_{-h/2}^{h/2} \sigma y dy = M_x + Py \tag{4.27b}$$

The integrals in the present case are solved by using the stress and strain diagrams in Figure 4.26. The distances d_1, e_1, and f_1 are determined as follows:

$$P = \sigma_y b d_1 + \frac{\sigma_y}{2} b e_1 - \frac{\sigma_1}{2} b f_1 \tag{4.27c}$$

$$M_{\text{int}} = \left[\sigma_y b d_1 \left(\frac{h}{2} - \frac{d_1}{2} \right) + \frac{\sigma_y b e_1}{2} \left(\frac{h}{2} - d_1 - \frac{e_1}{3} \right) + \frac{\sigma_1 b f_1}{2} \left(\frac{h}{2} - \frac{f_1}{3} \right) \right] \tag{4.27d}$$

$$d_1 + e_1 + f_1 = h \tag{4.27e}$$

From similar triangles in Figure 4.26a

$$\sigma_1 = \frac{\sigma_y f_1}{e_1} \tag{4.27f}$$

Divide both sides of Eq. (4.27c) by $A = bh$, the area of the cross-section of the rectangular section. Now substitute σ_1 from Eq. (4.27f), and $f_1 = h - d_1 - e_1$, from Eq. (4.27e) into Eq. (4.27c). After some calculation we get

$$e_1 = \frac{\sigma_y (h - d_1)^2}{2h(\sigma_y - \sigma_0)} \tag{4.27g}$$

where $\sigma_0 = P/A$ is the average stress. Similarly, substitute the values of f_1 and σ_1 from Eqs. (4.27e) and (4.27f) into Eq. (4.27d) and we get

$$M_{\text{int}} = \frac{b\sigma_y}{12e_1} [h^3 - 3hd_1^2 + 2d_1^3]$$

or

$$e_1 = \frac{b\sigma_y}{12M_{\text{int}}} [h^3 - 3hd_1^2 + 2d_1^3] \tag{4.27h}$$

Now equate the right-hand sides of Eqs. (4.27g) and (4.27h) and in solving we have

$$\frac{\sigma_y (h - d_1)^2}{2h(\sigma_y - \sigma_0)} = \frac{b\sigma_y}{12M_{\text{int}}} [(h + 2d_1)(h - d_1)^2]$$

$$d_1 = \frac{3M_{\text{int}}}{bh(\sigma_y - \sigma_0)} - \frac{h}{2} \tag{4.27i}$$

Substitute the d_1 obtained in Eq. (4.27i) into Eq. (4.27g) and we have

$$e_1 = \frac{\sigma_y}{2h(\sigma_y - \sigma_0)} \left[h - \left\{ \frac{3M_{\text{int}}}{bh(\sigma_0 - \sigma_y)} - \frac{h}{2} \right\} \right]^2$$

After some calculations we obtain

$$e_1 = \frac{9\sigma_y h}{8(\sigma_y - \sigma_0)^3} \left[\sigma_y - \sigma_0 - \frac{2M_{\text{int}}}{bh^2} \right]^2$$

or

$$e_1 = \frac{9\sigma_y \left[\frac{h}{2} \left(\frac{\sigma_y}{\sigma_0} - 1 \right) - \frac{M_{int}}{P} \right]^2}{2\sigma_0 h \left(\frac{\sigma_y}{\sigma_0} - 1 \right)^3} \tag{4.27j}$$

From Figure 4.26b, and the strength of materials [8], assuming the plane sections before bending remain plane after bending, we get

$$\varepsilon_y = \frac{e_1}{\rho}, \text{ or } \frac{\sigma_y}{E} = \frac{e_1}{\rho},$$

Therefore,

$$\frac{1}{\rho} = \frac{\sigma_y}{Ee_1} \tag{4.27k}$$

Substitute e_1 from Eq. (4.27j) into Eq. (4.27k) and obtain the inelastic moment curvature relation that should be used instead of Eq. (4.26c) when the stresses are above the elastic limit.

$$\frac{1}{\rho} = \frac{2\sigma_0 h \left(\frac{\sigma_y}{\sigma_0} - 1 \right)^3}{9E \left[\frac{h}{2} \left(\frac{\sigma_y}{\sigma_0} - 1 \right) - \frac{M_{int}}{P} \right]^2} \tag{4.27l}$$

Assume the deflection curve of the centroidal axis of the beam column is

$$y = \delta \sin \frac{\pi x}{L} \tag{4.27m}$$

where δ is the deflection of the beam column at the mid-span shown in Figure 4.25a.

$$\frac{d^2 y}{dx^2} = -\frac{\pi^2}{L^2} \delta \sin \frac{\pi x}{L}$$

$$\left. \frac{d^2 y}{dx^2} \right|_{x=\frac{L}{2}} = -\frac{\pi^2}{L^2} \delta \tag{4.27n}$$

Using Eqs. (4.26d), (4.27l), and (4.27n) we get

$$\frac{\pi^2}{L^2} \delta = \frac{2\sigma_0 h \left(\frac{\sigma_y}{\sigma_0} - 1 \right)^3}{9E \left[\frac{h}{2} \left(\frac{\sigma_y}{\sigma_0} - 1 \right) - \frac{M_{int}}{P} \right]^2}$$

Substitute M_{int} from Eq. (4.26b)

$$\delta \left[\frac{h}{2} \left(\frac{\sigma_y}{\sigma_0} - 1 \right) - \delta - \frac{M_x}{P} \right]^2 - \frac{2\sigma_0 h L^2}{9\pi^2 E} \left(\frac{\sigma_y}{\sigma_0} - 1 \right)^3 = 0 \tag{4.27o}$$

Euler's elastic buckling stress is given by

$$\sigma_e = \frac{\pi^2 EI}{L^2 A} = \frac{\pi^2 E \left(\frac{1}{12} bh^3\right)}{L^2 bh} = \frac{\pi^2 Eh^2}{12L^2}$$

Now Eq. (4.27o) can be written as

$$\frac{\delta}{h}\left[\frac{1}{2}\left(\frac{\sigma_y}{\sigma_0}-1\right)-\frac{\delta}{h}-\frac{M_x}{Ph}\right]^2 - \frac{\sigma_0}{54\sigma_e}\left(\frac{\sigma_y}{\sigma_0}-1\right)^3 = 0 \tag{4.27p}$$

Equation (4.27p) gives the load (in terms of average stress) versus mid-span deflection δ, in the inelastic range which can be drawn for the various values of M_x/P, provided the stress remains elastic on the tension side. The critical value of the average stress σ_0, can be obtained from the expression

$$\frac{d\sigma_0}{d\delta} = 0 \tag{4.27q}$$

The derivative of the expression in Eq. (4.27o) is taken with respect to δ and use Eq. (4.27q), which leads to

$$\left[\frac{h}{2}\left(\frac{\sigma_y}{\sigma_0}-1\right)-\delta-\frac{M_x}{P}\right]^2 - 2\delta\left[\frac{h}{2}\left(\frac{\sigma_y}{\sigma_0}-1\right)-\delta-\frac{M_x}{P}\right] = 0$$

or

$$\delta = \frac{1}{3}\left[\frac{h}{2}\left(\frac{\sigma_y}{\sigma_0}-1\right)-\frac{M_x}{P}\right] \tag{4.27r}$$

Substitute δ from Eq. (4.27r) into Eq. (4.27o) to obtain

$$\frac{1}{3}\left[\frac{h}{2}\left(\frac{\sigma_y}{\sigma_0}-1\right)-\frac{M_x}{P}\right]\left[\frac{h}{2}\left(\frac{\sigma_y}{\sigma_0}-1\right)-\frac{M_x}{P}-\frac{1}{3}\left\{\frac{h}{2}\left(\frac{\sigma_y}{\sigma_0}-1\right)-\frac{M_x}{P}\right\}\right]^2 - \frac{2\sigma_0 hL^2}{9\pi^2 E}\left(\frac{\sigma_y}{\sigma_0}-1\right)^3 = 0$$

$$\frac{1}{3}\left[\frac{h}{2}\left(\frac{\sigma_y}{\sigma_0}-1\right)-\frac{M_x}{P}\right]\frac{4}{9}\left[\frac{h}{2}\left(\frac{\sigma_y}{\sigma_0}-1\right)-\frac{M_x}{P}\right]^2 - \frac{2\sigma_0 hL^2}{9\pi^2 E}\left(\frac{\sigma_y}{\sigma_0}-1\right)^3 = 0$$

or

$$\sigma_c = \frac{\pi^2 Eh^2}{12L^2}\left[\frac{\frac{\sigma_y}{\sigma_c}-1-\frac{2M_x}{Ph}}{\frac{\sigma_y}{\sigma_c}-1}\right]^3 \tag{4.27s}$$

where the critical average stress $\sigma_c = \sigma_0$

$$\frac{\pi^2 Eh^2}{12L^2} = \frac{\pi^2 EI}{L^2 A} = \frac{\pi^2 E}{(L/r)^2}$$

or

$$\sigma_c = \frac{\pi^2 E}{(L/r)^2} \left[\frac{\frac{\sigma_y}{\sigma_c} - 1 - \frac{2M_x}{Ph}}{\frac{\sigma_y}{\sigma_c} - 1} \right]^3$$

(4.27t)

Equation (4.27t) gives the relation between the critical average stress and the slenderness ratio for any given value of M_x/P for the beam column. If there is no lateral load on the member, only the eccentric axial force is acting, then the parameter M_x/P can be substituted by the eccentricity e of the axial force. With the aid of Eq. (4.27t) the column curve giving the relation between the slenderness ratio, L/r, and the critical stress can be plotted in continuation of the Euler curve, provided the stress remains elastic on the tension side of the beam column.

4.10.2 Case 2: Yielding on Both the Compression and Tension Sides

In Case 2, the yield strength is reached in both the compression and tension sides of the member cross-section. It is assumed that the yield extends distances d_2 and f_2 in the compression and tension zones of the cross-section respectively, as shown in Figure 4.27a.

The procedure in Case 2 is similar to that in Case 1. The equations of equilibrium are used to find the distances d_2, e_2, and f_2 in Figure 4.27a.

$$P = \sigma_y b d_2 + \frac{\sigma_y}{2} b e_2 - \frac{\sigma_y}{2} b e_2 - \sigma_y b f_2$$

or

$$\sigma_0 = \frac{\sigma_y}{h}(d_2 - f_2)$$

(4.28a)

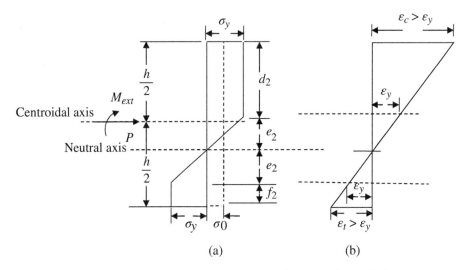

(a) (b)

Figure 4.27 Stress and strain distributions in Case 2 of an inelastic beam column: (a) Stress diagram; (b) Strain diagram.

where

$$\sigma_0 = \frac{P}{bh}.$$

$$M_{\text{int}} = \sigma_y b d_2 \left(\frac{h}{2} - \frac{d_2}{2} \right) + \sigma_y \frac{be_2}{2} \left(\frac{h}{2} - d_2 - \frac{e_2}{3} \right) + \sigma_y \frac{be_2}{2} \left(\frac{h}{2} - f_2 - \frac{e_2}{3} \right) + \sigma_y b f_2 \left(\frac{h}{2} - \frac{f_2}{2} \right)$$

(4.28b)

$$d_2 + 2e_2 + f_2 = h$$

(4.28c)

Equations (4.28a and 4.28c) give

$$\sigma_0 = \frac{\sigma_y}{h} [d_2 - h + d_2 + 2e_2]$$

$$e_2 = \frac{\sigma_0 h}{2\sigma_y} + \frac{h}{2} - d_2$$

(4.28d)

Equations (4.28c and 4.28d) give

$$f_2 = d_2 - \frac{\sigma_0 h}{\sigma_y}$$

(4.28e)

Simplifying Eq. (4.28b) we get

$$M_{\text{int}} = b\sigma_y \left[\frac{d_2 h}{2} - \frac{d_2^2}{2} + \frac{e_2 h}{2} - \frac{e_2 d_2}{2} - \frac{e_2^2}{3} - \frac{e_2 f_2}{2} + \frac{f_2 h}{2} - \frac{f_2^2}{2} \right]$$

(4.28f)

Substitute e_2 and f_2 from Eqs. (4.28d) and (4.28e) in Eq. (4.28f) and simplify

$$M_{\text{int}} = b\sigma_y \left[\frac{h^2}{6} - \frac{\sigma_0^2 h^2}{3\sigma_y^2} - \frac{d_2^2}{3} - \frac{\sigma_0 h^2}{6\sigma_y} + \frac{\sigma_0 h d_2}{3\sigma_y} + \frac{h d_2}{3} \right]$$

(4.28g)

or

$$d_2^2 - h d_2 \left(1 + \frac{\sigma_0}{\sigma_y} \right) - \left[\frac{h^2}{2} - \frac{\sigma_0^2 h^2}{\sigma_y^2} - \frac{\sigma_0 h^2}{2\sigma_y} - \frac{3M_{\text{int}}}{b\sigma_y} \right] = 0$$

(4.28h)

Equation (4.28h) is a quadratic equation in d_2, hence the roots are

$$d_2 = \frac{h \left(1 + \frac{\sigma_0}{\sigma_y} \right) \pm \sqrt{h^2 \left(1 + \frac{\sigma_0}{\sigma_y} \right)^2 + 4 \left[\frac{h^2}{2} - \frac{\sigma_0^2 h^2}{\sigma_y^2} - \frac{\sigma_0 h^2}{2\sigma_y} - \frac{3M_{\text{int}}}{b\sigma_y} \right]}}{2}$$

(4.28i)

or

$$d_2 = \frac{h}{2} \left(1 + \frac{\sigma_0}{\sigma_y} \right) - h \sqrt{\frac{3}{4} \left(1 - \frac{\sigma_0^2}{\sigma_y^2} \right) - \frac{3M_{\text{int}}}{bh^2 \sigma_y}}$$

(4.28j)

The negative sign is chosen before the term inside the square root so that we have e_2 a positive quantity, as shown later.

Substitute d_2 given in Eq. (4.28j) into Eqs. (4.28d) and (4.28e) and we obtain

$$e_2 = h\sqrt{\frac{3}{4}\left(1 - \frac{\sigma_0^2}{\sigma_y^2}\right) - \frac{3M_{int}}{bh^2\sigma_y}} \qquad (4.28k)$$

and

$$f_2 = \frac{h}{2}\left(1 - \frac{\sigma_0}{\sigma_y}\right) - h\sqrt{\frac{3}{4}\left(1 - \frac{\sigma_0^2}{\sigma_y^2}\right) - \frac{3M_{int}}{bh^2\sigma_y}} \qquad (4.28l)$$

$$\varepsilon_y = \frac{e_2}{\rho} \text{ or } \frac{\sigma_y}{E} = \frac{e_2}{\rho}$$

Therefore,

$$\frac{1}{\rho} = \frac{\sigma_y}{Ee_2} \qquad (4.28m)$$

Substitute e_2 from Eq. (4.28k) into Eq. (4.28m) and we have

$$\frac{1}{\rho} = \frac{\sigma_y}{Eh\sqrt{\frac{3}{4}\left(1 - \frac{\sigma_0^2}{\sigma_y^2}\right) - \frac{3M_{int}}{bh^2\sigma_y}}}$$

or

$$\frac{1}{\rho} = \sqrt{\frac{\frac{\sigma_y^3}{3E^2\sigma_0 h}}{\frac{h\sigma_y}{4\sigma_0}\left(1 - \frac{\sigma_0^2}{\sigma_y^2}\right) - \frac{M_{int}}{P}}} \qquad (4.28n)$$

Using Eqs. (4.26d), (4.27n), and (4.28n) we get

$$\frac{\pi^2\delta}{L} = \sqrt{\frac{\frac{\sigma_y^3}{3E^2\sigma_0 h}}{\frac{h\sigma_y}{4\sigma_0}\left(1 - \frac{\sigma_0^2}{\sigma_y^2}\right) - \frac{M_{int}}{P}}}$$

or

$$\delta\sqrt{\frac{h\sigma_y}{4\sigma_0}\left(1 - \frac{\sigma_0^2}{\sigma_y^2}\right)} - \delta - \frac{M_x}{P} - \frac{L^2\sigma_y}{\pi^2}\sqrt{\frac{\sigma_y}{3hE^2\sigma_0}} = 0 \qquad (4.28o)$$

Eq. (4.28n) gives the load (in terms of average stress) versus the mid-span deflection δ, relation in the inelastic range which can be drawn for the various values of M_x/P when the yield strength is reached on both the compression and tension sides of the cross- section. The critical value of the average stress σ_0, can be obtained from the expression

$$\frac{d\sigma_0}{d\delta} = 0 \qquad (4.27q)$$

The derivative of the expression in Eq. (4.28o) is taken with respect to δ and uses Eq. (4.27q), which leads to

$$\left[\frac{h\sigma_y}{4\sigma_0} \left(1 - \frac{\sigma_0^2}{\sigma_y^2}\right) - \delta - \frac{M_x}{P} \right]^{\frac{1}{2}} - \frac{\delta}{2} \left[\frac{h\sigma_y}{4\sigma_0} \left(1 - \frac{\sigma_0^2}{\sigma_y^2}\right) - \delta - \frac{M_x}{P} \right]^{-\frac{1}{2}} = 0$$

or

$$\delta = \frac{h\sigma_y}{6\sigma_0} \left(1 - \frac{\sigma_0^2}{\sigma_y^2}\right) - \frac{2M_x}{3P} \tag{4.28p}$$

Substitute δ from Eq. (4.28p) into Eq. (4.28o) to obtain

$$2 \left[\frac{h\sigma_y}{12\sigma_0} \left(1 - \frac{\sigma_0^2}{\sigma_y^2}\right) - \frac{M_x}{3P} \right]^{\frac{3}{2}} = \frac{L^2}{\pi^2} \left(\frac{\sigma_y^3}{3hE^2\sigma_0} \right)^{\frac{1}{2}}$$

$$\frac{4h^3}{12^3} \left[\frac{\sigma_y}{\sigma_0} - \frac{\sigma_0}{\sigma_y} - \frac{4M_x}{Ph} \right]^3 = \frac{L^4 \sigma_y^3}{3\pi^4 hE^2\sigma_0}$$

or

$$\sigma_c = \frac{12^2 L^4 \sigma_y^3}{\pi^4 h^4 E^2 \left[\frac{\sigma_y}{\sigma_c} - \frac{\sigma_c}{\sigma_y} - \frac{4M_x}{Ph} \right]^3} \tag{4.28q}$$

where the average critical stress $\sigma_c = \sigma_0$

$$\frac{\pi^2 E h^2}{12 L^2} = \frac{\pi^2 EI}{L^2 A} = \frac{\pi^2 E}{(L/r)^2}$$

or

$$\sigma_c = \frac{\left[\frac{(L/r)^2}{\pi^2 E} \right]^2 \sigma_y^3}{\left[\frac{\sigma_y}{\sigma_c} - \frac{\sigma_c}{\sigma_y} - \frac{4M_x}{Ph} \right]^3} \tag{4.28r}$$

Equation (4.28r) gives the relation between the critical average stress and the slenderness ratio for any given value of M_x/P for the beam column when the yield strength is reached on both the compression and tension sides of the cross-section. If there is no lateral load on the member, only the eccentric axial force is acting, then the parameter M_x/P can be substituted by the eccentricity e of the axial force.

If h is expressed by the core radius $s = h/6$ of the rectangular section in Eqs. (4.27t) and (4.28r), then the two equations can be used for non-rectangular cross-sections [7]. The core radius defines the core of a cross-section within which a compressive force can act on a short column without causing tensile stress in any extreme fiber. For an eccentrically loaded column

$$\sigma = \frac{P}{A} \pm \frac{Pec}{I} \qquad \text{or} \qquad \sigma = \frac{P}{A} \left(1 - \frac{ec}{r^2}\right)$$

For no tension in any extreme fiber, then the maximum value of $e = \frac{r^2}{c}$. The quantity, r^2/c, is called the core radius, where r is the radius of gyration of the cross-section about the centroidal axis parallel to the neutral axis, and c is the distance of the extreme fiber from the centroidal axis. For a rectangle, the core radius, $s = h/6$. Upon substituting the eccentricity ratio for a rectangle

$$\kappa = \frac{e}{s} = \frac{6e}{h} \tag{4.28s}$$

and $M_x/P = e$ into Eq. (4.27t) we have

$$\sigma_c = \frac{\pi^2 E}{(L/r)^2} \left[\frac{\frac{\sigma_y}{\sigma_c} - 1 - \frac{\kappa}{3}}{\frac{\sigma_y}{\sigma_c} - 1} \right]^3$$

or

$$\left(\frac{L}{r}\right)^2 = \frac{\pi^2 E}{\sigma_c} \left[\frac{3\left(\frac{\sigma_y}{\sigma_c} - 1\right) - \kappa}{3\left(\frac{\sigma_y}{\sigma_c} - 1\right)} \right]^3 \tag{4.28t}$$

Equation (4.28t) is valid for

$$\left(\frac{L}{r}\right)^2 - \frac{\pi^2 E \kappa^3}{9\sigma_y(3 - \kappa)} > 0, \text{Case 1. [7]}$$

Similarly, from Eq. (4.28r) we have

$$\sigma_c = \frac{\left[\frac{(L/r)^2}{\pi^2 E}\right]^2 \sigma_y^3}{\left[\frac{\sigma_y}{\sigma_c} - \frac{\sigma_c}{\sigma_y} - \frac{2\kappa}{3}\right]^3}$$

or

$$\left(\frac{L}{r}\right)^2 = \frac{\pi^2 E}{\sigma_y} \sqrt{\frac{\sigma_c}{\sigma_y} \left(\frac{\sigma_y}{\sigma_c} - \frac{\sigma_c}{\sigma_y} - \frac{2\kappa}{3}\right)^3} \tag{4.28u}$$

Equation (4.28u) is valid for

$$\left(\frac{L}{r}\right)^2 - \frac{\pi^2 E \kappa^3}{9\sigma_y(3 - \kappa)} < 0, \text{Case 2 [7].}$$

For large eccentricities $\kappa \geq 3$, Eq. (4.28u) is used because the stress distribution across the cross-section is that of case 2. Examples 4.3 and 4.4 are presented.

Example 4.3 Draw the column curves for the inelastic beam columns made of structural steel. Given data is:

Modulus of elasticity, $E = 29 \times 10^6$ psi (200, 000 MPa)

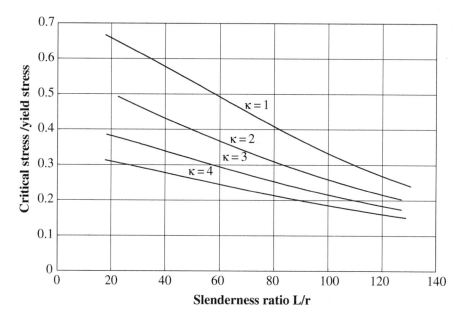

Figure 4.28 Column curves for the inelastic beam column.

Yield strength of steel, $\sigma_y = 50$ ksi (345 MPa)
Eccentricity ratio, $\kappa = 1, 2, 3,$ and 4.

Assume different values of critical stresses σ_c and apply Eq. (4.28t) for eccentricity ratios, $\kappa = 1$ and 2 to calculate the slenderness ratios L/r. Similarly apply Eq. (4.28u) for eccentricity ratios of $\kappa = 3$ and 4 to calculate the slenderness ratios L/r for different assumed values of critical stresses. The column curves in Figure 4.28 are obtained by this procedure.

Example 4.4 Draw the load deflection curve for a simply supported rectangular beam column subjected to an eccentric axial load. The member is made of structural steel with the following properties and dimensions:

Modulus of elasticity, $E = 29 \times 10^6$ psi (200, 000 MPa)
Yield strength of steel, $\sigma_y = 50$ ksi (345 MPa)
Eccentricity of the axial load, $e = 1.5$ in. (38 mm)
Length of the beam column – 150 in. (3.8 m)
Depth of the rectangular cross-section, $h = 4.5$ in. (114 mm)

Solution

$$\frac{e}{h} == \frac{1}{3}$$

Eccentricity ratio, $\kappa = \dfrac{6e}{h} = 2$

Radius of gyration, $r = \dfrac{h}{2\sqrt{3}}$

Slenderness ratio, $\dfrac{L}{r} = \dfrac{2\sqrt{3}L}{h} = 115.47$

Euler's buckling stress, $\sigma_e = \dfrac{\pi^2 E}{(L/r)^2} = 21.47$ ksi (148.04 MPa)

Within the elastic limit the combination of Eqs. (4.26b) and (4.26c) leads to

$$M_x + Py = -EI\dfrac{d^2y}{dx^2}$$

$$M_x + Py = EI\dfrac{\pi^2\delta}{L^2}, \quad \text{using Eq.(4.27n)}$$

Assume

$$M_x = Pe$$

Therefore,

$P(e + \delta) = \delta P_e, \quad$ where $P_e = \dfrac{\pi^2 EI}{L^2}$ is the Euler′s buckling load

Rearrange the terms and divide both sides by h

$$\dfrac{\delta}{h} = \dfrac{e}{h\left(\dfrac{P_e}{P} - 1\right)},$$

or

$$\dfrac{\delta}{h} = \dfrac{e}{h\left(\dfrac{\sigma_e}{\sigma_0} - 1\right)} \tag{4.29a}$$

where $\sigma_0 = P/A$ is the average axial stress on the member cross-section. The maximum stress within the elastic limit is given by

$$\sigma_{max} = \dfrac{P}{A} + \dfrac{Mc}{I} = \dfrac{P}{bh} + \dfrac{P(e + \delta)(h/2)}{bh^3/12}$$

or

$$\sigma_{max} = \sigma_0\left[1 + \dfrac{6(e + \delta)}{h}\right] \tag{4.29b}$$

Assume the values of σ_0, and find δ/h and σ_{max} from Eqs. (4.29a) and (4.29b) shown in Table 4.1 within the elastic limit. The maximum stress of 49.90 ksi (344.70 MPa), approximately the yield stress reaches at an average stress σ_0 of 10.30 ksi (71.07 MPa). Beyond this average stress Eq. (4.27p) for the inelastic analysis is used because the eccentricity ratio in the problem is $\kappa = 2$. For inelastic analyis it is better to assume δ/h quantities and calculate the average stress σ_0 from Eq. (4.27p).

Table 4.1 Load deflection relation for the beam column.

σ_0 ksi (MPa)	σ_0/σ_y	δ/h	σ_{max} ksi (MPa)
	Elastic analysis		
2 (13.80)	0.040	0.034	6.41 (44.23)
4 (27.60)	0.080	0.076	13.83 (95.44)
6 (41.40)	0.120	0.129	22.65 (156.31)
8 (55.20)	0.160	0.198	33.50 (231.17)
10.00 (69.00)	0.200	0.291	47.44 (327.31)
10.30 (71.07)	0.206	0.307	49.90 (344.27)

σ_0 ksi (MPa)	σ_0/σ_y	δ/h
	Inelastic analysis	
10.50 (72.45)	0.210	0.320
11.00 (75.90)	0.220	0.370
11.20 (77.28)	0.224	0.420
11.28 (77.80)	0.226	0.460
11.25 (77.63)	0.225	0.470
11.25 (77.63)	0.225	0.480
11.24 (77.56)	0.2248	0.490
11.23 (77.49)	0.2246	0.500
11.22 (77.42)	0.2244	05.10
11.20 (77.28)	0.2240	0.520
11.17 (77.07)	0.2234	0.530

The load deflection graph in terms of σ_0/σ_y and δ/h is shown in Figure 4.29. The first yield in the material occurs at $\sigma_0/\sigma_y = 0.206$, whereas the maximum stress in the material occurs at $\sigma_0/\sigma_y = 0.226$. The member is not able to resist further load after the maximum stress is reached.

The quantity $\sigma_0/\sigma_y = 0.225$ from the column curves in Figure 4.28 is for a slenderness ratio, $L/r = 115.47$ for the eccentricity ratio, $\kappa = 2$. A close result to 0.226 is obtained in this problem. The results shown are for the rectangular section made of elastic-perfectly plastic material. In cases where other shapes and materials are involved, numerical solutions are used to obtain maximum load. Galambos and Ketter [9] obtained numerical solution for steel I-beams for maximum load.

4.11 Design of Beam Columns

In beam columns consideration has to be given to additional moments called secondary moments which arise as a result of the axial force acting through the lateral displacements of the member. This is called $P-\delta$ effect, and these moments are to be added to the primary moments acting on the members due to applied end moments or transverse load. These secondary moments increase the stresses and deformations and should be considered in the design of beam columns. It is difficult to calculate the collapse load of beam columns theoretically. As shown in the previous sections, there are lengthy calculations in spite of the simplifying assumption of elastic-perfectly plastic material and if the section chosen was rectangular. For other cross-sections and materials it would be much more complex and often numerical solutions are used in such cases. To simplify the analysis the interaction equations are used in the design of beam columns. The interaction equations are graphs between M/M_u and P/P_u ratios as shown in Figure 4.30, where

P = Axial failure force in the presence of primary bending moment
P_u = Ultimate axial force (buckling load) at failure in the absence of primary bending moment
M = Maximum primary bending moment along with the axial force
M_u = Ultimate bending moment at failure (plastic moment if lateral torsional buckling does not occur) without the axial force

The simplest interaction equation is given by a straight line, as shown below:

$$\frac{P}{P_u} + \frac{M}{M_u} = 1 \qquad (4.30a)$$

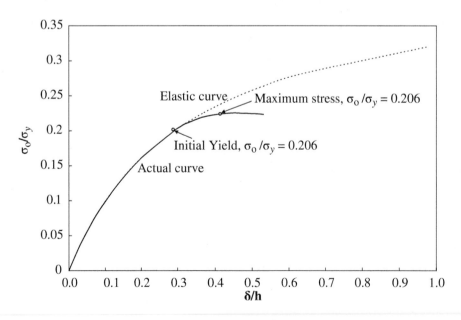

Figure 4.29 Load deflection relation for the beam column.

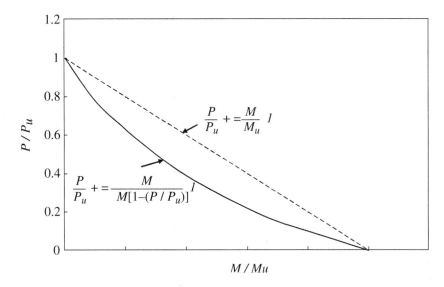

Figure 4.30 Interaction curve for a beam column.

It is given in the literature [9, 10] that the theoretical and experimental values fall below the straight line graph. Thus, the straight line formula of Eq. (4.30a) gives nonconservative values of strength for beam columns. This discrepancy occurs because only the primary bending moment, M, is included in Eq. (4.30a). The straight line formula is modified to include the secondary bending moment exerted by the axial force due to the lateral deflection of the member. If the primary bending moment, M, is multiplied by the amplification factor, $\frac{1}{1-(P/P_e)}$, to include the secondary moment as shown in Eq. (4.30b), then the references quoted before show the theoretical and experimental values fall closer to the curved line in Figure 4.30. It is also shown there that these interaction equations can be applied to different cross-sections and materials.

$$\frac{P}{P_u} + \frac{M}{M_u(1 - P/P_e)} = 1 \tag{4.30b}$$

Equations (4.3n and 4.5z) give the expressions for maximum moments when a concentrated force is acting at the middle of the span, and a uniformly distributed load acts on the entire span of simply supported beam columns respectively. These expressions can be written in the form

$$M_{\max} = M_0 \left[\frac{1 + \Psi(P/P_{ek})}{1 - (P/P_{ek})}\right] \tag{4.30c}$$

where M_{max} = maximum bending moment in the column including the secondary moment; M_0 = maximum primary bending moment in the column in the absence of axial force; P = actual axial force in the presence of bending moment; P_{ek} = critical axial force of the beam column considering the end conditions in the absence of primary bending moment; and Ψ = constant.

Table 4.2 Amplification factors Ψ and C_m.

Case	Ψ	C_m
	0	1.0
	−0.4	$1 - 0.4\frac{P}{P_{ek}}$
	−0.4	$1 - 0.4\frac{P}{P_{ek}}$
	−0.2	$1 - 0.2\frac{P}{P_{ek}}$
	−0.3	$1 - 0.3\frac{P}{P_{ek}}$
	−0.2	$1 - 0.2\frac{P}{P_{ek}}$

The values of Ψ have been adopted as −0.2 and 0 respectively for cases in the Sections 4.3.1 and 4.4 in the steel design by AISC [11]. Similarly the quantities Ψ are adopted for other cases of beam columns for their design as shown in Table 4.2.

4.11.1 Concept of Equivalent Moment and Factor C_m

In Eq. (4.12g) if M_B is the larger of the two end moments, then the term in the brackets can be considered as the amplification factor for beam columns subjected to two end moments and an axial force. The maximum moment thus obtained may lie within the span or outside the span. In the latter case, the maximum moment is then the larger of the two end moments.

$$M_{\max} = |M_B| \left[\sqrt{\frac{\left(\frac{M_A}{M_B}\right)^2 + 2\left(\frac{M_A}{M_B}\right)\cos kL + 1}{\sin^2 kL}} \right] \tag{4.12g}$$

For design purposes, the concept of an equivalent moment is used to simplify the calculations. The end moments M_A and M_B in the Figures 4.10 and 4.12 are replaced by two equal moments of magnitude M_{eqt} shown in Figure 4.31.

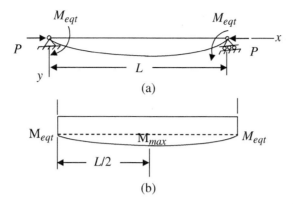

Figure 4.31 Beam column subjected to equivalent moment: (a) Equivalent moment; (b) Bending moment diagram.

M_{eqt} can be calculated by substituting $M_e = M_{eqt}$ in Eq. (4.10i), and equating the right-hand sides of Eqs. (4.10i) and (4.12g) as follows:

$$M_{eqt} \sec \frac{kL}{2} = |M_B| \left[\sqrt{\frac{\left(\frac{M_A}{M_B}\right)^2 + 2\left(\frac{M_A}{M_B}\right)\cos kL + 1}{\sin^2 kL}} \right]$$

or

$$M_{eqt} = |M_B| \left[\sqrt{\frac{\left(\frac{M_A}{M_B}\right)^2 + 2\left(\frac{M_A}{M_B}\right)\cos kL + 1}{2(1 - \cos kL)}} \right]$$

or

$$M_{eqt} = C_m |M_B| \tag{4.31a}$$

where C_m, the amplification factor using the equivalent moment concept, is given by

$$C_m = \sqrt{\frac{\left(\frac{M_A}{M_B}\right)^2 + 2\left(\frac{M_A}{M_B}\right)\cos kL + 1}{2(1 - \cos kL)}} \tag{4.31b}$$

The ratio M_A/M_B is taken as positive for a double curvature bending and is taken as negative for a single curvature bending as described before. Two simplified expressions for C_m have been given for design purposes. One of them as proposed by Austin [12] is

$$C_m = 0.6 - 0.4(M_A/M_B) \geq 0.4 \tag{4.31c}$$

The other expression given by Massonnet [13] is

$$C_m = \sqrt{0.3(M_A/M_B)^2 - 0.4(M_A/M_B) + 0.3} \qquad (4.31d)$$

In both the expressions for C_m, the ratio M_A/M_B is taken as positive for a double curvature bending and is taken as negative for a single curvature bending. It is shown by Chen and Lui [14] that the simplified values of C_m given by Eqs. (4.31c) and (4.31d) give a good approximation of the values given by Eq. (4.31b). The Austin equation is the simpler of the two, and has been adopted for design purposes when the beam columns are subjected to end moments and have no transverse loads on their span [11]. Now using Eq. (4.10i)

$$M_{\max} = M_{eqt} \sec \frac{kL}{2} \qquad (4.31e)$$

The substitution of Eq. (4.31a) into Eq. (4.31e) leads to

$$M_{\max} = C_m M_B \sec \frac{kL}{2}$$

or

$$M_{\max} \approx \frac{C_m}{1 - \dfrac{P}{P_{ek}}} M_B \qquad (4.31f)$$

Equation (4.31f) is derived for beam columns subjected to end moments, but it can be used for beam columns subjected to transverse loads by modifying the value of C_m. In cases where the primary bending moment is due to transverse loads, the comparison of Eqs. (4.30c) and (4.31f) leads to

$$C_m = 1 + \Psi(P/P_{ek}) \qquad (4.31g)$$

Table 4.2 gives the values of Ψ and C_m for different transverse loads and end supports for beam columns. Now Eq. (4.30b) can be written in the modified form incorporating the amplification factor C_m, as follows:

$$\frac{P}{P_u} + \frac{C_m M}{M_u(1 - P/P_e)} = 1 \qquad (4.31h)$$

If the bending occurs about two axes, x and y, then Eq. (4.31h) can be extended as follows:

$$\frac{P}{P_u} + \frac{C_{mx}M_x}{M_{ux}(1 - P/P_{ex})} + \frac{C_{my}M_y}{M_{uy}(1 - P/P_{ey})} = 1 \qquad (4.31i)$$

The variables in Eq. (4.31i) are the same as in Eq. (4.30b), where x and y subscripts relate to the respective axes.

4.11.2 AISC Design Criteria for Steel Beam Columns

4.11.2.1 Doubly and Singly Symmetric Members Subjected to Flexure and Compression

The AISC Steel Construction Manual 2017, Section H1, 15th edition [11] combines the Allowable Strength Design (ASD) and the Load Resistance Factor Design (LRFD) methods in the same manual. The interaction equations from the AISC manual for these members that are constrained to bend about a geometric axis (x and/or y) are given by

$$\frac{P_r}{P_c} + \frac{8}{9}\left(\frac{M_{rx}}{M_{cx}} + \frac{M_{ry}}{M_{cy}}\right) \leq 1.0 \qquad \text{for } \frac{P_r}{P_c} \geq 0.2 \tag{4.32a}$$

And

$$\frac{P_r}{2P_c} + \left(\frac{M_{rx}}{M_{cx}} + \frac{M_{ry}}{M_{cy}}\right) \leq 1.0 \qquad \text{for } \frac{P_r}{P_c} < 0.2 \tag{4.32b}$$

Where
P_r = required axial compressive strength, kips (N)
P_c = available axial compressive strength, kips (N)
M_r = required flexural strength, kip-in. (N-mm)
M_c = available flexural strength, kip-in. (N-mm)
x = subscript for strong axis
y = subscript for weak axis

For the LRFD design:

P_r = required axial compressive strength using LRFD load combinations (actual loads multiplied by load factors), kips (N)
$P_c = \phi_c P_n$ = design axial compressive strength, kips (N)
P_n = nominal compressive strength of the member in the absence of bending moment, kips (N)
M_r = required flexural strength using LRFD load combinations (actual loads multiplied by load factors), kip-in. (N-mm)
$M_c = \phi_b M_n$, design flexural strength of the member, kip-in. (N-mm)
M_n = nominal flexural strength of the member in the absence of axial force, kip-in. (N-mm)
ϕ_c = resistance factor for compression = 0.90
ϕ_b = resistance factor for flexure = 0.90

For the ASD design:

P_r = required axial compressive strength using ASD load combinations (actual loads), kips (N)
$P_c = P_n/\Omega_c$ = allowable axial compressive strength, kips (N)
M_r = required flexural strength using ASD load combinations (actual loads), kip-in. (N-mm)
$M_c = M_n/\Omega_b$ = allowable flexural strength, kip-in. (N-mm)
Ω_c = safety factor for compression = 1.67
Ω_b = safety factor for flexure = 1.67

4.11.2.2 Unsymmetric and Other Members Subject to Flexure and Axial Force

The interaction equation for shapes not covered by Section 4.11.2.1 is given by

$$\left|\frac{f_a}{F_a} + \frac{f_{bw}}{F_{bw}} + \frac{f_{bz}}{F_{bz}}\right| \leq 1.0 \tag{4.33a}$$

where f_a = required axial stress at the point of consideration, ksi (MPa)

F_a = available axial compressive stress at the point of consideration, ksi (MPa)

f_{bw}, f_{bz} = required flexural stresses at the point of consideration, ksi (MPa)

F_{bw}, F_{bz} = available flexural stresses at the point of consideration, ksi (MPa)

w = subscript for major principal axis

z = subscript for minor principal axis

Lower case stresses f, are computed by the elastic analysis for the applicable loads, including second-order effects. The upper case stresses F correspond to the limit state of yielding or buckling.

For the LRFD design:

f_a = required axial stress using LRFD load combinations, ksi (MPa)

$F_a = \phi_c F_{cr}$ = design axial compressive stress, ksi (MPa)

F_{cr} = flexural buckling stress [σ_{cr} by Eqs. (3.14h) or (3.14i)]

f_{bw}, f_{bz} = required flexural stresses at the point of consideration using LRFD load combinations, ksi (MPa)

$F_{bw}, F_{bz} = \frac{\phi_b M_n}{S}$ = design flexural stresses at the point of consideration, ksi (MPa)

S = Section modulus of the cross-section at the point of consideration

For the ASD design:

f_a = required axial stress using ASD load combinations, ksi (MPa)

$F_a = \frac{F_{cr}}{\Omega_c}$ = allowable axial compressive stress, ksi (MPa)

f_{bw}, f_{bz} = required flexural stresses at the point of consideration using ASD load combinations, ksi (MPa)

$F_{bw}, F_{bz} = \frac{M_n}{\Omega_b S}$ = allowable flexural stresses at the point of consideration, ksi (MPa)

The resistance factors, ϕ, and safety factors, Ω, are the same for Eqs. (4.32a), (4.32b), and (4.33a).

The required second-order axial strength, P_r, and flexural strength, M_r, are determined as follows:

$$M_r = B_1 M_{nt} + B_2 M_{lt} \tag{4.33b}$$

and

$$P_r = P_{nt} + B_2 P_{lt} \tag{4.33c}$$

where

$$B_1 = \frac{C_m}{1 - \alpha P_r / P_{e1}} \geq 1 \tag{4.33d}$$

For members loaded in axial compression, B_1 in Eq. (4.33d) is calculated by taking $P_r = P_{nt} + P_{lt}$. B_1 is the magnification factor to take into account second-order effects due to displacements between support points of beam columns, called $P - \delta$ effects. B_2 is the magnifier to account for second-order effects due to displacements of the support points, called $P - \Delta$ effects:

$$B_2 = \frac{1}{1 - \frac{\alpha \Sigma P_{nt}}{\Sigma P_{e2}}} \geq 1 \qquad (4.33e)$$

The various terms used in Eqs. (4.33b–4.33e) are:

M_{nt} = moment in the member calculated by using LRFD or ASD load combinations and first-order elastic analysis, assuming there is no lateral translation in the frame, kip-in. (N-mm)

M_{lt} = moment in the member calculated by first-order elastic analysis using LRFD or ASD load combinations caused by the lateral translation of the frame only, kip-in. (N-mm)

P_{nt} = axial compression forces in the member calculated by first-order analysis using LRFD or ASD load combinations, assuming there is no lateral translation in the frame, kips (N)

P_{lt} = axial compression forces in the member calculated by first-order analysis using LRFD or ASD load combinations, caused by the lateral translation of the frame only, kips (N)

α = a factor equal to 1.00 for LRFD design and 1.60 for ASD design

P_{e1} = elastic buckling resistance for the member in the plane of bending assuming zero sidesway
$= \frac{\pi^2 EI}{(K_1 L)^2}$, kips (N)

K_1 = effective length factor in the plane of bending assuming no lateral translation, taken as equal to 1.0, unless a smaller value is indicated by analysis

L = story height, in. (mm)

I = moment of inertia in the plane of bending, in.^{4} (mm^4)

E = modulus of elasticity of steel = 29 000 ksi (200 000 MPa)

C_m = modification coefficient as described before, assuming no lateral translation of the frame. Its value is taken as

(i) For beam columns not subjected to transverse loading between supports in the plane of bending,

$$C_m = 0.6 - 0.4(M_1/M_2)$$

where M_1 and M_2 are the smaller and larger moments at the ends of the unbraced length in the plane of bending, and are calculated from the first-order analysis. M_1/M_2 is positive for reverse or double curvature bending, negative for the single curvature.

(ii) For beam columns subjected to transverse loading between supports, the coefficient C_m is determined by analysis as shown in Table 4.2, or taken conservatively as 1.0 for all cases.

ΣP_{nt} = total vertical axial compressive forces supported by all the columns in a story including the gravity column loads, kips (N)

ΣP_{e2} = elastic buckling resistance for all the columns in a story $= \Sigma \frac{\pi^2 EI}{(K_2 L)^2}$, kips (N)

or $\Sigma P_{e2} = R_M \frac{\Sigma HL}{\Delta_H}$ (alternate expression), kips (N)

K_2 = effective length factors in the plane of bending based on the sidesway buckling analysis

$R_M = 1.0$ for braced frame systems and 0.85 for moment frame and combined systems

ΣH = total story shear due to lateral forces used to compute Δ_H, kips (N)

Δ_H = the first-order inter-story drift due to lateral forces, in. (mm)

The European [15], Canadian [16], and Australian [17] steel design codes calculate the cross-section strength, the member in-plane strength, and the member lateral-torsional strength separately. For a given ratio of bending and axial compression, the governing strength is taken as the least of these strengths [18, 19]. The AISC uses the same equations for all potential modes of failures. For comparison, equations from different codes for steel design are given below for cross-section strength. To make the comparison easier, the AISC notations are used in all equations.

4.11.3 Eurocode 3 (ECS, 1993) Design Criteria

For standard rolled I and H sections that are defined as class 1 or class 2, Eurocode 3 (EN 1993) [15] specifies the cross-section strength for strong axis bending as

$$\frac{P_u}{\phi P_y} + 0.9\frac{M_{ux}}{\phi M_{px}} \leq 1.0 \tag{4.34a}$$

and for weak axis bending as

$$\frac{M_{uy}}{\phi M_{py}} \leq 1.56\left(1 - \frac{P_u}{\phi P_y}\right)\left(\frac{P_u}{\phi P_y} + 0.6\right) \quad \text{for} \quad \frac{P_u}{P_y} > 0.2 \tag{4.34b}$$

and

$$\frac{M_{uy}}{\phi M_{py}} \leq 1.0 \quad \text{for} \quad \frac{P_u}{P_y} \leq 0.2 \tag{4.34c}$$

where

P_u = Design axial load on the member

P_y = Nominal yield load for the cross-section

M_{ux} = Maximum second-order elastic moment (amplified to account for the deflection of axial load) within span or at the ends of member about the x axis

M_{px} = Nominal plastic moment capacity of the cross-section about the x axis

M_{uy} = Maximum second-order elastic moment (amplified to account for deflection of axial load) within span or at the ends of member about the y axis

M_{py} = Nominal plastic moment capacity of the cross-section about the y axis

ϕ = Effective resistance factors = 0.91 = 1/1.1 for both axial compression and bending

4.11.4 Canadian Standards Association (CSA 1994 – CSA-S16.1)

For class 1 sections (suitable for plastic design) the cross-section strength for strong axis bending is given by [16]

$$\frac{P_u}{\phi P_y} + 0.85\frac{M_{ux}}{\phi M_{px}} \leq 1.0 \tag{4.35a}$$

and for weak axis bending the cross-section strength is

$$\frac{P_u}{\phi P_y} + 0.6\frac{M_{uy}}{\phi M_{py}} \le 1.0 \tag{4.35b}$$

Where P_u = Design axial load on the member as per CSA 1994 – CSA-S16.1

M_{ux}, M_{uy} = Maximum of second-order elastic moments about major and minor axes of cross-section at member ends

M_{px}, M_{py} = Plastic section strengths about major and minor axes of bending

ϕ = Resistance factor = 0.9 for both axial compression and bending

4.11.5 Australian Standard AS4100-1990

The cross-section strength for strong axis bending is given by [17]

$$\frac{P_u}{\phi P_y} + \frac{1}{1.18}\frac{M_{ux}}{\phi M_{px}} \le 1.0 \tag{4.36a}$$

The weak axis cross-section strength for an I shape is written as

$$\left(\frac{P_u}{\phi P_y}\right)^2 + \frac{1}{1.19}\frac{M_{uy}}{\phi(1.5S_yF_y)} \le 1.0 \tag{4.36b}$$

where

P_u = Design axial load on the member as per AS4100–1990

M_{ux}, M_{uy} = Maximum of second-order elastic moments about major and minor axes of cross-section in the member

M_{px}, M_{py} = Plastic section strengths about major and minor axes of bending

F_y = Yield stress

S_y = Elastic section modulus about y axis

ϕ = Resistance factor = 0.9 for both axial compression and bending

In the above equations the applied axial force and moment terms are calculated differently for various codes. The reader should consult these codes to know the full extent of application of these equations. Example 4.5 is presented.

Example 4.5 A W14 in. × 145 lb/ft (W360 mm × 216 kg/m) wide flange section of 50 ksi (345 MPa) steel is used as a beam column in a braced frame. It is bent in a single curvature by equal moments as shown in Figure 4.32. Determine if the section is satisfactory as per the AISC code.

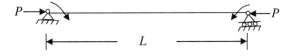

Figure 4.32 Beam column in a braced frame.

Given: $L_x = L_y = 14.0$ ft (4.27 m), P_D (Axial dead load) = 90 kips (405 kN), P_L (Axial live load) = 165 kips (743 kN).

The first-order moments are: M_{Dx} (Dead load moment) = 75 kip-ft (103 kN-m), M_{Lx} (Live load moment) = 135 kip-ft (185 kN-m), M_{Dy} (Dead load moment) = 45 kip-ft (62 kN-m), M_{Ly} (Live load moment) = 105 kip-ft (144 kN-m).

Properties of W 14 in. × 145 lb/ft (W 360 mm × 216 kg/m) are: $A = 42.7$ in.^{2} (27, 600 mm^2), $I_x = 1710$ in.^{4} (712×10^6 mm^4), $I_y = 677$ in.^{4} (283×10^6 mm^4), $r_x = 6.33$ in. (161 mm), $r_y = 3.98$ in. (101 mm), $z_x = 260$ in.^{3} (4260×10^3 mm^3), $z_y = 133$ in.^{3} (2180×10^3 mm^3).

Properties of steel are:

$F_y = 50$ ksi (345 MPa), $F_u = 65$ ksi (448.5 MPa), $E = 29 \times 10^6$ psi (200, 000 MPa).

For the braced frame, $K_x = K_y = 1.0$, $K_x L_x = K_y L_y = 14.0$ ft (4.27 m)

Solution (the LRFD Method)

$$P_r = 1.2 \ P_D + 1.6 \ P_L$$

$$P_r = 1.2(90) + 1.6(165) = 372 \text{ kips } [1.2 \ (405) + 1.6(743) = 1675 \text{ kN}]$$

$$M_{ntx} = 1.2 \ M_{Dx} + 1.6 \ M_{Lx}$$

$$M_{ntx} = 1.2(75) + 1.6(135) = 306 \text{ kip-ft}[1.2(103) + 1.6(185) = 420 \text{ kN-m}]$$

$$M_{nty} = 1.2 \ M_{Dy} + 1.6 \ M_{Ly}$$

$$M_{nty} = 1.2(45) + 1.6(105) = 222 \text{ kip-ft}[1.2(62) + 1.6(144) = 305 \text{ kN.m}]$$

$$\frac{K_y L_y}{r_y} = \frac{14 \times 12}{3.98} = 42.2 \quad \left[\frac{K_y L_y}{r_y} = \frac{4.27 \times 1000}{101} = 42.2 \right]$$

$$4.71 \sqrt{\frac{E}{F_y}} = 4.71 \sqrt{\frac{29 \times 10^6}{50,000}} = 113.4 \quad \left[4.71 \sqrt{\frac{E}{F_y}} = 4.71 \sqrt{\frac{200,000}{345}} = 113.4 \right]$$

$$\frac{K_y L_y}{r_y} = 42.2 < 4.71 \sqrt{\frac{E}{F_y}} = 113.4$$

Therefore,

$$F_{cr} = \left(0.658^{\frac{F_y}{F_e}} \right) F_y$$

where

$$F_e = \frac{\pi^2 E}{\left(\frac{KL}{r} \right)^2} = \frac{\pi^2 (29000)}{(42.2)^2} = 161.5 \text{ ksi } \left[F_e = \frac{\pi^2 (200,000)}{(42.2)^2} = 1114 \text{ MPa} \right] \text{ or}$$

$$F_{cr} = \left(0.658^{\frac{50}{161.5}} \right) (50) = 43.92 \text{ ksi } \left[F_{cr} = \left(0.658^{\frac{345}{1114}} \right) (345) = 303 \text{ MPa} \right]$$

$$P_n = F_{cr}A_g = (43.92)(42.7) = 1875 \text{ kips} \left[P_n = \frac{303(27600)}{1000} = 8363 \text{ kN} \right]$$

$$P_c = \phi_c P_n = 0.9(1875) = 1688 \text{ kips} [P_c = 0.9(8363) = 7527 \text{ kN}]$$

$$\frac{P_r}{P_c} = \frac{372}{1688} = 0.22 > 0.2 \left[\frac{P_r}{P_c} = \frac{1675}{7527} = 0.22 > 0.2 \right]$$

Therefore,

$$\frac{P_r}{P_c} + \frac{8}{9} \left(\frac{M_{rx}}{M_{cx}} + \frac{M_{ry}}{M_{cy}} \right) \leq 1.0$$

$$M_r = B_1 M_{nt} + B_2 M_{lt}, B_2 M_{lt} = 0 \text{ for braced frames.}$$
$$M_{rx} = B_{1x} M_{ntx}, \quad M_{ry} = B_{1y} M_{nty}$$

$$B_{1x} = \frac{C_{mx}}{1 - \alpha \frac{P_r}{P_{elx}}}, \quad B_{1y} = \frac{C_{my}}{1 - \alpha \frac{P_r}{P_{ely}}}, \quad \alpha = 1.00$$

$$P_{elx} = \frac{\pi^2 EI_x}{(K_x L_x)^2} = \frac{\pi^2(29000)(1710)}{(14x12)^2} = 17,341 \text{ kips} \left[P_{elx} = \frac{\pi^2(200,000)(712x10^6)}{(4.27x1000)^2 x 1000} = 77,082 \text{ kN} \right]$$

$$P_{ely} = \frac{\pi^2 EI_y}{(K_y L_y)^2} = \frac{\pi^2(29000)(677)}{(14x12)^2} = 6865 \text{ kips} \left[P_{ely} = \frac{\pi^2(200,000)(283x10^6)}{(4.27x1000)^2 x 1000} = 30,638 \text{ kN} \right]$$

$$C_{mx} = C_{my} = 0.6 - 0.4(-1) = 1.0$$

$$B_{1x} = \frac{1}{1 - (1)\frac{372}{17341}} = 1.02 \quad \left[B_{1x} = \frac{1}{1 - (1)\frac{1675}{77082}} = 1.02 \right]$$

$$B_{1y} = \frac{1}{1 - (1)\frac{372}{6865}} = 1.06 \quad \left[B_{1y} = \frac{1}{1 - (1)\frac{1675}{30638}} = 1.06 \right]$$

$$M_{rx} = 1.02(306) = 313 \text{ kip-ft } [M_{rx} = 1.02(420) = 429 \text{ kN-m}]$$
$$M_{ry} = 1.06(222) = 235 \text{ kip-ft } [M_{ry} = 1.06(305) = 323 \text{ kN-m})$$

The distance between braced points, $L_b = 14.0$ ft (4.27 m)
The limiting length for lateral-torsional buckling, $L_p = 14.1$ ft (4.3 m) for W 14×145 (W 360×216) wide flange section.
Therefore,

$$L_b = 14.0 \text{ ft } (4.27 \text{ m}) < L_p = 14.1 \text{ ft } (4.3 \text{ m})$$

$$M_{nx} = M_{px} = F_y Z_x = \frac{50(260)}{12} = 1083 \text{kip-ft} \left[M_{nx} = \frac{345(4260x10^3)}{1000x1000} = 1470 \text{kN-m} \right]$$

$$M_{cx} = \phi_b M_{nx} = 0.9(1083) = 975 \text{ kip-ft } [M_{cx} = 0.9 \ (1470) = 1323 \text{ kN-m}]$$

$$M_{ny} = M_{py} = F_y Z_y = \frac{50(133)}{12} = 554 \text{ kip-ft } \left[M_{ny} = \frac{345(2180 x 10^3)}{1000 x 1000} = 752 \text{ kN-m} \right]$$

$$M_{cy} = \phi_b M_{ny} = 0.9(554) = 499 \text{ kip-ft } [M_{cy} = 0.9(752) = 677 \text{ kN-m}]$$

$$\frac{P_r}{P_c} + \frac{8}{9} \left(\frac{M_{rx}}{M_{cx}} + \frac{M_{ry}}{M_{cy}} \right) \leq 1.0$$

$$\frac{372}{1688} + \frac{8}{9} \left(\frac{313}{975} + \frac{235}{499} \right) = 0.93 < 1.0 \quad \left[\frac{1675}{7527} + \frac{8}{9} \left(\frac{429}{1323} + \frac{323}{677} \right) = 0.93 < 1.0 \right]$$

The section is satisfactory.

Solution (the ASD Method)

$$P_r = P_D + P_L$$
$$P_r = 90 + 165 = 255 \text{ kips } [P_r = 405 + 743 = 1148 \text{ kN}]$$

$$M_{ntx} = M_{Dx} + M_{Lx}$$
$$M_{ntx} = 75 + 135 = 210 \text{ kip-ft } [M_{ntx} = 103 + 185 = 288 \text{ kN-m}]$$

$$M_{nty} = M_{Dy} + M_{Ly}$$
$$M_{nty} = 45 + 105 = 150 \text{ kip-ft } [M_{nty} = 62 + 144 = 206 \text{ kN-m}]$$

$$P_c = \frac{P_n}{\Omega_c} \ (P_n \text{ is the same as in LRFD method})$$

$$P_c = \frac{1875}{1.67} = 1123 \text{ kips } \left[P_c = \frac{8363}{1.67} = 5008 kN \right]$$

$$\frac{P_r}{P_c} = \frac{255}{1123} = 0.23 > 0.2 \quad \left[\frac{P_r}{P_c} = \frac{1148}{5008} = 0.227 > 0.2 \right]$$

Therefore,

$$\frac{P_r}{P_c} + \frac{8}{9} \left(\frac{M_{rx}}{M_{cx}} + \frac{M_{ry}}{M_{cy}} \right) \leq 1.0$$

$$B_{1x} = \frac{C_{mx}}{1 - \alpha \frac{P_r}{P_{elx}}}, \quad B_{1y} = \frac{C_{my}}{1 - \alpha \frac{P_r}{P_{ely}}}, \quad \alpha = 1.60$$

(C_{mx}, C_{my}, P_{elx} and P_{ely} are the same as in the LRFD method.)

$$B_{1x} = \frac{1}{1 - (1.6)\frac{255}{17341}} = 1.02 \quad \left[B_{1x} = \frac{1}{1 - (1.6)\frac{1148}{77082}} = 1.02 \right]$$

$$B_{1y} = \frac{1}{1 - (1.6)\frac{255}{6865}} = 1.06 \qquad \left[B_{1y} = \frac{1}{1 - (1.6)\frac{1148}{30638}} = 1.06 \right]$$

$M_r = B_1 M_{nt} + B_2 M_{lt}, \; B_2 M_{lt} = 0$ for braced frames.

$M_{rx} = B_{1x} M_{ntx}, \quad M_{ry} = B_{1y} \; M_{nty}$

$M_{rx} = 1.02(210) = 215$ kip-ft$[M_{rx} = 1.02(288) = 295$ kN-m$]$

$M_{ry} = 1.06(150) = 159$ kip-ft $[M_{ry} = 1.06(206) = 219$ kN-m$)$

$L_b = 14.0$ ft (4.27 m) $< L_p = 14.1$ ft (4.3 m) same as in the LRFD method.

$M_{nx} = M_{px} = 1083$ kip-ft $[M_{nx} = 1470$ kN-m$]$ same as in the LRFD method.

$$M_{cx} = \frac{M_{nx}}{\Omega_b} = \frac{1083}{1.67} = 649 \text{ kip-ft} \qquad \left[M_{cx} = \frac{1470}{1.67} = 880 \text{ kN-m} \right]$$

$M_{ny} = M_{py} = 554$ kip-ft $[M_{ny} = 752$ kN-m$]$ same as in the LRFD method.

$$M_{cy} = \frac{M_{ny}}{\Omega_b} = \frac{554}{1.67} = 332 \text{ kip-ft} \qquad \left[M_{cy} = \frac{752}{1.67} = 450 \text{ kN-m} \right]$$

$$\frac{P_r}{P_c} + \frac{8}{9}\left(\frac{M_{rx}}{M_{cx}} + \frac{M_{ry}}{M_{cy}} \right) \le 1.0$$

$$\frac{255}{1123} + \frac{8}{9}\left(\frac{215}{649} + \frac{159}{332} \right) = 0.95 < 1.0 \qquad \left[\frac{1148}{5008} + \frac{8}{9}\left(\frac{295}{880} + \frac{219}{450} \right) = 0.95 < 1.0 \right]$$

The section is satisfactory.

Problems

4.1 Find the deflection at a distance of x from the left support when a triangular load is acting as shown in Figure P4.1. Calculate the deflection and bending moment at the mid-span.

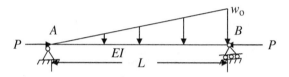

Figure P4.1

4.2 Find the deflection and bending moment at the mid-span of the beam with the load shown in Figure P4.2.

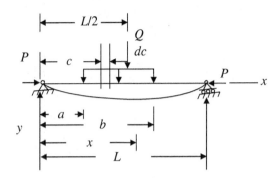

Figure P4.2

4.3 Compare the maximum moments (Figure P4.3) in the beam columns obtained for various values of P/P_{ek} from the exact equations in the text with those determined from the following expression:

$$M_{max} = M_0 \frac{1 + \Psi \frac{P}{P_{ek}}}{1 - \frac{P}{P_{ek}}}$$

Where P = axial force in the beam column, $P/P_{ek} = \pi^2 EI/(KL)^2$, $\psi = -0.4$

M_0 = Maximum moment in the beam column without the presence of axial load.

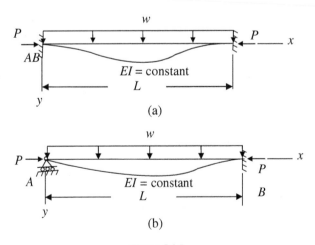

Figure P4.3

4.4 Find the critical load P for the continuous beam of Figure P.4.4 by the slope deflection method. The two spans are loaded by the uniformly distributed loads of w_1 and w_2. the modulus of elasticity of the entire beam is E.

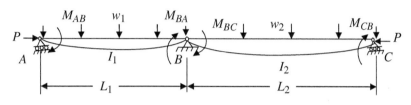

Figure P4.4

4.5 Show that the slope deflection coefficients for beam columns k_{ii} and k_{ij} are equal to 4 and 2 respectively if the axial force $P = 0$.

4.6 Find the maximum moment and its location for the beam column in Figure P4.6, when

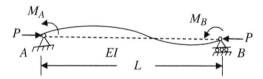

Figure P4.6

(a) $\frac{P}{P_e} = 0.2$, $\frac{M_A}{M_B} = 0$ (b) $\frac{P}{P_e} = 0.2$, $\frac{M_A}{M_B} = -0.2$ (c) $\frac{P}{P_e} = 0.4$, $\frac{M_A}{M_B} = 1.0$

4.7 A W14 in. × 68 lb/ft (W360 mm × 101 kg/m) wide flange section of 50 ksi (345 MPa) steel is used as a beam column in an unbraced frame. The axial load and moments obtained from the first-order analysis based on service loads are shown in Figure P4.7. The bending moments are about the x axis, the strong axis. The frame and the gravity loads are symmetric. Properties of W14 in. × 68 lb/ft (W360 mm × 101 kg/m) are:

$A = 20$ in.^{2} (12900 mm^2), $I_x = 722$ in.^{4} (302×10^6 mm^4),

$I_y = 121$ in.^{4} (50.6×10^4 mm^4), $r_x = 6.01$ in. (153 mm),

$r_y = 2.46$ in. (62.6 mm), $z_x = 115$ in.^{3} (1880×10^3 mm^3),

$z_y = 36.9$ in.^{3} (606×10^3 mm^3).

Properties of steel are: $\sigma_y = 50$ ksi (345 MPa), $F_u = 65$ ksi (448.5 MPa),

$E = 29 \times 10^6$ (200, 000 MPa).

Effective length factors are K_x (braced case) $= 0.85$, K_x (unbraced case) $= 1.2$, $K_y = 1.0$.

Determine if the section is satisfactory as per the AISC code.

$P_D = 95$ kips (423 kN), $P_L = 150$ kips (667 kN)

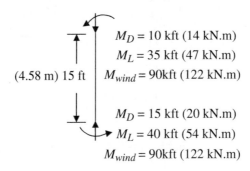

(4.58 m) 15 ft

$M_D = 10$ kft (14 kN.m)

$M_L = 35$ kft (47 kN.m)

$M_{wind} = 90$kft (122 kN.m)

$M_D = 15$ kft (20 kN.m)

$M_L = 40$ kft (54 kN.m)

$M_{wind} = 90$kft (122 kN.m)

Figure P4.7

References

1 CRC (2018). *Standard Mathematical Tables and Formulas*, 33e. Boca Raton, FL: CRC Press.

2 Kassimali, A. (2019). *Structural Analysis*, 5e. Boston, MA: Cengage Learning.

3 Timoshenko, S.P. and Gere, J.M. (1961). *Theory of Elastic Stability*, 2e. New York, NY: McGraw-Hill.

4 Nelson, J.K. and McCormack, J.C. (2003). *Structural Analysis: Using Classical and Matrix Methods*, 3e. Hoboken, NJ: Wiley.

5 Ježek, K. (1935) Näherungsberechnung der tragkraft exzentrisch gedrückter Stahlstäbe, *Der Stahlbau*, 8, 89.

6 Ježek, K. (1936) Die Tragfähigkeit axial gedrückter und auf Biegung beanspruchter Stahlstäbe, *Der Stahlbau*, 9, 12.

7 Bleich, F. and Bleich, H.H. (1952). *Buckling Strength of Metal Structures*. New York: McGraw-Hill.

8 Hibbeler, R.C. (2017). *Mechanics of Materials*, 10e. Hoboken, NJ: Pearson.

9 Galambos, T.V. and Ketter, R.L. (1959). Columns under combined bending and thrust. *Journal of Engineering Mechanics* 85 (2): 1–30.

10 Clark, J.W. (1955) Eccentrically loaded columns, Transactions of the ASCE, 120.

11 AISC (2017) *Steel Construction Manual*, 15e, AISC, Chicago, IL

12 Austin, W.J. (1961). Strength and design of metal beam-columns. *Journal of Structural Division* 87 (ST4): 1–32.

13 Massonnet, C. (1959). Stability considerations in the deign of steel columns. *Journal of Structural Division* 85 (September): 75–111.

14 Chen, W.F. and Lui, E.M. (1987). *Structural Stability Theory and Implementation*. Upper Saddle River, NJ: Prentice Hall.

15 ECS (1993). *Eurocode 3: Design of Steel Structures, EN 193–1-1*. Brussels, Belgium: European Committee for Standardization.

16 CSA (1994). *Limit State Design of Steel Structures, CAN/CSA-S16. 1-94*. Rexdale, Ontario, Canada: Canadian Standards Association.

17 SAA (1990). *Australian Standard: Steel Structures, AS 4100–1990*. North Sydney, New South Wales, Australia: Standards Association of Australia.

18 White, D.W. and Clarke, M. (1997). Design of beam columns in steel frames I: philosophies and procedures. *ASCE Journal of Structural Engineering* 123 (12): 1556–1564.

19 White, D.W. and Clarke, M. (1997). Design of beam columns in steel frames II: comparison of standards. *ASCE Journal of Structural Engineering* 123 (12): 1565–1564. 1574.

5

Frames

5.1 Introduction

So far we have considered the stability of members, i.e. columns, beams, and beam columns as individual members that have ideal boundary conditions, e.g. pinned, fixed, or free. In actual engineering systems these members exist as part of a larger framework, and their ends are elastically restrained by other members to which they are joined. It is usually assumed that, in a frame, members are connected by rigid joints, meaning all members connected at a joint rotate in the same direction by an equal amount. Under these conditions the member end restraints not only depend on the stiffness of members joining directly at a joint, but end restraints depend on the stiffness of all other members in the system. Therefore, we need to examine the stability of the whole frame. Another problem is that a member of a frame does not exactly behave like a column with spring supports at both ends, because the spring stiffness varies with the load. There can be different cases of loading in a frame. If the member of a frame are geometrically perfect and there is no primary moments present in the members, then there is no bending deformation and moment in the member until the critical load P_{cr} is reached. In this case, the problem can be solved as an eigenvalue problem as in the case of an individual column. On the other hand, if the members are geometrically imperfect or primary moments are present in the members due to eccentric loading or lateral loads, the frame will experience bending deformations from the instant the loads are applied and it requires second-order analysis considering geometric nonlinearity if the stresses are within the elastic limit. In this chapter we will consider frames with rigid joints whose member stresses are within the elastic limit at buckling.

5.2 Critical Loads by the Equilibrium Method

5.2.1 Portal Frame Without Sidesway

Consider a symmetrical frame fixed at the bottom in which sidesway is prevented, and it is loaded by point loads P as shown in the Figure 5.1a. The forces in the individual members of the frame are shown in Figure 5.1b. Considering the equilibrium of the free body diagram of the deflected shape of the left vertical column in Figure 5.1c, we have

$$P(-y) + (-EIy'') + \frac{M_A - M_B}{L_c}x - M_A = 0 \tag{5.1a}$$

or

$$y'' + k_c^2 y = \frac{M_A - M_B}{L_c}\frac{x}{EI_c} - \frac{M_A}{EI_c}, \text{ where } k_c^2 = \frac{P}{EI_c}$$

or

$$y'' + k_c^2 y = \frac{M_A k_c^2}{P}\left(\frac{x}{L_c} - 1\right) - \frac{M_B k_c^2}{P}\left(\frac{x}{L_c}\right) \tag{5.1b}$$

The general solution of Eq. (5.1b) is

$$y = A \sin k_c x + B \cos k_c x + \frac{M_A}{P}\left(\frac{x}{L_c} - 1\right) - \frac{M_B}{P}\left(\frac{x}{L_c}\right) \tag{5.1c}$$

$$y' = Ak_c \cos k_c x - Bk_c \sin k_c x + \frac{M_A}{PL_c} - \frac{M_B}{PL_c} \tag{5.1d}$$

The constants A and B are calculated by substituting boundary conditions in Eqs. (5.1c) and (5.1d).

$$y = 0 \text{ at } x = 0, \text{ therefore, } B = \frac{M_A}{P}$$

$$\text{and } y' = 0 \text{ at } x = 0, \text{ therefore, } A = -\frac{M_A - M_B}{PL_c k_c}$$

Substitute the values of A and B in Eq. (5.1c), and then

$$y = \frac{M_A}{P}\left(-\frac{1}{k_c L_c}\sin k_c x + \cos k_c x + \frac{x}{L_c} - 1\right) + \frac{M_B}{P}\left(\frac{1}{k_c L_c}\sin k_c x - \frac{x}{L_c}\right) \tag{5.1e}$$

There is no sidesway at top of the vertical column, therefore

$y = 0$ at $x = L_c$, and substituting in Eq. (5.1e) we have

$$M_A(k_c L_c \cos k_c L_c - \sin k_c L_c) + M_B(\sin k_c L_c - k_c L_c) = 0 \tag{5.1f}$$

Apply slope deflection equations to the horizontal member BC in Figure 5.1b. Neglecting the axial forces in the horizontal member BC, we have

$$M_B = \frac{EI_b}{L_b}(4\theta_B + 2\theta_C), \text{ Now } \theta_B = -\theta_c$$

Therefore,

$$M_B = \frac{2EI_b}{L_b}\theta_B \tag{5.1g}$$

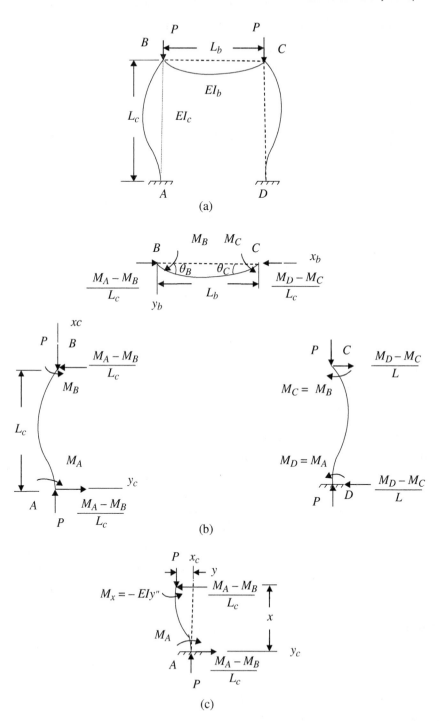

Figure 5.1 Symmetrical frame fixed at the base having no sidesway buckling: (a) No sidesway symmetrical buckling in a frame fixed at the base; (b) Forces in the horizontal and vertical members; (c) Free body diagram of the left vertical member.

At the joint B, compatibility of the slopes between the horizontal member BC and the vertical member AB gives

$$\theta_B = \frac{dy}{dx}\Big|_{x=l_c} \tag{5.1h}$$

Differentiating Eq. (5.1e)

$$\frac{dy}{dx} = \frac{M_A}{P}\left(-\frac{1}{L_c}\cos k_c x - k_c \sin k_c x + \frac{1}{L_c}\right) + \frac{M_B}{P}\left(\frac{1}{L_c}\cos k_c x - \frac{1}{L_c}\right) \tag{5.1i}$$

$$\frac{dy}{dx}\Big|_{x=l_c} = \frac{M_A}{P}\left(-\frac{1}{L_c}\cos k_c L_c - k_c \sin k_c L_c + \frac{1}{L_c}\right) + \frac{M_B}{P}\left(\frac{1}{L_c}\cos k_c L_c - \frac{1}{L_c}\right) \tag{5.1j}$$

Equations (5.1g), (5.1h), and (5.1j) provide

$$\frac{M_A}{P}\left(-\frac{1}{L_c}\cos k_c L_c - k_c \sin k_c L_c + \frac{1}{L_c}\right) + \frac{M_B}{P}\left(\frac{1}{L_c}\cos k_c L_c - \frac{1}{L_c}\right) = \frac{M_B L_b}{2EI_b}$$

or

$$M_A(-\cos k_c L_c - k_c L_c \sin k_c L_c + 1) + M_B\left(\cos k_c L_c - 1 - \frac{L_b L_c I_c k_c^2}{2I_b}\right) = 0 \tag{5.1k}$$

Equations (5.1f) and (5.1k) can be written as

$$\begin{bmatrix} k_c L_c \cos k_c L_c - \sin k_c L_c & \sin k_c L_c - k_c L_c \\ -\cos k_c L_c - k_c L_c \sin k_c L_c + 1 & \cos k_c L_c - 1 - \frac{L_b L_c I_c k_c^2}{2I_b} \end{bmatrix} \begin{Bmatrix} M_A \\ M_B \end{Bmatrix} = \begin{Bmatrix} 0 \\ 0 \end{Bmatrix} \tag{5.1l}$$

For a nontrivial solution of M_A and M_B, the determinant of the coefficient matrix in Eq. (5.1l) should be zero. Therefore,

$$\begin{vmatrix} k_c L_c \cos k_c L_c - \sin k_c L_c & \sin k_c L_c - k_c L_c \\ -\cos k_c L_c - k_c L_c \sin k_c L_c + 1 & \cos k_c L_c - 1 - \frac{L_b L_c I_c k_c^2}{2I_b} \end{vmatrix} = 0 \tag{5.1m}$$

or

$$2 - 2\cos k_c L_c - k_c L_c \sin k_c L_c + \frac{L_b I_c k_c}{2I_b}(\sin k_c L_c - k_c L_c \cos k_c L_c) = 0 \tag{5.1n}$$

Let

$$I_c = I_b = I, \text{ and } L_c = L_b = L$$

Then

$$\cos kL(4 + k^2 L^2) + kL \sin kL = 4 \tag{5.1o}$$

Solving the transcendental Eq. (5.1o) gives

$$k_c L = 5.02 \quad \text{or} \quad k_c^2 L^2 = (5.02)^2$$

Therefore,

$$P_{cr} = \frac{25.2EI}{L^2} \tag{5.1p}$$

5.2.1.1 Portal Frame Without Sidesway with Rigid or Extremely Flexible Beam

When the beam is rigid, it does not bend, and remains horizontal, as shown in Figure 5.2a. Since the joints between the columns and the beam are rigid, the columns cannot rotate at the upper ends. Since we are assuming no sidesway case, the columns cannot translate at their upper ends. Therefore, they behave as if they are fixed both at the top and the bottom. The critical load for such columns is given by Eq. (2.2f), i.e. $P_{cr} = \dfrac{4\pi^2 EI_c}{L_c^2} = 4P_e$, where P_e is the Euler buckling load, that is the critical load for the same dimensioned column as the present ones, whose both ends are pinned.

When the beam is extremely flexible, as shown in Figure 5.2b, it offers no resistance to rotation of the columns at their upper ends. The columns cannot translate at their upper ends because of no sidesway condition. The columns in this case behave as if they are fixed at the bottom and hinged at the top. The critical load for such columns is given by Eq. (2.5 h), i.e. $P_{cr} = \dfrac{20.19 EI_c}{L_c^2}$. In the practical cases the stiffness of the beams lies between the two extreme values shown in Figure 5.2. If $L_c = L$, and $I_c = I$, the critical load given by Eq. (5.1p) can be expressed as

$$\frac{20.19 EI}{L^2} < \frac{25.2 EI}{L^2} < \frac{4\pi^2 EI}{L^2} \tag{5.2a}$$

5.2.2 Portal Frame with Sidesway

If a frame is not braced, it may buckle in a sidesway mode. It is assumed there is no primary bending prior to buckling in the frame, and the material remains elastic during buckling and

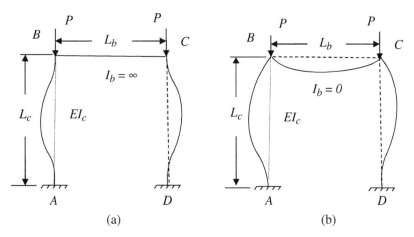

Figure 5.2 Frames fixed at the base having no sidesway with rigid and extremely flexible beams: (a) Frame with rigid beam; (b) Frame with extremely flexible beam.

the deformations are small. The frame is fixed at the bottom and is subjected to two axial forces each equal to P shown in Figure 5.3a.

The forces in the individual members of the frame are shown in Figure 5.3b neglecting the effect of the shear forces, V, coming on the vertical members from the equilibrium of the horizontal member. These shears are much smaller than the applied loads on the vertical columns. Considering the equilibrium of the free body diagram of deflected shape of the left vertical column in Figure 5.3c, we have

$$P \ y - M_A + EI_c y'' = 0$$

or

$$y'' + k_c^2 y = \frac{M_A}{EI_c}, \qquad (5.3a)$$

where

$$k_c^2 = \frac{P}{EI_c}.$$

The general solution of Eq. (5.3a) is

$$y = A \sin k_c x + B \cos k_c x + \frac{M_A}{P} \qquad (5.3b)$$

$$y' = A \cos k_c x - Bk_c \sin k_c x \qquad (5.3c)$$

The constants A and B are calculated by substituting boundary conditions in Eqs. (5.3b) and (5.3c).

$$y = 0 \text{ at } x = 0, \text{ therefore, } B = -\frac{M_A}{P}$$

and

$$y' = 0 \text{ at } x = 0, \text{ therefore, } A = 0$$

or

$$y = \frac{M_A}{P}(1 - \cos k_c x) \qquad (5.3d)$$

$$y = \Delta \text{ at } x = L_c$$

or

$$\Delta = \frac{M_A}{P}(1 - \cos k_c L_c) \qquad (5.3e)$$

From the moment equilibrium of the left column in Figure 5.3b we get

$$P\Delta = M_A + M_B \qquad (5.3f)$$

Substituting Δ from Eq. (5.3e) into Eq. (5.3f) we have

$$M_A \cos k_c L_c + M_B = 0 \qquad (5.3g)$$

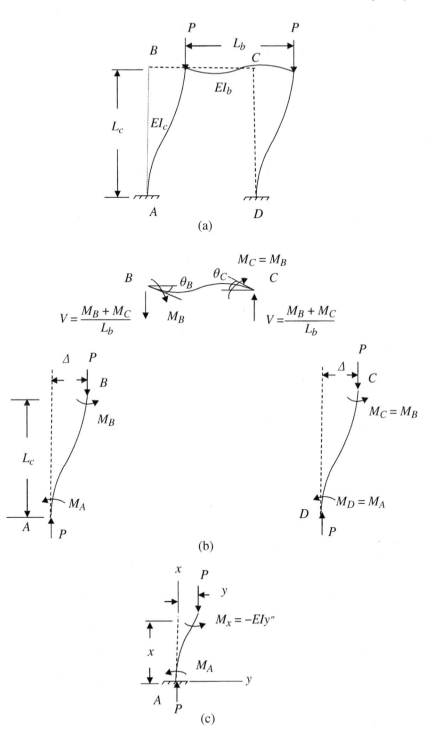

Figure 5.3 Portal frame fixed at the base with sidesway buckling: (a) Sidesway antisymmetrical buckling in a frame fixed at the base; (b) Forces in the horizontal and vertical members; (c) Free body diagram of the left vertical member.

Apply the slope deflection equations to the horizontal member BC in Figure 5.3b.

$$M_B = \frac{EI_b}{L_b}(4\theta_B + 2\theta_C), \text{ now } \theta_B = \theta_C$$

Therefore,

$$M_B = 6\frac{EI_b}{L_b}\theta_B \tag{5.3h}$$

At the joint B,

$$\left.\frac{dy}{dx}\right|_{x=L_C} = \theta_B \tag{5.3i}$$

Differentiating Eq. (5.3d)

$$\frac{dy}{dx} = \frac{M_A}{P}k_c \sin k_c x \tag{5.3j}$$

$$\left.\frac{dy}{dx}\right|_{x=L_C} = \frac{M_A}{P}k_c \sin k_c L_c \tag{5.3k}$$

Equations (5.3h), (5.3i), and (5.3k) give

$$\frac{M_A}{P}k_c \sin k_c L_c = \frac{M_B L_b}{6EI_b}, \quad \text{now } \frac{P}{EI_c} = k_c^2$$

Therefore,

$$\frac{6I_b}{k_c I_c}M_A \sin k_c L_c - M_B L_b = 0 \tag{5.3l}$$

Equations (5.3g) and (5.3l) can be written as

$$\begin{bmatrix} \cos k_c L_c & 1 \\ \dfrac{6I_b}{k_c I_c}\sin k_c L_c & -L_b \end{bmatrix}\begin{Bmatrix} M_A \\ M_B \end{Bmatrix} = \begin{Bmatrix} 0 \\ 0 \end{Bmatrix} \tag{5.3m}$$

For a nontrivial solution of M_A and M_B, the determinant of the coefficient matrix in Eq. (5.3m) should be zero. Therefore,

$$\begin{vmatrix} \cos k_c L_c & 1 \\ \dfrac{6I_b}{k_c I_c}\sin k_c L_c & -L_b \end{vmatrix} = 0 \tag{5.3n}$$

$$-L_b \cos k_c L_C - \frac{6I_b}{k_c I_c}\sin k_c L_c = 0$$

or

$$\frac{\tan k_c L_c}{k_c L_c} = -\frac{L_b}{6L_c}\frac{I_c}{I_b} \tag{5.3o}$$

If

$$I_c = I_b = I, \text{ and } L_c = L_b = L$$

$$\frac{\tan kL}{kL} = -\frac{1}{6} \tag{5.3p}$$

$$kL = 2.71 \quad \text{or} \quad k^2 L^2 = (2.71)^2$$

Therefore,

$$P_{cr} = \frac{7.34EI}{L^2} \tag{5.3q}$$

5.2.2.1 Portal Frame Having Sidesway with a Rigid or Extremely Flexible Beam

When the beam is rigid, it does not bend and remains horizontal as shown in Figure 5.4a. At the upper ends the vertical columns can translate but cannot rotate. Also, the inflexion points in the columns are at their mid-height. The critical load for the vertical columns is equal to that of a cantilever column whose length is half the column height in the frame. Therefore, the critical load is given by Eq. (2.4j), i.e. $P_{cr} = \dfrac{\pi^2 EI_c}{4\left(\frac{L_c}{2}\right)^2} = \dfrac{\pi^2 EI_c}{L_c^2} = P_e$, where P_e

is the Euler buckling load. If the beam is extremely flexible as in Figure 5.4b, the upper ends of the vertical columns behave as free ends. The entire vertical column behaves a fixed-free column of length, L_c. The critical load in this case is given by Eq. (2.4h), i.e. $P_{cr} = \dfrac{\pi^2 EI_c}{4L_c^2} = \dfrac{P_e}{4}$, where P_e is the Euler buckling load.

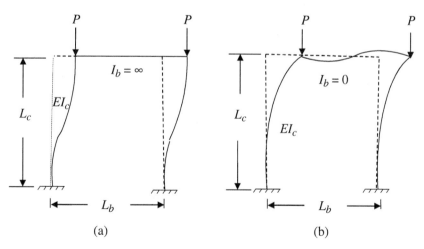

(a)   (b)

Figure 5.4 Frames fixed at the base having sideway with rigid and flexible beams: (a) Frame with rigid beam; (b) Frame with extremely flexible beam.

In the practical cases the stiffness of the beams lies between the two extreme values shown in Figure 5.4. If $L_c = L$, and $I_c = I$, the critical load given by Eq. (5.3q) can be expressed as

$$\frac{\pi^2 EI}{4L^2} < \frac{7.34EI}{L^2} < \frac{\pi^2 EI}{L^2} \tag{5.4a}$$

5.2.3 Frame with Prime Bending and Without Sidesway

In the previous cases buckling of the frames was studied when there were no transverse loads on the horizontal members, i.e. the beams. Let us consider a symmetrical frame with symmetrical loading as shown in Figure 5.5 so that there is no sidesway. The free body diagrams of the beam and the left column are given in Figures 5.5b and c. The reactions at A and D are P and H as shown in Figure 5.5a. The slope of the column at B is given by Eq. (4.8c), that is

$$\theta_B = -\frac{HL_c}{P}\left(k_c \cot k_c L_c - \frac{1}{L_c}\right) \tag{5.5a}$$

where

$$k_c = \sqrt{\frac{P}{EI_c}}$$

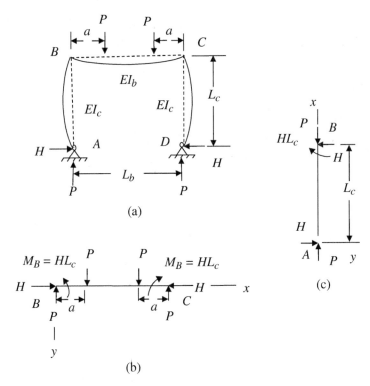

(a)

(b)

(c)

Figure 5.5 Symmetrical portal frame hinged at the base with prime bending and no sidesway: (a) Frame hinged at the base with transverse load and no sidesway; (b) Free body diagram of the horizontal member (beam); (c) Free body diagram of the vertical member (column).

The slope of the beam at B caused by two transverse forces is given by Eqs. (4.2w) and (4.2x) as follows:

$$\theta_B^1 = \frac{P \sin k_b(L_b - a)}{EI_b k_b^2 \sin k_b L_b} - \frac{P(L_b - a)}{EI_b L_b k_b^2} + \frac{P \sin k_b a}{EI_b k_b^2 \sin k_b L_b} - \frac{Pa}{EI_b L_b k_b^2}$$

where

$$k_b = \sqrt{\frac{H}{EI_b}}$$

or

$$\theta_B^1 = \frac{P}{H} \left(\frac{\sin k_b(L_b - a)}{\sin k_b L_b} + \frac{\sin k_b a}{\sin k_b L_b} - \frac{(L_b - a)}{L_b} - \frac{a}{L_b} \right)$$

or

$$\theta_B^1 = \frac{P}{H} \left[\sec \frac{k_b L_b}{2} \cos \left(\frac{k_b L_b}{2} - k_b a \right) - 1 \right] \tag{5.5b}$$

The slope of the beam at B caused by two end moments is given by Eq. (4.9b) as follows:

$$\theta_B^2 = \frac{HL_c}{H} \left(\frac{k_b \cos k_b L_b}{\sin k_b L_b} - \frac{1}{L_b} \right) - \frac{HL_c}{H} \left(\frac{k_b}{\sin k_b L_b} - \frac{1}{L_b} \right)$$

or

$$\theta_B^2 = -k_b L_c \tan \frac{k_b L_b}{2} \tag{5.5c}$$

Total slope of the beam at B is

$$\theta_B = \theta_B^1 + \theta_B^2$$

$$\theta_B = \frac{P}{H} \left[\sec \frac{k_b L_b}{2} \cos \left(\frac{k_b L_b}{2} - k_b a \right) - 1 \right] - k_b L_c \tan \frac{k_b L_b}{2} \tag{5.5d}$$

At the joint B

$$\theta_{B\ column} = \theta_{B\ beam} \tag{5.5e}$$

Substitute $\theta_{B\ column}$ and $\theta_{B\ beam}$ from Eqs. (5.5a) and (5.5d) respectively in Eq. (5.5e), and we get

$$k_b L_c \tan \frac{k_b L_b}{2} + \frac{H}{P}(1 - k_c L_c \cot k_c L_c) - \frac{P}{H} \left[\sec \frac{k_b L_b}{2} \cos \left(\frac{k_b L_b}{2} - k_b a \right) - 1 \right] = 0 \tag{5.5f}$$

Equation (5.5f) is the equation relating the horizontal force H to the applied transverse load P. Assume a case where $I_b = I_c = I$, $L_b = L_c = L$, and $a = L/3$, Eq. (5.5f) reduces to

$$k_b L \tan \frac{k_b L}{2} + \left(\frac{k_b L}{k_c L} \right)^2 (1 - k_c L \cot k_c L) - \left(\frac{k_c L}{k_b L} \right)^2 \left[\sec \frac{k_b L}{2} \cos \left(\frac{k_b L}{6} \right) - 1 \right] = 0 \tag{5.5g}$$

Assume $k_c L$, and find $k_b L$ by trial and error that satisfies Eq. (5.5g). Plot $k_c L = \sqrt{\frac{P}{EI}} L$ versus $k_b L = \sqrt{\frac{H}{EI}} L$ as shown in Figure 5.6. It can be seen in Figure 5.6 that as P increases, first H also

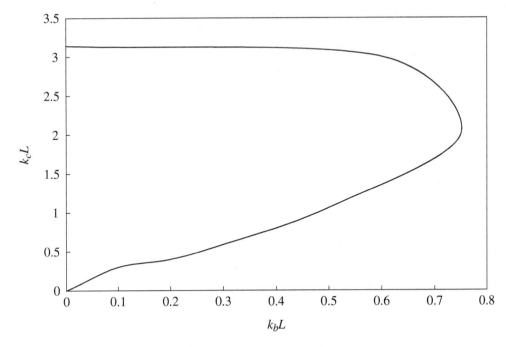

Figure 5.6 Plot of k_cL versus k_bL.

increases then decreases and becomes zero. It is because initially the frame behaves linearly, then as P increases further, the capacity of the column to sustain moment decreases, so the H also decreases. Finally, a stage is reached when the column is unable to resist any moment given by HL, and hence H also becomes zero [1]. The stage when the column is unable to resist any moment is characterized by the buckling of the frame and it occurs at $k_cL = \pi$. Therefore,

$$k_cL = \sqrt{\frac{P}{EI}}L = \pi$$

or

$$P_{cr} = \frac{\pi^2 EI}{L^2} \tag{5.5h}$$

5.3 Critical Loads by Slope Deflection Equations

5.3.1 Portal Frame Without Sidesway

The slope deflection equations for members subjected to both axial force and bending are given by Eq. (4.21e), where the constants in the coefficient matrix are given by Eqs. (4.21f) and (4.21g), and are given in Appendix A for various values of kL.

Consider the frame shown Figure 5.1a. The moments and the slopes in the frame are considered positive if they are clockwise in the slope deflection method. Positive moments and the

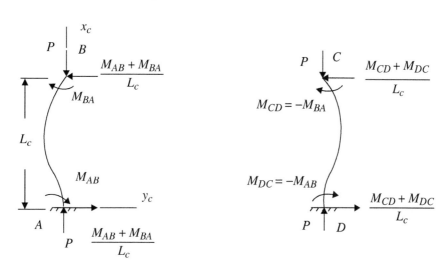

Figure 5.7 Portal frame fixed at the base without sidesway by slope deflection method.

corresponding forces in the members are shown in Figure 5.7. The slope deflection equation for the column AB can be written as

$$M_{BA} = \frac{EI_c}{L_c}(k_{11c}\theta_B + k_{12c}\theta_A) = \frac{EI_c}{L_c}(k_{11c}\theta_B) \tag{5.6a}$$

Because $\theta_A = 0$. The slope deflection equation for the horizontal beam BC can be written as

$$M_{BC} = \frac{EI_b}{L_b}(k_{11b}\theta_B + k_{12b}\theta_c)$$

Neglecting the axial force in the horizontal member BC, we have

$$M_{BC} = \frac{EI_b}{L_b}(4\theta_B + 2\theta_C), \text{ since } \theta_B = -\theta_C$$

$$M_{BC} = \frac{2EI_b}{L_b}\theta_B \tag{5.6b}$$

The subscripts c and b are used for the column and the beam respectively. For the joint equilibrium at B, we have

$$M_{BA} + M_{BC} = 0 \tag{5.6c}$$

or

$$\left[\frac{EI_c}{L_c}k_{11c} + \frac{2EI_b}{L_b} \right] \theta_B = 0, \qquad \theta_B \neq 0$$

Therefore,

$$\frac{EI_c}{L_c}k_{11c} + \frac{2EI_b}{L_b} = 0$$

Consider the case where $I_c = I_b = I$, and $L_c = L_b = L$, then

$$\frac{EI}{L}(k_{11c} + 2) = 0, \text{ or } k_{11c} = -2$$

From Appendix A,

$$kL = 5.02$$

or

$$(kL)^2 = 25.2$$

and

$$P_{cr} = \frac{25.2EI}{L^2} \tag{5.6d}$$

That is the same as in Eq. (5.1p).

5.3.2 Portal Frame with Sidesway

Now apply the slope deflection method to find the critical load for the frame shown in Fig. 5.3a.

Positive moments and the corresponding forces are shown in Figure 5.8. The slope deflection equation for the column AB is given by

$$M_{AB} = \frac{EI_c}{L_c} \left[k_{11c}\theta_A + k_{12c}\theta_B - (k_{11c} + k_{12c})\frac{\Delta}{L_c} \right]$$

$$M_{BA} = \frac{EI_c}{L} \left[k_{21c}\theta_A + k_{22c}\theta_B - (k_{21c} + k_{22c})\frac{\Delta}{L_c} \right]$$

$$\theta_A = 0,$$

therefore,

$$M_{AB} = \frac{EI_c}{L_c} \left[k_{12c}\theta_B - (k_{11c} + k_{12c})\frac{\Delta}{L_c} \right] \tag{5.7a}$$

$$M_{BA} = \frac{EI_c}{L_c} \left[k_{22c}\theta_B - (k_{21c} + k_{22c})\frac{\Delta}{L_c} \right] \tag{5.7b}$$

For the beam, the slope deflection equations are

$$M_{BC} = \frac{EI_b}{L_b}(k_{11b}\theta_B + k_{12b}\theta_C)$$

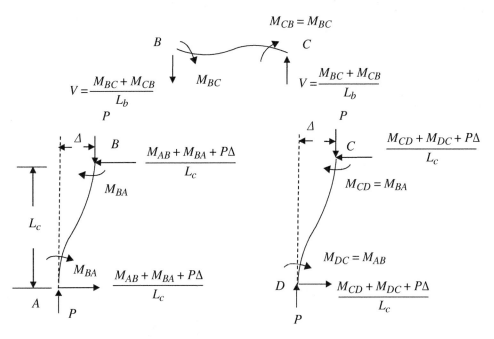

Figure 5.8 Portal frame fixed at the base with sidesway by slope deflection method.

Neglecting the axial force in the horizontal beam, we have

$$M_{BC} = \frac{EI_b}{L_b}(4\theta_B + 2\theta_C), \text{Since } \theta_B = \theta_C$$

$$M_{BC} = \frac{6EI_b}{L_b}\theta_B \tag{5.7c}$$

The subscripts b and c are used for the column and the beam respectively. For the joint equilibrium at B, we have

$$M_{BA} + M_{BC} = 0 \tag{5.7d}$$

$$\frac{EI_c}{L_c}\left[k_{22c}\theta_B - (k_{21c} + k_{22c})\frac{\Delta}{L_c}\right] + \frac{6EI_b}{L_b}\theta_B = 0$$

or

$$\left(k_{22c} + \frac{6I_b}{L_b}\frac{L_c}{I_c}\right)\theta_B - (k_{21c} + k_{22c})\frac{\Delta}{L_c} = 0 \tag{5.7e}$$

From the equilibrium of story shear in Figure 5.8, we get

$$\frac{M_{AB} + M_{BA} + P\Delta}{L_c} + \frac{M_{CD} + M_{DC} + P\Delta}{L_c} = 0$$

Now, $M_{AB} = M_{DC}$, and $M_{BA} = M_{CD}$ because the deformation of the frame is antisymmetric.

Therefore,

$$\frac{M_{AB} + M_{BA} + P\Delta}{L_c} = 0 \tag{5.7f}$$

Substitute the moments M_{AB} and M_{BA} in terms of θ_A, θ_B, Δ from Eqs. (5.7a) and (5.7b), and we obtain

$$\frac{1}{L_c}\left[\frac{EI_c}{L_c}\left\{k_{12c}\theta_B - (k_{11c} + k_{12c})\frac{\Delta}{L_c}\right\} + \frac{EI_c}{L_c}\left\{k_{22c}\theta_B - (k_{21c} + k_{22c})\frac{\Delta}{L_c}\right\} + P\Delta\right] = 0$$

The slope deflection coefficients, $k_{11c} = k_{22c}$, and $k_{12c} = k_{21c}$, therefore

$$-(k_{12c} + k_{11c})\,\theta_B + (2k_{11c} + 2k_{12c} - k_c^2 L_c^2)\frac{\Delta}{L_c} = 0 \tag{5.7g}$$

where

$$k_c^2 = \frac{P}{EI_c}.$$

Equations (5.7e) and (5.7g) can be written in the matrix form as

$$\begin{bmatrix} k_{11c} + \dfrac{6I_b\,L_c}{L_b\,I_c} & -(k_{12c} + k_{11c}) \\[4mm] -(k_{12c} + k_{11c}) & 2k_{11c} + 2k_{12c} - k_c^2 L_c^2 \end{bmatrix} \begin{Bmatrix} \theta_B \\[2mm] \dfrac{\Delta}{L_c} \end{Bmatrix} = \begin{Bmatrix} 0 \\[2mm] 0 \end{Bmatrix} \tag{5.7h}$$

Let $I_b = I_c = I$, and $L_b = L_c = L$, then Eq. (5.7h) becomes

$$\begin{bmatrix} k_{11} + 6 & -(k_{12} + k_{11}) \\[4mm] -(k_{12} + k_{11}) & 2k_{11} + 2k_{12} - k^2 L^2 \end{bmatrix} \begin{Bmatrix} \theta_B \\[2mm] \dfrac{\Delta}{L} \end{Bmatrix} = \begin{Bmatrix} 0 \\[2mm] 0 \end{Bmatrix} \tag{5.7i}$$

For a non-trivial solution for the quantities, θ_B and $\frac{\Delta}{L}$, the determinant of the coefficient matrix should be zero. So

$$\begin{vmatrix} k_{11} + 6 & -(k_{12} + k_{11}) \\[4mm] -(k_{12} + k_{11}) & 2k_{11} + 2k_{12} - k^2 L^2 \end{vmatrix} = 0 \tag{5.7j}$$

or

$$(k_{11} + 6)(2k_{11} + 2k_{12} - k^2 L^2) - (k_{12} + k_{11})^2 = 0$$

and

$$(k_{11}{}^2 + 12k_{11} - k^2 L^2 k_{11}) - (k_{12}{}^2 - 12k_{12} + 6k^2 L^2) = 0 \tag{5.7k}$$

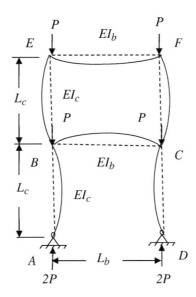

Figure 5.9 Two-story frame hinged at the base without sidesway by the slope deflection method.

Equation (5.7k) is a characteristic equation for the frame shown in Figure 5.3a. By trial and error and using Appendix A, the value of kL that satisfies Eq. (5.7k) is given by

$$kL = 2.71, \quad \text{and corresponding} \quad k_{11} = 2.9086, k_{12} = 2.3149$$
$$k^2 L^2 = 7.344$$

Therefore, the critical load for the frame is

$$P_{cr} = \frac{7.344 \, EI}{L^2}, \text{the same as in Eq. (5.3q)}. \tag{5.7l}$$

5.3.3 Two-Story Frame Without Sidesway

The slope deflection method is used to find the critical load for a two-story frame in Figure 5.9. The sidesway is prevented in the frame. It is symmetrical about the vertical center line of the frame, as shown in Figure 5.9. Therefore, we can consider only the left half to calculate the critical load for this frame. The slope deflection equations for the various members of the frame can be written as

$$M_{AB} = \frac{EI_c}{L_c}[k_{11AB}\theta_A + k_{12AB}\theta_B] \tag{5.8a}$$

$$M_{BA} = \frac{EI_c}{L_c}[k_{21AB}\theta_A + k_{22AB}\theta_B] \tag{5.8b}$$

$$M_{AB} = 0$$

because end A is a hinge support, therefore,

$$\theta_A = -\frac{k_{12AB}}{k_{11AB}}\theta_B, \text{ substituting in Eq. (5.8b)}$$

$$M_{BA} = \frac{EI_c}{L_c}\left[k_{22AB} - \frac{k_{12AB}^2}{k_{11AB}}\right]\theta_B \tag{5.8c}$$

$$M_{BE} = \frac{EI_c}{L_c}[k_{11BE}\theta_B + k_{12BE}\theta_E] \tag{5.8d}$$

$$M_{EB} = \frac{EI_c}{L_c}[k_{21BE}\theta_B + k_{22BE}\theta_E] \tag{5.8e}$$

$$M_{BC} = \frac{EI_b}{L_b}[k_{11BC}\theta_B + k_{12BC}\theta_C] \tag{5.8f}$$

$$M_{EF} = \frac{EI_b}{L_b}[k_{11EF}\theta_E + k_{12EF}\theta_F] \tag{5.8g}$$

Neglecting the axial force in the horizontal beams BC and EF, we have

$$k_{11BC} = k_{11EF} = 4, \text{ and } k_{12BC} = k_{12EF} = 2$$

Also, $\theta_B = -\theta_C$, and $\theta_E = -\theta_F$,

Substitute these values in Eqs. (5.8f) and (5.8g) to get

$$M_{BC} = \frac{EI_b}{L_b}[2\theta_B] \tag{5.8h}$$

$$M_{EF} = \frac{EI_b}{L_b}[2\theta_E] \tag{5.8i}$$

Moment equilibrium equations at the joints B and E are

$$M_{BA} + M_{BC} + M_{BE} = 0 \tag{5.8j}$$

$$M_{EB} + M_{EF} = 0 \tag{5.8k}$$

Substituting Eqs. (5.8c), (5.8d), (5.8e), (5.8h), and (5.8i), and $k_{11} = k_{22}$, $k_{12} = k_{21}$ into Eqs. (5.8j) and (5.8k) we have

$$\begin{bmatrix} \frac{EI_c}{L_c}\left(k_{11AB} - \frac{k_{12AB}^2}{k_{11AB}} + k_{11BE}\right) + \frac{2EI_b}{L_b} & \frac{EI_c}{L_c}k_{12BE} \\ \frac{EI_c}{L_c}k_{12BE} & \frac{EI_c}{L_c}k_{11BE} + 2\frac{EI_b}{L_b} \end{bmatrix}\begin{Bmatrix} \theta_B \\ \theta_E \end{Bmatrix} = \begin{Bmatrix} 0 \\ 0 \end{Bmatrix} \tag{5.8l}$$

Let $I_b = I_c = I$, and $L_b = L_c = L$. For a non-trivial solution for the quantities θ_B and θ_E, the determinant of the coefficient matrix in Eq. (5.8l) should be zero, so

$$\begin{vmatrix} k_{11AB} - \dfrac{k_{12AB}^2}{k_{11AB}} + k_{11BE} + 2 & k_{12BE} \\ \\ k_{12BE} & k_{11BE} + 2 \end{vmatrix} = 0 \tag{5.8m}$$

or

$$\left(k_{11AB} - \frac{k_{12AB}^2}{k_{11AB}} + k_{11BE} + 2 \right) (k_{11BE} + 2) - k_{12BE}^2 = 0 \tag{5.8n}$$

Therefore,

$$k_{11BE}(k_{11AB}^2 - k_{12AB}^2 + 4k_{11AB} + k_{11AB}k_{11BE}) + k_{11AB}(2k_{11AB} - k_{12BE}^2 + 4) - 2k_{12AB}^2 = 0 \tag{5.8o}$$

Now,

$$k_{BE}^2 = \frac{P}{EI} \quad \text{and} \quad k_{AB}^2 = \frac{2P}{EI}$$

Therefore,

$$k_{AB} = \sqrt{2}k_{BE}$$

Assume $k_{BE}L$ and calculate the corresponding, $k_{AB}L = \sqrt{2}k_{BE}L$, then find the slope deflection coefficients k_{11AB}, k_{12AB}, k_{11BE}, and k_{12BE} from Appendix A from the assumed and calculated values of $k_{BE}L$ and $k_{AB}L$. Alternately, the slope deflection coefficients can be obtained from Eqs. (4.8g), (4.8h), (4.21f), and (4.21g), where $u = \frac{kL}{2}$. Solve Eq. (5.8o) by trial and error to find the correct value of $k_{BE}L$. The trial and error procedure gives

$$k_{BE}L = 2.697 \tag{5.8p}$$

or

$$k_{BE}^2L^2 = 7.274 \quad \text{or} \quad \frac{PL^2}{EI} = 7.274$$

Therefore, the critical load for the two-story frame in Figure 5.9 is given by

$$P_{cr} = \frac{7.274EI}{L^2} \tag{5.8q}$$

5.3.4 Two-Bay Frame Without Sidesway

The slope deflection method is applied to the two-bay frame shown in Figure 5.10 to find the critical load.

Assume θ_A, θ_B, θ_C, θ_D, θ_E, and θ_F are the angles of rotation at the joints A, B, C, D, E, and F respectively. The slope deflection equations for the different members of the frame can be

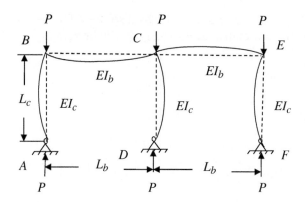

Figure 5.10 Two-bay frame hinged at the supports without sidesway by the slope deflection method.

written in terms of the slope deflection coefficients and moments, *M*, in different members as follows:

$$M_{AB} = \frac{EI_c}{L_c}[k_{11AB}\theta_A + k_{12AB}\theta_B]$$

$$M_{BA} = \frac{EI_c}{L_c}[k_{21AB}\theta_A + k_{22AB}\theta_B]$$

$$M_{AB} = 0, \text{ therefore}$$

$$\theta_A = -\frac{k_{12AB}}{k_{11AB}\theta_B}$$

or

$$M_{BA} = \frac{EI_c}{L_c}\left(k_{22AB} - \frac{k_{12AB}^2}{k_{11AB}}\right)\theta_B \tag{5.9a}$$

$$M_{BC} = \frac{EI_b}{L_b}(4\theta_B + 2\theta_C) \tag{5.9b}$$

$$M_{CB} = \frac{EI_b}{L_b}(2\theta_B + 4\theta_C) \tag{5.9c}$$

$$M_{CD} = \frac{EI_c}{L_c}(k_{11CD}\theta_C + k_{12CD}\theta_D)$$

$$M_{DC} = \frac{EI_c}{L_c}(k_{21CD}\theta_C + k_{22CD}\theta_D) = 0$$

Therefore,

$$\theta_D = -\frac{k_{21CD}}{k_{22CD}}\theta_C = -\frac{k_{12CD}}{k_{11CD}}\theta_C$$

$$M_{CD} = \frac{EI_c}{L_c}\left(k_{11CD}\theta_C - \frac{k_{12CD}^2}{k_{11CD}}\theta_C\right) \tag{5.9d}$$

$$M_{CE} = \frac{EI_b}{L_b}(4\theta_C + 2\theta_E) \tag{5.9e}$$

$$M_{EC} = \frac{EI_b}{L_b}(2\theta_C + 4\theta_E) \tag{5.9f}$$

$$M_{EF} = \frac{EI_c}{L_C}(k_{11EF}\theta_E + k_{12EF}\theta_F)$$

$$M_{FE} = \frac{EI_c}{L_c}(k_{12EF}\theta_E + k_{22EF}\theta_F) = 0$$

Therefore,

$$\theta_F = -\frac{k_{12EF}\theta_E}{k_{11EF}}$$

$$M_{EF} = \frac{EI_c}{L_c}\left(k_{11EF}\theta_E - \frac{k_{12EF}^2}{K_{11EF}}\theta_E\right) \tag{5.9g}$$

Moment equilibrium equations at the joints B, C, and E are given by

$$M_{BA} + M_{BC} = 0 \tag{5.9h}$$

$$M_{CB} + M_{CD} + M_{CE} = 0 \tag{5.9i}$$

$$M_{EC} + M_{EF} = 0 \tag{5.9j}$$

Substitute Eqs. (5.9a–5.9g) into Eqs. (5.9h)–(5.9j), also $k_{11} = k_{22}$, $k_{12} = k_{21}$, then we get

$$\begin{bmatrix} \frac{EI_c}{L_c}\left(k_{11AB} - \frac{k_{12AB}^2}{k_{11AB}}\right) + \frac{4EI_b}{L_b} & \frac{2EI_b}{L_b} & 0 \\[2em] \frac{2EI_b}{L_b} & \frac{EI_c}{L_c}\left(k_{11CD} - \frac{k_{12CD}^2}{k_{11CD}}\right) + \frac{8EI_b}{L_b} & \frac{2EI_b}{L_b} \\[2em] 0 & \frac{2EI_b}{L_b} & \frac{EI_c}{L_c}\left(k_{11EF} - \frac{k_{12EF}^2}{k_{11EF}}\right) + \frac{4EI_b}{L_b} \end{bmatrix}$$

$$\left\{\begin{matrix} \theta_B \\ \theta_C \\ \theta_E \end{matrix}\right\} = \left\{\begin{matrix} 0 \\ 0 \\ 0 \end{matrix}\right\} \tag{5.9k}$$

Let $I_b = I_c = I$ and $L_b = L_c = L$. For a nontrivial solution for the quantities θ_B, θ_C, and θ_E, the determinant of the coefficient matrix in Eq. (5.9k) should be zero, therefore

$$\begin{vmatrix} k_{11AB} - \dfrac{k_{12AB}^2}{k_{11AB}} + 4 & 2 & 0 \\[2ex] 2 & k_{11CD} - \dfrac{k_{12CD}^2}{k_{11CD}} + 8 & 2 \\[2ex] 0 & 2 & k_{11EF} - \dfrac{k_{12EF}^2}{k_{11EF}} + 4 \end{vmatrix} = 0 \tag{5.9l}$$

or

$$\left(k_{11AB} - \frac{k_{12AB}^2}{k_{11AB}} + 4 \right) \left[\left(k_{11CD} - \frac{k_{12CD}^2}{k_{11CD}} + 8 \right) \left(k_{11EF} - \frac{k_{12EF}^2}{k_{11EF}} + 4 \right) - 4 \right]$$

$$+ 2 \left[-2 \left(k_{11EF} - \frac{k_{12EF}^2}{k_{11EF}} + 4 \right) \right] = 0 \tag{5.9m}$$

Now,

$$k_{AB}^2 = k_{CD}^2 = k_{EF}^2 = \frac{P}{EI}$$

Assume a value of $k_{AB}L = k_{CD}L = k_{EF}L$, and then find the slope deflection coefficients k_{11AB}, k_{12AB}, k_{11CD}, and k_{12CD}, k_{11EF}, and k_{12EF} from Appendix A from the assumed values of $k_{AB}L$, $k_{CD}L$, and $k_{EF}L$. Alternately, the slope deflection coefficients can be obtained from Eqs. (4.8g), (4.8h), (4.21f), and (4.21g), where $u = \frac{kL}{2}$. Solve Eq. (5.9m) by trial and error to find the correct value of $k_{AB}L = k_{CD}L = k_{EF}L$. The trial and error procedure gives

$$k_{AB}L = k_{CD}L = k_{EF}L = 4.115 \tag{5.9n}$$

or

$$k_{AB}^2 L^2 = (4.115)^2 \quad \text{or} \quad \frac{PL^2}{EI} = 16.93$$

Therefore, the critical load for the two-bay frame in Figure 5.10 is given by

$$P_{cr} = \frac{16.93\,EI}{L^2} \tag{5.9o}$$

5.3.5 Frames with Prime Bending and Without Sidesway

5.3.5.1 Frame with Hinged Supports
The slope deflection method is used to find the critical load for the frame in Figure 5.5a. The frame and the transverse loads on the beam are symmetrical about the vertical center line of the frame. Therefore, we can consider only the left half of the frame to calculate the critical load. The slope deflection equations for the different members of the frame are

$$M_{AB} = \frac{EI_c}{L_c} [k_{11AB}\theta_A + k_{12AB}\theta_B] \tag{5.10a}$$

$$M_{BA} = \frac{EI_c}{L_c}[k_{21AB}\theta_A + k_{22AB}\theta_B] \tag{5.10b}$$

$M_{AB} = 0$, because end A is a hinge support, therefore,

$\theta_A = -\dfrac{k_{12AB}}{k_{11AB}}\theta_B$, substituting in Eq. (5.10b) leads to

$$M_{BA} = \frac{EI_c}{L_c}\left[k_{22AB} - \frac{k_{12AB}^2}{k_{11AB}}\right]\theta_B \tag{5.10c}$$

because $k_{12AB} = k_{21AB}$. For the beam

$$M_{BC} = \frac{EI_b}{L_b}(k_{11BC}\theta_B + k_{12BC}\theta_C) - \frac{Pa}{L_b}(L_b - a) \tag{5.10d}$$

Where $\frac{Pa}{L_b}(L_b - a)$ is the fixed end moment at B due to transverse load in the horizontal member BC, neglecting axial force. It is counterclockwise, so there is minus sign before the term as per the slope deflection convention.

Due to symmetry, $\theta_C = -\theta_B$

Therefore, Eq. (5.10d) becomes

$$M_{BC} = \frac{EI_b}{L_b}(k_{11BC} - k_{12BC})\,\theta_B - \frac{Pa}{L_b}(L_b - a) \tag{5.10e}$$

From Figures 5.5b and 5.5c we have

$$M_{BA} = HL_c, \text{ and } M_{BC} = -HL_c$$

From Eq. (5.10c)

$$\theta_{B\,column} = \frac{HL_c^2}{EI_c}\frac{1}{k_{22AB} - \dfrac{k_{12AB}^2}{k_{11AB}}} \tag{5.10f}$$

and Eq. (5.10e)

$$\theta_{B\,beam} = \frac{-HL_cL_b + Pa(L_b - a)}{EI_b(k_{11BC} - k_{12BC})} \tag{5.10g}$$

$$\theta_{B\,column} = \theta_{B\,beam}$$

$$\frac{HL_c^2}{EI_c}\frac{1}{k_{22AB} - \dfrac{k_{12AB}^2}{k_{11AB}}} = \frac{-HL_cL_b + Pa(L_b - a)}{EI_b(k_{11BC} - k_{12BC})} \tag{5.10h}$$

Equation (5.10h) is the equation relating the horizontal force H to the applied transverse load P. For the case $I_b = I_c = I$, $L_b = L_c = L$, and $a = L/3$, Eq. (5.10h) reduces to

$$\frac{\frac{HL^2}{EI}}{k_{22AB} - \frac{k_{12AB}^2}{k_{11AB}}} = \frac{-\frac{HL^2}{EI} + \frac{2}{9}\frac{PL^2}{EI}}{k_{11BC} - k_{12BC}} \tag{5.10i}$$

$$k_c = \sqrt{\frac{P}{EI_c}}, \text{and } k_b = \sqrt{\frac{H}{EI_b}}, \text{and } k_{11AB} = k_{22AB}$$

Therefore, Eq. (5.10i) can be written as

$$(k_b L)^2 \left(k_{11BC} - k_{12BC} + k_{11AB} - \frac{k_{12AB}^2}{k_{11AB}} \right) - \frac{2}{9}(k_c L)^2 \left(k_{11AB} - \frac{k_{12AB}^2}{k_{11AB}} \right) = 0 \tag{5.10j}$$

Assume $k_c L$ and find $k_b L$ that satisfies Eq. (5.10j) by trial and error. For each value of $k_c L$ and $k_b L$, find the slope deflection coefficients k_{11AB}, k_{12AB}, k_{11BC}, and k_{12BC} from Appendix A or from Eqs. (4.8g), (4.8h), (4.21f), and (4.21g). If we plot $k_c L = \sqrt{\frac{P}{EI}}L$ versus $k_b L = \sqrt{\frac{H}{EI}}L$ values that satisfy Eq. (5.10j), we will get Figure 5.6, and $k_c L = \pi$, as before. Therefore,

$$k_c L = \sqrt{\frac{P}{EI}}L = \pi$$

or

$$P_{cr} = \frac{\pi^2 EI}{L^2} \tag{5.10k}$$

Equations (5.5h) and (5.10k) are the same.

Alternately, at the joint B, the moment equilibrium equation is

$$M_{BA} + M_{BC} = 0$$

$$\frac{EI_c}{L_c}\left[k_{22AB} - \frac{k_{12AB}^2}{k_{11AB}} \right]\theta_B + \frac{EI_b}{L_b}(k_{11BC} - k_{12BC})\theta_B - \frac{Pa}{L_b}(L_b - a) = 0 \tag{5.10l}$$

For the case $I_b = I_c = I$, $L_b = L_c = L$, and $a = L/3$, the Eq. (5.10l) reduces to

$$\theta_B = \frac{\frac{2}{9}k_c^2 L^2}{k_{11AB} - \frac{k_{12AB}^2}{k_{11AB}} + k_{11BC} - k_{12BC}} \tag{5.10m}$$

where

$$k_c = \sqrt{\frac{P}{EI}}$$

Neglecting the axial force in the horizontal member, $k_{11BC} = 4$, and $k_{12BC} = 2$. Assume different values of $k_c L$, for each value find the slope deflection coefficients k_{11AB} and k_{12AB} in

Eq. (5.10m) from Appendix A or Eqs. (4.21f) and (4.21g). Find θ_B, for each assumed value of k_cL from Eq. (5.10m). Plot θ_B versus k_cL shown in Figure 5.11. The buckling of the frame occurs at $k_cL = 3.1416 = \pi$ as before where the curve tends to become horizontal, i.e. there is very large increase in θ_B with a small increase of k_cL. So $P_{cr} = \dfrac{\pi^2 EI}{L^2}$, which is the same as in Eq. (5.10k).

5.3.5.2 Frame with Fixed Supports

Consider a frame with fixed supports and loaded with transverse loads as shown in Figure 5.12. The slope deflection equations for the frame members are

$$M_{AB} = \frac{EI_c}{L_c}(k_{11AB}\theta_A + k_{12AB}\theta_B) \tag{5.11a}$$

$$M_{BA} = \frac{EI_c}{L_c}(k_{21AB}\theta_A + k_{22AB}\theta_B) \tag{5.11b}$$

Since $\theta_A = 0$

$$M_{BA} = \frac{EI_c}{L_c}(k_{22AB}\theta_B) \tag{5.11c}$$

$$M_{BC} = \frac{EI_b}{L_b}(k_{11BC}\theta_B + k_{12BC}\theta_C) - \frac{Pa}{L_b}(L_b - a) \tag{5.11d}$$

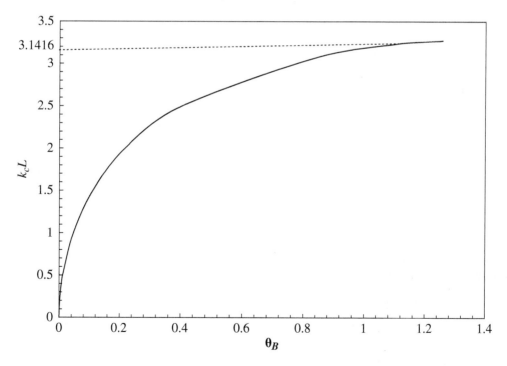

Figure 5.11 θ_B versus k_cL for a hinged frame with prime bending and no sidesway.

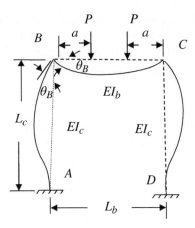

Figure 5.12 Frame fixed at the supports with transverse load and no sidesway.

where $\dfrac{Pa}{L_b}(L_b - a)$ is the fixed end moment at B due to transverse load in the horizontal member BC, neglecting axial force. It is counterclockwise, so there is a minus sign before the term as per the slope deflection convention.

$$\theta_C = -\theta_B$$

At the joint B, $M_{BA} + M_{BC} = 0$, therefore

$$\frac{EI_c}{L_c}(k_{22AB}\theta_B) + \frac{EI_b}{L_b}(k_{11BC} - k_{12BC})\theta_B - \frac{Pa}{L_b}(L_b - a) = 0 \tag{5.11e}$$

Assume $I_b = I_{c,} = I$, $L_b = L_c = L$, and $a = L/3$, then Eq. (5.11e) becomes

$$(k_{22AB} + k_{11BC} - k_{12Bc})\theta_B - \frac{2}{9}k_c^2 L^2 = 0 \tag{5.11f}$$

where

$$k_c = \sqrt{\frac{P}{EI}}$$

or

$$\theta_B = \frac{\frac{2}{9}k_c^2 L^2}{k_{22AB} + k_{11BC} - k_{12Bc}} \tag{5.11g}$$

Neglecting the axial force in the horizontal member

$$k_{11BC} = 4, \quad \text{and} \quad k_{12BC} = 2$$

or

$$\theta_B = \frac{\frac{2}{9}k_c^2 L^2}{k_{22AB} + 2} \tag{5.11h}$$

Assume different values of k_cL, and for each value find the slope deflection coefficients k_{22AB} in Eq. (5.11h) from Appendix A or Eqs. (4.21f) and (4.21g). Find θ_B, for each assumed value of k_cL from Eq. (5.11h). Plot θ_B versus k_cL shown in Figure 5.13. Buckling of the frame occurs at $k_cL = 4.67$ where the curve tends to become horizontal, i.e. there is a very large increase in θ_B with a small increase of k_cL. So

$$P_{cr} = \frac{21.81EI}{L^2} \tag{5.11i}$$

5.3.6 Frames with Prime Bending and Sidesway

Consider a hinged frame loaded with transverse loads on the horizontal member and allowed to sway horizontally as shown in Figure 5.14. The slope deflection equations for the members of the frame are

$$M_{AB} = \frac{EI_c}{L_c} \left[k_{11AB}\theta_A + k_{12AB}\theta_B - (k_{11AB} + k_{12AB})\frac{\Delta}{L_c} \right] \tag{5.12a}$$

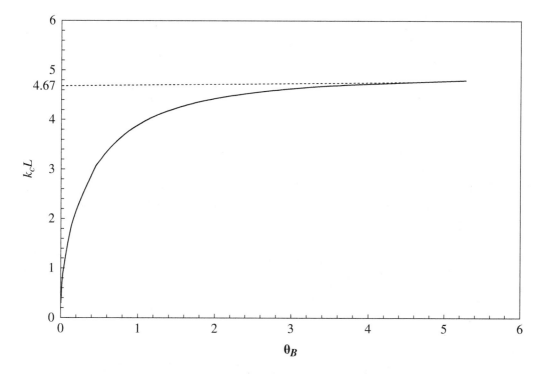

Figure 5.13 θ_B versus k_cL for a fixed frame with prime bending and no sidesway.

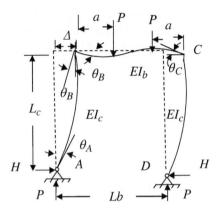

Figure 5.14 Frame hinged at the supports with transverse load and sidesway.

$$M_{BA} = \frac{EI_c}{L_c}\left[k_{21AB}\theta_A + k_{22AB}\theta_B - (k_{21AB} + k_{22AB})\frac{\Delta}{L_c}\right] \tag{5.12b}$$

$$M_{AB} = 0$$

$$0 = \frac{EI_c}{L_c}\left[k_{11AB}\theta_A + k_{12AB}\theta_B - (k_{11AB} + k_{12AB})\frac{\Delta}{L_c}\right]$$

$$\theta_A = \left(1 + \frac{k_{12AB}}{k_{11AB}}\right)\frac{\Delta}{L_c} - \frac{k_{12AB}}{k_{11AB}}\theta_B \tag{5.12c}$$

Substitute θ_A in Eq. (5.12b)

$$M_{BA} = \frac{EI_c}{L_c}\left[\left(k_{22AB} - \frac{k_{12AB}^2}{k_{11AB}}\right)\theta_B - \left(k_{22AB} - \frac{k_{12AB}^2}{k_{11AB}}\right)\frac{\Delta}{L_c}\right] \tag{5.12d}$$

$$M_{BC} = \frac{EI_b}{L_b}(k_{11BC}\theta_B + k_{12BC}\theta_C) - \frac{Pa}{L_b}(L_b - a) \tag{5.12e}$$

where $\frac{Pa}{L_b}(L_b - a)$ is the fixed end moment at B due to transverse load in the horizontal member BC, neglecting the axial force. It is counterclockwise, so there is a minus sign before the term as per the slope deflection convention.

Due to antisymmetry, $\theta_B = \theta_C$, therefore Eq. (5.12e) becomes

$$M_{BC} = \frac{EI_b}{L_b}(k_{11BC} + k_{12BC})\theta_B - \frac{Pa}{L_b}(L_b - a) \tag{5.12f}$$

For the joint equilibrium at B, we have

$$M_{BA} + M_{BC} = 0$$

$$\frac{EI_c}{L_c}\left[\left(k_{22AB} - \frac{k_{12AB}^2}{k_{11AB}}\right)\theta_B - \left(k_{22AB} - \frac{k_{12AB}^2}{k_{11AB}}\right)\frac{\Delta}{L_c}\right] + \frac{EI_b}{L_b}(k_{11BC} + k_{12BC})\theta_B - \frac{Pa}{L_b}(L_b - a) = 0$$

$$\tag{5.12g}$$

From the equilibrium of story shear in Figure 5.14, we have

$$\frac{M_{AB} + M_{BA} + P\Delta}{L_c} + \frac{M_{CD} + M_{DC} + P\Delta}{L_c} = 0 \tag{5.12h}$$

Ends A and D are hinges, so $M_{AB} = M_{DC} = 0$, and because of antisymmetry, $M_{BA} = M_{CD}$. Therefore, Eq. (5.12h) reduces to

$$\frac{M_{BA} + P\Delta}{L_c} = 0$$

or

$$\frac{EI_c}{L_c^2}\left[\left(k_{22AB} - \frac{k_{12AB}^2}{k_{11AB}}\right)\theta_B - \left(k_{22AB} - \frac{k_{12AB}^2}{k_{11AB}}\right)\frac{\Delta}{L_c}\right] + \frac{P\Delta}{L_c} = 0$$

$$\sqrt{\frac{P}{EI_c}} = k_c$$

or

$$\left(k_{22AB} - \frac{k_{12AB}^2}{k_{11AB}}\right)\theta_B - \left(k_{22AB} - \frac{k_{12AB}^2}{k_{11AB}} - k_c^2 L_c^2\right)\frac{\Delta}{L_c} = 0 \tag{5.12i}$$

For the case when $I_b = I_c = I$, $L_b = L_c = L$, Eq. (5.12g) gives

$$\left[\left(k_{22AB} - \frac{k_{12AB}^2}{k_{11AB}}\right) + (k_{11BC} + k_{12BC})\right]\theta_B - \left(k_{22AB} - \frac{k_{12AB}^2}{k_{11AB}}\right)\frac{\Delta}{L} - \frac{2}{9}k_c^2 L^2 = 0 \tag{5.12j}$$

Neglecting the axial force in the horizontal beam, $k_{11BC} = 4$ and $k_{12BC} = 2$, and Eq. (5.12j) becomes

$$\left(k_{22AB} - \frac{k_{12AB}^2}{k_{11AB}} + 6\right)\theta_B - \left(k_{22AB} - \frac{k_{12AB}^2}{k_{11AB}}\right)\frac{\Delta}{L} - \frac{2}{9}k_c^2 L^2 = 0 \tag{5.12k}$$

From Eq. (5.12i) we get

$$\left(k_{22AB} - \frac{k_{12AB}^2}{k_{11AB}}\right)\theta_B - \left(k_{22AB} - \frac{k_{12AB}^2}{k_{11AB}} - k_c^2 L^2\right)\frac{\Delta}{L} = 0 \tag{5.12l}$$

Equations (5.12k) and (5.12l) are solved simultaneously. From Eq. (5.12l) we have

$$\frac{\Delta}{L} = \frac{\left(k_{22AB} - \dfrac{k_{12AB}^2}{k_{11AB}}\right)\theta_B}{k_{22AB} - \dfrac{k_{12AB}^2}{k_{11AB}} - k_c^2 L^2} \tag{5.12m}$$

Substitute $\frac{\Delta}{L}$ from Eq. (5.12m) into Eq. (5.12k) and we have

$$\theta_B = \cfrac{\frac{2}{9}k_c^2 L^2}{k_{22AB} - \cfrac{k_{12AB}^2}{k_{11AB}} + 6 - \left(k_{22AB} - \cfrac{k_{12AB}^2}{k_{11AB}}\right)\left(\cfrac{k_{22AB} - \cfrac{k_{12AB}^2}{k_{11AB}}}{k_{22AB} - \cfrac{k_{12AB}^2}{k_{11AB}} - k_c^2 L^2}\right)} \tag{5.12n}$$

The slope deflection coefficients $k_{11AB} = k_{22AB}$.

Assume different values of $k_c L$, for each value, find the slope deflection coefficients k_{11AB} and k_{12AB} in Eq. (5.12n) from Appendix A or Eqs. (4.21f) and (4.21g). Find θ_B, for each assumed value of $k_c L$ from Eq. (5.12n). Plot θ_B versus $k_c L$ shown in Figure 5.15. Buckling of the frame occurs at $k_c L = 1.333$ where the curve tends to become horizontal, i.e. there is a very large increase in θ_B with a small increase of $k_c L$.

Or

$$\sqrt{\frac{P_{cr}}{EI}}L = 1.333$$

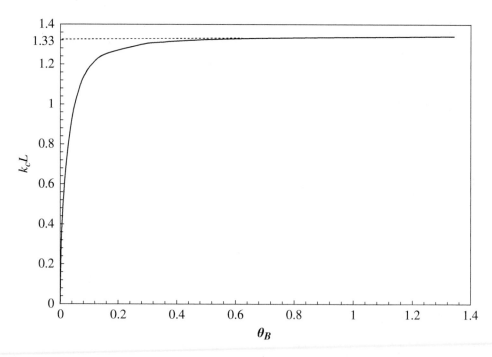

Figure 5.15 θ_B versus $k_c L$ for a hinged frame with prime bending and sidesway permitted.

Therefore,

$$P_{cr} = \frac{1.778EI}{L^2} \tag{5.12o}$$

which is the same value given by A. Kumar [1].

5.3.7 Box Frame Without Sidesway

Consider the buckling of a box frame that is symmetrical about the vertical center line of the frame. It is loaded as shown in Figure 5.16 such that there is no sidesway in the frame. The columns AB and CD are elastically restrained at the top and the bottom by the beams BC and AD.

The slope deflection equations for the members of the box frame are

$$M_{AB} = \frac{EI_c}{L_c}(k_{11AB}\theta_A + k_{12AB}\theta_B) \tag{5.13a}$$

$$M_{BA} = \frac{EI_c}{L_c}(k_{21AB}\theta_A + k_{22AB}\theta_B) \tag{5.13b}$$

Neglecting the horizontal force in the horizontal beams BC and AD

$$M_{BC} = \frac{EI_{b1}}{L_b}(4\theta_B + 2\theta_C) \tag{5.13c}$$

$$M_{AD} = \frac{EI_{b2}}{L_b}(4\theta_A + 2\theta_D) \tag{5.13d}$$

At the joint A, moment equilibrium gives

$$M_{AB} + M_{AD} = 0$$

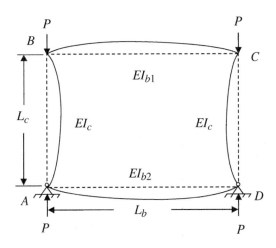

Figure 5.16 Symmetrical buckling without sidesway in a box frame.

$$\frac{EI_c}{L_c}(k_{11AB}\theta_A + k_{12AB}\theta_B) + \frac{EI_{b2}}{L_b}(4\theta_A + 2\theta_D) = 0 \tag{5.13e}$$

At the joint B, moment equilibrium gives

$$M_{BA} + M_{BC} = 0$$

$$\frac{EI_c}{L_c}(k_{21AB}\theta_A + k_{22AB}\theta_B) + \frac{EI_{b1}}{L_b}(4\theta_B + 2\theta_C) = 0 \tag{5.13f}$$

Due to symmetry about the vertical center line of the frame, we have $\theta_A = -\theta_D$, and $\theta_B = -\theta_C$, therefore

$$\frac{EI_c}{L_c}(k_{11AB}\theta_A + k_{12AB}\theta_B) + \frac{2EI_{b2}}{L_b}\theta_A = 0$$

$$\frac{EI_c}{L_c}(k_{21AB}\theta_A + k_{22AB}\theta_B) + \frac{2EI_{b1}}{L_b}\theta_B = 0$$

or

$$\begin{bmatrix} k_{11AB} + 2\frac{L_c}{L_b}\frac{I_{b2}}{I_c} & k_{12AB} \\ k_{21AB} & k_{22AB} + 2\frac{L_c}{L_b}\frac{I_{b1}}{I_c} \end{bmatrix} \left\{ \begin{matrix} \theta_A \\ \theta_B \end{matrix} \right\} = \left\{ \begin{matrix} 0 \\ 0 \end{matrix} \right\} \tag{5.13g}$$

For nontrivial solution of θ_A and θ_B we have

$$\begin{vmatrix} k_{11AB} + 2\frac{L_c}{L_b}\frac{I_{b2}}{I_c} & k_{12AB} \\ k_{21AB} & k_{22AB} + 2\frac{L_c}{L_b}\frac{I_{b1}}{I_c} \end{vmatrix} = 0 \tag{5.13h}$$

$$k_{11AB} = k_{22AB}, \text{ and } k_{12AB} = k_{21AB},$$

therefore,

$$\left(k_{11AB} + 2\frac{L_c}{L_b}\frac{I_{b2}}{I_c} \right) \left(k_{11AB} + 2\frac{L_c}{L_b}\frac{I_{b1}}{I_c} \right) - k_{12AB}^2 = 0 \tag{5.13i}$$

Assume various values of $k_c L_c$ for the columns and find corresponding quantities k_{11AB} and k_{12AB} for the given values of L_c, L_b, I_c, I_{b1}, and I_{b2} from Appendix A or Eqs. (4.21f) and (4.21g). The value of $k_c L_c$ that satisfies Eq. (5.13i) gives the critical load P for the frame.

For the case where $L_b = L_c = L$, and $I_c = I_{b1} = I_{b2} = I$, Eq. (5.13i) reduces to

$$k_{12AB} - k_{11AB} = 2 \tag{5.13j}$$

$k_c L = 4.0583$ satisfies Eq. (5.13j),

therefore,

$$k_c^2 L^2 = 16.47$$

or

$$P_{cr} = \frac{16.47EI}{L^2} \tag{5.13k}$$

Which is the same value given by N.G.R. Iyengar [2].

5.3.8 Multistory-Multibay Frames Without Sidesway

In practice, single story and single bay frames are not much used, instead multistory-multibay frames are used. Theoretically, critical loads for these frames under axial compression can also be found using the slope deflections equations as before, but it becomes very lengthy and complex to solve for frames that are of large size. A two-story and three-bay frame is solved for illustration purposes to show how the slope deflection method works for the multistory-multibay frames. The slope deflection equations are written for different members of the frame in Figure 5.17.

$$M_{BA} = \frac{EI_c}{L_c}\left(k_{22AB} - \frac{k_{12AB}^2}{k_{11AB}}\right)\theta_B \tag{5.14a}$$

$$M_{BC} = \frac{EI_c}{L_c}(k_{11BC}\theta_B + k_{12Bc}\theta_c) \tag{5.14b}$$

$$M_{CB} = \frac{EI_c}{L_c}(k_{21BC}\theta_B + k_{22BC}\theta_C) \tag{5.14c}$$

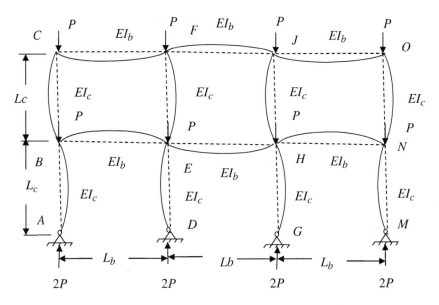

Figure 5.17 Symmetrical buckling of multistory multibay frame without sidesway.

Neglecting the axial force in the horizontal members, we get

$$M_{BE} = \frac{EI_b}{L_b}(4\theta_B + 2\theta_E) \tag{5.14d}$$

$$M_{EB} = \frac{EI_b}{L_b}(2\theta_B + 4\theta_E) \tag{5.14e}$$

$$M_{CF} = \frac{EI_b}{L_b}(4\theta_c + 2\theta_F) \tag{5.14f}$$

$$M_{FC} + \frac{EI_b}{L_b}(2\theta_C + 4\theta_F) \tag{5.14g}$$

$$M_{ED} = \frac{EI_c}{L_c}\left(k_{22DE} - \frac{k_{12DE}^2}{k_{11DE}}\right)\theta_E \tag{5.14h}$$

$$M_{EF} = \frac{EI_c}{L_c}(k_{11EF}\theta_E + k_{12EF}\theta_F) \tag{5.14i}$$

$$M_{FE} = \frac{EI_c}{L_c}(k_{21EF}\theta_E + k_{22EF}\theta_F) \tag{5.14j}$$

Neglecting the axial force in the horizontal members, we get

$$M_{EH} = \frac{EI_b}{L_b}(4\theta_E + 2\theta_H)$$

Because of symmetry about the vertical center line of the frame, $\theta_E = -\theta_H$, therefore

$$M_{EH} = \frac{EI_b}{L_b}(2\theta_E) \tag{5.14k}$$

$$M_{FJ} = \frac{EI_b}{L_b}(4\theta_F + 2\theta_J)$$

Because of symmetry about the vertical center line of the frame, $\theta_F = -\theta_J$, therefore,

$$M_{FJ} = \frac{EI_b}{L_b}(2\theta_F) \tag{5.14l}$$

From the moment equilibrium at the joints B, C, E, and F, we have

$$M_{BA} + M_{BC} + M_{BE} = 0 \tag{5.14m}$$

$$M_{CB} + M_{CF} = 0 \tag{5.14n}$$

$$M_{ED} + M_{EF} + M_{EB} + M_{EH} = 0 \tag{5.14o}$$

$$M_{FE} + M_{FC} + M_{FJ} = 0 \tag{5.14p}$$

Substitute for different moments from Eqs. (5.14a–5.14l) into Eqs. (5.14m)–(5.14p) to get

$$\frac{EI_c}{L_c}\left[\left(k_{22AB} - \frac{k_{12AB}^2}{k_{11AB}} + k_{11BC}\right)\theta_B + k_{12BC}\theta_c\right] + \frac{EI_b}{L_b}(4\theta_B + 2\theta_E) = 0 \tag{5.14q}$$

$$\frac{EI_c}{L_c}[k_{21BC}\theta_B + k_{22BC}\theta_c] + \frac{EI_b}{L_b}(4\theta_C + 2\theta_F) = 0 \tag{5.14r}$$

$$\frac{EI_c}{L_c}\left[\left(k_{22DE} - \frac{k_{12DE}^2}{k_{11DE}} + k_{11EF}\right)\theta_E + k_{12EF}\theta_F\right] + \frac{EI_b}{L_b}(2\theta_B + 6\theta_E) \tag{5.14s}$$

$$\frac{EI_c}{L_c}(k_{21EF}\theta_E + k_{22EF}\theta_F) + \frac{EI_b}{L_b}(2\theta_c + 6\theta_F) = 0 \tag{5.14t}$$

Equations (5.16q)–(5.16t) can be written as

$$\begin{bmatrix} \dfrac{EI_c}{L_c}\left(k_{22AB} - \dfrac{k_{12AB}^2}{k_{11AB}} + k_{11BC}\right) + \dfrac{4EI_b}{L_b} & \dfrac{EI_c}{L_c}k_{12BC} & \dfrac{2EI_b}{L_b} & 0 \\[3ex] \dfrac{EI_c}{L_c}k_{21BC} & \dfrac{EI_c}{L_c}(k_{22BC}) + \dfrac{4EI_b}{L_b} & 0 & \dfrac{2EI_b}{L_b} \\[3ex] \dfrac{2EI_b}{L_b} & 0 & \dfrac{EI_c}{L_c}\left(k_{22DE} - \dfrac{k_{12DE}^2}{k_{11DE}} + k_{11EF}\right) + \dfrac{6EI_b}{L_b} & \dfrac{EI_c}{L_c}k_{12EF} \\[3ex] 0 & \dfrac{2EI_b}{L_b} & \dfrac{EI_c}{L_c}k_{21EF} & \dfrac{EI_c}{L_c}k_{22EF} + \dfrac{6EI_b}{L_b} \end{bmatrix}$$

$$\times \begin{Bmatrix} \theta_B \\ \theta_C \\ \theta_E \\ \theta_F \end{Bmatrix} = \begin{Bmatrix} 0 \\ 0 \\ 0 \\ 0 \end{Bmatrix} \tag{5.14u}$$

For the case where $L_b = L_c = L$, and $I_b = I_c = I$, Eq. (5.14u) reduces to

$$\begin{vmatrix} \left(k_{22AB} - \dfrac{k_{12AB}^2}{k_{11AB}} + k_{11BC} + 4\right) & k_{12BC} & 2 & 0 \\[2mm] k_{21BC} & k_{22BC} + 4 & 0 & 2 \\[2mm] 2 & 0 & \left(k_{22DE} - \dfrac{k_{12DE}^2}{k_{11DE}} + k_{11EF} + 6\right) & k_{12EF} \\[2mm] 0 & 2 & k_{21EF} & k_{22EF} + 6 \end{vmatrix} \begin{Bmatrix} \theta_B \\ \theta_C \\ \theta_E \\ \theta_F \end{Bmatrix} = \begin{Bmatrix} 0 \\ 0 \\ 0 \\ 0 \end{Bmatrix}$$

$$(5.14v)$$

For a nontrivial solution of θ_B, θ_C, θ_E, and θ_F the determinant of the coefficient matrix should be zero, and we have

$$\begin{vmatrix} \left(k_{22AB} - \dfrac{k_{12AB}^2}{k_{11AB}} + k_{11BC} + 4\right) & k_{12BC} & 2 & 0 \\[2mm] k_{21BC} & k_{22BC} + 4 & 0 & 2 \\[2mm] 2 & 0 & \left(k_{22DE} - \dfrac{k_{12DE}^2}{k_{11DE}} + k_{11EF} + 6\right) & k_{12EF} \\[2mm] 0 & 2 & k_{21EF} & k_{22EF} + 6 \end{vmatrix} = 0$$

$$(5.14w)$$

Or $\left(k_{22AB} - \dfrac{k_{12AB}^2}{k_{11AB}} + k_{11BC} + 4\right)(k_{22BC} + 4)\left(k_{22DE} - \dfrac{k_{12DE}^2}{k_{11DE}} + k_{11EF} + 6\right)(k_{22EF} + 6)$

$-\left(k_{22AB} - \dfrac{k_{12AB}^2}{k_{11AB}} + k_{11BC} + 4\right)(k_{22BC} + 4)(k_{12EF})^2$

$-\left(k_{22AB} - \dfrac{k_{12AB}^2}{k_{11AB}} + k_{11BC} + 4\right)\left(k_{22DE} - \dfrac{k_{12DE}^2}{k_{11DE}} + k_{11EF} + 6\right)(4)$

$-k_{12BC}^2\left(k_{22DE} - \dfrac{k_{12DE}^2}{k_{11DE}} + k_{11EF} + 6\right)(k_{22EF} + 6)$

$+k_{12BC}^2 k_{12EF}^2 - 8k_{12BC}k_{12EF} - 4(k_{22BC} + 4)(k_{22EF} + 6) + 16 = 0 \qquad (5.14x)$

$$k_{BC}^2 = k_{EF}^2 = \frac{P}{EI}, \qquad\qquad k_{AB}^2 = k_{DE}^2 = \frac{2P}{EI}$$

$$k_{AB} = k_{DE} = \sqrt{2}k_{BC}$$

Assume $k_{BC}L = k_{EF}L$ and calculate the corresponding $k_{AB}L = k_{DE}L = \sqrt{2}k_{BC}L$. Find the slope deflection coefficients k_{11AB}, k_{12AB}, k_{11Bc}, k_{12BC}, $k_{11DE}, k_{12DE}, k_{11EF},$ and k_{12EF} from Appendix A from the assumed and calculated values of $k_{BC}L$, $k_{EF}L$, $k_{AB}L$, and $k_{DE}L$. Alternately, the slope deflection coefficients can be obtained from Eqs. (4.8g), (4.8h), (4.21f), and (4.21g), where $u = \frac{kL}{2}$. Equation (5.14x) is solved by trial and error to find the correct value of $k_{BC}L$. The trial and error procedure gives

$$k_{BC}L = 2.747$$

or

$$k_{BC}^2L^2 = 7.547 \qquad \text{or} \qquad \frac{PL^2}{EI} = 7.547$$

Therefore, the critical load for the two-story three-bay frame in Figure 5.17 is given by

$$P_{cr} = \frac{7.547EI}{L^2} \tag{5.14y}$$

5.4 Critical Loads by Matrix and Finite Element Methods

In the stability analysis of the frames the equilibrium is satisfied on the deformed shape of the structure. The first-order analysis in which the material is assumed as linearly elastic and the equilibrium is satisfied on the original shape of the structure is inadequate to measure the frame stability. To predict the stability limit of a frame, second-order analysis is performed. Second-order analysis can include both geometric and material nonlinearities. The analysis takes into consideration the effects of large deflections (geometrically nonlinear) and the inelastic material (material nonlinearity). Hence the inelastic second-order analysis is the most accurate method to find the loading strength of a structure. The load calculated from the elastic second-order analysis where only geometric nonlinearity is considered gives the elastic stability limit for the structure. A load-deflection graph is plotted where the curve approaches asymptotically the elastic stability limit load of the structure. Another approach is to calculate the elastic critical load by eigenvalue or bifurcation analysis where geometric nonlinearity is taken into account to form the structure stiffness matrix. The elastic analysis is valid only if stresses in a structure do not exceed the proportional limit at any time. In the second-order analysis both primary and secondary bending moments due to P-Δ (frame instability) and P-δ (member instability) effects are taken into account. The P-Δ moments are caused by the axial force acting though the relative lateral displacement of the two ends of a member such as caused by the sidesway of Δ in a frame [3]. The P-δ moments are caused by the axial force acting through the lateral displacement of a member as in the case of a frame where sidesway is prevented. The secondary moments due to P-δ are generated in the beam and the column of the frame due to the lateral deflection δ along the length of the column [4].

The matrix and finite element methods have gained popularity because of the extensive use of computers. Engineers can solve complex problems using these methods for which exact close form solutions are not available. In this method the element stiffness matrix which relates the element end forces to its end displacements is formed for every member of the frame.

The element stiffness matrices are then assembled into the structure stiffness matrix which relates the structural nodal forces to the structural nodal displacements. In the displacement method the element stiffness matrices are derived based on the general energy principles. In the finite element formulation, a displacement field is assumed for the element. The stiffness matrix relating the nodal forces to nodal displacements is derived by minimizing the total potential energy of the element.

5.4.1 Formation of the Element Stiffness Matrix

The principle of stationary total potential energy is used here to form the equilibrium equations for an element. Total potential energy Π for an element is given by

$$\Pi = U + V \tag{5.15a}$$

where U is the strain energy in the element, and V is the potential energy of the external forces on the element. From the principle of stationary total potential energy, equilibrium is obtained when the first variation of the total potential energy vanishes, that is

$$\delta\Pi = \delta(U + V) = \delta U + \delta V = 0 \tag{5.15b}$$

The potential energy of external forces on the element can be expressed as

$$V = -\int P_i \, dq_i \tag{5.15c}$$

P_i and dq_i are the external forces and the corresponding displacements. If the external forces and displacements in the element are expressed in terms of the n nodal forces $\{q\}$ and the corresponding nodal displacements $\{d\}$, then the potential energy of external forces is given by

$$V = -\sum_{i=1}^{n} q_i d_i = -\{d\}^T \{q\} \tag{5.15d}$$

The strain energy of the frame element is given by

$$U = \int_{Vol} \left(\int_{\varepsilon} \sigma \, d\varepsilon \right) dVol \tag{5.15e}$$

where σ and ε are the axial stress and the corresponding axial strain, and Vol is the volume of the element. Assuming linearly elastic material, from Hooke's law we have

$$\sigma = E\varepsilon$$

Substitution in Eq. (5.15e) and integrating we get

$$U = \frac{E}{2} \int_0^L \int_A (\varepsilon^2) \, dA \, dx \tag{5.15f}$$

where L is the length and A is the area of cross-section of the element. For nonlinear elastic analysis, the strain ε is given by

$$\varepsilon = u_{,x} + \frac{1}{2}u_{,x}^2 + \frac{1}{2}v_{,x}^2 - yv_{,xx} \tag{5.15g}$$

where u and v represent the axial and transverse displacements of the element respectively and subscripts show derivatives, for example

$$\frac{du}{dx} = u_x, \quad \frac{dv}{dx} = v_x, \quad \text{and} \quad \frac{d^2v}{dx^2} = v_{xx}$$

The first three terms in Eq. (5.15g) give the strain of the element at its centroid according to Green's strain tensor [5]. The last term gives the axial strain due to flexure in the fibers at a distance y from the centroid of the cross-section of the element. The second term u_x^2 is neglected in comparison to the first term u_x because the axial displacement is usually small. Substitution of Eq. (5.15g) into Eq. (5.15f) leads to

$$U = \frac{E}{2} \int_0^L \int_A \left(u_x + \frac{1}{2} v_x^2 - y v_{xx} \right)^2 dA \, dx \tag{5.15h}$$

$$U = \frac{E}{2} \int_0^L \int_A \left(u_x^2 + \frac{1}{4} v_x^4 + y^2 v_{xx}^2 + u_x v_x^2 - y v_x^2 v_{xx} - 2 y u_x v_{xx} \right) dx \, dA \tag{5.15i}$$

Neglect the term, $\frac{1}{4} v_x^4$ in comparison to other terms, and integrate with respect to the area A. Substitute $I = \int_A y^2 dA$, the second moment of inertia of the cross-section, and $\int_A y dA = 0$, the first moment of the area about the centroidal axis, which is zero. We get

$$U = \frac{E}{2} \int_0^L (A u_x^2 + I v_{xx}^2 + A u_x v_x^2) dx \tag{5.15j}$$

Substitute $EAu_x = P$, the axial force in the member. The first variation of the strain energy can be written as

$$\delta U = \int_0^L [EA(u_x)(\delta u_x) + EI(v_{xx})(\delta v_{xx}) + P(v_x)(\delta v_x)] \, dx \tag{5.15k}$$

The first variation of the potential energy obtained from Eq. (5.15d) is

$$\delta V = -\{\delta d\}^T \{q\} \tag{5.15l}$$

Apply the principle of stationary total potential energy, then from Eq. (5.15b) we get

$$\int_0^L [EA(u_x)(\delta u_x) + EI(v_{xx})(\delta v_{xx}) + P(v_x)(\delta v_x)] dx - \{\delta d\}^T \{q\} = 0 \tag{5.15m}$$

The displacement field for u and v in the element shown in Figure 5.18 is assumed as

$$u = a_0 + a_1 x \tag{5.15n}$$

$$v = b_0 + b_1 x + b_2 x^2 + b_3 x^3 \tag{5.15o}$$

The constants a_0, a_1, b_0, b_1, b_2, and b_3 are determined from the boundary conditions:

$$x = 0$$

$$u = d_1, \quad v = -d_2, \quad \text{and} \, v_x = d_3 \, \text{and}$$

$$x = L$$

$$u = d_4, \quad v = -d_5, \text{and} \, v_x = d_6$$

Substitute these boundary conditions in Eqs. (5.15n) and (5.15o) and we get

$$\left\{ \begin{matrix} u \\ v \end{matrix} \right\} = \left[\begin{matrix} \left(1 - \frac{x}{L}\right) & 0 & 0 & \frac{x}{L} & 0 & 0 \\ 0 & \left(-1 + \frac{3x^2}{L^2} - \frac{2x^3}{L^3}\right) & \left(x - \frac{2x^2}{L} + \frac{x^3}{L^2}\right) & 0 & \left(-\frac{3x^2}{L^2} + \frac{2x^3}{L^3}\right) & \left(-\frac{x^2}{L} + \frac{x^3}{L^2}\right) \end{matrix} \right]$$

$$\times \left\{ \begin{matrix} d_1 \\ d_2 \\ d_3 \\ d_4 \\ d_5 \\ d_6 \end{matrix} \right\} \tag{5.15p}$$

Eq. (5.15p) can be written as

$$\{u\} = [N]\{d\} \tag{5.15q}$$

where $\{u\}$ is the displacement function, $[N]$ is called the shape function, and $\{d\}$ represents the nodal displacements for the frame element. The total axial strain in the element can be represented by

$$\{\varepsilon\} = \{\varepsilon_0\} + \{\varepsilon_i\} \tag{5.15r}$$

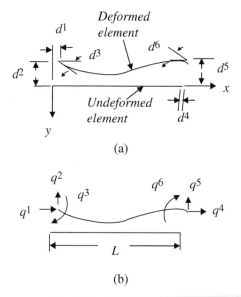

Figure 5.18 Beam element nodal displacements and forces: (a) Frame element displacements; (b) Frame element nodal forces.

where $\{\varepsilon_i\}$ is the initial strain due to axial force, and $\{\varepsilon_0\}$ linear elastic strain. The strain displacement relations for Eq. (5.15r) can be written as

$$\{\varepsilon\} = \begin{Bmatrix} u_{,x} \\ v_{,xx} \end{Bmatrix} + \{v_{,x}\} \tag{5.15s}$$

or

$$\{\varepsilon\} = [B_0]\{d\} + [B_i]\{d\} \tag{5.15t}$$

Eq. (5.15m) can be written as

$$\int_0^L [\,\delta u_{,x} \quad \delta v_{,xx}\,] \begin{bmatrix} EA & 0 \\ 0 & EI \end{bmatrix} \begin{Bmatrix} u_{,x} \\ v_{,xx} \end{Bmatrix} dx + P \int_0^L [\delta v_{,x}]^T [v_{,xx}]\, dx - \{\,\delta d\,\}^T \{\, q\,\} = 0 \tag{5.15u}$$

Substitute Eq. (5.15t) in Eq. (5.15u) and we have

$$\{\delta d\}^T \left\{ \int_0^L \left([B_0]^T \begin{bmatrix} EA & 0 \\ 0 & EI \end{bmatrix} [B_0] + P[B_i]^T[B_i] \right) dx \right\} \{d\} = \{\delta d\}^T \{q\} \tag{5.15v}$$

Equation (5.15v) is derived for any arbitrary increase in nodal displacements $\{\delta d\}$, therefore,

$$\left\{ \int_0^L \left([B_0]^T \begin{bmatrix} EA & 0 \\ 0 & EI \end{bmatrix} [B_0] + P[B_i]^T[B_i] \right) dx \right\} \{d\} = \{q\} \tag{5.15w}$$

or

$$\{[k_0] + P[k_P]\}\{d\} = \{q\} \tag{5.15x}$$

where the element stiffness matrix $\{k_0\}$ is the linear elastic or first order stiffness matrix. The $\{k_p\}$ is the element geometrical or initial stiffness matrix, it takes into account the effect of the axial load P on the stiffness of the flexural member. P is positive for a tensile axial force and is negative for a compressive axial force.

We can calculate the strain displacement relations coefficient matrices $[B_0]$ and $[B_i]$ from Eqs. (5.15p), (5.15s), and (5.15t) and are given as follows:

$$[B_0] = \begin{bmatrix} -\dfrac{1}{L} & 0 & 0 & \dfrac{1}{L} & 0 & 0 \\[2mm] 0 & \left(\dfrac{6}{L^2} - \dfrac{12x}{L^3}\right) & \left(-\dfrac{4}{L} + \dfrac{6x}{L^2}\right) & 0 & \left(-\dfrac{6}{L^2} + \dfrac{12x}{L^3}\right) & \left(-\dfrac{2}{L} + \dfrac{6x}{L^2}\right) \end{bmatrix}$$

and

$$[B_i] = \begin{bmatrix} 0 & \left(\dfrac{6x}{L^2} - \dfrac{6x^2}{L^3}\right) & \left(1 - \dfrac{4x}{L} + \dfrac{3x^2}{L^2}\right) & 0 & \left(-\dfrac{6x}{L^2} + \dfrac{6x^2}{L^3}\right) & \left(-\dfrac{2x}{L} + \dfrac{3x^2}{L^2}\right) \end{bmatrix}$$

Substitute $[B_0]$ in Eq. (5.15w) and we get

$$\int_0^L \left([B_0]^T \begin{bmatrix} EA & 0 \\ 0 & EI \end{bmatrix} [B_0] \right) \, dx = [k_0]$$

or

$$[k_0] = \begin{bmatrix} \dfrac{EA}{L} & 0 & 0 & -\dfrac{EA}{L} & 0 & 0 \\[2mm] 0 & \dfrac{12EI}{L^3} & -\dfrac{6EI}{L^2} & 0 & -\dfrac{12EI}{L^3} & -\dfrac{6EI}{L^2} \\[2mm] 0 & -\dfrac{6EI}{L^2} & \dfrac{4EI}{L} & 0 & \dfrac{6EI}{L^2} & \dfrac{2EI}{L} \\[2mm] -\dfrac{EA}{L} & 0 & 0 & \dfrac{EA}{L} & 0 & 0 \\[2mm] 0 & -\dfrac{12EI}{L^3} & \dfrac{6EI}{L^2} & 0 & \dfrac{12EI}{L^3} & \dfrac{6EI}{L^2} \\[2mm] 0 & -\dfrac{6EI}{L^2} & \dfrac{2EI}{L} & 0 & \dfrac{6EI}{L^2} & \dfrac{4EI}{L} \end{bmatrix} \tag{5.15y}$$

and

$$[k_P] = \begin{bmatrix} 0 & 0 & 0 & 0 & 0 & 0 \\[2mm] 0 & \dfrac{6}{5L} & -\dfrac{1}{10} & 0 & -\dfrac{6}{5L} & -\dfrac{1}{10} \\[2mm] 0 & -\dfrac{1}{10} & \dfrac{2L}{15} & 0 & \dfrac{1}{10} & -\dfrac{L}{30} \\[2mm] 0 & 0 & 0 & 0 & 0 & 0 \\[2mm] 0 & -\dfrac{6}{5L} & \dfrac{1}{10} & 0 & \dfrac{6}{5L} & \dfrac{1}{10} \\[2mm] 0 & -\dfrac{1}{10} & -\dfrac{L}{30} & 0 & \dfrac{1}{10} & \dfrac{2L}{15} \end{bmatrix} \tag{5.15z}$$

5.4.2 Formation of the Structure Stiffness Matrix

Prior to the assembly of element stiffness matrices into the structure stiffness matrix, it is necessary to transform the element nodal forces and displacements to global coordinates. This is accomplished by a transformation of coordinates. The element stiffness matrix is written in terms of element forces and deformations using element or local coordinates, whereas the structure stiffness matrix represents the relationship between structure forces and deformations in structure or global coordinates. Thus, element stiffness matrices are transformed into structure coordinates by transformation from element to global coordinates. The element and structure coordinates are shown in Figure 5.19 by the x, y and X, Y directions respectively. The out of plane z and Z axes coincide for a two-dimensional case of deformations and forces.

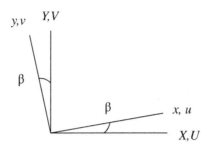

Figure 5.19 Global and local coordinates.

The transformation matrix T that transforms from element or local to structure or global coordinates is given by

$$\begin{Bmatrix} d_1 \\ d_2 \\ d_3 \\ d_4 \\ d_5 \\ d_6 \end{Bmatrix} = \begin{bmatrix} \cos\beta & \sin\beta & 0 & 0 & 0 & 0 \\ -\sin\beta & \cos\beta & 0 & 0 & 0 & 0 \\ 0 & 0 & 1 & 0 & 0 & 0 \\ 0 & 0 & 0 & \cos\beta & \sin\beta & 0 \\ 0 & 0 & 0 & -\sin\beta & \cos\beta & 0 \\ 0 & 0 & 0 & 0 & 0 & 1 \end{bmatrix} \begin{Bmatrix} D_1 \\ D_2 \\ D_3 \\ D_4 \\ D_5 \\ D_6 \end{Bmatrix} \tag{5.16a}$$

or

$$\{d\} = [T]\{D\} \tag{5.16b}$$

$\{d\}$ and $\{D\}$ are the element and structure nodal displacements matrices, respectively. The element nodal forces $\{q\}$ and structure nodal forces $\{Q\}$ transform in the same way as the deformations, that is

$$\{q\} = [T]\{Q\} \tag{5.16c}$$

The element nodal forces $\{q\}$ are related to the element nodal deformations $\{d\}$ by

$$[k]\{d\} = \{q\}$$

or

$$[k][T]\{D\} = [T]\{Q\}$$

or

$$[T]^{-1}[k][T]\{D\} = \{Q\}$$

Since $[T]$ is an orthogonal matrix, $[T]^{-1} = [T]^T$ or

$$[T]^T[k][T]\{D\} = \{Q\}$$

or

$$[K]\{D\} = \{Q\} \tag{5.16d}$$

Therefore,

$$[K] = [T]^T[k][T] \tag{5.16e}$$

An element stiffness matrix is transformed into the structure coordinates as shown in Eq. (5.16e) before assembling it into a structure stiffness matrix. By coordinate transformations, structure stiffness matrices can be formed from Eqs. (5.15y), (5.15z) and (5.16e) as follows:

$$[K_0] = [T]^T[k_0][T] \tag{5.16f}$$

and

$$[K_P] = [T]^T[k_P][T] \tag{5.16g}$$

5.4.3 In Span Loading

So far, we have considered concentrated nodal forces, but frame members are also subjected to distributed loads on their spans. The distributed loads are replaced by equivalent nodal forces by using the concept that the work of the distributed loads is equal to that of equivalent nodal forces for any arbitrary nodal displacements. Consider the distributed loading shown in Figure 5.20 on an element that consists of forces p_x and p_y along the span and perpendicular to the span of the element.

The work done by the distributed forces is given by

$$W = \int_0^L (p_x u + p_y v)\, dx \tag{5.17a}$$

or

$$W = \int_0^L [u\ v] \begin{Bmatrix} p_x \\ p_y \end{Bmatrix} dx \tag{5.17b}$$

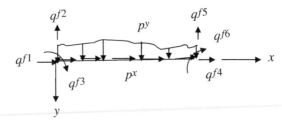

Figure 5.20 Distributed forces and equivalent nodal forces.

Substitution from Eq. (5.15q) into Eq. (5.17b) gives

$$W = \int_0^L \{d\}^T [N]^T \{p\} dx \tag{5.17c}$$

where

$$\{p\} = \begin{Bmatrix} p_x \\ p_y \end{Bmatrix} \tag{5.17d}$$

Work done by equivalent nodal forces $\{q_f\}$ is given by

$$W = \{d\}^T \{q_f\} \tag{5.17e}$$

Equate W in Eqs. (5.17c) and (5.17e), and we get

$$\{q_f\} = \begin{Bmatrix} q_{f1} \\ q_{f2} \\ q_{f3} \\ q_{f4} \\ q_{f5} \\ q_{f6} \end{Bmatrix} = \int_0^L [N]^T \{p\} dx \tag{5.17f}$$

The in span forces are accounted by introducing the equivalent nodal forces vector $\{q_f\}$ to the right side of Eq. (5.15x). Consider a simply supported beam column in Figure 5.21a subjected to the distributed force of $p_x = p$, and $p_y = w$. From Eq. (5.17f) we have

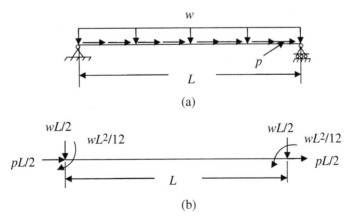

(a)

(b)

Figure 5.21 Beam with distributed load and equivalent nodal forces: (a) Beam with a distributed load; (b) The equivalent nodal forces.

$$\{q_f\} = \int_0^L \begin{bmatrix} \left(1-\dfrac{x}{L}\right) & 0 \\[2mm] 0 & \left(-1+\dfrac{3x^2}{L^2}-\dfrac{2x^3}{L^3}\right) \\[2mm] 0 & \left(x-\dfrac{2x^2}{L}+\dfrac{x^3}{L^2}\right) \\[2mm] \dfrac{x}{L} & 0 \\[2mm] 0 & \left(-\dfrac{3x^2}{L^2}+\dfrac{2x^3}{L^3}\right) \\[2mm] 0 & \left(-\dfrac{x^2}{L}+\dfrac{x^3}{L^2}\right) \end{bmatrix} \begin{Bmatrix} p \\ w \end{Bmatrix} dx = \begin{Bmatrix} \dfrac{pL}{2} \\[2mm] -\dfrac{wL}{2} \\[2mm] \dfrac{wL^2}{12} \\[2mm] \dfrac{pL}{2} \\[2mm] -\dfrac{wL}{2} \\[2mm] -\dfrac{wL^2}{12} \end{Bmatrix} \tag{5.17g}$$

The equivalent nodal forces are shown in Figure 5.21b and it is seen that the equivalent nodal forces are equal and opposite to the reactions in a fixed-fixed beam column under similar loading.

5.4.4 Buckling of a Frame Pinned at the Base and with Sidesway Permitted

Consider the frame loaded by two vertical forces each equal to P acting on the horizontal member shown in Figure 5.22a. Let each member consist of only single element, the positive element nodal forces and displacements are shown in Figure 5.22b. The members are assumed to be inextensible, that is the change in length due to axial force is neglected in the members. Therefore, each element has four degrees of freedom d_1, d_2, d_3, d_4 and the corresponding element forces q_1, q_2, q_3, q_4 shown in their positive sense in Figure 5.22b. So, the element stiffness matrices are 4×4 instead of 6×6 as shown in Eqs. (5.15y) and (5.15z). Each member is of length L and stiffness EI. The frame is hinged at the bottom and is allowed horizontal lateral movement at the top. The structure nodal forces and displacements are shown in Figure 5.22a in their positive sense. The structure and element coordinates are also shown in Figure 5.22. The element stiffness matrices for the column elements 1 and 3 are

$$[k_1] = [k_3] = \frac{EI}{L} \begin{bmatrix} \dfrac{12}{L^2} & -\dfrac{6}{L} & -\dfrac{12}{L^2} & -\dfrac{6}{L} \\[2mm] -\dfrac{6}{L} & 4 & \dfrac{6}{L} & 2 \\[2mm] -\dfrac{12}{L^2} & \dfrac{6}{L} & \dfrac{12}{L^2} & \dfrac{6}{L} \\[2mm] -\dfrac{6}{L} & 2 & \dfrac{6}{L} & 4 \end{bmatrix} - \frac{P}{L} \begin{bmatrix} \dfrac{6}{5} & -\dfrac{L}{10} & -\dfrac{6}{5} & -\dfrac{L}{10} \\[2mm] -\dfrac{L}{10} & \dfrac{2L^2}{15} & \dfrac{L}{10} & -\dfrac{L^2}{30} \\[2mm] -\dfrac{6}{5} & \dfrac{L}{10} & \dfrac{6}{5} & \dfrac{L}{10} \\[2mm] -\dfrac{L}{10} & -\dfrac{L^2}{30} & \dfrac{L}{10} & \dfrac{2L^2}{15} \end{bmatrix} \tag{5.18a}$$

or

$$[k_1] = [k_3] = \frac{EI}{L}[k_{01}] - \frac{P}{L}[k_{p1}] = \frac{EI}{L}[k_{03}] - \frac{P}{L}[k_{p3}] \tag{5.18b}$$

The element stiffness matrix for the beam element 2 is

$$[k_2] = \frac{EI}{L}\begin{bmatrix} \dfrac{12}{L^2} & -\dfrac{6}{L} & -\dfrac{12}{L^2} & -\dfrac{6}{L} \\[2mm] -\dfrac{6}{L} & 4 & \dfrac{6}{L} & 2 \\[2mm] -\dfrac{12}{L^2} & \dfrac{6}{L} & \dfrac{12}{L^2} & \dfrac{6}{L} \\[2mm] -\dfrac{6}{L} & 2 & \dfrac{6}{L} & 4 \end{bmatrix} = \frac{EI}{L}[k_{02}] \tag{5.18c}$$

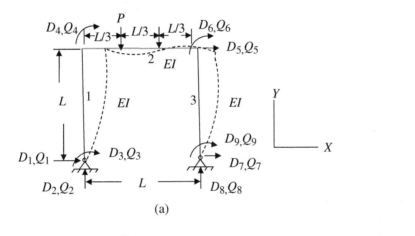

(a)

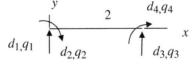

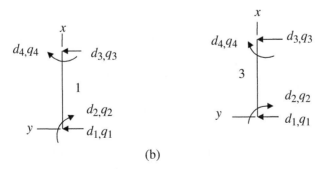

(b)

Figure 5.22 Structure and element forces and displacements: (a) Structure coordinates, degrees of freedom and forces; (b) Element coordinates, degrees of freedom and forces.

The element stiffness matrices are now converted into structure coordinates by using Eq. (5.16a). For element 1, $\beta = 90°$, substituting in Eq. (5.16a) we get

$$\begin{Bmatrix} d_1 \\ d_2 \\ d_3 \\ d_4 \end{Bmatrix} = \begin{bmatrix} -\sin\beta & \cos\beta & 0 & 0 & 0 \\ 0 & 0 & 1 & 0 & 0 \\ 0 & 0 & 0 & 0 & -\sin\beta \\ 0 & 0 & 0 & 1 & 0 \end{bmatrix} \begin{Bmatrix} D_1 \\ D_2 \\ D_3 \\ D_4 \\ D_5 \end{Bmatrix} \qquad (5.18\text{d})$$

or

$$[T_1] = \begin{bmatrix} -1 & 0 & 0 & 0 & 0 \\ 0 & 0 & 1 & 0 & 0 \\ 0 & 0 & 0 & 0 & -1 \\ 0 & 0 & 0 & 1 & 0 \end{bmatrix} \qquad (5.18\text{e})$$

For element 2, $\beta = 0$, substituting in Eq. (5.16a) we get

$$\begin{Bmatrix} d_1 \\ d_2 \\ d_3 \\ d_4 \end{Bmatrix} = \begin{bmatrix} 0 & 0 & 0 \\ 1 & 0 & 0 \\ 0 & -\sin\beta & 0 \\ 0 & 0 & 1 \end{bmatrix} \begin{Bmatrix} D_4 \\ D_5 \\ D_6 \end{Bmatrix} \qquad (5.18\text{f})$$

or

$$[T_2] = \begin{bmatrix} 0 & 0 & 0 \\ 1 & 0 & 0 \\ 0 & 0 & 0 \\ 0 & 0 & 1 \end{bmatrix} \qquad (5.18\text{g})$$

For element 3, $\beta = 90°$, substituting in Eq. (5.16a) we get

$$\begin{Bmatrix} d_1 \\ d_2 \\ d_3 \\ d_4 \end{Bmatrix} = \begin{bmatrix} -\sin\beta & \cos\beta & 0 & 0 & 0 \\ 0 & 0 & 1 & 0 & 0 \\ 0 & 0 & 0 & -\sin\beta & 0 \\ 0 & 0 & 0 & 0 & 1 \end{bmatrix} \begin{Bmatrix} D_7 \\ D_8 \\ D_9 \\ D_5 \\ D_6 \end{Bmatrix} \qquad (5.18\text{h})$$

or

$$[T_3] = \begin{bmatrix} -1 & 0 & 0 & 0 & 0 \\ 0 & 0 & 1 & 0 & 0 \\ 0 & 0 & 0 & -1 & 0 \\ 0 & 0 & 0 & 0 & 1 \end{bmatrix} \qquad (5.18\text{i})$$

The element stiffness matrices are now transformed into structure coordinates by using the following equation

$$[K] = [T]^T[k][T] \qquad (5.16\text{e})$$

For element 1

$$[K_1] = \frac{EI}{L}[T_1]^T[k_{01}][T_1] - \frac{P}{L}[T_1]^T[k_{p1}][T_1] \tag{5.18j}$$

or

$$[K_1] = \frac{EI}{L}
\begin{array}{ccccc}
D_1 & D_2 & D_3 & D_4 & D_5
\end{array}
\begin{bmatrix}
\dfrac{12}{L^2} & 0 & \dfrac{6}{L} & \dfrac{6}{L} & -\dfrac{12}{L^2} \\[6pt]
0 & 0 & 0 & 0 & 0 \\[6pt]
\dfrac{6}{L} & 0 & 4 & 2 & -\dfrac{6}{L} \\[6pt]
\dfrac{6}{L} & 0 & 2 & 4 & -\dfrac{6}{L} \\[6pt]
-\dfrac{12}{L^2} & 0 & -\dfrac{6}{L} & -\dfrac{6}{L} & \dfrac{12}{L^2}
\end{bmatrix}
- \frac{P}{L}
\begin{array}{ccccc}
D_1 & D_2 & D_3 & D_4 & D_5
\end{array}
\begin{bmatrix}
\dfrac{6}{5} & 0 & \dfrac{L}{10} & \dfrac{L}{10} & -\dfrac{6}{5} \\[6pt]
0 & 0 & 0 & 0 & 0 \\[6pt]
\dfrac{L}{10} & 0 & \dfrac{2L^2}{15} & -\dfrac{L^2}{30} & -\dfrac{L}{10} \\[6pt]
\dfrac{L}{10} & 0 & -\dfrac{L^2}{30} & \dfrac{2L^2}{15} & -\dfrac{L}{10} \\[6pt]
-\dfrac{6}{5} & 0 & -\dfrac{L}{10} & -\dfrac{L}{10} & \dfrac{6}{5}
\end{bmatrix}
\begin{array}{l}
D_1 \\ D_2 \\ D_3 \\ D_4 \\ D_5
\end{array}
\tag{5.18k}$$

For element 2

$$[K_2] = \frac{EI}{L}[T_2]^T[k_{02}][T_2] \tag{5.18l}$$

or

$$[K_2] = \frac{EI}{L}
\begin{array}{ccc}
D_4 & D_5 & D_6
\end{array}
\begin{bmatrix}
4 & 0 & 2 \\
0 & 0 & 0 \\
2 & 0 & 4
\end{bmatrix}
\begin{array}{l}
D_4 \\ D_5 \\ D_6
\end{array}
\tag{5.18m}$$

For element 3

$$[K_3] = \frac{EI}{L}[T_3]^T[k_{02}][T_3] - \frac{P}{L}[T_3]^T[k_{p3}][T_3] \tag{5.18n}$$

or

$$[K_3] = \frac{EI}{L}
\begin{array}{ccccc}
D_7 & D_8 & D_9 & D_5 & D_6
\end{array}
\begin{bmatrix}
\dfrac{12}{L^2} & 0 & \dfrac{6}{L} & -\dfrac{12}{L^2} & \dfrac{6}{L} \\[6pt]
0 & 0 & 0 & 0 & 0 \\[6pt]
\dfrac{6}{L} & 0 & 4 & -\dfrac{6}{L} & 2 \\[6pt]
-\dfrac{12}{L^2} & 0 & -\dfrac{6}{L} & \dfrac{12}{L^2} & -\dfrac{6}{L} \\[6pt]
\dfrac{6}{L} & 0 & 2 & -\dfrac{6}{L} & 4
\end{bmatrix}
- \frac{P}{L}
\begin{array}{ccccc}
D_7 & D_8 & D_9 & D_5 & D_6
\end{array}
\begin{bmatrix}
\dfrac{6}{5} & 0 & \dfrac{L}{10} & -\dfrac{6}{5} & \dfrac{L}{10} \\[6pt]
0 & 0 & 0 & 0 & 0 \\[6pt]
\dfrac{L}{10} & 0 & \dfrac{2L^2}{15} & -\dfrac{L}{10} & -\dfrac{L^2}{30} \\[6pt]
-\dfrac{6}{5} & 0 & -\dfrac{L}{10} & \dfrac{6}{5} & -\dfrac{L}{10} \\[6pt]
\dfrac{L}{10} & 0 & -\dfrac{L^2}{30} & -\dfrac{L}{10} & \dfrac{2L^2}{15}
\end{bmatrix}
\begin{array}{l}
D_7 \\ D_8 \\ D_9 \\ D_5 \\ D_6
\end{array}$$

$$\tag{5.18o}$$

Three element matrices in structure coordinates given by Eqs. (5.18k), (5.18m), and (5.18o) are assembled to give the structure stiffness matrix. The procedure to assemble a structure stiffness matrix can be seen in a text book on structural analysis using the matrix methods [6]. The total structure stiffness matrix can be written as

$$[K] = \frac{EI}{L}[K_0] - \frac{P}{L}[K_P]$$

(5.18p)

$$[K_o] = \begin{bmatrix}
\frac{12}{L^2} & 0 & \frac{6}{L} & \frac{6}{L} & -\frac{12}{L^2} & 0 & 0 & 0 & 0 \\
0 & 0 & 0 & 0 & 0 & 0 & 0 & 0 & 0 \\
\frac{6}{L} & 0 & 4 & 2 & -\frac{6}{L} & 0 & 0 & 0 & 0 \\
\frac{6}{L} & 0 & 2 & 4+4 & -\frac{6}{L} & 2 & 0 & 0 & 0 \\
-\frac{12}{L^2} & 0 & -\frac{6}{L} & -\frac{6}{L} & \frac{12}{L^2}+\frac{12}{L^2} & -\frac{6}{L} & -\frac{12}{L^2} & 0 & -\frac{6}{L} \\
0 & 0 & 0 & 2 & -\frac{6}{L} & 4+4 & \frac{6}{L} & 0 & 2 \\
0 & 0 & 0 & 0 & -\frac{12}{L^2} & \frac{6}{L} & \frac{12}{L^2} & 0 & \frac{6}{L} \\
0 & 0 & 0 & 0 & 0 & 0 & 0 & 0 & 0 \\
0 & 0 & 0 & 0 & -\frac{6}{L} & 2 & \frac{6}{L} & 0 & 4
\end{bmatrix}
\begin{matrix} D_1 \\ D_2 \\ D_3 \\ D_4 \\ D_5 \\ D_6 \\ D_7 \\ D_8 \\ D_9 \end{matrix}$$

(columns: $D_1\ D_2\ D_3\ D_4\ D_5\ D_6\ D_7\ D_8\ D_9$)

(5.18q)

$$[K_P] = \begin{bmatrix}
\frac{6}{5} & 0 & \frac{L}{10} & \frac{L}{10} & -\frac{6}{5} & 0 & 0 & 0 & 0 \\
0 & 0 & 0 & 0 & 0 & 0 & 0 & 0 & 0 \\
\frac{L}{10} & 0 & \frac{2L^2}{15} & -\frac{L^2}{30} & -\frac{L}{10} & 0 & 0 & 0 & 0 \\
\frac{L}{10} & 0 & -\frac{L^2}{30} & \frac{2L^2}{15} & -\frac{L}{10} & 0 & 0 & 0 & 0 \\
-\frac{6}{5} & 0 & -\frac{L}{10} & -\frac{L}{10} & \frac{6}{5}+\frac{6}{5} & -\frac{L}{10} & -\frac{6}{5} & 0 & -\frac{L}{10} \\
0 & 0 & 0 & 0 & -\frac{L}{10} & \frac{2L^2}{15} & \frac{L}{10} & 0 & -\frac{L^2}{30} \\
0 & 0 & 0 & 0 & -\frac{6}{5} & \frac{L}{10} & \frac{6}{5} & 0 & \frac{L}{10} \\
0 & 0 & 0 & 0 & 0 & 0 & 0 & 0 & 0 \\
0 & 0 & 0 & 0 & -\frac{L}{10} & -\frac{L^2}{30} & \frac{L}{10} & 0 & \frac{2L^2}{15}
\end{bmatrix}
\begin{matrix} D_1 \\ D_2 \\ D_3 \\ D_4 \\ D_5 \\ D_6 \\ D_7 \\ D_8 \\ D_9 \end{matrix}$$

(columns: $D_1\ D_2\ D_3\ D_4\ D_5\ D_6\ D_7\ D_8\ D_9$)

(5.18r)

The supports at the bottom of the frame are hinged, so the horizontal and vertical displacements and the corresponding structure degrees of freedoms D_1, D_2, D_7, and D_8 are zero there. Eliminate the rows and columns corresponding to these degrees of freedom from the structure stiffness matrix to form the reduced structure stiffness matrix given by

$$[K] = \frac{EI}{L} \begin{bmatrix} & D_3 & D_4 & D_5 & D_6 & D_9 \\ 4 & 2 & -\dfrac{6}{L} & 0 & 0 \\ 2 & 8 & -\dfrac{6}{L} & 2 & 0 \\ -\dfrac{6}{L} & -\dfrac{6}{L} & \dfrac{24}{L^2} & -\dfrac{6}{L} & -\dfrac{6}{L} \\ 0 & 2 & -\dfrac{6}{L} & 8 & 2 \\ 0 & 0 & -\dfrac{6}{L} & 2 & 4 \end{bmatrix} - \frac{P}{L} \begin{bmatrix} & D_3 & D_4 & D_5 & D_6 & D_9 \\ \dfrac{2L^2}{15} & -\dfrac{L^2}{30} & -\dfrac{L}{10} & 0 & 0 \\ -\dfrac{L^2}{30} & \dfrac{2L^2}{15} & -\dfrac{L}{10} & 0 & 0 \\ -\dfrac{L}{10} & -\dfrac{L}{10} & \dfrac{12}{5} & -\dfrac{L}{10} & -\dfrac{L}{10} \\ 0 & 0 & -\dfrac{L}{10} & \dfrac{2L^2}{15} & -\dfrac{L^2}{30} \\ 0 & 0 & -\dfrac{L}{10} & -\dfrac{L^2}{30} & \dfrac{2L^2}{15} \end{bmatrix} \begin{matrix} D_3 \\ D_4 \\ D_5 \\ D_6 \\ D_9 \end{matrix}$$

$$(5.18s)$$

The nodal forces from Eq. (5.17g) are given by

$$\begin{Bmatrix} Q_3 \\ Q_4 \\ Q_5 \\ Q_6 \\ Q_9 \end{Bmatrix} = \begin{Bmatrix} 0 \\ \dfrac{2PL}{9} \\ 0 \\ -\dfrac{2PL}{9} \\ 0 \end{Bmatrix} \qquad (5.18t)$$

where $2PL/9$ is the fixed end moment magnitude on both ends of the horizontal member due to load in Figure 5.22. The clockwise moment is taken as positive and the counterclockwise moment as negative. The equilibrium equation for the frame in the deformed position is given by

$$[K]\{D\} = \{Q\} \qquad (5.18u)$$

where $[K]$ and $\{Q\}$ are given by Eqs. (5.18s) and (5.18t) respectively, and $\{D\}$ is given by

$$\{D\} = \begin{Bmatrix} D3 \\ D4 \\ D5 \\ D6 \\ D9 \end{Bmatrix} \qquad (5.18v)$$

The deflected shape of the frame is antisymmetric, therefore, $D_3 = D_9$, and $D_4 = D_6$. The stiffness matrix in Eq. (5.18s) can be further reduced to

$$[K] = \frac{EI}{L} \begin{bmatrix} 4 & 2 & -\dfrac{6}{L} \\ 2 & 10 & -\dfrac{6}{L} \\ -\dfrac{12}{L} & -\dfrac{12}{L} & \dfrac{24}{L^2} \end{bmatrix} - \frac{P}{L} \begin{bmatrix} \dfrac{2L^2}{15} & -\dfrac{L^2}{30} & -\dfrac{L}{10} \\ -\dfrac{L^2}{30} & \dfrac{2L^2}{15} & -\dfrac{L}{10} \\ -\dfrac{L}{5} & -\dfrac{L}{5} & \dfrac{12}{5} \end{bmatrix} \tag{5.18t}$$

The structure nodal displacements and the corresponding structure forces are related by

$$\left\{ \frac{EI}{L} \begin{bmatrix} 4 & 2 & -\dfrac{6}{L} \\ 2 & 10 & -\dfrac{6}{L} \\ -\dfrac{12}{L} & -\dfrac{12}{L} & \dfrac{24}{L^2} \end{bmatrix} - \frac{P}{L} \begin{bmatrix} \dfrac{2L^2}{15} & -\dfrac{L^2}{30} & -\dfrac{L}{10} \\ -\dfrac{L^2}{30} & \dfrac{2L^2}{15} & -\dfrac{L}{10} \\ -\dfrac{L}{5} & -\dfrac{L}{5} & \dfrac{12}{5} \end{bmatrix} \right\} \begin{Bmatrix} D_3 \\ D_4 \\ D_5 \end{Bmatrix} = \begin{Bmatrix} 0 \\ \dfrac{2PL}{9} \\ 0 \end{Bmatrix} \tag{5.18u}$$

At the critical load, there is large increase in $\{D\}$ for a small increase in $\{Q\}$ because the bending stiffness of the member is considerably reduced. We can write Eq. (5.18u) as

$$\{D\} = [K]^{-1}\{Q\} \tag{5.18v}$$

The inverse of the stiffness matrix, $\{K\}^{-1}$, is determined by dividing the adjoint matrix of $[K]$ by its determinant. At the critical load found by bifurcation theory, the bending stiffness of the structure vanishes, so the determinant of $[K]$ should be zero. This leads to an eigenvalue problem, the solution of which gives the critical load as the minimum eigenvalue. The structure stiffness matrix from Eq. (5.18t) can be written as

$$[K] = \begin{bmatrix} \dfrac{4EI}{L} - \dfrac{2PL}{15} & \dfrac{2EI}{L} + \dfrac{PL}{30} & -\dfrac{6EI}{L^2} + \dfrac{P}{10} \\ \dfrac{2EI}{L} + \dfrac{PL}{30} & \dfrac{10EI}{L} - \dfrac{2PL}{15} & -\dfrac{6EI}{L^2} + \dfrac{P}{10} \\ -\dfrac{12EI}{L^2} + \dfrac{P}{5} & -\dfrac{12EI}{L^2} + \dfrac{P}{5} & \dfrac{24EI}{L^3} - \dfrac{12P}{5L} \end{bmatrix} \tag{5.18w}$$

The determinant of the structure stiffness matrix in Eq. (5.18w) is equated to zero, thus

$$|K| = \frac{EI}{L} \begin{vmatrix} 4 - 4\lambda & 2 + \lambda & \dfrac{-6 + 3\lambda}{L} \\ 2 + \lambda & 10 - 4\lambda & \dfrac{-6 + 3\lambda}{L} \\ \dfrac{-12 + 6\lambda}{L} & \dfrac{-12 + 6\lambda}{L} & \dfrac{24 - 72\lambda}{L^2} \end{vmatrix} = 0, \text{ where } \lambda = \frac{PL^2}{30EI}$$

or

$$25\lambda^3 - 105\lambda^2 + 72\lambda - 4 = 0 \tag{5.18x}$$

or

$$\lambda = 0.0609$$

or

$$P_{cr} = \frac{1.827EI}{L^2} \tag{5.18y}$$

That is almost the same as $P_{cr} = \dfrac{1.778EI}{L^2}$ in Eq. (5.12o).

5.4.5 Nonlinear Geometric or Large Deflection Analysis (Second-Order Elastic Analysis)

The elastic critical load for the frame can be found by conducting the geometric nonlinear analysis where the loads on the frame are increased in steps and the stiffness matrix is updated at the end of each step to account for the deformation that occurred in the previous step. Thus, the solution of a nonlinear problem is obtained by a series of linear analyses. In one of the approaches the load is applied in a number of increments and within each increment an iterative technique is applied to obtain the solution. It is assumed that during a particular iteration cycle the structure behaves linearly. The deformed configuration of the structure at the end of each cycle is used to analyze the structure in the next cycle. A second-order analysis described here will generate a load-deflection curve, the peak point on the curve where the tangent is horizontal gives the critical load. The incremental-iterative procedure is described below.

1. The element stiffness matrices are formulated by Eqs. (5.15x), (5.15y), and (5.15z) in the element coordinate systems, then they are converted to the structure coordinate system and assembled to form the structure stiffness matrix. For the first load increment and first iteration, the stiffness matrix is formed on the basis of undeformed geometry.
2. To describe the procedure in general, let us say that the equilibrium and kinematic states of the frame are known at the end of load step (*i*-1), and it is desired to find the state of the structure at the end of load step *i*.
3. The incremental displacements at the end of first iteration of the load step *i* are given by

$$[K_i^1]\{\Delta D_i^1\} = \{\Delta Q_i\} \tag{5.19a}$$

or

$$\{\Delta D_i^1\} = [K_i^1]^{-1}\{\Delta Q_i\} \tag{5.19b}$$

where

$[K_i^1]$ = Stiffness matrix formed for the first iteration of load step *i*, and based on the equilibrium and
kinematic states at the end of load step (*i*-1)

$\{\Delta D_i^1\}$ = Incremental structural nodal displacement vector calculated in the first iteration of the load
step *i*

$\{\Delta Q_i\}$ = Load increment at the load step *i*

4. The structural nodal displacement vector is now given by

$$\{D_i^1\} = \{D_{i-1}\} + \{\Delta D_i^1\} \tag{5.19c}$$

where
$\{D_i^1\}$ = Structural nodal displacement vector after first iteration of load step i
$\{D_{i-1}\}$ = Structure nodal displacement vector at the beginning of load step i
In Figure 5.23a the equilibrium iterations are performed by the Newton-Raphson method and are shown schematically for a one degree of freedom system.

5. Extract the element end displacement vector $\{d_i\}$ from $\{D_i^1\}$ for each element.

6. Calculate the element axial displacement u and element end rotations θ_A and θ_B from Figure 5.24 as follows:

$$\theta_A = \alpha + d_3 - \beta \tag{5.19d}$$

$$\theta_B = \alpha + d_6 - \beta \tag{5.19e}$$

$$u = L_f - L_o \tag{5.19f}$$

where
$\begin{bmatrix} d_1 & d_2 & d_3 & d_4 & d_5 & d_6 \end{bmatrix}^T$ = Element end displacement vector
α = Initial slope of the underformed element with reference to the horizontal
β = Slope of the deformed element with reference to the horizontal
L_o = Initial length of the element in the undeformed position
L_f = Length of the element in deformed position
u = Axial deformation in the element

$$\beta = \tan^{-1} \frac{d_5 + L_o \sin \alpha - d_2}{d_4 + L_o \cos \alpha - d_1} \tag{5.19g}$$

$$L_f = \sqrt{(d_4 + L_o \cos \alpha - d_1)^2 + (d_5 + L_o \sin \alpha - d_2)^2} \tag{5.19h}$$

7. Calculate the axial force P, and end moments M_A and M_B for each element from the slope deflection equations [3]

$$M_A = \frac{EI}{L_f}(s_{ii}\theta_A + s_{ij}\theta_B) + M_{FA} \tag{5.19i}$$

$$M_B = \frac{EI}{L_f}(s_{ij}\theta_A + s_{jj}\theta_B) + M_{FB} \tag{5.19j}$$

$$P = \frac{EA}{L_f}u \tag{5.19k}$$

where EI = Stiffness of the element; A is the area of cross-section of the element; s_{ii} etc. are the slope deflection coefficients for a beam column from Appendix A; M_{FA} and M_{FB} are the fixed end moments at the ends A and B of the element respectively, which are zero if there is no in span load in the element.

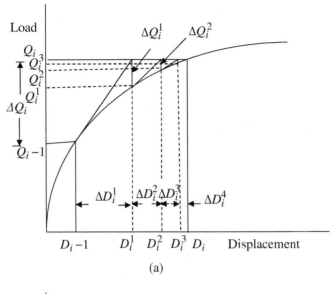

(a)

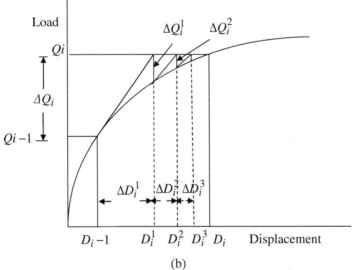

(b)

Figure 5.23 Second-order frame analysis with the Newton-Raphson iteration method: (a) Newton-Raphson method; (b) Modified Newton-Raphson method.

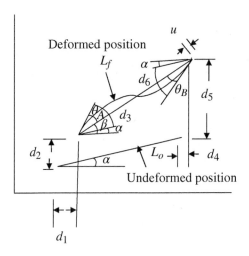

Figure 5.24 Large deflection analysis kinematics relations.

8. For each element evaluate the end forces $\{q\}$ in the structure coordinates from the internal forces P, M_A, and M_B as shown in Figure 5.25. These forces are given by Eq. (5.19l).

$$
\begin{Bmatrix} q_1 \\ q_2 \\ q_3 \\ q_4 \\ q_5 \\ q_6 \end{Bmatrix} = \begin{bmatrix} -\cos\beta & \dfrac{\sin\beta}{L_f} & \dfrac{\sin\beta}{L_f} \\[2mm] -\sin\beta & -\dfrac{\cos\beta}{L_f} & -\dfrac{\cos\beta}{L_f} \\[2mm] 0 & 1 & 0 \\[2mm] \cos\beta & -\dfrac{\sin\beta}{L_f} & -\dfrac{\sin\beta}{L_f} \\[2mm] \sin\beta & \dfrac{\cos\beta}{L_f} & \dfrac{\cos\beta}{L_f} \\[2mm] 0 & 0 & 1 \end{bmatrix} \begin{Bmatrix} P \\ M_A \\ M_B \end{Bmatrix}
$$

(5.19l)

In Eq. (5.19l) clockwise moments are positive and the axial tension is taken as positive. The element end forces in structure coordinates are calculated for all the elements and assembled to get the internal force vector, $\{Q_i^1\}$, at the end of first iteration of the load step i as shown in Figure 5.23a.

9. The external force vector $\{Q_i\}$ at the end of load step i is

$$\{Q_i\} = \{Q_{i-1}\} + \Delta Q_i$$

(5.19m)

10. The difference between the external force vector and the internal force vector gives the unbalanced force ΔQ_i^1 after the first iteration of load step i as

$$\{\Delta Q_i^1\} = \{Q_i\} - \{Q_i^1\}$$

(5.19n)

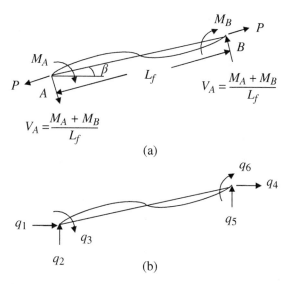

Figure 5.25 Internal forces in the element: (a) Axial force and the end moments in the element; (b) Internal nodal forces in the element in the structure coordinates.

11. Using the current value of the axial force, update the stiffness matrix $[k]$ for each element and form the updated structure stiffness matrix $[K_i^2]$ for the second iteration of load step i. Calculate the incremental displacement vector $\{\Delta D_i^2\}$ as follows:

$$\{\Delta D_i^2\} = [K_i^2]^{-1}\{\Delta Q_i^1\} \tag{5.19o}$$

The formation and inversion of the stiffness matrix require considerable computer time, sometimes the same structure stiffness matrix is used for all iterations in a load increment. The stiffness matrix is changed at the end of each load step or when there is difficulty in convergence during iteration within a load step. This procedure is called modified Newton-Raphson method, and is shown schematically for one degree of freedom system in Figure 5.23b.

12. Form the updated structure nodal displacement vector $\{D_i^2\}$ as follows:

$$\{D_i^2\} = \{D_{i-1}\} + \sum_{j=1}^{2} \Delta D_i^j \tag{5.19p}$$

13. Extract the element end displacement vector $\{d_i\}$ from $\{D_i^2\}$ for each element. Find new values of u, θ_A, θ_B; and then find P, M_A, and M_B for the second iteration of the load step i as was found for first iteration for all the elements.

14. Find the new element end forces in the structure coordinates and form the structure internal force vector $\{Q_i^2\}$ for the second iteration of the load step i as was done in step 8.

15. The new unbalanced force at the end of second iteration of load step i is given by

$$\{\Delta Q_i^2\} = \{Q_i\} - \{Q_i^2\} \tag{5.19q}$$

16. Repeat the process from steps 11–15 until the solution converges. The convergence is said to occur when the unbalanced force $\{\Delta Q_i^j\}$ at the end of the j-th iteration is negligible, i.e. it falls within the allowable tolerance.

17. The structure nodal displacements at the end of load step i if the convergence occurs after n number of iterations, are given by

$$\{D_i\} = \{D_{i-1}\} + \sum_{j=1}^{n} \Delta D_i^j$$

18. Now increase the load by another load step ΔQ_{i+1}, and repeat the procedure from steps 3–17 for the load step $(i+1)$.

19. This way, a complete nonlinear load displacement curve can be drawn and the critical load of the frame can be found.

5.5 Design of Frame Members

In the design of frames, the stability of the entire frame is to be considered, including the stability of individual members. The critical load for the frame can be found by carrying out a stability analysis of the entire frame. From this the loads in the individual compression members at the time of buckling of the frame can be found. To simplify the procedure for the design of individual compression members, a simple approximate approach is used. In this approach the effective lengths of various compression members in the framed structure are found by the Julian and Lawrence (1959) [7] method. The effective length is found by considering the effect of members that frame directly into the compression member, whereas the effect of members not directly connected to the compression member is neglected. The method is based on the following assumptions:

1. All members in the frame are prismatic and their behavior is elastic.
2. Axial compressive force in the beams is neglected.
3. All joints are rigid.
4. The stiffness parameter $L\sqrt{\frac{P}{EI}}$ is constant for all the columns in a story.
5. All columns buckle simultaneously in a story.
6. At a joint the restraint provided by beams is distributed to the columns above and below the joint in proportion to their $\frac{I}{L}$ values.
7. For braced frames, rotations at the opposite ends of the restraining beams are equal in magnitude and opposite in direction (single curvature bending in beams).
8. For unbraced frames, rotations at the opposite ends of the restraining beams are equal in magnitude and direction (double curvature bending in beams). The inflection point is in the middle of a beam span if only lateral loads are acting.

The Julian and Lawrence (1959) [7] approach produces alignment charts and is described for braced and unbraced frames.

5.5.1 Braced Frames (Sidesway Inhibited)

It is intended to find the effective length factor K for the column BC in a braced frame shown partly in Figure 5.26. When the frame buckles, the moments are developed at the ends of the members. These moments for the columns AB, BC, and CD in Figure 5.26 are given by the slope deflection equations as follows:

$$(M_B)_{AB} = \left(\frac{EI}{L}\right)_{AB}(k_{ii}\theta_B + k_{ij}\theta_C) \tag{5.20a}$$

$$(M_B)_{BC} = \left(\frac{EI}{L}\right)_{BC}(k_{ii}\theta_B + k_{ij}\theta_C) \tag{5.20b}$$

$$(M_C)_{BC} = \left(\frac{EI}{L}\right)_{BC}(k_{ij}\theta_B + k_{jj}\theta_C) \tag{5.20c}$$

$$(M_C)_{CD} = \left(\frac{EI}{L}\right)_{CD}(k_{ii}\theta_C + k_{ij}\theta_B) \tag{5.20d}$$

where E is the modulus of elasticity of the material of columns, I and L are the moment of inertia and unbraced length of the columns in the plane of buckling.

For horizontal beams, the axial force is neglected, therefore, $k_{ii} = k_{jj} = 4$, and $k_{ij} = k_{ji} = 2$.

The slope deflection equations for the horizontal beams BE, BF, CG, and CH in Figure 5.26 are as follows:

$$(M_B)_{BE} = \left(\frac{EI}{L}\right)_{BE}(4\theta_B - 2\theta_B) = \left(\frac{2EI}{L}\right)_{BE}\theta_B \tag{5.20e}$$

$$(M_B)_{BF} = \left(\frac{EI}{L}\right)_{BF}(4\theta_B - 2\theta_B) = \left(\frac{2EI}{L}\right)_{BF}\theta_B \tag{5.20f}$$

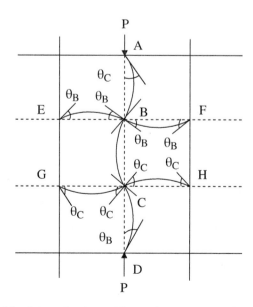

Figure 5.26 Column in a braced frame along with adjoining members.

$$(M_C)_{CG} = \left(\frac{EI}{L}\right)_{CG}(4\theta_C - 2\theta_C) = \left(\frac{2EI}{L}\right)_{CG}\theta_C \tag{5.20g}$$

$$(M_C)_{CH} = \left(\frac{EI}{L}\right)_{CH}(4\theta_C - 2\theta_C) = \left(\frac{2EI}{L}\right)_{CH}\theta_C \tag{5.20h}$$

where E is the modulus of elasticity of the material of beams, I and L are the moment of inertia and unbraced length of the beams in the plane of buckling.

Joint equilibrium at B gives

$$M_{BA} + M_{BC} + M_{BE} + M_{BF} = 0 \tag{5.20i}$$

Substituting from Eqs. (5.20a), (5.20e), and (5.20f) we get

$$M_{BC} = -\left(\frac{EI}{L}\right)_{AB}(k_{ii}\theta_B + k_{ij}\theta_C) - \left(\frac{2EI}{L}\right)_{BE}\theta_B - \left(\frac{2EI}{L}\right)_{BF}\theta_B \tag{5.20j}$$

Equating the right-hand terms in Eqs. (5.20b) and (5.20j) we have

$$\left[2\frac{\sum\limits_B\left(\frac{EI}{L}\right)_b}{\sum\limits_B\left(\frac{EI}{L}\right)_c} + k_{ii}\right]\theta_B + k_{ij}\theta_C = 0 \tag{5.20k}$$

where

$$\sum_B\left(\frac{EI}{L}\right)_b = \left(\frac{EI}{L}\right)_{BE} + \left(\frac{EI}{L}\right)_{BF} = \text{Sum of the stiffness of beams meeting at the joint } B, \text{ and}$$

$$\sum_B\left(\frac{EI}{L}\right)_c = \left(\frac{EI}{L}\right)_{AB} + \left(\frac{EI}{L}\right)_{BC} = \text{Sum of the stiffness of columns meeting at the joint } B.$$

Let

$$G_B = \frac{\sum\limits_B\left(\frac{EI}{L}\right)_c}{\sum\limits_B\left(\frac{EI}{L}\right)_b} \tag{5.20l}$$

Joint equilibrium at C gives

$$M_{CB} + M_{CD} + M_{CG} + M_{CH} = 0 \tag{5.20m}$$

Substituting from Eqs. (5.20d), (5.20g), and (5.20h), we get

$$M_{CB} = -\left(\frac{EI}{L}\right)_{CD}(k_{ii}\theta_C + k_{ij}\theta_B) - \left(\frac{2EI}{L}\right)_{CG}\theta_C - \left(\frac{2EI}{L}\right)_{CH}\theta_C \tag{5.20n}$$

Equating the right-hand terms in Eqs. (5.20c) and (5.20n), we have

$$k_{ij}\theta_B + \left[2\frac{\sum\limits_C\left(\frac{EI}{L}\right)_b}{\sum\limits_C\left(\frac{EI}{L}\right)_c} + k_{ii}\right]\theta_C = 0 \tag{5.20o}$$

where

$$\sum_C \left(\frac{EI}{L}\right)_b = \left(\frac{EI}{L}\right)_{CG} + \left(\frac{EI}{L}\right)_{CH} = \text{Sum of the stiffness of beams meeting at the joint } C, \text{and}$$

$$\sum_C \left(\frac{EI}{L}\right)_c = \left(\frac{EI}{L}\right)_{CB} + \left(\frac{EI}{L}\right)_{CD} = \text{Sum of the stiffness of columns meeting at the joint } C.$$

Let

$$G_C = \frac{\sum_C \left(\frac{EI}{L}\right)_c}{\sum_C \left(\frac{EI}{L}\right)_b} \tag{5.20p}$$

Equations (5.20k) and (5.20o) can be written as

$$\left(k_{ii} + \frac{2}{G_B}\right)\theta_B + k_{ij}\theta_C = 0$$

$$k_{ij}\theta_B + \left(k_{ii} + \frac{2}{G_C}\right)\theta_C = 0$$

or

$$\begin{bmatrix} k_{ii} + \dfrac{2}{G_B} & k_{ij} \\ k_{ij} & k_{ii} + \dfrac{2}{G_C} \end{bmatrix} \begin{Bmatrix} \theta_B \\ \theta_C \end{Bmatrix} = \begin{Bmatrix} 0 \\ 0 \end{Bmatrix} \tag{5.20q}$$

For a nontrivial solution the determinant of the coefficient matrix should be zero. Therefore,

$$\begin{vmatrix} k_{ii} + \dfrac{2}{G_B} & k_{ij} \\ k_{ij} & k_{ii} + \dfrac{2}{G_C} \end{vmatrix} = 0 \tag{5.20r}$$

The slope deflection coefficients are given before by Eqs. (4.21f) and (4.21g) as

$$k_{ii} = k_{jj} = \frac{12\psi(u)}{4\psi^2(u) - \phi^2(u)} \tag{4.21f}$$

$$k_{ij} = k_{ji} = \frac{6\phi(u)}{4\psi^2(u) - \phi^2(u)} \tag{4.21g}$$

and the parameters $\psi(u)$ and $\phi(u)$ are given before by Eqs. (4.8g) and (4.8h) as

$$\psi(u) = \frac{3}{2u}\left[\frac{1}{2u} - \frac{1}{\tan 2u}\right] \tag{4.8g}$$

$$\phi(u) = \frac{3}{u}\left[\frac{1}{\sin 2u} - \frac{1}{2u}\right] \tag{4.8h}$$

where

$$u = \frac{kL}{2}$$

or

$$2u = kL = \sqrt{\frac{P}{EI}}L = \sqrt{\frac{PL^2}{EI\pi^2}}\,\pi = \pi\sqrt{\frac{P}{P_e}} = \frac{\pi}{K}$$

Because at buckling, $P = \dfrac{\pi^2 EI}{(KL)^2}$, where K is the effective length factor for the column under consideration.

Therefore,

$$k_{ii} = \frac{\left(\dfrac{\tan\dfrac{\pi}{K} - \dfrac{\pi}{K}}{\dfrac{\pi^2}{K^2}\tan\dfrac{\pi}{K}}\right)}{\left(\dfrac{\tan\dfrac{\pi}{K} - \dfrac{\pi}{K}}{\dfrac{\pi^2}{K^2}\tan\dfrac{\pi}{K}}\right)^2 - \left(\dfrac{\dfrac{\pi}{K} - \sin\dfrac{\pi}{K}}{\dfrac{\pi^2}{K^2}\sin\dfrac{\pi}{K}}\right)^2}$$

(5.20s)

or

$$k_{ii} = \frac{\dfrac{\pi}{K}\sin\dfrac{\pi}{K} - \left(\dfrac{\pi}{K}\right)^2\cos\dfrac{\pi}{K}}{2 - 2\cos\dfrac{\pi}{K} - \dfrac{\pi}{K}\sin\dfrac{\pi}{K}}$$

(5.20t)

Similarly,

$$k_{ij} = \frac{\left(\dfrac{\pi}{K}\right)^2 - \dfrac{\pi}{K}\sin\dfrac{\pi}{K}}{2 - 2\cos\dfrac{\pi}{K} - \dfrac{\pi}{K}\sin\dfrac{\pi}{K}}$$

(5.20u)

Equation (5.20r) can be written as

$$k_{ii}^2 + 2k_{ii}\frac{G_B + G_C}{G_B G_C} + \frac{4}{G_B G_C} - k_{ij}^2 = 0$$

(5.20v)

Substitute Eqs. (5.20t) and (5.20u) into Eq. (5.20v), and we obtain

$$\left(\frac{\dfrac{\pi}{K}\sin\dfrac{\pi}{K} - \left(\dfrac{\pi}{K}\right)^2\cos\dfrac{\pi}{K}}{2 - 2\cos\dfrac{\pi}{K} - \dfrac{\pi}{K}\sin\dfrac{\pi}{K}}\right)^2 + 2\left(\frac{\dfrac{\pi}{K}\sin\dfrac{\pi}{K} - \left(\dfrac{\pi}{K}\right)^2\cos\dfrac{\pi}{K}}{2 - 2\cos\dfrac{\pi}{K} - \dfrac{\pi}{K}\sin\dfrac{\pi}{K}}\right)\frac{G_B + G_C}{G_B G_C}$$

$$+ \frac{4}{G_B G_C} - \left(\frac{\left(\dfrac{\pi}{K}\right)^2 - \dfrac{\pi}{K}\sin\dfrac{\pi}{K}}{2 - 2\cos\dfrac{\pi}{K} - \dfrac{\pi}{K}\sin\dfrac{\pi}{K}}\right)^2 = 0$$

(5.20w)

or

$$\frac{\left(\frac{\pi}{K}\right)^3 \sin\frac{\pi}{K}}{2 - 2\cos\frac{\pi}{K} - \frac{\pi}{K}\sin\frac{\pi}{K}} + 2\left(\frac{\frac{\pi}{K}\sin\frac{\pi}{K} - \left(\frac{\pi}{K}\right)^2\cos\frac{\pi}{K}}{2 - 2\cos\frac{\pi}{K} - \frac{\pi}{K}\sin\frac{\pi}{K}}\right)\frac{G_B + G_C}{G_B G_C} + \frac{4}{G_B G_C} = 0$$

or

$$\frac{G_B G_C}{4}\left(\frac{\pi}{K}\right)^2 + \frac{1}{2}\left(1 - \frac{\left(\frac{\pi}{K}\right)^2\cos\frac{\pi}{K}}{\frac{\pi}{K}\sin\frac{\pi}{K}}\right)(G_B + G_C) + \frac{2}{\frac{\pi}{K}\sin\frac{\pi}{K}} - \frac{2\cos\frac{\pi}{K}}{\frac{\pi}{K}\sin\frac{\pi}{K}} - 1 = 0$$

or

$$\frac{G_B G_C}{4}\left(\frac{\pi}{K}\right)^2 + \left(\frac{G_B + G_C}{2}\right)\left(1 - \frac{\frac{\pi}{K}}{\tan\frac{\pi}{K}}\right) + \frac{2}{\frac{\pi}{K}}\tan\left(\frac{\pi}{2K}\right) - 1 = 0 \tag{5.20x}$$

Equation (5.20x) is shown in the form of a nomograph in Figure 5.27, where the effective length factor values K for a column vary from 0.5 to 1.0. If the column end beams are much stiffer than the column, then the K factors are lower. $K = 0.5$ corresponds to the fixed end condition and $K = 1.0$ means the end of the column is pin connected. The use of the nomograph is recommended by the American Institute of Steel Construction (AISC) [8]. The subscripts A and B refer to the joints at the two ends of the column section being considered. G is defined assuming the modulus of elasticity E to be constant for all the columns and beams as follows:

$$G = \frac{\sum(I_c/L_c)}{\sum(I_g/L_g)} \tag{5.20y}$$

where Σ indicates a summation of all members rigidly connected to the joint and lying on the plane in which buckling of the column is being considered. I_c is the moment of inertia and L_c is the unsupported length of a column section, whereas I_g is the moment of inertia and L_g is the unsupported length of a girder or other restraining member. I_c and I_g are taken about axes perpendicular to the plane of buckling being considered.

The Structural Stability Research Council (SSRC) makes several recommendations concerning the use of the nomograph:

For pinned columns at the footing, theoretically, $G = \infty$, but it could be taken $= 10$ for practical designs. For rigid connections of columns to a properly designed footing, theoretically, G approaches zero, but it may be taken $= 1.0$ as no connection is perfectly rigid.

For the design of a column in a frame, compute G at each end of the column and label the values G_A and G_B as desired. Then draw a straight line on the nomograph between the G_A and G_B values and read K where the line intersects the center K scale. For example, if $G_A = 0.8$, $G_B = 2.0$, then $K = 0.8$ as shown by the intersection of dotted line with the center K scale.

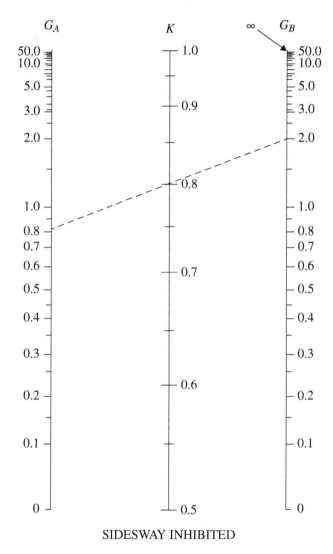

Figure 5.27 Nomograph for effective length factor K of columns in braced frames.

5.5.2 Unbraced Frames (Sidesway Not Inhibited)

Now we examine the effective length factor K for laterally unbraced frames. In these frames the two ends of columns move laterally relative to each other, i.e. the columns are subjected to sidesway. We wish to determine the effective length factor K for the column BC shown in Figure 5.28. The method is based on the same assumptions that were used for the braced frames except that in the assumption "7" rotations at the opposite ends of the restraining beams are equal in magnitude and same in direction (double curvature bending in horizontal beams).

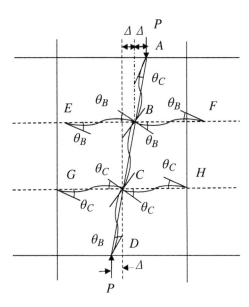

Figure 5.28 Column in unbraced frame along with adjoining members.

The end moments for the columns *AB*, *BC*, and *CD* are given by the slope deflection equations as follows:

$$(M_B)_{AB} = \left(\frac{EI}{L}\right)_{AB}\left[k_{ii}\theta_B + k_{ij}\theta_C - (k_{ii} + k_{ij})\frac{\Delta}{L_{AB}}\right] \tag{5.21a}$$

$$(M_B)_{BC} = \left(\frac{EI}{L}\right)_{BC}\left[k_{ii}\theta_B + k_{ij}\theta_C - (k_{ii} + k_{ij})\frac{\Delta}{L_{BC}}\right] \tag{5.21b}$$

$$(M_C)_{BC} = \left(\frac{EI}{L}\right)_{BC}\left[k_{ij}\theta_B + k_{jj}\theta_C - (k_{jj} + k_{ij})\frac{\Delta}{L_{BC}}\right] \tag{5.21c}$$

$$(M_C)_{CD} = \left(\frac{EI}{L}\right)_{CD}\left[k_{ii}\theta_C + k_{ij}\theta_B - (k_{ii} + k_{ij})\frac{\Delta}{L_{CD}}\right] \tag{5.21d}$$

where E is the modulus of elasticity of the materials of columns, I and L are the moment of inertia and unbraced length of the columns in the plane of buckling.

For horizontal beams, the axial force is neglected, therefore, $k_{ii} = k_{jj} = 4$, and $k_{ij} = k_{ji} = 2$.

The slope deflection equations for the horizontal beams BE, BF, CG, and CH in Figure 5.28 are as follows:

$$(M_B)_{BE} = \left(\frac{EI}{L}\right)_{BE}(4\theta_B + 2\theta_B) = \left(\frac{6EI}{L}\right)_{BE}\theta_B \tag{5.21e}$$

$$(M_B)_{BF} = \left(\frac{EI}{L}\right)_{BF}(4\theta_B + 2\theta_B) = \left(\frac{6EI}{L}\right)_{BF}\theta_B \tag{5.21f}$$

$$(M_C)_{CG} = \left(\frac{EI}{L}\right)_{CG}(4\theta_C + 2\theta_C) = \left(\frac{6EI}{L}\right)_{CG}\theta_C \tag{5.21g}$$

$$(M_C)_{CH} = \left(\frac{EI}{L}\right)_{CH}(4\theta_C + 2\theta_C) = \left(\frac{6EI}{L}\right)_{CH}\theta_C \tag{5.21h}$$

where E is the modulus of elasticity of the material of beams, I and L are the moment of inertia and unbraced length of the beams in the plane of buckling.

Joint equilibrium at B gives

$$M_{BA} + M_{BC} + M_{BE} + M_{BF} = 0 \tag{5.21i}$$

Substitute Eqs. (5.21a), (5.21e), and (5.21f) into Eq. (5.21i) to get

$$M_{BC} = -\left(\frac{EI}{L}\right)_{AB}\left[k_{ii}\theta_B + k_{ij}\theta_C - (k_{ii} + k_{ij})\frac{\Delta}{L_{AB}}\right] - \left(\frac{6EI}{L}\right)_{BE}\theta_B - \left(\frac{6EI}{L}\right)_{BF}\theta_B \tag{5.21j}$$

Equate the right-hand sides in Eqs. (5.21b) and (5.21j), and assume $L_{AB} = L_{BC}$, to get

$$\left[6\frac{\sum_B\left(\frac{EI}{L}\right)_b}{\sum_B\left(\frac{EI}{L}\right)_c} + k_{ii}\right]\theta_B + k_{ij}\theta_C - (k_{ii} + k_{ij})\frac{\Delta}{L_{AB}} = 0 \tag{5.21k}$$

where

$$\sum_B\left(\frac{EI}{L}\right)_b = \left(\frac{EI}{L}\right)_{BE} + \left(\frac{EI}{L}\right)_{BF} = \text{Sum of the stiffness of beams meeting at the joint } B, \text{ and}$$

$$\sum_B\left(\frac{EI}{L}\right)_c = \left(\frac{EI}{L}\right)_{AB} + \left(\frac{EI}{L}\right)_{BC} = \text{Sum of the stiffness of columns meeting at the joint } B.$$

Let

$$G_B = \frac{\sum_B\left(\frac{EI}{L}\right)_c}{\sum_B\left(\frac{EI}{L}\right)_b} \tag{5.21l}$$

Joint equilibrium at C gives

$$M_{CB} + M_{CD} + M_{CG} + M_{CH} = 0 \tag{5.21m}$$

Substituting from Eqs. (5.21d), (5.21g), and (5.21h) we get

$$M_{CB} = -\left(\frac{EI}{L}\right)_{CD}\left[k_{ii}\theta_C + k_{ij}\theta_B - (k_{ii} + k_{ij})\frac{\Delta}{L_{CD}}\right] - \left(\frac{6EI}{L}\right)_{CG}\theta_C - \left(\frac{6EI}{L}\right)_{CH}\theta_C \tag{5.21n}$$

Equate the right-hand terms in Eqs. (5.21c) and (5.21n), and assume $L_{BC} = L_{CD}$, to get

$$k_{ij}\theta_B + \left[6\frac{\sum_C\left(\frac{EI}{L}\right)_b}{\sum_C\left(\frac{EI}{L}\right)_c} + k_{ii}\right]\theta_C - (k_{ii} + k_{ij})\frac{\Delta}{L_{BC}} = 0 \tag{5.21o}$$

where

$$\sum_C \left(\frac{EI}{L}\right)_b = \left(\frac{EI}{L}\right)_{CG} + \left(\frac{EI}{L}\right)_{CH} = \text{Sum of the stiffness of beams meeting at the joint } C, \text{and}$$

$$\sum_C \left(\frac{EI}{L}\right)_c = \left(\frac{EI}{L}\right)_{CB} + \left(\frac{EI}{L}\right)_{CD} = \text{Sum of the stiffness of columns meeting at the joint } C.$$

$$G_C = \frac{\sum\limits_C \left(\frac{EI}{L}\right)_c}{\sum\limits_C \left(\frac{EI}{L}\right)_b} \tag{5.20p}$$

Member equilibrium of column BC in Figure 5.29 gives

$$M_{BC} + M_{CB} + P\Delta - VL_{BC} = 0$$

Since there is no external horizontal force present, $V = 0.$ or

$$M_{BC} + M_{CB} + P\Delta = 0 \tag{5.21p}$$

Equation (5.21b) can be written as

$$\frac{M_{BC}}{\left(\frac{EI}{L}\right)_{BC}} = k_{ii}\theta_B + k_{ij}\theta_C - (k_{ii} + k_{ij})\frac{\Delta}{L_{BC}}$$

and substitute in Eq. (5.21j) to get

$$M_{BC} = -\left(\frac{EI}{L}\right)_{AB}\left[\frac{M_{BC}}{\left(\frac{EI}{L}\right)_{BC}}\right] - 6\left[\left(\frac{EI}{L}\right)_{BE} + \left(\frac{EI}{L}\right)_{BF}\right]\theta_B$$

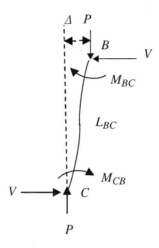

Figure 5.29 Member equilibrium for column *BC*.

or

$$M_{BC} = -6 \frac{\sum\limits_{B} \left(\frac{EI}{L}\right)_b}{\sum\limits_{B} \left(\frac{EI}{L}\right)_c} \left(\frac{EI}{L}\right)_{BC} \theta_B \tag{5.21q}$$

Similarly considering equations at the joint C, we get

$$M_{CB} = -6 \frac{\sum\limits_{C} \left(\frac{EI}{L}\right)_b}{\sum\limits_{6} \left(\frac{EI}{L}\right)_c} \left(\frac{EI}{L}\right)_{BC} \theta_C \tag{5.21r}$$

$$\sqrt{\frac{P}{EI}} = k$$

Therefore,

$$P\Delta = \left(\frac{EI}{L}\right)_{BC} (kL)_{BC}^2 \frac{\Delta}{L_{BC}} \tag{5.21s}$$

Substitute Eqs. (5.21q), (5.21r), and (5.21s) into Eq. (5.21p), and we obtain

$$-6 \frac{\sum\limits_{B} \left(\frac{EI}{L}\right)_b}{\sum\limits_{B} \left(\frac{EI}{L}\right)_c} \left(\frac{EI}{L}\right)_{BC} \theta_B - 6 \frac{\sum\limits_{C} \left(\frac{EI}{L}\right)_b}{\sum\limits_{6} \left(\frac{EI}{L}\right)_c} \left(\frac{EI}{L}\right)_{BC} \theta_C + \left(\frac{EI}{L}\right)_{BC} (kL)_{BC}^2 \frac{\Delta}{L_{BC}} = 0$$

or

$$-\frac{6}{G_B} \theta_B - \frac{6}{G_C} \theta_C + (kL)_{BC}^2 \frac{\Delta}{L_{BC}} = 0 \tag{5.21t}$$

Equations (5.21k), (5.21o), and (5.21t) can be written in matrix form as

$$\begin{bmatrix} \left(k_{ii} + \dfrac{6}{G_B}\right) & k_{ij} & -(k_{ii} + k_{ij}) \\[2ex] k_{ij} & \left(k_{ii} + \dfrac{6}{G_C}\right) & -(k_{ii} + k_{ij}) \\[2ex] -\dfrac{6}{G_B} & -\dfrac{6}{G_C BC} & (kL)_{BC}^2 \end{bmatrix} \begin{Bmatrix} \theta_B \\[1ex] \theta_C \\[1ex] \dfrac{\Delta}{L_{BC}} \end{Bmatrix} = \begin{Bmatrix} 0 \\ 0 \\ 0 \end{Bmatrix} \tag{5.21u}$$

For the nontrivial solution for θ_B, θ_C, and $\frac{\Delta}{L_{BC}}$ the determinant of the coefficient matrix should be zero, or

$$\begin{vmatrix} \left(k_{ii} + \dfrac{6}{G_B}\right) & k_{ij} & -(k_{ii} + k_{ij}) \\[2ex] k_{ij} & \left(k_{ii} + \dfrac{6}{G_C}\right) & -(k_{ii} + k_{ij}) \\[2ex] -\dfrac{6}{G_B} & -\dfrac{6}{G_C} & (kL)_{BC}^2 \end{vmatrix} = 0 \tag{5.21v}$$

or

$$\left(k_{ii} + \frac{6}{G_B}\right)\left[\left(k_{ii} + \frac{6}{G_C}\right)(kL)^2_{BC} - \frac{6}{G_C}(k_{ii} + k_{ij})\right] + k_{ij}\left[\frac{6}{G_B}(k_{ii} + k_{ij}) - k_{ij}(kL)^2_{BC}\right]$$

$$- (k_{ii} + k_{ij})\left[-\frac{6}{G_C}k_{ij} + \frac{6}{G_B}\left(k_{ii} + \frac{6}{G_C}\right)\right] = 0$$

After some algebraic calculations we get

$$6(k^2_{ij} - k^2_{ii})(G_B + G_C) - 72(k_{ii} + k_{ij}) + [(k^2_{ii} - k^2_{ij})G_B G_C + 6k_{ii}(G_B + G_C) + 36](kL)^2_{BC} = 0$$

$$(5.21\text{w})$$

Substitute k_{ii} and k_{ij} from Eqs. (5.20t) and (5.20u) into Eq. (5.21w), and we get

$$-\frac{6\left(\frac{\pi}{K}\right)^3 \sin\frac{\pi}{K}}{2 - 2\cos\frac{\pi}{K} - \frac{\pi}{K}\sin\frac{\pi}{K}}(G_B + G_C) - \frac{72\left(\frac{\pi}{K}\right)^2\left(1 - \cos\frac{\pi}{K}\right)}{2 - 2\cos\frac{\pi}{K} - \frac{\pi}{K}\sin\frac{\pi}{K}}$$

$$+\left[\frac{\left(\frac{\pi}{K}\right)^3 \sin\frac{\pi}{K}}{2 - 2\cos\frac{\pi}{K} - \frac{\pi}{K}\sin\frac{\pi}{K}}(G_B + G_C) + \frac{6\left(\frac{\pi}{K}\right)\sin\frac{\pi}{K} - 6\left(\frac{\pi}{K}\right)^2\cos\frac{\pi}{K}}{2 - 2\cos\frac{\pi}{K} - \frac{\pi}{K}\sin\frac{\pi}{K}}(G_B + G_C) + 36\right](kL)^2_{BC} = 0$$

$$(5.21\text{x})$$

After some algebraic calculations we obtain

$$G_B G_C\left(\frac{\pi}{K}\right)^2 \tan\frac{\pi}{K} - 6\frac{\pi}{K}(G_B + G_C) - 36\tan\frac{\pi}{K} = 0$$

or

$$\frac{G_B G_C\left(\frac{\pi}{K}\right)^2 - 36}{6(G_B + G_C)} - \frac{\frac{\pi}{K}}{\tan\frac{\pi}{K}} = 0 \qquad (5.21\text{y})$$

Equation (5.21y) is shown in the form of a nomograph by AISC [8] in Figure 5.30, where the effective length factors K vary from 1.0 to above 20.0.

As before, A and B refer to the two ends of the column cross-section under consideration. Similarly, G is defined assuming the modulus of elasticity E to be constant for all the columns and beams as follows:

$$G = \frac{\Sigma(I_c/L_c)}{\Sigma(I_g/L_g)} \qquad (5.20\text{y})$$

The nomograph of the unbraced frames in Figure 5.30 is used in a similar way as described for the nomograph of the braced frames in Figure 5.27.

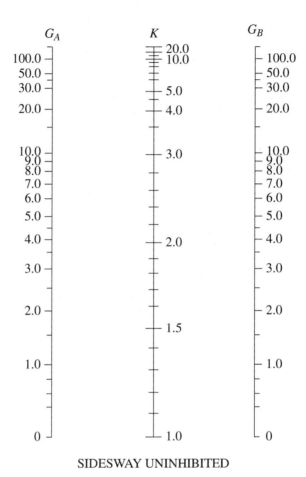

SIDESWAY UNINHIBITED

Figure 5.30 Nomograph for effective length factor K of columns in unbraced frame.

5.5.3 Inelastic Buckling of Frames

The nomographs were made for elastic buckling of beam columns, i.e. the stresses are in the elastic region. If stresses in the beam columns are in the inelastic region at buckling, the end restraint parameter G is modified. The G value should be corrected by multiplying the elastic G value by the stiffness reduction factor (SRF) given by J.A. Yura [9] as

$$\text{Stiffness reduction factor (SRF)} = \frac{E_T}{E} \tag{5.22a}$$

where E_T is the tangent modulus of the column material. Therefore,

$$G_{inelastic} = G_{elastic}\frac{E_T}{E} \tag{5.22b}$$

Where $G_{inelastic}$ = End restraint parameter for inelastic buckling of columns

and $G_{elastic}$ = End restrain parameter for elastic buckling of columns

The G values for inelastic columns are smaller than those of elastic columns and therefore smaller K factors are obtained for inelastic columns than those for elastic columns. It is conservative to base the column design on elastic K factors.

Example 5.1

Calculate the critical load for the columns AB and BE in the frame of Figure 5.9 by using the nomograph in Figure 5.27. Assume $I_b = I_c = I$, and $L_b = L_c = L$. Compare the result with that obtained   in Section 5.3.3.

Solution

From Eq. (5.20y) we have

$$G = \frac{\Sigma(I_c/L_c)}{\Sigma(I_g/L_g)}$$

Column BE

$$G_E = \frac{\dfrac{I_{EB}}{L_{EB}}}{\dfrac{I_{EF}}{L_{EF}}} = \frac{\dfrac{I}{L}}{\dfrac{I}{L}} = 1.0, \quad G_B = \frac{\dfrac{I_{BA}}{L_{BA}} + \dfrac{I_{BE}}{L_{BE}}}{\dfrac{I_{BC}}{L_{BC}}} = \frac{\dfrac{I}{L} + \dfrac{I}{L}}{\dfrac{I}{L}} = 2.0$$

Draw a straight line between $G_A = 1.0$ and $G_B = 2.0$ on the nomograph in Figure 5.27, it would intersect at $K = 0.82$. Therefore,

$$P_{cr_{BE}} = \frac{\pi^2 EI}{(0.82L)^2} = \frac{14.68EI}{L^2}$$

Column AB

$$G_B = 2.0, \quad G_A = \infty \quad \text{(Hinge)}$$

And $K = 0.92$ is obtained as before from the nomograph in Figure 5.27. The force on the column $AB = 2P$. Therefore,

$$P_{cr_{AB}} = \frac{1}{2} \frac{\pi^2 EI}{(0.92L)^2} = \frac{5.83EI}{L^2}$$

The critical load for the column AB governs the maximum critical load the frame can take. Therefore, the critical load for the frame in Figure 5.9 is given by

$$P_{cr} = \frac{5.83EI}{L^2}$$

Equation (5.8q) gives $P_{cr} = \dfrac{7.274EI}{L^2}$ in Section 5.3.3. Hence, the nomograph method gives a conservative value for the frame critical load.

Example 5.2

Design the columns in the two-story, two-bay frame by considering the dead load, live load, and wind load. Given: Dead load = 1.2 k/ft (17.52 kN/m), Live load = 2.00 k/ft (29.2 kN/m), the wind load is taken as concentrated on the floor levels shown in Figure 5.31.

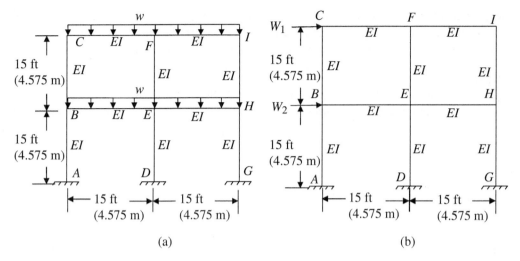

Figure 5.31 Two-story, two-bay frame's geometry and loading: (a) Gravity load on the frame; (b) Wind load on the frame.

Solution

The classical method of moment distribution is used in this example to perform the first-order analysis for the frame. First, the stiffness factors for the various members are calculated and are given by $K = \frac{4EI}{L}$. In this case, the stiffness factors for different members are the same because they have same length and cross-sectional area. Then the distribution factors (DF) are calculated at the different joints.

Gravity loading
Consider the dead load of $w = 1.2$ k/ft (17.52 kN/m) in Figure 5.31a. The fixed end moments (FEM) at the ends of the horizontal beam members BE, EH, CF, and FI are given by

$$\frac{wL^2}{12} = \frac{1.2(15)^2}{12} = 22.5 \text{ k.ft} \quad \left[\frac{17.52(4.575)^2}{12} = 30.56 \text{ kN.m}\right]$$

The clockwise moments are taken positive and counterclockwise moments are taken as negative.

The moment distribution steps are shown in Table 5.1.

Wind load
The wind load is assumed to be concentrated horizontal forces of $W_1 = 3$ kips (13.5 kN), and $W_2 = 6$ kips (27.0 kN) shown in Figure 5.31b. Assume fixed end moments of -100 k.ft. (-137 kN.m) and calculate the end moments at the joints shown in Table 5.2.

From the free body diagrams of the columns calculate the end reaction shown in Figure 5.32. The sum of the reactions at A, D, and G is given by

$$7.77 \text{ k} + 9.91 \text{ k} + 7.77 \text{ k} = 25.45 \text{ k} \, [34.97 \text{ kN} + 44.6 \text{ kN} + 34.97 \text{ kN} = 114.54 \text{ kN}]$$

Table 5.1 Moment distribution for gravity load.

Joint	A	B			C		D	E			
Member	AB	BA	BC	BE	CB	CF	DE	ED	EB	EH	EF
DF		$\frac{1}{3}$	$\frac{1}{3}$	$\frac{1}{3}$	$\frac{1}{2}$	$\frac{1}{2}$		$\frac{1}{4}$	$\frac{1}{4}$	$\frac{1}{4}$	$\frac{1}{4}$
FEM k.ft. (kN.m)				−22.5 (−30.6)		−22.5 (−30.6)			22.5 (30.6)	−22.5 (−30.6)	
DIST.		7.5 (10.2)	7.5 (10.2)	7.5 (10.2)	11.25 (15.3)	11.25 (15.3)					
CO	3.75 (5.1)		5.62 (7.65)		3.75 (5.1)				3.75 (5.1)	−3.75 (−5.1)	
DIST.		−1.87 (−2.55)	−1.87 (−2.55)	−1.88 (−2.55)	−1.88 (−2.55)	−1.87 (−2.55)					
CO	−0.94 (−1.28)		−0.94 (−1.28)		−0.94 (−1.28)				−0.94 (−1.28)	0.94 (1.28)	
DIST.		0.31 (0.43)	0.31 (0.43)	0.32 (0.42)	0.47 (0.64)	0.47 (0.64)					
Final Moment	2.81 (3.82)	5.94 (8.08)	10.62 (14.45)	−16.56 (−22.5)	12.65 (17.2)	−12.65 (−17.2)			25.31 (34.4)	−25.31 (−34.4)	

Joint	F			G	H			I	
Member	FE	FC	FI	GH	HG	HE	HI	IF	IH
DF	$\frac{1}{3}$	$\frac{1}{3}$	$\frac{1}{3}$		$\frac{1}{3}$	$\frac{1}{3}$	$\frac{1}{3}$	$\frac{1}{2}$	$\frac{1}{2}$
FEM k.ft. (kN.m)		22.5 (30.6)	−22.5 (−30.6)			22.5 (30.6)		22.5 (30.5)	
DIST					−7.5 (−10.2)	−7.5 (−10.2)	−7.5 (−10.2)	−11.25 (−15.25)	−11.25 (−15.25)
CO		5.62 (7.65)	−5.62 (−7.65)	−3.75 (−5.1)			−5.62 (−7.65)		−3.75 (−5.1)
DIST					1.87 (2.55)	1.87 (2.55)	1.88 (2.55)	1.88 (2.55)	1.87 (2.55)
CO		−0.94 (−1.28)	0.94 (1.28)	0.94 (1.28)			0.94 (1.28)		0.94 (1.28)
DIST					−0.31 (−0.43)	−0.31 (−0.43)	−0.32 (−0.42)	−0.47 (−0.64)	−0.47 (−0.64)
Final Moment		27.18 (36.97)	−27.18 (−36.97)	−2.81 (3.82)	−5.94 (−8.08)	16.56 (22.5)	−10.62 (−14.45)	12.66 (17.16)	−12.66 (−17.16)

Table 5.2 Moment distribution for wind loads.

Joint	A	B			C		D	E			
Member	AB	BA	BC	BE	CB	CF	DE	ED	EB	EH	EF
DF	$\frac{1}{3}$	$\frac{1}{3}$	$\frac{1}{3}$	$\frac{1}{3}$	$\frac{1}{2}$	$\frac{1}{2}$		$\frac{1}{4}$	$\frac{1}{4}$	$\frac{1}{4}$	$\frac{1}{4}$
FEM k.ft. (kN.m)	−100 (−137)	−100 (−137)	−100 (−137)		−100 (−137)		−100 (−137)	−100 (−137)			−100 (−137)
DIST.		66.67 (91.5)	66.67 (91.5)	66.67 (91.5)	50 (68.6)	50 (68.6)		50 (68.6)	50 (68.6)	50 (68.6)	50 (68.6)
CO	33.34 (45.8)		25 (34.3)	25 (34.3)	33.34 (45.8)	16.67 (22.9)	25 (34.3)		33.33 (45.8)	33.33 (45.8)	16.67 (22.9)
DIST.		−16.67 (−22.9)	−16.67 (−22.9)	−16.67 (−22.9)	−25 (−34.4)	−25 (−34.4)		−20.83 (−28.6)	−20.83 (−28.6)	−20.83 (−28.6)	−20.83 (−28.6)
CO	−8.34 (−11.5)		−12.5 (−17.2)	−10.42 (−14.3)	−8.34 (−11.5)	−12.5 (−17.2)	−10.42 (−14.3)		−8.34 (−11.5)	−8.34 (−11.5)	−12.5 (−17.2)
DIST.		7.64 (10.50)	7.64 (10.50)	7.64 (10.50)	10.42 (14.4)	10.42 (14.4)		7.3 (10.1)	7.3 (10.1)	7.3 (10.1)	7.3 (10.1)
CO	3.82 (5.3)		5.21 (7.2)	3.65 (5.1)	3.82 (5.3)	5.91 (8.1)	3.65 (5.1)		3.82 (5.3)	3.82 (5.3)	5.91 (8.1)
D IST.		−2.95 (−4.1)	−2.95 (−4.1)	−2.96 (−4.1)	−4.87 (−6.7)	−4.86 (−6.7)		−3.39 (−4.68)	−3.39 (−4.68)	−3.39 (−4.68)	−3.38 (−4.67)
Final Moment	−71.18 (−97.4)	−45.31 (−62.0)	−27.6 (−37.8)	72.9 (99.8)	−40.63 (−55.5)	40.64 (55.6)	−81.77 (−111.9)	−66.92 (−91.6)	61.89 (84.7)	61.89 (84.7)	−56.84 (−77.8)
Corrected Final Moment	−25.17 (−34.5)	−16.02 (−21.9)	−4.78 (−6.5)	20.8 (28.4)	−7.04 (−9.61)	7.04 (9.61)	−28.92 (−39.6)	−23.67 (−32.4)	16.76 (23.0)	16.76 (23.0)	−9.85 (−13.5)

Joint	F			G	H			I	
Member	FE	FC	FI	GH	HG	HE	HI	IF	IH
DF	$\frac{1}{3}$	$\frac{1}{3}$	$\frac{1}{3}$		$\frac{1}{3}$	$\frac{1}{3}$	$\frac{1}{3}$	$\frac{1}{2}$	$\frac{1}{2}$
FEM k.ft. (kN.M)	−100 (−137)			−100 (−137)	−100 (−137)		−100 (−137)		−100 (−137)
DIST.	33.33 (45.8)	33.33 (45.8)	33.33 (45.8)		66.67 (91.5)	66.67 (91.5)	66.67 (91.5)	50 (68.6)	50 (68.6)
CO	25 (34.3)	25 (34.3)	25 (34.3)	33.34 (45.8)		25 (34.3)	25 (34.3)	16.67 (22.9)	33.34 (45.8)

(Continued)

Table 5.2 (Continued)

Joint	F			G	H			I	
Member	FE	FC	FI	GH	HG	HE	HI	IF	IH
DIST.	−25	−25	−25		−16.67	−16.67	−16.67	−25	−25
	(−34.3)	(−34.3)	(−34.3)		(−22.9)	(−22.9)	(−22.9)	(−34.3)	(−34.3)
CO	−10.42	−12.5	−12.5	−8.34		−10.42	−12.5	−12.5	−8.34
	(−14.3)	(−17.2)	(−17.2)	(−11.5)		(−14.4)	(−17.2)	(−17.2)	(−11.5)
DIST.	11.81	11.81	11.81		7.64	7.64	7.64	10.42	10.42
	(16.3)	(16.3)	(16.3)		(10.5)	(10.5)	(10.5)	(14.3)	(14.3)
CO	3.65	5.41	5.21	3.82		3.65	5.41	5.91	3.82
	(5.0)	(7.4)	(7.2)	(5.3)		(5.0)	(7.4)	(8.1)	(5.3)
DIST.	−4.76	−4.76	−4.75		−3.02	−3.02	−3.02	−4.87	−4.86
	(−6.6)	(−6.6)	(−6.4)		(−4.2)	(−4.1)	(−4.1)	(−6.7)	(−6.7)
Final Moment	−66.39	33.31	33.31	−71.18	−45.31	72.84	−27.60	40.63	−40.63
	(−90.8)	(45.4)	(45.4)	(−97.4)	(−62.0)	(99.8)	(−37.8)	(55.7)	(−55.5)
Corrected Final Moment	−11.51	5.76	5.76	−25.17	−16.05	20.81	−4.76	7.04	−7.04
	(−15.7)	(7.9)	(7.9)	(−34.5)	(−21.9)	(28.4)	(−6.5)	(9.6)	(−9.6)

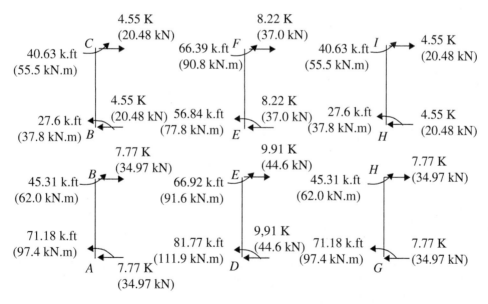

Figure 5.32 End moments and end reactions in columns due to −100 k.ft. (−137 kN.m) end moments applied in Table 5.2.

where this sum should be equal to 9 K (40.5 kN), the bottom story horizontal shear due to service wind force. The corresponding end moments in the columns of the bottom story can be determined by proportion. Therefore, the end moments developed in the bottom story columns by the service wind load in Figure 5.31b are given by

$$M_{AB} = \frac{71.18x9}{25.45} = 25.17 \, \text{k.ft} \quad \left[\frac{97.4x40.5}{114.54} = 30.48 \, \text{k.Nm}\right]$$

$$M_{BA} = \frac{45.31x9}{25.45} = 16.02 \, \text{k.ft} \quad \left[\frac{62x40.5}{114.54} = 21.95 \, \text{kN.m}\right]$$

$$M_{DE} = \frac{81.77x9}{25.4} = 28.92 \, \text{k.ft} \quad \left[\frac{111.9x40.5}{114.54} = 39.61 \, \text{kN.m}\right]$$

$$M_{ED} = \frac{66.92x9}{25.45} = 23.67 \, \text{k.ft} \quad \left[\frac{91.6x40.5}{114.54} = 32.43 \, \text{kN.m}\right]$$

$$M_{GH} = \frac{71.18x9}{25.45} = 25.17 \, \text{k.ft} \quad \left[\frac{97.4x40.5}{114.54} = 34.48 \, \text{kN.m}\right]$$

$$M_{HG} = \frac{45.31x9}{25.45} = 16.05 \, \text{k.ft} \quad \left[\frac{62x40.5}{114.54} = 21.95 \, \text{kN.m}\right]$$

The sum of the reactions at B, E, and H is given by

$$4.55 \, \text{k} + 8.22 \, \text{k} + 4.55 \, \text{k} = 17.32 \, \text{k} \, [20.48 \, \text{kN} + 37 \, \text{kN} + 20.48 \, \text{kN} = 77.96 \, \text{kN}]$$

where this sum should be equal to 3 K (13.5 kN), the top story horizontal shear due to service wind force. The corresponding end moments in the columns of the top story can be determined by proportion. Therefore, the end moments developed in the top story columns by the service wind load in Figure 5.31b are given by

$$M_{BC} = \frac{27.6x3}{17.31} = 4.78 \, \text{k.ft} \quad \left[\frac{37.8x13.5}{77.96} = 6.55 \, \text{kN.m}\right]$$

$$M_{CB} = \frac{40.63x3}{17.31} = 7.04 \, \text{k.ft} \quad \left[\frac{55.5x13.5}{77.96} = 9.61 \, \text{kN.m}\right]$$

$$M_{EF} = \frac{56.84x3}{17.31} = 9.85 \, \text{k.ft} \quad \left[\frac{77.8x13.5}{77.96} = 13.47 \, \text{kN.m}\right]$$

$$M_{FE} = \frac{66.39x3}{17.31} = 11.51 \, \text{k.ft} \quad \left[\frac{90.8x13.5}{77.96} = 15.72 \, \text{kN.m}\right]$$

$$M_{HI} = \frac{27.60x3}{17.31} = 4.76 \, \text{k.ft} \quad \left[\frac{37.8x13.5}{77.96} = 6.55 \, \text{kN.m}\right]$$

$$M_{IH} = \frac{40.63x3}{17.31} = 7.04 \text{k.ft} \quad \left[\frac{55.65x13.5}{77.96} = 9.64 \text{kN.m}\right]$$

The calculated end moments for the wind load are shown as Corrected Final Moments in Table 5.2. Assume $1.2D + 1.6L$ is the most severe load combination [10] for the gravity load in the load resistance factor design, where D is the dead load and L is the live load. For this load combination the uniformly distributed gravity load acting on the horizontal beams of the frame is

$$1.2D + 1.6L = 1.2 \times 1.2 + 1.6 \times 2.0 = 4.64 \, \text{k/ft} \, [1.2 \times 17.52 + 1.6 \times 29.2 = 67.74 \, \text{kN/m}]$$

The end moments calculated in Table 5.1 for a gravity load of 1.2 k/ft. (17.52 kN/m) are multiplied by $\frac{4.64}{1.2} = 3.87$ $\left[\frac{67.74}{17.52} = 3.87\right]$ to get the end moments for the gravity load case of $1.2D + 1.6L$. Now draw the free body diagrams of the frame members and find end reactions for the load case $1.2D + 1.6L$ in Figure 5.33.

The span moments ($wL^2/8$ at the midspan) on the horizontal members are now superimposed on the end moments shown in the Figure 5.33. The bending moment diagrams for the various members of the frame are shown in Figure 5.34 due to the gravity load equal to $1.2D + 1.6L$.

Assume $1.2D + L + 1.0\,W$ is the most severe load combination [10] in the presence of wind load for the load resistance factor design, where W is the wind load. The bending moment diagrams for the wind load on the frame are shown in Figure 5.35, from Table 5.2.

The free body diagrams of the frame members and the end reactions, for the wind load are shown in Figure 5.36. The symbols used in this example are explained in the Section 4.12.2. For the design of columns, the following load combinations are considered:

Load case − $1.2\,D + 1.6\,L$

The moments M_{nt} and the axial forces P_{nt} are given below from Figure 5.33 for various columns:

Columns *BC* and *HI*	$M_{nt} = 48.91$ k.ft (67.13 kN.m), $P_{nt} = 31.05$ k (139.73 kN)
Columns *AB* and *GH*	$M_{nt} = 22.97$ k.ft (31.53 kN.m), $P_{nt} = 63.59$ k (286.16 kN)
Column *FE*	$M_{nt} = 0$, $P_{nt} = 77.1$ k (346.96 kN)
Column *DE*	$M_{nt} = 0$, $P_{nt} = 151.22$ k (680.5 kN)

where the subscript *nt* stands for no translation.

Load case − $1.2\,D + L + W$

The moments M_{nt} and the axial forces P_{nt} are calculated from Figure 5.33 by multiplying the quantities there by the ratio $\frac{1.2D + L}{1.2D + 1.6L} = \frac{1.2(1.2) + (2)}{1.2(1.2) + 1.6(2)} = 0.741$ $\left[\frac{1.2(17.52) + 29.2}{1.2(17.52) + 1.6(29.2)}\right]$ $= 0.741$ for various columns and are as follows:

Columns *BC* and *HI*	$M_{nt} = 36.26$ k.ft (49.76 kN.m), $P_{nt} = 23.02$ k (103.58 kN)
Columns *AB* and *GH*	$M_{nt} = 17.03$ k.ft ($22.97 \times 0.741 = 17.03$ k.ft)
	[23.37 kN.m ($31.53 \times 0.741 = 23.37$ kN.m)]
	$P_{nt} = 47.14$ k ($63.59 \times 0.741 = 47.14$ k)
	[212.13 kN ($286.16 \times 0.741 = 212.13$ kN)]
Column *FE*	$M_{nt} = 0$, $P_{nt} = 57.13$ k (257.1 kN)
Column *DE*	$M_{nt} = 0$, $P_{nt} = 112.11$ k (504.45 kN)

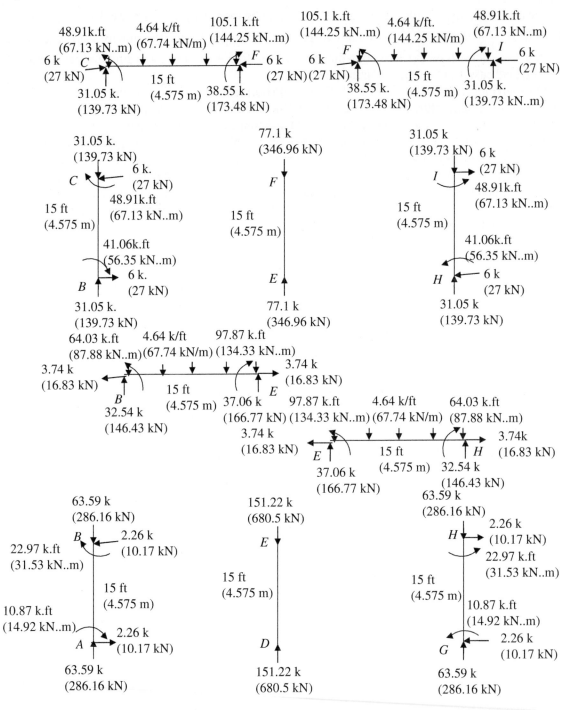

Figure 5.33 Free body diagrams for the frame under gravity load.

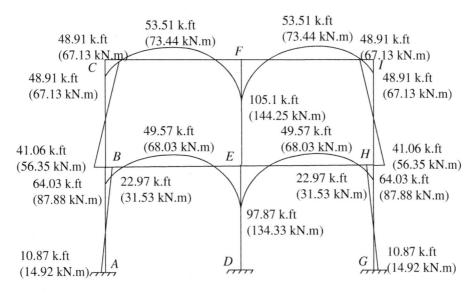

Figure 5.34 Bending moment diagrams for the gravity load from the first-order analysis.

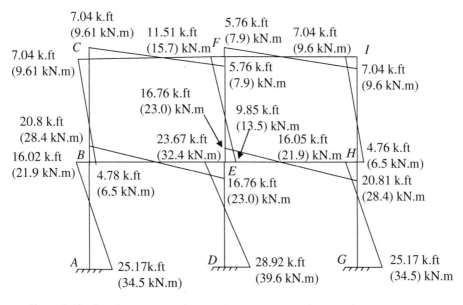

Figure 5.35 Bending moment diagrams for the wind load from the first-order analysis.

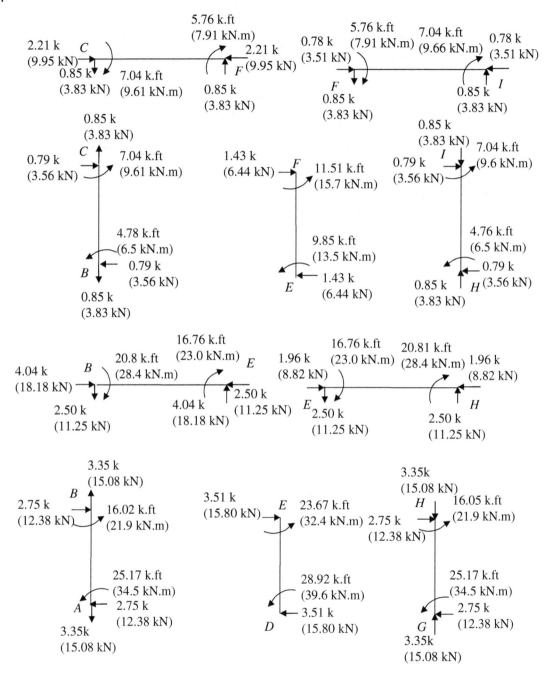

Figure 5.36 Free body diagrams for the frame under wind load.

The moments M_{lt} and the axial forces P_{lt} are given below from Figure 5.36 for various columns:

Column BC $M_{lt} = 7.04$ k.ft (9.61 kN.m), $P_{lt} = 0.85$ k (3.83 kN) Tension

Column HI $M_{lt} = 7.04$ k.ft (9.61 kN.m), $P_{lt} = 0.85$ k (3.83 kN) Compression

Column AB $M_{lt} = 25.17$ k.ft (34.5 kN.m), $P_{lt} = 3.35$ k (15.08 kN) Tension

Column GH $M_{lt} = 25.17$ k.ft (34.5 kN.m), $P_{lt} = 3.35$ k (15.08 kN) Compression

Column FE $M_{lt} = 11.51$ k.ft (15.7 kN.m), $P_{lt} = 0$

Column DE $M_{lt} = 28.92$ k.ft (39.6 kN.m), $P_{lt} = 0$

where the subscript lt stands for lateral translation.

Assume the frame consists of members as shown in Figure 5.37.

Design of column GH by AISC LRFD method [8]

Gravity load $-$ $1.2D + 1.6L$

P_{nt} (Axial load with no lateral translation of the frame) $= 63.59$ k (286.16 kN)

$M_{nt} =$ (Moment with no lateral translation of the frame) $= 22.97$ k.ft (31.53 kN.m)

Section properties of the wide flange $W\,10 \times 30$ ($W\,250 \times 44.8$) are:
Nominal weight $W = 30$ lb/ft (0.439 kN/m); Area $A_g = 8.84$ in.^{2} (5720 mm^2); depth $d = 10.5$ in. (266 mm); Area moment of inertia about x axis $I_{xx} = 170$ in.^{4} (71.1 $\times 10^6$ mm^4); Radius of gyration about x axis $r_{xx} = 4.38$ in. (111 mm); Area moment of inertia about y axis $I_{yy} = 16.7$ in.^{4} (7.03 $\times 10^6$ mm^4); Radius of gyration about y axis $r_{yy} = 1.37$ in. (35.1 mm).

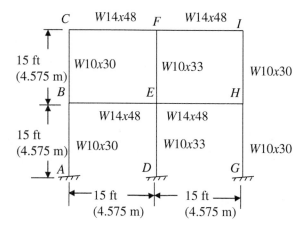

Figure 5.37 Two-story, two-bay frame.

Section properties of the wide flange $W\,10 \times 33$ ($W\,250 \times 49.1$) are:

$W = 33$ lb/ft (0.482 kN/m); $A_g = 9.71$ in.^{2} (6250 mm^2); $d = 9.73$ in.(247 mm);
$I_{xx} = 171$ in.^{4} (70.6 $\times\,10^6$ mm^4); $r_{xx} = 4.19$ in. (106 mm); $I_{yy} = 36.6$ in.^{4} (15.1 $\times\,10^6$ mm^4);
$r_{yy} = 1.94$ in. (49.2 mm).

Section properties of the wide flange $W\,14 \times 48$ ($W\,360 \times 72$) are:

$W = 48$ lb/ft (0.706 kN/m); $A_g = 14.1$ in.^{2} (9110 mm^2); $d = 13.8$ in.(350 mm);
$I_{xx} = 484$ in.^{4} (201 $\times\,10^6$ mm^4); $r_{xx} = 5.85$ in. (149 mm); $I_{yy} = 51.4$ in.^{4} (21.4 $\times\,10^6$ mm^4);
$r_{yy} = 1.91$ in. (48.5 mm).

The material properties are:
Yield strength $F_y = 50$ ksi (345 MPa); Ultimate strength $F_u = 65$ ksi (448.5 MPa); Modulus of elasticity $E = 29 \times 10^6$ psi (200,000 MPa). Assume $W10 \times 30$ ($W250 \times 44.8$) is used for the member GH along with other members as shown in Figure 5.37. End G is a fixed end, so $G_G = 1.0$.

$$G_H = \frac{\Sigma(I_c/L_c)}{\Sigma(I_g/L_g)} = \frac{\dfrac{170}{15} + \dfrac{170}{15}}{\dfrac{484}{15}} = 0.7 \qquad \left(G_H = \frac{\dfrac{71.1 \times 10^6}{4575} + \dfrac{71.1 \times 10^6}{4575}}{\dfrac{201 \times 10^6}{4575}} = 0.7 \right)$$

Draw a straight line between $G_G = 1.0$, and $G_H = 0.7$ on the non-sway nomograph in Figure 5.27, it would intersect at $K_{x1} = 0.75$. Similarly, from the sidesway uninhibited nomograph in Figure 5.30, $K_{x2} = 1.28$. Assume for the member GH, $K_y = 1.0$.

$$\frac{K_y L_y}{r_{yy}} = \frac{1 \times 15 \times 12}{1.37} = 131.39 \qquad \left[\frac{K_y L_y}{r_{yy}} = \frac{1 \times 4575}{35.1} = 131.39 \right]$$

$$4.71\sqrt{\frac{E}{F_y}} = 4.71\sqrt{\frac{29 \times 10^6}{50,000}} = 113.4 < 131.39 \qquad \left[4.71\sqrt{\frac{E}{F_y}} = 4.71\sqrt{\frac{200,000}{345}} = 113.4 < 131.39 \right]$$

Hence the column will have elastic buckling, and the critical stress F_{cr} is given by

$$F_{cr} = 0.877\, F_e$$

where the Euler buckling stress F_e is obtained from

$$F_e = \frac{\pi^2 E}{\left(\frac{K_y L_y}{r_{yy}}\right)^2} = \frac{\pi^2(29{,}000)}{(131.39)^2} = 16.58 \text{ ksi} \quad \left[F_e = \frac{\pi^2(200{,}000)}{(131.39)^2} = 114.34 \text{ MPa}\right]$$

$$F_{cr} = 0.877 \times 16.58 = 14.54 \text{ ksi} \quad [F_{cr} = 0.877 \times 114.34 = 100.28 \text{ MPa}]$$

$$P_n = F_{cr} A_g = 14.54 \times 8.84 = 128.53 \text{ kips} \quad \left[P_n = \frac{100.28(5720)}{1000} = 573.60 \text{ kN}\right]$$

$$P_c = \phi P_n = 0.9(128.53) = 115.68 \text{ kips} \quad [P_c = 0.9(573.60) = 516.24 \text{ kN}]$$

$$\frac{P_r}{P_c} = \frac{63.59}{115.68} = 0.55 > 0.2 \quad \left[\frac{P_r}{P_c} = \frac{286.16}{516.24} = 0.55 > 0.2\right]$$

$$\frac{P_r}{P_c} + \frac{8}{9}\left(\frac{M_{rx}}{M_{cx}} + \frac{M_{ry}}{M_{cy}}\right) \le 1.0$$

$$M_r = B_{1x} M_{nt} + B_2 M_{lt}$$

For gravity loads, $M_{lt} = 0$, therefore $M_{rx} = B_{1x} M_{ntx}$

$$B_{1x} = \frac{C_{mx}}{1 - \alpha \frac{P_r}{P_{elx}}} \ge 1, \quad \alpha = 1.0 \text{ for LRFD design.}$$

$$C_{mx} = 0.6 - 0.4\left(\frac{M_{1x}}{M_{2x}}\right)$$

where M_{1x} and M_{2x} are the smaller and larger of the end moments in the member *GH*.
or

$$C_{mx} = 0.6 - 0.4\left(\frac{10.87}{22.97}\right) = 0.411 > 0.4 \quad \left[C_{mx} = 0.6 - 0.4\left(\frac{14.92}{31.53}\right) = 0.411 > 0.4\right]$$

Therefore, $C_{mx} = 0.411$

$$P_{elx} = \frac{\pi^2 E I_x}{(K_{x1} L_x)^2} = \frac{\pi^2(29000)(170)}{(0.75x15x12)^2} = 2669.80 \text{ kips}$$

$$\left[P_{elx} = \frac{\pi^2(200000)(71.1x10^6)}{(0.75x4.575x1000)^2 x1000} = 11920.50 \text{ kN}\right]$$

$$B_{1x} = \frac{0.411}{1 - (1)\frac{63.59}{2669.80}} = 0.421 \ge 1 \quad \left[B_{1x} = \frac{0.411}{1 - (1)\frac{286.16}{11920.50}} = 0.421 \ge 1\right]$$

Therefore, $B_{1x} = 1.0$, and $M_{rx} = 1\,(22.97) = 22.97 \text{ k.ft} \quad [M_{rx} = 1\,(31.53) = 31.53 \text{ kN. m}]$

L_b(unbraced length for lateral displacement) $= 15 \text{ ft } (4.574 \text{ m})$

From the AISC manual [8] $L_p = 4.84 \text{ ft } (1.48 \text{ m})$, and $L_r = 16.1 \text{ ft } (4.91 \text{ m})$ for the W10 $\times$ 30 (250 $\times$ 44.8) section:

where

L_p = Limiting laterally unbraced length for compact sections to achieve their full plastic moment, M_p, capacity, and

L_r = Limiting laterally unbraced length for the limit state of inelastic lateral-torsional buckling. If the unbraced length is more than L_r, the section will buckle elastically.

$$L_p = 4.84 \text{ ft}(1.48 \text{ m}) < L_b = 15 \text{ ft } (4.575 \text{ m}) < L_r = 16.1 \text{ ft}(4.91 \text{ m})$$

Therefore, from the AISC manual [8], $\phi_b M_n = M_{cx} = 90.5$ k. ft [124.21 kN. m]

$$\frac{P_r}{P_c} + \frac{8}{9}\frac{M_{rx}}{M_{cx}} = \frac{63.59}{115.68} + \frac{8}{9}\left(\frac{22.97}{90.5}\right) = 0.55 + 0.226 = 0.776 < 1.0$$

$$\left[\frac{P_r}{P_c} + \frac{8}{9}\frac{M_{rx}}{M_{cx}} = \frac{286.16}{516.24} + \frac{8}{9}\left(\frac{31.53}{124.21}\right) = 0.55 + 0.226 = 0.776 < 1.0\right]$$

Gravity and wind load combined for Column $GH - 1.2D + L + W$

P_{nt} (Axial load with no lateral translation of the frame) = 47.14 k (212.13 kN) Compression

M_{ntx} (Moment with no lateral translation of the frame) = 17.03 k.ft. (23.37 kN.m)

P_{lt} (Axial load in the member caused by the lateral translation of the frame only by wind load) = 3.35 k (15.08 kN) Compression

M_{ltx} (Moment in the member caused by the lateral translation of the frame only by wind load) = 25.17 k.ft. (34.5 kN.m)

$$P_r = P_{nt} + B_2 P_{lt}$$
$$M_{rx} = B_{1x}M_{ntx} + B_2 M_{ltx}$$
$$B_{1x} = \frac{C_{mx}}{1 - \dfrac{\alpha P_r}{P_{elx}}} \geq 1$$

To find B_{1x}

$$P_r = P_{nt} + P_{lt} = 47.14 + 3.35 = 50.49 \text{ k} \quad [212.13 + 15.08 = 227.21 \text{ kN}]$$
$$M_{1x} = 10.87(0.741) = 8.05 \text{ k.ft} \quad [14.92 \,(0.741) = 11.06 \text{ kN.m}]$$
$$C_{mx} = 0.6 - 0.4\left(\frac{8.05}{17.03}\right) = 0.411 \quad \left[C_{mx} = 0.6 - 0.4\left(\frac{11.06}{23.37}\right) = 0.411\right]$$

The P_{elx} for W10 × 30 (W250 × 49.1) is the same as for gravity load calculated before and is equal to

$$P_{elx} = 2669.80 \text{ k} \quad (11920.50 \text{ kN})$$

$$B_{1x} = \frac{0.411}{1 - \dfrac{1(50.49)}{2269.80}} = 0.42 < 1.0 \quad \left[B_{1x} = \frac{0.411}{1 - \dfrac{1(227.21)}{11920.50}} = 0.42 < 1.0\right]$$

Therefore,

$$B_{1x} = 1.0$$

$$\Sigma P_{nt} = (63.59 + 151.22 + 63.59)0.741 - 3.35 + 3.35 = 206.29 \text{ k}$$

$$[(286.16 + 680.5 + 286.16)0.741 - 15.08 + 15.08 = 928.34 \text{ kN}]$$

$$P_{e2x} = \frac{\pi^2 EI}{(K_{x2}L)^2} = \frac{\pi^2(29000)(170)}{(1.28x15x12)^2} = 916.6k$$

$$\left[P_{e2x} = \frac{\pi^2(200,000)(71.1x10^6)}{(1.28x4.575x1000)^2(1000)} = 4092.58 \text{ kN}\right]$$

$$\Sigma P_{e2x} = 3(916.6) = 2749.8 \text{ k} \quad [3(4092.58) = 12,277.74 \text{ kN}]$$

$$B_{2x} = \frac{1}{1 - \dfrac{\alpha \Sigma P_{nt}}{\Sigma P_{e2x}}} \geq 1$$

$$B_{2x} = \frac{1}{1 - \dfrac{1(206.29)}{2749.8}} = 1.08 \quad \left[B_{2x} = \frac{1}{1 - \dfrac{1(928.34)}{12277.74}} = 1.08\right]$$

$$P_r = P_{nt} + B_{2x}P_{ltx} = 47.14 + 1.08(3.35) = 50.76 \text{ k} \quad [212.3 + 1.08(15.08) = 228.42 \text{ kN}]$$

$$M_{rx} = B_{1x}M_{ntx} + B_{2x}M_{ltx}$$

$$M_{rx} = 1(17.03) + 1.08(25.17) = 44.21 \text{ k.ft} \quad [M_{rx} = 1(23.37) + 1.08(34.5) = 60.63 \text{ kN.m}]$$

$$\frac{K_{x2}L_x}{r_{xx}} = \frac{1.28x15x12}{4.38} = 52.60 \quad \left[\frac{K_xL_x}{r_{xx}} = \frac{1.28x4.575x1000}{111} = 52.76\right]$$

$$< \frac{K_yL_y}{r_{yy}} = \frac{1x15x12}{1.37} = 131.39 \quad \left[\frac{K_yL_y}{r_{yy}} = \frac{1x4.575x1000}{35.1} = 130.34\right]$$

Therefore, $P_c = 115.68 \text{ k}$ (516.24 kN) as before.

$$\frac{P_r}{P_c} = \frac{50.76}{115.68} = 0.439 > 0.2 \quad \left[\frac{P_r}{P_c} = \frac{228.42}{516.24} = 0.442 > 0.2\right]$$

$$\frac{P_r}{P_c} + \frac{8}{9}\left(\frac{M_{rx}}{M_{cx}} + \frac{M_{ry}}{M_{cy}}\right) \leq 1.0, M_{ry} = 0$$

therefore,

$$\frac{P_r}{P_c} + \frac{8}{9}\left(\frac{M_{rx}}{M_{cx}}\right) = 0.439 + \frac{8}{9}\left(\frac{44.21}{90.5}\right) = 0.873 < 1.0 \text{ O.K.}$$

$$\left[\frac{P_r}{P_c} + \frac{8}{9}\left(\frac{M_{rx}}{M_{cx}}\right) = 0.442 + \frac{8}{9}\left(\frac{60.63}{124.21}\right) = 0.876 < 1.0 \text{ O.K.}\right]$$

The section is satisfactory.

<u>Design of column *GH* by the AISC ASD method [8]</u>

Gravity load – $D + L$

$D + L = 1.2 + 2.0 = 3.2\,\text{k/ft}$ [$17.52 + 29.2 = 46.72\,\text{kN/m}$]

$$\frac{3.2}{1.2(1.2) + 1.6(2.0)} = \frac{3.2}{4.64} = 0.69 \quad \left[\frac{46.72}{1.2(17.52) + 1.6(29.2)} = \frac{46.72}{67.74} = 0.69 \right]$$

$P_r = P_D + P_L = P_{nt} = 63.59 \times 0.69 = 43.88\,\text{k}$ [$286.16 \times 0.69 = 197.45\,\text{kN}$]

$M_{nt} = 22.97 \times 0.69 = 15.85\,\text{k.ft}$ [$31.53 \times 0.69 = 21.76\,\text{kN.M}$]

$$P_c = \frac{P_n}{\Omega_c} \; (P_n \text{ is the same as in LRFD design})$$

where

Ω_c = safety factor for compression = 1.67

or

$$P_c = \frac{128.53}{1.67} = 76.96\,\text{k} \quad \left[P_c = \frac{573.6}{1.67} = 343.47\,\text{kN} \right]$$

$$\frac{P_r}{P_c} = \frac{43.88}{76.96} = 0.57 > 0.2 \quad \left[\frac{P_r}{P_c} = \frac{197.45}{343.47} = 0.57 > 0.2 \right]$$

$$\frac{P_r}{P_c} + \frac{8}{9} \left(\frac{M_{rx}}{M_{cx}} + \frac{M_{ry}}{M_{cy}} \right) \leq 1.0$$

$M_r = B_1 M_{nt} + B_2 M_{lt}$

For gravity loads, $M_{lt} = 0$, therefore $M_{rx} = B_{1x} M_{ntx}$

$$B_{1x} = \frac{C_{mx}}{1 - \alpha \dfrac{P_r}{P_{elx}}}, \quad \alpha = 1.6 \text{ for ASD design.}$$

$$C_{mx} = 0.6 - 0.4 \left(\frac{M_{1x}}{M_{2x}} \right)$$

$M_{1x} = 10.87 \times 0.69 = 7.5\,\text{k.ft}$ [$14.92 \times 0.69 = 10.29\,\text{kN.m}$]

$M_{2x} = 22.97 \times 0.69 = 15.85\,\text{k.ft}$ [$31.53 \times 0.69 = 21.76\,\text{kN.m}$]

$$C_{mx} = 0.6 - 0.4 \left(\frac{7.5}{15.85} \right) = 0.411 > 0.4 \quad \left[C_{mx} = 0.6 - 0.4 \left(\frac{10.29}{21.76} \right) = 0.411 > 0.4 \right]$$

Therefore,

$$C_{mx} = 0.411$$

$$B_{1x} = \frac{C_{mx}}{1 - \alpha \dfrac{P_r}{P_{elx}}}$$

P_{elx} is the same as for the LRFD design.

or

$$B_{1x} = \frac{0.411}{1 - 1.6\dfrac{43.88}{2669.80}} = 0.422 < 1.0 \quad \left[B_{1x} = \frac{0.411}{1 - 1.6\dfrac{197.45}{11920.50}} = 0.422 < 1.0 \right]$$

Therefore,

$$B_{1x} = 1.0$$

$$M_{rx} = B_{1x}M_{ntx} = 1.0(15.85) = 15.85 \text{ k.ft} \quad [1.0(21.76) = 21.76 \text{ kN.m}]$$

Nominal moment of resistance from the AISC manual [8]

$$M_n = \frac{90.5}{0.9} = 100.56 \text{ k.ft} \quad \left[M_n = \frac{124.21}{0.9} = 138.01 \text{ kN.m} \right]$$

$$M_c = \frac{M_n}{\Omega_b}$$

Ω_b = Safety factor for flexure = 1.67

$$M_{cx} = \frac{100.56}{1.67} = 60.22 \text{ k.ft} \quad \left[M_{cx} = \frac{138.01}{1.67} = 82.64 \text{ kN.m} \right]$$

$$\frac{P_r}{P_c} + \frac{8}{9}\left(\frac{M_{rx}}{M_{cx}}\right) = \frac{43.88}{76.96} + \frac{8}{9}\left(\frac{15.85}{60.22}\right) = 0.57+).234 = 0.804 < 1.0 \text{ O.K.}$$

or

$$\left[\frac{P_r}{P_c} + \frac{8}{9}\left(\frac{M_{rx}}{M_{cx}}\right) = \frac{197.45}{343.47} + \frac{8}{9}\left(\frac{21.76}{82.64}\right) = 0.57 + 0.234 = 0.804 < 1.0 \text{ O.K.} \right]$$

Gravity and wind load combined for Column $GH - D + 0.75\ L + 0.75\ (0.6W)$

The load combination is given by

$$D + 0.75L + 0.45W$$

or

$$D + 0.75L = 1.2 + 0.75 \times 2 = 2.7 \text{ k/ft} \quad [17.52 + 0.75 \times 29.2 = 39.42 \text{ kN/m}]$$

$$\frac{2.7}{4.64} = 0.582 \quad \left[\frac{39.42}{67.74} = 0.582 \right]$$

$$P_{nt} = 63.59 \times 0.582 = 37\,k \quad [286.16 \times 0.582 = 166.55\,kN]$$

$$M_{nt} = 22.97 \times 0.582 = 13.37\,k.ft\,[31.53 \times 0.582 = 18.35\,kN.m]$$

$$P_{lt} = 3.35 \times 0.45 = 1.51\,k \quad [15.08 \times 0.45 = 6.79\,kN]$$

$$M_{lt} = 16.05 \times 0.45 = 7.22\,k.ft \quad [21.9 \times 0.45 = 9.86\,kN.m]$$

$$P_r = P_{nt} + B_2 P_{lt}$$

$$M_{rx} = B_{1x} M_{ntx} + B_2 M_{ltx}$$

$$B_{1x} = \frac{C_{mx}}{1 - \dfrac{\alpha P_r}{P_{elx}}} \geq 1$$

To find B_{1x}

$$P_r = P_{nt} + P_{lt} = 37 + 1.51 = 38.51\,k \quad [166.55 + 6.79 = 173.34\,kN]$$
$$M_{1x} = 10.87\,(0.582) = 6.33\,k.ft \quad [14.92\,(0.582) = 8.68\,kN.m]$$
$$M_{2x} = 22.97(0.582) = 13.37\,k.ft \quad [31.53(0.582) = 18.35\,kN.m]$$
$$C_{mx} = 0.6 - 0.4\left(\frac{6.33}{13.37}\right) = 0.411 > 0.4 \quad \left[C_{mx} = 0.6 - 0.4\left(\frac{8.68}{18.35}\right) = 0.411 > 0.4\right]$$

The P_{elx} for W10 × 30 (W250 × 49.1) is the same as for the gravity load calculated before for the LRFD method, and is equal to

$$P_{elx} = 2669.80\,k(11920.50\,kN)$$

$$B_{1x} = \frac{0.411}{1 - \dfrac{1.6(38.51)}{2269.80}} = 0.422 < 1.0 \quad \left[B_{1x} = \frac{0.411}{1 - \dfrac{1.6(173.34)}{11920.50}} = 0.421 < 1.0\right]$$

Therefore,

$$B_{1x} = 1.0$$

$$\Sigma P_{nt} = (63.59 + 151.22 + 63.59)0.582 - (3.35 - 3.35)0.45 = 162.03\,k$$
$$[(286.16 + 680.5 + 286.16)0.582 - (15.08 - 15.08)0.45 = 729.14\,kN]$$

$$P_{e2x} = \frac{\pi^2 EI}{(K_{x2}L)^2} = \frac{\pi^2(29000)(170)}{(1.28x15x12)^2} = 916.6k$$

$$\left[P_{e2x} = \frac{\pi^2(200,000)(71.1x10^6)}{(1.28x4.575x1000)^2(1000)} = 4092.58kN \right]$$

$\Sigma P_{e2x} = 3\,(916.6) = 2749.8$ k $\,[3(4092.58) = 12,277.74$ kN], same as in the LRFD method

$$B_{2x} = \frac{1}{1 - \dfrac{\alpha \Sigma P_{nt}}{\Sigma P_{e2x}}} \geq 1$$

$$B_{2x} = \frac{1}{1 - \dfrac{1.6(162.03)}{2749.8}} = 1.10 \quad \left[B_{2x} = \frac{1}{1 - \dfrac{1.6(729.14)}{12277.74}} = 1.10 \right]$$

$P_r = P_{nt} + B_{2x}P_{ltx} = 37 + 1.10(1.51) = 38.66$ k $[166.55 + 1.10(6.79) = 174.02$ kN$]$

$M_{rx} = B_{1x}M_{ntx} + B_{2x}M_{ltx}$

$M_{rx} = 1(13.37) + 1.10(11.33) = 25.83$ k.ft $\,[M_{rx} = 1(18.35) + 1.10(15.53) = 35.43$ kN.m$]$

$$\frac{K_{x2}L_x}{r_{xx}} = \frac{1.28x15x12}{4.38} = 52.60 \quad \left[\frac{K_xL_x}{r_{xx}} = \frac{1.28x4.575x1000}{111} = 52.76 \right]$$

$$< \frac{K_yL_y}{r_{yy}} = \frac{1x15x12}{1.37} = 131.39 \quad \left[\frac{K_yL_y}{r_{yy}} = \frac{1x4.575x1000}{35.1} = 130.34 \right]$$

Therefore, $P_c = 76.96$ k $\,(343.47$ kN$)$ as for the ASD design before

$M_{cx} = 60.22$ k.ft $[82.64$ N.m$]$ from the AISC manual [8]

$$\frac{P_r}{P_c} = \frac{38.66}{76.96} = 0.502 > 0.2 \quad \left[\frac{P_r}{P_c} = \frac{174.02}{343.47} = 0.507 > 0.2 \right]$$

$$\frac{P_r}{P_c} + \frac{8}{9}\left(\frac{M_{rx}}{M_{cx}} + \frac{M_{ry}}{M_{cy}} \right) \leq 1.0$$

$M_{ry} = 0,$

therefore,

$$\frac{P_r}{P_c} + \frac{8}{9}\left(\frac{M_{rx}}{M_{cx}} \right) = \frac{38.66}{76.96} + \frac{8}{9}\left(\frac{25.83}{60.22} \right) = 0.502 + 0.381 = 0.883 < 1.0 \text{ O.K.}$$

$$\left[\frac{P_r}{P_c} + \frac{8}{9}\left(\frac{M_{rx}}{M_{cx}} \right) = \frac{174.02}{343.47} + \frac{8}{9}\left(\frac{35.43}{82.64} \right) = 0.507 + 0.381 = 0.888 < 1.0 \text{ O.K.} \right]$$

The section is satisfactory.

Similarly, the other members of the frame can be designed.

Problems

5.1 Find the buckling load of the frame in Figure P5.1 by the equilibrium method when there is no sidesway.

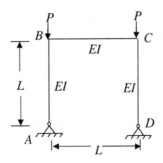

Figure P5.1

5.2 Find the buckling load of the frame in Figure P5.1 by the equilibrium method when there is sidesway.

5.3 Find the buckling load of the frame in Figure P5.3 by the slope deflection method when there is no sidesway. EI = Constant.

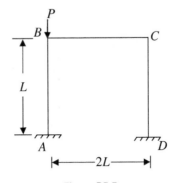

Figure P5.3

5.4 Find the buckling load of the frame in Figure P5.3 by the slope deflection method when the sidesway is permitted. EI = Constant.

5.5 Find the critical load of the frame in Figure P5.5 by the matrix method.

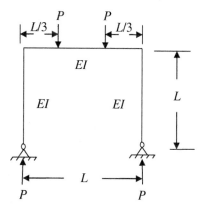

Figure P5.5

5.6 For the frame in Figure P5.6, find the difference in G factor for both sway inhibited and sway not inhibited cases, when
   (a) the ends of the beams are pinned
   (b) the ends of the beams are fixed.

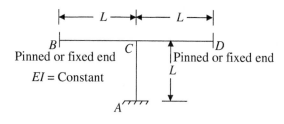

Figure P5.6

References

1 Kumar, A. (1998). *Stability of Structures*. New Delhi, India: Allied Publishers Ltd.
2 Iyengar, N.G.R. (2007). *Elastic Stability of Structural Elements*. Delhi, India: Macmillan India, Ltd.
3 Chen, W.F. and Lui, E.M. (1991). *Structural Design of Steel Frames*. Boca Raton, FL: CRC Press.
4 Ziemian, R.D. (ed.) (2010). *Guide to Stability Design Criteria for Metal Structures*, 6e. Hoboken, NJ: Wiley.
5 Fung, Y.C. (1965). *Foundations of Solid Mechanics*. Englewood Cliffs, NJ: Prentice Hall.

6 Weaver, W. and Gere, J.M. (1990). *Matrix Analysis of Framed Structures*, 3e. New York: Van Nostrand Reinhold.

7 Julian, O.G. and Lawrence, L.S. (1959) Notes on J and L nomograms for determination of effective lengths. Unpublished report.

8 AISC (2017) Steel Construction Manual, Fifteenth Edition, American Institute of Steel Construction (AISC), Chicago, IL.

9 Yura, J.A. (1971). The effective length of columns in unbraced frames. *Engineering Journal* 8 (2): 37–42.

10 ASCE (2016) Standard ASCE/SEI-16. *Minimum Design Loads and Associated Criteria for Buildings and Other Structures*. Reston, VA: American Society of Civil Engineers.

6

Torsional Buckling and Lateral Buckling of Beams

6.1   Introduction

So far we have assumed that buckling of members occurs due to bending only, where the bending deformation takes place in the plane of one of the principal axes and there is no rotation. This occurs if the bending rigidity of a member EI is smaller than the torsion rigidity GJ. Closed thin-walled and solid sections usually fail in this type of failure. However, beams and columns can also fail by buckling due to torsion alone or by a combined torsion-flexure mode. This type of failure occurs in sections where torsion rigidity is low in comparison to the flexural rigidity. Open thin-walled sections made of narrow rectangles, such as I, channel, T, angle, etc. sections usually have low torsion rigidity and are thus likely to buckle through torsion. A pure torsion buckling can occur in a section with two axes of symmetry where the centroidal and shear center axes coincide. For other sections, such as angle and channel sections, where the centroidal and shear center axes do not coincide, combined torsion-flexure mode of buckling occurs. The torsion-flexural buckling also occurs in transversely loaded beams, when the compression flange becomes unstable and wants to buckle laterally but the tension flange is stable and straight. We will consider in this chapter both axially loaded columns and transversally loaded beams for different buckling modes.

6.2   Pure Torsion of Thin-Walled Cross-Sections

In the case of a circular cross-section, there is no warping when a twisting moment is applied. When a twisting moment is applied to a non-circular structural member, the cross-section warps in the axial direction in addition to twisting. If the noncircular member is twisted by couples at its ends, acting in planes normal to the axis of the member, and the ends are free to warp, we have the case of pure torsion as shown in Figure 6.1a. In this case the transverse sections that were plane before twisting are no longer plane after deformation. Since the warping deformation is the same for all cross-sections and it takes place without causing any axial strain/stress, shear stresses are the only stresses produced. The distribution of the shear stresses depends on the shape of the cross-section and is the same for all sections along the length of the member. These shear stresses entirely resist the applied torque similar to those present in

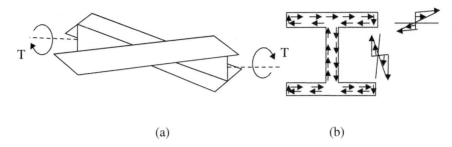

(a)   (b)

Figure 6.1 Twisting of an I-section free to warp: (a) I-section under uniform torsion; (b) Shear stress distribution.

a circular section. The shear stresses are called St. Venant shear stresses, and the associated torque is called St. Venant or pure torsion, T_{sv}.

For pure torsion, the relation between the angle of twist, β, and the applied torque, $T = T_{sv}$, is given by [1].

$$T_{sv} = GJ\frac{d\beta}{dz} \tag{6.1a}$$

where G is the shear modulus of elasticity of the material, J is the torsion constant of the cross-section, and z is the distance along the longitudinal axis of the member. The quantity, $\dfrac{d\beta}{dz}$, is the angle of twist per unit length, and is constant for pure torsion over the length of the member. For thin-walled open sections, it can be assumed that the shearing stress at any point in an element is parallel to the tangent to the middle line at that point in that element. The magnitude of the shear stresses is proportional to the distance from that middle line shown in Figure 6.1b. The torsion constant J for the cross-section is given by

$$J = \frac{1}{3}\sum_{i=1}^{n} b_i t_i^3 \tag{6.1b}$$

where b_i and t_i are the length of the middle line and thickness, respectively, of any element, i, of the cross-section. n is the number of elements in the cross-section. Eq. (6.1b) is valid if b_i/t_i is more than 10, if it is smaller than 10, a correction factor is used [2]. The values of J, for various shapes are given in Appendix B. For a thin-walled closed section, a membrane analogy [2] can be used to find the torsion constant J as follows:

$$J = \frac{4A^2}{\oint \frac{dl}{t}} \tag{6.1c}$$

where A is the area enclosed by the middle line, l is the peripheral mean length, and t is the thickness of the section. For sections made of thin rectangular elements, the maximum shear stress due to pure torsion, T_{sv} is given by Boresi and Schmidt [2].

$$\tau_{max} = \frac{2T_{sv}h_{max}}{J} \tag{6.1d}$$

where h_{max} is the maximum value of h_i.

6.3 Non-uniform Torsion of Thin-Walled Open Cross-Sections

6.3.1 I-section

The non-uniform torsion occurs if any cross-sections of the beam are not free to warp or the torsion moment varies along the length of the member. In such cases, the warping due to torsion varies along the bar and there will be axial compressive or tension stresses in the cross-section in addition to the shear stresses. Consider the case of a symmetrical I section cantilever beam. One end of the beam is fixed where there is no warping allowed, and the other end is free where the cross-section can warp. In this case a constraint on warping results in a differential bending of the flanges, one flange bends to the right and another to the left, as shown in Figure 6.2a. The applied torque T is resisted partially by the shear stresses due to pure torsion, as shown before, and partially by the bending resistance of the flanges. The total applied torque is given by

$$T = T_{sv} + T_w \tag{6.2a}$$

where T_{sv} is the St. Venant's torsion, and T_w is the warping torsion resistance due to the bending of the flanges. If M_f denotes the bending moment in one flange due to the bending of flanges, and V_f is the corresponding shear force shown in Figure 6.2b, then by beam theory

$$V_f = -\frac{dM_f}{dz} \tag{6.2b}$$

The minus sign in Eq. (6.2b) is present because the bending moment M_f decreases as the distance z along the beam increases. The shear forces V_f are acting in the opposite direction because the flanges bend in the opposite directions. The couple formed by these two shear forces is the value of the warping torsion given as

$$T_w = V_f(h) \tag{6.2c}$$

or from Eq. (6.2b),

$$T_w = -\frac{dM_f}{dz}(h) \tag{6.2d}$$

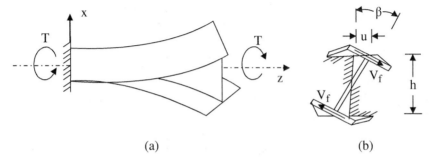

(a)　　　　　　　　　　　(b)

Figure 6.2 Noncircular section under the action of applied torque: (a) Flange bending due to warping; (b) Twisting of I-section.

Let u be the lateral displacement of the upper flange and β is the angle of twist at any distance z, then

$$u = \beta \frac{h}{2} \tag{6.2e}$$

The bending moment in the upper flange is given by

$$M_f = EI_f \frac{d^2 u}{dz^2} \tag{6.2f}$$

Equations (6.2d)–(6.2f) are combined to give

$$T_w = -\frac{EI_f h^2}{2} \frac{d^3 \beta}{dz^3}$$

or

$$T_w = -EC_w \frac{d^3 \beta}{dz^3} \tag{6.2g}$$

where I_f = moment of inertia of one flange about its strong axis, E = modulus of elasticity of the material of the cross-section, and $C_w = \dfrac{I_f h^2}{2}$ is called the warping constant of the I-section. In the case of non-uniform torsion, the total resistance of an I-section is obtained from Eqs. (6.1a) and (6.2g) as follows:

$$T = T_{sv} + T_w = GJ \frac{d\beta}{dz} - EC_w \frac{d^3 \beta}{dz^3} \tag{6.2h}$$

The angle of twist β can be found from the integration of Eq. (6.2h), if T is the known function of z. Now, when β is known, the St. Venant's and warping torsion resistances, T_{sv} and T_w, can be obtained. The stresses in the beam can be calculated from these two torsion components.

6.3.2 General Thin-Walled Open Cross-Sections

The warping constant calculated before for an I-section is a property of a cross-section and is different for different cross-sections. The warping constant C_w for a general open thin-walled section in Figure 6.3 is derived here with the help of material available in the literature [3]. The quantity w_s is called the warping function and is expressed as

$$w_s = \int_0^s rds \tag{6.3a}$$

where r = distance from the tangent at the middle line of the cross-section at any point P around the cross-section to the shear center O, it is taken as positive if a vector along the tangent and pointing toward the increasing s direction acts counterclockwise about the axis of rotation. s = distance along the middle line from one end of the cross-section. The quantity, $\bar{w}_s$, called the average value of w_s over the entire cross-section is written as

$$\bar{w}_s = \frac{1}{m} \int_0^m w_s ds \tag{6.3b}$$

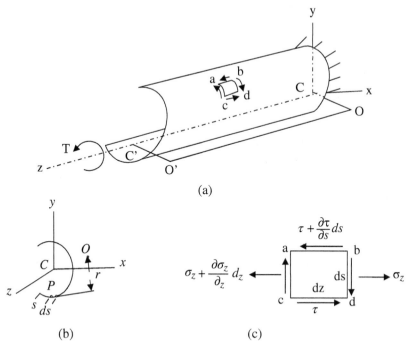

Figure 6.3 Thin-walled open section bar under non-uniform torsion: (a) Thin wall open section bar; (b) Thin walled open cross-section; (c) Element from the bar wall.

where m is the total length of the middle line of the cross-section. The quantity w, called the warping displacement [3], denotes the displacement in the z direction (along the longitudinal axis of the member) and is given by

$$w = (\bar{w}_s - w_s)\frac{d\beta}{dz} \tag{6.3c}$$

where $\dfrac{d\beta}{dz}$ is the angle of twist per unit length, and is variable along the length of the member in the non-uniform torsion. The axial strain at a point in the member is given by

$$\varepsilon_z = \frac{\partial w}{\partial z}$$

or

$$\varepsilon_z = (\bar{w}_s - w_s)\frac{d^2\beta}{dz^2} \tag{6.3d}$$

because $\bar{w}_s$ and w_s are independent of z. The normal stress produced by non-uniform torsion by applying Hooke's law is

$$\sigma_z = E\varepsilon_z = E(\bar{w}_s - w_s)\frac{d^2\beta}{dz^2} \tag{6.3e}$$

The sum of the forces on a cross-section in the z direction should be zero when it is subjected to torsion only, i.e. $\Sigma F_z = 0$. Therefore,

$$\int_0^m \sigma_z t ds = E\frac{d^2\beta}{dz^2} \int_0^m (\bar{w}_s - w_s)t ds = 0$$

or

$$\int_0^m (\bar{w}_s - w_s)t ds = 0 \tag{6.3f}$$

Assume a thin bar of open arbitrary shape shown in Figure 6.3a is built in at one end and is subjected to a torque T at the free end. The line OO' is the shear center axis, and it remains straight during torsion [3]. The cross-sections of the bar rotate about the axis OO'. Let β be the angle of rotation of any cross-section as before. The normal stresses σ_z produce shearing stresses as shown in Figure 6.3c. Let us consider an element $abcd$ cut out from the wall of the member shown in Figure 6.3b. It is assumed that the shearing stresses τ, are constant over the thickness t and are parallel to the tangent to the middle surface of the cross-section. Assume t, the thickness of the cross-section, may vary with s but it is independent of z. In the z direction, $\Sigma F_z = 0$ in Figure 6.3b, therefore

$$\sigma_z t ds - \left(\sigma_z + \frac{\partial \sigma_z}{\partial z}dz\right)t ds + \tau t dz - \left(\tau t + \frac{\partial(\tau t)}{\partial s}ds\right)dz = 0$$

or

$$\frac{\partial \sigma_z}{\partial z}t ds dz + \frac{\partial(\tau t)}{\partial s}ds dz = 0$$

or

$$\frac{\partial(\tau t)}{\partial s} = -t\frac{\partial \sigma_z}{\partial z} \tag{6.3g}$$

Substituting σ_z from Eq. (6.3e) for σ_z in Eq. (6.3g) gives

$$\frac{\partial(\tau t)}{\partial s} = -Et(\bar{w}_s - w_s)\frac{d^3\beta}{dz^3}$$

Integrating both sides with respect to s, substitute $\tau = 0$ for $s = 0$, and since β is independent of s, we get

$$\tau t = -E\frac{d^3\beta}{dz^3}\int_0^s (\bar{w}_s - w_s)t ds \tag{6.3h}$$

The warping torsion resistance T_w is calculated by summing the moments of the elemental shear forces about the shear center. Therefore,

$$T_w = \int_0^m r\tau t\, ds$$

Substituting τt from Eq. (6.3h) we get

$$T_w = -E\frac{d^3\beta}{dz^3}\int_0^m \left[\int_0^s (\bar{w}_s - w_s)t\, ds\right] r\, ds \tag{6.3i}$$

From Eq. (6.3a) we have

$$w_s = \int_0^s r \, ds$$

or

$$r \, ds = d(w_s)$$

We can write the following because $\bar{w}_s$ is independent of s,

$$r \, ds = -\frac{d}{ds}(\bar{w}_s - w_s) \, ds$$

The integrant in Eq. (6.3i) can be expressed as

$$\int_0^m \left[\int_0^s (\bar{w}_s - w_s)t \, ds \right] r \, ds = -\int_0^m \left[\int_0^s (\bar{w}_s - w_s)t \, ds \right] \frac{d(\bar{w}_s - w_s)}{ds} \, ds$$

Integration of the right-hand side by parts and using Eq. (6.3f) gives

$$\int_0^m \left[\int_0^s (\bar{w}_s - w_s)t \, ds \right] r \, ds = \int_0^m (\bar{w}_s - w_s)^2 t \, ds$$

Substitute in Eq.(6.3i) and we get

$$T_w = -E\frac{d^3\beta}{dz^3} \int_0^m (\bar{w}_s - w_s)^2 t \, ds$$

or

$$T_w = -EC_w \frac{d^3\beta}{dz^3} \tag{6.3j}$$

where C_w is the warping constant for the thin-walled open cross-section of general shape, and is given by

$$C_w = \int_0^m (\bar{w}_s - w_s)^2 t \, ds \tag{6.3k}$$

For non-uniform torsion, the total torsion resistance of a thin-walled open cross-section of general shape, Eq. (6.2h) is still applicable, i.e.

$$T = T_{sv} + T_w = GJ\frac{d\beta}{dz} - EC_w \frac{d^3\beta}{dz^3} \tag{6.2h}$$

where the warping constant of a general shape, thin-walled cross-section is given by Eq. (6.3k), and the torsion constant J is given by Eq. (6.1b). Equation (6.2h) is applicable to any thin-walled open section. If T is known as a function of z, the angle of twist β is found by solving the differential Eq. (6.2h). When β is known, torques T_{sv} and T_w can be found from Eqs. (6.1a) and (6.3j), respectively. The stresses due to St. Venant's torque T_{sv} are calculated in the same way as for pure torsion. The normal and shear stresses produced by warping torsion T_w are obtained from Eqs. (6.3e) and (6.3h), respectively.

6.3.3 Warping Constant C_w of a Channel Section

Consider a channel section and assume that the rotation occurs about the longitudinal axis passing through the shear center O in Figure 6.4. The warping constant is given by

$$C_w = \int_0^m (\bar{w}_s - w_s)^2 t\, ds \tag{6.3k}$$

The warping function w_s is given from Eq. (6.3a) as

$$w_s = \int_0^s r\, ds$$

For the element EF, $0 \le s \le b$, the distance $r = \frac{h}{2}$, and it is positive

$$w_s = \int_0^s \frac{h}{2}\, ds = \frac{sh}{2} \tag{6.4a}$$

For the element FG, $b \le s \le b+h$, the distance $r = e$, and it is negative

$$w_s = \frac{bh}{2} - \int_b^s e\, ds = \frac{bh}{2} + be - se \tag{6.4b}$$

For the element GH, $b+h \le s \le 2b+h$, the distance $r = \frac{h}{2}$, and it is positive

$$w_s = \frac{bh}{2} - he + \int_{b+h}^s \frac{h}{2}\, ds = \frac{bh}{2} - he + \frac{h}{2}(s - b - h) = -he - \frac{h^2}{2} + \frac{hs}{2} \tag{6.4c}$$

The average value of the warping function $\bar{w}_s$ is given from Eq. (6.3h) as

$$\bar{w}_s = \frac{1}{m} \int_0^m w_s\, ds$$

$$\text{or} \quad \bar{w}_s = \frac{1}{m} \left[\int_0^b \frac{sh}{2}\, ds + \int_b^{b+h} \left(\frac{bh}{2} + be - se \right) ds + \int_{b+h}^{2b+h} \left(-he - \frac{h^2}{2} + \frac{hs}{2} \right) ds \right]$$

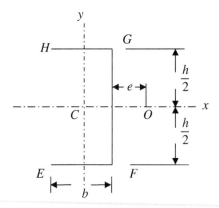

Figure 6.4 Channel cross-section.

or $\quad \bar{w}_s = \dfrac{1}{m} \left[\dfrac{h}{2}(b - e)(2b + h) \right]$

$\qquad m = b + h + b = 2b + h$

or $\quad \bar{w}_s = \dfrac{h}{2}(b - e)$ $\qquad\qquad\qquad\qquad\qquad$ (6.4d)

Substitute Eqs. (6.4a)–(6.4d) into Eq. (6.3c), and we get

$$w = \left(\dfrac{h}{2} \right)(b - e - s)\dfrac{d\beta}{dz} \qquad\qquad 0 \le s \le b \qquad\qquad (6.4e)$$

$$w = e\left(s - b - \dfrac{h}{2} \right)\dfrac{d\beta}{dz} \qquad\qquad b \le s \le b + h \qquad\qquad (6.4f)$$

$$w = \dfrac{h}{2}(b + e + h - s)\dfrac{d\beta}{dz} \qquad\qquad b + h \le s \le 2b + h \qquad\qquad (6.4g)$$

The warping constant C_w is given, as before, by

$$C_w = \int_0^m (\bar{w}_s - w_s)^2 t \, ds \qquad\qquad\qquad\qquad (6.3k)$$

Substitute Eqs. (6.4a)–(6.4d) into Eq. (6.3k), and we obtain

$$C_w = \int_0^b \dfrac{h^2}{4}(b - e - s)^2 t \, ds + \int_b^{b+h} e^2\left(-b - \dfrac{h}{2} + s \right)^2 t \, ds + \int_{b+h}^{2b+h} \dfrac{h^2}{4}(b + e + h - s)^2 t \, ds$$

$$(6.4h)$$

Find the squares of the quantities in the parentheses and integrate with respect to s, noting that b, e, h, and t are constant. The integration which is simple and lengthy gives

$$C_w = \dfrac{h^2 t}{4}\left(2e^2 b - 2eb^2 + \dfrac{2}{3}b^3 \right) + \dfrac{e^2 t h^3}{12}$$

or

$$C_w = \dfrac{h^2 t}{12}[e^2 h + 2b^3 - 6eb(b - e)] \qquad\qquad (6.4i)$$

The distance e for the channel section in Figure 6.4 whose flange thickness t_f and web thickness t_w are constants is given in Appendix B as

$$e = \dfrac{3b^2 t_f}{6bt_f + ht_w} \qquad\qquad\qquad\qquad (6.4j)$$

If

$$t_f = t_w = t,$$

then

$$e = \dfrac{3b^2}{6b + h} \qquad\qquad\qquad\qquad (6.4k)$$

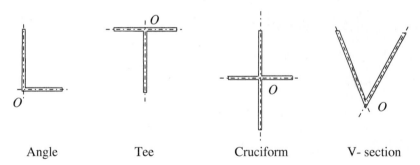

Figure 6.5 Cross-section having zero warping constant ($C_w = 0$).

Substitute e from Eq. (6.4k) into Eq. (6.4j), we get the warping constant for the channel section given by

$$C_w = \frac{h^2 t b^3}{12}\left(\frac{3b + 2h}{6b + h}\right) \tag{6.4l}$$

Similarly, the warping constant can be found for various other sections. The torsion constants J and Cw, and the shear center for different thin-walled open cross-section are shown in Appendix B. For cross-sections that consist of thin rectangular elements that intersect at a common point, such as those shown in Figure 6.5, if the rotation is taken about an axis passing through the shear center O, the warping constant is zero. The warping constant is zero because the distance r in Eq. (6.3a) is zero for all the points lying on the center lines of different elements. This means these sections do not warp when subjected to torsion moments.

6.4 Torsional Buckling of Columns

A doubly symmetric bar with a thin-walled cross-section when subjected to an axial compressive force may have pure torsion buckling, while its longitudinal axis remains straight. This type of buckling is important for columns that have short length and wide flanges. The pure torsional buckling of doubly symmetric bars whose center of gravity and shear center coincide, such as symmetric I-section, a Z-section, or cruciform section is considered here. Consider a doubly symmetric cruciform section having four identical flanges of width b, and thickness t. The length of the column is L as shown in Figure 6.6.

The column cross-section has x and y axes of symmetry, and z is the longitudinal axis. Pure torsional buckling may occur under compression and the column may take the buckled shape shown in Figure 6.6a. In the buckled shape, each flange has rotated about the z axis, whereas the longitudinal z axis remains straight. Assume the column is pin-supported at the top and the bottom. We consider the equilibrium of the deflected shape of the column to find the buckling load. Initially the column is straight and is subjected to an axial force. When the axial force reaches its critical value, the column can be in equilibrium in a slightly deflected shape shown in Figure 6.6. Now, in addition to the initial uniform compressive stresses, the bending stresses

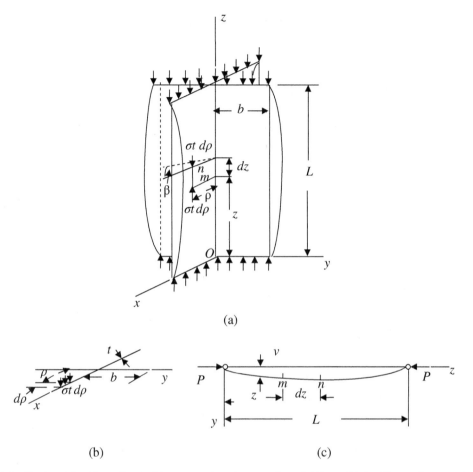

(a)

(b)   (c)

Figure 6.6 Torsional buckling of a doubly symmetric column under axial force: (a) Cruciform column under compressive load; (b) Cross-section of the element *mn*; (c) A section of the bent column.

are superposed on the cross-section. The differential equation of the deflected longitudinal fiber *mn* at a distance ρ from the z axis is found from Eq. (4.1f) in Chapter 4, and can be rewritten as

$$EI_x\frac{d^4v}{dz^4} + P\frac{d^2v}{dz^2} = \frac{w(z)}{EI} \tag{6.5a}$$

Taking $w(z) = 0$, therefore,

$$EI_x\frac{d^4v}{dz^4} = -P\frac{d^2v}{dz^2} \tag{6.5b}$$

From Eqs. (2.1a and 2.1b) we have

$$M = EI\frac{d^2v}{dz^2}$$

or

$$\frac{dM}{dz} = EI\frac{d^3v}{dz^3}$$

From *Mechanics of Materials* [4].

$$\frac{dM}{dz} = V, \quad \frac{dV}{dz} = w(z)$$

where M and V are the bending moment and shear force at a section of a beam loaded by lateral load of intensity $w(z)$ per unit length. Therefore,

$$V = EI\frac{d^3v}{dz^3}$$

or

$$\frac{dV}{dz} = EI\frac{d^4v}{dz^4}$$

or

$$w(z) = EI\frac{d^4v}{dz^4} \tag{6.5c}$$

Comparing Eqs. (6.5b) and (6.5c), we can say that the column is subjected to a fictitious lateral load intensity of $-P\dfrac{d^2v}{dz^2}$. The lateral load is responsible for the torsion of the column about the z axis. The cross-section area of the fiber mn is $t\,d\rho$, shown in Figure 6.6b. The deflection of the element mn in the y direction due to torsion buckling is given by

$$v = \rho\beta \tag{6.5d}$$

where β is the small angle of twist of the cross-section due to torsion buckling. It is assumed the cross-section does not distort, its shape does not change, and it only rotates. The vertical compressive force in the element mn at its ends is given by $\sigma t\,d\rho$, where $\sigma = \dfrac{P}{A_c}$ is the initial compressive stress in the column, and A_c is the cross-sectional area. The equivalent lateral load intensity $w(z)$ due to the compressive force is found from Eq. (6.5b) and we get

$$w(z) = -(\sigma t d\rho)\frac{d^2v}{dz^2} \tag{6.5e}$$

Substitute Eq. (6.5d) into Eq. (6.5e) and we have

$$w(z) = -\sigma t(d\rho)\rho\frac{d^2\beta}{dz^2} \tag{6.5f}$$

The fictitious lateral force $w(z)$ acts on the column at a distance ρ from the z axis. The torsional moment of the force $w(z)$ acting on the element mn in Figure 6.6c about the z axis is equal to

$$-\sigma t\rho d\rho\frac{d^2\beta}{dz^2}(dz)(\rho) = -\sigma\frac{d^2\beta}{dz^2}dzt\rho^2 d\rho \tag{6.5g}$$

Summing up the moments for the entire cross-section at a distance z along the length of the column, we get the torsion acting on the element mn of the buckled bar. This torsion is given by

$$m_z dz = -\sigma \frac{d^2\beta}{dz^2} dz \int_A t\rho^2 d\rho = -\sigma \frac{d^2\beta}{dz^2} I_0 dz$$

where I_0 is the polar moment of inertia of the entire cross-section about the shear center O. The entire cross-section in this case consists of four flanges in Figure 6.6. In this derivation, the centroid and the shear center of the cross-section coincide. The torque generated per unit length of the column m_z is obtained from

$$m_z = -\sigma \frac{d^2\beta}{dz^2} I_0 \tag{6.5h}$$

Equation (6.5h) is valid for any cross-section where the centroid and the shear center coincide. We can use Eq. (6.2h) derived for non-uniform torsion in a thin-walled open cross-section to find the torsional buckling load.

$$T = GJ \frac{d\beta}{dz} - EC_w \frac{d^3\beta}{dz^3} \tag{6.2h}$$

Differentiate Eq. (6.2h) with respect to z, and we have

$$\frac{dT}{dz} = GJ \frac{d^2\beta}{dz^2} - EC_w \frac{d^4\beta}{dz^4} \tag{6.5i}$$

Assuming the counterclockwise torsional moment is positive, then considering the equilibrium of the element of length dz in Figure 6.7, we have

$$-T + m_z dz + T + dT = 0$$

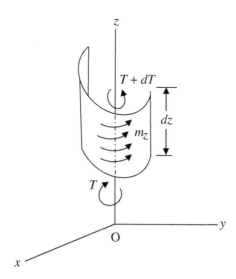

Figure 6.7 Torsional buckling of an open thin-walled section.

or

$$m_z = -\frac{dT}{dz} \tag{6.5j}$$

Combine Eqs. (6.5i) and (6.5j) and we get

$$m_z = -GJ\frac{d^2\beta}{dz^2} + EC_w\frac{d^4\beta}{dz^4} \tag{6.5k}$$

Substitute Eq. (6.5h) into Eq. (6.5k) and we obtain

$$EC_w\frac{d^4\beta}{dz^4} - (GJ - \sigma I_o)\frac{d^2\beta}{dz^2} = 0 \tag{6.5l}$$

Equation (6.5l) is the governing differential equation for the pure torsional buckling of a thin-walled open cross-section column as long as the shear center and centroid coincide. The critical compressive stress σ_z and the critical load can be calculated from Eq. (6.5l). In addition to the torsional mode, the column may have pure flexural modes in the XZ and YZ planes in Figure 6.6. The buckling load will be the minimum of the three loads corresponding to the three modes.

6.5 Torsional Buckling Load

6.5.1 Thin-Walled Open Sections with Rectangular Elements Intersecting at a Point

For the column shown in Figure 6.6, $C_w = 0$, therefore, for torsional buckling of a column of the cruciform section, Eq. (6.5l) gives

$$GJ - \sigma I_o = 0$$

or

$$\sigma_{cr} = \frac{GJ}{I_o} \tag{6.6a}$$

where the torsional constant of the cross-section from Eq. (6.1b) is given by

$$J = \frac{1}{3}\sum_1^4 b_i t_i^3 = \frac{1}{3}(4bt^3) = \frac{4}{3}bt^3$$

and the polar moment of inertia of the cruciform section about the z axis passing through the shear center o is equal to

$$I_o = \frac{4}{3}tb^3$$

Therefore,

$$\sigma_{cr} = \frac{G\left(\frac{4}{3}bt^3\right)}{\frac{4}{3}(tb^3)} = \frac{Gt^2}{b^2} \tag{6.6b}$$

6.5.2 Thin-Walled Open Doubly Symmetric Sections

For column cross-sections where the shear center and centroid coincide, while the warping constant C_w is not equal to zero, the critical stress and the critical load can be found from Eq. (6.5l) as follows:

$$\frac{d^4\beta}{dz^4} + \lambda^2\frac{d^2\beta}{dz^2} = 0 \tag{6.7a}$$

where

$$\lambda^2 = \frac{\sigma I_o - GJ}{EC_w} \tag{6.7b}$$

The solution of Eq. (6.7a) is given by (see Section 2.5)

$$\beta = A\sin\lambda z + B\cos\lambda z + Cz + D \tag{6.7c}$$

6.5.2.1 Pinned-pinned Columns

For the compressed pinned-pinned column, the ends of the column cannot rotate about the z axis, but are free to warp, the boundary conditions are:

$$\beta = 0 \text{ and } \frac{d^2\beta}{dz^2} = 0 \text{ (because the bending moment is zero) at } z = 0$$

$$\beta = 0 \text{ and } \frac{d^2\beta}{dz^2} = 0 \text{ (because the bending moment is zero) at } z = L$$

Apply the boundary conditions at $z = 0$ to Eq. (6.7c) and we have

$$B + D = 0$$

$$\frac{d^2\beta}{dz^2} = -\lambda^2 A\sin\lambda z - \lambda^2 B\cos\lambda z$$

or

$$B = 0 \text{ and } D = 0$$

Now, apply the boundary conditions at $z = L$ to Eq. (6.7c) and we get

$$\beta = A\sin\lambda L + CL = 0$$

$$\frac{d^2\beta}{dz^2} = -\lambda^2 A\sin\lambda L = 0$$

or

$$A\sin\lambda L = 0 \text{ and } C = 0$$

Therefore, $\sin\lambda L = 0,$ or $\lambda L = n\pi, n = 1, 2, 3, \text{----}$

Substitute λ into Eq. (6.7b) to obtain

$$\frac{n^2\pi^2}{L^2} = \frac{\sigma I_o - GJ}{EC_w}$$

or

$$\sigma = \frac{1}{I_o}\left(GJ + \frac{n^2\pi^2}{L^2}EC_w\right)$$

the smallest value of $n = 1$, which corresponds to the critical stress given by

$$\sigma_{cr} = \frac{1}{I_o}\left(GJ + \frac{\pi^2}{L^2}EC_w\right)$$

The critical load for this case for pure torsion is obtained from

$$P_{cr_z} = \frac{1}{r_o^2}\left(GJ + \frac{\pi^2}{L^2}EC_w\right) \tag{6.8a}$$

where r_o is the polar radius of gyration of the cross-section with respect to the shear center. The column may buckle because of lateral bending about the x and y axes given by Euler's formula as follows:

$$P_{cr_x} = \frac{\pi^2 EI_x}{L^2}$$

$$P_{cr_y} = \frac{\pi^2 EI_y}{L^2}$$

where I_x and I_y are the moment of inertia of the cross-section about the x and y axes. The buckling load is taken as the minimum of P_{cr_x}, P_{cr_y}, and P_{cr_z}. The three buckling loads are independent of each other.

The buckling shape at the critical stress is obtained from

$$\beta = A\sin\frac{\pi z}{L} \tag{6.8b}$$

Example 6.1 Find the torsional buckling load of a 20 ft (6.1 m) long column that is simply supported at its ends, whose cross-section is a W18×65 (in. × lb./ft) or W460×97 (mm × kg/m) wide flange shape, according to the American Institute of Steel Construction manual [5]. The area of cross-section is $A = 19.1$ in.^{2} (12 300 mm^2). The moment of inertia about the x and y axes are: $I_x = 1070$ in.^{4} (445 × 10^6 mm^4), and $I_y = 54.8$ in.^{4} (22.8 × 10^6 mm^4). The radius of gyration about the x and y axes are: $r_x = 7.49$ in. (190 mm), and $r_y = 1.69$ in. (43.1 mm). The torsional constant $J = 2.73$ in.^{4} (1130 × 10^{3}mm^4), and the warping constant $C_w = 4240$ in.^{6} (1140 × 10^{9}mm^6). The modulus of elasticity of steel $E = 29 \times 10^6$ psi (200 000 MPa), and the modulus of rigidity $G = 11 \times 10^6$ psi (75 000 MPa).

$$r_o = \sqrt{r_x^2 + r_y^2} = \sqrt{(7.49)^2 + (1.69)^2} = 7.68 \text{ in. } (\sqrt{(190)^2 + (43.1)^2} = 194.83 \text{ mm})$$

The pure torsional buckling load from Eq. (6.5f) is

$$P_{cr_z} = \frac{1}{(7.68)^2}\left(11x10^6(2.73) + \frac{\pi^2}{(20x12)^2}(29x10^6)(4240)\right)\frac{1}{1000} = 866.34 \text{ kips}$$

$$\left(\frac{1}{(194.83)^2}\left[75000(1130x10^3) + \frac{\pi^2}{(6.1x1000)^2}(200000)(1140x10^9)\right]\frac{1}{1000} = 3825.86 \text{ kN}\right)$$

The buckling load because of bending about the x axis is given by Euler's formula

$$P_{cr_x} = \frac{\pi^2 EI_x}{L^2} = \frac{\pi^2(29x10^6)(1070)}{(20x12)^2}x\frac{1}{1000} = 5316.91 \text{ kips}$$

$$\left(\frac{\pi^2(200000)(445x10^6)}{(6.1x1000)^2}x\frac{1}{1000} = 23606.42 \text{ kN}\right)$$

The buckling load because of bending about the y axis is given by Euler's formula

$$P_{cr_y} = \frac{\pi^2 EI_y}{L^2} = \frac{\pi^2(29x10^6)(54.8)}{(20x12)^2}x\frac{1}{1000} = 272.31 \text{ kips}$$

$$\left(\frac{\pi^2(200000)(22.8x10^6)}{(6.1x1000)^2}x\frac{1}{1000} = 1209.50 \text{ kN}\right)$$

The lowest of the three values gives the buckling load for the column

$$P_{cr} = 272.31 \text{ kips } (1209.50 \text{ kN})$$

For the doubly symmetric, hot rolled steel sections, the flexural buckling load about the weak axis is usually the lowest of the three buckling loads. Hence, these sections are designed for the flexural load and we neglect the pure torsional buckling load.

6.5.2.2 Fixed-fixed Columns

For the fixed-fixed columns, the ends of the column cannot rotate about the z axis and cannot warp. The boundary conditions are given by

$$\beta = 0 \quad \text{and} \quad \frac{d\beta}{dz} = 0 \text{ (From Eq.6.3c, because } w = 0) \text{ at } z = 0$$

$$\beta = 0 \quad \text{and} \quad \frac{d\beta}{dz} = 0 \text{ (From Eq.6.3c, because } w = 0) \text{ at } z = L$$

Apply the boundary conditions at $z = 0$ to Eq. (6.5e) and we have

$$B + D = 0 \tag{6.9a}$$

$$\frac{d\beta}{dz} = A\lambda \cos \lambda z - B\lambda \sin \lambda z + C$$

or

$$A\lambda + C = 0 \tag{6.9b}$$

Now, apply the boundary conditions at $z = L$ to Eq. (6.7c) and we get

$$A \sin \lambda L + B \cos \lambda L + CL + D = 0, \text{ and}$$

$$A\lambda Cos\lambda L - B\lambda \sin \lambda L + C = 0$$

Substitute C and D in terms of A and B from above to get

$$\begin{bmatrix} \sin \lambda L - \lambda L & \cos \lambda L - 1 \\ \lambda(\cos \lambda L - 1) & -\lambda \sin \lambda L \end{bmatrix} \begin{Bmatrix} A \\ B \end{Bmatrix} = \begin{Bmatrix} 0 \\ 0 \end{Bmatrix} \tag{6.9c}$$

For a non-trivial solution, the determinant of the coefficient matrix in Eq. (6.9c) should be zero

$$\begin{vmatrix} \sin \lambda L - \lambda L & \cos \lambda L - 1 \\ \lambda(\cos \lambda L - 1) & -\lambda \sin \lambda L \end{vmatrix} = 0$$

or

$$\sin \frac{\lambda L}{2} \left(\lambda L \cos \frac{\lambda L}{2} - 2 \sin \frac{\lambda L}{2} \right) = 0 \qquad (6.9\text{d})$$

or

$$\sin \frac{\lambda L}{2} = 0,$$

therefore,

$$\frac{\lambda L}{2} = n\pi$$

$$\lambda L = 2n\pi, \quad n = 1, 2, 3 - - - -$$

The minimum value of λ occurs when $n = 1$, therefore

$$\lambda L = 2\pi$$

Also, from Eq. (6.5k) we have

$$\tan \frac{\lambda L}{2} = \frac{\lambda L}{2}$$

or

$$\lambda L = 8.987$$

The minimum critical stress is found from

$$\lambda L = 2\pi$$

or $\lambda^2 = \frac{4\pi^2}{L^2}$. Using Eq. (6.5d) we obtain

$$\frac{4\pi^2}{L^2} = \frac{\sigma I_o - GJ}{EC_w}$$

The compressive critical stress is given by

$$\sigma_{cr} = \frac{1}{I_o} \left(GJ + \frac{4\pi^2}{L^2} EC_w \right)$$

The critical load for this case for pure torsion is obtained from

$$P_{cr} = \frac{1}{r_o^2} \left(GJ + \frac{4\pi^2}{L^2} EC_w \right) \qquad (6.9\text{e})$$

where r_o is the polar radius of gyration of the cross-section with respect to the shear center. The column may buckle because of lateral bending about the x and y axes given by Euler's formula as follows:

$$P_{cr_x} = \frac{4\pi^2 E I_x}{L^2}$$

$$P_{cr_y} = \frac{4\pi^2 E I_y}{L^2}$$

where I_x and I_y are the moment of inertia of the cross-section about the x and y axes. The buckling load is taken as the minimum of P_{cr_x}, P_{cr_y}, and P_{cr_z}. The three buckling loads are independent of each other.

Equations (6.9a) and (6.9b) give

$$B = -D \text{ and } A = C = 0$$

Therefore, using Eq. (6.7c), the buckled shape is of the form is given below

$$\beta = B\left(\cos\frac{2\pi z}{L} - 1\right) \tag{6.9f}$$

Example 6.2 Find the torsional buckling load for the column in the Example 6.1 if both ends are fixed.

$$r_o = \sqrt{r_x^2 + r_y^2} = \sqrt{(7.49)^2 + (1.69)^2} = 7.68 \text{ in. } (\sqrt{(190)^2 + (43.1)^2} = 194.83 \text{ mm})$$

The pure torsional buckling load from Eq. (6.5l) is

$$P_{cr_z} = \frac{1}{(7.68)^2}\left(11\mathrm{x}10^6(2.73) + \frac{4\mathrm{x}\pi^2}{(20\mathrm{x}12)^2}(29\mathrm{x}10^6)(4240)\right)\frac{1}{1000} = 1937.96 \text{ kips}$$

$$\left(\frac{1}{(194.83)^2}\left[75000(1130\mathrm{x}10^3) + \frac{4\mathrm{x}\pi^2}{(6.1\mathrm{x}1000)^2}(200000)(1140\mathrm{x}10^9)\right]\frac{1}{1000} = 8605.38 \text{ kN}\right)$$

The buckling load because of bending about the x axis is given by Euler's formula

$$P_{cr_x} = \frac{4\pi^2 E I_x}{L^2} = \frac{4\mathrm{x}\pi^2(29\mathrm{x}10^6)(1070)}{(20\mathrm{x}12)^2}\mathrm{x}\frac{1}{1000} = 21267.63 \text{ kips}$$

$$\left(\frac{4\mathrm{x}\pi^2(200000)(445\mathrm{x}10^6)}{(6.1\mathrm{x}1000)^2}\mathrm{x}\frac{1}{1000} = 94425.67 \text{ kN}\right)$$

The buckling load because of bending about the y axis is given by Euler's formula

$$P_{cr_y} = \frac{4\pi^2 E I_y}{L^2} = \frac{4\mathrm{x}\pi^2(29\mathrm{x}10^6)(54.8)}{(20\mathrm{x}12)^2}\mathrm{x}\frac{1}{1000} = 1089.24 \text{ kips}$$

$$\left(\frac{4\mathrm{x}\pi^2(200000)(22.8\mathrm{x}10^6)}{(6.1\mathrm{x}1000)^2}\mathrm{x}\frac{1}{1000} = 4838.00 \text{ kN}\right)$$

The lowest of the three values gives the buckling load for the column

$$P_{cr} = 1089.24 \text{ kips } (4838.00 \text{ kN})$$

For the doubly symmetric, hot rolled steel sections, the flexural buckling load about the weak axis is usually the lowest of the three buckling loads. Hence these sections are designed for the flexural load and we neglect the pure torsional buckling load.

6.6   Torsional Flexural Buckling

In general, the buckling failure usually occurs in thin-walled open cross-sections due to a combination of torsion and bending. Consider an unsymmetrical section shown in Figure 6.8 to

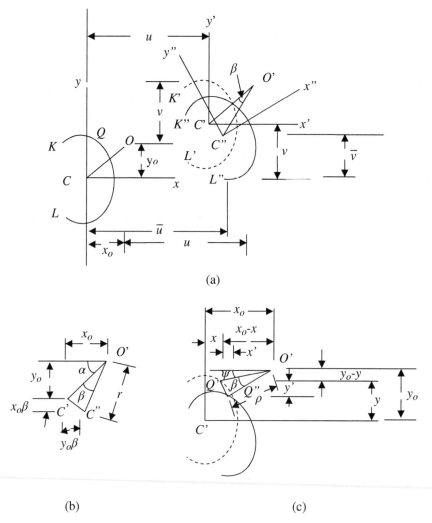

Figure 6.8   Torsional flexural buckling of a thin-walled section: (a) Displacement of section during torsional flexural buckling; (b) Displacement of centroid; (c) Displacement of a point on the section.

study this type of buckling. Let x and y be the principal axes of the cross-section passing through its centroid C, and x_o and y_o be the coordinates of the shear center O. The shear center of a cross-section is the point through which if the lateral forces pass, there will be bending without twisting. Also, the shear center is the center of rotation of the section in pure torsion. It is assumed that during buckling the deformation consists of bending about the two axes x and y, and a rotation of the cross-section about the shear center. The shear center displaces by u and v in the x and y directions respectively during buckling. In addition, during buckling, there is a rotation of the cross-section about the shear center O by an angle β about the longitudinal axis z of the member passing through the shear center. It is assumed the geometric shape of the cross-section in the xy plane remains constant during buckling. The lengths OC, $O'C'$, and $O'C''$ in Figure 6.8a shown as r in Figure 6.8b remain constant.

The point O moves to O' and the point C moves to C', and the cross-section moves from KL to $K'L'$ during the translation of the cross-section in Figure 6.8a. The cross-section moves to $K''L''$ and the centroid moves to C'' in Figure 6.8a during the rotation of the section about the shear center.

The displacement $\bar{u}$ of the centroid C to C'' during buckling in the x direction in Figures 6.8a and b is given by

$$\bar{u} = x_o + u - r\cos(\alpha + \beta)$$

or

$$\bar{u} = x_o + u - r\cos\alpha\cos\beta + r\sin\alpha\sin\beta = x_o + u - x_o\cos\beta + y_o\sin\beta$$

For small deformations, $\cos\beta \approx 1$ and $\sin\beta \approx \beta$, hence

$$\bar{u} = u + y_o\beta \tag{6.10a}$$

The displacement $\bar{v}$ of the centroid C to C'' during buckling in the y direction in Figures 6.8a and b is obtained from

$$\bar{v} = y_o + v - r\sin(\alpha + \beta)$$

$$\text{or} \quad \bar{v} = y_o + v - r\sin\alpha\cos\beta - r\cos\alpha\sin\beta = y_o + v - y_o\cos\beta - x_o\sin\beta$$

$$\text{or} \quad \bar{v} = v - x_o\beta \tag{6.10b}$$

The differential equations for the deflection curve of the longitudinal axis passing through the shear center from the theory of bending are

$$M_y = -EI_y\frac{d^2u}{dz^2} = P(u + y_o\beta)$$

$$\text{or} \quad EI_y\frac{d^2u}{dz^2} + P(u + y_o\beta) = 0 \tag{6.10c}$$

$$M_x = -EI_x\frac{d^2v}{dz^2} = P(v - x_o\beta)$$

$$\text{or} \quad EI_x\frac{d^2v}{dz^2} + P(v - x_o\beta) = 0 \tag{6.10d}$$

where P is the axial compressive force. A point Q on the cross-section moves to Q' during the translation of the cross-section, and then it moves to Q'' during the rotation of the section about the shear center in Figure 6.8c. The displacement of the point Q to Q'' during buckling in the x direction in Figures 6.8a and 6.8c is given by

$$u_Q = u + x' = u + Q'Q'' \sin \psi = u + \rho\beta \sin \psi$$

$$\text{or} \quad u_Q = u + (y_o - y)\beta \tag{6.10e}$$

The displacement of the point Q to Q'' during buckling in the y direction in Figures 6.8a and 6.8c is obtained from

$$v_Q = v - y' = v - Q'Q'' \cos \psi = v - \rho\beta \cos \psi$$

or

$$v_Q = v - (x_o - x)\beta \tag{6.10f}$$

Consider an element of length dz and cross-section $t\,ds$ at point Q having coordinates x, y in the plane of the cross-section. The compressive force in the element is $\sigma t ds$. Using Eq. (6.5b), the forces in the x and y direction per unit length are respectively

$$-(\sigma t ds)\frac{d^2 u_Q}{dz^2} = -(\sigma t ds)\frac{d^2}{dz^2}[u + (y_o - y)\beta] \tag{6.10g}$$

and

$$-(\sigma t ds)\frac{d^2 v_Q}{dz^2} = -(\sigma t ds)\frac{d^2}{dz^2}[v - (x_o - x)\beta] \tag{6.10h}$$

The moment of the forces given by Eqs. (6.10g) and (6.10h) about the longitudinal axis z passing through the shear center gives the torque per unit length for the elementary strip of the bar as

$$dm_z = (\sigma t ds)(x_o - x)\left[\frac{d^2 v}{dz^2} - (x_o - x)\frac{d^2\beta}{dz^2}\right] - (\sigma t ds)(y_o - y)\left[\frac{d^2 u}{dz^2} + (y_o - y)\frac{d^2\beta}{dz^2}\right] \tag{6.10i}$$

Integrate over the entire cross-section area A_c, and we get the torque per unit length of the bar as follows:

$$m_z = \int_{A_c} dm_z = \sigma x_o \frac{d^2 v}{dz^2}\int_{A_c} t ds - \sigma\frac{d^2 v}{dz^2}\int_{A_c} xt ds - \sigma\frac{d^2\beta}{dz^2}\int_{A_c}(x_o - x)^2 t ds -$$

$$\sigma y_o \frac{d^2 u}{dz^2}\int_{A_c} t ds + \sigma\frac{d^2 u}{dz^2}\int_{A_c} yt ds - \sigma\frac{d^2\beta}{dz^2}\int_{A_c}(y_o - y)^2 t ds \tag{6.10j}$$

Now

$$\sigma\int_{A_c} t ds = P, \qquad \int_{A_c} xt ds = 0, \qquad \int_{A_c} yt ds = 0$$

$$\int_{A_c} x^2 t ds = I_y, \qquad \int_{A_c} y^2 t ds = I_x$$

Therefore,

$$m_z = Px_o\frac{d^2 v}{dz^2} - Px_o^2\frac{d^2\beta}{dz^2} - \sigma I_y\frac{d^2\beta}{dz^2} - Py_o\frac{d^2 u}{dz^2} - Py_o^2\frac{d^2\beta}{dz^2} - \sigma I_x\frac{d^2\beta}{dz^2} \tag{6.10k}$$

or

$$m_z = P \left[x_o \frac{d^2v}{dz^2} - y_o \frac{d^2u}{dz^2} \right] - P \left[x_o^2 + y_o^2 + \frac{I_x}{A_c} + \frac{I_y}{A_c} \right] \frac{d^2\beta}{dz^2}$$

or

$$m_z = P \left[x_o \frac{d^2v}{dz^2} - y_o \frac{d^2u}{dz^2} \right] - \frac{PI_o}{A_c} \frac{d^2\beta}{dz^2} \tag{6.10l}$$

where

$$I_o = I_x + I_y + A_c(x_o^2 + y_o^2)$$

I_x and I_y are the moments of inertia of the section about the principal axes of the cross-section passing through the centroid C. I_o is the polar moment of inertia of the cross-section about the longitudinal axis passing through the shear center O. For non-uniform torsion from Eq. (6.5k) we have

$$m_z = EC_w \frac{d^4\beta}{dz^4} - GJ \frac{d^2\beta}{dz^2} \tag{6.5k}$$

Combine Eqs. (6.5k and 6.10l) to get

$$EC_w \frac{d^4\beta}{dz^4} - \left(GJ - \frac{I_o P}{A_c} \right) \frac{d^2\beta}{dz^2} - Px_o \frac{d^2v}{dz^2} + Py_o \frac{d^2u}{dz^2} = 0 \tag{6.10m}$$

Equations (6.10c), (6.10d), and (6.10m) are the three simultaneous differential equations for buckling by bending and torsion, and can be used to find the critical loads. The angle of rotation β appears in all the equations and these equations are coupled. In general, torsional buckling and bending of the axis occur simultaneously.

6.6.1 Pinned-pinned Columns

The ends of the column are free to warp and they can rotate about the x and y axes, but they cannot rotate about the z axis. The ends cannot deflect in the x and y directions also. The boundary conditions are

$$u = v = \beta = 0 \text{ at } z = 0$$

$$u = v = \beta = 0 \text{ at } z = L$$

$$\frac{d^2u}{dz^2} = \frac{d^2v}{dz^2} = \frac{d^2\beta}{dz^2} = 0 \text{ (Because the bending moment is zero) at } z = 0$$

$$\frac{d^2u}{dz^2} = \frac{d^2v}{dz^2} = \frac{d^2\beta}{dz^2} = 0 \text{ (Because the bending moment is zero) at } z = L$$

Assume the solution of Eqs. (6.10c, 6.10d, and 6.10m) as

$$u = A \sin \frac{\pi z}{L} \tag{6.11a}$$

$$v = B \sin \frac{\pi z}{L} \tag{6.11b}$$

$$\beta = C \sin \frac{\pi z}{L} \tag{6.11c}$$

The above solution satisfies the boundary conditions. Equations (6.11a)–(6.11c) are substituted into Eqs. (6.10c, 6.10d, and 6.10m) and we obtain

$$\left(P - \frac{\pi^2 EI_y}{L^2}\right) A + Py_o C = 0$$

$$\left(P - \frac{\pi^2 EI_x}{L^2}\right) B - Px_o C = 0$$

$$Py_o A - Px_o B - \left(EC_w \frac{\pi^2}{L^2} + GJ - \frac{I_o}{A_c} P\right) C = 0$$

Let

$$\frac{\pi^2 EI_x}{L^2} = P_x, \quad \frac{\pi^2 EI_y}{L^2} = P_y, \quad \text{and} \quad P_\phi = \frac{A_c}{I_o}\left(GJ + \frac{\pi^2 EC_w}{L^2}\right) \tag{6.11d}$$

where P_x and P_y are the Euler buckling loads about the x and y axes, and P_ϕ is the critical load for pure torsional buckling (Eq. 6.8a).

$$A(P - P_y) + CPy_o = 0 \tag{6.11e}$$

$$B(P - P_x) - CPx_o = 0 \tag{6.11f}$$

$$APy_o - BPx_o + C\frac{I_o}{A_c}(P - P_\phi) = 0 \tag{6.11g}$$

Equations (6.11e)–(6.11g) can be written in matrix form as

$$\begin{bmatrix} (P - P_y) & 0 & Py_o \\ 0 & (P - P_x) & -Px_o \\ Py_o & -Px_o & \frac{I_o}{A_c}(P - P_\phi) \end{bmatrix} \begin{Bmatrix} A \\ B \\ C \end{Bmatrix} = \begin{Bmatrix} 0 \\ 0 \\ 0 \end{Bmatrix} \tag{6.11h}$$

The determinant of the coefficient matrix should be zero for a nontrivial solution of Eqs. (6.11h). Hence, we get

$$\begin{vmatrix} (P - P_y) & 0 & Py_o \\ 0 & (P - P_x) & -Px_o \\ Py_0 & -Px_o & \frac{I_o}{A_c}(P - P_\phi) \end{vmatrix} = 0 \tag{6.11i}$$

The characteristic equation is given by

$$\frac{I_o}{A_c}(P - P_y)(P - P_x)(P - P_\phi) - (P - P_y)P^2 x_o^2 - (P - P_x)P^2 y_o^2 = 0 \tag{6.11j}$$

or

$$\left(\frac{I_o}{A_c} - x_o^2 - y_o^2\right) P^3 + \frac{I_o}{A_c}\left(P_y \frac{A_c x_o^2}{I_o} + P_x \frac{A_c y_o^2}{I_o} - P_x - P_y - P_\phi\right) P^2$$

$$+ \frac{PI_o}{A_c}(P_x P_y + P_x P_\phi + P_y P_\phi) - \frac{I_o}{A_c} P_x P_y P_\phi = 0$$

or

$$\frac{(I_x + I_y)}{I_o} P^3 + \left[\frac{A_c}{I_o}(P_x y_o^2 + P_y x_o^2) - (P_x + P_y + P_\phi) \right] P^2$$

$$+ (P_x P_y + P_x P_\phi + P_y P_\phi)P - P_x P_y P_\phi = 0 \tag{6.11k}$$

If we know the properties of the column material and the cross-section, we can calculate the coefficients in Eq. (6.11k). The solution of the cubic equation in P will give three roots, i.e. the three values of P, and the smallest value will be the critical load. If a column section is symmetrical about two axes, the shear center coincides with the centroid, and we have $x_o = y_o = 0$. Then, the Eq. (6.11j) becomes

$$(P - P_y)(P - P_x)(P - P_\phi) = 0 \tag{6.11l}$$

The three roots of this equation are

$$P = P_y = \frac{\pi^2 EI_y}{L}$$

$$P = P_x = \frac{\pi^2 EI_x}{L}$$

$$P = P_\phi = \frac{A_c}{I_o}\left(GJ + \frac{\pi^2 EC_w}{L^2} \right)$$

This is the same as the previous case of pure torsional buckling. In this case, as stated before, there are two pure flexural buckling modes about the x and y axes, and the third is the pure torsional buckling. The least of the three loads, P_x, P_y, and P_ϕ gives the critical buckling load. The flexural Euler buckling load about the weak axis is almost the least of the three values for the hot rolled steel, wide flange sections. Hence, the flexural buckling load is usually used for the design of these cross-sections.

6.6.2 Fixed-fixed Columns

If the ends of the column are built-in, the boundary conditions are:

$$u = v = \beta = 0 \qquad \text{at } z = 0 \text{ and } z = L$$

$$\text{and} \quad \frac{du}{dz} = \frac{dv}{dz} = \frac{d\beta}{dz} = 0 \qquad \text{at } z = 0 \text{ and } z = L$$

There will be fixed-end moments during buckling that have to be considered as shown in Eq. (2.2b). Hence, add the fixed end moments in Eqs. (6.10c and 6.10d) and we obtain

$$EI_y \frac{d^2u}{dz^2} + P(u + y_o\beta) = EI_y\left(\frac{d^2u}{dz^2} \right)_{z=0} \tag{6.12a}$$

$$EI_x \frac{d^2v}{dz^2} + P(v - x_o\beta) = EI_x\left(\frac{d^2v}{dz^2} \right)_{z=0} \tag{6.12b}$$

The governing Eq. (6.10m) was developed by considering an element between two sections, hence, it is not affected by the end conditions of a column.

$$EC_w \frac{d^4\beta}{dz^4} - \left(GJ - \frac{I_oP}{A_c}\right)\frac{d^2\beta}{dz^2} - Px_o\frac{d^2v}{dz^2} + Py_o\frac{d^2u}{dz^2} = 0 \qquad (6.10m)$$

Equations (6.12a), (6.12b), and (6.10m) are the three governing equations of the torsional flexural buckling of columns with built-in ends. Assume the solution of these three equations is

$$u = A\left(1 - \cos\frac{2\pi z}{L}\right) \qquad (6.12c)$$

$$v = B\left(1 - \cos\frac{2\pi z}{L}\right) \qquad (6.12d)$$

$$\beta = C\left(1 - \cos\frac{2\pi z}{L}\right) \qquad (6.12e)$$

The assumed solution satisfies the boundary conditions. Substitute Eqs. (6.12c)–(6.12e) into Eqs. (6.12a, 6.12b, and 6.10m) and we get the same equation Eq. (6.11h) as before

$$\begin{bmatrix} (P - P_y) & 0 & Py_o \\ 0 & (P - P_x) & -Px_o \\ Py_o & -Px_o & \frac{I_o}{A_c}(P - P_\phi) \end{bmatrix}\begin{Bmatrix} A \\ B \\ C \end{Bmatrix} = \begin{Bmatrix} 0 \\ 0 \\ 0 \end{Bmatrix} \qquad (6.12f)$$

where in Eq. (6.12f)

$$P_x = \frac{4\pi^2 EI_x}{L^2}, \quad P_y = \frac{4\pi^2 EI_y}{L^2}, \quad \text{and } P_\phi = \frac{A_c}{I_o}\left(GJ + \frac{4\pi^2 EC_w}{L^2}\right)$$

The determinant of the coefficient matrix should be zero for a non-trivial solution of Eqs. (6.12f), and we get, as before, the characteristic equation to find the critical loads.

$$\frac{I_o}{A_c}(P - P_y)(P - P_x)(P - P_\phi) - (P - P_y)P^2x_o^2 - (P - P_x)P^2y_o^2 = 0 \qquad (6.12g)$$

Or, as before

$$\frac{(I_x + I_y)}{I_o}P^3 + \left[\frac{A_c}{I_o}(P_xy_o^2 + P_yx_o^2) - (P_x + P_y + P_\phi)\right]P^2 + (P_xP_y + P_xP_\phi + P_yP_\phi)P - P_xP_yP_\phi = 0 \qquad (6.12h)$$

The cubic characteristic equation for calculating the critical loads for built-in columns is the same as for simply supported ends, just use $\frac{4\pi^2}{L^2}$ in place of $\frac{\pi^2}{L^2}$. The calculations for the doubly symmetric sections where the Shear center and the centroid coincide are also similar except for the stated change.

6.6.3 Singly Symmetric Sections

When a cross-section is symmetric about one axis, such as a channel and T-section, the shear center lies on the axis of symmetry. Let us assume that the x axis is the axis of

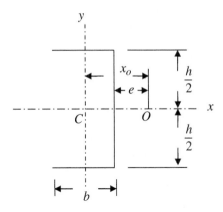

Figure 6.9 Singly symmetric section.

symmetry. In this case, $y_o = 0$, as shown in Figure 6.9. Equations (6.10c), (6.10d), and (6.10m) become

$$EI_y \frac{d^2u}{dz^2} + Pu = 0 \tag{6.13a}$$

$$EI_x \frac{d^2v}{dz^2} + P(v - x_o\beta) = 0 \tag{6.13b}$$

$$EC_w \frac{d^4\beta}{dz^4} - \left(GJ - \frac{I_oP}{A_c}\right)\frac{d^2\beta}{dz^2} - Px_o\frac{d^2v}{dz^2} = 0 \tag{6.13c}$$

Equation (6.13a) does not contain β and the buckling given by the equation in the plane of symmetry is independent of torsion. Equations (6.13b) and (6.13c) contain β and hence the buckling perpendicular to the plane of symmetry is coupled with torsion.

6.6.3.1 Pinned-pinned Columns
Since the ends are hinged, the column can rotate about x and y axes but cannot deflect in these directions. In addition, the column is free to warp but cannot rotate about the z axis. The boundary conditions are:

$$u = v = \beta = 0 \qquad \text{at } z = 0 \text{ and } z = L \tag{6.14a}$$

$$\frac{d^2u}{dz^2} = \frac{d^2v}{dz^2} = \frac{d^2\beta}{dz^2} = 0 \qquad \text{at } z = 0 \text{ and } z = L \tag{6.14b}$$

The shape functions satisfying the boundary conditions as before are given by Eqs. (6.11a)–(6.11c) as follows:

$$u = A\sin\frac{\pi z}{L} \tag{6.11a}$$

$$v = B\sin\frac{\pi z}{L} \tag{6.11b}$$

$$\beta = C\sin\frac{\pi z}{L} \tag{6.11c}$$

Equation (6.13a) can be solved independently by substituting in this equation Eq. (6.11a). The buckling load in this case is the flexural buckling load given by Euler's formula as

$$P_{cr1} = \frac{\pi^2 EI_y}{L^2} \tag{6.14c}$$

The shape functions in Eqs. (6.11b and 6.11c) represent the first mode shape of buckling by flexure and torsion. These shape functions are substituted into governing differential equations given by Eqs. (6.13b) and (6.13c), and we get

$$\begin{bmatrix} (P - P_x) & -Px_o \\ -Px_o & \frac{I_o}{A_c}(P - P_\phi) \end{bmatrix} \begin{Bmatrix} B \\ C \end{Bmatrix} = \begin{Bmatrix} 0 \\ 0 \end{Bmatrix} \tag{6.14d}$$

The various terms in Eq. (6.14d) have been defined before. The determinant of the coefficient matrix should be zero for a nontrivial solution of Eqs. (6.14d). Hence we get

$$\begin{vmatrix} (P - P_x) & -Px_o \\ -Px_o & \frac{I_o}{A_c}(P - P_\phi) \end{vmatrix} = 0 \tag{6.14e}$$

$$\frac{I_o}{A_c}(P - P_x)(P - P_\phi) - P^2 x_o^2 = 0$$

Hence, the quadratic characteristic equation is given by

$$P^2 \left(1 - \frac{A_c x_o^2}{I_o}\right) - P(P_x + P_\phi) + P_x P_\phi = 0 \tag{6.14f}$$

The two roots of the quadratic characteristic equation are

$$P_{cr2,\, cr3} = \frac{(P_x + P_\phi) \pm \sqrt{(P_x + P_\phi)^2 - 4P_x P_\phi \left(1 - \frac{A_c x_o^2}{I_o}\right)}}{2\left(1 - \frac{A_c x_o^2}{I_o}\right)}$$

or

$$P_{cr2,cr3} = \frac{1}{2k}\left[(P_x + P_\phi) \pm \sqrt{(P_x + P_\phi)^2 - 4P_x P_\phi k}\right] \tag{6.14g}$$

where

$$k = \left(1 - \frac{A_c x_o^2}{I_o}\right), \text{ and } I_o = Ax_o^2 + I_x + I_y$$

I_x and I_y are the moments of inertia of the section about the principal axes of the cross-section passing through the centroid C. I_o is the polar moment of inertia of the cross-section about the longitudinal axis passing through the shear center O.

The two roots, P_{cr2} and P_{cr3}, correspond to a combination of bending and twisting, i.e. torsional flexural buckling. The smaller root is smaller than P_x or P_ϕ, while the larger root is larger than P_x or P_ϕ. For sections with one axis of symmetry, the critical load is the smaller of the two roots of Eq. (6.14g) corresponding to the torsional flexural buckling or the Euler load, P_{cr1}, corresponding to the bending in the plane of symmetry. Thus, sections that are symmetric about one principal axis can fail either by flexural buckling in the plane of symmetry, or by torsional flexural buckling in the plane perpendicular to the plane of symmetry. The non-symmetric sections, on the other hand, always fail by torsional flexural buckling.

Example 6.3 Find the torsional flexural buckling load of a 25 ft (4.57 m) column that is simply supported at its ends, where the cross-section is an equal L $6\times6''\times3/4''$ (152 mm × 152 mm × 19 mm) , according to the AISC manual [5]. The area of the cross-section is $A_c = 8.46$ in.^{2} (5420 mm^2). The moment of inertia about the x and y axes are: $I_x = 28.1$ in.^{4} (11.6×10^6 mm^4), and $I_y = 28.1$ in.^{4} (11.6×10^6 mm^4). The radius of gyration about the x, y, and z axes is: $r_x = 1.82$ in. (46.3 mm), $r_y = 1.82$ in. (46.3 mm), and $r_z = 1.17$ in. (29.7 mm). the torsional constant $J = 1.61$ in.^{4} (629×10^3 mm^4), $C_w = 4.17$ in.^{6} (1.1×10^9 mm^6), $r_o = 3.24$ in. (82.4 mm), $\tan \alpha = 1$, $\bar{x} = \bar{y} = 1.77$ in. (44.9 mm).

u and z are the major and minor principal axes for the equal angle section (Figure 6.10). C is the centroid and O (at the intersection of the center lines of two legs of the angle) is the shear center for the cross-section. Some of the physical properties of the angle section [6] are calculated below

$$I_z = A\,r_z^2 = 8.44\,(1.17)^2 = 11.6\ \text{in.}^4 \quad [5420\,(29.7)^2 = 4.78 \times 10^6\,\text{mm}^4]$$

$$I_u = I_x + I_y - I_z = 28.1 + 28.1 - 11.6 = 44.6\ \text{in.}^4$$

$$[11.6 \times 10^6 + 11.6 \times 10^6 - 4.78 \times 10^6 = 18.42 \times 10^6\,\text{mm}^4]$$

$$x_o = y_o = \bar{x} - \frac{t}{2} = 1.77 - \frac{3}{8} = 1.395\ \text{in.}\quad \left[44.9 - \frac{19}{2} = 35.4\ \text{mm}\right]$$

$$u_o = x_o \cos \alpha + y_o \sin \alpha = 1.395 \times 0.707 + 1.395 \times 0.707 = 1.973\ \text{in.}$$

$$[35.4 \times 0.707 + 35.4 \times 0.707 = 50.06\ \text{mm}]$$

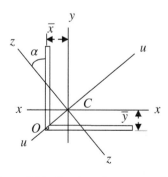

Figure 6.10 Equal angle cross-section.

Shear center O lies at the intersection of the two legs of the angle, $z_o = 0$.

$$I_o = I_u + I_z + Au_o^2 + Az_o^2$$
$$I_o = 44.6 + 11.6 + 8.46\,(1.973)^2 + 0 = 89.13 \text{ in.}^4$$
$$[11.6 \times 10^6 + 11.6 \times 10^6 + 5420\,(50.06)^2 + 0 = 36.78 \times 10^6 \text{mm}^4]$$

$$P_{cr1} = P_z = \frac{\pi^2 EI_z}{L^2} = \frac{\pi^2 x 29 x 10^6 x 11.6}{(15x12)^2} x \frac{1}{1000} = 102.47 \text{kips} \left[\frac{\pi^2 x 200000 x 4.78 x 10^6}{(4.57 x 1000)^2} x \frac{1}{1000} = 451.78 \text{ kN} \right]$$

$$P_u = \frac{\pi^2 EI_u}{L^2} = \frac{\pi^2 x 29 x 10^6 x 44.6}{(15x12)^2} x \frac{1}{1000} = 393.99 \text{ kips} \left[\frac{\pi^2 x 200000 x 18.42 x 10^6}{(4.57 x 1000)^2} x \frac{1}{1000} = 1740.95 \text{ kN} \right]$$

$$P_\phi = \frac{A_c}{I_o} \left(GJ + \frac{\pi^2 EC_w}{L^2} \right) = \frac{8.46}{89.13} \left(11 x 10^6 x 1.61 + \frac{\pi^2 x 29 x 10^6 x 4.17}{(15x12)^2} \right) x \frac{1}{1000} = 1684.49 \text{ kips}$$

$$\left[\frac{5420}{36.78 x 10^6} \left(75000 x 629 x 10^3 + \frac{\pi^2 x 200000 x 1.1 x 10^9}{(4.57 x 1000)^2} \right) x \frac{1}{1000} = 6966.67 \text{ kN} \right]$$

$$k = 1 - \frac{A_c x_o^2}{I_o} = 1 - \frac{8.46 x 1.395^2}{89.13} = 0.815 \left[1 - \frac{5420 x 35.4^2}{36.78 x 10^6} = 0.815 \right]$$

$$P_{cr2,cr3} = \frac{1}{2k} \left[(P_u + P_\phi) \pm \sqrt{(P_u + P_\phi)^2 - 4 P_u P_\phi k} \right]$$

$$P_{cr2,cr3} = \frac{1}{2 x 0.815} \left[(393.99 + 1684.49) \pm \sqrt{(393.99 + 1684.49)^2 - 4(393.99)(1684.49)(0.815)} \right]$$

$$P_{cr2} = 2176.08 \text{ kips and } P_{cr3} = 373.22 \text{ kips}$$

$$P_{cr2,cr3} = \frac{1}{2 x 0.815} \left[(1740.95 + 6966.67) \pm \sqrt{(1740.95 + 6966.67)^2 - 4(1740.95)(6966.67)(0.815)} \right]$$

$$P_{cr2} = 9037.54 \text{ kN} \qquad \text{and} \qquad P_{cr3} = 1645.33 \text{ kN}$$

The critical buckling load for the simply supported single angle column is given by the minimum of three values P_{cr1}, P_{cr2}, P_{cr3}, hence

$$P_{cr} = 102.47 \text{ kips (451.78 kN)}.$$

6.6.3.2 Fixed-fixed Columns

In this case the governing equations of buckling are obtained from Eqs. (6.12a), (6.12b), and (6.10m) by substituting $y_o = 0$. The governing equations of buckling are

$$EI_y \frac{d^2 u}{dz^2} + Pu = EI_y \left(\frac{d^2 u}{dz^2} \right)_{z=0} \tag{6.15a}$$

$$EI_x \frac{d^2 v}{dz^2} + P(v - x_o \beta) = EI_x \left(\frac{d^2 v}{dz^2} \right)_{z=0} \tag{6.15b}$$

$$EC_w \frac{d^4 \beta}{dz^4} - \left(GJ - \frac{I_o P}{A_c} \right) \frac{d^2 \beta}{dz^2} - P x_o \frac{d^2 v}{dz^2} = 0 \tag{6.15c}$$

The boundary conditions are

$$u = v = \beta = 0 \qquad \text{at } z = 0 \text{ and } z = L \tag{6.15d}$$

$$\frac{du}{dz} = \frac{dv}{dz} = \frac{d\beta}{dz} = 0 \qquad \text{at } z = 0 \text{ and } z = L \tag{6.15e}$$

Assume the solution of these three equations is the same as given by Eqs. (6.12c)–(6.12e).

$$u = A\left(1 - \cos\frac{2\pi z}{L}\right) \tag{6.12c}$$

$$v = B\left(1 - \cos\frac{2\pi z}{L}\right) \tag{6.12d}$$

$$\beta = C\left(1 - \cos\frac{2\pi z}{L}\right) \tag{6.12e}$$

Equation (6.15a) can be solved independently by substituting in this equation Eq. (6.12c). The buckling load in this case is the flexural buckling load given by Euler's formula as

$$P_{cr1} = \frac{4\pi^2 EI_y}{L^2} \tag{6.15f}$$

The shape functions in Eqs. (6.12d and 6.12e) represent the first mode shape of buckling by flexure and torsion. These shape functions are substituted into governing differential equations given by Eqs. (6.15b) and (6.15c), and we get Eq. (6.14f) as the quadratic characteristic equation which is the same as given for a singly symmetric pinned-pinned column

$$P^2\left(1 - \frac{A_c x_o^2}{I_o}\right) - P(P_x + P_\phi) + P_x P_\phi = 0 \tag{6.14f}$$

where

$$P_x = \frac{4\pi^2 EI_x}{L^2} \text{ and } P_\phi = \frac{A_c}{I_o}\left(GJ + \frac{4\pi^2 EC_w}{L^2}\right) \tag{6.15g}$$

The two roots of the quadratic Eq. (6.14f) are given by the same Eq. (6.14g) as given before for a singly symmetric pinned-pinned column

$$P_{cr2,cr3} = \frac{1}{2k}[(P_x + P_\phi) \pm \sqrt{(P_x + P_\phi)^2 - 4P_x P_\phi k}] \tag{6.14g}$$

where $k = \left(1 - \frac{A_c x_o^2}{I_o}\right)$, and $I_o = Ax_o^2 + I_x + I_y$, as before, for pinned-pinned columns. The smaller of the roots given by Eq. (6.14g) or the Euler buckling load given by Eq. (6.15f) in the plane of symmetry gives the critical buckling load for the column.

Example 6.4 Find the torsional flexural buckling load of a 15 ft (4.57 m) column that is built-in at its ends, where the cross-section is MC 10×41.1 (in. × lb./ft) or MC 250×61.2 (mm × kg/m), miscellaneous channel, according to the AISC manual [5]. The area of cross-section is $A_c = 12.1$ in.^{2} (7780 mm^2), $\bar{x} = 1.09$ in. (27.7 mm), $e_o = 0.864$ in. (22.1 mm)

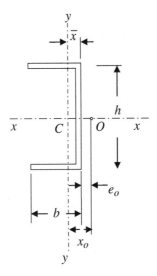

Figure 6.11 Channel cross-section.

(Figure 6.11). The moment of inertia about the x and y axes is: $I_x = 158$ in.^{4} (64.8×10^6 mm^4), and $I_y = 15.8$ in.^{4} (6.65×10^6 mm^4). The radius of gyration about the x and y axes is: $r_x = 3.61$ in. (91.3 mm), and $r_y = 1.14$ in. (29.2 mm). The torsional constant $J = 2.27$ in.^{4} (942×10^3 mm^4), $C_w = 270$ in.^{6} (84.6×10^9 mm^6), $r_o = 4.26$ in. (108 mm).

$$I_o = I_x + I_y + A_c(x_o{}^2 + y_o{}^2)$$

$$x_o = \bar{x} + e_o = 1.09 + 0.864 = 1.954 \text{ in.} [27.7 + 22.1 = 49.8 \text{ mm}]$$

$$I_o = 158 + 15.8 + 12.1\,(1.954^2 + 0) = 220 \text{ in.}^4$$
$$[64.8 \times 10^6 + 6.65 \times 10^6 + 7780(49.8)^2 + 0 = 90.74 \times 10^6 \text{ mm}^4]$$

$$k = 1 - \frac{A_c x_0^2}{I_o} = 1 - \frac{12.1(1.954)^2}{220} = 0.79 \quad \left[1 - \frac{7780(49.8)^2}{90.74 \text{x} 10^6} = 0.79 \right]$$

$$P_{cr1} = P_y = \frac{4\pi^2 EI_y}{L^2} = \frac{4(\pi^2)(29 \times 10^6)(15.8)}{(15 \times 12)^2} \times \frac{1}{1000} = 558.32 \text{ kips}$$

$$\left[\frac{4(\pi^2)(200000)(6.5 \times 10^6)}{(4.57 \times 1000)^2} \times \frac{1}{1000} = 2514.08 \text{ kN} \right]$$

$$P_x = \frac{4\pi^2 EI_x}{L^2} = \frac{4(\pi^2)(29 \times 10^6)(158)}{(15 \times 12)^2} \times \frac{1}{1000} = 5583.04 \text{ kips}$$

$$\left[\frac{4(\pi^2)(200000)(64.8 \times 10^6)}{(4.57 \times 1000)^2} \times \frac{1}{1000} = 24,498.08 \text{ kN} \right]$$

$$P_\phi = \frac{A_c}{I_o}\left(GJ + \frac{4\pi^2 EC_w}{L^2}\right) = \frac{12.1}{220}\left(11 \times 10^6(2.27) + \frac{4(\pi^2)(29 \times 10^6)(270)}{(15 \times 12)^2}\right) \times \frac{1}{1000} = 1898.08 \text{ kips}$$

$$\left[\frac{7780}{90.74 \times 10^6}\left(75000(942 \times 10^3) + \frac{4(\pi^2)(200000)(84.6 \times 10^9)}{(4.57 \times 1000)^2}\right) \times \frac{1}{1000} = 8799.75 \text{ kN}\right]$$

$$P_{cr2,cr3} = \frac{1}{2k}[(P_x + P_\phi) \pm \sqrt{(P_x + P_\phi)^2 - 4P_x P_\phi k}]$$

$$P_{cr2,cr3} = \frac{1}{2 \times 0.79}[(5583.04 + 1898.08) \pm \sqrt{(5583.04 + 1898.08)^2 - 4(5583.04)(1898.08)(0.79)}]$$

$$P_{cr2} = 7735.74 \text{ kips} \quad \text{and} \quad P_{cr3} = 1734.27 \text{ kips}$$

$$\left[P_{cr2,cr3} = \frac{1}{2 \times 0.79}[(24, 498.08 + 8799.75) \pm \sqrt{(24, 498.08 + 8799.75)^2 - 4(24, 498.08)(8799.75)(0.79)}]\right]$$

$$[P_{cr2} = 34161.04 \text{ kN} \quad \text{and} \quad P_{cr3} = 7989.23 \text{ kN}]$$

The critical buckling load for the built-in column whose cross-section is single miscellaneous channel is given by the minimum of three values $P_{cr1}, P_{cr2}, P_{cr3}$, hence

$$P_{cr} = 558.32 \text{ kips } [2514.08 \text{ kN}].$$

6.7 Torsional Flexural Buckling: The Energy Approach

The energy method can be used to find the torsional flexural buckling load for columns as an alternative to the differential equations method. This method can be used to obtain an approximate solution though there is the possibility to converge to almost an exact solution. This method was used in Chapter 1 and some of the same principles are used here. We find the total potential energy, Π, of a system given by

$$\Pi = U + V \tag{6.16a}$$

where U is the strain energy and V is the potential of the external loads.

6.7.1 Strain Energy of Torsional Flexural Buckling

The strain energy is stored due to bending about the x and y axes and due to torsion about the z axis during buckling in a column. Strain energy due to torsion is stored due to shear stress developed in St. Venant's torsion and the longitudinal stresses developed during warping torsion. The strain energy stored in a member due to longitudinal stresses associated with bending moments M_x and M_y about the x and y axes respectively is given by

$$U_b = \frac{1}{2}\int_0^L \int_A \sigma\varepsilon dA\, dz \tag{6.17a}$$

within the elastic limit, $\sigma = E\varepsilon$

Hence,

$$U_b = \frac{1}{2} \int_0^L \int_A \frac{\sigma^2}{E} dA \, dz$$

Using the flexure formula from *Mechanics of Materials* [4], the strain energy of bending due to moments M_x and M_y about the x and y axes respectively is obtained from

$$U_b = \frac{1}{2} \int_0^L \frac{M_x^2 dz}{EI_x} + \frac{1}{2} \int_0^L \frac{M_y^2 dz}{EI_y} \tag{6.17b}$$

where I_x and I_y are the area moment of inertias about the x and y axes of the member cross-section. Use Eqs. (2.1a and 2.1b) to get

$$U_b = \frac{1}{2} \int_0^L EI_x \left(\frac{d^2v}{dz^2}\right)^2 dz + \frac{1}{2} \int_0^L EI_y \left(\frac{d^2u}{dz}\right)^2 dz \tag{6.17c}$$

where u and v are the displacements in the x and y directions.

Strain energy stored in a member due to shear stresses associated with St. Venant's torsion is given by

$$U_{sv} = \frac{1}{2} \int_0^L \int_A \tau\gamma dA \, dz \tag{6.17d}$$

within elastic limit $\tau = G\gamma$

$$U_{sv} = \frac{1}{2} \int_0^L \int_A \frac{\tau^2}{G} dA \, dz$$

Hence,

$$\tau = \frac{T_{sv}}{J} r, \text{ and from Eq. (6.1a) } T_{sv} = GJ\frac{d\beta}{dz}$$

or

$$U_{sv} = \frac{1}{2} \int_0^L GJ \left(\frac{d\beta}{dz}\right)^2 dz \tag{6.17e}$$

where

$$J = \int_A r^2 dA$$

Strain energy due to longitudinal stresses associated with the warping torsion is obtained from

$$U_w = \frac{1}{2} \int_0^L \int_A \sigma\varepsilon dA \, dz \tag{6.17f}$$

We can proceed by considering the example of an I-section where the bending moment in the flange is given by

$$M_f = EI_f \frac{d^2u}{dz^2} \tag{6.17g}$$

The strain energy due to bending in one flange associated with warping torsion is

$$U_w = \frac{1}{2} \int_0^L EI_f \left(\frac{d^2u}{dz^2} \right)^2 dz \qquad (6.17h)$$

$$u = \beta \frac{h}{2}$$
$$\frac{d^2u}{dz^2} = \frac{h}{2} \frac{d^2\beta}{dz^2} \qquad (6.17i)$$

or

$$U_w = \frac{1}{4} \int_0^L EC_w \left(\frac{d^2\beta}{dz^2} \right)^2 dz$$

where

$$C_w = \frac{1}{2} I_f h^2 \qquad (6.17j)$$

The strain energy of bending due to warping in both flanges is

$$U_w = \frac{1}{2} \int_0^L EC_w \left(\frac{d^2\beta}{dz^2} \right)^2 dz \qquad (6.17k)$$

Alternate approach

In this approach a general thin-walled open cross-section is considered. Use Eqs. (6.17f, 6.3d, and 6.3e) to get

$$U_w = \frac{1}{2} \int_0^L \int_A E(\bar{w}_s - w_s)^2 \left(\frac{d^2\beta}{dz^2} \right)^2 dA\, dz \qquad (6.17l)$$

$$U_w = \frac{1}{2} \int_0^L EC_w \left(\frac{d^2\beta}{dz^2} \right)^2 dz \qquad (6.17m)$$

where

$$C_w = \int_A (\bar{w}_s - w_s)^2 dA \qquad (6.17n)$$

Equations (6.17k) and (6.17m) are the same.

$$U = U_b + U_{sv} + U_w \qquad (6.17o)$$

or

$$U = \frac{1}{2} \int_0^L EI_x \left(\frac{d^2v}{dz^2} \right)^2 dz + \frac{1}{2} \int_0^L EI_y \left(\frac{d^2u}{dz} \right)^2 dz + \frac{1}{2} \int_0^L GJ \left(\frac{d\beta}{dz} \right)^2 dz + \frac{1}{2} \int_0^L EC_w \left(\frac{d^2\beta}{dz^2} \right)^2 dz \qquad (6.17p)$$

Assume a pinned-pinned column and substitute the assumed deformations u, v, and β from Eqs. (6.11a)–(6.11c) in the strain energy expression (6.17p) to get

$$U = \frac{\pi^2}{4L} \left[A^2 \frac{\pi^2 EI_y}{L^2} + B^2 \frac{\pi^2 EI_x}{L^2} + C^2 \left(GJ + \frac{\pi^2 EC_w}{L^2} \right) \right] \qquad (6.17q)$$

where

$$\int_0^L \sin^2 \frac{\pi z}{L} dz = \int_0^L \cos^2 \frac{\pi z}{L} dz = \frac{L}{2}$$

6.7.2 Potential Energy of External Loads in Torsional Flexural Buckling

When an external load is applied to the column, it deforms and the external force P moves a distance of ΔL as shown in Figure 6.12a. The potential energy of the external forces is obtained from the negative of the products of the forces and the corresponding displacements in the directions of the forces. The potential energy of the external forces is obtained from

$$V = -\int \sigma dA \,(\Delta L) \tag{6.18a}$$

where σ is the axial stress in the column cross-section, and dA is the elementary area of the cross-section.

$$\Delta L = S - L \tag{6.18b}$$

Consider an element dz in the fiber AB parallel to the z axis in Figure 6.12b. The deformed length, ds, of the element is obtained as follows:

$$ds^2 = (du')^2 + (dv')^2 + (dz)^2 \tag{6.18c}$$

$$ds = \left[1 + \left(\frac{du'}{dz} \right)^2 + \left(\frac{dv'}{dz} \right)^2 \right]^{\frac{1}{2}} dz \tag{6.18d}$$

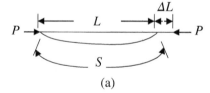

(a)

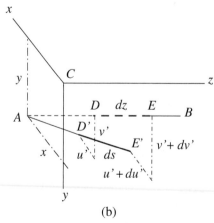

(b)

Figure 6.12 Torsional flexural buckling of a column: (a) Deformed shape of the column; (b) Deformation of an element.

The quantity in the brackets can be written in a series form by using the binomial theorem [7]. The theorem states

$$(1+x)^n = 1 + nx + \frac{n(n-1)}{2!}x^2 + \frac{n(n-1)(n-2)}{3!}x^3 + ------$$

If x is small, then higher powers of x can be neglected and the binomial expansion can be written as

$$(1+x)^n \approx 1 + nx$$

Equation (6.18d) can be written by using the binomial theorem [7] as follows:

$$\int_0^S ds = \int_0^L \left[1 + \frac{1}{2}\left(\frac{du'}{dz}\right)^2 + \frac{1}{2}\left(\frac{dv'}{dz}\right)^2 + ---- \right] dz$$

or

$$\Delta L = S - L = \frac{1}{2}\int_0^L \left[\left(\frac{du'}{dz}\right)^2 + \left(\frac{dv'}{dz}\right)^2 \right] dz \tag{6.18e}$$

Equations (6.10e) and (6.10f) give

$$u' = u + (y_o - y)\beta \tag{6.18f}$$

$$v' = v - (x_o - x)\beta \tag{6.18g}$$

$$\frac{du'}{dz} = \frac{du}{dz} + y_o\frac{d\beta}{dz} - y\frac{d\beta}{dz} \tag{6.18h}$$

$$\frac{dv'}{dz} = \frac{dv}{dz} - x_o\frac{d\beta}{dz} + x\frac{d\beta}{dz} \tag{6.18i}$$

Substitute Eqs. (6.18h and 6.18i) into Eq. (6.18e) and we have

$$\Delta L = \frac{1}{2}\int_0^L \left[\left(\frac{du}{dz}\right)^2 + \left(\frac{dv}{dz}\right)^2 + (x^2 + y^2 + x_o^2 + y_o^2)\left(\frac{d\beta}{dz}\right)^2 \right.$$
$$\left. -2(xx_o + yy_o)\left(\frac{d\beta}{dz}\right)^2 + 2(y_o - y)\frac{du}{dz}\frac{d\beta}{dz} - 2(x_o - x)\frac{dv}{dz}\frac{d\beta}{dz} \right] dz \tag{6.18j}$$

Substitute Eq. (6.18j) into Eq. (6.18a) to get

$$V = -\frac{1}{2}\int_0^L \int_A \sigma\left[\left(\frac{du}{dz}\right)^2 + \left(\frac{dv}{dz}\right)^2 + (x^2 + y^2 + x_o^2 + y_o^2)\left(\frac{d\beta}{dz}\right)^2 - 2(xx_o + yy_o)\left(\frac{d\beta}{dz}\right)^2 \right.$$
$$\left. + 2(y_0 - y)\frac{du}{dz}\frac{d\beta}{dz} - 2(x_0 - x)\frac{dv}{dz}\frac{d\beta}{dz} \right] dAdz$$

$$V = -\frac{1}{2}\int_0^L \sigma\left[\left(\frac{du}{dz}\right)^2 A + \left(\frac{dv}{dz}\right)^2 A + I_0\left(\frac{d\beta}{dz}\right)^2 + 2y_0A\frac{du}{dz}\frac{d\beta}{dz} - 2x_0A\frac{dv}{dz}\frac{d\beta}{dz} \right] dz$$

where

$$\int_A x dA = 0, \quad \int_A y dA = 0, \quad \int_A x^2 dA = I_y, \quad \int_A y^2 dA = I_x$$

and

$$I_0 = I_x + I_y + A(x_0^2 + y_0^2), \quad \sigma A = P$$

Hence,

$$V = -\frac{P}{2} \int_0^L \left[\left(\frac{du}{dz}\right)^2 + \left(\frac{dv}{dz}\right)^2 + \frac{I_0}{A}\left(\frac{d\beta}{dz}\right)^2 \right.$$
$$\left. + 2y_0 \left(\frac{du}{dz}\right)\left(\frac{d\beta}{dz}\right) - 2x_0 \left(\frac{dv}{dz}\right)\left(\frac{d\beta}{dz}\right) \right] dz$$

$$(6.18k)$$

Assume u, v, and β given by Eqs. (6.11a)–(6.11c) for a pinned-pinned column and substitute in Eq. (6.18k) to get the expression for the potential energy of the external forces

$$V = -\frac{P\pi^2}{4L}\left[A^2 + B^2 + \frac{I_0}{A_c}C^2 - 2BCx_0 + 2ACy_0\right] \tag{6.18l}$$

The total potential energy of the column is

$$\Pi = U + V$$
$$\Pi = \frac{\pi^2}{4L}\left[A^2 \frac{\pi^2 EI_y}{L^2} + B^2 \frac{\pi^2 EI_x}{L^2} + C^2\left(GJ + \frac{\pi^2 EC_w}{L^2}\right)\right]$$
$$- \frac{P\pi^2}{4L}\left[A^2 + B^2 + \frac{I_0}{A_c}C^2 - 2BCx_0 + 2ACy_0\right]$$

Substitute

$$P_x = \frac{\pi^2 EI_x}{L^2}, P_y = \frac{\pi^2 EI_y}{L^2}, \text{and } P_\phi = \frac{A_c}{I_0}\left(GJ + \frac{\pi^2 EC_w}{L^2}\right) \text{ as before to give}$$

$$\Pi = \frac{\pi^2}{4L}\left[A^2(P_y - P) + B^2(P_x - P) + \frac{C^2 I_0}{A_c}(P_\phi - P) - 2ACPy_0 + 2BCPx_0\right] \tag{6.18m}$$

The expression in Eq. (6.18m) gives the total potential energy in the column in a slightly deformed position. The critical load is the axial compressive force at which the system is in equilibrium in a slightly deflected shape. Hence, the total potential energy, Π, has a stationary value at the critical load. The total potential energy is a function of three variables A, B, C. Its stationary value can be achieved if the first variation of the total potential energy with respect to the three variables is separately equal to zero.

$$\frac{\partial \Pi}{\partial A} = A(P_y - P) - CPy_0 = 0 \tag{6.18n}$$

$$\frac{\partial \Pi}{\partial B} = B(P_x - P) + CPx_o = 0 \tag{6.18o}$$

$$\frac{\partial \Pi}{\partial C} = -APy_o + BPx_o + C\frac{I_o}{A_c}(P_\phi - P) \tag{6.18p}$$

Equations (6.18n)–(6.18p) can be written in matrix form as

$$\begin{bmatrix} (P_y - P) & 0 & -Py_o \\ 0 & (P_x - P) & Px_o \\ -Py_o & Px_o & \frac{I_o}{A_c}(P_\phi - P) \end{bmatrix} \begin{Bmatrix} A \\ B \\ C \end{Bmatrix} = \begin{Bmatrix} 0 \\ 0 \\ 0 \end{Bmatrix} \tag{6.18q}$$

For a nontrivial solution, the determinant of the coefficient matrix should be zero.

$$\begin{vmatrix} (P_y - P) & 0 & -Py_o \\ 0 & (P_x - P) & Px_o \\ -Py_o & Px_o & \frac{I_o}{A_c}(P_\phi - P) \end{vmatrix} = 0 \tag{6.18r}$$

or

$$\frac{I_o}{A_c}(P_y - P)(P_x - P)(P_\phi - P) - (P_y - P)P^2x_o^2 - (P_x - P)P^2y_o^2 = 0 \tag{6.18s}$$

Hence, we get the same expression for finding the critical load for a pinned-pinned column by the energy method as was obtained from the deformed shape equilibrium approach in Eq. (6.11j). For sections that are symmetrical about both the x and y axes, $x_o = y_o = 0$, and Eq. (6.18s) reduces to

$$(P_y - P)(P_x - P)(P_\phi - P) = 0$$

This is the same expression as Eq. (6.11l). For sections that are symmetric about one axis only, say, the x axis, the shear center lies on the axis of symmetry, the x axis. In this case, $y_o = 0$, and Eq. (6.18s) reduces to

$$\frac{I_o}{A_c}(P_y - P)(P_x - P)(P_\phi - P) - (P_y - P)P^2x_o^2 = 0$$

or

$$(P_y - P)\left[\frac{I_o}{A_c}(P_x - P)(P_\phi - P) - P^2x_o^2\right] = 0 \tag{6.18t}$$

The expression in Eq. (6.18t) is satisfied if

$$(P_y - P) = 0$$

or

$$P_{cr1} = P_y = \frac{\pi^2 EI_y}{L^2},$$

which is the same as in Eq. (6.14c) and

$$\frac{I_o}{A_c}(P_x - P)(P_\phi - P) - P^2x_o^2 = 0$$

or

$$P^2\left(1 - \frac{A_c x_o^2}{I_o}\right) - P(P_x + P_\phi) + P_x P_\phi = 0 \tag{6.18u}$$

This is the same expression as Eq. (6.14f). Hence, the solutions obtained from the energy approach and from the equilibrium of the deformed shape are the same. The energy approach will give the same solution as before for a fixed-fixed column if we use $\frac{4\pi^2}{L^2}$ in place of $\frac{\pi^2}{L^2}$.

6.8 Lateral Buckling of Beams

When a beam is loaded by a transverse load in the plane of the web, it may buckle sideways at a certain critical load if it is not provided adequate lateral support, as shown in Figure 6.13. If the flexural rigidity of the beam in the plane of the web is much larger than its lateral rigidity, then the lateral buckling is important in the design of the beam. In this case, the beam may buckle and collapse long before the bending stresses due to the transverse load reach the yield point. If the transverse loads remain below a certain value, the structure is in stable equilibrium, i.e. it will come back to its plane configuration after it is slightly twisted and buckled laterally if the force causing the displacement ceases to act. When the loads increase, a loading is reached when the displaced equilibrium position of the beam is also possible in addition to its equilibrium plane configuration. The plane configuration is now not stable, and the lowest load at which this change occurs is called the critical load for lateral buckling of the beam.

6.8.1 Lateral Buckling of Simply Supported, Narrow Rectangular Beams in Pure Bending

Consider a narrow rectangular, simply supported beam in pure bending as shown in Figure 6.14. Simply supported means that the ends of the beam are free to rotate about the principal axes x and y, but the beam cannot rotate about the principal axis z. Assume the beam is elastic and the applied moments act in the yz plane, the plane of maximum rigidity. It is also assumed that a small lateral deflection occurs under the action of the moments M_x, and the geometry of the cross-section does not change during buckling. We form the

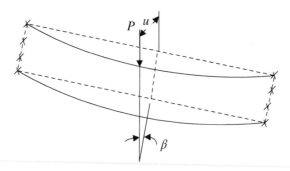

Figure 6.13 Lateral buckling of beams.

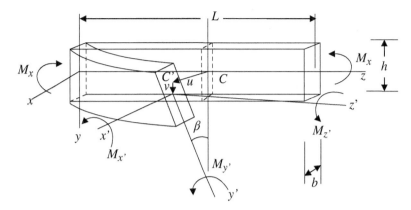

Figure 6.14 Lateral buckling of simply supported, narrow rectangular beam under pure bending.

differential equations of equilibrium for the deformed beam and from there we can calculate the critical load.

The fixed set of axes x, y, and z are chosen as per the right-hand screw rule, and a second set of axes x', y', and z' are chosen relative to the deformed shape of the member. The x' and y' axes coincide with the principal directions of the cross-section, and the z' axis is in the direction of the tangent to the deformed axis of the member after buckling. The deformation of the beam is defined by the u and v components of the displacement of the centroid of the cross-section in the x and y directions respectively, and the angle of rotation β of the cross-section about the z axis. The u and v are taken as positive in the positive directions of the x and y axes respectively, and β is the positive rotation about the z axis according to the right-hand screw rule sign convention. So, all three motions are positive in Figure 6.14.

The external moment M_x can be resolved into internal moment components $M_{x'}$, $M_{y'}$, and $M_{z'}$ by considering the equilibrium of the free body diagram of the deflected shape of the beam In Figure 6.14. The positive direction of these internal moments is shown in Figure 6.14. The internal moment components are calculated as follows:

In the xy plane, the angle between the x and the x' axis is the angle of rotation β in Figure 6.15a. The moment components about the x' and y' axes are given by

$$M_{x'} = M_x \quad \cos \beta = M_x \ (\cos \beta \approx 1 \text{ for small angle of rotation } \beta) \tag{6.19a}$$

And

$$M_{y'} = -M_X \sin \beta = -M_x\beta \ (\sin \ \beta \approx \beta \text{ for small angle of rotation } \beta) \tag{6.19b}$$

In the xz plane, the angle between the z and z' axes is equal to $\dfrac{du}{dz}$ in Figure 6.15b. The moment components about the x' and z' axes are

$$M_{z'} = M_x \sin \frac{du}{dz} = M_x\frac{du}{dz} \tag{6.19c}$$

$M_{x'} = M_x \cos \frac{du}{dz} = M_x$ as before for small angles.

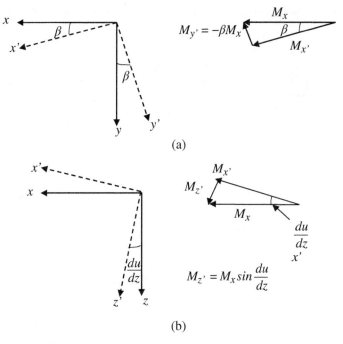

Figure 6.15 Internal moment components: (a) Moments in the *xy* plane; (b) Moments in the *xz* plane.

The differential equations of beam bending are

$$M_{x'} = -EI_x \frac{d^2 v}{dz^2} \tag{6.19d}$$

$$M_{y'} = EI_y \frac{d^2 u}{dz^2} \tag{6.19e}$$

For small angle of twist β, it is assumed that the curvature and moment of inertia in the $x'z'$ and $y'z'$ planes may be taken the same as their corresponding values in the xz and yz planes respectively. The negative sign in Eq. (6.19d) shows that for positive moment M_x shown in Figure 6.14 the corresponding curvature $\frac{d^2 v}{dz^2}$ in $y' - z'$ plane is negative. The differential equation for the twisting moment of the buckled member is taken from Eq. (6.2h), where $C_w = 0$, because, for a narrow rectangle, warping restraint given by C_w can be neglected. The equation for the twisting moment for the buckled bar is

$$M_{z'} = M_x \frac{du}{dz} = GJ \frac{d\beta}{dz} \tag{6.19f}$$

Substitute Eqs. (6.19a)–(6.19c) into Eqs. (6.19d)–(6.19f) to get

$$M_x + EI_x \frac{d^2 v}{dz^2} = 0 \tag{6.19g}$$

$$\beta M_x + EI_y \frac{d^2u}{dz^2} = 0 \tag{6.19h}$$

$$GJ\frac{d\beta}{dz} - M_x\frac{du}{dz} = 0 \tag{6.19i}$$

Equation (6.19g) contains only variable v, hence this equation is not coupled with Eqs. (6.19h and 6.19i) and can be solved independently for bending in the yz plane. Equations (6.19h) and (6.19i) that describe the lateral buckling and the twisting contain variables u and β, and are thus coupled. These equations have to be solved simultaneously to get the lateral buckling load. If we differentiate Eq. (6.19i) with respect to z, and substitute for the term, $\dfrac{d^2u}{dz^2}$, from this expression into Eq. (6.19h), we have

$$\frac{d^2\beta}{dz^2} + \frac{M_x^2\beta}{EI_yGJ} = 0$$

or

$$\frac{d^2\beta}{dz^2} + k^2\beta = 0 \tag{6.19j}$$

where

$$\frac{M_x^2}{EI_yGJ} = k^2 \tag{6.19k}$$

The solution of Eq. (6.19j) is

$$\beta = A\sin kz + B\cos kz \tag{6.19l}$$

The ends of the beam are simply supported, hence, the deflections and the bending moments are zero there. The boundary conditions are

$$u = v = \frac{d^2u}{dz^2} = \frac{d^2v}{dz^2} = 0 \text{ at } z = 0 \text{ and } z = L \tag{6.19m}$$

The ends of the member do not rotate about the z axis, but are free to warp. Thus,

$$\beta = 0 \text{ at } z = 0 \text{ and } z = L \tag{6.19n}$$

Substitute $\beta = 0$ at $z = 0$ and at $z = L$, into Eq. (6.19l) and we get

$$0 = 0 + B(1), \text{ or } B = 0, \text{ and}$$
$$0 = A\sin kL$$

A cannot be zero because we are considering a deformed shape, so

$$\text{Sin } kL = 0$$

or

$$kL = n\pi, \quad n = 0, 1, 2, 3, \text{---------------}$$

the least value of $kL = \pi$, or $\quad k^2L^2 = \pi^2$ or

$$\frac{M_x^2 L^2}{EI_y GJ} = \pi^2$$

The critical moment is given by

$$M_{cr} = \frac{\pi}{L}\sqrt{EI_y GJ} \tag{6.19o}$$

The maximum critical stress because of lateral buckling is

$$\sigma_{cr} = \frac{M_{cr}(h/2)}{I_x} = \frac{\pi}{L}\sqrt{EI_y GJ}\frac{h/2}{I_x}$$

$$J = \frac{hb^3}{3}, I_x = \frac{bh^3}{12}$$

or

$$\sigma_{cr} = \frac{\pi\sqrt{GE}}{L/b}\sqrt{\frac{I_y}{I_x}} \tag{6.19p}$$

The critical stress is proportional to the ratio I_y/I_x and is inversely proportional to L/b. So, the lateral buckling is important for long, narrow, and deep beams.

6.8.2 Lateral Buckling of Simply Supported I Beams in Pure Bending

Consider a simply supported I beam in pure bending as shown in Figure 6.16. This problem is similar to the problem in Section 6.8.1 and there is no change in the loading and the support

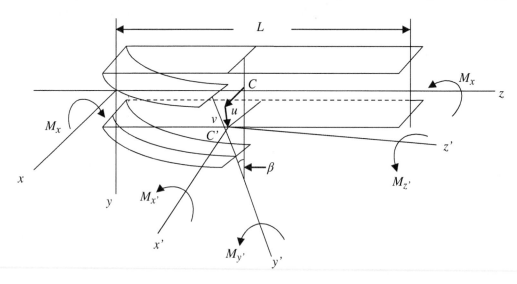

Figure 6.16 Lateral buckling of simply supported I beam under pure bending.

conditions. Equations (6.19a)–(6.19e), (6.19g), and (6.19h) apply here. The only equation different in this case is Eq. (6.19f) that is modified because in an I section there are both St. Venant's and warping torsions. Hence the torsion equation in this case is written as

$$M_{z'} = M_x \frac{du}{dz} = GJ \frac{d\beta}{dz} - EC_w \frac{d^3\beta}{dz^3} \tag{6.20a}$$

The governing differential equations for the I-section under pure bending are

$$M_x + EI \frac{d^2v}{dz^2} = 0 \tag{6.19g}$$

$$\beta M_x + EI \frac{d^2u}{dz^2} = 0 \tag{6.19h}$$

$$GJ \frac{d\beta}{dz} - EC_w \frac{d^3\beta}{dz^3} - M_x \frac{du}{dz} = 0 \tag{6.20b}$$

The bending in the vertical plane yz is described by Eq. (6.19g) and the equation is independent of lateral buckling. The lateral torsional buckling is defined by the coupled Eqs. (6.19h and 6.20b). Differentiate Eq. (6.20b) with respect to z and substitute for the term $\frac{d^2u}{dz^2}$ into Eq. (6.19h) to get

$$\frac{d^4\beta}{dz^4} - \frac{GJ}{EC_w} \frac{d^2\beta}{dz^2} - \frac{\beta M_x^2}{E^2 I_y C_w} = 0 \tag{6.20c}$$

Substitute

$$a = \frac{GJ}{2EC_w}, \quad b = \frac{M_x^2}{E^2 I_y C_w} \tag{6.20d}$$

or

$$\frac{d^4\beta}{dz^4} - 2a \frac{d^2\beta}{dz^2} - b\beta = 0 \tag{6.20e}$$

Equation (6.20e) is a linear differential equation of the fourth order and has constant coefficients. The general solution of Eq. (6.20e) is

$$\beta = A \sin mz + B \cos mz + Ce^{nz} + De^{-nz} \tag{6.20f}$$

where m and n are positive, real quantities [3] are given by

$$m = \sqrt{-a + \sqrt{a^2 + b}}, \quad n = \sqrt{a + \sqrt{a^2 + b}} \tag{6.20g}$$

Assume the ends of the member do not rotate about the z axis, but are free to warp. Hence

$$\beta = \frac{d^2\beta}{dz^2} = 0 \text{ at } z = 0 \text{ and } z = L \tag{6.20h}$$

$$\frac{d^2\beta}{dz^2} = -Am^2 \sin mz - Bm^2 \cos mz + Cn^2 e^{nz} + Dn^2 e^{-nz} \tag{6.20i}$$

By setting the boundary conditions at $z = 0$ in Eqs. (6.20f and 6.20i) we get

$$B + C + D = 0 \tag{6.20j}$$

$$-Bm^2 + Cn^2 + Dn^2 = 0 \tag{6.20k}$$

From Eqs. (6.20j and 6.20k) we obtain

$$-B(m^2 + n^2) = 0$$

Therefore,

$$B = 0, \quad C = -D \tag{6.20l}$$

Substitute Eq. (6.20l) into Eq. (6.20f) to get

$$\beta = A \sin mz - 2D \sinh nz \tag{6.20m}$$

$$\frac{d^2\beta}{dz^2} = -Am^2 \sin mz - 2Dn^2 \sinh nz \tag{6.20n}$$

Substitute the boundary conditions at $z = L$ in Eqs. (6.20m and 6.20n), and we have

$$A \sin mL - 2D \sinh nL = 0 \tag{6.20o}$$

$$Am^2 \sin mL + 2Dn^2 \sinh nL = 0 \tag{6.20p}$$

or

$$\begin{bmatrix} \sin mL & -2\sinh nL \\ m^2 \sin mL & 2n^2 \sinh nL \end{bmatrix} \begin{Bmatrix} A \\ D \end{Bmatrix} = \begin{Bmatrix} 0 \\ 0 \end{Bmatrix} \tag{6.20q}$$

For a nontrivial solution, the determinant of Eq. (6.20q) should be zero

$$\begin{vmatrix} \sin mL & -2\sinh nL \\ m^2 \sin mL & 2n^2 \sinh nL \end{vmatrix} = 0$$

or

$$\sin mL(n^2 \sinh L + m^2 \sinh nL) = 0$$

Since m and n are positive nonzero quantities, we obtain

$$\sin mL = 0 \tag{6.20r}$$

From Eqs. (6.20o and 6.20p), we obtain $D = 0$. Hence the buckled shape from Eq. (6.20f) is given by

$$\beta = A \sin mz \tag{6.20s}$$

The values of m satisfying Eq. (6.20r) are

$$mL = n\pi, \quad n = 1, 2, 3 \text{-----}$$

The smallest value of m is given by

$$m = \frac{\pi}{L}$$

(6.20t)

Using Eq. (6.20g) we have

$$-a + \sqrt{a^2 + b} = \frac{\pi^2}{L^2}$$

Substitute for a and b from Eq. (6.20d) and we get

$$M_x^2 = \frac{\pi^2}{L^2} \left(EI_y GJ + \frac{\pi^2}{L^2} E^2 I_y C_w \right)$$

The critical moment is given by

$$M_{cr} = \frac{\pi}{L} \sqrt{EI_y GJ \left(1 + \frac{\pi^2}{L^2} \frac{EC_w}{GJ} \right)}$$

(6.20u)

or

$$M_{cr} = \alpha \frac{\sqrt{EI_y GJ}}{L}$$

(6.20v)

where

$$\alpha = \pi \sqrt{1 + \frac{\pi^2}{L^2} \frac{EC_w}{GJ}}$$

(6.20w)

6.8.3 Lateral Buckling of Simply Supported I Beams: Concentrated Load at the Mid-Span

If a simply supported beam is bent in the yz plane by a concentrated load P applied at the centroid of the mid-span cross-section in Figure 6.17, lateral buckling could occur when the load reaches a critical value. A set of axes x, y parallel to the principal axes of the cross-section, and the z along the longitudinal axis are taken with the origin at the centroid of the cross-section at mid-span. It is assumed that the beam can rotate with respect to the principal axes of the cross-section that are parallel to the x and y axes, but the supports restrict the rotation about the z axis. Hence lateral buckling of the beam is accompanied by twisting of the beam.

Consider the cross-section in the free body diagram of the deflected shape of the beam at a distance z from the origin at the mid-span in Figure 6.17. The moments on this cross-section due to external forces about the axes passing through the centroid of the cross-section and parallel to the $x, y,$ and z axes are

$$-\frac{P}{2} \left(\frac{L}{2} + z \right) + Pz + M_x = 0$$

or

$$M_x = \frac{P}{2} \left(\frac{L}{2} - z \right)$$

(6.21a)

$$M_y = 0$$

(6.21b)

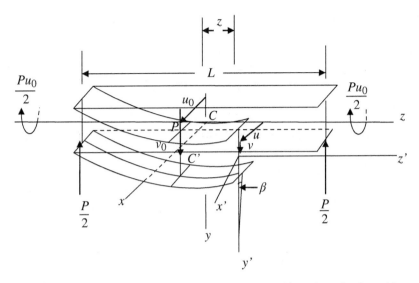

Figure 6.17 Lateral buckling of simply supported I beam subjected to a load at mid-span.

$$-\frac{Pu_0}{2} + \frac{Pu}{2} + P(u_0 - u) + M_z = 0$$

$$M_z = -\frac{P}{2}(u_0 - u) \tag{6.21c}$$

where u_0 is the lateral deflection of the centroid of the mid-span section and u is the lateral deflection at any cross-section at a distance z from the mid-span in Figure 6.17. The internal moment components in the directions of axes x', y', and z' at the cross-section are obtained from Figure 6.18 as follows:

$$M_{x'} = M_x - M_z\frac{du}{dz} = \frac{P}{2}\left(\frac{L}{2} - z\right) + \frac{P}{2}(u_0 - u)\frac{du}{dz} \tag{6.21d}$$

$$M_{y'} = -\beta\frac{P}{2}\left(\frac{L}{2} - z\right) + \frac{P}{2}(u_0 - u)\frac{dv}{dz} \tag{6.21e}$$

$$M_{z'} = -\frac{P}{2}(u_0 - u) + \frac{P}{2}\left(\frac{L}{2} - z\right)\frac{du}{dz} \tag{6.21f}$$

The internal resisting moments are expressed as

$$M_{x'} = -EI_x\frac{d^2v}{dz^2} \tag{6.21g}$$

$$M_{y'} = EI_y\frac{d^2u}{dz^2} \tag{6.21h}$$

There is a minus sign on the right-hand side of Eq. (6.21g) because positive moment produces negative curvature as per the right-hand screw rule. Substitute the moments from Eqs. (6.21d

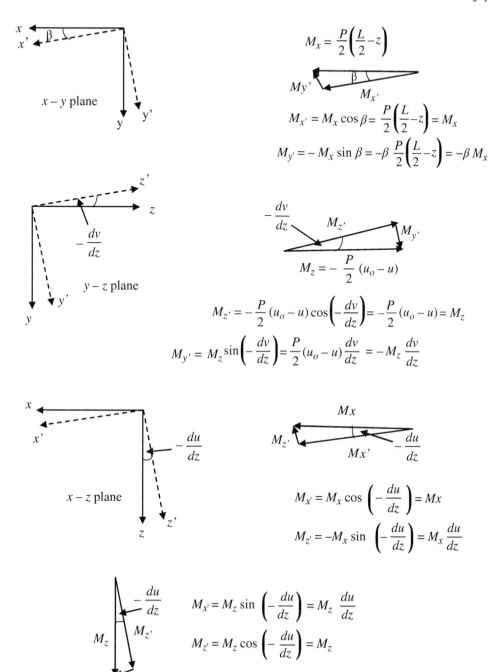

$$M_x = \frac{P}{2}\left(\frac{L}{2} - z\right)$$

$$M_{x'} = M_x \cos\beta = \frac{P}{2}\left(\frac{L}{2} - z\right) = M_x$$

$$M_{y'} = -M_x \sin\beta = -\beta \frac{P}{2}\left(\frac{L}{2} - z\right) = -\beta M_x$$

$$M_z = -\frac{P}{2}(u_o - u)$$

$$M_{z'} = -\frac{P}{2}(u_o - u)\cos\left(-\frac{dv}{dz}\right) = -\frac{P}{2}(u_o - u) = M_z$$

$$M_{y'} = M_z \sin\left(-\frac{dv}{dz}\right) = \frac{P}{2}(u_o - u)\frac{dv}{dz} = -M_z\frac{dv}{dz}$$

$$M_{x'} = M_x \cos\left(-\frac{du}{dz}\right) = Mx$$

$$M_{z'} = -M_x \sin\left(-\frac{du}{dz}\right) = M_x\frac{du}{dz}$$

$$M_{x'} = M_z \sin\left(-\frac{du}{dz}\right) = M_z\frac{du}{dz}$$

$$M_{z'} = M_z \cos\left(-\frac{du}{dz}\right) = M_z$$

Figure 6.18 Internal moment components during lateral buckling.

and 6.21e) into Eqs. (6.21g and 6.21h) respectively to get

$$EI_x \frac{d^2v}{dz^2} + \frac{P}{2}\left(\frac{L}{2} - z\right) = 0 \tag{6.21i}$$

$$EI_y \frac{d^2u}{dz^2} + \beta\frac{P}{2}\left(\frac{L}{2} - z\right) = 0 \tag{6.21j}$$

The terms containing the multiplication of du/dz and $(u_o - u)$, the multiplication of dv/dz and $(u_o - u)$, have been neglected in Eq. (6.21d) and Eq. (6.21e), since these quantities are small. The internal torsional resisting moment is given by

$$T = T_{sv} + T_w = GJ\frac{d\beta}{dz} - EC_w\frac{d^3\beta}{dz^3}$$

$$M_{z'} = T \tag{6.2h}$$

From Eqs. (6.21f and 6.2h) we get

$$GJ\frac{d\beta}{dz} - EC_w\frac{d^3\beta}{dz^3} + \frac{P}{2}(u_o - u) - \frac{P}{2}\left(\frac{L}{2} - z\right)\frac{du}{dz} = 0 \tag{6.21k}$$

Equation (6.21i) describes the buckling behavior in the yz plane and is not coupled with lateral buckling. We are interested in the lateral buckling here, that is described by Eqs. (6.21j and 6.21k). Eliminate u between Eqs. (6.21j and 6.21k) by differentiating Eq. (6.21k) with respect to z, note that $du_o/dz = 0$ and ignore du/dz because it is small, and substitute d^2u/dz^2 from Eq. (6.21j) to get

$$GJ\frac{d^2\beta}{dz^2} - EC_w\frac{d^4\beta}{dz^4} + \frac{P^2\beta}{4EI_y}\left(\frac{L}{2} - z\right)^2 = 0 \tag{6.21l}$$

The solution of this differential equation is obtained by the method of infinite series by Timoshenko and Gere [3] and given as

$$P_{cr} = \alpha\frac{\sqrt{EI_yGJ}}{L^2} \tag{6.21m}$$

The values of α given in the reference are plotted in Figure 6.19 for the concentrated mid-span load acting on the upper flange, centroid, and the lower flange of the beam. When the load acts at the upper flange, the external torsional moment is the maximum given by

$$M_z = -\frac{P}{2}\left(u_o + \frac{\beta_o h}{2} - u\right) \tag{6.21n}$$

If the load acts at the lower flange, the external moment is the least given by

$$M_z = -\frac{P}{2}\left(u_o - \frac{\beta_o h}{2} - u\right) \tag{6.21o}$$

where β_o is the rotation of the mid-span cross-section. When the load acts at the centroid of the cross-section, the torsional moment is given by Eq. (6.21c). Hence, the critical force P_{cr} is the least when the load acts on the upper flange and the critical force is the maximum when it acts on the lower flange of an I section, as shown in Figure 6.19.

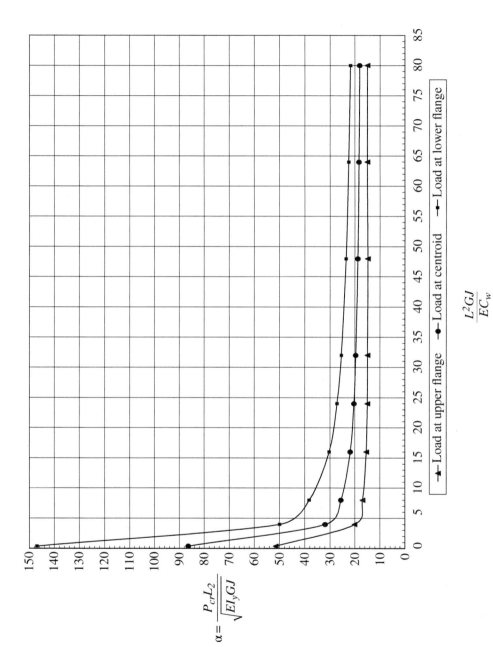

$$\alpha = \frac{P_{cr}L_2}{\sqrt{EI_yGJ}}$$

$$\frac{L^2GJ}{EC_w}$$

─▲─ Load at upper flange ─●─ Load at centroid ─◀─ Load at lower flange

─■─ Load at upper flange ─■─ Load at lower flange

Figure 6.19 Lateral buckling of a simply supported I beam subjected to mid-span load.

6.8.4 Lateral Buckling of Cantilever I Beams: Concentrated Load at the Free End

Consider a cantilever beam subjected to a concentrated force P applied at the centroid of the end cross-section. The deflected shape of the beam in the yz plane becomes unstable as the force P is increased and lateral buckling occurs, as shown in Figure 6.20. A set of axes x, y parallel to the principal axes of the cross-section, and the z axis along the longitudinal axis of the beam are taken with the origin at the centroid of the cross-section at the fixed end. Consider the equilibrium of the deflected shape of the beam. Now, consider the cross-section a-a at a distance z from the origin O at the fixed end. Considering the equilibrium of the deformed beam, the moments of the external force P at the section a-a with respect to the axes passing through the centroid of the cross-section and parallel to the x, y, and z axes are

$$PL - Pz + M_x = 0$$
$$M_x = -P(L - z) \tag{6.22a}$$
$$M_y = 0 \tag{6.22b}$$
$$-Pu_0 + Pu + M_z = 0$$
$$M_z = P(u_0 - u) \tag{6.22c}$$

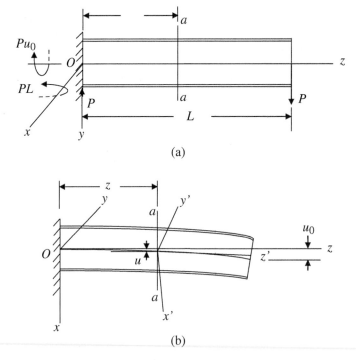

Figure 6.20 Lateral buckling of I beam acted upon by a concentrated load at the free end: (a) Cantilever beam with a concentrated force P at the free end; (b) Deflected top plan view.

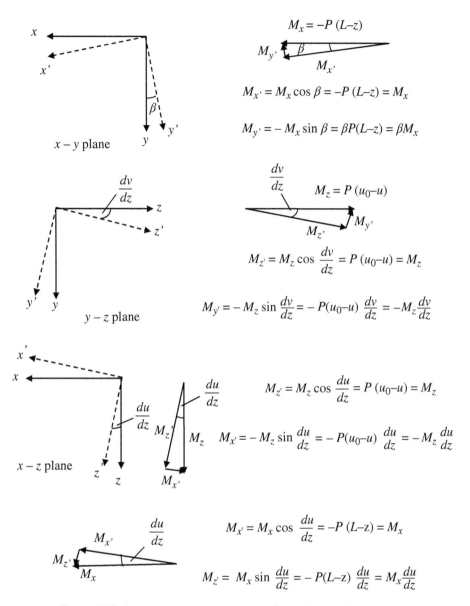

Figure 6.21 Internal moment components of a buckled cantilever beam.

The quantity u_0 is the deflection of the centroid of the cross-section at the free end of the beam in the x direction, taken as positive in the positive x direction, whereas u is the deflection of the centroid of the cross-section at a distance z in the x direction. The internal moment components at the section a-a in the directions of axes x', y', and z' in Figure 6.20 from Figure 6.21 are as follows:

$$M_{x'} = M_x - M_z \frac{du}{dz} = -P(L - z) - P(u_0 - u)\frac{du}{dz} \tag{6.22d}$$

$$M_{y'} = -\beta M_x - M_z \frac{dv}{dz} = \beta P(L - z) - P(u_0 - u)\frac{dv}{dz} \tag{6.22e}$$

$$M_{z'} = M_z + M_x \frac{du}{dz} = P(u_0 - u) - P(L - z)\frac{du}{dz} \tag{6.22f}$$

The internal moment components are given by Eqs. (6.21g and 6.21h)

$$M_{x'} = -EI_x \frac{d^2v}{dz^2} \tag{6.21g}$$

$$M_{y'} = EI_y \frac{d^2u}{dz^2} \tag{6.21h}$$

Substitute the moments given by Eqs. (6.22d and 6.22e) into Eqs. (6.21g and 6.21h), and we obtain

$$EI_x \frac{d^2v}{dz^2} - P(L - z) = 0 \tag{6.22g}$$

$$EI_y \frac{d^2u}{dz^2} - \beta P(L - z) = 0 \tag{6.22h}$$

The terms containing the multiplication of du/dz and $(u_o - u)$, the multiplication of dv/dz and $(u_o - u)$, have been neglected in Eq. (6.22d) and Eq. (6.22e), since these quantities are small. The internal torsional resisting moment is given by

$$T = T_{sv} + T_w = GJ\frac{d\beta}{dz} - EC_w \frac{d^3\beta}{dz^3}$$

$$M_{z'} = T \tag{6.2h}$$

Combine Eqs. (6.22f and 6.2h) to give

$$GJ\frac{d\beta}{dz} - EC_w \frac{d^3\beta}{dz^3} - P(u_0 - u) + P(L - z)\frac{du}{dz} = 0 \tag{6.22i}$$

Eliminate u between Eqs. (6.22h and 6.22i) by differentiating Eq. (6.22i) with respect to z, note $du_0/dz = 0$ and ignore du/dz because it is small. Now substitute d^2u/dz^2 from Eq. (6.22h) to obtain

$$GJ\frac{d^2\beta}{dz^2} - EC_w \frac{d^4\beta}{dz^4} + \frac{P^2}{EI_y}(L - z)^2\beta = 0 \tag{6.22j}$$

or

$$\frac{d^4\beta}{dz^4} - \frac{GJ}{EC_w}\frac{d^2\beta}{dz^2} + \frac{P^2}{E^2 I_y C_w}(L - z)^2\beta = 0 \tag{6.22k}$$

The solution of this differential equation is shown by Timoshenko and Gere [3] to be of the form

$$P_{cr} = \alpha\frac{\sqrt{EI_y GJ}}{L^2} \tag{6.22l}$$

where for large values of the ratio $\frac{GJL^2}{EC_w}$, approximate value of α is given by the equation

$$\alpha = \frac{4.013}{\left(1 - \sqrt{\dfrac{EC_w}{GJL^2}}\right)^2} \tag{6.22m}$$

6.8.4.1 Lateral Buckling of Cantilever Narrow Rectangular Beams: Concentrated Load at the Free End

If the beam cross-section is a narrow rectangle, the warping constant C_w can be neglected, the differential Eq. (6.22j) for the calculation of the angle of twist β becomes

$$\frac{d^2\beta}{dz^2} + \frac{P^2}{EI_yGJ}(L-z)^2\beta = 0 \tag{6.23a}$$

The solution of Eq. (6.23a) is shown by Timoshenko and Gere [3] to be

$$P_{cr} = \frac{4.013}{L^2}\sqrt{EI_yGJ} \tag{6.23b}$$

6.8.5 Lateral Buckling of Narrow Rectangular Cantilever Beams Acted on by Uniform Moment

Let us study a cantilever beam of rectangular cross-section acted on by uniform moment by solving the differential equations of equilibrium for the deformed beam. Consider the equilibrium of the deformed shape of the beam shown in Figure 6.22. The internal moment components at the section a-a with respect to the axes passing through the centroid of the cross-section and parallel to the x, y, and z axes are

$$M_x = -M \tag{6.24a}$$

$$M_y = 0 \tag{6.24b}$$

$$M_z = 0 \tag{6.24c}$$

The internal moment components at the section a-a in the directions of axes x', y', and z' can be obtained as before from Figure 6.21 and Eqs. (6.22d)–(6.22f) as follows:

$$M_{x'} = M_x = -M \tag{6.24d}$$

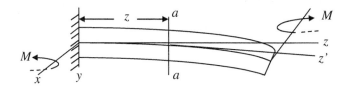

Figure 6.22 Lateral buckling of cantilever beam under uniform moment.

$$M_{y'} = -\beta M_x = \beta M \tag{6.24e}$$

$$M_{z'} = -M\frac{du}{dz} \tag{6.24f}$$

Use Eqs. (6.21g, 6.21h, 6.24d, and 6.24e) to get

$$EI_x\frac{d^2v}{dz^2} - M = 0 \tag{6.24g}$$

$$EI_y\frac{d^2u}{dz^2} - \beta M = 0 \tag{6.24h}$$

For a narrow rectangular beam, the warping constant C_w can be neglected, hence the internal torsional resisting moment is given by

$$T = GJ\frac{d\beta}{dz}$$

$$M_{z'} = T \tag{6.24i}$$

Combine Eqs. (6.24f and 6.24i) to obtain

$$GJ\frac{d\beta}{dz} + M\frac{du}{dz} = 0 \tag{6.24j}$$

Eliminate u between Eqs. (6.24h and 6.24j) by differentiating Eq. (6.24j) with respect to z, and substitute $\dfrac{d^2u}{dz^2}$ from Eq. (6.24h) to get

$$GJ\frac{d^2\beta}{dz^2} + \beta\frac{M^2}{EI_y} = 0 \tag{6.24k}$$

Substitute $\dfrac{M^2}{EI_yGJ} = k^2$, and we have

$$\frac{d^2\beta}{dz^2} + k^2\beta = 0 \tag{6.24l}$$

The solution of Eq. (6.24l) is given by

$$\beta = A\sin kz + B\cos kz \tag{6.24m}$$

The boundary conditions are:

$$\beta = 0 \text{ at } z = 0, \text{ and } \frac{d\beta}{dz} = 0 \text{ at } z = L \tag{6.24n}$$

Substitute the boundary conditions in Eq.(6.24n) to obtain

$$B = 0, \text{ and } \cos kL = 0$$

Hence,

$$kL = \frac{n\pi}{2}, \quad n = 1, 3, 5, --$$

The smallest load is obtained when $n = 1$, therefore

$$kL = \frac{\pi}{2}$$

or

$$\frac{M^2}{EI_y GJ} = \frac{\pi^2}{4L^2}$$

Therefore,

$$M_{cr} = \frac{\pi}{2L}\sqrt{EI_y GJ} \tag{6.24o}$$

6.9 The Energy Method

6.9.1 Lateral Buckling of Simply Supported I Beams: Concentrated Load at the Mid-Span

The lateral buckling load of a simply supported I beam in Figure 6.17 subjected to a concentrated load at the mid-span can also be found by using the energy method. It is assumed the concentrated load is applied at the centroid of the cross-section. When the beam buckles laterally, the strain energy consists of energy due to lateral buckling because of the bending moment about the y axis, and due to the twisting moment about the longitudinal z axis. The x and y axes are the principal axes of the cross-section, whereas the z axis is the centroidal axis. The strain energy of bending in the plane of the beam is comparatively small, hence it can be neglected. The total strain energy, U, is given by

$$U = U_b + U_{sv} + U_w \tag{6.25a}$$

where U_b, U_{sv}, and U_w are the strain energies due to lateral bending, St. Venant's torsion, and warping torsion, respectively. Use Eqs. (6.17b, 6.17e, and 6.17k) to get

$$U = \frac{1}{2}EI_y \int_0^L \left(\frac{d^2 u}{dz^2}\right)^2 dz + \frac{1}{2}GJ \int_0^L \left(\frac{d\beta}{dz}\right)^2 dz + \frac{1}{2}EC_w \int_0^L \left(\frac{d^2\beta}{dz^2}\right)^2 dz \tag{6.25b}$$

The potential energy of the external load will consist of the negative of the product of the force P and the vertical distance the force moves during lateral buckling. When the load P is applied, the member bends in the vertical plane and moves from position 1 to 2 as shown in Figure 6.23a. Then the member moves from position 2 to 3 due to lateral buckling. The displacement from 2 to 3 consists of a horizontal displacement u_0 and a vertical movement w_0. In Eq. (6.25b) the strain energy caused by u is considered, whereas the strain energy due to w, being small, is neglected [1]. If the rigidity of the beam is very large in the plane of the web in comparison to the rigidity in the lateral direction, the assumption is sufficiently accurate for practical purposes. We are concerned here about the potential energy due to lateral buckling. The vertical movement of the force P during lateral buckling is w_0. Consider an element of length dz in Figure 6.23b located at a distance z from the mid span of the beam. The horizontal

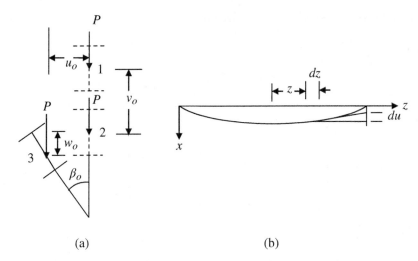

(a)   (b)

Figure 6.23 Lateral buckling of I beam subjected to concentrated load at the center of the beam span: (a) Different positions of the beam deformation (b) Horizontal deviation of an element during lateral buckling.

deviation du due to lateral buckling at the right support between the two tangents at the two ends of the element [1] is given by the moment area theorem as

$$-du = \frac{M_{y'}}{EI_y} dz \left(\frac{L}{2} - z\right)$$

Vertical displacement, dw, corresponding to du is

$$dw = \beta du = -\frac{M_{y'}}{EI_y} \beta \left(\frac{L}{2} - z\right) dz$$

where β is the angle of twist about z axis. Hence, the vertical displacement at the mid-span is

$$w_o = -\int_0^{L/2} \frac{M_{y'}}{EI_y} \beta \left(\frac{L}{2} - z\right) dz$$

From Eq. (6.21e), neglecting the second term as before because it is small, we have

$$M_{y'} = -\beta M_x = -\beta \frac{P}{2} \left(\frac{L}{2} - z\right) \tag{6.25c}$$

Also,

$$M_{y'} = EI_y \frac{d^2 u}{dz^2} \tag{6.21h}$$

Combine Eqs. (6.21h and 6.25c) to get

$$\frac{d^2 u}{dz^2} = \frac{M_{y'}}{EI_y} = -\frac{P\beta}{2EI_y} \left(\frac{L}{2} - z\right) \tag{6.25d}$$

Therefore,

$$w_o = \int_0^{L/2} \frac{P\beta^2}{2EI_y} \left(\frac{L}{2} - z\right)^2 dz \tag{6.25e}$$

The potential energy, V, due to the external force P is given by

$$V = -Pw_o = -\int_0^{L/2} \frac{P^2\beta^2}{2EI_y} \left(\frac{L}{2} - z\right)^2 dz \tag{6.25f}$$

The total potential energy due to lateral buckling is

$$\Pi = U + V$$

$$\Pi = \frac{1}{2}EI_y \int_0^L \left(\frac{d^2u}{dz^2}\right)^2 dz + \frac{1}{2}GJ \int_0^L \left(\frac{d\beta}{dz}\right)^2 dz + \frac{1}{2}EC_w \int_0^L \left(\frac{d^2\beta}{dz^2}\right)^2 dz - -\int_0^{L/2} \frac{P^2\beta^2}{2EI_y} \left(\frac{L}{2} - z\right)^2 dz \tag{6.25g}$$

Substitute Eq. (6.25d) into Eq. (6.25g) and we have

$$\Pi = -\frac{P^2}{4EI_y} \int_0^{L/2} \beta^2 \left(\frac{L}{2} - z\right)^2 dz + \frac{1}{2}GJ \int_0^L \left(\frac{d\beta}{dz}\right)^2 dz + \frac{1}{2}EC_w \int_0^L \left(\frac{d^2\beta}{dz^2}\right)^2 dz \tag{6.25h}$$

Assume

$$\beta = A\cos\frac{\pi z}{L} \tag{6.25i}$$

The expression in Eq. (6.25i) satisfies the boundary conditions that the ends of the member do not rotate about the z axis, but are free to warp. Hence,

$$\beta = \frac{d^2\beta}{dz^2} = 0 \text{ at } z = -\frac{L}{2} \text{ and } z = \frac{L}{2} \tag{6.25j}$$

Substitute Eq. (6.25i) in Eq. (6.25h) and we get

$$\Pi = -\frac{P^2A^2}{4EI_y} \int_0^{L/2} \left(\frac{L}{2} - z\right)^2 \cos^2\frac{\pi z}{L}dz + \frac{GJA^2\pi^2}{2L^2} \int_0^L \sin^2\frac{\pi z}{L}dz + \frac{EC_wA^2\pi^4}{2L^4} \int_0^L \cos^2\frac{\pi z}{L}dz \tag{6.25k}$$

In Eq. (6.25k) the definite integrals have the following values:

$$\int_0^{L/2} \left(\frac{L}{2} - z\right)^2 \cos^2\frac{\pi z}{L}dz = \frac{L^3}{8\pi^2}\left(\frac{\pi^2}{6} + 1\right)$$

$$\int_0^L \sin^2\frac{\pi z}{L}dz = \int_0^L \cos^2\frac{\pi z}{L}dz = \frac{L}{2} \tag{6.25l}$$

Substitute Eqs. (6.25l) into Eq. (6.25k) to get

$$\Pi = -\frac{P^2A^2L^3}{32EI_y\pi^2}\left(\frac{\pi^2}{6} + 1\right) + \frac{GJA^2\pi^2}{4L} + \frac{EC_wA^2\pi^4}{4L^3} \tag{6.25m}$$

Table 6.1 Values of α for simply supported I beams with concentrated load at the mid-span acting through the centroid of the beam cross-section.

$\dfrac{L^2 GJ}{EC_w}$	0.4	4	8	16	24	32	48	64	80
α	86.97	31.96	25.65	21.83	20.39	19.63	18.85	18.44	18.19

At the critical load, the neutral equilibrium is possible, therefore $\dfrac{d\Pi}{dA} = 0$. Hence,

$$\frac{d\Pi}{dA} = \frac{A}{2}\left[-\frac{P^2 L^3}{8EI_y \pi^2}\left(\frac{\pi^2}{6}+1\right) + \frac{GJ\pi^2}{L} + \frac{EC_w \pi^4}{L^3}\right] = 0$$

Hence, the critical load P_{cr} is given by

$$P_{cr} = \frac{4\pi^2}{L^2}\sqrt{\frac{3EI_y}{\pi^2+6}\left(GJ + \frac{EC_w \pi^2}{L^2}\right)} \qquad (6.25n)$$

Equation (6.25n) can be rewritten as

$$P_{cr} = 4\pi^2 \sqrt{\frac{3}{\pi^2+6}}\frac{\sqrt{EI_y GJ}}{L^2}\sqrt{1 + \frac{EC_w \pi^2}{L^2 GJ}} \qquad (6.25o)$$

or

$$P_{cr} = \alpha\frac{\sqrt{EI_y GJ}}{L^2} \qquad (6.25p)$$

where

$$\alpha = 17.1647\sqrt{1 + \frac{EC_w \pi^2}{L^2 GJ}} \qquad (6.25q)$$

The values of α given by Eq. (6.25q) are given in Table 6.1., and these values compare well with those given in Figure 6.19.

6.9.1.1 Lateral Buckling of Simply Supported, Narrow Rectangular Beams: Concentrated Load at the Mid-Span

For a narrow rectangular beam, the warping restraint given by C_w can be neglected, therefore, the critical load given by Eq. (6.25n) can be written as

$$P_{cr} = \frac{4\pi^2}{L^2}\sqrt{\frac{3EI_y GJ}{\pi^2+6}} \qquad (6.26a)$$

or

$$P_{cr} = \frac{17.1647\sqrt{EI_y GJ}}{L^2} \qquad (6.26b)$$

The constant, 17.1647, obtained here by the energy method is very close to the exact solution given by Timoshenko and Gere [3], where the constant is 16.94.

6.9.2 Lateral Buckling of Simply Supported I Beams: Uniformly Distributed Load

It is assumed the uniform load is applied at the centroid of the cross-section. The energy method described for the case of a concentrated load at the mid-span can be used for the beam loaded with a uniform load of q per unit length in Figure 6.24. The total strain energy for the beam is given by Eq. (6.25b) as before

$$U = \frac{1}{2}EI_y \int_0^L \left(\frac{d^2u}{dz^2}\right)^2 dz + \frac{1}{2}GJ \int_0^L \left(\frac{d\beta}{dz}\right)^2 dz + \frac{1}{2}EC_w \int_0^L \left(\frac{d^2\beta}{dz^2}\right)^2 dz \qquad (6.25b)$$

The potential energy of applied load is calculated by the same procedure as given for the simply supported beam loaded with a concentrated load at the mid-span. The external force is uniform, so to calculate the potential energy, the elemental load, $q(dz)$, acting on the element of length dz is multiplied by the vertical deflection, w, under the element and the negative sign of the product is taken. The product is then integrated over the span of the beam to get the total potential energy of the external uniform load. Consider an element dz located at a distance z from the mid-span of the beam in Figure 6.24. The horizontal deviation du due to lateral bending between the tangents drawn at the two ends of the element at the right support is given by the moment area theorem as before as

$$-du = \frac{M_{y'}}{EI_y} dz \left(\frac{L}{2} - z\right)$$

and the vertical displacement corresponding to du is

$$dw = \beta du = -\frac{M_{y'}}{EI_y}\beta \left(\frac{L}{2} - z\right) dz$$

The tangent to the deflected curve of the beam at A is horizontal so the vertical deviation of point B from the tangent at A gives the vertical displacement of the mid-span by moment area theorem. Hence the vertical displacement w_0 at the center of the span is given by

$$w_o = -\int_0^{L/2} \frac{M_{y'}}{EI_y}\beta \left(\frac{L}{2} - z\right) dz$$

$$M_{y'} = -\beta M_x = -\beta \left[\frac{qL}{2}\left(\frac{L}{2} - z\right) - \frac{q}{2}\left(\frac{L}{2} - z\right)^2\right]$$

or

$$M_{y'} = -\beta \frac{q}{8}(L^2 - 4z^2) \qquad (6.27a)$$

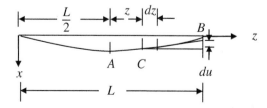

Figure 6.24 Horizontal deviation of an element due to lateral buckling.

or

$$w_0 = \frac{q}{8EI_y} \int_0^{L/2} \left(\frac{L^3}{2} - L^2 z - 2Lz^2 + 4z^3 \right) \beta^2 dz \tag{6.27b}$$

Assume

$$\beta = A \cos \frac{\pi z}{L} \tag{6.27c}$$

The β assumed above satisfies the boundary conditions at the ends of the beam. Therefore,

$$w_0 = \frac{qA^2}{8EI_y} \int_0^{L/2} \left(\frac{L^3}{2} - L^2 z - 2Lz^2 + 4z^3 \right) \cos^2 \frac{\pi z}{L} \tag{6.27d}$$

In Eq. (6.27d) the definite integrals have the following values:

$$\int_0^{L/2} \cos^2 \frac{\pi z}{L} dz = \frac{L}{4}$$

$$\int_0^{L/2} z \cos^2 \frac{\pi z}{L} dz = \frac{L^2}{16} - \frac{L^2}{4\pi^2}$$

$$\int_0^{L/2} z^2 \cos^2 \frac{\pi z}{L} dz = \frac{L^3}{48} - \frac{L^3}{8\pi^2}$$

$$\int_0^{L/2} z^3 \cos^2 \frac{\pi z}{L} dz = \frac{L^4}{128} - \frac{3L^4}{32\pi^2} + \frac{3L^4}{8\pi^4} \tag{6.27e}$$

Substitute the values of definite integrals from Eq. (6.27e) in Eq. (6.27d) to get

$$w_0 = \frac{qA^2}{8EI_y} \left[\frac{L^3}{2} \left(\frac{L}{4} \right) - L^2 \left(\frac{L^2}{16} - \frac{L^2}{4\pi^2} \right) - 2L \left(\frac{L^3}{48} - \frac{L^3}{8\pi^2} \right) + 4 \left(\frac{L^4}{128} - \frac{3L^4}{32\pi^2} + \frac{3L^4}{8\pi^4} \right) \right]$$

or

$$w_0 = \frac{qA^2}{8EI_y} \left(\frac{5L^4}{96} + \frac{L^4}{8\pi^2} + \frac{3L^4}{2\pi^4} \right) \tag{6.27f}$$

Now consider an element of length ds at a distance s from the mid-span within the distance z as shown in Figure 6.25. The horizontal deviation du_s due to lateral bending between the tangents drawn at the two ends of length ds at the point D in Figure 6.25b is given by

$$-du_s = \frac{M_{y'}}{EI_y} (z - s)ds$$

and the vertical displacement corresponding to du_s is

$$dw = \beta du = -\frac{M_{y'}}{EI_y} \beta (z - s)ds$$

The moment M_x at D is obtained from

$$M_x = \frac{qL}{2} \left(\frac{L}{2} - s \right) - \frac{q}{2} \left(\frac{L}{2} - s \right)^2$$

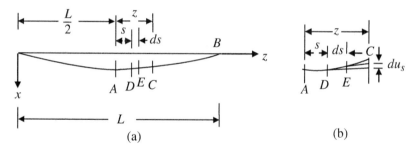

Figure 6.25 Horizontal deviation of element of length ds: (a) Element of length ds within the length z; (b) Enlarged view of AC.

Hence,

$$M_{y'} = -\beta M_x = -\beta \frac{q}{8}(L^2 - 4s^2) \tag{6.27g}$$

Similar to the vertical displacement at mid-span, the vertical deviation of point C from the tangent at mid-span A in the deflected beam is expressed as

$$t_{C/A} = -\int_0^z \frac{M_{y'}}{EI_y}\beta(z - s)ds$$

or

$$t_{C/A} = \frac{q}{8EI_y}\int_0^z \beta^2(L^2 - 4s^2)(z - s)ds \tag{6.27h}$$

Assume

$$\beta = A\cos\frac{\pi s}{L}$$

It satisfies the boundary conditions at the ends of the beam segment AC. Substitute the assumed β in Eq. (6.27h) to obtain

$$t_{C/A} = \frac{qA^2}{8EI_y}\int_0^z (L^2 z - L^2 s - 4zs^2 + 4s^3)\cos^2\frac{\pi s}{L}ds \tag{6.27i}$$

In Eq. (6.27i) the definite integrals have the following values:

$$\int_0^z \cos^2\frac{\pi s}{L}ds = \frac{z}{2} + \frac{L}{4\pi}\sin\frac{2\pi z}{L}$$

$$\int_0^z s\cos^2\frac{\pi s}{L}ds = \frac{z^2}{4} + \frac{Lz}{4\pi}\sin\frac{2\pi z}{L} + \frac{L^2}{8\pi^2}\cos\frac{2\pi z}{L} - \frac{L^2}{8\pi^2}$$

$$\int_0^z s^3\cos^2\frac{\pi s}{L}ds = \frac{z^4}{8} + \left(\frac{3L^2 z^2}{8\pi^2} - \frac{3L^4}{16\pi^4}\right)\cos\frac{2\pi z}{L} + \left(\frac{Lz^3}{4\pi} - \frac{3L^3 z}{8\pi^3}\right)\sin\frac{2\pi z}{L} + \frac{3L^4}{16\pi^4} \tag{6.27j}$$

Substitute Eqs. (6.27j) into Eq. (6.27i) to obtain

$$t_{C/A} = \frac{qA^2}{8EI_y}\left[\frac{L^2 z^2}{4} - \frac{z^4}{6} + \frac{L^4}{8\pi^2} + \frac{3L^4}{4\pi^4} - \left(\frac{zL^3}{\pi^3}\right)\sin\frac{2\pi z}{l} + \left(-\frac{L^4}{8\pi^2} + \frac{L^2 z^2}{2\pi^2} - \frac{3L^4}{4\pi^4}\right)\cos\frac{2\pi z}{L}\right]$$

The vertical displacement at C at a distance z from the mid-span is equal to

$$w = w_0 - t_{C/A}$$

or

$$w = \frac{qA^2}{8EI_y} \left[\frac{5L^4}{96} + \frac{3L^4}{4\pi^4} - \frac{L^2z^2}{4} + \frac{z^4}{6} + \left(\frac{zL^3}{\pi^3} \right) \sin \frac{2\pi z}{L} - \left(-\frac{L^4}{8\pi^2} + \frac{L^2z^2}{2\pi^2} - \frac{3L^4}{4\pi^4} \right) \cos \frac{2\pi z}{L} \right]$$

$$(6.27k)$$

The potential energy, dV, due to the external uniform load $(q \, dz)$ over the span length dz is given by

$$dV = -q dz \, (w)$$

Hence, the total potential energy V of the uniform load is obtained from

$$V = -\frac{q^2A^2}{4EI_y} \int_0^{L/2} \left[\frac{5L^4}{96} + \frac{3L^4}{4\pi^4} - \frac{L^2z^2}{4} + \frac{z^4}{6} + \left(\frac{zL^3}{\pi^3} \right) \sin \frac{2\pi z}{L} - \left(-\frac{L^4}{8\pi^2} + \frac{L^2z^2}{2\pi^2} - \frac{3L^4}{4\pi^4} \right) \cos \frac{2\pi z}{L} \right] dz$$

$$(6.27l)$$

In Eq. (6.17l) the definite integrals have the following values:

$$\int_0^{L/2} z \sin \frac{2\pi z}{L} dz = \frac{L^2}{4\pi}$$

$$\int_0^{L/2} z^3 \sin \frac{2\pi z}{L} dz = \frac{L^4}{16\pi} - \frac{3L^4}{8\pi^3}$$

$$\int_0^{L/2} \cos \frac{2\pi z}{L} dz = 0$$

$$\int_0^{L/2} z^2 \cos \frac{2\pi z}{L} dz = -\frac{L^3}{4\pi^2} \qquad (6.27m)$$

Substitute Eqs. (6.27m) into Eq.(6.27l) to get

$$V = -\frac{q^2A^2L^5}{16EI_y} \left(\frac{1}{15} + \frac{3}{\pi^4} \right) \qquad (6.27n)$$

The total potential energy due to lateral buckling is

$$\Pi = U + V$$

Combine Eqs. (6.25b and 6.27n) and we have

$$\Pi = \frac{1}{2}EI_y \int_0^L \left(\frac{d^2u}{dz^2} \right)^2 dz + \frac{1}{2}GJ \int_0^L \left(\frac{d\beta}{dz} \right)^2 dz + \frac{1}{2}EC_w \int_0^L \left(\frac{d^2\beta}{dz^2} \right)^2 dz - \frac{q^2A^2L^5}{16EI_y} \left(\frac{1}{15} + \frac{3}{\pi^4} \right)$$

$$(6.27o)$$

$$M_{y'} = EI_y \frac{d^2u}{dz^2} \qquad (6.21h)$$

Combine Eqs. (6.21h and 6.27a) to have

$$\frac{d^2u}{dz^2} = \frac{M_{y'}}{EI_y} = -\frac{\beta q}{8EI_y}(L^2 - 4z^2) \tag{6.27p}$$

Substitute Eqs. (6.27c and 6.27p) to obtain

$$\frac{1}{2}EI_y \int_0^L \left(\frac{d^2u}{dz^2}\right)^2 dz = \frac{q^2A^2}{64EI_y} \int_0^{L/2} (L^4 - 8L^2z^2 + 16z^4)\cos^2\frac{\pi z}{L} dz \tag{6.27q}$$

In Eq. (6.27q) the definite integral has the following value:

$$\int_0^{L/2} z^4 \cos^2\frac{\pi z}{L} dz = \frac{L^5}{8}\left(\frac{1}{40} - \frac{1}{2\pi^2} + \frac{3}{\pi^4}\right) \tag{6.27r}$$

Substitute Eqs. (6.27e and 6.27r) into Eq. (6.27q) to get

$$\frac{1}{2}EI_y \int_0^{L/2} \left(\frac{d^2u}{dz^2}\right)^2 dz = \frac{q^2A^2L^5}{16EI_y}\left(\frac{1}{30} + \frac{3}{2\pi^4}\right) \tag{6.27s}$$

The following expressions were derived before (to derive Eq. (6.25m))

$$\frac{1}{2}GJ \int_0^{L/2} \left(\frac{d\beta}{dz}\right)^2 dz = \frac{GJA^2\pi^2}{4L} \tag{6.27t}$$

$$\frac{1}{2}EC_w \int_0^L \left(\frac{d^2\beta}{dz^2}\right)^2 dz = \frac{EC_wA^2\pi^4}{4L^3} \tag{6.27u}$$

Substitute Eqs. (6.27s)–(6.27u) into Eq. (6.27o) and we have total potential energy equal to

$$\Pi = \frac{q^2A^2L^5}{16EI_y}\left(-\frac{1}{30} - \frac{3}{2\pi^4}\right) + \frac{GJA^2\pi^2}{4L} + \frac{EC_wA^2\pi^4}{4L^3} \tag{6.27v}$$

At the critical load, the neutral equilibrium is possible, therefore, $\dfrac{d\Pi}{dA} = 0$. Hence,

$$\frac{d\Pi}{dA} = \frac{A}{2}\left[\frac{q^2L^5}{4EI_y}\left(-\frac{1}{30} - \frac{3}{2\pi^4}\right) + \frac{GJ\pi^2}{L} + \frac{EC_w\pi^4}{L^3}\right] = 0$$

$$q^2 = \frac{4EI_y\pi^4}{L^5\left(\dfrac{\pi^4 + 45}{30}\right)}\left(\frac{GJ\pi^2}{L} + \frac{EC_w\pi^4}{L^3}\right)$$

$$(qL)_{cr} = 2\pi^3\left(\sqrt{\frac{30}{\pi^4 + 45}}\right)\frac{\sqrt{EI_yGJ}}{L^2}\sqrt{1 + \frac{EC_w\pi^2}{GJL^2}} \tag{6.27w}$$

or

$$(qL)_{cr} = \alpha\frac{\sqrt{EI_yGJ}}{L^2} \tag{6.27x}$$

Table 6.2 Values of α for simply supported I beams with uniformly distributed load acting through the centroid of the beam cross-section.

$\dfrac{L^2 GJ}{EC_w}$	0.4	4	8	16	24	32	48	64	80
α	144.21	52.99	42.54	36.19	33.81	32.55	31.25	3.58	30.16

where

$$\alpha = 28.46\sqrt{1 + \frac{EC_w \pi^2}{GJL^2}} \qquad (6.27\text{y})$$

The values of α given by Eq. (6.27y) are given in Table 6.2.

6.9.2.1 Lateral Buckling of Simply Supported, Narrow Rectangular Beams: Uniformly Distributed Load along the Centroidal Axis

For a narrow rectangle the warping constant C_w can be assumed to be zero, therefore the critical load given by Eq. (6.27w) can be written as

$$(qL)_{cr} = 2\pi^3 \left(\sqrt{\frac{30}{\pi^4 + 45}} \right) \frac{\sqrt{EI_y GJ}}{L^2}$$

$$(qL)_{cr} = 28.46 \frac{\sqrt{EI_y GJ}}{L^2} \qquad (6.28\text{a})$$

The value of α is given by $28.3 \dfrac{\sqrt{EI_y GJ}}{L^2}$ in Timoshenko and Gere's book [3], which is obtained from the exact solution. The critical load obtained from Eq. (6.28a) is only 0.57% different from the exact solution. The solution here is not exact because only one term is taken in the series to calculate the β value.

6.9.3 Lateral Buckling of Cantilever Rectangular Beams: Concentrated Load at the Free End

The energy method is used here to calculate the lateral buckling load of a cantilever rectangular beam acted upon by a concentrated force P at the centroid of the cross-section at the free end, shown in Figure 6.20. The total strain energy of the beam is given by Eq. (6.25b) in which the term containing warping constant is neglected.

$$U = \frac{1}{2}EI_y \int_0^L \left(\frac{d^2 u}{dz^2} \right)^2 dz + \frac{1}{2}GJ \int_0^L \left(\frac{d\beta}{dz} \right)^2 dz \qquad (6.29\text{a})$$

Consider an element AB of length dz at a distance z from the fixed support in Figure 6.26. The horizontal deviation du at the free end due to lateral buckling between the tangents drawn at the two ends of the element is given below by the moment area theorem as before

$$du = \frac{M_{y'}}{EI_y} dz(L - z) \qquad (6.29\text{b})$$

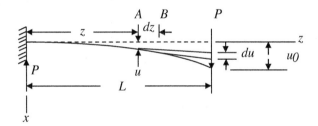

Figure 6.26   Top plan view of cantilever beam.

The vertical displacement corresponding to du in Figure 6.26 is given by

$$dw = \beta du = \beta \frac{M_{y'}}{EI_y}(L - z)dz \tag{6.29c}$$

Hence, the vertical displacement w_0 at the free end of the cantilever is given by

$$w_0 = \int_0^L \frac{\beta M_{y'}}{EI_y}(L - z)dz \tag{6.29d}$$

From Eq. (6.22e), neglecting the second term as before because it is small, we have

$$M_{y'} = -\beta M_x = \beta P(L - z) \tag{6.29e}$$

$$M_{y'} = EI_y \frac{d^2u}{dz^2} \tag{6.21h}$$

Combine Eqs. (6.21h and 6.29e) to obtain

$$\frac{d^2u}{dz^2} = \frac{M_{y'}}{EI_y} = \frac{\beta P(L - z)}{EI_y} \tag{6.29f}$$

Therefore,

$$w_0 = \int_0^L \frac{P\beta^2(L - z)^2}{EI_y}dz \tag{6.29g}$$

The potential energy, V, due to the external force P is given by

$$V = -Pw_0 = -\int_0^L \frac{P^2\beta^2(L - z)^2}{EI_y}dz \tag{6.29h}$$

Total potential energy due to lateral buckling is

$$\Pi = U + V$$

$$\Pi = \frac{1}{2}EI_y \int_0^L \left(\frac{d^2u}{dz^2}\right)^2 dz + \frac{1}{2}GJ \int_0^L \left(\frac{d\beta}{dz}\right)^2 dz - \int_0^L \frac{P^2\beta^2(L - z)^2}{EI_y}dz \tag{6.29i}$$

Substitute Eq. (6.29f) into Eq. (6.29i) and we have

$$\Pi = -\frac{1}{2} \int_0^L \frac{P^2 \beta^2 (L-z)^2}{EI_y} dz + \frac{1}{2} GJ \int_0^L \left(\frac{d\beta}{dz}\right)^2 dz \tag{6.29j}$$

Assume [8]

$$\beta = \frac{\beta_L (2Lz - z^2)}{L^2} \tag{6.29k}$$

where β_L is the rotation of the cross-section at the free end.

The expression in Eq. (6.29k) satisfies the boundary conditions

$$\beta = 0 \text{ at } z = 0 \text{ (No rotation at the fixed end)} \tag{6.29l}$$

and

$$\frac{d\beta}{dz} = 0 \text{ at } z = L \left(\text{St.Venant's torsion, } T_{sv} = GJ \frac{d\beta}{dz} = 0, \text{ at the free end} \right) \tag{6.29m}$$

Substitute Eq. (6.29k) into Eq. (6.29j) to get

$$\Pi = -\frac{P^2}{2EI_y} \int_0^L \left(\frac{\beta_L}{L^2}(2Lz - z^2)\right)^2 (L-z)^2 dz + \frac{1}{2} \int_0^L GJ \left(\frac{\beta_L}{L^2}(2L - 2z)\right)^2 dz \tag{6.29n}$$

After integrating Eq. (6.29n), we have

$$\Pi = -\frac{4P^2 \beta_L^2 L^3}{105 EI_y} + \frac{2GJ\beta_L^2}{3L} \tag{6.29o}$$

The critical load is obtained by considering the neutral equilibrium. Therefore,

$$\frac{d\Pi}{d\beta_L} = 0$$

or

$$\frac{d\Pi}{d\beta_L} = -\beta_L \left(\frac{-8P^2 L^3}{105 EI_y} + \frac{4GJ}{3L} \right) = 0$$

Hence, the critical load is given by

$$P_{cr} = \frac{4.183}{L^2} \sqrt{EI_y GJ} \tag{6.29p}$$

The exact solution for this problem given in Timoshenko and Gere's book [3] is given by $P_{cr} = \frac{4.013}{L^2} \sqrt{EI_y GJ}$. The solution given by Eq. (6.29p) is 4.24% different from the exact solution.

6.10 Beams with Different Support and Loading Conditions

We have studied simply supported beams under the action of uniform moments, concentrated load at mid-span, and uniformly distributed load by using differential equations or energy approach. We also developed the solutions for cantilever beams subjected to uniform moment and concentrated load at the free end. Now it is intended to find general expression for the lateral buckling of beams with different support and loading conditions.

6.10.1 Different Support Conditions

The critical moment expressions for cantilever beams and simply supported beams acted on by uniform moments are as follows:

$$M_{cr} = \frac{\pi}{L}\sqrt{EI_yGJ} \text{ (Simply supported narrow rectangular beam)} \tag{6.19o}$$

$$M_{cr} = \frac{\pi}{L}\sqrt{EI_yGJ\left(1 + \frac{\pi^2}{L^2}\frac{EC_w}{GJ}\right)} \text{ (Simply supported I − beam)} \tag{6.20u}$$

$$M_{cr} = \frac{\pi}{2L}\sqrt{EI_yGJ} \text{ (Cantilever narrow rectangular beam)} \tag{6.24o}$$

The fixed beam in Figure 6.27 can rotate about the principal axis x, but cannot rotate about the principal axis y at the ends. The applied moments act in the y–z plane, the plane of maximum rigidity. Hence the beam is simply supported when bending in the y–z plane. The beam is acted on by a uniform moment in the y–z plane. The critical moment can be calculated by assuming that there will be inflection points at the distances of $L/4$ from the ends. The middle portion of the beam of length $L/2$ acts similar to the simply supported beam shown in Figure 6.16. Hence, the critical moments can be found from Eqs. (6.19o and 6.20u) by substituting $L/2$ for L, and we obtain

$$M_{cr} = \frac{2\pi}{L}\sqrt{EI_yGJ} \text{ (Narrow rectangle beam fixed at the ends)} \tag{6.30a}$$

$$M_{cr} = \frac{2\pi}{L}\sqrt{EI_yGJ\left(1 + \frac{4\pi^2}{L^2}\frac{EC_w}{GJ}\right)} \text{ (I − beam fixed at the ends)} \tag{6.30b}$$

The critical moments given by Eqs. (6.19o, 6.20u, 6.30a, and 6.30b) suggest that we can find the critical moments for different end conditions by using the concept of effective length, We can write a general equation for a beam acted on by uniform moments as follows;

$$M_{cr} = \frac{\pi}{kL}\sqrt{EI_yGJ\left[1 + \frac{\pi^2}{(kL)^2}\frac{EC_w}{GJ}\right]} \tag{6.30c}$$

In Eq. (6.30c), kL, is the effective length of the beam. For a simply supported beam, $kL = L$, the actual length of the beam; for a cantilever, $kL = 2L$; for a beam with one end simply supported and the other fixed, $kL = 0.70L$; and for the fixed beam, $kL = L/2$. Equation (6.30c) can be used

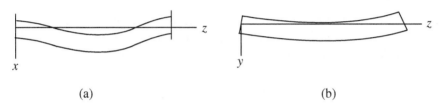

(a)   (b)

Figure 6.27 Lateral buckling of fixed-fixed beams: (a) Warping and lateral bending prevented at both ends in the x-z plane; (b) Bending in the y-z plane permitted at both ends.

for narrow rectangles, by taking $C_w = 0$. Galambos [9], Nethercot [10], and Vlasov [11] have listed effective length factors for other support conditions.

6.10.2 Different Loading Conditions

The critical moment for a simply supported beam subjected to uniform moments is given by Eq. (6.20u), as shown below:

$$M_{cr} = \frac{\pi}{L}\sqrt{EI_yGJ\left(1 + \frac{\pi^2}{L^2}\frac{EC_w}{GJ}\right)} \tag{6.20u}$$

Whereas the same beam acted on by a concentrated load at mid-span the critical load is given by Eq. (6.25o). as shown below:

$$P_{cr} = 4\pi^2\sqrt{\frac{3}{\pi^2 + 6}}\frac{\sqrt{EI_yGJ}}{L^2}\sqrt{1 + \frac{EC_w\pi^2}{L^2GJ}} \tag{6.25o}$$

At the critical load P_{cr}, the critical moment M_{cr} will be at the center of the span that can be obtained from Eq. (6.25o) as

$$M_{cr} = M_{max} = \frac{P_{cr}L}{4} = 1.37\frac{\pi}{L}\sqrt{EI_yGJ\left(1 + \frac{\pi^2EC_w}{L^2GJ}\right)} \tag{6.31a}$$

The critical load for a simply supported beam under the action of uniformly distributed load is given by Eq. (6.27w), as shown below:

$$(qL)_{cr} = 2\pi^3\left(\sqrt{\frac{30}{\pi^4 + 45}}\right)\frac{\sqrt{EI_yGJ}}{L^2}\sqrt{1 + \frac{\pi^2EC_w}{L^2GJ}} \tag{6.27w}$$

The critical moment M_{cr} will be at the center of the span obtained from

$$M_{cr} = M_{max} = \frac{q_{cr}L^2}{8} = 1.13\frac{\pi}{L}\sqrt{EI_yGJ\left(1 + \frac{\pi^2EC_w}{L^2GJ}\right)} \tag{6.31b}$$

If we use the criterion of critical moment instead of critical load, then we may write a general expression for different loading conditions to give the critical moment as follows:

$$M_{cr} = C\frac{\pi}{L}\sqrt{EI_yGJ\left(1 + \frac{\pi^2EC_w}{L^2GJ}\right)} \tag{6.31c}$$

Where the constant C varies with different load conditions, it is being equal to 1.0 for uniform moment, 1.37 for concentrated load at the mid span, and 1.13 for the uniformly distributed load.

If we combine Eqs. (6.30c and 6.31c), the critical moment expression can be written as

$$M_{cr} = C\frac{\pi}{kL}\sqrt{EI_yGJ\left(1 + \frac{\pi^2EC_w}{(kL)^2GJ}\right)} \tag{6.31d}$$

The values of constant C are given by Clark and Hill [12] corresponding to different loading conditions.

6.10.2.1 Beams with Unequal Moments

If a beam is subjected to unequal moments at its ends, moments in the beam vary along the span. In such cases the governing differential equations will have variable coefficients and solutions to such problems are difficult. For the purpose of design, Salvadori [13] suggested a solution given below:

$$M_{cr} = C_b \frac{\pi}{L} \sqrt{EI_y GJ \left(1 + \frac{\pi^2 EC_w}{L^2 GJ}\right)} \tag{6.32a}$$

where C_b is given by

$$C_b = 1.75 + 1.05 \left(\frac{M_A}{M_B}\right) + 0.3 \left(\frac{M_A}{M_B}\right)^2 \leq 2.3 \tag{6.32b}$$

In Eq. (6.32b), M_A is the smaller of the two end moments, and M_B is the larger moment. The ratio $\left(\frac{M_A}{M_B}\right)$ is positive for a double curvature bending shown in Figure 4.21, and the ratio is negative for a single curvature bending shown in Figure 4.24. If the loading is of another form, an expression given by Kirby and Nethercot [14] can be used to calculate C_b.

$$C_b = \frac{12}{3(M_1/M_{max}) + 4(M_2/M_{max}) + 3(M_3/M_{max}) + 2} \tag{6.32c}$$

where M_1, M_2, and M_3 are the moments at the quarter point, mid-point, and three-quarter point of the beam respectively, and the moment M_{max} is the maximum moment in the beam. Other cases of loading and boundary conditions, including continuous beams, are given by Chen and Lui [8].

6.11 Design for Torsional and Lateral Buckling

In the design of beams, Eq. (6.32a) is many times simplified by neglecting St. Venant's torsion or the warping term inside the square root sign. For members that have long unbraced spans, the warping term can be omitted because the warping term decreases with the span increase, whereas the St. Venant's torsion resistance does not. For thin-walled members, the St. Venant's torsion resistance is proportional to the cube of the thickness in Eq. (6.1b), hence it can be neglected in comparison to the warping term.

6.11.1 AISC Design Criteria for Steel Beams

According to AISC [5], if a beam can remain stable until it reaches its plastic moment capacity M_p, the nominal flexural strength, M_n, of the beam is given by

$$M_n = M_p = F_y Z \tag{6.33a}$$

where F_y is the yield strength of steel, and Z is the plastic section modulus. This depends on the cross-sectional geometry, and the lateral unbraced length of the beam. The AISC specifications define two types of buckling in the beam cross-sections: local buckling or the overall lateral buckling. A beam may fail by flange local buckling, web local buckling, or lateral torsional buckling. Any of these failures can be either in the elastic or inelastic range. The strength corresponding to these modes is calculated and the smallest value is taken.

6.11.1.1 Local Buckling

Steel beam cross-sections are classified as compact, noncompact, or slender-element sections. For a section to be qualified as compact, its flanges must be continuously connected to the web or webs, and the width to thickness ratio of its compression elements must be within the limiting ratio λ_p. The elements supported on only one side parallel to the direction of the compressive force are called unstiffened, such as the flanges of I-shaped members and tees, legs of angles and flanges of channels. The elements supported along two edges parallel to the direction of the compressive force are called stiffened, such as webs of rolled steel sections. The classification of the shapes is given in Table 6.3. For flanges of I-shaped members and tees, the width b is one half the flange width b_f. For legs of angles and flanges of channels, the width b is the full nominal dimension. For webs of rolled sections, the width is h is the clear distance between the flanges less the fillet or corner radius at each flange.

For noncompact sections, local buckling may occur before the plastic moment capacity is reached. The nominal moment capacity M_n for doubly symmetric I shapes with compact webs and noncompact or slender flanges bent about their major axis is calculated as below.

Compression flange local buckling

For sections with noncompact flanges, buckling is inelastic, and

$$M_n = M_p - (M_p - 0.7F_yS_x)\left(\frac{\lambda - \lambda_p}{\lambda_r - \lambda_p}\right) \tag{6.34a}$$

For sections with slender flanges the buckling is elastic, and

$$M_n = \frac{0.9Ek_cS_x}{\lambda^2} \tag{6.34b}$$

Table 6.3 Width thickness ratios for compression elements in flexure.

Element	λ	λ_p	λ_r
Flanges of rolled I shapes	$\dfrac{b_f}{2t_f}$	$0.38\sqrt{\dfrac{E}{F_y}}$	$1.0\sqrt{\dfrac{E}{F_y}}$
Flanges of channels	$\dfrac{b_f}{t_f}$	$0.38\sqrt{\dfrac{E}{F_y}}$	$1.0\sqrt{\dfrac{E}{F_y}}$
Webs of doubly symmetric I shapes and channels	$\dfrac{h}{t_w}$	$3.76\sqrt{\dfrac{E}{F_y}}$	$5.70\sqrt{\dfrac{E}{F_y}}$

where λ = width to thickness ratio; t_f = thickness of the element; E = modulus of elasticity of material
If $\lambda \leq \lambda_p$, and the flange is continuously connected to the web, the shape is compact.
$\lambda_p < \lambda \leq \lambda_r$, the flange of the shape is noncompact.
$\lambda > \lambda_r$, the flange of the shape is slender.

where $\lambda = \dfrac{b_f}{2t_f}$; λ_p = limiting width to thickness ratio for a compact flange in Table 6.3; and λ_r = limiting width to thickness ratio for a noncompact flange in Table 6.3.

$k_c = \dfrac{4}{\sqrt{h/t_w}}$, its value is not taken less than 0.35 and not more than 0.76 for calculation purposes; and S_x = elastic section modulus taken about x axis.

The nominal moment capacity for doubly symmetric I shapes with noncompact or slender webs bent about their major axis is calculated as below.

<u>Compression flange local buckling</u>

$$M_n = R_{pg}F_{cr}S_{xc} \tag{6.34c}$$

where S_{xc} = elastic section modulus referred to the compression flange; and the bending strength reduction factor R_{pg} is expressed as

$$R_{pg} = 1 - \frac{a_w}{1200 + 300a_w}\left(\frac{h_c}{t_w} - 5.7\sqrt{\frac{E}{F_y}}\right) \leq 1.0 \tag{6.34d}$$

and

$$a_w = \frac{h_c t_w}{b_{fc}t_{fc}} \leq 1.0 \tag{6.34e}$$

where h_c = two times the depth of the web in compression; t_w = thickness of the web; b_{fc} = compression flange width; t_{fc} = compression flange thickness.

For sections with noncompact flanges

$$F_{cr} = F_y - (0.3F_y)\left(\frac{\lambda - \lambda_p}{\lambda_r - \lambda_p}\right) \tag{6.34f}$$

For sections with slender flange sections

$$F_{cr} = \frac{0.9Ek_c}{\left(\dfrac{b_f}{2t_f}\right)^2} \tag{6.34g}$$

$k_c = \dfrac{4}{\sqrt{h/t_w}}$, it should not be taken to be less than 0.35 nor greater than 0.76.

6.11.1.2 Lateral Torsional Buckling

Lateral torsional buckling can be prevented by lateral supporting or bracing the beam at close intervals. The moment capacity of a beam depends upon the compactness of the cross-section and the lateral unbraced length L_b, the distance between points of lateral support, or bracing. For doubly symmetric compact I shapes and channels bent about their major axis the moment capacity depends on the unbraced length and is given as follows:

When $L_b \leq L_p$

where L_p = the maximum laterally unbraced length for which a beam can reach its plastic moment capacity, M_p, and the lateral torsional buckling will not occur, as before:

$$M_n = M_p = F_y Z \tag{6.33a}$$

When $L_p < L_b \leq L_r$, the moment capacity is based on the inelastic lateral torsional buckling given as

$$M_n = C_b \left[M_p - (M_p - 0.7F_y S_x) \left(\frac{L_b - L_p}{L_r - L_p} \right) \right] \leq M_p \tag{6.35a}$$

where L_r = unbraced length at which elastic torsional buckling will occur
When $L_b > L_r$, the moment capacity is based on elastic lateral torsional buckling given as

$$M_n = F_{cr} S_x \leq M_p \tag{6.35b}$$

$$F_{cr} = \frac{C_b \pi^2 E}{\left(\dfrac{L_b}{r_{ts}} \right)^2} \sqrt{1 + 0.078 \frac{Jc}{S_x h_0} \left(\frac{L_b}{r_{ts}} \right)^2} \tag{6.35c}$$

where

C_b = lateral torsional buckling modification factor for nonuniform moment over the unbraced length L_b, and is obtained from

$$C_b = \frac{12.5 M_{max}}{2.5 M_{max} + 3 M_A + 4 M_B + 3 M_C} R_m \leq 3.0 \tag{6.35d}$$

Note: Equation (6.35d) is similar to Eq. (6.32c) shown before.
where

M_{max} = absolute value of maximum moment in the unbraced segment
M_A = absolute value of moment at quarter point of the unbraced segment
M_B = absolute value of moment at centerline of the unbraced segment
M_C = absolute value of moment at three-quarter point of the unbraced segment
R_m = cross-section mono-symmetry parameter
   = 1.0 for doubly symmetric members
   = 1.0 for singly symmetric members subjected to single curvature bending
   $= 0.5 + 2 \left(\dfrac{I_{yc}}{I_y} \right)^2$ for singly symmetric members subjected to reverse curvature bending
I_y = moment of inertia of the cross-section about the principal y axis
I_{yc} = moment of inertia about the y axis referred to the compression flange, or if reverse curvature bending,
   referred to the smaller flange
C_b is permitted to be taken as 1.0 for all cases, which is its value for uniform bending.
J = torsional constant
S_x = elastic section modulus about the x axis
h_0 = distance between the flange centroids

$c = 1$ for doubly symmetric I shapes
and for a channel

$$c = \frac{h_0}{2}\sqrt{\frac{I_y}{C_w}} \tag{6.35e}$$

$$r_{ts}^2 = \frac{\sqrt{I_y C_w}}{S_x} \tag{6.35f}$$

For doubly symmetric I shapes with rectangular flanges, the warping constant $C_w = \frac{I_y h_0^2}{4}$.

Note: For non-uniform moment acting on a beam, the critical elastic lateral torsional buckling moment is given by Eq. (6.32a) derived before.

$$M_n = M_{cr} = C_b \frac{\pi}{L}\sqrt{EI_y GJ + \frac{\pi^2 E^2 I_y C_w}{L^2}} \tag{6.32a}$$

$$G = \frac{E}{2(1+v)}$$

For steel, $v = 0.3$, hence $G = 0.385\,E$. Substitute $L = L_b$, and C_w from Eq. (6.35f) into Eq. (6.32a) to get

$$M_n = C_b \frac{\pi^2 E}{\left(\frac{L_b}{r_{ts}}\right)^2}\sqrt{\frac{0.385 I_y J}{\pi^2}\frac{L_b^2}{r_{ts}^4} + S_x^2}$$

Substitute the value of c from Eq. (6.35e), and simplifying further we obtain

$$F_{cr} = \frac{M_n}{S_x} = \frac{C_b \pi^2 E}{\left(\frac{L_b}{r_{ts}}\right)^2}\sqrt{1 + 0.078\frac{Jc}{S_x h_0}\left(\frac{L}{r_{ts}}\right)^2} \tag{6.35c}$$

which is the same as Eq. (6.35c) given before in the AISC [5] design criteria. It shows the critical elastic lateral torsional buckling moment used in the code is that obtained from the theory of elastic stability shown before. The inelastic lateral torsional buckling moment given by Eq. (6.35a) is the linear interpolation between the plastic moment M_P at $L_b = L_P$ and the elastic lateral torsional buckling moment M_r at $L_b = L_r$ shown in Figure 6.28.
   The lengths L_p and L_r are obtained as follows:

$$L_p = 1.76 r_y \sqrt{\frac{E}{F_y}} \tag{6.35g}$$

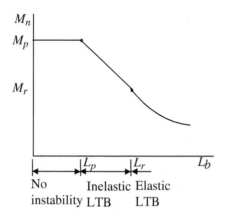

Figure 6.28 Unbraced length L_b versus nominal moment capacity M_n for compact sections.

where r_y is the radius of gyration of a cross-section about the y axis, and

$$L_r = 1.95 r_{ts} \frac{E}{0.7F_y} \sqrt{\frac{Jc}{S_x h_0}} \sqrt{1 + \sqrt{1 + 6.76\left(\frac{0.7F_y}{E}\frac{S_x h_0}{Jc}\right)}} \qquad (6.35h)$$

If the square root term in Eq. (6.35c) is taken as 1, according to AISC manual [5], then Eq. (6.35h) becomes

$$L_r = \pi r_{ts} \sqrt{\frac{E}{0.7F_y}} \qquad (6.35i)$$

For a doubly symmetric I shape, warping constant, $C_w = \dfrac{I_y h_0^2}{4}$, hence

$$r_{ts}^2 = \frac{I_y h_0}{2 S_x} \qquad (6.35j)$$

The r_{ts} may be taken as per AISC manual [5], equal to the radius of gyration of the compression flange plus one-sixth of the web

$$r_{ts} = \frac{b_f}{\sqrt{12\left(1 + \frac{1}{6}\frac{h t_w}{b_f t_f}\right)}} \qquad (6.35k)$$

Example 6.5

Find the design and the allowable strength of a W18 in.× 65 lb./ft (W460 mm × 97 kg/m) wide flange section [5] of 50 ksi (345 MPa) steel. It is subjected to a uniformly distributed load and the bending is about its strong axis on a 40 ft (12.2 m) simply supported span. Assume the lateral bracing as follows:

(a) The beam is laterally braced at 4 ft (1.22 m) intervals.
(b) The beam is laterally braced at the ends and the third points ($L_b = 13.33$ ft or 4.07 m).
(c) The beam is laterally braced at the supports only ($L_b = 40$ ft or 12.2 m).

Cross-section properties of W18in.×65 lb./ft (W460 mm × 97 kg/m) wide flange section are:

$d = 18.4$ in. (466 mm), $t_w = 0.45$ in. (11.4 mm), $h_0 = 17.6$ in. (447 mm), $I_x = 1070$ in.^{4} (445 × 10^{6}mm^4), $S_x = 117$ in.^{3} (1910 × 10^{3}mm^3), $Z_x = 133$ in.^{3} (2180 × 10^{3}mm^3), $r_y = 1.69$ in. (43.1 mm), $J = 2.73$ in.^{4} (1130 × 10^{3}mm^4), $C_w = 4240$ in.^{6} (1140 × 10^{9}mm^6), $\frac{b_f}{2t_f} = 5.06(5.08)$, $\frac{h}{t_w} = 35.7$ (35.8), $r_{ts} = 2.03$ in. (51.56 mm), $c = 1$.

Properties of steel are: $F_y = 50$ ksi (345 MPa), $F_u = 65$ ksi (448.5 MPa), $E = 29 \times 10^6$ psi psi (200 000 MPa).

Solution
Equations used are from the AISC Steel Construction Manual [5].

$$L_p = 1.76r_y\sqrt{\frac{E}{F_y}} = 1.76(1.69)\sqrt{\frac{29000}{50}} = 71.64 \text{ in.} = 5.97ft$$

$$\left[1.76(43.1)\sqrt{\frac{200000}{345}} = 1826.40 \text{ mm} = 1.83 \text{ m}\right]$$

$$L_r = 1.95r_{ts}\frac{E}{0.7F_y}\sqrt{\frac{Jc}{S_xh_0}}\sqrt{1 + \sqrt{1 + 6.76\left(\frac{0.7F_y}{E}\frac{S_xh_0}{Jc}\right)}} \tag{6.35h}$$

or

$$L_r = 1.95(2.03)\frac{29000}{0.7(50)}\sqrt{\frac{2.73(1)}{117(17.60)}}\sqrt{1 + \sqrt{1 + 6.76\left(\frac{0.7x50}{29000}\frac{117x17.60}{2.73x1}\right)}}$$

$$= 228.93 \text{ in.} = 19.08 \text{ ft}$$

$$\left[1.95(51.56)\frac{200000}{0.7(345)}\sqrt{\frac{1130x10^3(1)}{1910x10^3(447)}}\sqrt{1 + \sqrt{1 + 6.76\left(\frac{0.7x345}{200000}\frac{1910x10^3x447}{1130x10^3x1}\right)}}\right.$$
$$\left. = 5808.81mm = 5.81 \text{ m}\right]$$

Flange compactness

$$\lambda_p = 0.38\sqrt{\frac{E}{F_y}} = 0.38\sqrt{\frac{29000}{50}} = 9.15 \left[0.38\sqrt{\frac{200000}{345}} = 9.15\right]$$

$$\lambda = \frac{b_f}{2t_f} = 5.06 < \lambda_p = 9.15 \,[\lambda = 5.08 < \lambda_p = 9.15]$$

Hence, the flanges are compact.

Web compactness

$$\lambda_p = 3.76\sqrt{\frac{E}{F_y}} = 3.76\sqrt{\frac{29000}{50}} = 90.55 \left[3.76\sqrt{\frac{200000}{345}} = 90.53\right]$$

$$\lambda = \frac{h}{t_w} = 35.7 < \lambda_p = 90.55 \,[\lambda = 35.8 < \lambda_p = 90.53]$$

Hence, the web is compact. The section is compact because both the flanges and the web are compact.

(a) The unbraced length $L_b = 4$ ft $< L_p = 5.97$ ft $[L_b = 1.22$ m $< L_p = 1.83$ m]
Therefore, the beam is adequately braced, and its moment strength is governed by the plastic moment M_p of the beam.

$$M_n = M_p = F_y Z_x = \frac{50(133)}{12} = 554.17\text{kft} \left[\frac{345(2180\text{x}10^3)}{1000\text{x}1000} = 752.10 \text{ kN.m}\right]$$

Design moment strength from the LRFD method is obtained from

$$M_u = \phi_b M_n = 0.90(554.17) = 498.75 \text{ kft } [0.90(752.10) = 676.89 \text{ kN.m}]$$

Allowable moment from the ASD method is calculated as

$$M_a = \frac{M_n}{\Omega_b} = \frac{554.17}{1.67} = 331.84 \text{ kft } \left[\frac{752.10}{1.67} = 450.36 \text{ kN.m}\right]$$

(b) $L_p = 5.97$ ft $< L_b = 13.33$ ft $< L_r = 19.08$ ft $[L_p = 1.22$ m $< L_b = 4.07$ m $< L_r = 5.81$ m]

Therefore, the nominal moment capacity M_n of the beam is controlled by the inelastic lateral torsional buckling.

$$M_n = C_b \left[M_p - (M_p - 0.7F_y S_x)\left(\frac{L_b - L_p}{L_r - L_p}\right)\right] \leq M_p \tag{6.35a}$$

$$C_b = \frac{12.5M_{max}}{2.5M_{max} + 3M_A + 4M_B + 3M_C} R_m \leq 3.0 \tag{6.35d}$$

$R_m = 1.0$ for doubly symmetric shapes
Bending moment at a distance x shown in Figure 6.29 is expressed as

$$M = \frac{wL}{2}x - \frac{wx^2}{2} = \frac{w}{2}(Lx - x^2)$$

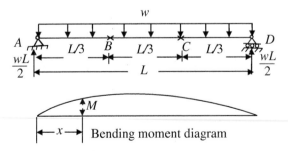

Figure 6.29 Simply supported beam laterally braced at third points.

For the end segments AB and CD

$$M_{max} = \frac{w}{2}\left(L\left(\frac{L}{3}\right) - \frac{L^2}{9}\right) = \frac{wL^2}{9}$$

$$M_A = \frac{w}{2}\left(L\left(\frac{L}{12}\right) - \frac{L^2}{144}\right) = \frac{11wL^2}{288}$$

$$M_B = \frac{w}{2}\left(L\left(\frac{L}{6}\right) - \frac{L^2}{36}\right) = \frac{5wL^2}{72}$$

$$M_C = \frac{w}{2}\left(L\left(\frac{L}{4}\right) - \frac{L^2}{16}\right) = \frac{3wL^2}{32}$$

$$C_b = \frac{12.5(1.00)}{2.5(1.00) + 3(0.344) + 4(0.625) + 3(0.844)} = 1.46$$

Similarly, for the center segment BC

$$C_b = \frac{12.5(1.00)}{2.5(1.00) + 3(0.972) + 4(1.00) + 3(0.972)} = 1.01$$

Hence, the center segment controls the design
The nominal moment strength is obtained from Eq. (6.21i)

$$M_n = 1.01\left\{554.17 - \left(554.17 - \frac{0.7x50x117}{12}\right)\left(\frac{13.33 - 5.97}{19.08 - 5.97}\right)\right\} = 439.17\,\text{kft}$$

$$\left[1.01\left\{752.10 - \left(752.10 - \frac{0.7x345x1910x10^3}{1000x1000}\right)\left(\frac{4.07 - 1.83}{5.81 - 1.83}\right)\right\} = 594.30\,\text{kN.m}\right]$$

Design moment strength from the LRFD method is obtained from

$$M_u = \phi_b M_n = 0.90(439.17) = 395.25\,\text{kft}\,[0.90(594.30) = 534.87\,\text{kN.m}]$$

Allowable moment from the ASD method is calculated as

$$M_a = \frac{M_n}{\Omega_b} = \frac{439.17}{1.67} = 262.98\,\text{kft}\,\left[\frac{594.30}{1.67} = 355.87\,\text{kN.m}\right]$$

(c) $L_b = 40\,\text{ft} > L_r = 19.08\,\text{ft}\,[L_b = 12.20\,\text{m} > L_r = 5.81\,\text{m}\,]$
Therefore, the nominal moment capacity M_n of the beam is controlled by the elastic lateral torsional buckling.

$$M_n = F_{cr}S_x \le M_p \tag{6.35b}$$

$$F_{cr} = \frac{C_b \pi^2 E}{\left(\frac{L_b}{r_{ts}}\right)^2}\sqrt{1 + 0.078\frac{Jc}{S_x h_0}\left(\frac{L_b}{r_{ts}}\right)^2} \tag{6.35c}$$

As before

$$C_b = \frac{12.5(1.00)}{2.5(1.00) + 3(0.75) + 4(1.00) + 3(0.75)} = 1.14$$

$$F_{cr} = \frac{1.14x\pi^2 x29000}{\left(\frac{40x12}{2.03}\right)^2} \sqrt{1 + 0.078\frac{2.73x1}{117x17.6}\left(\frac{40x12}{2.03}\right)^2} = 15.20 \text{ ksi}$$

$$\left[\frac{1.14x\pi^2 x200000}{\left(\frac{12.20x1000}{51.56}\right)^2}\sqrt{1 + 0.078\frac{1130x10^3 x1}{1910x10^3 x447}\left(\frac{12.20x1000}{51.56}\right)^2} = 104.65 \text{ MPa}\right]$$

$$M_n = \frac{15.20x117}{12} = 148.20 \text{ kft} \left[\frac{104.65x1910x10^3}{1000x1000} = 199.88 \text{ kN.m}\right]$$

Design moment strength from the LRFD method is obtained from

$$M_u = \phi_b M_n = 0.90(148.20) = 133.38\text{kft} \ [0.90(199.88) = 179.82 \text{ kN.m}]$$

Allowable moment from the ASD method is calculated as

$$M_a = \frac{M_n}{\Omega_b} = \frac{148.20}{1.67} = 88.74\text{kft} \left[\frac{199.88}{1.67} = 119.69 \text{ kN.m}\right]$$

Example 6.6
Find the design and the allowable strength of a W14 in.× 90 lb./ft (W360 mm × 134 kg/m) wide flange section [5] of 50 ksi (345 MPa) steel. It is subjected to a uniformly distributed load and the bending is about its strong axis on a 40 ft (12.2 m) simply supported span. Assume the lateral bracing as follows:

(a) The beam is laterally braced at 4 ft (1.22 m) intervals.
(b) The beam is laterally braced at the ends and the third points ($L_b = 13.33$ ft or 4.07 m).

Cross-section properties of W14 in.×90 lb./ft (W360 mm × 134 kg/m) wide flange section are:

$d = 14.0$ in. (356 mm), $t_w = 0.44$ in. (11.2 mm), $h_0 = 13.3$ in. (337.82 mm), $I_x = 999$ in.^{4} (415×10^6 mm^4), $S_x = 143$ in.^{3} (2330×10^3 mm^3), $Z_x = 157$ in.^{3} (2560×10^3 mm^3), $r_y = 3.70$ in. (94.0 mm), $J = 4.06$ in.^{4} (1680×10^3 mm^4), $C_w = 16,000$ in.^{6} (4310×10^9 mm^6), $\frac{b_f}{2t_f} = 10.2(10.3)$, $\frac{h}{t_w} = 25.9$ (25.9), $r_{ts} = 4.11$ in. (104.39 mm), $c = 1$.

Properties of steel are: $F_y = 50$ ksi (345 MPa), $F_u = 65$ ksi (448.5 MPa), $E = 29 \times 10^6$ psi (200 000 MPa).

Solution
Equations used are from the AISC Steel Construction Manual [5].

$$L_p = 1.76r_y\sqrt{\frac{E}{F_y}} = 1.76(3.70)\sqrt{\frac{29000}{50}} = 156.83 \text{ in.} = 13.07\text{ft}$$

$$\left[1.76(94.0)\sqrt{\frac{200000}{345}} = 3983.33 \text{ mm} = 3.98 \text{ m}\right]$$

$$L_r = 1.95r_{ts}\frac{E}{0.7F_y}\sqrt{\frac{Jc}{S_x h_0}}\sqrt{1 + \sqrt{1 + 6.76\left(\frac{0.7F_y}{E}\frac{S_x h_0}{Jc}\right)}} \tag{6.35h}$$

or

$$L_r = 1.95(4.11)\frac{29000}{0.7(50)}\sqrt{\frac{4.06(1)}{143(13.3)}}\sqrt{1 + \sqrt{1 + 6.76\left(\frac{0.7 \times 50}{29000} \times \frac{143 \times 13.3}{4.06 \times 1}\right)}}$$

$$= 548.49 \text{ in.} = 45.71 \text{ ft}$$

$$\left[1.95(104.39)\frac{200000}{0.7(345)}\sqrt{\frac{1680 \times 10^3(1)}{2330 \times 10^3(337.82)}}\sqrt{1 + \sqrt{1 + 6.76\left(\frac{0.7 \times 345}{200000}\frac{2330 \times 10^3 \times 337.82}{1680 \times 10^3 \times 1}\right)}}\right.$$

$$\left. = 13924.35\text{mm} = 13.92 \text{ m}\right]$$

Flange compactness

$$\lambda_p = 0.38\sqrt{\frac{E}{F_y}} = 0.38\sqrt{\frac{29000}{50}} = 9.15 \quad \left[0.38\sqrt{\frac{200000}{345}} = 9.15\right]$$

$$\lambda_r = 1.0\sqrt{\frac{E}{F_y}} = 1.0\sqrt{\frac{29000}{50}} = 24.08 \quad \left[1.0\sqrt{\frac{200,000}{345}} = 24.08\right]$$

$$\lambda_p = 9.15 < \lambda = 10.2 < \lambda_r = 24.08 \, [\lambda_p = 9.15 < \lambda = 10.3 < \lambda_r = 24.08]$$

Hence, the flanges are noncompact.

Web compactness

$$\lambda_p = 3.76\sqrt{\frac{E}{F_y}} = 3.76\sqrt{\frac{29000}{50}} = 90.55 \quad \left[3.76\sqrt{\frac{200000}{345}} = 90.53\right]$$

$$\lambda = \frac{h}{t_w} = 25.9 < \lambda_p = 90.55 \, [\lambda = 25.9 < \lambda_p = 90.53]$$

Hence, the web is compact. The flanges are noncompact, hence W14×90 (W360×134) is designed as a noncompact section.

(a) The unbraced length $L_b = 4$ ft $< L_p = 13.07$ ft $[L_b = 1.22$ m $< L_p = 3.98$ m]
Therefore, the beam is adequately braced, and its moment strength is governed by the inelastic flange local buckling.

$$M_p = F_y Z_x = \frac{50(157)}{12} = 654.17\text{kft} \quad \left[\frac{345(2560 \times 10^3)}{1000 \times 1000} = 883.20 \text{ kN.m}\right]$$

The nominal moment strength is obtained from

$$M_n = M_p - (M_p - 0.7F_y S_x)\left(\frac{\lambda - \lambda_p}{\lambda_r - \lambda_p}\right) \tag{6.34a}$$

$$M_n = 654.17 - \left(654.17 - \frac{0.7 \times 50 \times 143}{12}\right)\left(\frac{10.2 - 9.15}{24.08 - 9.15}\right) = 637.50 \text{ kft}$$

$$\left[883.2 - \left(883.2 - \frac{0.7 \times 345 \times 2330 \times 10^3}{1000 \times 1000}\right)\left(\frac{10.3 - 9.15}{24.08 - 9.15}\right) = 860.66 \text{ kN.m}\right]$$

The design strength from the LRFD method is given as

$$M_u = \phi_b M_n = 0.90(637.50) = 573.75 \text{ kft } [0.90(860.66) = 774.59 \text{ kN.m}]$$

Allowable moment from the ASD method is calculated as

$$M_a = \frac{M_n}{\Omega_b} = \frac{637.50}{1.67} = 381.74 \text{ kft } \left[\frac{860.66}{1.67} = 515.37 \text{ kN.m}\right]$$

(b) $L_p = 13.07 \text{ ft} < L_b = 13.33 \text{ ft} < L_r = 45.71 \text{ ft } [L_p = 3.98 \text{ m} < L_b = 4.07 \text{ m} < L_r = 13.92 \text{ m}]$
The nominal moment capacity M_n of the beam is controlled by either the flange local buckling or the inelastic lateral torsional buckling. From the solution of part (a), the flexural nominal strength for flange local buckling is

$$M_n = 637.50 \text{ kft } [860.66 \text{ kN.m}]$$

For the inelastic lateral torsional buckling the flexural nominal strength is

$$M_n = C_b \left[M_p - (M_p - 0.7F_y S_x)\left(\frac{L_b - L_p}{L_r - L_p}\right)\right] \leq M_p \tag{6.35a}$$

$C_b = 1.01$ from Example 6.5

$$M_n = 1.01\left[654.17 - \left(654.17 - \frac{0.75x50x143}{12}\right)\left(\frac{13.33 - 13.07}{45.71 - 13.07}\right)\right] = 659.04 \text{ kft} > 637.50 \text{ kft}$$

$$\left[1.01\left\{883.20 - \left(883.20 - \frac{0.75x345x2330x10^3}{1000x1000}\right)\left(\frac{4.07 - 3.98}{13.92 - 3.98}\right)\right\} = 889.47 \text{ kN.m} > 860.66 \text{ kN.m}\right]$$

Hence, the nominal flexural strength is controlled by the flange local buckling and is given by

$$M_n = 637.50 \text{ kft } [860.66 \text{ kN.m}]$$

Design moment strength for the load and resistance factor design (LRFD) method is

$$M_u = 0.9(637.50) = 573.75 \text{ kft } [0.9(860.66) = 774.59 \text{ kN.m}]$$

Allowable moment from the ASD method is

$$M_a = \frac{637.50}{1.67} = 381.74 \text{ kft } \left[\frac{860.66}{1.67} = 515.37 \text{ kN.m}\right]$$

Example 6.7

Solve for the flexural strength of the beam in Example 6.6 if the beam is supported on a 50 ft (15.25 m) simple span. Also, the beam is laterally braced at the supports only ($L_b = 50$ ft or 15.25 m).

Solution

Equations used are from the AISC Steel Construction Manual [5].

$L_b = 50 \text{ ft } [15.25] \text{ m} > L_r = 45.71 \text{ ft } [13.94 \text{ m}]$
$C_b = 1.14$ (from Example 6.5)

The nominal moment capacity M_n of the beam is controlled by either the flange local buckling or the elastic lateral torsional buckling. From the solution of Example 6.11.2 (part a), the flexural nominal strength for flange local buckling is

$$M_n = 637.50 \text{ kft } [860.66 \text{ kN.m}]$$

For the elastic lateral torsional buckling, the flexural nominal strength is

$$M_n = F_{cr}S_x \leq M_p \tag{6.35b}$$

$$F_{cr} = \frac{C_b\pi^2 E}{\left(\frac{L_b}{r_{ts}}\right)^2}\sqrt{1 + 0.078\frac{Jc}{S_x h_0}\left(\frac{L_b}{r_{ts}}\right)^2} \tag{6.35c}$$

$$F_{cr} = \frac{1.14x\pi^2 x29000}{\left(\frac{50x12}{4.11}\right)^2}\sqrt{1 + 0.078\frac{4.06x1}{143x13.3}\left(\frac{50x12}{4.11}\right)^2} = 32.65 \text{ ksi}$$

$$\left[\frac{1.14x\pi^2 x200,000}{\left(\frac{15.25x1000}{104.39}\right)^2}\sqrt{1 + 0.078\frac{1680x10^3 x1}{2330x10^3 x337.82}\left(\frac{15.25x1000}{104.39}\right)^2} = 224.99 \text{ MPa}\right]$$

The nominal flexural strength is given by

$$M_n = \frac{32.65(143)}{12} = 389.08 \text{ kft} < 637.50 \text{ kft}$$

$$\left[\frac{224.99 \times 2330 \times 10^3}{1000 \times 1000} = 524.23 \text{ kN.m} < 860.66 \text{ kN.m}\right]$$

Hence, the nominal flexural strength is controlled by the elastic lateral torsional buckling and is given by

$$M_n = 389.08 \text{ kft } [524.23 \text{ kN.m}]$$

Design moment strength for the LRFD method is

$$M_u = 0.9(389.08) = 350.17 \text{ kft } [0.9\,(524.23) = 471.81 \text{ kN.m}]$$

Allowable moment from the ASD method is

$$M_a = \frac{389.08}{1.67} = 232.98 \text{ kft } \left[\frac{524.33}{1.67} = 313.91 \text{ kN.m}\right]$$

Problems

6.1 Find the torsional buckling load for the column in Example 6.1 if one end is fixed and the other is simply supported.

6.2 Find the torsional buckling load of a 20 ft (6.1 m) cantilever column whose cross-section is the same as in Example 6.4.

6.3 Find the lateral buckling load of a simple supported I section in Example 6.1 by the energy method if the concentrated load is applied at the mid-span and the load acts at the
(a) upper flange
(b) centroid of the cross-section
(c) lower flange

6.4 Use Eq. (6.20g) to calculate the coefficient C_b for different loading conditions.
(a) Simple supported beam subjected to the concentrated load at the mid-span in Figure P6.4a.
(b) Simple supported beam subjected to the uniform increasing load in Figure P6.4b.

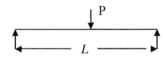

Figure P6.4 (a)

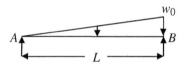

Figure P6.4 (b)

6.5 Derive the differential equation governing the lateral torsional buckling load of a cantilever I beam subjected to uniform distributed load shown in Figure P6.5.

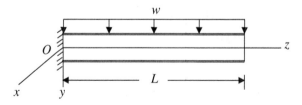

Figure P6.5 Cantilever beam.

6.6 Derive the differential equation governing the lateral torsional buckling load of a simple supported I beam subjected to uniform distributed load shown in Figure P6.6.

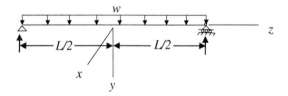

Figure P6.6 Simple supported beam.

6.7 Calculate the design and the allowable strength of a W18 in.×65 lb./ft (W460 mm × 97 kg/m) wide flange section of 50 ksi (345 MPa) steel. It is subjected to a uniformly distributed load and the bending is about its strong axis on a 40 ft (12.2 m) beam fixed at both end supports. Assume the lateral bracing as follows:
(a) The beam is laterally braced at 4 ft (1.22 m) intervals.
(b) The beam is laterally braced at the ends and the third points ($L_b = 13.33$ ft or 4.07 m).
(c) The beam is laterally braced at the supports only ($L_b = 40$ ft or 12.2 m).
Cross-section properties of W18in.×65 lb./ft (W460 mm × 97 kg/m) wide flange section are:
$d = 18.4$ in. (466 mm), $t_w = 0.45$ in. (11.4 mm), $h_0 = 17.6$ in. (447 mm), $I_x = 1070$ in.^{4} (445×10^6 mm^4), $S_x = 117$ in.^{3} (1910×10^3 mm^3), $Z_x = 133$ in.^{3} (2180×10^3 mm^3), $r_y = 1.69$ in. (43.1 mm), $J = 2.73$ in.^{4} (1130×10^3 mm^4), $C_w = 4240$ in.^{6} (1140×10^9 mm^6), $\dfrac{b_f}{2t_f} = 5.06(5.08)$, $\dfrac{h}{t_w} = 35.7$ (35.8), $r_{ts} = 2.03$ in. (51.56 mm), $c = 1$.
Properties of steel are: $F_y = 50$ ksi (345 MPa), $F_u = 65$ ksi (448.5 MPa), $E = 29 \times 10^6$ psi (200 000 MPa), $G = 11 \times 10^6$ psi (75 000 MPa).

References

1 Chajes, A. (1974). *Principles of Structural Stability Theory*. Englewood Cliffs, NJ: Prentice Hall, Inc.
2 Boresi, A.P. and Schmidt, R.J. (2003). *Advanced Mechanics of Materials*, 6e. New York: Wiley.
3 Timoshenko, S.P. and Gere, J.M. (1961). *Theory of Elastic Stability*, 2e. New York: McGraw-Hill.
4 Hibbeler, R.C. (2017). *Mechanics of Materials*, 10e. Hoboken, NJ: Pearson.
5 AISC (2017). *Steel Construction Manual*, 15e. Chicago, IL: AISC.
6 Ziemian, R.D. (ed.) (2010). *Guide to Stability Design Criteria for Metal Structures*, 6e. Hoboken, NJ: Wiley.
7 CRC (2018). *Standard Mathematical Tables and Formulas*, 33e. Boca Raton, FL: CRC Press.

8 Chen, W.F. and Lui, E.M. (1987). *Structural Stability Theory and Implementation*. Upper Saddle River, NJ: Prentice Hall.

9 Galambos, T.V. (1968). *Structural Members and Frames*. Englewood Cliffs, NJ: Prentice Hall, Inc.

10 Nethercot, D.A. (1983). Elastic lateral buckling of beams in beams and beam columns: stability and strength. In: *Beams and Beam Columns* (ed. R. Narayanan), 1–33. Barking, Essex, England: Applied Science Publishers.

11 Vlasov, V.Z. (1961). *Thin Walled Elastic Beams* (trans. Y. Schechtman). Jerusalem: Israel Program for Scientific Translation.

12 Clark, J.W. and Hill, H.N. (1960). Lateral buckling of beams. *Journal of the Structural Division* 86 (ST7): 175–196.

13 Salvadori, M.G. (1955). Lateral buckling of I-beams. *Transactions of ASCE* 120: 1165.

14 Kirby, P.A. and Nethercot, D.A. (1979). *Design for Structural Stability*. New York: Wiley.

7

Buckling of Plates

7.1 Introduction

Thin plates are considered as two-dimensional members and the bending takes place in two planes, whereas columns were one-dimensional members and the bending could be assumed in one plane. The deflections and bending moments in the columns were functions of one independent variable, whereas in plates these quantities are dependent on two independent variables. Hence, the governing equations in the columns were ordinary differential equations. In the case of plates, the behavior is governed by partial differential equations. Another difference between the behavior of columns and plates is that the columns cannot take additional forces beyond their buckling loads, and the critical load is the failure load for columns. Plates have additional capacity to resist forces beyond their buckling load and their failure load is much higher. The failure load for the plates can be found by considering their post-buckling behavior. Plates are used as elements of wide flange, channel, angles, and other column cross-sections. In such cases, the buckling of plates is responsible for the local buckling of the column cross-sections. In addition, the plates are used as flat surfaces in buildings, bridges, aircraft wings, plate girder webs, ship hulls, etc. Plates are of different shapes, and sizes, e.g. circular and rectangular plates, thin and thick plates. Plates are supported by different types of supports.

7.2 Theory of Plate Bending

It is customary to take the X and Y axes along the edges of the plate and the Z axis perpendicular and downwards as shown in Figure 7.1. If the plate is subjected to small deflections, the plane midway between the top and the bottom surfaces of the plate called the middle surface is the XY plane. Consider the bending of a rectangular plate by distributed load $q(x, y)$ as shown in Figure 7.1a that acts perpendicular to the middle surface of the plate. Assume h is the thickness of the plate and is small in comparison to the other dimensions of the plate.

The following assumptions are made in the bending theory of thin plates:

1. The lines normal to the middle surface prior to bending remain normal and straight to the middle surface during bending.

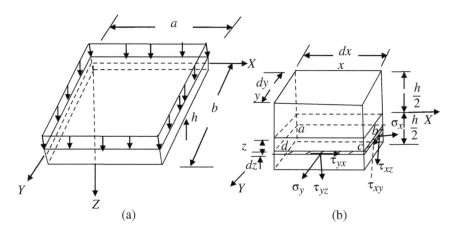

Figure 7.1 Plate coordinates and differential element: (a) Coordinates of a plate; (b) Differential element of plate.

2. The normal stress σ_z and the strain ε_z are negligible and hence the transverse deflection of points in the plate varies with their x and y coordinates but is independent of the coordinate z, i.e. $w = w(x,y)$.
3. It is assumed there is no strain in the middle surface of the plate caused by bending if the transverse deflections of the plate are small in comparison to its thickness, and is therefore the neutral surface.

In addition, it is assumed here that the material of the plate is homogeneous and isotropic and obeys Hooke's law. Hence, the plate can be treated as a two-dimensional stress problem and we get linear differential equations with constant coefficients to describe the behavior of the plate.

An element of the plate of length dx and dy and thickness h is shown in Figure 7.1b. The strains of an elementary lamina $abcd$ at a distance z from the neutral surface and of thickness dz are given by

$$\varepsilon_x = \frac{z}{\rho_x}, \qquad \varepsilon_y = \frac{z}{\rho_y} \tag{7.1a}$$

where $1/\rho_x$ and $1/\rho_y$ are the curvatures of the neutral surface in the XZ and YZ planes. The stress strain relations of an elastic plane stress case are

$$\sigma_x = \frac{E}{1 - v^2}(\varepsilon_x + v\varepsilon_y) \tag{7.1b}$$

$$\sigma_y = \frac{E}{1 - v^2}(\varepsilon_y + v\varepsilon_x) \tag{7.1c}$$

$$\tau_{xy} = \frac{E}{2(1 + v)}\gamma_{xy} \tag{7.1d}$$

E is the modulus of elasticity and v is the Poisson's ratio of the plate material. σ_x, σ_y, and ε_x, ε_y are the stresses and strains at a point in the lamina $abcd$ in the X and Y directions respectively.

The thin element *abcd* at a distance *z* below the middle surface undergoes displacements shown in Figure 7.2a during bending. The strain of fiber *ab* in the *X* direction is given by

$$\varepsilon_x = \frac{a'b' - ab}{ab} = \frac{dx + \left(u_b + \frac{\partial u_b}{\partial x}dx\right) - u_b - dx}{dx}$$

or

$$\varepsilon_x = \frac{\partial u_b}{\partial x} \tag{7.1e}$$

Similarly, the strain of the fiber *ad* in the *y* direction is

$$\varepsilon_y = \frac{\partial v_b}{\partial y} \tag{7.1f}$$

The shear strain is given by the change in the angle *dab* and is

$$\gamma_{xy} = \frac{\partial u_b}{\partial y} + \frac{\partial v_b}{\partial x} \tag{7.1g}$$

It is assumed that the plane sections remain plane during bending (during transverse displacement *w*), the displacements *u* and *v* of the point *e* above a distance *z* from the middle surface in Figure 7.2b are given by

$$u_b = -z\frac{\partial w}{\partial x} \tag{7.1h}$$

Similarly,

$$v_b = -z\frac{\partial w}{\partial y} \tag{7.1i}$$

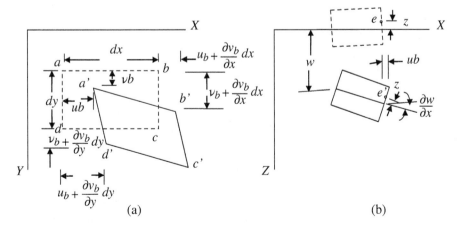

Figure 7.2 Displacements in the plate due to bending: (a) Displacements in the *X – Y* plane; (b) Displacements in the *X – Z* plane.

The negative sign indicates that for negative z (point e in Figure 7.2b) and positive slope $\left(\frac{\partial w}{\partial x}\right)$, displacements are positive. Substitute Eqs. (7.1h) and (7.1i) into Eqs. (7.1e)–(7.1g) to get

$$\varepsilon_x = -z\frac{\partial^2 w}{\partial x^2} \tag{7.1j}$$

$$\varepsilon_y = -z\frac{\partial^2 w}{\partial y^2} \tag{7.1k}$$

$$\gamma_{xy} = -2z\frac{\partial^2 w}{\partial x \partial y} \tag{7.1l}$$

For small deflections of the plate, the curvatures of the middle surface of the plate are given by [1]:

$$\frac{1}{\rho_x} = -\frac{\partial^2 w}{\partial x^2}, \quad \frac{1}{\rho_y} = -\frac{\partial^2 w}{\partial y^2}$$

The quantity, $1/\rho_{xy}$, is called the twist of the middle surface with respect to the X and Y axes, and is given by [1]:

$$\frac{1}{\rho_{xy}} = \frac{\partial^2 w}{\partial x \partial y}$$

Substituting Eqs. (7.1j)–(7.1l) into Eqs. (7.1b)–(7.1d) we get

$$\sigma_x = -\frac{Ez}{1-v^2}\left(\frac{\partial^2 w}{\partial x^2} + v\frac{\partial^2 w}{\partial y^2}\right) \tag{7.1m}$$

$$\sigma_y = -\frac{Ez}{1-v^2}\left(\frac{\partial^2 w}{\partial y^2} + v\frac{\partial^2 w}{\partial x^2}\right) \tag{7.1n}$$

$$\tau_{xy} = -\frac{Ez}{1+v}\frac{\partial^2 w}{\partial x \partial y} \tag{7.1o}$$

The stresses are positive when they are in the direction shown in Figure 7.1b. The normal and shear stresses distributed over the lateral sides of the element in Figure 7.1b produce the following moments and vertical shear forces:

$$M_x = \int_{-\frac{h}{2}}^{\frac{h}{2}} \sigma_x z\,dz \quad M_y = \int_{-\frac{h}{2}}^{\frac{h}{2}} \sigma_y z\,dz \tag{7.1p}$$

$$M_{xy} = -\int_{-\frac{h}{2}}^{\frac{h}{2}} \tau_{xy} z\,dz \quad M_{yx} = \int_{-\frac{h}{2}}^{\frac{h}{2}} \tau_{yx} z\,dz \tag{7.1q}$$

$$Q_x = \int_{-\frac{h}{2}}^{\frac{h}{2}} \tau_{xz}\,dz \quad Q_y = \int_{-\frac{h}{2}}^{\frac{h}{2}} \tau_{yz}\,dz$$

where M_x and M_y are the bending moments per unit length acting on the lateral sides of the element parallel to the Y and X axes respectively. These are considered positive when they produce compression at the top and tension at the bottom surfaces of the plate. The moments M_{xy} and M_{yx} are the twisting moments per unit length on the lateral sides of the element. The negative sign in Eq. (7.1q) shows that M_{xy} is negative when z and τ_{xy} are positive. Q_x and Q_y are the vertical shear forces per unit length.

The bending moments M_x, M_y, the twisting moments M_{xy}, M_{yx}, and vertical shear forces are positive as shown in Figure 7.3. In addition, the distributed load $q(x,y)$ acts perpendicular to the middle surface of the plate. Since $\tau_{xy} = \tau_{yx}$, hence $M_{yx} = -M_{xy}$. Substituting Eqs. (7.1m)–(7.1o) into Eqs. (7.1p) and (7.1q) we have

$$M_x = -D\left(\frac{\partial^2 w}{\partial x^2} + v\frac{\partial^2 w}{\partial y^2}\right) \tag{7.1r}$$

$$M_y = -D\left(\frac{\partial^2 w}{\partial y^2} + v\frac{\partial^2 w}{\partial x^2}\right) \tag{7.1s}$$

$$M_{xy} = -M_{yx} = D(1-v)\frac{\partial^2 w}{\partial x \partial y} \tag{7.1t}$$

where $D = \dfrac{Eh^3}{12(1-v^2)}$ is called the flexural rigidity of the plate per unit width.

The plate element in Figure 7.3 is in equilibrium in which all forces are in the Z direction and the moments are about the X and Y axes. Therefore, we have to consider equations of equilibrium in the Z direction for forces, and equations of equilibrium for moments about the X and Y directions, for a total of three equations of equilibrium. Hence,

$$\Sigma F_Z = 0$$

$$-Q_x dy + \left(Q_x + \frac{\partial Q_x}{\partial x}dx\right)dy - Q_y dx + \left(Q_y + \frac{\partial Q_y}{\partial y}dy\right)dx + q\,dx\,dy = 0$$

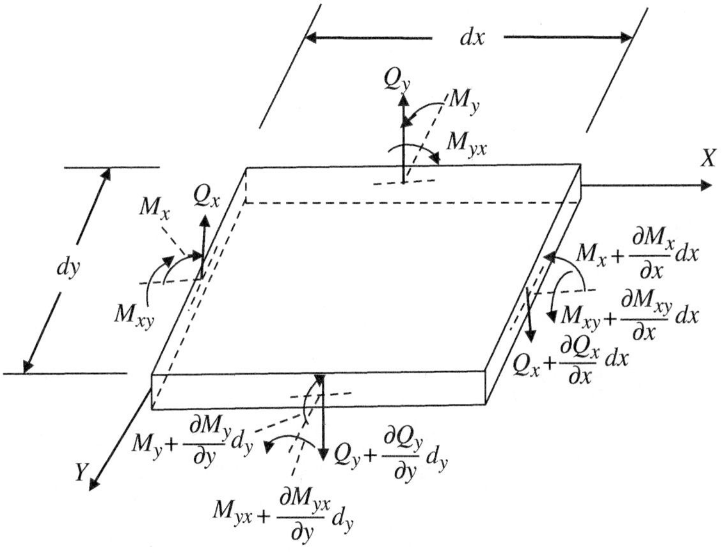

Figure 7.3 Moments and shears on a plate element.

or

$$\frac{\partial Q_x}{\partial x} + \frac{\partial Q_y}{\partial y} + q = 0 \tag{7.1u}$$

$\Sigma M_X = 0$

$$M_y dx - \left(M_y + \frac{\partial M_y}{\partial y} dy\right) dx + \left(Q_y + \frac{\partial Q_y}{\partial y} dy\right) dxdy - Q_x dy \frac{dy}{2} + \left(Q_x + \frac{\partial Q_x}{\partial x} dx\right) dy \frac{dy}{2}$$

$$+ qdxdy\frac{dy}{2} - M_{xy} dy + \left(M_{xy} + \frac{\partial M_{xy}}{\partial x} dx\right) dy = 0$$

Neglecting higher-order terms, we have

$$\frac{\partial M_{xy}}{\partial x} - \frac{\partial M_y}{\partial y} + Q_y = 0 \tag{7.1v}$$

$\Sigma M_Y = 0$

$$- M_x dy + \left(M_x + \frac{\partial M_x}{\partial x} dx\right) dy - - M_{yx} dx + \left(M_{yx} + \frac{\partial M_{yx}}{\partial y} dy\right) dx - \left(Q_x + \frac{\partial Q_x}{\partial x} dx\right) dydx$$

$$+ Q_y dx \frac{dx}{2} - \left(Q_y + \frac{\partial Q_y}{\partial y} dy\right) dx \frac{dx}{2} - qdxdy\frac{dx}{2} = 0$$

Neglecting higher-order terms, we have

$$\frac{\partial M_{yx}}{\partial y} + \frac{\partial M_x}{\partial x} - Q_x = 0 \tag{7.1w}$$

Differentiate Q_x and Q_y from Eqs. (7.1v) and (7.1w), and substitute in Eq. (7.1u) to get

$$\frac{\partial^2 M_{yx}}{\partial x \partial y} + \frac{\partial^2 M_x}{\partial x^2} - \frac{\partial^2 M_{xy}}{\partial x \partial y} + \frac{\partial^2 M_y}{\partial y^2} + q = 0$$

$$M_{xy} = -M_{yx}$$

Hence,

$$\frac{\partial^2 M_x}{\partial x^2} + \frac{\partial^2 M_y}{\partial y^2} - 2\frac{\partial^2 M_{xy}}{\partial x \partial y} = -q \tag{7.1x}$$

Substitute Eqs. (7.1r)–(7.1t) into Eq. (7.1x) to obtain

$$\frac{\partial^4 w}{\partial x^4} + 2\frac{\partial^4 w}{\partial x^2 \partial y^2} + \frac{\partial^4 w}{\partial y^4} = \frac{q}{D} \tag{7.1y}$$

The deflection of a plate can be determined by the integration of Eq. (7.1y) for a given load q and the given boundary conditions. The bending and twisting moments are then calculated from Eqs. (7.1r)–(7.1t). The shearing forces are obtained from the Eqs. (7.1v) and (7.1w) as follows:

$$Q_x = -D\frac{\partial}{\partial x}\left(\frac{\partial^2 w}{\partial x^2} + \frac{\partial^2 w}{\partial y^2}\right), \quad \text{and} \quad Q_y = -D\frac{\partial}{\partial y}\left(\frac{\partial^2 w}{\partial x^2} + \frac{\partial^2 w}{\partial y^2}\right) \tag{7.1z}$$

7.3 Buckling of Thin Plates

The stability problems in plates can be investigated by assuming that the plate slightly buckles under the action of forces applied in the middle surface. The critical forces are obtained from the magnitudes of these forces that keep the plate in the buckled shape. The equations of equilibrium for the plate are formed in a slightly deformed position to find the critical in-plane load for a flat plate. An element of a laterally bent plate is acted on by two sets of forces. One set is the applied in-plane forces, and the second set consists of moments and shears resulting from the transverse bending of the plate. The equations of equilibrium for these two sets of forces are considered separately and then combined [2].

7.3.1 In-plane Forces

Consider a plate subjected to axial and shear forces in the middle plane of the plate. A small element cut from the plate by the planes XZ and YZ and acted on by in-plane forces N_x, N_y, N_{xy}, and N_{yx} per unit length is shown in Figure 7.4. The in-plane forces are due to the in-plane loads because the middle surface strains caused by the bending of the plate are neglected.

The sum of the forces in the X and Y axes give the following equilibrium equations:

$$-N_x dy + \left(N_x + \frac{\partial N_x}{\partial x} dx\right) dy - N_{yx} dx + \left(N_{yx} + \frac{\partial N_{yx}}{\partial y} dy\right) dx = 0$$

$$\frac{\partial N_x}{\partial x} + \frac{\partial N_{yx}}{\partial y} = 0 \tag{7.2a}$$

$$-N_y dx + \left(N_y + \frac{\partial N_y}{\partial y} dy\right) dx - N_{xy} dy + \left(N_{xy} + \frac{\partial N_{xy}}{\partial x} dx\right) dy$$

$$\frac{\partial N_y}{\partial y} + \frac{\partial N_{xy}}{\partial x} = 0 \tag{7.2b}$$

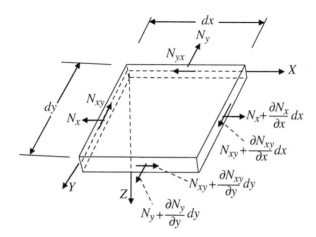

Figure 7.4 In-plane forces on a plate element.

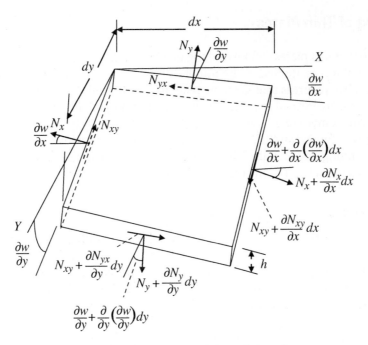

Figure 7.5 In-plane forces on a deformed plate element.

The sum of the moments of the forces about the x, y, and z axes shown in Figure 7.4 are identically zero. To consider the sum of the components of the forces in Figure 7.4 in the Z direction, we have to consider the deflection of the plate shown in Figure 7.5.

The components of the normal forces N_x in the Z direction due to the curvature of the plate in the XZ plane are

$$-N_x dy \frac{\partial w}{\partial x} + \left(N_x + \frac{\partial N_x}{\partial x} dx \right) \left(\frac{\partial w}{\partial x} + \frac{\partial^2 w}{\partial x^2} dx \right) dy$$

Neglecting higher-order terms, we have

$$N_x \frac{\partial^2 w}{\partial x^2} dxdy + \frac{\partial N_x}{\partial x} \frac{\partial w}{\partial x} dxdy \tag{7.2c}$$

The components of the normal forces N_y in the Z direction due to the curvature of the plate in the YZ plane are

$$-N_y dx \frac{\partial w}{\partial y} + \left(N_y + \frac{\partial N_y}{\partial y} dy \right) \left(\frac{\partial w}{\partial y} + \frac{\partial^2 w}{\partial y^2} dy \right) dx$$

Neglecting higher-order terms, we have

$$N_y \frac{\partial^2 w}{\partial y^2} dxdy + \frac{\partial N_y}{\partial y} \frac{\partial w}{\partial y} dxdy \tag{7.2d}$$

The z component of the shear forces N_{xy} is

$$-N_{xy}dy\frac{\partial w}{\partial y} + \left(N_{xy} + \frac{\partial N_{xy}}{\partial x}dx\right) dy \left(\frac{\partial w}{\partial y} + \frac{\partial^2 w}{\partial x \partial y}dx\right)$$

Neglecting higher-order terms, we have

$$N_{xy}\frac{\partial^2 w}{\partial x \partial y}dxdy + \frac{\partial N_{xy}}{\partial x}\frac{\partial w}{\partial y}dxdy \tag{7.2e}$$

The z component of the shear forces N_{yx} is

$$-N_{yx}dx\frac{\partial w}{\partial x} + \left(N_{yx} + \frac{\partial N_{yx}}{\partial y}dy\right) dx \left(\frac{\partial w}{\partial x} + \frac{\partial^2 w}{\partial x \partial y}dy\right)$$

Neglecting higher-order terms, we have

$$N_{yx}\frac{\partial^2 w}{\partial x \partial y}dxdy + \frac{\partial N_{yx}}{\partial y}\frac{\partial w}{\partial x}dxdy \tag{7.2f}$$

Adding expressions (7.2c)–(7.2f) and using Eqs. (7.2a) and (7.2b), the sum of the z components of all the normal and shear forces after substituting $N_{xy} = N_{yx}$ is given by

$$N_x\frac{\partial^2 w}{\partial x^2}dxdy + N_y\frac{\partial^2 w}{\partial y^2}dxdy + 2N_{xy}\frac{\partial^2 w}{\partial x \partial y}dxdy \tag{7.2g}$$

In addition to the in-plane forces in Figure 7.4, the differential element of a slightly bent plate will have moments and shear forces shown in Figure 7.3. The sum of the components of shear forces in the z directions is

$$-Q_xdy + \left(Q_x + \frac{\partial Q_x}{\partial x}dx\right) dy - Q_ydx + \left(Q_y + \frac{\partial Q_y}{\partial y}dy\right) dx$$

and the expression becomes

$$\left(\frac{\partial Q_x}{\partial x} + \frac{\partial Q_y}{\partial y}\right) dxdy \tag{7.2h}$$

Add the z components given by Eqs. (7.2g) and (7.2h) to get the equilibrium equation of forces in the plate in z direction as

$$\left(\frac{\partial Q_x}{\partial x} + \frac{\partial Q_y}{\partial y}\right) dxdy + N_x\frac{\partial^2 w}{\partial x^2}dxdy + N_y\frac{\partial^2 w}{\partial y^2}dxdy + 2N_{xy}\frac{\partial^2 w}{\partial x \partial y}dxdy = 0 \tag{7.2i}$$

Using Eq. (7.1z) we get the following equation

$$\frac{\partial^4 w}{\partial x^4} + 2\frac{\partial^4 w}{\partial x^2 \partial y^2} + \frac{\partial^4 w}{\partial y^4} = \frac{1}{D}\left(N_x\frac{\partial^2 w}{\partial x^2} + N_y\frac{\partial^2 w}{\partial y^2} + 2N_{xy}\frac{\partial^2 w}{\partial x \partial y}\right) \tag{7.2j}$$

Equation (7.2j) is the equation of buckling of a rectangular thin plate under the action of in-plane forces N_x, N_y, and N_{xy}. The in-plane forces N_x and N_y are positive for tensile loads in Eq. (7.2j), whereas in buckling these forces are negative. Equation (7.2j) is a fourth-order partial differential equation in x and y. It requires eight boundary conditions, four in the x and four in the y direction to get a unique solution.

7.4 Boundary Conditions

The boundary conditions for rectangular plates with different support conditions are considered here.

7.4.1 Simply Supported Edge

If the edge $x = 0$ of the plate in Figure 7.1a is simply supported, the deflection $w = 0$ at this edge. This edge can rotate about the Y axis freely, hence there is no bending moment M_x along this edge. Therefore,

$$(w)_{x=0} = 0, \quad \text{and using Eq.(7.1r),} \quad \left(\frac{\partial^2 w}{\partial x^2} + v \frac{\partial^2 w}{\partial y^2} \right)_{x=0} = 0 \tag{7.3a}$$

7.4.2 Built-in Edge

If the edge $x = 0$ of the plate in Figure 7.1a, is built-in, the deflection along this edge is zero. The edge cannot rotate about the Y axis, and the tangent to the deflected surface along this edge coincides with the initial middle plane of the plate. Hence,

$$(w)_{x=0} = 0, \quad \text{and} \quad \left(\frac{\partial w}{\partial x} \right)_{x=0} = 0 \tag{7.4}$$

7.4.3 Free Edge

If $x = a$ is the free edge of the plate in Figure 7.6, then

$$(M_x)_{x=a} = 0, \quad (M_{xy})_{x=a} = 0, \quad \text{and} \quad (Q_x)_{x=a} = 0 \tag{7.5a}$$

Kirchoff's theory of an elastic rod in 1859 [3] proved that these three boundary conditions can be reduced to two boundary conditions that are sufficient to solve the problem for w completely. The twisting moment of $M_{xy} \, dy$ acting on an element of length dy at the edge $x = a$ can be

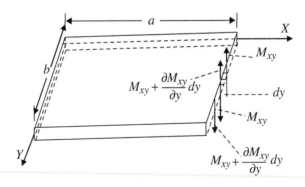

Figure 7.6 Plate boundary conditions at the free edge.

replaced by two vertical forces of M_{xy} acting at dy distance apart shown in Figure 7.6. This does not change the bending of the plate and produces only local changes.

The distribution of twisting moments M_{xy} is statically equivalent to the distribution of vertical shear forces of $-\left(\dfrac{\partial M_{xy}}{\partial y}\right)_{x=a}$ per unit length. Hence, the net vertical shear force at the edge $x = a$ is given by $Q_x - \dfrac{\partial M_{xy}}{\partial y}$ per unit length. Therefore, at the free edge

$$(M_x)_{x=a} = 0 \quad \text{and} \quad \left(Q_x - \frac{\partial M_{xy}}{\partial y}\right)_{x=a} = 0 \tag{7.5b}$$

Substitute Eqs. (7.1r), (7.1t), and (7.1z) for M_x, M_{xy}, and Q_x respectively in Eq. (7.5b) to get

$$\left(\frac{\partial^2 w}{\partial x^2} + v\frac{\partial^2 w}{\partial y^2}\right)_{x=a} = 0 \tag{7.5c}$$

and

$$-D\left(\frac{\partial^3 w}{\partial x^3} + \frac{\partial^3 w}{\partial x \partial y^2}\right) - D(1 - v)\frac{\partial^3 w}{\partial x \partial y^2} = 0$$

or

$$\left[\frac{\partial^3 w}{\partial x^3} + (2 - v)\frac{\partial^3 w}{\partial x \partial y^2}\right]_{x=a} = 0 \tag{7.5d}$$

The boundary conditions at the free edge, $x = a$, are given by Eqs. (7.5c) and (7.5d).

7.4.4 Elastically Supported and Elastically Built-in Edge

If the edge $x = a$ of a rectangular plate is rigidly joined with a supporting beam shown in Figure 7.7, the deflection at the edge is equal to the deflection in the beam. Also, the rotation of the edge is equal to the twist of the beam.

The force per unit length from the plate on the beam from Eq. (7.5d) is

$$-\left(Q_x - \frac{\partial M_{xy}}{\partial y}\right)_{x=a} = D\left(\frac{\partial^3 w}{\partial x^3} + (2 - v)\frac{\partial^3 w}{\partial x \partial y^2}\right)_{x=a} \tag{7.6a}$$

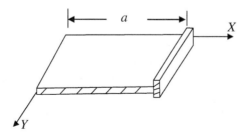

Figure 7.7 Elastically supported edge with a beam.

The differential equation for the beam from Eq. (4.1d) is

$$EI\frac{\partial^4 w}{\partial y^4} + P\frac{\partial^2 w}{\partial y^2} = q$$

where q is the force per unit length on the beam. If the axial force P is zero, then using Eq. (7.6a) we have

$$EI\left(\frac{\partial^4 w}{\partial y^4}\right)_{x=a} = D\left(\frac{\partial^3 w}{\partial x^3} + (2-v)\frac{\partial^3 w}{\partial x \partial y^2}\right)_{x=a} \tag{7.6b}$$

where EI is the flexural stiffness of the beam. The angle of rotation of any cross-section of the beam is counterclockwise and hence is taken as $-\left(\dfrac{\partial w}{\partial x}\right)_{x=a}$. The rate of change of this slope along the length of the beam is $-\left(\dfrac{\partial^2 w}{\partial x \partial y}\right)_{x=a}$. The twisting moment in the beam is given by $-GJ\left(\dfrac{\partial^2 w}{\partial x \partial y}\right)_{x=a}$, where GJ is the torsional stiffness of the beam. G is the shear modulus of the beam material, and J is the torsion constant of the beam cross-section. For the equilibrium of the edge, the twisting moment per unit length of the beam and the bending moment, M_x, per unit length in the plate are related as follows:

$$-GJ\frac{\partial}{\partial y}\left(\frac{\partial^2 w}{\partial x \partial y}\right)_{x=a} + (M_x)_{x=a} = 0$$

Substitute M_x from Eq. (7.1r) to have

$$GJ\left(\frac{\partial^3 w}{\partial x \partial y^2}\right)_{x=a} + D\left(\frac{\partial^2 w}{\partial x^2} + v\frac{\partial^2 w}{\partial y^2}\right)_{x=a} = 0 \tag{7.6c}$$

Equations (7.6b) and (7.6c) are the boundary conditions at the elastically supported edge $x = a$, of the plate that is supported by the beam.

7.5 Buckling of Rectangular Plates Uniformly Compressed in One Direction

7.5.1 Buckling of Rectangular Plates with Simply Supported Edges

Consider a simply supported rectangular plate with sides of a and b subjected to a compressive force of N_x per unit length acting on the edges $x = 0$ and $x = a$, shown in Figure 7.8. Assuming the edges of the plate can move in the plane of the plate, hence there is no additional in-plane force, i.e. $N_y = 0$ and $N_{xy} = 0$.

The governing equation of plate buckling, Eq. (7.2j) is written as

$$\frac{\partial^4 w}{\partial x^4} + 2\frac{\partial^4 w}{\partial x^2 \partial y^2} + \frac{\partial^4 w}{\partial y^4} + \frac{N_x}{D}\frac{\partial^2 w}{\partial x^2} = 0 \tag{7.7a}$$

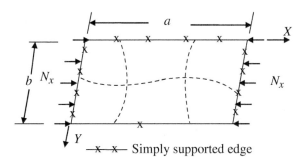

Figure 7.8 Simply supported plate under compressive axial force.

Equation (7.7a) is written by taking N_x as negative in Eq. (7.2j) because it is a compressive force. Since all edges are simply supported, the boundary conditions are given by Eq. (7.3a) as

$$w = 0, \quad \text{and} \quad \frac{\partial^2 w}{\partial x^2} + v\frac{\partial^2 w}{\partial y^2} = 0 \quad \text{at } x = 0 \text{ and } x = a \tag{7.7b}$$

and

$$w = 0, \quad \text{and} \quad \frac{\partial^2 w}{\partial y^2} + v\frac{\partial^2 w}{\partial x^2} = 0 \quad \text{at } y = 0 \text{ and } y = b \tag{7.7c}$$

The lateral deflection $w = 0$ along all four sides, hence

$$\frac{\partial^2 w}{\partial y^2} = 0 \text{ at } x = 0 \text{ and } x = a \tag{7.7d}$$

and

$$\frac{\partial^2 w}{\partial x^2} = 0 \text{ at } y = 0 \text{ and } y = b \tag{7.7e}$$

Substitute Eqs. (7.7d) and (7.7e) into Eqs. (7.7b) and (7.7c) and we have

$$w = 0, \text{and} \quad \frac{\partial^2 w}{\partial x^2} = 0 \text{ at } x = 0 \text{ and } x = a \tag{7.7f}$$

$$w = 0, \text{and} \quad \frac{\partial^2 w}{\partial y^2} = 0 \text{ at } y = 0 \text{ and } y = b \tag{7.7g}$$

We are considering small deformations of the plate, hence the bending strains are only considered and the in-plane strains due to bending are neglected. Therefore, only the boundary conditions related to transverse deformations are needed. For simply supported plates, the deflection surface can be assumed as

$$w = \sum_{m=1}^{\infty} \sum_{n=1}^{\infty} A_{mn} \sin \frac{m\pi x}{a} \sin \frac{n\pi y}{b} m = 1, 2, 3 ----, n = 1, 2, 3, --- \tag{7.7h}$$

where m and n define the number of half waves that the plate buckles in the x and y directions respectively. A_{mn} gives the amplitudes of the mode shapes or shape functions. The double trigonometric series in Eq. (7.7h) satisfies boundary conditions given by Eqs. (7.7f) and (7.7g). The substitution of w and its derivatives from Eqs. (7.7h) into Eq. (7.7a) gives

$$\sum_{m=1}^{\infty}\sum_{n=1}^{\infty} A_{mn} \left[\frac{m^4 \pi^4}{a^4} + \frac{2m^2 n^2 \pi^4}{a^2 b^2} + \frac{n^4 \pi^4}{b^4} - \frac{N_x}{D} \frac{m^2 \pi^2}{a^2} \right] \sin \frac{m \pi x}{a} \sin \frac{n \pi y}{b} = 0 \qquad (7.7i)$$

The terms in the brackets in Eq. (7.7i) consist of the sum of an infinite number of independent functions. The only way such a sum can be zero is when the coefficient of each of the terms is equal to zero. Hence,

$$A_{mn} \left[\pi^4 \left(\frac{m^2}{a^2} + \frac{n^2}{b^2} \right)^2 - \frac{N_x}{D} \frac{m^2 \pi^2}{a^2} \right] = 0 \qquad (7.7j)$$

Equation (7.7j) is satisfied if either $A_{mn} = 0$, or the term in the brackets is zero. If $A_{mn} = 0$, then the plate does not deform for all loads, and this is the trivial solution. The nontrivial solution is obtained by equating the term in the brackets equal to zero and we have

$$N_x = \frac{Da^2 \pi^2}{m^2} \left(\frac{m^2}{a^2} + \frac{n^2}{b^2} \right)^2$$

or

$$N_x = \frac{D\pi^2}{b^2} \left(\frac{mb}{a} + \frac{n^2 a}{mb} \right)^2 \qquad (7.7k)$$

We wish to know the lowest value of N_x called the critical load at which the equilibrium of the plate can change from the plane to a bent shape. The values of m and n, the number of half-waves that will minimize N_x is to be found. In Eq. (7.7k) as n increases N_x increases, therefore $n = 1$ gives the smallest value of N_x. This shows that the plate buckles in a single half sine wave in the y direction. The number of half sine waves in the x direction corresponding to the minimum value of N_x is found by taking the derivative of the expression in Eq. (7.7k) with respect to m at $n = 1$, and equating it to zero.

$$\frac{dN_x}{dm} = \frac{2D\pi^2}{b^2} \left(\frac{mb}{a} + \frac{a}{mb} \right) \left(\frac{b}{a} - \frac{a}{bm^2} \right) = 0$$

This gives

$$m = \frac{a}{b} \qquad (7.7l)$$

Substitute Eqs. (7.7l) into (7.7k) and we obtain

$$(N_x)_{cr} = \frac{4D\pi^2}{b^2} \qquad (7.7m)$$

Thus, the simply supported plate buckles with one half-wave in the y direction, and $m = a/b$ half-waves in the x direction. This means a/b is a whole number and Eq. (7.7m) gives the critical load only when a/b is a whole number. For plates in this category, the buckling pattern consists of one half-wave in the Y direction and a/b half-waves in the X direction. Hence, the plate buckles into a/b square waves. If a/b is not a whole number, we can write Eq. (7.7k) as

$$(N_x)_{cr} = \frac{kD\pi^2}{b^2} \tag{7.7n}$$

where

$$k = \left(\frac{mb}{a} + \frac{n^2 a}{mb} \right)^2 \tag{7.7o}$$

The value of k depends on the aspect ratio of the plate a/b and on the m and n values, the number of half-waves into which the plate buckles. The lowest value of N_x occurs when $n = 1$ as before. That is the plate buckles with one wave in the Y direction. In the X direction we have to find the variation of k versus a/b for various values of m, with $n = 1$ from Eq. (7.7o), so that N_x is the lowest value. The variation of k for various values of m with $n = 1$ from Eq. (7.7o) is plotted in Figure 7.9.

The curves showing variation of k for $m = 1, 2, 3, 4,$ and 5 are shown in Figure 7.9. We are interested in the lowest value of k for a particular aspect ratio a/b to get the minimum N_x. The solid lines in Figure 7.9 give the critical values of buckling load coefficient k as a function of a/b along with the number of half-waves into which the plate will buckle. For all plates of a/b ratio less than $\sqrt{2}$, k values are obtained from the curve drawn for $m = 1$ in Figure 7.9. Hence these plates buckle into a single half-wave in the x direction. For plates of $\sqrt{2} < a/b < \sqrt{6}$, the critical value of k is given by the curve for $m = 2$ in Figure 7.9, and these plates buckle into two half-waves in the x direction at buckling. Similarly, for the plates with $\sqrt{6} < a/b < \sqrt{12}$ the

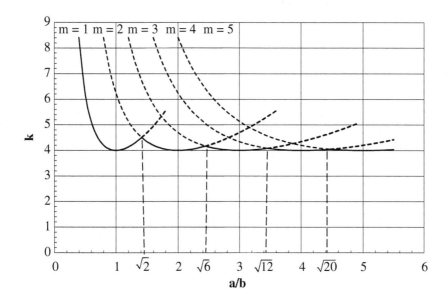

Figure 7.9 Variation of buckling load coefficient k for uniaxially compressed plate.

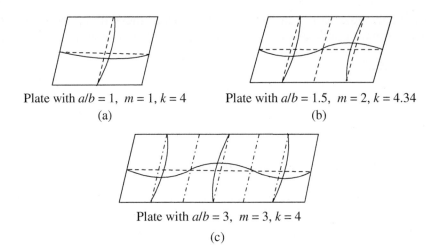

Plate with $a/b = 1$, $m = 1$, $k = 4$
(a)

Plate with $a/b = 1.5$, $m = 2$, $k = 4.34$
(b)

Plate with $a/b = 3$, $m = 3$, $k = 4$
(c)

Figure 7.10 Buckling modes of plates with different dimensions.

curve for $m = 3$, and for the plates of $\sqrt{12} < a/b < \sqrt{20}$ the curve for $m = 4$ give the critical values of k to calculate the minimum buckling force N_x. The buckling modes for plates of a/b ratios of 1, 1.5, and 3 are given in Figure 7.10.

The transition from m to $m+1$ half-waves occurs when the two corresponding curves have equal k values. Thus, from Eq. (7.7o)

$$\frac{mb}{a} + \frac{a}{mb} = \frac{(m+1)b}{a} + \frac{a}{(m+1)b} \quad \text{or} \quad \frac{a}{mb} = \frac{b}{a} + \frac{a}{(m+1)b} \tag{7.7p}$$

Let $\dfrac{a}{b} = p$, Eq. (7.7p) reduces to

$$\frac{p}{m} = \frac{1}{p} + \frac{p}{m+1} \quad \text{or} \quad p = \frac{a}{b} = \sqrt{m(m+1)} \tag{7.7q}$$

For $m = 1$, there is a transition from one to two half-waves at the aspect ratio of $a/b = \sqrt{2}$. For $m = 2$, the transition from two to three half-waves occurs at $a/b = \sqrt{6}$. Similarly, the transition from three to four half-waves occurs at $a/b = \sqrt{12}$, and from four to five half-waves occurs at $a/b = \sqrt{20}$. For $a/b > 4$, the buckling coefficient k varies very little from 4, and can be approximately taken as 4.0, as seen in Figure 7.9. The critical compressive stress from Eq. (7.7n) is

$$\sigma_{cr} = \frac{N_x}{h} = \frac{kD\pi^2}{b^2 h} \tag{7.7r}$$

where $D = \dfrac{Eh^3}{12(1 - v^2)}$, b is the width of the plate, and h is the thickness of the plate. Thus,

$$\sigma_{cr} = \frac{k\pi^2 E}{12(1 - v^2)} \frac{1}{(b/h)^2} \tag{7.7s}$$

The buckling load for a plate given by Eq. (7.7r) is similar to the expression for a column given by

$$\sigma_{cr} = \frac{\pi^2 E}{(KL/r)^2} \tag{7.7t}$$

where KL is the effective length of the column that depends on the boundary conditions. The plate and column buckling strengths are proportional to the modulus of elasticity of the material E, and inversely proportional to $(b/h)^2$ in the plate, and the slenderness ratio squared, $(L/r)^2$, in the column. Thus, the critical stress in a column depends on its length, whereas the critical stress in a plate depends on its width, and is independent of its length.

7.5.2 Buckling of Rectangular Plates with Other Boundary Conditions

We now consider plates that are simply supported on the loading edges ($x = 0$ and $x = a$), on which a compressive force of N_x per unit length is applied. The other two edges have various edge conditions (Figure 7.11). The governing equation of plate buckling is

$$\frac{\partial^4 w}{\partial x^4} + 2\frac{\partial^4 w}{\partial x^2 \partial y^2} + \frac{\partial^4 w}{\partial y^4} + \frac{N_x}{D}\frac{\partial^2 w}{\partial x^2} = 0 \tag{7.8a}$$

Assume that under the action of compressive force N_x the plate buckles in m sinusoidal half-waves in the x direction. Hence, we take the solution of Eq. (7.8a) in the form

$$w = f(y)\sin\frac{m\pi x}{a} \tag{7.8b}$$

where $f(y)$ is a function of y only. Equation (7.8b) satisfies the boundary conditions

$$w = 0 \quad \text{and} \quad \frac{\partial^2 w}{\partial x^2} + v\frac{\partial^2 w}{\partial y^2} = 0$$

at the simply supported edges, $x = 0$ and $x = a$. Taking the derivatives of (Eq. 7.8b) we have

$$\frac{\partial^2 w}{\partial x^2} = -f(y)\frac{m^2\pi^2}{a^2}\sin\frac{m\pi x}{a}, \quad \frac{\partial^4 w}{\partial x^2 \partial y^2} = -\frac{d^2 f}{dy^2}\frac{m^2\pi^2}{a^2}\sin\frac{m\pi x}{a},$$

$$\frac{\partial^4 w}{\partial x^4} = f(y)\frac{m^4\pi^4}{a^4}\sin\frac{m\pi x}{a}, \quad \text{and} \quad \frac{\partial^4 w}{\partial y^4} = \frac{d^4 f}{dy^4}\sin\frac{m\pi x}{a}$$

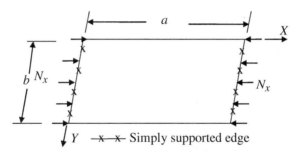

Figure 7.11 Plates under compressive force with different edges.

Substitute these derivatives into Eq. (7.8a) to have the following ordinary differential equation because $\sin \dfrac{m\pi x}{a}$ is not zero:

$$\frac{d^4 f}{dy^4} - \frac{2m^2\pi^2}{a^2}\frac{d^2 f}{dy^2} + \left(\frac{m^4\pi^4}{a^4} - \frac{N_x}{D}\frac{m^2\pi^2}{a^2}\right) f = 0 \tag{7.8c}$$

The general solution of Eq. (7.8c) can be written as

$$f(y) = A_1 \sinh \alpha y + A_2 \cosh \alpha y + A_3 \sin \beta y + A_4 \cos \beta y \tag{7.8d}$$

where

$$\alpha = \sqrt{\frac{m^2\pi^2}{a^2} + \sqrt{\frac{N_x}{D}\frac{m^2\pi^2}{a^2}}}, \qquad \beta = \sqrt{-\frac{m^2\pi^2}{a^2} + \sqrt{\frac{N_x}{D}\frac{m^2\pi^2}{a^2}}} \tag{7.8e}$$

Due to some constraints on the edges $y = 0$ and $y = b$, we always have $\dfrac{N_x}{D} > \dfrac{m^2\pi^2}{a^2}$ [3]. The constants A_1, A_2, A_3, and A_4 are determined from the edge conditions at $y = 0$ and $y = b$.

7.5.3 Loading Edges Simply Supported, the Side $y = 0$ Is Clamped, and the Side $y = b$ Is Free

The plate and the loading are shown in Figure 7.12. The boundary conditions in this case are given by

$$w = 0, \qquad \frac{\partial w}{\partial y} = 0 \qquad \text{at } y = 0 \tag{7.9a}$$

$$\frac{\partial^2 w}{\partial y^2} + v\frac{\partial^2 w}{\partial x^2} = 0, \qquad \frac{\partial^3 w}{\partial y^3} + (2 - v)\frac{\partial^3 w}{\partial x^2 \partial y} = 0 \quad \text{at } y = b \tag{7.9b}$$

The governing equation here is Eq. (7.8c), as in Section 7.5.2:

$$\frac{d^4 f}{dy^4} - \frac{2m^2\pi^2}{a^2}\frac{d^2 f}{dy^2} + \left(\frac{m^4\pi^4}{a^4} - \frac{N_x}{D}\frac{m^2\pi^2}{a^2}\right) f = 0 \tag{7.8c}$$

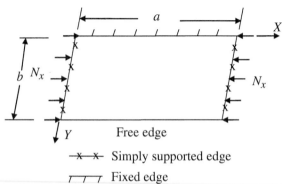

Free edge
—×—×— Simply supported edge
⊓⊓⊓ Fixed edge

Figure 7.12 Plate with loading edges simply supported, side $y = 0$ clamped, and $y = b$ free.

The general solution of Eq. (7.8c) is

$$f(y) = A_1 \sinh \alpha y + A_2 \cosh \alpha y + A_3 \sin \beta y + A_4 \cos \beta y \tag{7.8d}$$

where

$$\alpha = \sqrt{\frac{m^2 \pi^2}{a^2} + \sqrt{\frac{N_x}{D} \frac{m^2 \pi^2}{a^2}}}, \quad \beta = \sqrt{-\frac{m^2 \pi^2}{a^2} + \sqrt{\frac{N_x}{D} \frac{m^2 \pi^2}{a^2}}} \tag{7.8e}$$

Use Eqs. (7.8b) and (7.9a) and (7.9b) to get the boundary conditions in terms of f, as follows:

$$f(y) = 0 \quad \text{at } y = 0 \tag{7.9c}$$

$$\frac{df}{dy} = 0 \quad \text{at } y = 0 \tag{7.9d}$$

$$\frac{d^2 f}{dy^2} - \frac{m^2 \pi^2 v}{a^2} f(y) = 0 \quad \text{at } y = b \tag{7.9e}$$

$$\frac{d^3 f}{dy^3} - (2 - v)\frac{m^2 \pi^2}{a^2} \frac{df}{dy} = 0 \quad \text{at } y = b \tag{7.9f}$$

Using Eqs. (7.9c)–(7.9f) and (7.8d), we get the following algebraic equations:

$$A_2 + A_4 = 0 \tag{7.9g}$$

$$\alpha A_1 + \beta A_3 = 0 \tag{7.9h}$$

$$\alpha^2 A_1 \sinh \alpha b + \alpha^2 A_2 \cosh \alpha b - \beta^2 A_3 \sin \beta b - \beta^2 A_4 \cos \beta b$$
$$-\frac{m^2 \pi^2 v}{a^2} \left[A_1 \sinh \alpha b + A_2 \cosh \alpha b + A_3 \sin \beta b + A_4 \cos \beta b \right] = 0 \tag{7.9i}$$

$$\alpha^3 A_1 \cosh \alpha b + \alpha^3 A_2 \sinh \alpha b - \beta^3 A_3 \cos \beta b + \beta^3 A_4 \sin \beta b$$
$$-(2 - v)\frac{m^2 \pi^2}{a^2} \left[\alpha A_1 \cosh \alpha b + \alpha A_2 \sinh \alpha b + \beta A_3 \cos \beta b - \beta A_4 \sin \beta b \right] = 0 \tag{7.9j}$$

Equations (7.9i) and (7.9j) can be simplified using Eqs. (7.9g) and (7.9h) and written as

$$A_1 \left[\sinh \alpha b \left(\alpha^2 - \frac{m^2 \pi^2 v}{a^2} \right) + \left(\beta^2 + \frac{m^2 \pi^2 v}{a^2} \right) \frac{\alpha}{\beta} \sin \beta b \right]$$
$$+ A_2 \left[\cosh \alpha b \left(\alpha^2 - \frac{m^2 \pi^2 v}{a^2} \right) + \left(\beta^2 + \frac{m^2 \pi^2 v}{a^2} \right) \cos \beta b \right] = 0 \tag{7.9k}$$

and

$$A_1 \left[\alpha \cosh \alpha b \left\{ \alpha^2 - (2 - v)\frac{m^2 \pi^2}{a^2} \right\} + \alpha \cos \beta b \left\{ \beta^2 + (2 - v)\frac{m^2 \pi^2}{a^2} \right\} \right]$$
$$+ A_2 \left[\alpha \sinh \alpha b \left\{ \alpha^2 - (2 - v)\frac{m^2 \pi^2}{a^2} \right\} - \beta \sin \beta b \left\{ \beta^2 + (2 - v)\frac{m^2 \pi^2}{a^2} \right\} \right] = 0 \tag{7.9l}$$

Equation (7.8e) can be written as

$$\alpha^2 = \frac{m^2\pi^2}{a^2} + \frac{m\pi}{a}\sqrt{\frac{N_x}{D}}, \qquad \beta^2 = -\frac{m^2\pi^2}{a^2} + \frac{m\pi}{a}\sqrt{\frac{N_x}{D}}$$

or

$$\alpha^2 = \beta^2 + \frac{2m^2\pi^2}{a^2}$$

Substitute

$$s = \alpha^2 - \frac{m^2\pi^2 v}{a^2} = \beta^2 + \frac{2m^2\pi^2}{a^2} - \frac{m^2\pi^2 v}{a^2}$$

or

$$s = \alpha^2 - \frac{m^2\pi^2 v}{a^2} = \beta^2 + (2-v)\frac{m^2\pi^2}{a^2} \tag{7.9m}$$

Substitute

$$t = \beta^2 + \frac{m^2\pi^2 v}{a^2} = \alpha^2 - \frac{2m^2\pi^2}{a^2} + \frac{m^2\pi^2 v}{a^2}$$

or

$$t = \beta^2 + \frac{m^2\pi^2 v}{a^2} = \alpha^2 - (2-v)\frac{m^2\pi^2}{a^2} \tag{7.9n}$$

Make the substitutions of Eqs. (7.9m) and (7.9n), and we can write Eqs. (7.9k) and (7.9l) as follows:

$$\begin{bmatrix} s\sinh\alpha b + t\dfrac{\alpha}{\beta}\sin\beta b & s\cosh\alpha b + t\cos\beta b \\ \alpha t\cosh\alpha b + s\alpha\cos\beta b & \alpha t\sinh\alpha b - \beta s\sin\beta b \end{bmatrix} \begin{Bmatrix} A_1 \\ A_2 \end{Bmatrix} = \begin{Bmatrix} 0 \\ 0 \end{Bmatrix} \tag{7.9o}$$

For a nontrivial solution for A_1 and A_2, the determinant of the coefficient matrix should be zero. Hence,

$$\begin{vmatrix} s\sinh\alpha b + t\dfrac{\alpha}{\beta}\sin\beta b & s\cosh\alpha b + t\cos\beta b \\ \alpha t\cosh\alpha b + s\alpha\cos\beta b & \alpha t\sinh\alpha b - \beta s\sin\beta b \end{vmatrix} = 0 \tag{7.9p}$$

or

$$\left(s\sinh\alpha b + t\frac{\alpha}{\beta}\sin\beta b \right)(\alpha t\sinh\alpha b - \beta s\sin\beta b)$$
$$-(s\cosh\alpha b + t\cos\beta b)(\alpha t\cosh\alpha b + \alpha s\cos\beta b) = 0$$

The critical compressive force can be obtained from the following transcendental characteristic equation:

$$2\alpha\beta st + \alpha\beta(s^2 + t^2)\cosh\alpha b\cos\beta b - (\alpha^2 t^2 - \beta^2 s^2)\sinh\alpha b\sin\beta b = 0$$

or

$$2st + (s^2 + t^2)\cosh \alpha b \cos \beta b - \left(\frac{\alpha}{\beta}t^2 - \frac{\beta}{\alpha}s^2\right)\sinh \alpha b \sin \beta b = 0$$

or

$$2 + \left(\frac{s}{t} + \frac{t}{s}\right)\cos h\alpha b \cos \beta b - \left(\frac{\alpha}{\beta}\frac{t}{s} - \frac{\beta}{\alpha}\frac{s}{t}\right)\sinh \alpha b \sin \beta b = 0 \tag{7.9q}$$

Equation (7.9q) can be used to calculate the critical value of N_x if the dimensions of the plate and the material properties are known, since α and β contain N_x. The magnitude of critical N_x can be expressed as

$$(N_x)_{cr} = \frac{kD\pi^2}{b^2} \tag{7.7n}$$

and

$$\sigma_{cr} = \frac{(N_x)_{cr}}{h} = \frac{k\pi^2 E}{12(1 - v^2)} \frac{1}{(b/h)^2} \tag{7.7s}$$

Example 7.1   Determine the critical stress for a simply supported plate on the edges $x = 0$ and $x = a$, and clamped on the side $y = 0$ and free on the side $y = b$. The plate is loaded in the x direction on the simply supported edges as shown in Figure 7.12. Given: the modulus of elasticity, $E = 29 \times 10^6$ psi (200 000 MPa), Poisson's ratio, $v = 0.3$, $a/b = 1.0$, $h/b = 0.01$.
   Find the value of k that satisfies the transcendental Eq. (7.9q) for different a/b ratios.
   Let $\frac{a}{b} = 1.0$. By trial and error the minimum value of $k = 1.65$ is found at $m = 1$. The calculations here show that $k = 1.65$ satisfies Eq. (7.9q) at $m = 1$.

$$\alpha b = \frac{m\pi}{a/b}\sqrt{1 + \frac{a}{mb}\sqrt{k}} = \frac{1(\pi)}{1}\sqrt{1 + 1\sqrt{1.65}} = 4.748$$

$$sb^2 = (\alpha b)^2 - v\frac{m^2\pi^2}{(a/b)^2} = (4.748)^2 - 0.3\frac{1(\pi)^2}{1} = 19.583$$

$$\beta b = \frac{m\pi}{a/b}\sqrt{-1 + \frac{a}{mb}\sqrt{k}} = \frac{1(\pi)}{1}\sqrt{-1 + 1\sqrt{1.65}} = 1.676$$

$$tb^2 = \beta^2 b^2 + v\frac{m^2\pi^2}{(a/b)^2} = (1.676)^2 + 0.3\frac{1(\pi)^2}{1} = 5.770$$

$$\frac{s}{t} = \frac{19.583}{5.770} = 3.394, \quad \frac{t}{s} = 0.295$$

$$\frac{\alpha}{\beta} = \frac{4.748}{1.676} = 2.833, \quad \frac{\beta}{\alpha} = 0.353$$

$$\cosh \alpha b = \cosh 4.748 = 57.68, \quad \sinh \alpha b = \sinh 4.748 = 57.67$$

$$\cos \beta b = \cos 1.676 = -0.105, \quad \sin \beta b = \sin 1.676 = 0.994$$

From Eq. (7.9q) we have

$$2 + (3.394 + 0.295)(57.68)(-0.105) - (2.833x0.295 - 0.353x3.394)(57.67)(0.994)$$
$$= 0.429 \approx 0$$

Therefore, $k = 1.65$ satisfies Eq. (7.9q).
From Eq. (7.7s)

$$\sigma_{cr} = \frac{k\pi^2 E}{12(1 - v^2)} \frac{1}{(b/h)^2} \tag{7.7s}$$

$$\sigma_{cr} = \frac{1.65(\pi)^2 x29x10^6}{12(1 - 0.3^2)}(0.01)^2 = 4324.73 psi \quad \left[\frac{1.65(\pi)^2 x200,000}{12(1 - 0.3^2)}(0.01)^2 = 29.83 MPa \right]$$

Find the value of k that satisfies the transcendental Eq. (7.9q) for different a/b ratios as shown in Example 7.5.1. Then find N_x and critical stress σ_{cr} from Eqs. (7.7n) and (7.7s) respectively knowing the material properties and dimensions of the plate. The critical values of k along with m values for various a/b ratios are given in Table 7.1. For $a/b = 2.8$, 3.0, 4.0, and 5.0, the minimum value of k occurs for $m = 2$, a phenomenon similar to the results shown in Figure 7.9 for the simply supported plates.

Table 7.1 Values of the factor k for an axially compressed plate when two loaded sides are simply supported, the side $y = 0$ is clamped and the side $y = b$ is free (Figure 7.12).

a/b	m	k
0.2	1	25.21
0.4	1	6.65
0.6	1	3.28
0.8	1	2.15
1.0	1	1.65
1.2	1	1.43
1.4	1	1.32
1.6	1	1.28
1.8	1	1.29
2.0	1	1.34
2.2	1	1.40
2.4	1	1.49
2.6	1	1.60
2.8	2	1.31
3.0	2	1.29
4.0	2	1.34
5.0	2	1.55

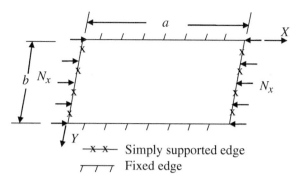

Figure 7.13 Plate with loading edges simply supported, sides $y = 0$ and $y = b$ clamped.

7.5.4 Loading Edges Simply Supported and the Sides $y = 0$ and $y = b$ Are Clamped

The plate and the loading are shown in Figure 7.13. The boundary conditions in this case are given by

$$w = 0 \qquad \text{at } y = 0 \text{ and } y = b \tag{7.10a}$$

$$\frac{\partial w}{\partial y} = 0 \qquad \text{at } y = 0 \text{ and } y = b \tag{7.10b}$$

We can transform the boundary conditions given by Eqs. (7.10a) and (7.10b) in terms of $f(y)$ using Eq. (7.8b). Substitute $f(y)$ and its derivatives from Eq. (7.8d) into Eqs. (7.10a) and (7.10b) to get

$$A_2 + A_4 = 0 \tag{7.10c}$$

$$\alpha A_1 + \beta A_3 = 0 \tag{7.10d}$$

$$A_1 \sinh \alpha b + A_2 \cosh \alpha b + A_3 \sin \beta b + A_4 \cos \beta b = 0 \tag{7.10e}$$

$$A_1 \alpha \cosh \alpha b + A_2 \alpha \sinh \alpha b + A_3 \beta \cos \beta b - A_4 \beta \sin \beta b = 0 \tag{7.10f}$$

Equations (7.10c) and (7.10d) give

$$A_4 = -A_2 \qquad \text{and} \qquad A_3 = -\left(\frac{\alpha}{\beta}\right) A_1$$

Now substitute A_3 and A_4 in Eqs. (7.10e) and (7.10f) to get

$$\begin{bmatrix} \sinh \alpha b - \dfrac{\alpha}{\beta} \sin \beta b & \cosh \alpha b - \cos \beta b \\ \alpha \cosh \alpha b - \alpha \cos \beta b & \alpha \sinh \alpha b + \beta \sin \beta b \end{bmatrix} \begin{Bmatrix} A_1 \\ A_2 \end{Bmatrix} = \begin{Bmatrix} 0 \\ 0 \end{Bmatrix} \tag{7.10g}$$

For a nontrivial solution the determinant of the coefficient matrix in Eq. (7.10g) should vanish

$$\begin{vmatrix} \sinh \alpha b - \dfrac{\alpha}{\beta} \sin \beta b & \cosh \alpha b - \cos \beta b \\ \alpha \cosh \alpha b - \alpha \cos \beta b & \alpha \sinh \alpha b + \beta \sin \beta b \end{vmatrix} = 0 \tag{7.10h}$$

or

$$\left(\sinh \alpha b - \frac{\alpha}{\beta} \sin \beta b \right) (\alpha \sinh \alpha b + \beta \sin \beta b)$$
$$-(\cosh \alpha b - \cos \beta b)(\alpha \cosh \alpha b - \alpha \cos \beta b) = 0$$

The critical value of the compressive forces can be obtained from the following transcendental characteristic equation:

$$2(1 - \cosh \alpha b \cos \beta b) = \left(\frac{\beta}{\alpha} - \frac{\alpha}{\beta} \right) \sinh \alpha b \sin \beta b \qquad (7.10i)$$

where α and β are given by Eq. (7.8e). The critical load is again given by Eq. (7.7n). Substitute the critical load N_x given by Eq. (7.7n) in Eq. (7.8e) to get

$$\alpha b = \frac{m\pi}{a/b} \sqrt{1 + \frac{a}{mb} \sqrt{k}}, \quad \text{and} \quad \beta b = \frac{m\pi}{a/b} \sqrt{-1 + \frac{a}{mb} \sqrt{k}} \qquad (7.10j)$$

The critical load of N_x per unit length is given by Eq. (7.7k) as before

$$N_x = \frac{D\pi^2}{b^2} \left(\frac{mb}{a} + \frac{n^2 a}{mb} \right)^2 \qquad (7.7k)$$

or

$$N_x = \frac{D\pi^2}{a^2} \left(m + \frac{1}{m} \frac{a^2}{b^2} \right)^2 \text{ for } n = 1 \qquad (7.10k)$$

The quantity outside the parentheses in Eq. (7.10k) represents Euler's buckling load for a strip of unit width and length a. The quantity in the parentheses determines in what proportion the stability of a plate is greater than the stability of an isolated column strip, and it depends on the ratio a/b and the quantity m, the number of half-waves into which the plate buckles. If $a < b$, the second term in the parentheses of Eq. (7.10k) is always smaller than the first, and the minimum value of the expression given by Eq. (7.10k) is when $m = 1$. If we write the critical load as

$$(N_x)_{cr} = \frac{kD\pi^2}{b^2} \qquad (7.7n)$$

For $a/b \leq 1$, the minimum value of k occurs for $m = 1$. When the $a/b > 1$, the values of $m = 2$, 3, - -- give the minimum values of k similar to the results shown in Figure 7.9 for the simply supported plates.

Find the value of k that satisfies transcendental Eq. (7.10i). Once the k is known, the critical force N_x can be found from Eq. (7.7n). The critical stress can be found from Eq. (7.7s) knowing the material properties and the dimensions of the plate. The variation of k with various values of a/b is given in Table 7.2 along with m values.

Table 7.2 Values of k for axially compressed plate when two sides are simply supported and the other two are clamped in Figure 7.13.

a/b	m	k
0.2	1	27.46
0.3	1	13.91
0.4	1	9.45
0.5	1	7.69
0.6	1	7.05
0.7	1	7.00
0.8	1	7.29
0.9	1	7.83
1.0	1	8.60
$\sqrt{2} = 1.414$	2	7.02
2.0	2	8.60
$\sqrt{6} = 2.45$	3	7.38
3.0	3	8.62
$\sqrt{12} = 3.46$	4	7.64
$\sqrt{20} = 4.47$	5	7.82

Example 7.2 The smallest value of k is 7.00 in Table 7.2. If $E = 29 \times 10^6$ psi (200 000 MPa), $v = 0.3$, $h/b = 0.01$, then the critical stress is given by

$$\sigma_{cr} = \frac{k\pi^2 E}{12(1 - v^2)} \frac{1}{\left(b/h\right)^2} \tag{7.7s}$$

or

$$\sigma_{cr} = \frac{7(\pi)^2 x29x10^6}{12(1 - 0.3^2)}(0.01)^2 = 18,347.34 \text{ psi} \left[\frac{7(\pi^2)x200,000}{12(1 - 0.3^2)}(0.01)^2 = 126.53 \text{ MPa}\right]$$

7.5.5 Loading Edges Simply Supported, the Side $y = 0$ Is Simply Supported, and the Side $y = b$ Is Free

The plate and the loading are shown in Figure 7.14. The boundary conditions in this case are given by

$$w = 0, \quad \frac{\partial^2 w}{\partial y^2} + v\frac{\partial^2 w}{\partial x^2} = 0 \text{ at } y = 0 \tag{7.11a}$$

$$\frac{\partial^2 w}{\partial y^2} + v\frac{\partial^2 w}{\partial x^2} = 0, \quad \frac{\partial^3 w}{\partial y^3} + (2 - v)\frac{\partial^3 w}{\partial x^2 \partial y} = 0 \text{ at } y = b \tag{7.11b}$$

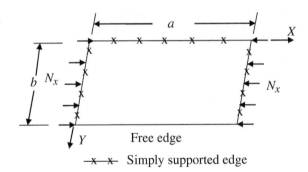

Figure 7.14 Plate with loading edges simply supported, side $y = 0$ simply supported, and $y = b$ free.

The governing equation for the plate is Eq. (7.8c) as in Section 7.5.2. The general solution is given by Eq. (7.8d)

$$f(y) = A_1 \sinh \alpha y + A_2 \cosh \alpha y + A_3 \sin \beta y + A_4 \cos \beta y \tag{7.8d}$$

where α and β are given by Eq. (7.8e).

From Eq. (7.8b) we have,

$$w = f(y) \sin \frac{m\pi x}{a} \tag{7.8b}$$

Use the boundary conditions in Eqs. (7.11a) and (7.11b) and (7.8b) to get the boundary conditions in terms of f as follows:

$$f(y) = 0 \text{ at } y = 0 \tag{7.11c}$$

$$\frac{d^2 f}{dy^2} - \frac{m^2 \pi^2 v}{a^2} f(y) = 0 \text{ at } y = 0 \tag{7.11d}$$

$$\frac{d^2 f}{dy^2} - \frac{m^2 \pi^2 v}{a^2} f(y) = 0 \text{ at } y = b \tag{7.11e}$$

$$\frac{d^3 f}{dy^3} - (2 - v)\frac{m^2 \pi^2}{a^2}\frac{df}{dy} = 0 \text{ at } y = b \tag{7.11f}$$

Use Eqs. (7.11c) and (7.11d) and (7.8d) to get

$$A_2 = 0 \quad \text{and} \quad A_4 = 0 \tag{7.11g}$$

Hence Eq. (7.8d) can be written as

$$f(y) = A_1 \sinh \alpha y + A_3 \sin \beta y \tag{7.11h}$$

Substitute Eq. (7.11h) into Eqs. (7.11e) and (7.11f) to obtain

$$A_1 \left[\sinh \alpha b \left(\alpha^2 - \frac{m^2 \pi^2 v}{a^2} \right) \right] - A_3 \left[\sin \beta b \left(\beta^2 + \frac{m^2 \pi^2 v}{a^2} \right) \right] = 0 \tag{7.11i}$$

$$A_1 \left[\alpha \cosh \alpha b \left\{ \alpha^2 - (2 - v)\frac{m^2 \pi^2}{a^2} \right\} \right] - A_3 \left[\beta \cos \beta b \left\{ \beta^2 + (2 - v)\frac{m^2 \pi^2}{a^2} \right\} \right] = 0 \tag{7.11j}$$

or

$$
\begin{bmatrix}
\sinh \alpha b \left(\alpha^2 - \dfrac{m^2 \pi^2 \nu}{a^2} \right) & -\sin \beta b \left(\beta^2 + \dfrac{m^2 \pi^2 \nu}{a^2} \right) \\[4mm]
\alpha \cosh \alpha b \left\{ \alpha^2 - (2 - \nu)\dfrac{m^2 \pi^2}{a^2} \right\} & -\beta \cos \beta b \left\{ \beta^2 + (2 - \nu)\dfrac{m^2 \pi^2}{a^2} \right\}
\end{bmatrix}
\begin{Bmatrix} A_1 \\[4mm] A_3 \end{Bmatrix}
=
\begin{Bmatrix} 0 \\[4mm] 0 \end{Bmatrix}
$$

(7.11k)

For a nontrivial solution for A_1 and A_3, the determinant of the coefficient matrix in Eq. (7.11k) should be zero. Hence

$$
\begin{vmatrix}
\sinh \alpha b \left(\alpha^2 - \dfrac{m^2 \pi^2 \nu}{a^2} \right) & -\sin \beta b \left(\beta^2 + \dfrac{m^2 \pi^2 \nu}{a^2} \right) \\[4mm]
\alpha \cosh \alpha b \left\{ \alpha^2 - (2 - \nu)\dfrac{m^2 \pi^2}{a^2} \right\} & -\beta \cos \beta b \left\{ \beta^2 + (2 - \nu)\dfrac{m^2 \pi^2}{a^2} \right\}
\end{vmatrix}
= 0
$$

(7.11l)

or

$$
\alpha \left(\beta^2 + \frac{m^2 \pi^2 \nu}{a^2} \right)^2 \tan \beta b = \beta \left(\alpha^2 - \frac{m^2 \pi^2 \nu}{a^2} \right)^2 \tanh \alpha b
$$

(7.11m)

The critical compressive force is obtained from the characteristic Eq. (7.11m). The quantities αb and βb are given by Eq. 7.10j.

$$
\alpha b = \frac{m\pi}{a/b} \sqrt{1 + \frac{a}{mb}\sqrt{k}}, \text{ and } \beta b = \frac{m\pi}{a/b} \sqrt{-1 + \frac{a}{mb}\sqrt{k}}
$$

(7.10j)

Equation (7.11m) can be written as

$$
\frac{\alpha}{\beta} \frac{t^2}{s^2} \tan \beta b = \tanh \alpha b
$$

(7.11n)

where

$$
s = \alpha^2 - \frac{m^2 \pi^2 \nu}{a^2} = \beta^2 + (2 - \nu)\frac{m^2 \pi^2}{a^2}
$$

(7.9m)

and

$$
t = \beta^2 + \frac{m^2 \pi^2 \nu}{a^2} = \alpha^2 - (2 - \nu)\frac{m^2 \pi^2}{a^2}
$$

(7.9n)

Find the value of k that satisfies Eq. (7.11n). The critical force N_x can be found from Eq. (7.7n) knowing k. The critical stress σ_{cr} is found from Eq. (7.7s) if the material properties and the dimensions of the plate are known. The critical values of k along with m for various a/b ratios are given in Table 7.3. All minimum values of k occur at $m = 1$, meaning the buckled plate consists of only one half-wave.

Table 7.3 Factor k for axially compressed plate when two loaded sides are simply supported, the side $y = 0$ is simply supported and the side $y = b$ is free (Figure 7.14).

a/b	m	k
0.2	1	25.15
0.4	1	6.58
0.6	1	3.15
0.8	1	1.95
1.0	1	1.43
1.2	1	1.13
1.4	1	0.95
1.6	1	0.83
1.8	1	0.75
2.0	1	0.69
2.2	1	0.63
2.4	1	0.60
2.6	1	0.58
2.8	1	0.56
3.0	1	0.55
4.0	1	0.49
5.0	1	0.47

Example 7.3 Determine the critical stress for a simply supported plate on the edges $x = 0$ and $x = a$, and simply supported on the side $y = 0$ and free on the side $y = b$. The plate is loaded in the x direction on the simply supported edges as shown in Figure 7.14. Given: the modulus of elasticity, $E = 29 \times 10^6$ psi (200 000 MPa), Poisson's ratio, $v = 0.3$, $a/b = 3.0$, $h/b = 0.01$.

Take $m = 1$ (for $a/b = 3.0$, $m = 1$ gives the minimum critical load shown in Table 7.3), and assume $k = 0.55$.

$$\alpha b = \frac{m\pi}{a/b}\sqrt{1 + \frac{a}{mb}\sqrt{k}} = \frac{1(\pi)}{3}\sqrt{1 + 3\sqrt{0.55}} = 1.8805$$

$$sb^2 = (\alpha b)^2 - v\frac{m^2\pi^2}{(a/b)^2} = (1.8805)^2 - 0.3\frac{1(\pi)^2}{(3)^2} = 3.2073$$

$$\beta b = \frac{m\pi}{a/b}\sqrt{-1 + \frac{a}{mb}\sqrt{k}} = \frac{1(\pi)}{3}\sqrt{-1 + 3\sqrt{0.55}} = 1.1589$$

$$tb^2 = \beta^2 b^2 + v\frac{m^2\pi^2}{(a/b)^2} = (1.1589)^2 + 0.3\frac{1(\pi)^2}{(3)^2} = 1.6720$$

$$\frac{\alpha}{\beta} = \frac{1.8805}{1.1589} = 1.6227, \quad \frac{t^2}{s^2} = \left(\frac{1.6720}{3.2073}\right)^2 = 0.2718$$

A_4

From Eq. (7.11n) we have

$$1.6227(0.2718)(2.2889) - 0.9545 = 0.055$$

Therefore,

$k = 0.55$

From Eq. (7.7s)

$$\sigma_{cr} = \frac{k\pi^2 E}{12(1 - v^2)} \frac{1}{(b/h)^2} \tag{7.7s}$$

$$\sigma_{cr} = \frac{0.55(\pi)^2 x 29x10^6}{12(1 - 0.3^2)}(0.01)^2 = 1441.57 \text{ psi} \quad \left[\frac{0.55(\pi)^2 x 200,000}{12(1 - 0.3^2)}(0.01)^2 = 9.94 \text{ MPa}\right]$$

7.5.6 Loading Edges Simply Supported, the Side $y = 0$ Is Elastically Built-in and the Side $y = b$ Is Free

The plate and the loading are shown in Figure 7.15. In many practical cases, the constraint at a support will be that between a simply supported edge and the built-in edge. We consider that the upper edge of the plate is elastically built in, and the bending moments that develop at this edge during buckling of the plate are proportional to the rotation of the edge. The restraining support may consist of a flange shown in Figure 7.15.

The boundary conditions at the upper edge, $y = 0$ are

$$w = 0 \tag{7.12a}$$

For the second boundary condition, the continuity between the restraining flange and the web plate are considered. For continuity, it is necessary that the angle of rotation at the edge of the buckling plate is equal to the angle of rotation at the edge of the restraining support.

The angle of rotation of the flange during buckling $= \left(\dfrac{\partial w}{\partial y}\right)_{y=0}$

The rate of change of the slope along the upper edge $= \left(\dfrac{\partial^2 w}{\partial y \partial x}\right)_{y=0}$

Hence, the twisting moment at the upper edge at any section along the x axis $= GJ\left(\dfrac{\partial^2 w}{\partial x \partial y}\right)_{y=0}$

where GJ is the torsional rigidity of the flange. The rate of change of the torsional moment is

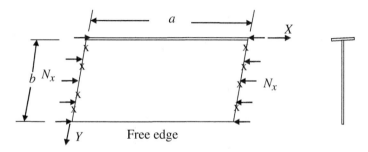

Figure 7.15 Plate with loading edges simply supported, side $y = 0$ elastically built in, and $y = b$ free.

equal to the bending moment M_y per unit length at the upper edge of the web plate. The signs of the torsional and bending moments are same, hence

$$GJ\left(\frac{\partial^3 w}{\partial x^2 \partial y}\right)_{y=0} = (M_y)_{y=0}$$

Using Eq. (7.1s) we have

$$GJ\left(\frac{\partial^3 w}{\partial x^2 \partial y}\right)_{y=0} + D\left(\frac{\partial^2 w}{\partial y^2} + v\frac{\partial^2 w}{\partial x^2}\right)_{y=0} = 0 \tag{7.12b}$$

From Eq. (7.8b) we have

$$w = f(y)\sin\frac{m\pi x}{a}$$

$$\frac{\partial^2 w}{\partial x^2} = -f(y)\frac{m^2\pi^2}{a^2}\sin\frac{m\pi x}{a} = -\frac{m^2\pi^2}{a^2}w$$

$$\frac{\partial^3 w}{\partial x^2 \partial y} = -\frac{m^2\pi^2}{a^2}\frac{\partial w}{\partial y}$$

Since $w = 0$ at $y = 0$, we have, $\dfrac{\partial^2 w}{\partial x^2} = 0$. From Eq. (7.12b) we have

$$GJ\frac{m^2\pi^2}{a^2}\frac{\partial w}{\partial y} = D\frac{\partial^2 w}{\partial y^2} \tag{7.12c}$$

Equations (7.12a) and (7.12c) give the two boundary conditions at the elastically built in edge at $y = 0$. At the free edge, $y = b$ the boundary conditions are given by Eq. (7.11b)

$$\frac{\partial^2 w}{\partial y^2} + v\frac{\partial^2 w}{\partial x^2} = 0, \qquad \frac{\partial^3 w}{\partial y^3} + (2-v)\frac{\partial^3 w}{\partial x^2 \partial y} = 0 \tag{7.11b}$$

The general solution of the governing differential equation for the plate is given by Eq. (7.8d)

$$f(y) = A_1 \sinh \alpha y + A_2 \cosh \alpha y + A_3 \sin \beta y + A_4 \cos \beta y \tag{7.8d}$$

Substitute Eqs. (7.8b) and (7.8d) in the boundary conditions at $y = 0$ given by Eqs. (7.12a) and (7.12c) to give

$$A_2 + A_4 = 0 \tag{7.12d}$$

and

$$GJ\frac{m^2\pi^2}{a^2}(A_1\alpha + A_3\beta) = D(A_2\alpha^2 - A_4\beta^2)$$

Let

$$r = GJ\frac{m^2\pi^2}{a^2 D}$$

or

$$A_1\alpha + A_3\beta = \frac{A_2\alpha^2 - A_4\beta^2}{r} \tag{7.12e}$$

Substitute Eqs. (7.8b) and (7.8d) in the boundary conditions at $y = b$ given by Eq. (7.11b) to get

$$A_1\alpha^2 \sinh \alpha b + A_2\alpha^2 \cosh \alpha b - A_3\beta^2 \sin \beta b - A_4\beta^2 \cos \beta b$$

$$-\frac{m^2\pi^2 v}{a^2}(A_1 \sinh \alpha b + A_2 \cosh \alpha b + A_3 \sin \beta b + A_4 \cos \beta b) = 0 \qquad (7.12f)$$

$$A_1\alpha^3 \cosh \alpha b + A_2\alpha^3 \sinh \alpha b - A_3\beta^3 \cos \beta b + A_4\beta^3 \sin \beta b$$

$$-(2 - v)\frac{m^2\pi^2}{a^2}(A_1\alpha \cosh \alpha b + A_2\alpha \sinh \alpha b + A_3\beta \cos \beta b - A_4\beta \sin \beta b) = 0 \qquad (7.12g)$$

Use Eqs. (7.12d) and (7.12e) to eliminate A_3 and A_4 as follows:

$$A_4 = -A_2 \qquad (7.12h)$$

and

$$A_3 = \left[-A_1\alpha + A_2 \left(\frac{\alpha^2 + \beta^2}{r}\right)\right]\frac{1}{\beta} \qquad (7.12i)$$

A_3 and A_4 are substituted from Eqs. (7.12h) and (7.12i) in the Eqs.(7.12f) and (7.12g) and we have

$$A_1 \left[s \sinh \alpha b + \frac{\alpha}{\beta}t \sin \beta b\right] + A_2 \left[s \cosh \alpha b + t \cos \beta b - t \left(\frac{\alpha^2 + \beta^2}{\beta r}\right) \sin \beta b\right] = 0 \qquad (7.12j)$$

$$A_1 [t\alpha \cosh \alpha b + s\alpha \cos \beta b] + A_2 \left[t\alpha \sinh \alpha b - s\beta \sin \beta b - s \left(\frac{\alpha^2 + \beta^2}{r}\right) \cos \beta b\right] = 0 \qquad (7.12k)$$

where s and t are given by Eqs. (7.9m) and (7.9n).

If the torsional rigidity of the flange is very large, then $1/r = 0$, and Eqs. (7.12j) and (7.12k) reduce to those of one end fixed and one end free, Eqs. (7.9k) and (7.9l). Equations (7.12j) and (7.12k) are written as

$$\begin{bmatrix} s \sinh \alpha b + \dfrac{\alpha}{\beta}t \sin \beta b & s \cosh \alpha b + t \cos \beta b - t \left(\dfrac{\alpha^2 + \beta^2}{\beta r}\right) \sin \beta b \\ t\alpha \cosh \alpha b + s\alpha \cos \beta b & t\alpha \sinh \alpha b - s\beta \sin \beta b - s \left(\dfrac{\alpha^2 + \beta^2}{r}\right) \cos \beta b \end{bmatrix} \begin{Bmatrix} A_1 \\ A_2 \end{Bmatrix} = \begin{Bmatrix} 0 \\ 0 \end{Bmatrix} \qquad (7.12l)$$

For the nontrivial solution of A_1 and A_2 the determinant of the coefficient matrix in Eq. (7.12l) should be zero. Hence

$$\begin{vmatrix} s \sinh \alpha b + \dfrac{\alpha}{\beta}t \sin \beta b & s \cosh \alpha b + t \cos \beta b - t \left(\dfrac{\alpha^2 + \beta^2}{\beta r}\right) \sin \beta b \\ t\alpha \cosh \alpha b + s\alpha \cos \beta b & t\alpha \sinh \alpha b - s\beta \sin \beta b - s \left(\dfrac{\alpha^2 + \beta^2}{r}\right) \cos \beta b \end{vmatrix} = 0 \qquad (7.12m)$$

or

$$\frac{2st}{\cos \beta b \cosh \alpha b} + (t^2 + s^2) + \frac{1}{\alpha \beta}(s^2 \beta^2 - t^2 \alpha^2)\tan \beta b \tanh \alpha b$$

$$- \left(\frac{\alpha^2 + \beta^2}{r\alpha}\right)\frac{\alpha}{\beta}\left(t^2 \tan \beta b - s^2\frac{\beta}{\alpha}\tanh \alpha b\right) = 0 \tag{7.12n}$$

or

$$\frac{2}{\cos \beta b \cosh \alpha b} + \left(\frac{t}{s} + \frac{s}{t}\right) + \left(\frac{\beta}{\alpha}\frac{s}{t} - \frac{\alpha}{\beta}\frac{t}{s}\right)\tan \beta b \tanh \alpha b$$

$$- \frac{\alpha}{\beta}\left(\frac{\alpha b}{rb} + \frac{\beta^2 b^2}{(rb)(\alpha b)}\right)\left(\frac{t}{s}\tan \beta b - \frac{s}{t}\frac{\beta}{\alpha}\tanh \alpha b\right) = 0 \tag{7.12o}$$

Find the values of k that satisfy Eq. (7.12o). the critical force N_x can be found from Eq. (7.7n) knowing k. The critical stress is found from Eq. (7.7s) if the material properties and the dimensions of the plate are known. The critical values of k along with m for various a/b ratios are given in Table 7.4. It is seen that with the increase of r, the factor k increases. For $rb = 10$, the values of k approach the values given in the Table 7.1 for the clamped edge.

Table 7.4 Factor k for axially compressed plate when the two loaded sides are simply supported, the side $y = 0$ is elastically built in and the side $y = b$ is free (Figure 7.15).

a/b	k (m)	
	rb = 1	rb = 10
0.2	25.17(1)	25.18(1)
0.4	6.59(1)	6.62(1)
0.6	3.16(1)	3.22(1)
0.8	2.0(1)	2.06(1)
1.0	1.44(1)	1.55(1)
1.2	1.15(1)	1.30(1)
1.4	0.98(1)	1.17(1)
1.6	0.88(1)	1.12(1)
1.8	0.83(1)	1.10(1)
2.0	0.78(1)	1.12(1)
2.2	0.76(1)	1.15(1)
2.4	0.75(1)	1.21(1)
2.6	0.74(1)	1.27(1)
2.8	0.76(1)	1.17(2)
3.0	0.77(1)	1.14(2)
4.0	0.78(2)	1.12(2)
5.0	0.75(2)	1.24(2)

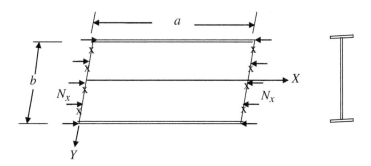

Figure 7.16 Plate with loading edges simply supported, and elastically built in at $y = \pm b/2$.

7.5.7 Loading Edges Simply Supported, the Sides $y = \pm b/2$ Are Elastically Built-in

Consider the buckling of a vertical web partially constrained by the flanges. The plate and the loading are shown in Figure 7.16. The governing equation of plate buckling is

$$\frac{\partial^4 w}{\partial x^4} + 2\frac{\partial^4 w}{\partial x^2 \partial y^2} + \frac{\partial^4 w}{\partial y^4} + \frac{N_x}{D}\frac{\partial^2 w}{\partial x^2} = 0 \tag{7.8a}$$

Assume

$$w = f(y)\sin\frac{m\pi x}{a} \tag{7.8b}$$

Hence,

$$\frac{d^4 f}{dy^4} - \frac{2m^2\pi^2}{a^2}\frac{d^2 f}{dy^2} + \left(\frac{m^4\pi^4}{a^4} - \frac{N_x}{D}\frac{m^2\pi^2}{a^2}\right)f = 0 \tag{7.8c}$$

The general solution of Eq. (7.8c) can be written as

$$f(y) = A_1 \sinh \alpha y + A_2 \cosh \alpha y + A_3 \sin \beta y + A_4 \cos \beta y \tag{7.8d}$$

where

$$\alpha = \sqrt{\frac{m^2\pi^2}{a^2} + \sqrt{\frac{N_x}{D}\frac{m^2\pi^2}{a^2}}}, \quad \beta = \sqrt{-\frac{m^2\pi^2}{a^2} + \sqrt{\frac{N_x}{D}\frac{m^2\pi^2}{a^2}}} \tag{7.8e}$$

The deflection of the plate is symmetrical function of y for the coordinates shown in Figure 7.16. Therefore, $A_1 = A_3 = 0$, and Eq. (7.8d) reduces to

$$f(y) = A_2 \cosh \alpha y + A_4 \cos \beta y \tag{7.13a}$$

$$\frac{df}{dy} = A_2 \alpha \sinh \alpha y - A_4 \beta \sin \beta y \tag{7.13b}$$

$$\frac{d^2 f}{dy^2} = A_2 \alpha^2 \cosh \alpha y - A_4 \beta^2 \cos \beta y \tag{7.13c}$$

Let GJ be the torsional rigidity of the flanges at $y = \pm b/2$.

The boundary conditions at the upper edge, $y = -b/2$ are similar to the plate in Section 7.5.6, that is

$$w = 0 \tag{7.12a}$$

and

$$GJ\frac{m^2\pi^2}{a^2}\frac{\partial w}{\partial y} = D\frac{\partial^2 w}{\partial y^2} \tag{7.12c}$$

Equations (7.12a) and (7.12c) represent the boundary conditions at $y = -b/2$.
Substitute Eqs. (7.8b) and (7.13a) in (7.12a) to obtain at $y = -b/2$

$$A_2 \cosh\frac{\alpha b}{2} + A_4 \cos\frac{\beta b}{2} = 0 \tag{7.13d}$$

From Eq. (7.8b) we have

$$\frac{\partial w}{\partial y} = \frac{df}{dy}\sin\frac{m\pi x}{a}$$

or

$$\frac{\partial w}{\partial y} = (A_2\alpha\sinh\alpha y - A_4\beta\sin\beta y)\sin\frac{m\pi x}{a} \tag{7.13e}$$

$$\frac{\partial^2 w}{\partial y^2} = \frac{d^2 f}{dy^2}\sin\frac{m\pi x}{a}$$

or

$$\frac{\partial^2 w}{\partial y^2} = (A_2\alpha^2\cosh\alpha y - A_4\beta^2\cos\beta y)\sin\frac{m\pi x}{a} \tag{7.13f}$$

Substitute Eqs. (7.13e) and (7.13f) and $y = -b/2$ into Eq. (7.12c) to obtain

$$A_2\left(GJ\frac{m^2\pi^2}{a^2 D}\alpha\sinh\frac{\alpha b}{2} + \alpha^2\cosh\frac{\alpha b}{2}\right) - A_4\left(GJ\frac{m^2\pi^2}{a^2 D}\beta\sin\frac{\beta b}{2} + \beta^2\cos\frac{\beta b}{2}\right) = 0$$

Let

$$r = GJ\frac{m^2\pi^2}{a^2 D}$$

$$A_2\left(\alpha\sinh\frac{\alpha b}{2} + \frac{\alpha^2}{r}\cosh\frac{\alpha b}{2}\right) - A_4\left(\beta\sin\frac{\beta b}{2} + \frac{\beta^2}{r}\cos\frac{\beta b}{2}\right) = 0 \tag{7.13g}$$

The boundary conditions at the lower edge $y = b/2$ are given by

$$w = 0 \tag{7.13h}$$

The angle of rotation of the flange during buckling $= -\left(\dfrac{\partial w}{\partial y}\right)_{y=b/2}$

The rate of change of the slope along the lower edge $= -\left(\dfrac{\partial^2 w}{\partial y\partial x}\right)_{y=b/2}$

Hence, the twisting moment at the lower edge at any section along the x axis $=$
$-GJ\left(\dfrac{\partial^2 w}{\partial x\partial y}\right)_{y=b/2}$

The rate of change of the torsional moment is equal to the bending moment M_y per unit length at the lower edge of the web plate. The signs of the torsional and bending moments are same, hence

$$-GJ\left(\frac{\partial^3 w}{\partial x^2 \partial y}\right)_{y=b/2} = (M_y)_{y=b/2}$$

or

$$-GJ\left(\frac{\partial^3 w}{\partial x^2 \partial y}\right)_{y=b/2} + D\left(\frac{\partial^2 w}{\partial y^2} + v\frac{\partial^2 w}{\partial x^2}\right)_{y=b/2} = 0 \tag{7.13i}$$

Since $w = 0$ at $y = b/2$, therefore, $\dfrac{\partial^2 w}{\partial x^2} = 0$. From Eq. (7.13i) we have

$$GJ\frac{m^2\pi^2}{a^2}\frac{\partial w}{\partial y} = -D\frac{\partial^2 w}{\partial y^2} \tag{7.13j}$$

Equations (7.13h) and (7.13j) give the two boundary conditions at the elastically built in edge at $y = b/2$.

Substitute Eqs. (7.8b) and (7.13a) into (7.13h) to obtain at $y = b/2$.

$$A_2 \cosh\frac{\alpha b}{2} + A_4 \cos\frac{\beta b}{2} = 0 \tag{7.13k}$$

Substitute Eqs. (7.13e) and (7.13f) and $y = b/2$ into Eq. (7.13j) to obtain

$$GJ\frac{m^2\pi^2}{a^2}\left(A_2\alpha \sinh\frac{\alpha b}{2} - A_4\beta \sin\frac{\beta b}{2}\right) = -D\left(A_2\alpha^2 \cosh\frac{\alpha b}{2} - A_4\beta^2 \cos\frac{\beta b}{2}\right)$$

or

$$A_2\left(\alpha \sinh\frac{\alpha b}{2} + \frac{\alpha^2}{r} \cosh\frac{\alpha b}{2}\right) - A_4\left(\beta \sin\frac{\beta b}{2} + \frac{\beta^2}{r} \cos\frac{\beta b}{2}\right) = 0 \tag{7.13l}$$

Equations (7.13d) and (7.13k), and (7.13g) and (7.13l) are identical respectively, hence Eqs. (7.13d) and (7.13g) can be written as

$$\begin{bmatrix} \cosh\dfrac{\alpha b}{2} & \cos\dfrac{\beta b}{2} \\ \alpha \sinh\dfrac{\alpha b}{2} + \dfrac{\alpha^2}{r} \cosh\dfrac{\alpha b}{2} & -\beta \sin\dfrac{\beta b}{2} - \dfrac{\beta^2}{r} \cos\dfrac{\beta b}{2} \end{bmatrix} \begin{Bmatrix} A_2 \\ A_4 \end{Bmatrix} = \begin{Bmatrix} 0 \\ 0 \end{Bmatrix} \tag{7.13m}$$

For a nontrivial solution for A_2 and A_4 the determinant of the coefficient matrix in Eq. (7.13m) should be zero. Hence,

$$\begin{vmatrix} \cosh\dfrac{\alpha b}{2} & \cos\dfrac{\beta b}{2} \\ \alpha \sinh\dfrac{\alpha b}{2} + \dfrac{\alpha^2}{r} \cosh\dfrac{\alpha b}{2} & -\beta \sin\dfrac{\beta b}{2} - \dfrac{\beta^2}{r} \cos\dfrac{\beta b}{2} \end{vmatrix} = 0 \tag{7.13n}$$

The characteristic equation is obtained by expanding the determinant of Eq. (7.13n) as

$$\alpha \tanh\frac{\alpha b}{2} + \beta \tan\frac{\beta b}{2} + \frac{\alpha^2 + \beta^2}{r} = 0 \tag{7.13o}$$

Equation (7.13o) can be solved to obtain the critical value of N_x. The above solution was given by Bryan in 1891 [4].

If the plate is simply supported at $y = \pm b/2$, we have $r = GJ\dfrac{m^2\pi^2}{a^2 D} = 0$, and

$$\alpha \tanh \frac{\alpha b}{2} + \beta \tan \frac{\beta b}{2} = \infty$$

$\tanh \dfrac{\alpha b}{2}$ is a finite value, therefore,

$$\tan \frac{\beta b}{2} = \infty, \text{ or } \quad \frac{\beta b}{2} = \frac{\pi}{2}$$

Substitute the value of β from Eq. (7.8e) to obtain

$$\frac{m\pi b}{2a}\sqrt{-1 + \frac{a}{m\pi}\sqrt{\frac{N_x}{D}}} = \frac{\pi}{2}$$

$$\frac{a}{m\pi}\sqrt{\frac{N_x}{D}} = \frac{a^2}{m^2 b^2} + 1$$

or

$$N_x = \frac{D\pi^2}{b^2}\left(\frac{mb}{a} + \frac{a}{mb}\right)^2$$

This is the same as Eq. (7.7k) for $n = 1$.

7.5.7.1 Loading Edges Simply Supported, the Sides $y = \pm b/2$ Are Elastically Restrained by Rotational Springs

The deflection is a symmetric function of y for the coordinate shown in Figure 7.16. Hence, $A_1 = A_3 = 0$ and

$$f(y) = A_2 \cosh \alpha y + A_4 \cos \beta y \tag{7.14}$$

Let c be the rotational spring constant per unit length. The angles of rotation of the plate during buckling are shown in Figure 7.17.

The boundary conditions are as follows:

$$w = 0, \qquad M_y = c\frac{\partial w}{\partial y}, \quad \text{at } y = -\frac{b}{2}$$

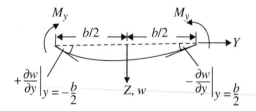

Figure 7.17 Plate with loading edges simply supported, and elastically restrained at $y = \pm b/2$.

and

$$w = 0, \quad M_y = c\left(-\frac{\partial w}{\partial y}\right) \quad \text{at } y = \frac{b}{2}$$

The determinant to find the coefficients A_2 and A_4 can be calculated by the same procedure as in Section 7.5.7, and is as follows:

$$\begin{vmatrix} \cosh\dfrac{\alpha b}{2} & \cos\dfrac{\beta b}{2} \\[2ex] \alpha\sinh\dfrac{\alpha b}{2} + \dfrac{D}{c}\alpha^2\cosh\dfrac{\alpha b}{2} & -\beta\sin\dfrac{\beta b}{2} - \dfrac{D}{c}\beta^2\cos\dfrac{\beta b}{2} \end{vmatrix} = 0$$

7.5.7.2 Loading Edges Simply Supported, the Sides $y = 0$ and $y = b$ Are Elastically Built with Different Flange Sizes

The plate and the loading are shown in Figure 7.18. Let GJ_1 and GJ_2 be the torsional rigidities of the flanges at $y = 0$ and $y = b$ respectively. The boundary conditions at the upper edge $y = 0$ are given by Eqs. (7.12a) and (7.12c), that are similar to the plate in Section 7.5.6

$$w = 0 \tag{7.12a}$$

and

$$GJ_1\frac{m^2\pi^2}{a^2}\frac{\partial w}{\partial y} = D\frac{\partial^2 w}{\partial y^2} \tag{7.12c}$$

At the lower edge, $y = b$, the boundary conditions are given by

$$w = 0 \tag{7.15a}$$

The angle of rotation of the flange during buckling $= -\left(\dfrac{\partial w}{\partial y}\right)_{y=b}$

The rate of change of the slope along the lower edge $= -\left(\dfrac{\partial^2 w}{\partial y\partial x}\right)_{y=b}$

Hence, the twisting moment at the lower edge at any section along the x axis $= -GJ_2\left(\dfrac{\partial^2 w}{\partial x\partial y}\right)_{y=b}$

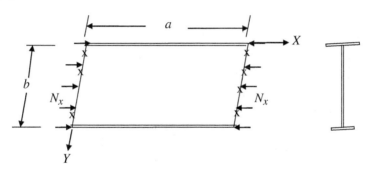

Figure 7.18 Plate with loading edges simply supported, and elastically built in at $y = 0$ and $y = b$.

The rate of change of the torsional moment is equal to the bending moment M_y per unit length at the lower edge of the web plate. The signs of the torsional and bending moments are same, hence,

$$-GJ_2\left(\frac{\partial^3 w}{\partial x^2 \partial y}\right)_{y=b} = (M_y)_{y=b}$$

or

$$-GJ_2\left(\frac{\partial^3 w}{\partial x^2 \partial y}\right)_{y=b} + D\left(\frac{\partial^2 w}{\partial y^2} + v\frac{\partial^2 w}{\partial x^2}\right)_{y=b} = 0 \tag{7.15b}$$

From Eq. (7.8b), we have

$$w = f(y)\sin\frac{m\pi x}{a}$$

$$\frac{\partial^2 w}{\partial x^2} = -f(y)\frac{m^2\pi^2}{a^2}\sin\frac{m\pi x}{a} = -\frac{m^2\pi^2}{a^2}w$$

$$\frac{\partial^3 w}{\partial x^2 \partial y} = -\frac{m^2\pi^2}{a^2}\frac{\partial w}{\partial y}$$

Since $w = 0$ at $y = b$, we have, $\dfrac{\partial^2 w}{\partial x^2} = 0$. Use Eq. (7.15b) to obtain

$$GJ_2\frac{m^2\pi^2}{a^2}\frac{\partial w}{\partial y} = -D\frac{\partial^2 w}{\partial y^2} \tag{7.15c}$$

Equations (7.15a) and (7.15c) give the two boundary conditions at the elastically built-in edge at $y = b$. The general solution of the governing differential equation for the plate is given by

$$f(y) = A_1\sinh\alpha y + A_2\cosh\alpha y + A_3\sin\beta y + A_4\cos\beta y \tag{7.8d}$$

Substitute Eqs. (7.8b) and (7.8d) in the boundary conditions at $y = 0$ given by Eqs. (7.12a) and (7.12c) to obtain

$$A_2 + A_4 = 0 \tag{7.15d}$$

and

$$GJ_1\frac{m^2\pi^2}{a^2}(A_1\alpha + A_3\beta) = D(A_2\alpha^2 - A_4\beta^2)$$

Let

$$r_1 = GJ_1\frac{m^2\pi^2}{a^2D}$$

or

$$A_1\alpha + A_3\beta = \frac{A_2\alpha^2 - A_4\beta^2}{r_1}$$

or

$$A_3 = -A_1\frac{\alpha}{\beta} + \frac{A_2}{r_1\beta}(\alpha^2 + \beta^2) \tag{7.15e}$$

Substitute Eqs. (7.8b) and (7.8d) in the boundary conditions at $y = b$ given by Eqs. (7.15a) and (7.15c) to get

$$A_1 \sinh \alpha b + A_2 \cosh \alpha b + A_3 \sin \beta b + A_4 \cos \beta b = 0 \qquad (7.15f)$$

$$GJ_2 \frac{m^2 \pi^2}{a^2}(A_1 \alpha \cosh \alpha b + A_2 \alpha \sinh \alpha b + A_3 \beta \cos \beta b - A_4 \beta \sin \beta b)$$
$$= -D(A_1 \alpha^2 \sinh \alpha b + A_2 \alpha^2 \cosh \alpha b - A_3 \beta^2 \sin \beta b - A_4 \beta^2 \cos \beta b) \qquad (7.15g)$$

Let

$$r_2 = GJ_2 \frac{m^2 \pi^2}{a^2 D}$$

Use Eq. (7.15d) and (7.15e) to eliminate A_3 and A_4 from Eqs. (7.15f) and (7.15g) to have

$$A_1 \left[\sinh \alpha b - \frac{\alpha}{\beta} \sin \beta b \right] + A_2 \left[\cosh \alpha b + \frac{1}{r_1 \beta}(\alpha^2 + \beta^2) \sin \beta b - \cos \beta b \right] = 0 \qquad (7.15h)$$

$$A_1 \left[\alpha \cosh \alpha b - \alpha \cos \beta b + \frac{1}{r_2}(\alpha^2 \sinh \alpha b + \alpha\beta \sin \beta b) \right]$$
$$+ A_2 \left[\begin{array}{l} \alpha \sinh \alpha b + \frac{1}{r_1}(\alpha^2 + \beta^2) \cos \beta b + \beta \sin \beta b + \frac{1}{r_2}\alpha^2 \cosh \alpha b \\ -\frac{1}{r_1 r_2}\beta(\alpha^2 + \beta^2) \sin \beta b + \frac{1}{r_2}\beta^2 \cos \beta b \end{array} \right] = 0 \qquad (7.15i)$$

or

$$\begin{bmatrix} \sinh \alpha b - \frac{\alpha}{\beta}\sin\beta b & \cosh\alpha b + \frac{1}{r_1\beta}(\alpha^2+\beta^2)\sin\beta b - \cos\beta b \\[2mm] \left\{ \begin{array}{l} \alpha\cosh\alpha b - \alpha\cos\beta b \\[1mm] +\frac{1}{r_2}(\alpha^2\sinh\alpha b + \alpha\beta\sin\beta b) \end{array}\right\} & \left\{ \begin{array}{l} \alpha\sinh\alpha b + \frac{1}{r_1}(\alpha^2+\beta^2)\cos\beta b \\[1mm] +\beta\sin\beta b + \frac{1}{r_2}\alpha^2\cosh\alpha b \\[1mm] -\frac{1}{r_1 r_2}\beta(\alpha^2+\beta^2)\sin\beta b + \frac{1}{r_2}\beta^2\cos\beta b \end{array}\right\} \end{bmatrix} \begin{Bmatrix} A_1 \\ A_2 \end{Bmatrix} = \begin{Bmatrix} 0 \\ 0 \end{Bmatrix}$$

$$(7.15j)$$

For a nontrivial solution for A_1 and A_2 the determinant of the coefficient matrix in Eq. (7.15j) should be zero. Hence,

$$\begin{vmatrix} \sinh\alpha b - \frac{\alpha}{\beta}\sin\beta b & \cosh\alpha b + \frac{1}{r_1\beta}(\alpha^2+\beta^2)\sin\beta b - \cos\beta b \\[2mm] \left\{ \begin{array}{l} \alpha\cosh\alpha b - \alpha\cos\beta b \\[1mm] +\frac{1}{r_2}(\alpha^2\sinh\alpha b + \alpha\beta\sin\beta b) \end{array}\right\} & \left\{ \begin{array}{l} \alpha\sinh\alpha b + \frac{1}{r_1}(\alpha^2+\beta^2)\cos\beta b \\[1mm] +\beta\sin\beta b + \frac{1}{r_2}\alpha^2\cosh\alpha b \\[1mm] -\frac{1}{r_1 r_2}\beta(\alpha^2+\beta^2)\sin\beta b + \frac{1}{r_2}\beta^2\cos\beta b \end{array}\right\} \end{vmatrix} = 0 \qquad (7.15k)$$

The characteristic equation is obtained by expanding the determinant of Eq. (7.15k) and solved to obtain the critical value of N_x.

7.5.8 Loading Edges Simply Supported, the Sides $y = 0$ and $y = b$ Are Supported by Elastic Beams

The plate is simply supported at the edges $x = 0$ and $x = a$, and the plate is supported by two equal elastic beams along the edges $y = \pm b/2$ as shown in Figure 7.19. The plate is free to rotate along the edges $y = \pm b/2$ during buckling, but the deflections of the plate at these edges are resisted by the elastic beams. Since the plate is free to rotate, the boundary condition is

$$M_y = -D\left(\frac{\partial^2 w}{\partial y^2} + v\frac{\partial^2 w}{\partial x^2}\right) = 0$$

or

$$\left(\frac{\partial^2 w}{\partial y^2} + v\frac{\partial^2 w}{\partial x^2}\right) = 0 \text{ at } y = \pm b/2 \tag{7.16a}$$

To obtain the second boundary condition we consider the beams are simply supported at the ends, and the modulus of elasticity of the beams and plate materials is the same. The differential equation for the beam deflection is given by Eq. (4.1d) as

$$EI\frac{\partial^4 w}{\partial x^4} + P\frac{\partial^2 w}{\partial x^2} = q \tag{7.16b}$$

where P is the axial compressive force on the beam, and q is the intensity of the lateral load transmitted from the plate to the beam. At

$$y = \frac{b}{2}$$

from Figure 7.3 and Eq. (7.5b) we have

$$q = -\left(Q_y + \frac{\partial M_{yx}}{\partial x}\right) \tag{7.16c}$$

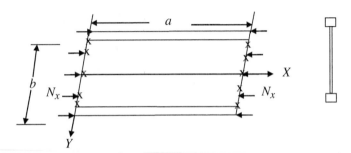

Figure 7.19 Plate with loading edges simply supported, and supported by beams at $y = 0$ and $y = b$.

Substitute Eqs. (7.1t) and (7.1z) into Eq. (7.16c) to obtain

$$q = D\left[\frac{\partial^3 w}{\partial y^3} + (2 - v)\frac{\partial^3 w}{\partial x^2 \partial y}\right] \qquad (7.16d)$$

At

$$y = -\frac{b}{2}$$

$$q = Q_y + \frac{\partial M_{yx}}{\partial x} \qquad (7.16e)$$

Substitute Eqs. (7.1t) and (7.1z) into Eq. (7.16e) to obtain

$$q = -D\left[\frac{\partial^3 w}{\partial y^3} + (2 - v)\frac{\partial^3 w}{\partial x^2 \partial y}\right] \qquad (7.16f)$$

Substitute the q values in Eqs. (7.16d) and (7.16f) in the beam deflection Eq. (7.16b) to have the following boundary conditions:

$$EI\frac{\partial^4 w}{\partial x^4} = D\left[\frac{\partial^3 w}{\partial y^3} + (2 - v)\frac{\partial^3 w}{\partial x^2 \partial y}\right] - P\frac{\partial^2 w}{\partial x^2} \quad \text{at } y = \frac{b}{2} \qquad (7.16g)$$

$$EI\frac{\partial^4 w}{\partial x^4} = -D\left[\frac{\partial^3 w}{\partial y^3} + (2 - v)\frac{\partial^3 w}{\partial x^2 \partial y}\right] - P\frac{\partial^2 w}{\partial x^2} \quad \text{at } y = -\frac{b}{2} \qquad (7.16h)$$

The deflection of the plate is given by

$$w = f(y)\sin\frac{m\pi x}{a} \qquad (7.8b)$$

The deflection of the plate is a symmetrical function of y for the coordinates shown in Figure 7.19. Hence, $A_1 = A_3 = 0$, and, as before, Eq. (7.8d) reduces to

$$f(y) = A_2 \cosh \alpha y + A_4 \cos \beta y \qquad (7.14a)$$

Substitute derivatives of w and $f(y)$ in Eq. (7.16a) to get $y = \pm\frac{b}{2}$

$$A_2\left(\alpha^2 - \frac{m^2\pi^2 v}{a^2}\right)\cosh\frac{\alpha b}{2} - A_4\left(\beta^2 + \frac{m^2\pi^2 v}{a^2}\right)\cos\frac{\beta b}{2} = 0 \qquad (7.16i)$$

Similarly, substitute derivatives of w and $f(y)$ in Eqs. (7.16g) and (7.16h) that will yield the same expression for $y = \pm\frac{b}{2}$ as follows:

$$EI\frac{m^4\pi^4}{a^4}\left[A_2\cosh\frac{\alpha b}{2} + A_4\cos\frac{\beta b}{2}\right]\sin\frac{m\pi x}{a}$$

$$= D\left[A_2\alpha^3\sinh\frac{\alpha b}{2} + A_4\beta^3\sin\frac{\beta b}{2} + (2 - v)\left(-\frac{m^2\pi^2}{a^2}\right)\right.$$

$$\times\left(A_2\alpha\sinh\frac{\alpha b}{2} - A_4\beta\sin\frac{\beta b}{2}\right)\right]\sin\frac{m\pi x}{a}$$

$$+ P\left(\frac{m^2\pi^2}{a^2}\right)\left[A_2\cosh\frac{\alpha b}{2} + A_4\cos\frac{\beta b}{2}\right]\sin\frac{m\pi x}{a} \qquad (7.16j)$$

or

$$
\begin{bmatrix}
\left(\alpha^2 - \dfrac{m^2\pi^2 v}{a^2}\right)\cosh\dfrac{\alpha b}{2} & -\left(\beta^2 + \dfrac{m^2\pi^2 v}{a^2}\right)\cos\dfrac{\beta b}{2} \\[2em]
\begin{aligned}
& EI\dfrac{m^4\pi^4}{a^4}\cosh\dfrac{\alpha b}{2} - D\alpha^3\sin h\dfrac{\alpha b}{2} \\
& +D(2-v)\dfrac{m^2\pi^2}{a^2}\alpha\sinh\dfrac{\alpha b}{2} \\
& -P\dfrac{m^2\pi^2}{a^2}\cosh\dfrac{\alpha b}{2}
\end{aligned}
&
\begin{aligned}
& EI\dfrac{m^4\pi^4}{a^4}\cos\dfrac{\beta b}{2} - D\beta^3\sin\dfrac{\beta b}{2} \\
& -D(2-v)\dfrac{m^2\pi^2}{a^2}\beta\sin\dfrac{\beta b}{2} \\
& -P\dfrac{m^2\pi^2}{a^2}\cos\dfrac{\beta b}{2}
\end{aligned}
\end{bmatrix}
\begin{Bmatrix} A_2 \\ A_4 \end{Bmatrix}
=
\begin{Bmatrix} 0 \\ 0 \end{Bmatrix}
$$

(7.16k)

The determinant of the coefficient matrix will be zero to get nontrivial solution for A_2 and A_4. Hence,

$$
\begin{vmatrix}
\left(\alpha^2 - \dfrac{m^2\pi^2 v}{a^2}\right)\cosh\dfrac{\alpha b}{2} & -\left(\beta^2 + \dfrac{m^2\pi^2 v}{a^2}\right)\cos\dfrac{\beta b}{2} \\[2em]
\begin{aligned}
& EI\dfrac{m^4\pi^4}{a^4}\cosh\dfrac{\alpha b}{2} - D\alpha^3\sinh\dfrac{\alpha b}{2} \\
& +D(2-v)\dfrac{m^2\pi^2}{a^2}\alpha\sinh\dfrac{\alpha b}{2} \\
& -P\dfrac{m^2\pi^2}{a^2}\cosh\dfrac{\alpha b}{2}
\end{aligned}
&
\begin{aligned}
& EI\dfrac{m^4\pi^4}{a^4}\cos\dfrac{\beta b}{2} - D\beta^3\sin\dfrac{\beta b}{2} \\
& -D(2-v)\dfrac{m^2\pi^2}{a^2}\beta\sin\dfrac{\beta b}{2} \\
& -P\dfrac{m^2\pi^2}{a^2}\cos\dfrac{\beta b}{2}
\end{aligned}
\end{vmatrix}
= 0
$$

(7.16l)

or

$$
\left(\alpha^2 - \frac{m^2\pi^2 v}{a^2}\right)\cosh\frac{\alpha b}{2}\left[EI\frac{m^4\pi^4}{a^4}\cos\frac{\beta b}{2} - D\beta^3\sin\frac{\beta b}{2}\right.
$$

$$
\left. -D(2-v)\frac{m^2\pi^2}{a^2}\beta\sin\frac{\beta b}{2} - P\frac{m^2\pi^2}{a^2}\cos\frac{\beta b}{2}\right]
$$

$$
+\left(\beta^2 + \frac{m^2\pi^2 v}{a^2}\right)\cos\frac{\beta b}{2}
$$

$$
\times\left[
\begin{aligned}
& EI\frac{m^4\pi^4}{a^4}\cosh\frac{\alpha b}{2} - D\alpha^3\sin h\frac{\alpha b}{2} + D(2-v)\frac{m^2\pi^2}{a^2}\alpha\sinh\frac{\alpha b}{2} \\
& -P\frac{m^2\pi^2}{a^2}\cosh\frac{\alpha b}{2}
\end{aligned}
\right] = 0
$$

or

$$\beta \tan \frac{\beta b}{2} D \left[-\alpha^2 \beta^2 \frac{a^2}{m^2 \pi^2} - 2\alpha^2 + v(\alpha^2 + \beta^2) + (2-v)\frac{vm^2\pi^2}{a^2} \right] \frac{m^2\pi^2}{a^2}$$

$$+ \alpha \tanh \frac{\alpha b}{2} D \left[-\alpha^2 \beta^2 \frac{a^2}{m^2 \pi^2} + 2\beta^2 - v(\alpha^2 + \beta^2) + (2-v)\frac{vm^2\pi^2}{a^2} \right] \frac{m^2\pi^2}{a^2}$$

$$= \frac{m^2\pi^2}{a^2}(\alpha^2 + \beta^2)\left(P - \frac{m^2\pi^2 EI}{a^2} \right) \tag{7.16m}$$

From Eq. (7.8e) we have

$$\alpha^2 + \beta^2 = \frac{2m\pi}{a}\sqrt{\frac{N_x}{D}} \quad \text{and} \quad \alpha^2\beta^2 = \frac{m^2\pi^2}{a^2}\frac{N_x}{D} - \frac{m^4\pi^4}{a^4}$$

Substituting in Eq. (7.16m), we get

$$\beta \tan \frac{\beta b}{2} \frac{m^2\pi^2}{a^2} \left[-\frac{N_x}{D}\frac{a^2}{m^2\pi^2} - 1 - \frac{2a}{m\pi}\sqrt{\frac{N_x}{D}}(1-v) + v(2-v) \right]$$

$$+ \alpha \tanh \frac{\alpha b}{2} \frac{m^2\pi^2}{a^2} \left[-\frac{N_x}{D}\frac{a^2}{m^2\pi^2} - 1 + \frac{2a}{m\pi}\sqrt{\frac{N_x}{D}}(1-v) + v(2-v) \right]$$

$$= \frac{2m\pi}{a}\sqrt{\frac{N_x}{D}}\left(\frac{P}{D} - \frac{m^2\pi^2}{a^2}\frac{EI}{D} \right) \tag{7.16n}$$

or

$$\beta\left(1 - v + \frac{a}{m\pi}\sqrt{\frac{N_x}{D}} \right)^2 \tan \frac{\beta b}{2} + \alpha\left(1 - v - \frac{a}{m\pi}\sqrt{\frac{N_x}{D}} \right)^2 \tanh \frac{\alpha b}{2}$$

$$= \frac{2m\pi}{a}\sqrt{\frac{N_x}{D}}\left(\frac{EI}{D} - \frac{a^2}{m^2\pi^2}\frac{P}{D} \right) \tag{7.16o}$$

From Eq. (7.7n)

$$N_x = \frac{kD\pi^2}{b^2}$$

$$\frac{P}{D} = \frac{AN_x}{hD} = \frac{A}{h}\frac{k\pi^2}{b^2}$$

where h is the thickness of the plate, b is the width of the plate, and A is the area of cross-section of one beam.

Substituting for N_x and P/D in Eq. (7.16o), we have

$$\beta b\left(1 - v + \frac{a}{m\pi}\sqrt{k}\frac{\pi}{b} \right)^2 \tan \frac{\beta b}{2} + \alpha b\left(1 - v - \frac{a}{m\pi}\sqrt{k}\frac{\pi}{b} \right)^2 \tanh \frac{\alpha b}{2}$$

$$= \frac{2m\pi^2}{\frac{a}{b}}\sqrt{k}\left(\frac{EI}{bD} - \frac{a^2 k}{m^2 b^2}\frac{A}{bh} \right) \tag{7.16p}$$

where EI = flexural rigidity of one beam, bD = flexural rigidity of the plate, and bh = area of cross-section of the plate. αb and βb are given by

$$\alpha b = \frac{m\pi}{a/b}\sqrt{1 + \frac{a}{mb}\sqrt{k}} \quad \text{and} \quad \beta b = \frac{m\pi}{a/b}\sqrt{-1 + \frac{a}{mb}\sqrt{k}}$$

The procedure for calculation is as follows:

1. Calculate $\dfrac{EI}{bD}$ and $\dfrac{A}{bh}$ using beam and plate dimensions.
2. Assume m, the number of half-waves.
3. For a particular value of $\frac{a}{b}$ ratio, find the k that satisfies the transcendental Eq. (7.16p). The critical stress σ_{cr} for the plate is given by

$$\sigma_{cr} = \frac{k\pi^2 E}{12(1 - v^2)} \frac{1}{(b/h)^2} \tag{7.7s}$$

7.6 The Energy Method

The principle of stationary potential energy can also be used in the case of plates as it was used in Section 2.11 for columns. This method is useful where the solution of the differential Eq. (7.2j) is not known. For plates with stiffeners, non-uniform plates where the plate thickness varies with x and y but is symmetrical with respect to the z axis, and the plates are subjected to varying edge forces, the energy method can be used to approximate a solution of the critical load. We assume the plate is stressed by forces acting in the middle of the plate and undergoes small bending consistent with the boundary conditions. It is assumed the middle plane of the plate does not stretch, hence the strain energy due to bending is only considered. For thin plates where the thickness h is less than 1/10 of plate lateral dimensions, it is assumed that the stresses σ_z, τ_{xz}, and τ_{yz} are negligible. Thus, the plate is in-plane stress containing internal stresses of σ_x, σy, and τ_{xy}.

7.6.1 Strain Energy Due to Bending in Plates

The strain energy stored in the plate is given by

$$U = \frac{1}{2}\int_{-\frac{h}{2}}^{\frac{h}{2}}\int_0^b \int_0^a (\sigma_x \, \varepsilon_x + \sigma_y \, \varepsilon_y + \tau_{xy} \, \gamma_{xy}) \, dx \, dy \, dz \tag{7.17a}$$

For the plane stress, the constitutive equations are

$$\sigma_x = \frac{E}{1 - v^2}(\varepsilon_x + v\varepsilon_y) \tag{7.1b}$$

$$\sigma_y = \frac{E}{1 - v^2}(\varepsilon_y + v\varepsilon_x) \tag{7.1c}$$

$$\tau_{xy} = \frac{E}{2(1 + v)}\gamma_{xy} \tag{7.1d}$$

Substitute Eqs. (7.1j)–(7.1o) into Eq. (7.17a) to obtain

$$U = \frac{1}{2} \int_{-\frac{h}{2}}^{\frac{h}{2}} \int_0^b \int_0^a \left[\frac{Ez}{1-v^2} \left(\frac{\partial^2 w}{\partial x^2} + v \frac{\partial^2 w}{\partial y^2} \right) \left(z \frac{\partial^2 w}{\partial x^2} \right) + \frac{Ez}{1-v^2} \left(\frac{\partial^2 w}{\partial y^2} + v \frac{\partial^2 w}{\partial x^2} \right) \left(z \frac{\partial^2 w}{\partial y^2} \right) \right.$$
$$\left. + \frac{Ez}{1+v} \frac{\partial^2 w}{\partial x \partial y} \left(2z \frac{\partial^2 w}{\partial x \partial y} \right) \right] dx\,dy\,dz \tag{7.17b}$$

$$U = \frac{D}{2} \int_0^b \int_0^a \left[\left(\frac{\partial^2 w}{\partial x^2} \right)^2 + \left(\frac{\partial^2 w}{\partial y^2} \right)^2 + 2v \left(\frac{\partial^2 w}{\partial x^2} \right) \left(\frac{\partial^2 w}{\partial y^2} \right) + 2(1-v) \left(\frac{\partial^2 w}{\partial x \partial y} \right)^2 \right] dx\,dy \tag{7.17c}$$

Equation (7.17c) is the equation of strain energy due to bending only in a thin plate.

7.6.2 Potential Energy of the External Forces in Plates

7.6.2.1 Potential Energy Due to N_x and N_y

Consider a strip of width dy in Figure 7.20. The load acting on this strip is $N_x dy$. The potential energy for any strip of width dy is given by

$$dv_1 = -N_x dy (\Delta L) \tag{7.18a}$$

where ΔL is the change in length of the strip due to bending. The change in length of the strip can be obtained from Eq. (2.28e) as

$$\Delta L = \frac{1}{2} \int_0^a \left(\frac{\partial w}{\partial x} \right)^2 dx$$
$$dv_1 = -(N_x dy) \left[\frac{1}{2} \int_0^a \left(\frac{\partial w}{\partial x} \right)^2 dx \right] \tag{7.18b}$$

There is no factor of $1/2$ present in calculating the external work because $N_x\,dy$ remains constant during the shortening of this strip due to bending. The total potential energy of the entire plate

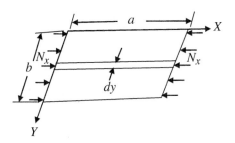

Figure 7.20 Plate subjected to in-plane external force.

is obtained by adding the potential energies of all such strips or by integrating with respect to y. Hence,

$$V_1 = -\frac{1}{2} \int_0^a \int_0^b N_x \left(\frac{\partial w}{\partial x} \right)^2 dx dy \tag{7.18c}$$

Similarly considering a strip of width dx, we can write the potential energy of the plate due to N_y as

$$V_2 = -\frac{1}{2} \int_0^a \int_0^b N_y \left(\frac{\partial w}{\partial y} \right)^2 dx dy \tag{7.18d}$$

7.6.2.2 Potential Energy Due to N_{xy}

To find the shear strain due to displacement w, consider two small elements OA and OB of lengths dx and dy respectively in the directions of axes X and Y shown in Figure 7.21. The elements move to $O'A'$ and $O'B'$ due to displacement w. The difference between the angles $AOB \left(\frac{\pi}{2} \right)$ and $A'O'B'$ is the shear strain, γ, due to w. Hence, angle $A'O'B' = \frac{\pi}{2} - \gamma$. From the triangle $A'O'B'$

$$(A'B')^2 = (O'A')^2 + (O'B')^2 - 2(O'A')(O'B') \cos\left(\frac{\pi}{2} - \gamma \right) \tag{7.19a}$$

$$\cos\left(\frac{\pi}{2} - \gamma \right) = \sin\gamma = \gamma \text{ for small values of } \gamma.$$

$$(O'A')^2 = dx^2 + \left(\frac{\partial w}{\partial x} dx \right)^2$$

$$(O'B')^2 = dy^2 + \left(\frac{\partial w}{\partial y} dy \right)^2$$

$$(A'B')^2 = dx^2 + \left(\frac{\partial w}{\partial x} dx \right)^2 + dy^2 + \left(\frac{\partial w}{\partial y} dy \right)^2 - 2(dx)(dy)\gamma \tag{7.19b}$$

Also

$$(A'B')^2 = dx^2 + dy^2 + \left(\frac{\partial w}{\partial y} dy - \frac{\partial w}{\partial x} dx \right)^2 \tag{7.19c}$$

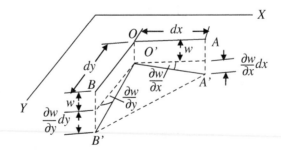

Figure 7.21 Shear strain corresponding to deflection w.

Equations (7.19b) and (7.19c) give

$$\gamma = \frac{\partial w}{\partial x}\frac{\partial w}{\partial y}$$

Hence, the potential energy of the plate due to N_{xy} is given by

$$V_3 = -\frac{1}{2}\int_0^b\int_0^a 2N_{xy}\left(\frac{\partial w}{\partial x}\right)\left(\frac{\partial w}{\partial y}\right)dxdy \tag{7.19d}$$

The factor 2 is to account for the contribution of N_{yx}.

7.6.3 Rectangular Plate Subjected to Uniaxial Compressive Force and Fixed on All Edges

The plate and the loading are shown in Figure 7.22 and the boundary conditions are

$$w = \frac{\partial w}{\partial x} = 0 \ \text{ at } x = 0, a \tag{7.20a}$$

$$w = \frac{\partial w}{\partial y} = 0 \ \text{ at } y = 0, b \tag{7.20b}$$

The edges of the plate are free to move in the XY plane. The boundary conditions are satisfied if w is assumed as

$$w = A\left(1 - \cos\frac{2\pi x}{a}\right)\left(1 - \cos\frac{2\pi y}{b}\right)$$

$$\frac{\partial w}{\partial x} = A\left(\frac{2\pi}{a}\right)\left(\sin\frac{2\pi x}{a}\right)\left(1 - \cos\frac{2\pi y}{b}\right)$$

$$\frac{\partial^2 w}{\partial x^2} = A\frac{4\pi^2}{a^2}\cos\frac{2\pi x}{a}\left(1 - \cos\frac{2\pi y}{b}\right)$$

$$\frac{\partial w}{\partial y} = A\left(\frac{2\pi}{b}\right)\left(1 - \cos\frac{2\pi x}{a}\right)\sin\frac{2\pi y}{b}$$

$$\frac{\partial^2 w}{\partial y^2} = A\left(\frac{4\pi^2}{b^2}\right)\left(1 - \cos\frac{2\pi x}{a}\right)\cos\frac{2\pi y}{b}$$

$$\frac{\partial^2 w}{\partial x\partial y} = A\left(\frac{4\pi^2}{ab}\right)\sin\frac{2\pi x}{a}\sin\frac{2\pi y}{b} \tag{7.20c}$$

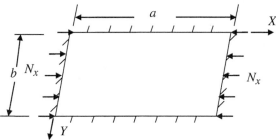

Figure 7.22 Fixed rectangular plate under compressive forces.

From Eq. (7.17c), we have

$$U = \frac{D}{2} \int_0^b \int_0^a \left[\left(\frac{\partial^2 w}{\partial x^2} \right)^2 + \left(\frac{\partial^2 w}{\partial y^2} \right)^2 + 2v \left(\frac{\partial^2 w}{\partial x^2} \right) \left(\frac{\partial^2 w}{\partial y^2} \right) + 2(1-v) \left(\frac{\partial^2 w}{\partial x \partial y} \right)^2 \right] dxdy$$

or

$$U = \left(\frac{D}{2} \right) \left(\frac{16\pi^4 A^2}{a^2 b^2} \right) \left[\int_0^b \int_0^a \frac{b^2}{a^2} \cos^2 \frac{2\pi x}{a} \left(1 + \cos^2 \frac{2\pi y}{b} - 2 \cos \frac{2\pi y}{b} \right) \right.$$

$$+ \frac{a^2}{b^2} \cos^2 \frac{2\pi y}{b} \left(1 + \cos^2 \frac{2\pi x}{a} - 2 \cos \frac{2\pi x}{a} \right)$$

$$+ 2v \left(\cos \frac{2\pi x}{a} - \cos^2 \frac{2\pi x}{a} \right) \left(\cos \frac{2\pi y}{b} - \cos^2 \frac{2\pi y}{b} \right)$$

$$\left. + 2(1-v) \left(\sin^2 \frac{2\pi x}{a} \right) \left(\sin^2 \frac{2\pi y}{b} \right) \right] dxdy \tag{7.20d}$$

We have the following definite integrals

$$\int_0^a \sin^2 \frac{2\pi x}{a} dx = \frac{a}{2}, \int_0^b \sin^2 \frac{2\pi y}{b} dy = \frac{b}{2} \tag{7.20e}$$

$$\int_0^a \cos^2 \frac{2\pi x}{a} dx = \frac{a}{2}, \int_0^b \cos^2 \frac{2\pi y}{b} dy = \frac{b}{2} \tag{7.20f}$$

$$\int_0^a \cos^2 \frac{2\pi x}{a} dx = 0, \int_0^b \cos^2 \frac{2\pi y}{b} dy = 0 \tag{7.20g}$$

Use the definite integrals in Eqs. (7.20e)–(7.20g) to obtain

$$U = \frac{2D\pi^4 A^2}{a^2 \beta} (3 + 2\beta^2 + 3\beta^4) \tag{7.20h}$$

where the aspect ratio $\frac{a}{b}$ of the plate $= \beta$.

The potential energy of the external force N_x is given by Eq. (7.18c) as

$$V = -\frac{1}{2} \int_0^b \int_0^a \left[N_x \left(\frac{\partial w}{\partial x} \right)^2 \right] dxdy \tag{7.18c}$$

or

$$V = -\frac{1}{2} \int_0^b \int_0^a N_x \left[A^2 \frac{4\pi^2}{a^2} \left(\sin^2 \frac{2\pi x}{a} \right) \left(1 + \cos^2 \frac{2\pi y}{b} - 2 \cos \frac{2\pi y}{b} \right) \right] dxdy \tag{7.20i}$$

Use the definite integrals given by Eqs. (7.20e)–(7.20g), to get

$$V = -\frac{3N_x \pi^2 A^2 b}{2a} \tag{7.20j}$$

The total potential energy of the plate is obtained from Eqs. (7.20h) and (7.20i) as follows:

$$\Pi = U + V = \frac{2D\pi^4 A^2}{a^2 \beta}(3 + 2\beta^2 + 3\beta^4) - \frac{3}{2}\frac{N_x \pi^2 A^2}{\beta} \tag{7.20k}$$

Using the principle of stationary potential energy and the Rayleigh-Ritz method, take the variation of Π with respect to A and equate it to zero. Thus, we get

$$\frac{d\Pi}{dA} = \frac{4D\pi^4 A}{a^2 \beta}(3 + 2\beta^2 + 3\beta^4) - \frac{3N_x \pi^2 A}{\beta} = 0$$

or

$$(N_x)_{cr} = \frac{4D\pi^2 a^2}{3}\left(\frac{3}{a^4} + \frac{3}{b^4} + \frac{2}{a^2 b^2}\right) \tag{7.20l}$$

For a square plate, $\dfrac{a}{b} = \beta = 1$, and from Eq. (7.20l) we have

$$(N_x)_{cr} = \frac{32D\pi^2}{3a^2} = 10.67\frac{D\pi^2}{a^2} \tag{7.20m}$$

Levy [5] obtained an exact solution of this problem by using infinite series for w as

$$(N_x)_{cr} = 10.07\frac{D\pi^2}{a^2} \tag{7.20n}$$

Error in the one-term solution of Eq. (7.20m) is equal to $[(10.67-10.07)/10.07] \times 100 = 5.96\%$. The energy method gives an upper bound to the exact value.

7.6.4 A Rectangular Plate with Clamped Edges under Compressive Pressure in Two Perpendicular Directions

The plate and the loading are shown in Figure 7.23 and the boundary conditions are

$$w = \frac{\partial w}{\partial x} = 0 \text{ at } x = 0, a \tag{7.20a}$$

$$w = \frac{\partial w}{\partial y} = 0 \text{ at } y = 0, b \tag{7.20b}$$

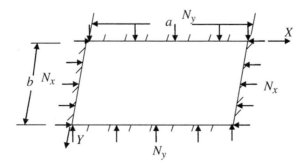

Figure 7.23   Fixed rectangular plate under compressive forces in two perpendicular directions.

The edges of the plate are free to move in the XY plane. The boundary conditions are satisfied if w is assumed as

$$w = A\left(1 - \cos\frac{2\pi x}{a}\right)\left(1 - \cos\frac{2\pi y}{b}\right) \tag{7.20c}$$

The strain energy for the plate is the same as in the case in Section 7.6.3 and is as follows:

$$U = \frac{2D\pi^4 A^2}{a^2\beta}(3 + 2\beta^2 + 3\beta^4) \tag{7.20h}$$

From Eqs. (7.18c) and (7.18d) the work done by the compressive forces during the buckling of the plate is given by

$$V = -\frac{1}{2}N_x\int_0^b\int_0^a\left(\frac{\partial w}{\partial x}\right)^2 dx\,dy - \frac{1}{2}N_y\int_0^b\int_0^a\left(\frac{\partial w}{\partial y}\right)^2 dx\,dy \tag{7.21a}$$

or

$$V = -\frac{1}{2}\int_0^b\int_0^a N_x\left[A^2\frac{4\pi^2}{a^2}\left(\sin^2\frac{2\pi x}{a}\right)\left(1 + \cos^2\frac{2\pi y}{b} - 2\cos\frac{2\pi y}{b}\right)\right]dx\,dy$$
$$-\frac{1}{2}\int_0^b\int_0^a N_y\left[A^2\frac{4\pi^2}{b^2}\left(\sin^2\frac{2\pi y}{b}\right)\left(1 + \cos^2\frac{2\pi x}{a} - 2\cos\frac{2\pi x}{a}\right)\right]dx\,dy \tag{7.21b}$$

or

$$V = -\frac{3\pi^2 A^2 b}{2a}\left(N_x + \frac{a^2}{b^2}N_y\right) \tag{7.21c}$$

The total potential energy Π of the plate is obtained by adding the strain and potential energies from Eqs. (7.20h) and (7.21c) as follows:

$$\Pi = U + V = \frac{2D\pi^4 A^2}{a^2\beta}(3 + 2\beta^2 + 3\beta^4) - \frac{3\pi^2 A^2}{2\beta}(N_x + \beta^2 N_y) \tag{7.21d}$$

where the aspect ratio $\dfrac{a}{b}$ of the plate $= \beta$.

Using the principle of stationary potential energy and the Rayleigh-Ritz method, take the variation of Π with respect to A and equate it to zero. Thus, we get

$$\frac{d\Pi}{dA} = \frac{4D\pi^4 A}{a^2\beta}(3 + 2\beta^2 + 3\beta^4) - \frac{3\pi^2 A}{\beta}(N_x + \beta^2 N_y) = 0$$

or

$$\left(N_x + \frac{a^2}{b^2}N_y\right)_{cr} = \frac{4D\pi^2 a^2}{3}\left(\frac{3}{a^4} + \frac{3}{b^4} + \frac{2}{a^2 b^2}\right) \tag{7.21e}$$

For a square plate, $a/b = 1$, thus

$$(N_x + N_y)_{cr} = \frac{32D\pi^2}{3a^2} \tag{7.21f}$$

If $N_y = 0$, we get the same result as given by Eq. (7.20m) in Section 7.6.3. If $N_x = N_y = N_{cr}$, we get

$$N_{cr} = \frac{16D\pi^2}{a^2} = 5.34\frac{D\pi^2}{a^2} \tag{7.21g}$$

7.6.5 Buckling of Simply Supported Rectangular Plates Under the Action of Shear Forces

Consider a simply supported plate on all sides subjected to shear forces N_{xy} as shown in Figure 7.24. The energy method is used to calculate the distributed shear force N_{xy} per unit length along the edges at which buckling takes place. The boundary conditions are satisfied if we take the double series as follows:

$$w = \sum_{m=1}^{\infty} \sum_{n=1}^{\infty} A_{mn} \sin \frac{m\pi x}{a} \sin \frac{n\pi y}{b} \quad m = 1, 2, 3\text{-----}, n = 1, 2, 3\text{---} \tag{7.22a}$$

The strain energy expression is the same as before

$$U = \frac{D}{2} \int_0^b \int_0^a \left[\left(\frac{\partial^2 w}{\partial x^2} \right)^2 + \left(\frac{\partial^2 w}{\partial y^2} \right)^2 + 2v \left(\frac{\partial^2 w}{\partial x^2} \right) \left(\frac{\partial^2 w}{\partial y^2} \right) + 2(1-v) \left(\frac{\partial^2 w}{\partial x \partial y} \right)^2 \right] dx dy \tag{7.22b}$$

The derivatives of w needed to calculate the strain energy are

$$\frac{\partial w}{\partial x} = \sum_{m=1}^{\infty} \sum_{n=1}^{\infty} A_{mn} \frac{m\pi}{a} \cos \frac{m\pi x}{a} \sin \frac{n\pi y}{b}$$

$$\frac{\partial^2 w}{\partial x^2} = -\sum_{m=1}^{\infty} \sum_{n=1}^{\infty} A_{mn} \frac{m^2\pi^2}{a^2} \sin \frac{m\pi x}{a} \sin \frac{n\pi y}{b}$$

$$\frac{\partial w}{\partial y} = \sum_{m=1}^{\infty} \sum_{n=1}^{\infty} A_{mn} \frac{n\pi}{b} \sin \frac{m\pi x}{a} \cos \frac{n\pi y}{b}$$

$$\frac{\partial^2 w}{\partial y^2} = -\sum_{m=1}^{\infty} \sum_{n=1}^{\infty} A_{mn} \frac{n^2\pi^2}{b^2} \sin \frac{m\pi x}{a} \sin \frac{n\pi y}{b}$$

$$\frac{\partial^2 w}{\partial x \partial y} = \sum_{m=1}^{\infty} \sum_{n=1}^{\infty} A_{mn} \frac{mn\pi^2}{ab} \cos \frac{m\pi x}{a} \cos \frac{n\pi y}{b} \tag{7.22c}$$

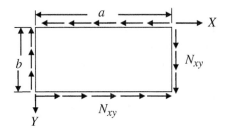

Figure 7.24 Simply supported plate subjected to pure shear.

The following definite integrals are also needed to calculate the strain energy

$$\int_0^b \int_0^a \sin^2 \frac{m\pi x}{a} \sin^2 \frac{n\pi y}{b} dxdy = \frac{ab}{4} \tag{7.22d}$$

$$\int_0^b \int_0^a \cos^2 \frac{m\pi x}{a} \cos^2 \frac{n\pi y}{b} dxdy = \frac{ab}{4}$$

Substitute derivatives of w given by Eqs. (7.22c) into Eq. (7.22b) and make use of the definite integrals in Eq. (7.22d) to have

$$U = \frac{abD}{8} \sum_{m=1}^{\infty} \sum_{n=1}^{\infty} A_{mn}^2 \left(\frac{m^4 \pi^4}{a^4} + \frac{n^4 \pi^4}{b^4} + 2\frac{m^2 n^2 \pi^4}{a^2 b^2} \right)$$

or

$$U = \frac{\pi^4 abD}{8} \sum_{m=1}^{\infty} \sum_{n=1}^{\infty} A_{mn}^2 \left(\frac{m^2}{a^2} + \frac{n^2}{b^2} \right)^2 \tag{7.22e}$$

From Eq. (7.19d) the potential energy of the external shear N_{xy} is given by

$$V = -\frac{1}{2} \int_0^b 2N_{xy} \left(\frac{\partial w}{\partial x} \right) \left(\frac{\partial w}{\partial y} \right) dxdy \tag{7.19d}$$

Substitute the derivatives of w from Eqs. (7.22c) in (7.19d) to get

$$V = -N_{xy} \int_0^b \int_0^a \sum_{m=1}^{\infty} \sum_{n=1}^{\infty} \sum_{r=1}^{\infty} \sum_{s=1}^{\infty} \frac{ms\pi^2}{ab} \left(A_{mn} A_{rs} \cos \frac{m\pi x}{a} \sin \frac{n\pi y}{b} \sin \frac{r\pi x}{a} \cos \frac{s\pi y}{b} \right) dxdy \tag{7.22f}$$

The following definite integrals are also needed to calculate the potential energy

$$\int_0^a \sin \frac{m\pi x}{a} \cos \frac{r\pi x}{a} dx = 0 \qquad \text{if } m \pm r \text{ is an even number}$$

$$\int_0^a \sin \frac{m\pi x}{a} \cos \frac{r\pi x}{a} dx = \frac{2a}{\pi} \frac{m}{m^2 - r^2} \qquad \text{if } m \pm r \text{ is an odd number} \tag{7.22g}$$

$$V = -N_{xy} \sum_{m=1}^{\infty} \sum_{n=1}^{\infty} \sum_{r=1}^{\infty} \sum_{s=1}^{\infty} \frac{ms\pi^2}{ab} \left\{ A_{mn} A_{rs} \left(\frac{2a}{\pi} \frac{r}{r^2 - m^2} \right) \left(\frac{2b}{\pi} \frac{n}{n^2 - s^2} \right) \right\}$$

$$V = -4N_{xy} \sum_{m=1}^{\infty} \sum_{n=1}^{\infty} \sum_{r=1}^{\infty} \sum_{s=1}^{\infty} A_{mn} A_{rs} \frac{mnrs}{(m^2 - r^2)(s^2 - n^2)} \tag{7.22h}$$

In Eq. (7.22h) the integers m, n, r, s have such values that $m \pm r$ and $n \pm s$ are odd numbers. The total potential energy in the plate is

$$\Pi = U + V = \frac{\pi^4 abD}{8} \sum_{m=1}^{\infty} \sum_{n=1}^{\infty} A_{mn}^2 \left(\frac{m^2}{a^2} + \frac{n^2}{b^2} \right)^2 - 4N_{xy} \sum_{m=1}^{\infty} \sum_{n=1}^{\infty} \sum_{r=1}^{\infty} \sum_{s=1}^{\infty} A_{mn} A_{rs} \frac{mnrs}{(m^2 - r^2)(s^2 - n^2)} \tag{7.22i}$$

In these equations the combinations of $m \pm r$ and $n \pm s$ are odd numbers from Eq. (7.22h). Let us assume m, n, r, s can each have only the values of 1 or 2. The possible combinations of m, n, r, s are as follows:

```
m n r s
1 1 2 2
1 2 2 1
2 1 1 2
2 2 1 1
```

With the above combinations of $m, n, r,$ and s the total potential energy in the plate is given by

$$\Pi = \frac{\pi^4 abD}{8}\left[A_{11}^2\left(\frac{1}{a^2}+\frac{1}{b^2}\right)^2 + A_{12}^2\left(\frac{1}{a^2}+\frac{4}{b^2}\right)^2 + A_{21}^2\left(\frac{4}{a^2}+\frac{1}{b^2}\right)^2 + A_{22}^2\left(\frac{4}{a^2}+\frac{4}{b^2}\right)^2\right]$$

$$- 4N_{xy}\left[A_{11}A_{22}\frac{(1)(1)(2)(2)}{(1-4)(4-1)} + A_{12}A_{21}\frac{(1)(2)(2)(1)}{(1-4)(1-4)}\right.$$

$$\left. + A_{21}A_{12}\frac{(2)(1)(1)(2)}{(4-1)(4-1)} + A_{22}A_{11}\frac{(2)(2)(1)(1)}{(4-1)(1-4)}\right]$$

or

$$\Pi = \frac{\pi^4 abD}{8}\left[A_{11}^2\left(\frac{1}{a^2}+\frac{1}{b^2}\right)^2 + A_{12}^2\left(\frac{1}{a^2}+\frac{4}{b^2}\right)^2 + A_{21}^2\left(\frac{4}{a^2}+\frac{1}{b^2}\right)^2 + A_{22}^2\left(\frac{4}{a^2}+\frac{4}{b^2}\right)^2\right]$$

$$+ \frac{32}{9}N_{xy}(A_{11}A_{22} - A_{12}A_{21}) \tag{7.22j}$$

Using the principle of stationary potential energy and the Rayleigh-Ritz method, we are interested in finding the constants A_{mn} and A_{rs} such that the total potential energy Π is a minimum. Take the derivatives of Π with respect to each of these coefficients and equate them to zero and we get a system of homogeneous linear equations.

$$\frac{d\Pi}{dA_{11}} = \frac{\pi^4 abD}{4}\left[A_{11}\left(\frac{1}{a^2}+\frac{1}{b^2}\right)^2\right] + \frac{32}{9}N_{xy}(A_{22}) = 0$$

$$\frac{d\Pi}{dA_{12}} = \frac{\pi^4 abD}{4}\left[A_{12}\left(\frac{1}{a^2}+\frac{4}{b^2}\right)^2\right] - \frac{32}{9}N_{xy}(A_{21}) = 0$$

$$\frac{d\Pi}{dA_{21}} = \frac{\pi^4 abD}{4}\left[A_{21}\left(\frac{4}{a^2}+\frac{1}{b^2}\right)^2\right] - \frac{32}{9}N_{xy}(A_{12}) = 0$$

$$\frac{d\Pi}{dA_{22}} = \frac{\pi^4 abD}{4}\left[A_{22}\left(\frac{4}{a^2}+\frac{4}{b^2}\right)^2\right] + \frac{32}{9}N_{xy}(A_{11}) = 0 \tag{7.22k}$$

Take $\dfrac{\pi^4 abD}{4}\left(\dfrac{1}{a^2}+\dfrac{1}{b^2}\right)^2 = \dfrac{\pi^4 D}{4\beta^3 b^2}(1+\beta^2)^2,$ where $\beta = \dfrac{a}{b}$. Similarly, we can calculate other coefficients in terms of β.

Since $(m + r)$ and $(n + s)$ are both odd, $(m + r + n + s)$ is even. If $(m + n)$ is even, $(r + s)$ must be even, and if $(m + n)$ is odd, $(r + s)$ must be odd. Thus, $(i + j)$ is either even or odd in the A_{ij} coefficients in each of the linear homogeneous Eq. (7.22k) and can be subdivided into two sets of equations that can be solved independently. In one set of equations $(i + j)$ is even and in the other set $(i + j)$ is odd. The equations where $(i + j)$ is even, the buckling mode is symmetric and gives maximum deflection at the center of the plate, i.e. $\left(\dfrac{\partial w}{\partial x} = \dfrac{\partial w}{\partial y} = 0 \right)$. In the equations where $(i + j)$ is odd, the buckling mode is antisymmetric and requires zero deflection at the center of the plate, i.e. $w = 0$. One of these modes give the minimum value of N_{xy}, the critical shear force at which buckling occurs. The two sets of equations corresponding to $(i + j)$ even and odd are respectively as follows:

$$\begin{bmatrix} \dfrac{\delta(1 + \beta^2)^2}{4\beta^3} & \dfrac{32}{9} N_{xy} \\[2ex] \dfrac{32}{9} N_{xy} & \dfrac{16\delta(1 + \beta^2)^2}{4\beta^3} \end{bmatrix} \left\{ \begin{matrix} A_{11} \\ A_{22} \end{matrix} \right\} = \left\{ \begin{matrix} 0 \\ 0 \end{matrix} \right\} \tag{7.22l}$$

and

$$\begin{bmatrix} \dfrac{\delta(1 + 4\beta^2)^2}{4\beta^3} & -\dfrac{32}{9} N_{xy} \\[2ex] -\dfrac{32}{9} N_{xy} & \dfrac{\delta(4 + \beta^2)^2}{4\beta^3} \end{bmatrix} \left\{ \begin{matrix} A_{12} \\ A_{21} \end{matrix} \right\} = \left\{ \begin{matrix} 0 \\ 0 \end{matrix} \right\} \tag{7.22m}$$

where

$$\delta = \frac{\pi^4 D}{b^2}.$$

For nontrivial solutions of A_{ij} in the Eqs. (7.22l) and (7.22m), the determinants of the coefficient matrices should be zero. Hence, from Eq. (7.22l)

$$\left(\frac{\delta(1 + \beta^2)^2}{4\beta^3} \right) \left(\frac{16\delta(1 + \beta^2)^2}{4\beta^3} \right) - \left(\frac{32}{9} \right)^2 N_{xy}^2 = 0$$

or

$$N_{xy} = \pm \frac{9}{32} \delta \left(\frac{1}{\beta^3} + \frac{2}{\beta} + \beta \right) \tag{7.22n}$$

For a square plate, $\dfrac{a}{b} = \beta = 1$, and

$$N_{xy} = \pm \frac{9}{8} \delta \tag{7.22o}$$

The plus and minus signs indicate that the direction of the shear force does not affect the critical value of shear force.

$$N_{xy} = \frac{9}{8}\frac{\pi^4 D}{b^2} = 11.10\frac{\pi^2 D}{b^2} = k\frac{\pi^2 D}{b^2} \tag{7.22p}$$

or

$$k = 11.10 \tag{7.22q}$$

From Eq. (7.22m) we have

$$\left(\frac{\delta(1 + 4\beta^2)^2}{4\beta^3}\right)\left(\frac{\delta(4 + \beta^2)^2}{4\beta^3}\right) - \left(\frac{32}{9}\right)^2 N_{xy}^2 = 0 \tag{7.22r}$$

or

$$N_{xy} = \pm\frac{9}{128}\delta\left(\frac{4}{\beta^3} + \frac{17}{\beta} + 4\beta\right) \tag{7.22s}$$

For a square plate, $\beta = 1$ leads to

$$N_{xy} = 17.35\frac{\pi^2 D}{b^2} = k\frac{\pi^2 D}{b^2}$$

or

$$k = 17.35$$

Hence, the minimum

$$N_{xy} = 11.10\frac{\pi^2 D}{b^2} \tag{7.22t}$$

The solution given by Eq. (7.22t) is an approximate solution. The more accurate solution [3] is obtained by taking more terms in Eq. (7.22k), and for a square plate the critical shear stress is given by the expression

$$\tau_{cr} = k\frac{\pi^2 D}{b^2 h} = 9.34\frac{\pi^2 D}{b^2 h}$$

The error in τ_{cr} by taking $m = n = 1,2$ obtained from Eq. (7.22t) $= \frac{11.10 - 9.34}{9.34}x100 = 18.84\%$.

Changes in the coefficient k by taking more and more terms are given in Table 7.5, whereas Table 7.6 gives values of k calculated by [6] taking $m = n = 10$ in the series for $w(x, y)$.

For $a/b < 2$, $i + j = $ even gives the smallest value of $(N_{xy})_{cr}$. For larger plates both sets of equations should be considered.

7.6.6 Buckling of Simply Supported Rectangular Plates Under Combined Bending and Compression

Consider a simply supported plate subjected to in-plane distributed forces on sides $x = 0$ and $x = a$ that are varying with respect to y. The intensity of the force per unit length of the edge is given by

$$N_x = N_0\left(1 - \frac{\alpha y}{b}\right) \tag{7.23a}$$

Table 7.5 Effect of the number of terms in the series for $w(x, y)$ on k.

$\beta = \dfrac{a}{b}$	$m = n = 2$		$m = n = 3$		$m = n = 4$	
	$i+j$ = even	$i+j$ = odd	$i+j$ = even	$i+j$ = odd	$i+j$ = even	$i+j$ = odd
1.0	11.10	17.35	9.42	11.73	9.40	11.65
1.2	9.56	14.75	8.00	9.87	8.05	9.76
1.4	8.86	13.32	7.37	8.68	7.35	8.50
1.5	8.68	12.85	7.15	8.24	7.07	8.03

Table 7.6 Values of k for $m = n = 10$.

β	$i+j$ = even	$i+j$ = odd
1.0	9.35	11.63
1.2	8.00	9.70
1.5	7.07	7.97
2.0	6.59	6.61
2.5	6.29	6.06
3.0	6.04	5.89
4.0	5.67	5.77

where N_0 is the intensity of compressive force at $y = 0$ and α is a numerical factor, by changing it, we get different types of forces on the plate. When $\alpha = 0$, we get a uniformly distributed compressive load, for $\alpha = 2$ the case of pure bending is obtained. If $\alpha < 2$, we get a combination of an axial compressive force and bending moment as shown in Figure 7.25.

The deflection of the simply supported plate can be taken as

$$w = \sin\frac{m\pi x}{a} \sum_{n=1}^{\infty} A_{mn} \sin\frac{n\pi y}{b} \tag{7.23b}$$

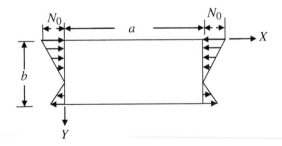

Figure 7.25 Rectangular plate subjected to combined bending and compression.

This is equivalent to the assumption that the buckled plate is subdivided along the X axis in m half-waves. The plate can be considered as simply supported between two nodal lines containing one half-wave. The strain energy of bending due to deflection given by Eq. (7.23b) can be expressed from Eq. (7.22e) as

$$U = \frac{\pi^4 abD}{8} \sum_{n=1}^{\infty} A_{mn}^2 \left(\frac{m^2}{a^2} + \frac{n^2}{b^2} \right)^2 \tag{7.23c}$$

The potential energy due to applied forces is

$$V = -\frac{1}{2} \int_0^b \int_0^a N_x \left(\frac{\partial w}{\partial x} \right)^2 dxdy \tag{7.23d}$$

$$\frac{\partial w}{\partial x} = \frac{m\pi}{a} \cos \frac{m\pi x}{a} \sum_{n=1}^{\infty} A_{mn} \sin \frac{\pi y}{b} \tag{7.23e}$$

$$V = -\frac{1}{2} \int_0^b \int_0^a \left[N_0 \left(1 - \frac{\alpha y}{b} \right) \frac{m^2 \pi^2}{a^2} \cos^2 \frac{m\pi x}{a} \left(\sum_{n=1}^{\infty} \sum_{s=1}^{\infty} A_{mn} A_{ms} \sin \frac{n\pi y}{b} \sin \frac{s\pi y}{b} \right) \right] dxdy \tag{7.23f}$$

The following definite integrals are needed to calculate the potential energy

$$\int_0^b y \sin \frac{n\pi y}{b} \sin \frac{s\pi y}{b} dy = \frac{b^2}{4} \qquad \text{for } n = s$$

$$\int_0^b y \sin \frac{n\pi y}{b} \sin \frac{s\pi y}{b} dy = 0 \qquad \text{for } n \neq s \text{ and } n \pm s \text{ an even number}$$

$$\int_0^b y \sin \frac{n\pi y}{b} \sin \frac{s\pi y}{b} dy = -\frac{4b^2}{\pi^2} \frac{ns}{(n^2 - s^2)^2} \qquad \text{for } n \neq s \text{ and } n \pm s \text{ an odd number}$$

$$\tag{7.23g}$$

$$V = -\frac{N_0 \, ab}{2 \quad 4} \sum_{n=1}^{\infty} \frac{m^2 \pi^2}{a^2} A_{mn}^2 + \frac{N_0 \alpha \, a \, m^2 \pi^2}{2b \quad 2 \quad a^2} \left[\frac{b^2}{4} \sum_{n=1}^{\infty} A_{mn}^2 - \frac{8b^2}{\pi^2} \sum_{n=1}^{\infty} \sum_{s} \frac{A_{mn} A_{ms} ns}{(n^2 - s^2)^2} \right] \tag{7.23h}$$

The total potential energy Π can be written by combining the strain and potential energies from Eqs. (7.23c) and (7.23h) as

$$\Pi = U + V = \frac{\pi^4 abD}{8} \sum_{n=1}^{\infty} A_{mn}^2 \left(\frac{m^2}{a^2} + \frac{n^2}{b^2} \right)^2 - \frac{N_0 ab}{8} \sum_{n=1}^{\infty} \frac{m^2 \pi^2}{a^2} A_{mn}^2$$

$$+ \frac{N_0 \alpha b}{16a} m^2 \pi^2 \sum_{n=1}^{\infty} A_{mn}^2 - \frac{2N_0 \alpha b}{a} m^2 \sum_{n=1}^{\infty} \sum_{s} \frac{A_{mn} A_{ms} ns}{(n^2 - s^2)^2} \tag{7.23i}$$

where s is taken such that $n \pm s$ is an odd number. Take derivatives of Π with respect to $A_{mn}(n = 1, 2, - - -)$, and we get a system of linear homogeneous equations given below:

$$\frac{\pi^4 abD}{4} A_{mn} \left(\frac{m^2}{a^2} + \frac{n^2}{b^2} \right)^2 - \frac{N_0 ab}{4} \frac{m^2 \pi^2}{a^2} A_{mn} + \frac{N_0 \alpha b}{8a} m^2 \pi^2 A_{mn} - \frac{2N_0 \alpha b}{a} m^2 \sum_{s} \frac{A_{ms} ns}{(n^2 - s^2)^2} = 0$$

or

$$\pi^4 D A_{mn}\left(\frac{m^2}{a^2}+\frac{n^2}{b^2}\right)^2 - N_0\frac{m^2\pi^2}{a^2}A_{mn} + \frac{N_0\alpha}{2a^2}m^2\pi^2 A_{mn} - \frac{8N_0\alpha}{a^2}m^2\sum_s^{\infty}\frac{A_{ms}ns}{(n^2-s^2)^2} = 0$$

(7.23j)

In Section 7.5.1 containing simply supported plates under axial compression load, it is seen that these plates buckle into equal half-waves with nodal lines perpendicular to the X-axis. We can consider each buckle a plate simply supported on its four edges. Assume $m = 1$ in Eq. (7.23j) and we obtain a system of the following equations:

$$A_{1n}\left[\left(1+n^2\frac{a^2}{b^2}\right)^2 - \frac{N_0 a^2}{\pi^2 D}\left(1-\frac{\alpha}{2}\right)\right] - \frac{8a^2 N_0\alpha}{\pi^4 D}\sum_s^{\infty}\frac{A_{1s}ns}{(n^2-s^2)^2} = 0$$

(7.23k)

where $n \pm s$ is an odd number. Equation (7.23k) are linear homogeneous equations in A_{11}, A_{12}, ... For a nontrivial solution, the determinants of the coefficient matrices should be zero. If only the coefficient A_{11} is considered, and others are taken as zero, we obtain one term approximate solution given below:

$$A_{11}\left[\left(1+\frac{a^2}{b^2}\right)^2 - \frac{N_0 a^2}{\pi^2 D}\left(1-\frac{\alpha}{2}\right)\right] = 0$$

The critical force $(N_0)_{cr}$ per unit length is obtained from

$$(N_0)_{cr} = \left(\frac{b}{a}+\frac{a}{b}\right)^2\frac{\pi^2 D}{b^2}\frac{1}{1-\dfrac{\alpha}{2}}$$

(7.23l)

The critical stress σ_{cr} is defined by, $\sigma_{cr} = \dfrac{(N_0)_{cr}}{h}$, where h is the thickness of the plate. Hence,

$$\sigma_{cr} = \left(\frac{b}{a}+\frac{a}{b}\right)^2\frac{\pi^2 D}{b^2 h}\frac{1}{1-\dfrac{\alpha}{2}}$$

(7.23m)

For uniform compression, $\alpha = 0$, Eq. (7.23m) reduces to

$$\sigma_{cr} = \left(\frac{b}{a}+\frac{a}{b}\right)^2\frac{\pi^2 D}{b^2 h}$$

(7.23n)

It is the same expression for uniformly compressed plate given by Eq. (7.7k) for $m = 1$, $n = 1$. This solution, however, is not accurate for higher values of $a/b = \alpha$. To obtain a more accurate solution, more terms should be considered in the set of equations given by Eq. (7.23k). Consider a two-term solution where $n = 1, 2$ and take two terms A_{11} and A_{12}.

$n = 1$ and $s = 2$, so that $n + s$ is odd, we get from Eq. (7.23k)

$$A_{11}\left[\left(1+\frac{a^2}{b^2}\right)^2 - \frac{N_0 a^2}{\pi^2 D}\left(1-\frac{\alpha}{2}\right)\right] - \frac{8a^2 N_0\alpha}{\pi^4 D}A_{12}\frac{(1)(2)}{(1^2-2^2)^2} = 0$$

(7.23o)

$n = 2$ and $s = 1$, so that $n + s$ is odd, we get from Eq. (7.23k)

$$A_{12}\left[\left(1+\frac{4a^2}{b^2}\right)^2 - \frac{N_0 a^2}{\pi^2 D}\left(1-\frac{\alpha}{2}\right)\right] - \frac{8a^2 N_0\alpha}{\pi^4 D}A_{11}\frac{(2)(1)}{(2^2-1^2)^2} = 0$$

(7.23p)

Equations (7.23o) and (7.23p) can be written in the matrix form as

$$
\left[
\begin{array}{cc}
\left(1+\dfrac{a^2}{b^2}\right)^2 - \dfrac{N_0 a^2}{\pi^2 D}\left(1-\dfrac{\alpha}{2}\right) & -\dfrac{8a^2 N_0 \alpha}{\pi^4 D}\dfrac{2}{9} \\[3mm]
-\dfrac{8a^2 N_0 \alpha}{\pi^4 D}\dfrac{2}{9} & \left(1+\dfrac{4a^2}{b^2}\right)^2 - \dfrac{N_0 a^2}{\pi^2 D}\left(1-\dfrac{\alpha}{2}\right)
\end{array}
\right]
\begin{Bmatrix} A_{11} \\[2mm] A_{12} \end{Bmatrix}
=
\begin{Bmatrix} 0 \\[2mm] 0 \end{Bmatrix}
\tag{7.23q}
$$

For a nontrivial solution of A_{11} and A_{12}, the determinant of the coefficient matrix is zero, hence,

$$
\left[\left(1+\dfrac{a^2}{b^2}\right)^2 - \dfrac{N_0 a^2}{\pi^2 D}\left(1-\dfrac{\alpha}{2}\right)\right]\left[\left(1+\dfrac{4a^2}{b^2}\right)^2 - \dfrac{N_0 a^2}{\pi^2 D}\left(1-\dfrac{\alpha}{2}\right)\right] - \left(\dfrac{8a^2 N_0 \alpha}{\pi^4 D}\dfrac{2}{9}\right)^2 = 0
$$

$$
\left(\dfrac{N_0 a^2}{\pi^2 D}\right)^2\left[\left(1-\dfrac{\alpha}{2}\right)^2 - \left(\dfrac{16\alpha}{9\pi^2}\right)^2\right] - \dfrac{N_0 a^2}{\pi^2 D}\left(1-\dfrac{\alpha}{2}\right)\left[\left(1+\dfrac{a^2}{b^2}\right)^2 + \left(1+\dfrac{4a^2}{b^2}\right)\right]
$$

$$
+ \left(1+\dfrac{a^2}{b^2}\right)^2\left(1+\dfrac{4a^2}{b^2}\right)^2 = 0
\tag{7.23r}
$$

For pure bending, $\alpha = 2$, for square plate $a/b = 1$, and substitute in Eq. (7.23r) to get

$$
\left(\dfrac{N_0 a^2}{\pi^2 D}\right)\left(\dfrac{32}{9\pi^2}\right) = 10
$$

or

$$
(N_0)_{cr} = \dfrac{45}{16}\dfrac{\pi^4 D}{b^2}
\tag{7.23s}
$$

and

$$
\sigma_{cr} = \dfrac{45}{16}\dfrac{\pi^4 D}{b^2 h} = k\dfrac{\pi^2 D}{b^2 h}
\tag{7.23t}
$$

where

$$
k = \dfrac{45\pi^2}{16} = 27.76
\tag{7.23u}
$$

The exact value of $k = 25.6$ given by Timoshenko and Gere [3]. The error in the two-term solution is

$$
\text{Error} = \dfrac{27.76 - 25.6}{25.6}x100 = 8.44\%
$$

To increase the accuracy for $\alpha = 2$, let us take a three-term solution where $n = 1, 2, 3$ and consider three terms A_{11}, A_{12}, A_{13}.

Take $n = 1$ and $s = 2$, so that $n + s$ is odd, we get from Eq. (7.23k)

$$
\left(1+\dfrac{a^2}{b^2}\right)^2 A_{11} - \dfrac{16 N_0 a^2}{\pi^4 D}\left(\dfrac{2}{9}A_{12}\right) = 0
\tag{7.23v}
$$

Take $n = 2$ and $s = 1, 3$, so that $n + s$ is odd, we get from Eq. (7.23k)

$$A_{12}\left(1 + \frac{4a^2}{b^2}\right)^2 - \frac{16N_0 a^2}{\pi^4 D}\left(A_{11}\frac{2x1}{(4-1)^2} + A_{13}\frac{2x3}{(4-9)^2}\right) = 0 \qquad (7.23w)$$

Take $n = 3$ and $s = 2$, so that $n + s$ is odd, we get from Eq. (7.23k)

$$A_{13}\left(1 + \frac{9a^2}{b^2}\right)^2 - \frac{16N_0 a^2}{\pi^4 D}\left(A_{12}\frac{3x2}{(9-4)^2}\right) = 0 \qquad (7.23x)$$

Assume $\dfrac{32N_0 a^2}{\pi^4 D} = \lambda$ and $a/b = 1$. Eqs. (7.23v)–(7.23x) can be written as

$$\begin{bmatrix} 4 & -\dfrac{\lambda}{9} & 0 \\[2mm] -\dfrac{\lambda}{9} & 25 & -\dfrac{3}{25}\lambda \\[2mm] 0 & -\dfrac{3}{25}\lambda & 100 \end{bmatrix} \begin{Bmatrix} A_{11} \\[2mm] A_{12} \\[2mm] A_{13} \end{Bmatrix} = \begin{Bmatrix} 0 \\[2mm] 0 \\[2mm] 0 \end{Bmatrix} \qquad (7.23y)$$

For a nontrivial solution of A_{11}, A_{12}, A_{13}, the determinant of the coefficient matrix should be zero. Hence,

$$4\left[2500 - \left(\frac{3}{25}\right)^2\lambda^2\right] - \frac{\lambda}{9}\left(\frac{100\lambda}{9}\right) = 0$$

$$\lambda = \frac{32N_0 a^2}{\pi^4 D} = 87.97$$

$$(N_0)_{cr} = \frac{87.97}{32}\frac{\pi^4 D}{b^2} \quad \text{or} \quad \sigma_{cr} = \frac{87.97}{32}\frac{\pi^4 D}{b^2 h} = k\frac{\pi^2 D}{b^2 h}$$

where

$$k = \frac{87.97\pi^2}{32} = 27.13 \qquad (7.23z)$$

$$\text{Error} = \frac{27.13 - 25.6}{25.6}x100 = 5.98\%$$

Similarly, by taking more equations of the system of equations given by Eq. (7.23k), a higher accuracy for the solution of k can be obtained.

7.6.7 Buckling of Plates with Stiffeners

We have seen the critical stress for normal and shear forces is proportional to the flexural rigidity of the plate. For given boundary conditions and the aspect ratio a/b, the critical stress is proportional to h^2/b^2, where h is the thickness and b is the width of the plate. This shows that the stability of a plate can be improved by increasing the thickness of the plate, but it will increase the weight of the plate. Another way of increasing the stability is to reduce the width of a plate. This is accomplished by providing longitudinal stiffeners. The stiffeners are rigidly connected to a plate. Hence, at common points stiffeners deflect and twist similar to the plate. Longitudinal stiffeners carry a portion of the compressive force as well as subdivide the plate into smaller panels. Stiffeners can be provided in the transverse direction also. We saw before that simply supported rectangular plates that are compressed in one direction buckle in a number of half-waves that correspond to the width of the plate. Thus, these stiffeners have to be placed at much closer distances that are less than the width of the plate to increase the stability of the plate. The Rayleigh-Ritz method can be used to find the critical load for the plates reinforced with stiffeners by adding the strain energy stored in the stiffeners and taking into account the potential energy of the external forces in the stiffeners. We assume that the deflections are small in the stiffened plates and the material is elastic and isotropic.

7.6.8 Simply Supported Rectangular Plates with Longitudinal Stiffeners

Consider a plate with a longitudinal stiffener placed at a distance c_i from the edge $y = 0$ shown in Figure 7.26. Using the energy method, the strain and potential energies are contributed by both the plate and the stringer. Assume the deflected shape of the plate during buckling as a double trigonometric series that satisfies the boundary conditions

$$w = \sum_{m=1}^{\infty} \sum_{n=1}^{\infty} A_{mn} \sin \frac{m\pi x}{a} \sin \frac{n\pi y}{b} \tag{7.24a}$$

The strain energy of bending of the plate is given by Eq. (7.22e)

$$U = \frac{\pi^4 abD}{8} \sum_{m=1}^{\infty} \sum_{n=1}^{\infty} A_{mn}^2 \left(\frac{m^2}{a^2} + \frac{n^2}{b^2} \right)^2 \tag{7.22e}$$

The strain energy of an ith stiffener placed at a distance c_i from the edge $y = 0$ is obtained from

$$U_i = \frac{EI_i}{2} \int_0^a \left(\frac{\partial^2 w}{\partial x^2} \right)^2_{y = c_i} dx + GJ_i \int_0^a \left(\frac{\partial^2 w}{\partial x \partial y} \right)^2_{y = c_i} \tag{7.24b}$$

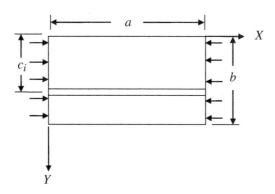

Figure 7.26 Simply supported rectangular plate with longitudinal stiffener.

EI_i is the flexural rigidity of the rib and GJ_i is its torsional rigidity. By neglecting the strain energy of twist of the stiffener we get

$$U_i = \frac{EI_i}{2} \int_0^a \sum_{m=1}^{\infty} \left(A_{m1} \frac{m^2 \pi^2}{a^2} \sin \frac{m\pi x}{a} \sin \frac{\pi y}{b} + A_{m2} \frac{m^2 \pi^2}{a^2} \sin \frac{m\pi x}{a} \sin \frac{2\pi y}{b} + --- \right)^2_{y=c_i} dx$$

The definite integrals have the values given below

$$\int_0^a \sin^2 \frac{m\pi x}{a} dx = \int_0^a \cos^2 \frac{m\pi x}{a} dx = \frac{a}{2}$$

$$\int_0^a \sin \frac{m\pi x}{a} \sin \frac{n\pi y}{b} dx = 0, \quad m \neq n$$

We get

$$U_i = \frac{\pi^4 EI_i}{4a^3} \sum_{m=1}^{\infty} m^4 \left(A_{m1} \sin \frac{\pi c_i}{b} + A_{m2} \sin \frac{2\pi c_i}{b} + -- \right)^2 \tag{7.24c}$$

The potential energy of the force N_x on the plate during buckling is given by

$$V = -\frac{N_x}{2} \int_0^b \int_0^a \left(\frac{\partial w}{\partial x} \right)^2 dxdy$$

or

$$V = -\frac{N_x}{2} \int_0^b \int_0^a \left(\sum_{m=1}^{\infty} \sum_{n=1}^{\infty} A_{mn} \frac{m\pi}{a} \cos \frac{m\pi x}{a} \sin \frac{n\pi y}{b} \right)^2 dxdy$$

or

$$V = -\frac{N_x}{2} \frac{ab}{4} \sum_{m=1}^{\infty} \sum_{n=1}^{\infty} A_{mn}^2 \frac{m^2 \pi^2}{a^2} \tag{7.24d}$$

The potential energy of the compressive force P_i acting on a stiffener i during buckling is

$$V_i = -\frac{P_i}{2} \int_0^a \left(\frac{\partial w}{\partial x} \right)^2_{y=c_i} dx$$

or

$$V_i = -\frac{P_i}{2} \int_0^a \sum_{m=1}^{\infty} \left(A_{m1} \frac{m\pi}{a} \cos \frac{m\pi x}{a} \sin \frac{\pi y}{b} + A_{m2} \frac{m\pi}{a} \cos \frac{m\pi x}{a} \sin \frac{2\pi y}{b} + -- \right)^2_{y=c_i} dx$$

$$V_i = -\frac{P_i \pi^2}{4a} \sum_{m=1}^{\infty} m^2 \left(A_{m1} \sin \frac{\pi c_i}{b} + A_{m2} \sin \frac{2\pi c_i}{b} + -- \right)^2 \tag{7.24e}$$

The total potential energy of the plate and stiffeners is obtained from

$$\Pi = U + \sum_{i=1}^{r} U_i + V + \sum_{i=1}^{r} V_i \tag{7.24f}$$

where r is the number of longitudinal stiffeners provided at distances $c_1, c_2, \ldots c_r$ from the edge $y = 0$.

$$\Pi = \left(\frac{\pi^2 D}{8a^2}\right)\left(\frac{\pi^2 b}{a}\right)\sum_{m=1}^{\infty}\sum_{n=1}^{\infty} A_{mn}^2 (m^2 + \beta^2 n^2)^2 + \left(\frac{\pi^2 D}{8a^2}\right)\left(\frac{2\pi^2 b}{a}\right)$$

$$\sum_{i=1}^{r} \gamma_i \sum_{m=1}^{\infty} m^4 \left(\sum_{n=1}^{\infty} A_{mn} \sin\frac{n\pi c_i}{b}\right)^2$$

$$-\left(\frac{N_x}{8}\right)\left(\frac{\pi^2 b}{a}\right)\sum_{m=1}^{\infty}\sum_{n=1}^{\infty} m^2 A_{mn}^2 - \left(\frac{N_x}{8}\right)\left(\frac{2\pi^2 b}{a}\right)\sum_{i=1}^{r}\delta_i\sum_{m=1}^{\infty} m^2\left(\sum_{n=1}^{\infty} A_{mn}\sin\frac{n\pi c_i}{b}\right)^2$$

$$(7.24g)$$

where

$$\frac{a}{b} = \beta, \qquad \frac{EI_i}{bD} = \gamma_i, \qquad \frac{P_i}{bN_x} = \frac{A_i}{bh} = \delta_i$$

and bh = Cross-sectional area of the plate
A_i = Cross-sectional area of one stiffener

The minimum buckling load for the plate, N_x, can be obtained by taking the derivatives of the total potential energy, Π, with respect to the coefficients A_{mn} and equating them to zero. We get a set of homogeneous linear algebraic equations as follows:

$$\frac{\pi^2 D}{b^2\beta^2}\left[A_{mn}(m^2 + n^2\beta^2)^2 + 2\sum_{i=1}^{r}\gamma_i\sin\frac{n\pi c_i}{b}m^4\sum_{s=1}^{\infty}A_{ms}\sin\frac{s\pi c_i}{b}\right]$$

$$-N_x\left[m^2 A_{mn} + 2\sum_{i=1}^{r}\delta_i\sin\frac{n\pi c_i}{b}m^2\sum_{s=1}^{\infty}A_{ms}\sin\frac{s\pi c_i}{b}\right] = 0 \qquad (7.24h)$$

Assume there is only one longitudinal stiffener provided at $c_i = \dfrac{b}{2}$. We saw before that for a simply supported plate, minimum buckling load occurs when the plate buckles into one half sine wave in the X direction, i.e. $m = 1$. Then we can write equations for different values of n, the number of half-waves in the Y direction, from Eq. (7.24h) in the following form.

Take $n = 1$ to get

$$\frac{\pi^2 D}{b^2\beta^2}\left[A_{11}(1 + \beta^2)^2 + 2\gamma\right.$$

$$\times\left\{\sin\frac{\pi}{2}(1)^4\left(A_{11}\sin\frac{\pi}{2} + A_{12}\sin\pi + A_{13}\sin\frac{3\pi}{2} + A_{14}\sin 2\pi + A_{15}\sin\frac{5\pi}{2} + -\right)\right\}\right]$$

$$-N_x\left[(1)^2 A_{11} + 2\delta\right.$$

$$\times\left[\left\{\sin\frac{\pi}{2}(1)^2\left(A_{11}\sin\frac{\pi}{2} + A_{12}\sin\pi + A_{13}\sin\frac{3\pi}{2} + A_{14}\sin 2\pi + A_{15}\sin\frac{5\pi}{2} + -\right)\right\}\right] = 0$$

or

$$\frac{\pi^2 D}{b^2 \beta^2}[A_{11}(1+\beta^2)^2 + 2\gamma(A_{11} - A_{13} + A_{15} - -)] - N_x[A_{11} + 2\delta(A_{11} - A_{13} + A_{15} - -)] = 0$$

(7.24i)

Similarly, take $n = 2$ to have

$$\frac{\pi^2 D}{b^2 \beta^2} A_{12}(1 + 4\beta^2)^2 - N_x A_{12} = 0$$

(7.24j)

For $n = 3$

$$\frac{\pi^2 D}{b^2 \beta^2}[A_{13}(1+9\beta^2)^2 - 2\gamma(A_{11} - A_{13} + A_{15} - -)] - N_x[A_{13} - 2\delta(A_{11} - A_{13} + A_{15} - -)] = 0$$

(7.24k)

For $n = 4$

$$\frac{\pi^2 D}{b^2 \beta^2} A_{14}(1 + 16\beta^2)^2 - N_x A_{14} = 0$$

(7.24l)

Equations for other values of n can be written the same way. For even values of $n = 2, 4, \ldots$, etc. the stiffener does not contribute to the buckling load. In these cases, a nodal line coincides with the stiffener, and the stiffener remains straight during the buckling of the plate. To connect the buckling load with the flexural rigidity of the stiffener, the equations having $n = 1, 3, 5 \ldots$, etc. should be considered.

The first approximation to the N_{cr} can be obtained by taking only the A_{11} coefficient and assuming others zero. This way, we obtain from Eq. (7.24i)

$$\frac{\pi^2 D}{b^2 \beta^2}[A_{11}(1+\beta^2)^2 + 2\gamma A_{11}] - N_x(1 + 2\delta)A_{11} = 0$$

or

$$N_{cr} = \frac{\pi^2 D}{b^2 \beta^2} \frac{(1+\beta^2)^2 + 2\gamma}{(1+2\delta)}$$

(7.24m)

The second approximation of N_{cr} is obtained by considering the coefficients A_{11} and A_{13} in Eqs. (7.24i) and (7.24k), as follows;

$$\frac{\pi^2 D}{b^2 \beta^2}[A_{11}(1+\beta^2)^2 + 2\gamma(A_{11} - A_{13})] - N_x[A_{11} + 2\delta(A_{11} - A_{13})] = 0$$

$$\frac{\pi^2 D}{b^2 \beta^2}[A_{13}(1+9\beta^2)^2 - 2\gamma(A_{11} - A_{13})] - N_x[-2\delta A_{11} + A_{13}(1 + 2\delta)] = 0$$

or

$$\begin{bmatrix} \dfrac{c}{\beta^2} - k(1+2\delta) & -\dfrac{2\gamma}{\beta^2} + 2k\delta \\[3mm] -\dfrac{2\gamma}{\beta^2} + 2k\delta & \dfrac{d}{\beta^2} - k(1+2\delta) \end{bmatrix} \begin{Bmatrix} A_{11} \\[3mm] A_{13} \end{Bmatrix} = \begin{Bmatrix} 0 \\[3mm] 0 \end{Bmatrix}$$

(7.24n)

where

$$(1 + \beta^2)^2 + 2\gamma = c, \quad (1 + 9\beta^2)^2 + 2\gamma = d, \quad \text{and} \quad \frac{N_x b^2}{\pi^2 D} = k$$

The determinant of the coefficient matrix in Eq. (7.24n) should be zero for nontrivial solution for the A_{11} and A_{13} coefficients. Hence,

$$\begin{vmatrix} \dfrac{c}{\beta^2} - k(1 + 2\delta) & -\dfrac{2\gamma}{\beta^2} + 2k\delta \\ -\dfrac{2\gamma}{\beta^2} + 2k\delta & \dfrac{d}{\beta^2} - k(1 + 2\delta) \end{vmatrix} = 0$$

or

$$\left[\frac{c}{\beta^2} - k(1 + 2\delta) \right] \left[\frac{d}{\beta^2} - k(1 + 2\delta) \right] - \left[-\frac{2\gamma}{\beta^2} + 2k\delta \right]^2 = 0$$

or

$$(k\beta^2)^2(1 + 4\delta) - k\beta^2[(1 + 2\delta)(c + d) - 8\gamma\delta] + cd - 4\gamma^2 = 0 \tag{7.24o}$$

For various values of β, γ, and δ, the value of k can be found from Eq. (7.24o), then

$$N_{cr} = k\frac{\pi^2 D}{b^2} \tag{7.24p}$$

The critical stress is given by

$$\sigma_{cr} = k\frac{\pi^2 D}{b^2 h} \tag{7.24q}$$

7.6.8.1 Plates with Two Longitudinal Stiffeners Dividing the Width of the Plate

Suppose there are two equal longitudinal stiffeners dividing the width of the plate in three equal parts. The problem can be analyzed the same way as before. Assume the deflection of the plate is symmetrical about the middle longitudinal axis ($y = b/2$), and the stiffeners are placed at the distances of $c_1 = b/3$ and $c_2 = 2b/3$. Take only one term of A_{11}, then from Eq. (7.24h) by taking $m = n = 1$ we can write

$$\frac{\pi^2 D}{b^2 \beta^2} \left[A_{11}(1 + \beta^2)^2 + 2\left\{ \gamma_1 \sin\frac{\pi}{3} A_{11} \sin\frac{\pi}{3} + \gamma_2 \sin\frac{2\pi}{3} A_{11} \sin\frac{2\pi}{3} \right\} \right]$$

$$-N_x \left[A_{11} + 2\left\{ \delta_1 \sin\frac{\pi}{3} A_{11} \sin\frac{\pi}{3} + \delta_2 \sin\frac{2\pi}{3} A_{11} \sin\frac{2\pi}{3} \right\} \right] = 0 \tag{7.25a}$$

or

$$\frac{\pi^2 D}{b^2 \beta^2} \left[A_{11}(1 + \beta^2)^2 + 2\gamma\left(\frac{\sqrt{3}}{2}\right)^2 + 2\gamma\left(\frac{\sqrt{3}}{2}\right)^2 \right] A_{11} - N_x$$

$$\times \left[1 + 2\delta\left(\frac{\sqrt{3}}{2}\right)^2 + 2\delta\left(\frac{\sqrt{3}}{2}\right)^2 \right] A_{11} = 0$$

The first approximation of the critical load is given by

$$N_{cr} = \frac{\pi^2 D}{b^2 \beta^2} \frac{(1+\beta^2)^2 + 3\gamma}{1 + 3\delta} \tag{7.25b}$$

$$\sigma_{cr} = \frac{\pi^2 D}{b^2 \beta^2 h} \frac{(1+\beta^2)^2 + 3\gamma}{1 + 3\delta} \tag{7.25c}$$

If there are more than two stiffeners, the critical stress from the first approximation is obtained from Eq. (7.24h) as

$$\sigma_{cr} = \frac{\pi^2 D}{b^2 \beta^2 h} \frac{(1+\beta^2)^2 + 2\sum\limits_{i=1}^{r} \gamma_i \sin^2 \frac{\pi c_i}{b}}{1 + 2\sum\limits_{i=1}^{r} \delta_i \sin^2 \frac{\pi c_i}{b}} \tag{7.25d}$$

7.6.9 Simply Supported Rectangular Compressed Plate with Transverse Stiffeners

Suppose the plate is simply supported on all sides and is uniaxially compressed as shown in Figure 7.27. Assume there is a transverse stiffener at a distance $x = c_i$ from the edge $x = 0$. In this the strain energy results from the bending of the plate and the stringer/stringers. The potential energy is contributed by the shortening of the plate. The stringer/stringers do not contribute to the potential energy because these are placed in the Y direction and the load is applied in the X direction. We assume that the plate forms one-half sine wave in the Y direction, hence the deflected shape of the plate during buckling is assumed as

$$w = \sum_{m=1}^{\infty} A_m \sin \frac{m\pi x}{a} \sin \frac{\pi y}{b} \tag{7.26a}$$

where $n = 1$.

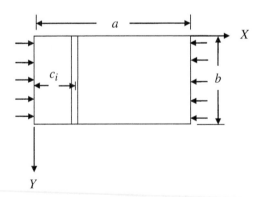

Figure 7.27 Simply supported rectangular plate with transverse stiffener.

The strain energy of the plate by using Eq. (7.22e) is obtained from

$$U = \frac{\pi^4 D}{8\beta a^2} \sum_{m=1}^{\infty} A_m^2 (m^2 + \beta^2)^2 \tag{7.26b}$$

The strain energy in the stiffener is given by

$$U_i = \frac{EI_i}{2} \int_0^b \left(\frac{\partial^2 w}{\partial y^2} \right)_{x=c_i}^2 dy \tag{7.26c}$$

or

$$U_i = \frac{EI_i}{2} \int_0^b \sum_{m=1}^{\infty} \left(A_m \frac{\pi^2}{b^2} \sin \frac{m\pi c_i}{a} \sin \frac{\pi y}{b} \right)^2 dy$$

or

$$U_i = \frac{\pi^4 EI_i}{4b^3} \sum_{m=1}^{\infty} A_m^2 \sin^2 \frac{m\pi c_i}{a} \tag{7.26d}$$

The potential energy of the plate is given by

$$V = -\frac{N_x}{2} \int_0^b \int_0^a \left(\frac{\partial w}{\partial x} \right)^2 dxdy$$

or

$$V = -\frac{N_x}{2} \int_0^b \int_0^a \left(\sum_{m=1}^{\infty} A_m \frac{m\pi}{a} \cos \frac{m\pi x}{a} \sin \frac{\pi y}{b} \right)^2 dxdy$$

or

$$V = -\frac{\pi^2 N_x}{8\beta} \sum_{m=1}^{\infty} A_m^2 m^2 \tag{7.26e}$$

The total potential energy is

$$\Pi = U + \sum_{i=1}^{r} U_i + V \tag{7.26f}$$

where r is the number of transverse stiffeners provided at distances $c_1, c_2, \text{------} c_r$ from the edge $x = 0$.

$$\Pi = \frac{\pi^2 D}{\beta^2 b^2} \frac{\pi^2}{8\beta} \sum_{m=1}^{\infty} A_m^2 (m^2 + \beta^2)^2 + \left(\frac{\pi^2 D}{\beta^2 b^2} \right) \left(\frac{\pi^2}{8\beta} \right) 2\beta^3 \sum_{i=1}^{r} \gamma_1 \sum_{m=1}^{\infty} A_m^2 \sin^2 \frac{m\pi c_i}{a}$$

$$-\frac{\pi^2 N_x}{8\beta} \sum_{m=1}^{\infty} A_m^2 m^2 \tag{7.26g}$$

where

$$\beta = \frac{a}{b}, \quad \frac{EI_i}{bD} = \gamma_i.$$

The minimum buckling load for the plate, N_x, can be obtained by taking the derivatives of the total potential energy, Π, with respect to the coefficients A_m and equating them to zero. We get a set of homogeneous linear algebraic equations. Hence,

$$\frac{d\Pi}{dA_m} = 0$$

$$\frac{\pi^2 D}{b^2} \left[A_m (m^2 + \beta^2)^2 + 2\beta^3 \sum_{i=1}^{r} \gamma_i \sin \frac{m\pi c_i}{a} \left(A_1 \sin \frac{\pi c_i}{a} + A_2 \sin \frac{2\pi c_i}{a} + -- \right) \right] - N_x \beta^2 m^2 A_m = 0$$

(7.26h)

If there is only one transverse stiffener in the middle, i.e. $c_1 = a/2$, and assume $m = 1$, the number of half sine waves along the X direction in Eq. (7.26a). We obtain the first approximation for critical force N_x, from Eq. (7.26h) by taking A_1 as the only nonzero coefficient

$$\frac{\pi^2 D}{b^2} [A_1 (1 + \beta^2)^2 + 2\beta^3 \gamma A_1] - N_x \beta^2 A_1 = 0$$

or

$$N_{cr} = \frac{\pi^2 D}{b^2 \beta^2} [(1 + \beta^2)^2 + 2\gamma \beta^3]$$

(7.26i)

and

$$\sigma_{cr} = \frac{\pi^2 D}{b^2 h \beta^2} [(1 + \beta^2)^2 + 2\gamma \beta^3]$$

(7.26j)

If we assume A_1, A_2, and A_3 different than zero by taking $m = 1, 2,$ and 3, we get from Eq. (7.26h) three homogeneous linear equations.

$m = 1$

$$\frac{\pi^2 D}{b^2} \left[A_1 (1 + \beta^2)^2 + 2\beta^3 \gamma \sin \frac{\pi}{2} \left(A_1 \sin \frac{\pi}{2} + A_2 \sin \pi + A_3 \sin \frac{3\pi}{2} + -- \right) \right] - A_1 N_x \beta^2 = 0$$

$m = 2$

$$\frac{\pi^2 D}{b^2} \left[A_2 (4 + \beta^2)^2 + 2\beta^3 \gamma \sin \pi \left(A_1 \sin \frac{\pi}{2} + A_2 \sin \pi + A_3 \sin \frac{3\pi}{2} + -- \right) \right] - 4A_2 N_x \beta^2 = 0$$

$m = 3$

$$\frac{\pi^2 D}{b^2} \left[A_3 (9 + \beta^2)^2 + 2\beta^3 \gamma \sin \frac{3\pi}{2} \left(A_1 \sin \frac{\pi}{2} + A_2 \sin \pi + A_3 \sin \frac{3\pi}{2} + -- \right) \right] - 9A_3 N_x \beta^2 = 0$$

or

$$A_1 \left[(1 + \beta^2)^2 + 2\beta^3 \gamma - \frac{N_x b^2 \beta^2}{\pi^2 D} \right] - 2\beta^3 \gamma A_3 = 0$$

(7.26k)

$$A_2\left[(4+\beta^2)^2 - \frac{4N_x b^2 \beta^2}{\pi^2 D}\right] = 0 \tag{7.26l}$$

$$A_1[-2\beta^3\gamma] + A_3\left[(9+\beta^2)^2 + 2\beta^3\gamma - \frac{9N_x b^2 \beta^2}{\pi^2 D}\right] = 0 \tag{7.26m}$$

If we consider Eqs. (7.26k) and (7.26m), we can get a better approximation of the critical force N_{cr}.

$$\begin{bmatrix} e - k\beta^2 & -2\beta^3\gamma \\ -2\beta^3\gamma & f - 9k\beta^2 \end{bmatrix} \begin{Bmatrix} A_1 \\ A_3 \end{Bmatrix} = \begin{Bmatrix} 0 \\ 0 \end{Bmatrix} \tag{7.26n}$$

where

$$(1+\beta^2)^2 + 2\beta^3\gamma = e, \quad (9+\beta^2)^2 + 2\beta^3\gamma = f, \quad \text{and} \quad \frac{N_x b^2}{\pi^2 D} = k$$

For nontrivial solution of the coefficients A_1 and A_3 the determinant of the coefficient matrix is zero, thus

$$\begin{vmatrix} e - k\beta^2 & -2\beta^3\gamma \\ -2\beta^3\gamma & f - 9k\beta^2 \end{vmatrix} = 0$$

or

$$(e - k\beta^2)(f - 9k\beta^2) - 4\beta^6\gamma^2 = 0$$

or

$$9k^2\beta^4 - k\beta^2(9e+f) + ef - 4\beta^6\gamma^2 = 0 \tag{7.26o}$$

For various values of β and γ, the value of k can be found from Eq. (7.26o). Then N_{cr} and σ_{cr} can be found from Eqs. (7.24p) and (7.24q).

If γ is increased, the plate will buckle in two half sine-waves and the stiffener becomes the nodal line in the buckled plate, i.e. the stiffener remains straight during the buckling of plate. In this case, the critical force is obtained from Eq. (7.26l).

$$N_{cr} = \frac{\pi^2 D}{b^2} \frac{(4+\beta^2)^2}{4\beta^2} \tag{7.26p}$$

For a square plate, $\beta = 1$, hence

$$N_{cr} = 6.25\frac{\pi^2 D}{b^2}, \quad k = 6.25$$

whereas for the unstiffened square plate, $k = 4$ given in Figure 7.10a. The effect of the stiffener on the critical force depends on a/b ratio of the plate.

For $a/b = 1.41$, Eq. (7.26p) gives for a stiffened plate

$$k = \frac{(4 + \beta^2)^2}{4\beta^2} = \frac{(4 + 1.41^2)^2}{4(1.41)^2} = 4.51$$

For $a/b = \sqrt{2} = 1.41$, Eq. (7.7o) and Figure 7.9 give for an unstiffened plate

$$k = \left(\frac{mb}{a} + \frac{n^2 a}{mb}\right)^2 = \left(\frac{1}{1.41} + 1.41\right)^2 = 4.49$$

where $m = 1$ or 2 from Figure 7.9 for the minimum value of k.

7.6.10 Simply Supported Rectangular Plate with Stiffeners in Both the Longitudinal and Transverse Directions

If there are a number of equal and equidistant stiffeners, we can consider the stiffened plate as a plate having two different flexural rigidities in the two perpendicular directions [1]. It is considered an orthotropic plate and the moment curvature relations are obtained from

$$M_x = -\frac{(EI)_x}{1 - \nu_{xy}\nu_{yx}}\left(\frac{\partial^2 w}{\partial x^2} + \nu_{yx}\frac{\partial^2 w}{\partial y^2}\right) \tag{7.27a}$$

$$M_y = -\frac{(EI)_y}{1 - \nu_{xy}\nu_{yx}}\left(\frac{\partial^2 w}{\partial y^2} + \nu_{xy}\frac{\partial^2 w}{\partial x^2}\right) \tag{7.27b}$$

$$M_{xy} = 2(GI)_{xy}\frac{\partial^2 w}{\partial x \partial y} = 2D_{xy}\frac{\partial^2 w}{\partial x \partial y} \tag{7.27c}$$

where D_{xy} is the torsional rigidity of the plate, ν_{xy} and ν_{yx} are Poisson's ratios in the X and Y directions.

Combine Eqs. (7.1v), (7.1w), and (7.2i) and we have

$$\frac{\partial^2 M_x}{\partial x^2} - 2\frac{\partial^2 M_{xy}}{\partial x \partial y} + \frac{\partial^2 M_y}{\partial y^2} + N_x\frac{\partial^2 w}{\partial x^2} + N_y\frac{\partial^2 w}{\partial y^2} + 2N_{xy}\frac{\partial^2 w}{\partial x \partial y} = 0 \tag{7.27d}$$

Substitute Eqs. (7.27a)–(7.27c) into Eq. (7.27d), assume that the plate is subjected to a uniform compression in the X direction ($N_y = N_{xy} = 0$), then we can write

$$D_x\frac{\partial^4 w}{\partial x^4} + 2H\frac{\partial^4 w}{\partial x^2 \partial y^2} + D_y\frac{\partial^4 w}{\partial y^4} = N_x\frac{\partial^2 w}{\partial x^2} \tag{7.27e}$$

where

$$D_x = \frac{E_x h^3}{12(1 - \nu_{xy}\nu_{yx})}, \quad D_y = \frac{E_y h^3}{12(1 - \nu_{xy}\nu_{yx})}, \quad D_{xy} = \frac{Gh^3}{12} = \frac{E}{2(1 + \nu)}\frac{h^3}{12} \tag{7.27f}$$

$$2H = D_x\nu_{yx} + D_y\nu_{xy} + 4D_{xy}$$

$$D_x\nu_{yx} = D_y\nu_{xy}$$

$$H = D_x\nu_{yx} + 2D_{xy} \tag{7.27g}$$

where D_x and D_y are the flexural rigidities of the plate corresponding to the bending moments M_x and M_y respectively. Assume the plate buckles into one half sine wave in the X and Y directions, then

$$w = A \sin \frac{\pi x}{a} \sin \frac{\pi y}{b} \qquad (7.27h)$$

Substitute w and its derivatives in Eq. (7.27e) and we get

$$\left(D_x \frac{\pi^4}{a^4} + 2H \frac{\pi^4}{a^2 b^2} + D_y \frac{\pi^4}{b^4} - N_x \frac{\pi^2}{a^2} \right) A \sin \frac{\pi x}{a} \sin \frac{\pi y}{b} = 0 \qquad (7.27i)$$

or

$$N_x = \frac{\pi^2}{b^2} \left(D_x \frac{b^2}{a^2} + 2H + D_y \frac{a^2}{b^2} \right)$$

or

$$N_x = \frac{\pi^2}{b^2} \left(\frac{D_x}{\beta^2} + 2H + D_y \beta^2 \right) \qquad (7.27j)$$

where $\beta = a/b$. The minimum value of critical force $(N_x)_{cr}$ is obtained by taking the derivative of N_x with respect to β and equating it to zero. Thus

$$\frac{dN_x}{d\beta} = \frac{\pi^2}{b^2} \left(-\frac{2D_x}{\beta^3} + 2D_y \beta \right) = 0$$

or

$$D_x = D_y \beta^4, \text{ and } \beta = \left(\frac{D_x}{D_y} \right)^{\frac{1}{4}} \qquad (7.27k)$$

Substitute β from Eqs. (7.27k) into Eq. (7.27j) to obtain

$$(N_x)_{cr} = \frac{2\pi^2}{b^2} \left(\sqrt{D_x D_y} + H \right) \qquad (7.27l)$$

The critical compressive force is given by Eq. (7.27l). For an isotropic plate, Eq. (7.27g) gives

$$H = v \frac{Eh^3}{12(1 - v^2)} + 2 \frac{E}{2(1 + v)} \frac{h^3}{12} = \frac{Eh^3}{12(1 - v^2)} = D$$

Hence for an isotropic plate, $D_x = D_y = H = D$ and from Eq. (7.27l)

$$(N_x)_{cr} = \frac{4\pi^2 D}{b^2},$$

The same result as obtained in Eq. (7.7m).

7.7 Buckling of Circular Plates

We have so far analyzed the stability of rectangular plates. Many times circular plates are used in practice, e.g. pressure vessels, containers, foundations, etc. In the analysis of circular plates, it is simpler to employ polar coordinates. It is assumed that the material is linearly elastic and isotropic. Consider a circular plate of uniform cross-section that is subjected to a uniform compressive force N_r distributed around the edge of the plate as shown in Figure 7.28. The symmetric mode of deflection about the axis passing through the center of the plate and perpendicular to the plate is considered. The deflected surface of the plate is a surface of revolution, where ϕ is the angle between the axis of revolution and normal to the deflected plate at any point. Let O be the origin of coordinates at the center of the un-deflected plate. Let us take r to be the radial distance of a point in the middle plane of the plate, θ is the horizontal angle, and the z coordinate is perpendicular to the original plate surface before deflection.

The maximum slope of the deflection surface at any point is $- dw/dr$. For small deflections the curvatures of the middle surface of the plate in the diametral section rz and in the perpendicular direction are respectively given by [1]:

$$\frac{1}{r_n} = -\frac{d^2w}{dr^2} = \frac{d\phi}{dr} \tag{7.28a}$$

$$\frac{1}{r_t} = -\frac{1}{r}\frac{dw}{dr} = \frac{\phi}{r} \tag{7.28b}$$

For pure bending

$$M_r = D\left(\frac{1}{r_n} + v\frac{1}{r_t}\right) \tag{7.28c}$$

and

$$M_\theta = D\left(\frac{1}{r_t} + v\frac{1}{r_n}\right) \tag{7.28d}$$

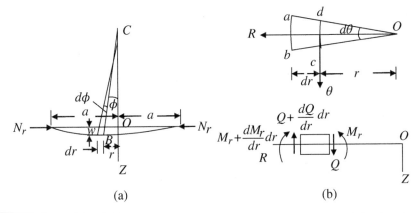

(a)   (b)

Figure 7.28 Circular plate subjected to compressive force at the edge: (a) Buckling of circular plate; (b) Internal forces on an element.

Substitute the curvature values given by Eqs. (7.28a) and (7.28b) in the moment curvature relations for pure bending, assuming these also hold during buckling of the plate subjected to pure compression

$$M_r = -D\left(\frac{d^2w}{dr^2} + \frac{v}{r}\frac{dw}{dr}\right) = D\left(\frac{d\phi}{dr} + v\frac{\phi}{r}\right) \tag{7.28e}$$

and

$$M_\theta = -D\left(\frac{1}{r}\frac{dw}{dr} + v\frac{d^2w}{dr^2}\right) = D\left(\frac{\phi}{r} + v\frac{d\phi}{dr}\right) \tag{7.28f}$$

where M_r and M_θ are the bending moments per unit length on the circumferential and radial sections respectively shown in Figure 7.28. Q and $Q + \frac{dQ}{dr}dr$ are the shear forces per unit length on the circumferential sections cd and ab respectively in Figure 7.28. Ignore the small difference in the shear forces. Consider the moment equilibrium of the element $abcd$ about θ axis to get

$$\left(M_r + \frac{dM_r}{dr}dr\right)(r + dr)d\theta - M_r r d\theta - M_\theta dr d\theta + Qr dr d\theta = 0$$

Neglecting small quantities,

$$M_r + r\frac{dM}{dr} - M_\theta + Qr = 0 \tag{7.28g}$$

Substitute for M_r and M_θ from Eqs. (7.28e) and (7.28f) into Eq. (7.28g) to have

$$\frac{d^2\phi}{dr^2} + \frac{1}{r}\frac{d\phi}{dr} - \frac{\phi}{r^2} = -\frac{Q}{D} \tag{7.28h}$$

or

$$\frac{d^3w}{dr^3} + \frac{1}{r}\frac{d^2w}{dr^2} - \frac{1}{r^2}\frac{dw}{dr} = \frac{Q}{D} \tag{7.28i}$$

The shear force Q as a component of the distributed force N_r around the plate edge as shown in Figure 7.29 is written

$$Q = N_r\phi \tag{7.28j}$$

Let

$$\frac{N_r}{D} = \lambda^2 \tag{7.28k}$$

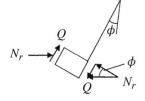

Figure 7.29 Plate element showing shear and compressive force.

Substitute for Q and N_r/D in Eq. (7.28h) to obtain

$$r^2 \frac{d^2\phi}{dr^2} + r\frac{d\phi}{dr} + (\lambda^2 r^2 - 1)\phi = 0 \tag{7.28l}$$

Let $u = \lambda r$, thus, $\frac{du}{dr} = \lambda$, hence

$$\frac{d\phi}{dr} = \frac{d\phi}{du}\frac{du}{dr} = \lambda\frac{d\phi}{du}, \quad \text{and} \quad r\frac{d\phi}{dr} = u\frac{d\phi}{du}$$

Similarly,

$$r^2 \frac{d^2\phi}{dr^2} = u^2 \frac{d^2\phi}{du^2}$$

Now, we can write Eq. (7.28l) in terms of the new variable, u, and its derivatives

$$u^2 \frac{d^2\phi}{du^2} + u\frac{d\phi}{du} + (u^2 - 1)\phi = 0 \tag{7.28m}$$

It is a Bessel differential equation of order 1 for real variable u. If instead of $(u^2 - 1)$, the term was $(u^2 - n)$, the Bessel differential equation would be classified of order n [7]. The general solution of Eq. (7.28m) is [3]

$$\phi = C_1 J_1(u) + C_2 Y_1(u) \tag{7.28n}$$

where $J_1(u)$ and $Y_1(u)$ are Bessel functions of the first order of the first and second kinds, respectively. The coefficients C_1 and C_2 are obtained from the boundary conditions.

7.7.1 Clamped Plate

For a circular plate that is fixed around its edge, the boundary conditions are:

1. At the center of the plate ($r = u = 0$), and the angle $\phi = 0$ because of the symmetry of the deflected plate. Since $Y_1(0) = -\infty$ [8] as u approaches zero, we should take $C_2 = 0$. Hence,

$$\phi = C_1 J_1(u) \tag{7.29a}$$

2. The slope at the clamped edge of the plate is zero, i.e. $\phi|_{r=a} = 0$. Hence from Eq. (7.29a) we have

$$C_1 J_1(\lambda a) = 0$$

or

$$J_1(\lambda a) = 0 \tag{7.29b}$$

From the table of function $J_1(u)$ [8], the smallest root of Eq. (7.29b) is

$$\lambda a = 3.832$$

Substitute for λ in Eq. (7.28k) to get

$$(N_r)_{cr} = \frac{14.684D}{a^2} \qquad (7.29c)$$

For a strip of unit width and length equal to the diameter of the plate having clamped ends, the critical compressive force is

$$(N_r)_{cr} = \frac{\pi^2 D}{a^2}$$

It shows the buckling capacity of a circular plate is 48.78% higher than for a strip of unit width with clamped ends.

7.7.2 Simply Supported Plate

For a circular plate that is simply supported around its edge, the boundary conditions are

1. From the symmetry of deflection at the center of the plate we have

$$C_2 = 0$$

or

$$\phi = C_1 J_1(u) \qquad (7.30a)$$

2. The bending moment along the edge is zero, hence,

$$M_r = \left(\frac{d\phi}{dr} + v\frac{\phi}{r} \right) = 0$$

$$\frac{d\phi}{dr} = \lambda\frac{d\phi}{du} = C_1\lambda\frac{dJ_1(u)}{du}$$

or

$$C_1\lambda\left[\frac{dJ_1(u)}{du} + v\frac{J_1(u)}{u} \right]_{r=a} = 0 \qquad (7.30b)$$

The derivative of $J_1(u)$ is written as [7]

$$\frac{d}{du}[uJ_1(u)] = uJ_0(u)$$

or

$$\frac{dJ_1}{du} = J_0(u) - \frac{J_1(u)}{u} \qquad (7.30c)$$

From Eqs. (7.30b) and (7.30c) we obtain

$$C_1\lambda\left[J_0(u) - \frac{J_1(u)}{u} + v\frac{J_1(u)}{u} \right]_{r=a} = 0 \qquad (7.30d)$$

where J_0 is the Bessel function of the zero order:

$$\lambda J_0(u) - \frac{J_1(u)}{a} + v\frac{J_1(u)}{a} = 0$$

Assume the Poisson's ratio $v = 0.3$ and we get

$$\lambda a J_0(\lambda a) - 0.7 J_1(\lambda a) = 0 \tag{7.30e}$$

The smallest root of Eq. (7.30e) by using tables of functions J_0 and J_1 is found to be, $\lambda a = 2.05$. Substitute for λ in Eq. (7.28k) to get

$$(N_r)_{cr} = \frac{4.2025D}{a^2} \tag{7.30f}$$

The critical force for the clamped circular plate in Eq. (7.29c) is 3.494 times the critical force for the simply supported circular plate given by Eq. (7.30f).

7.8 The Finite Difference Method

The finite difference method is a numerical technique to solve differential equations. The method reduces an infinite degree of freedom problem to one with finite degrees of freedom. In this method, a differential equation is replaced by a set of equivalent algebraic difference equations that are easier to solve. This is achieved by replacing each derivative of a function at a point by an algebraic expression consisting of the value of the function at the reference and neighboring points. Hence, the accuracy of the solution is increased if the points are closer to one another because the algebraic approximations of the derivatives are more accurate. The number of simultaneous equations solved increase as the number of points increase. This method has the disadvantage that it gives the values of the function at discrete points only instead of an analytical expression that can be used for the entire system. If an analytical expression for the function is needed, it is obtained by fitting a curve to the discrete values obtained. More discussion on this technique is given by Salvadori and Baron [9], Ralston [10], and Szilard [11].

Let us consider a function $f(x)$ shown in Figure 7.30 whose values are known at the reference point $x = i$ and at other evenly spaced points to the left and the right of it. The first derivative of the function $f(x)$ at the point $x = i$ can be written as

$$\left.\frac{df}{dx}\right|_{x=i} = f'(x_i) \cong \Delta f_i = \frac{f(x_{i+h}) - f(x_i)}{h} \tag{7.31a}$$

where $f(x_i)$ and $f(x_{i+h})$ are the values of the function $f(x)$ at $x = i$ and $x = i + h$, and h is the interval at which the function's values are known. The accuracy of differentiation given by Δf_i is called the first difference. It increases as the interval h is reduced. The differentiation in Eq. (7.31a) is called the forward difference.

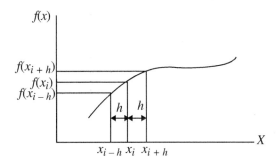

Figure 7.30 Grid points for the central finite difference method.

Similarly, we can write

$$\frac{df}{dx}\bigg|_{x=i} = f'(x_i) \cong \Delta f_i = \frac{f(x_i) - f(x_{i-h})}{h} \tag{7.31b}$$

where $f(x_{i-h})$ is the value of the function at $x = i - h$. The differentiation in Eq. (7.31b) is called the backward difference. If we use the values of the function $f(x)$ on both sides of $x = i$, the differentiation can be written as

$$\frac{df}{dx}\bigg|_{x=i} = f'(x_i) \cong \Delta f_i = \frac{f(x_{i+h}) - f(x_{i-h})}{2h} \tag{7.31c}$$

The differentiation in Eq. (7.31c) is called the central difference, and it is the most accurate of the three differentiations for an interval h. The higher derivatives of the function $f(x)$ are derived here for the central differences only. The second derivative, $\frac{d^2 f}{dx^2}$, is approximately expressed by the second difference, $\Delta^2 f_i$, by taking the difference of the first difference. Hence,

$$\frac{d^2 f}{dx^2}\bigg|_{x=i} = f''(x) \cong \Delta^2 f_i = \Delta(\Delta f_i) = \frac{\Delta[f(x_{i+h/2}) - f(x_{i-h/2})]}{h} = \frac{\Delta f(x_{i+h/2}) - \Delta f(x_{i-h/2})}{h}$$

$$\frac{d^2 f}{dx^2}\bigg|_{x=i} \cong \Delta^2 f_i = \frac{\dfrac{f(x_{i+h}) - f(x_i)}{h} - \dfrac{f(x_i) - f(x_{i-h})}{h}}{h} = \frac{f(x_{i+h}) - 2f(x_i) + f(x_{i-h})}{h^2} \tag{7.31d}$$

Similarly, the third derivative is approximately expressed by the third difference as follows:

$$\frac{d^3 f}{dx3}\bigg|_{x=i} = f'''(x) \cong \Delta^3 f_i = \Delta^2(\Delta f_i) = \frac{\Delta^2 f(x_{i+h}) - \Delta^2 f(x_{i-h})}{2h}$$

$$\frac{d^3 f}{dx^3}\bigg|_{x=i} \cong \Delta^3 f_i = \frac{\dfrac{f(x_{i+2h}) - 2f(x_{i+h}) + f(x_i)}{h^2} - \dfrac{f(x_i) - 2f(x_{i-h}) + f(x_{i-2h})}{h^2}}{2h}$$

$$= \frac{f(x_{i+2h}) - 2f(x_{i+h}) + 2f(x_{i-h}) - f(x_{i-2h})}{2h^3} \tag{7.31e}$$

The fourth derivative is approximately expressed by the fourth difference as

$$\frac{d^4f}{dx^4}\bigg|_{x=i} \cong \Delta^4 f_i = \Delta^2(\Delta^2 f_i) = \frac{\Delta^2 f(x_{i+h}) - 2\Delta^2 f(x_i) + \Delta^2 f(x_{i-h})}{h^2}$$

$$\Delta^4 f_i = \frac{\dfrac{f(x_{i+2h}) - 2f(x_{i+h}) + f(x_i)}{h^2} - 2\left[\dfrac{f(x_{i+h}) - 2f(x_i) + f(x_{i-h})}{h^2}\right]}{h^2}$$
$$+ \dfrac{f(x_i) - 2f(x_{i-h}) + f(x_{i-2h})}{h^2}$$

$$\frac{d^4 f}{dx^4}\bigg|_{x=i} \cong \cong \Delta^4 f_i = -\frac{f(x_{i+2h}) - 4f(x_{i+h}) + 6f(x_i) - 4f(x_{i-h}) + f(x_{i-2h})}{h^4} \tag{7.31f}$$

To solve plate problems it is necessary to obtain the expressions for partial derivatives in terms of differences. Consider a plate that is represented by a number of discrete points as shown in Figure 7.31. The points are equally spaced at distances of h and k in the X and Y directions respectively. Equations (7.31d) and (7.31f) can be used to write the second and fourth difference expressions representing the second- and fourth-order partial derivatives as

$$\left(\frac{\partial^2 w}{\partial x^2}\right)_{i,j} = \frac{w_{i+h,j} - 2w_{i,j} + w_{i-h,j}}{h^2} \tag{7.31g}$$

$$\left(\frac{\partial^4 w}{\partial x^4}\right)_{i,j} = \frac{w_{i+2h,j} - 4w_{i+h,j} + 6w_{i,j} - 4w_{i-h,j} + w_{i-2h,j}}{h^4} \tag{7.31h}$$

$$\left(\frac{\partial^2 w}{\partial y^2}\right)_{i,j} = \frac{w_{i,j+k} - 2w_{i,j} + w_{i,j-k}}{k^2} \tag{7.31i}$$

$$\left(\frac{\partial^4 w}{\partial y^4}\right)_{i,j} = \frac{w_{i,j+2k} - 4w_{i,j+k} + 6w_{i,j} - 4w_{i,j-k} + w_{i,j-2k}}{k^4} \tag{7.31j}$$

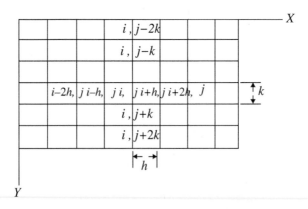

Figure 7.31 Finite difference grid points in two dimensions.

$$\left(\frac{\partial^4 w}{\partial x^2 \partial y^2}\right)_{i,j} = \frac{\partial^2}{\partial x^2}\left(\frac{\partial^2 w}{\partial y^2}\right)_{i,j} = \frac{1}{h^2}\left[\left(\frac{\partial^2 w}{\partial y^2}\right)_{i+h,j} - 2\left(\frac{\partial^2 w}{\partial y^2}\right)_{i,j} + \left(\frac{\partial^2 w}{\partial y^2}\right)_{i-h,j}\right]$$

$$\left(\frac{\partial^4 w}{\partial x^2 \partial y^2}\right)_{i,j} = \frac{1}{h^2 k^2}[(w_{i+h,j+k} - 2w_{i+h,j} + w_{i+h,j-k}) - 2(w_{i,j+k} - 2w_{i,j} + w_{i,j-k})$$

$$+ (w_{i-h,j+k} - 2w_{i-h,j} + w_{i-h,j-k})] \quad (7.31k)$$

7.8.1 Critical Load for a Simply Supported Plate Subjected to Biaxial Loading

Consider a plate that is simply supported on all sides acted on by distributed compressive forces of $N_x = N_y = N$ shown in Figure 7.32.

The governing equation of plate buckling, Eq. (7.2j) is written as

$$\frac{\partial^4 w}{\partial x^4} + 2\frac{\partial^4 w}{\partial x^2 \partial y^2} + \frac{\partial^4 w}{\partial y^4} + \frac{N}{D}\left(\frac{\partial^2 w}{\partial x^2} + \frac{\partial^2 w}{\partial y^2}\right) = 0 \quad (7.32a)$$

The boundary conditions for a simply supported plate are

$$w = \frac{\partial^2 w}{\partial x^2} = 0 \quad \text{at } x = 0, a \quad (7.32b)$$

$$w = \frac{\partial^2 w}{\partial y^2} = 0 \quad \text{at } y = 0, b \quad (7.32c)$$

Divide the plate into a mesh of $m \times n$ size. The distances of points in the x and y directions are $h = a/m$ and $k = b/n$ respectively, as shown in Figure 7.33.

Consider a point, $w(i, j)$, on the edges $x = 0$ and $x = a$ then

$$w(i,j) = 0 \text{ for } i = 0 \text{ and } j = 0,1,2, \text{---}n \quad (7.32d)$$

$$w(i,j) = 0 \text{ for } i = m \text{ and } j = 0,1,2, \text{---}n \quad (7.32e)$$

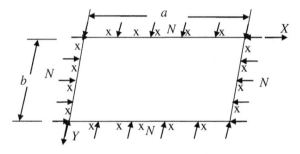

Figure 7.32 Simply supported plate with a biaxial loading.

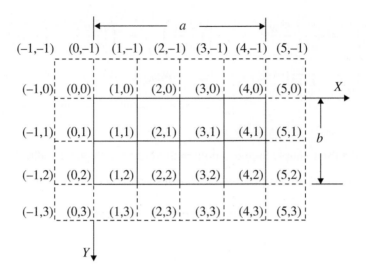

Figure 7.33 Finite difference mesh for $m = 4$ and $n = 2$.

Now

$$\left(\frac{\partial^2 w}{\partial x^2}\right)_{i,j} = \frac{w_{i+h,j} - 2w_{i,j} + w_{i-h,j}}{h^2} = 0,$$

hence,

$$w_{i+h,j} - 2w_{i,j} + w_{i-h,j} = 0$$

Using Eqs. (7.32d) and (7.32e) we have

$$w_{i+h,j} = -w_{i-h,j} \tag{7.32f}$$

at

$$i = 0, \text{ and } j = 0, 1, 2, \text{---}n$$

and
$$i = m, \text{ and } j = 0, 1, 2, \text{- - - }n$$

Similarly, if we consider a point $w(i, j)$ on the edges $y = 0$ and $y = b$, we can show using Eqs. (7.32c) and (7.31i) that

$$w_{i,j+k} = -w_{i,j-k} \tag{7.32g}$$

at $j = 0$, and $i = 0, 1, 2, \text{- - - }m$
and
$$j = n, \text{ and } i = 0, 1, 2, \text{- - - }m$$

This shows that for a simply supported edge, the deflection at a point immediately outside the boundary is negative of the deflection at the corresponding point inside the boundary. We write for the governing Eq. (7.32a) the difference equation at the internal point (1,1) as

$$
\frac{1}{h^4}\left[6w_{1,1} - 4w_{2,1} - 4w_{0,1} + w_{3,1} + w_{-1,1}\right] + \frac{1}{k^4}\left[6w_{1,1} - 4w_{1,2} - 4w_{1,0} + w_{1,-1} + w_{1,3}\right]
$$

$$
+\frac{1}{h^2k^2}\left[8w_{1,1} - 4w_{2,1} - 4w_{0,1} - 4w_{1,2} - 4w_{1,0} + 2w_{2,2} + 2w_{2,0} + 2w_{0,2} + 2w_{0,0}\right]
$$

$$
+\frac{N}{D}\left[\frac{w_{2,1} - 2w_{1,1} + w_{0,1}}{h^2} + \frac{w_{1,2} - 2w_{1,1} + w_{1,0}}{k^2}\right] = 0
$$

Substitute $w_{0,1} = w_{1,2} = w_{1,0} = w_{2,2} = w_{2,0} = w_{0,2} = w_{0,0} = 0$ at the boundaries to obtain

$$
\frac{1}{h^4}\left[6w_{1,1} - 4w_{2,1} + w_{3,1} + w_{-1,1}\right] + \frac{1}{k^4}\left[6w_{1,1} + w_{1,-1} + w_{1,3}\right] + \frac{1}{h^2k^2}\left[8w_{1,1} - 4w_{2,1}\right]
$$

$$
+\frac{N}{D}\left[\frac{w_{2,1} - 2w_{1,1}}{h^2} + \frac{-2w_{1,1}}{k^2}\right] = 0
$$

Substitute $w_{3,1} = w_{1,1}$ for symmetrical deflection of the plate and for simply supported boundary conditions $w_{-1,1} = w_{1,-1} = w_{1,3} = -w_{1,1}$ to get

$$
\frac{1}{h^4}\left[6w_{1,1} - 4w_{2,1}\right] + \frac{1}{k^4}\left[4w_{1,1}\right] + \frac{1}{h^2k^2}\left[8w_{1,1} - 4w_{2,1}\right] + \frac{N}{D}\left[\frac{w_{2,1} - 2w_{1,1}}{h^2} + \frac{-2w_{1,1}}{k^2}\right] = 0
$$

$$
\frac{w_{1,1}}{h^4k^4}\left[4h^4 + 6k^4 + 8h^2k^2 - \frac{2Nh^2k^4}{D} - \frac{2Nh^4k^2}{D}\right] - \frac{w_{2,1}}{h^4k^4}\left[4k^4 + 4h^2k^2 - \frac{Nh^2k^4}{D}\right] = 0
$$

$$
(7.32h)
$$

Similarly, the difference equation at the internal point (2,1) is

$$
\frac{1}{h^4}\left[6w_{2,1} - 4w_{3,1} - 4w_{1,1} + w_{4,1} + w_{0,1}\right] + \frac{1}{k^4}\left[6w_{2,1} - 4w_{2,2} - 4w_{2,0} + w_{2,-1} + w_{2,3}\right]
$$

$$
+\frac{1}{h^2k^2}\left[8w_{2,1} - 4w_{3,1} - 4w_{1,1} - 4w_{2,2} - 4w_{2,0} + 2w_{3,2} + 2w_{3,0} + 2w_{1,2} + 2w_{1,0}\right]
$$

$$
+\frac{N}{D}\left[\frac{w_{3,1} - 2w_{2,1} + w_{1,1}}{h^2} + \frac{w_{2,2} - 2w_{2,1} + w_{2,0}}{k^2}\right] = 0
$$

Substitute $w_{0,1} = w_{4,1} = w_{2,2} = w_{2,0} = w_{3,2} = w_{3,0} = w_{1,2} = w_{1,0} = 0$ at the boundaries to obtain

$$
\frac{1}{h^4}\left[6w_{2,1} - 4w_{3,1} - 4w_{1,1}\right] + \frac{1}{k^4}\left[6w_{2,1} + w_{2,-1} + w_{2,3}\right] + \frac{1}{h^2k^2}\left[8w_{2,1} - 4w_{3,1} - 4w_{1,1}\right]
$$

$$
+\frac{N}{D}\left[\frac{w_{3,1} - 2w_{2,1} + w_{1,1}}{h^2} + \frac{-2w_{2,1}}{k^2}\right] = 0
$$

Substitute $w_{3,1} = w_{1,1}$ for the symmetrical deflection of the plate and for the simply supported boundary conditions $w_{2,-1} = w_{2,3} = -w_{2,1}, w_{1,3} = -w_{1,1}$ to get

$$
\frac{w_{1,1}}{h^4k^4}\left[-8k^4 - 8h^2k^2 + \frac{2Nh^2k^4}{D}\right] + \frac{w_{2,1}}{h^4k^4}\left[4h^4 + 6k^4 + 8h^2k^2 - \frac{2Nh^2k^4}{D} - \frac{2Nh^4k^2}{D}\right] = 0
$$

$$
(7.32i)
$$

Equations (7.32h) and (7.32i) can be written in the form

$$
\begin{bmatrix} \left(\begin{array}{c} 4h^4 + 6k^4 + 8h^2k^2 \\ -\dfrac{2Nh^2k^4}{D} - \dfrac{2Nh^4k^2}{D} \end{array} \right) & -\left(4k^4 + 4h^2k^2 - \dfrac{Nh^2k^4}{D} \right) \\[6mm] -\left(8k^4 + 8h^2k^2 - \dfrac{2Nh^2k^4}{D} \right) & \left(\begin{array}{c} 4h^4 + 6k^4 + 8h^2k^2 \\ -\dfrac{2Nh^2k^4}{D} - \dfrac{2Nh^4k^2}{D} \end{array} \right) \end{bmatrix} \begin{Bmatrix} w_{1,1} \\[4mm] w_{2,1} \end{Bmatrix} = \begin{Bmatrix} 0 \\[4mm] 0 \end{Bmatrix} = 0 \qquad (7.32\text{j})
$$

Substitute

$$
4h^4 + 6k^4 + 8h^2k^2 - \frac{2Nh^2k^4}{D} - \frac{2Nh^4k^2}{D} = c
$$

and

$$
4k^4 + 4h^2k^2 - \frac{Nh^2k^4}{D} = d
$$

Hence, Eq. (7.32j) can be written as

$$
\begin{bmatrix} c & -d \\ -2d & c \end{bmatrix} \begin{Bmatrix} w_{1,1} \\ w_{2,1} \end{Bmatrix} = \begin{Bmatrix} 0 \\ 0 \end{Bmatrix} \qquad (7.32\text{k})
$$

For a nontrivial solution for $w_{1,1}$ and $w_{2,1}$, the determinant of the coefficient matrix in Eq. (7.32k) should be zero, thus

$$
c^2 - 2d^2 = 0 \qquad (7.32\text{l})
$$

For $\dfrac{a}{b} = 1$

$$
h = \frac{a}{4}, \quad k = \frac{b}{2} = \frac{a}{2}, \quad k = 2h
$$

$$
c = 132h^4 - \frac{40N}{D}h^6
$$

$$
d = 80h^4 - \frac{16Nh^6}{D}
$$

Substitute c and d in Eq. (7.32l) to get

$$
N_{cr} = 1.76\frac{\pi^2 D}{a^2} \qquad (7.32\text{m})
$$

The exact solution is $2\pi^2 D/a^2$, given in [3]. Hence, the approximate solution for 4×2 mesh size by the finite difference method differs by 12% [8].

For $\dfrac{a}{b} = 0.5$

$$
h = \frac{a}{4}, \quad k = \frac{b}{2} = a, \quad k = 4h
$$

$$c = 1668h^4 - \frac{544Nh^6}{D}$$

$$d = 1088h^4 - 256\frac{Nh^6}{D}$$

Substitute c and d in Eq. (7.32l) to obtain

$$N_{cr} = 1.15\frac{\pi^2 D}{a^2} \tag{7.32n}$$

The exact solution is $1.25\pi^2 D/a^2$, given in [3]. So the approximate solution for 4×2 mesh size by the finite difference method differs by 8%. The accuracy of the finite difference solution can be increased by decreasing the mesh size.

7.9 The Finite Element Method

The finite element method used in Section 5.4 for frames can also be applied to solve plate buckling problems. A plate can be discretized into a number of elements, each connected to the other at the nodal points. A plate is a two-dimensional structure, hence triangular, rectangular and other shapes of elements can be used to model the structure. In this method the structure is divided into number of finite elements that are connected at the joints called nodes. A continuum is thus discretized into a finite degree of freedom system, the greater the number of elements, the higher the accuracy of analysis one can expect. The unknowns in this method are the displacements and their derivatives at the nodes. A displacement function can be expressed as a polynomial whose coefficients are determined from the boundary conditions. The function should satisfy the continuity requirements within the element and should satisfy the compatibility conditions between the adjacent elements and should include rigid body displacements. By choosing such a function we can expect convergence of the solution. For more on the finite element method, books on the subject by Zienkiewicz [12] and Desai [13] can be recommended. A plate divided into a number of rectangular elements and the rectangular plate bending element are shown in Figures 7.34a and 7.34b respectively.

The plate element has three degrees of freedom at each node. The vertical deflection w_i and rotations about the x and y axes are shown at node i in Figure 7.34c. The negative sign in $\frac{\partial w_i}{\partial x}$ indicates that if positive displacement dw_i at a distance dx from the node i is considered, then the rotation about Y axis at the node i is in the negative direction of the Y axis. Therefore, the plate element has 12° of freedom and the shape function can be a 12-term polynomial. The shape function is given by

$$w(x,y) = \alpha_1 + \alpha_2 x + \alpha_3 y + \alpha_4 x^2 + \alpha_5 xy + \alpha_6 y^2 + \alpha_7 x^3 + \alpha_8 x^2 y + \alpha_9 xy^2 + \alpha_{10} y^3 + \alpha_{11} x^3 y + \alpha_{12} xy^3 \tag{7.33a}$$

or

$$w = [\psi]\{\alpha\} \tag{7.33b}$$

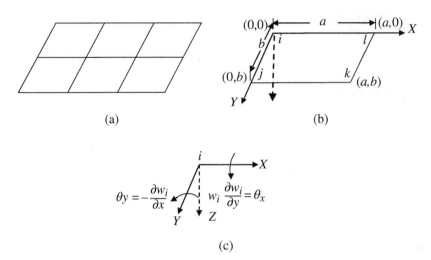

Figure 7.34 Finite element modeling of a plate: (a) Rectangular plate' (b) Rectangular plate bending element; (c) Nodal displacements.

where

$$[\psi] = \begin{bmatrix} 1 & x & y & x^2 & xy & y^2 & x^3 & x^2y & xy^2 & y^3 & x^3y & xy^3 \end{bmatrix} \tag{7.33c}$$

and

$$\{\alpha\} = \begin{bmatrix} \alpha_1 & \alpha_2 & \alpha_3 & \alpha_4 & \alpha_5 & \alpha_6 & \alpha_7 & \alpha_8 & \alpha_9 & \alpha_{10} & \alpha_{11} & \alpha_{12} \end{bmatrix}^T \tag{7.33d}$$

The conditions at node i are

$$w = w_i, \quad -\frac{\partial w}{\partial x} = -\frac{\partial w_i}{\partial x}, \quad \frac{\partial w}{\partial y} = \frac{\partial w_i}{\partial y}, \quad i = 1, 2, 3, 4 \tag{7.33e}$$

Use Eqs. (7.33a) and (7.33e) to write

$$[A]\{\alpha\} = \{\delta\}_e \tag{7.33f}$$

where

$$[A] = \begin{bmatrix}
1 & 0 & 0 & 0 & 0 & 0 & 0 & 0 & 0 & 0 & 0 & 0 \\
0 & 0 & 1 & 0 & 0 & 0 & 0 & 0 & 0 & 0 & 0 & 0 \\
0 & -1 & 0 & 0 & 0 & 0 & 0 & 0 & 0 & 0 & 0 & 0 \\
1 & 0 & b & 0 & 0 & b^2 & 0 & 0 & 0 & b^3 & 0 & 0 \\
0 & 0 & 1 & 0 & 0 & 2b & 0 & 0 & 0 & 3b^2 & 0 & 0 \\
0 & -1 & 0 & 0 & -b & 0 & 0 & 0 & -b^2 & 0 & 0 & -b^3 \\
1 & a & b & a^2 & ab & b^2 & a^3 & a^2b & ab^2 & b^3 & a^3b & ab^3 \\
0 & 0 & 1 & 0 & a & 2b & 0 & a^2 & 2ab & 3b^2 & a^3 & 3ab^2 \\
0 & -1 & 0 & -2a & -b & 0 & -3a^2 & -2ab & -b^2 & 0 & -3a^2b & -b^3 \\
1 & a & 0 & a^2 & 0 & 0 & a^3 & 0 & 0 & 0 & 0 & 0 \\
0 & 0 & 1 & 0 & a & 0 & 0 & a^2 & 0 & 0 & a^3 & 0 \\
0 & -1 & 0 & -2a & 0 & 0 & -3a^2 & 0 & 0 & 0 & 0 & 0
\end{bmatrix} \tag{7.33g}$$

and $\{\delta\}_e$ is the nodal displacement vector for an element that can be written as

$$\{\delta\}_e = \begin{Bmatrix} \delta_1 \\ \delta_2 \\ \delta_3 \\ \delta_4 \end{Bmatrix} \tag{7.33h}$$

where

$$\{\delta_1\} = \begin{Bmatrix} w_i \\ \dfrac{\partial w_i}{\partial y} \\ -\dfrac{\partial w_i}{\partial x} \end{Bmatrix}, i = 1, 2, 3, 4, \tag{7.33i}$$

From Eq. (7.33f)

$$\{\alpha\} = [A]^{-1}\{\delta\}_e \tag{7.33j}$$

and from Eq. (7.33g)

$$w = [\psi][A]^{-1}\{\delta\}_e \tag{7.33k}$$

The bending moments and the torsional moments acting on a $dx \times dy$ rectangular differential plate element are given by

$$\{M\} = \begin{Bmatrix} M_x \\ M_y \\ M_{xy} \end{Bmatrix} \tag{7.33l}$$

The corresponding curvatures and the twists are obtained from

$$\{\phi\} = \begin{Bmatrix} -\dfrac{\partial^2 w}{\partial x^2} \\ -\dfrac{\partial^2 w}{\partial y^2} \\ 2\dfrac{\partial^2 w}{\partial x \partial y} \end{Bmatrix} \tag{7.33m}$$

Equation (7.33m) can be written as

$$\{\phi\} = [Q]\{\alpha\} \tag{7.33n}$$

where $[Q]$ is obtained from Eq. (7.33a) by differentiating twice as follows:

$$[Q] = \begin{bmatrix} 0 & 0 & 0 & -2 & 0 & 0 & -6x & -2y & 0 & 0 & -6xy & 0 \\ 0 & 0 & 0 & 0 & 0 & -2 & 0 & 0 & -2x & -6y & 0 & -6xy \\ 0 & 0 & 0 & 0 & 2 & 0 & 0 & 4x & 4y & 0 & 6x^2 & 6y^2 \end{bmatrix} \tag{7.33o}$$

From Eqs. (7.33j) and (7.33n) we have

$$\{\phi\} = [Q][A]^{-1}\{\delta\}_e$$

or

$$\{\phi\} = [B]\{\delta\}_e \tag{7.33p}$$

where

$$[B] = [Q][A]^{-/1}$$

For a linear elastic material, the moment curvature relations are given by Eqs. (7.1r)–(7.1t), so we can write

$$[M] = [D]\{\phi\} \tag{7.33q}$$

where

$$[D] = \frac{Eh^3}{12(1 - v^2)}\begin{bmatrix} 1 & v & 0 \\ v & 1 & 0 \\ 0 & 0 & \left(\dfrac{1 - v}{2}\right) \end{bmatrix} \tag{7.33r}$$

E and v are the modulus of elasticity and Poisson's ratio of the plate material respectively, and h is the thickness of the plate. The principle of stationary potential energy is used as was applied in Section 5.4 to formulate the stiffness matrix for the plate element. The strain energy of bending for the differential plate element is given by

$$dU = \frac{1}{2}\{\phi\}^T[M]\,dxdy$$

The strain energy for the entire plate element is obtained from

$$U = \frac{1}{2}\int_0^b \int_0^a \{\phi\}^T[M]\,dxdy$$

or

$$U = \frac{1}{2}\int_0^b \int_0^a \{\phi\}^T[D]\{\phi\}\,dxdy$$

or

$$U = \frac{1}{2}\{\delta\}_e^T \int_0^b \int_0^a [B]^T[D][B]\{\delta\}_e\,dxdy \tag{7.33s}$$

The external work is done by the nodal forces $\{q\}_e$ consisting of two moments and a lateral force at each node, corresponding to the nodal degrees of freedom $\{\delta\}_e$, and by the in-plane forces $[N]$. For an element, the nodal forces are

$$\{q\}_e = \begin{Bmatrix} q_1 \\ q_2 \\ q_3 \\ q_4 \end{Bmatrix} \tag{7.33t}$$

where

$$\{q_i\} = \begin{Bmatrix} Q_i \\ M_{yi} \\ M_{xi} \end{Bmatrix}$$

The relation between the nodal forces and the nodal displacements is given by

$$\{q\}_e = [k]_e \{\delta\}_e \tag{7.33u}$$

where $[k]_e$ is the element stiffness matrix. The external work due to nodal forces is

$$W_q = \frac{1}{2} \{\delta\}_e^T \{q\}_e$$

or

$$W_q = \frac{1}{2} \{\delta\}_e^T [k]_e \{\delta\}_e$$

The external work done by the in-plane forces is obtained from Eqs. (7.18c), (7.18d), and (7.20f)

$$W_N = \frac{1}{2} \int_0^b \int_0^a \{\beta\}^T [N] \{\beta\} dxdy$$

where

$$\{\beta\} = \begin{Bmatrix} \dfrac{\partial w}{\partial x} \\ \dfrac{\partial w}{\partial y} \end{Bmatrix}$$

$$\{\beta\} = [P]\{\alpha\} \quad \text{and} \quad [N] = \begin{bmatrix} N_x & N_{xy} \\ N_{xy} & N_y \end{bmatrix}$$

where $[P]$ is obtained from Eq. (7.33a) by differentiating once as follows:

$$[P] = \begin{bmatrix} 0 & 1 & 0 & 2x & y & 0 & 3x^2 & 2xy & y^2 & 0 & 3x^2y & y^3 \\ 0 & 0 & 1 & 0 & x & 2y & 0 & x^2 & 2xy & 3y^2 & x^3 & 3xy^2 \end{bmatrix}$$

$$\{\beta\} = [P][A]^{-1}\{\delta\}_e = [C]\{\delta\}_e$$

where

$$[C] = [P][A]^{-1}$$

Hence,

$$W_N = \frac{1}{2} \int_0^b \int_0^a \{\delta\}_e^T [C]^T [N][C]\{\delta\}_e dxdy$$

The total external work

$$W = W_q + W_N = \frac{1}{2}\{\delta\}_e^T [k]_e \{\delta\}_e + \frac{1}{2}\{\delta\}_e^T \int_0^b \int_0^a [C]^T [N][C]\{\delta\}_e dxdy$$

Equate the total external work to the strain energy and we get

$$\frac{1}{2}\{\delta\}_e^T [k]_e \{\delta\}_e + \frac{1}{2}\{\delta\}_e^T \int_0^b \int_0^a [C]^T [N] [C] \{\delta\}_e dxdy = \frac{1}{2}\{\delta\}_e^T \int_0^b \int_0^a [B]^T [D] [B] \{\delta\}_e dxdy$$

or

$$[k]_e \{\delta\}_e = \left\{ \int_0^b \int_0^a [B]^T [D] [B] \, dxdy - \int_0^b \int_0^a [C]^T [N] [C] \, dxdy \right\} \{\delta\}_e$$

Therefore, the stiffness matrix of a plate element is given by

$$[k]_e = \int_0^b \int_0^a [B]^T [D] [B] \, dxdy - \int_0^b \int_0^a [C]^T [N] [C] \, dxdy \qquad (7.33v)$$

or

$$[k]_e = [k]_e^1 - [k]_e^2 \qquad (7.33w)$$

Thus, the stiffness matrix for the plate element contains two matrices, the first one, $[k]_e^1$, is the pure bending matrix, and the second, $[k]_e^2$, is called the geometric or initial stress stiffness matrix that gives the influence of in-plane forces on the bending stiffness. The numerical values of the stiffness matrices $[k]_e^1$ and $[k]_e^2$ require the calculation of $[A]$, $[B]$, and $[C]$ matrices and then solve the integrals in Eq. (7.33v). We get the element stiffness matrix in Eq. (7.33w) in the local coordinates of the element shown in Figure 7.34. These element stiffness matrices are converted into a global coordinate system and then combined to give the structure stiffness matrix. The structure stiffness matrix is then reduced by applying the boundary conditions. When a uniform compressive force of N_x is applied on a plate, the critical loading is obtained from the equations

$$\{[K^1] - N_x [K^2]\} \{\delta\} = 0 \qquad (7.33x)$$

where $[K^1]$ and $[K^2]$ are the structure bending and geometric or initial stress stiffness matrices, and $\{\delta\}$ is the structure nodal displacement vector. For a nontrivial solution, the determinant is $|\{[K^1] - N_x[K^2]\}| = 0$, which gives the characteristic equation whose smallest root called the lowest eigenvalue gives the critical load. If we are dealing with a large number of equations, then the lowest eigenvalue is obtained by iteration. Equation (7.33x) is pre-multiplied by $[K^2]^{-1}$ before conducting iteration as follows:

$$\left\{ [K^2]^{-1} [K^1] - N_x [K^2]^{-1} [K^2] \right\} \{\delta\} = 0$$

or

$$\left\{ [K^2]^{-1} [K^1] - N_x [I] \right\} \{\delta\} = 0 \qquad (7.33y)$$

where I is the identity matrix. By iteration the highest eigenvalue is obtained first from the matrix in Eq. (7.33y). The matrix in Eq. (7.33y) is inverted to give

$$\left\{ [K1]^{-1} [K^2] - \frac{1}{N_x} [I] \right\} \{\delta\} = 0 \qquad (7.33z)$$

From the iteration process of the matrix in Eq. (7.33z) we find first the highest value of $1/N_x$, which will be the lowest value for N_x. For more details of the procedure to find the critical load for plates under compressive loading by finite element method, books on the subject mentioned above in this chapter can be referred to.

7.10 Large Deflection Theory of Plates

The in-plane equilibrium equations for the small displacement theory of plates are given by

$$\frac{\partial N_x}{\partial x} + \frac{\partial N_{yx}}{\partial y} = 0 \tag{7.2a}$$

$$\frac{\partial N_y}{\partial y} + \frac{\partial N_{xy}}{\partial x} = 0 \tag{7.2b}$$

The equilibrium equation in the z direction is

$$D\left(\frac{\partial^4 w}{\partial x^4} + 2\frac{\partial^4 w}{\partial x^2 \partial y^2} + \frac{\partial^4 w}{\partial y^4}\right) - N_x\frac{\partial^2 w}{\partial x^2} - N_y\frac{\partial^2 w}{\partial y^2} - 2N_{xy}\frac{\partial^2 w}{\partial x \partial y} = 0 \tag{7.2j}$$

In these equations the strains in the middle surface of the plate due to bending are neglected, therefore, the quantities N_x, N_y, and N_{xy} are constant, whereas in the large deflection analyses these quantities are variable unknown forces as a function of x and y. They represent variable membrane forces in the middle surface of the plate due to bending, which were neglected in the small deflection theory, and the constant applied edge loads. There are four unknown variables in the above nonlinear equations, three in-plane forces N_x, N_y, N_{xy}, and vertical deflection w. Hence, one additional equation is needed to solve the problem. The additional equation is obtained by considering the strain-displacement relations of the middle surface of the plate.

At any point in the plate the total displacement consists of two parts, one due to in-plane forces that is constant over the thickness of the plate, and the other due to bending that varies from zero at the middle surface to maximum at the outer surfaces. The total displacements u and v at a point in the plate are given by

$$u = u_0 + u_b$$
$$v = v_0 + v_b$$

where u_0 and v_0 are the middle surface displacements, u_b and v_b are the displacements due to bending relative to the middle surface. The strain ε_z is neglected, hence w is constant over the thickness of the plate. Consider an element AB of the middle surface of the plate shown in Figure 7.35. The element AB goes to position $A'B'$ due to buckling of the plate. The elongation of the element due to in-plane displacement u_0 is given by $\frac{\partial u_0}{\partial x}dx$. The change in the length of the element due to w displacement from Chapter 2 (see Eq. 2.28e) is given by $\frac{1}{2}\left(\frac{\partial w}{\partial x}\right)^2 dx$.

The total middle surface strain in the X direction is

$$\varepsilon_{x0} = \frac{\partial u_0}{\partial x} + \frac{1}{2}\left(\frac{\partial w}{\partial x}\right)^2 \tag{7.34a}$$

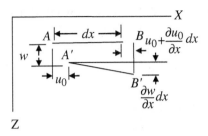

Figure 7.35 Axial displacements in plate large deflection analysis.

Similarly, the middle surface strain in the Y direction is

$$\varepsilon_{y0} = \frac{\partial v_0}{\partial y} + \frac{1}{2}\left(\frac{\partial w}{\partial y}\right)^2 \tag{7.34b}$$

The middle surface shear strain that is the change in the angle between the two perpendicular lines OA and OB occurs due to the buckling of the plate shown in Figure 7.36. It consists of two parts, one due to the plane displacements u_0 and v_0 and the other due to the transverse displacement w.

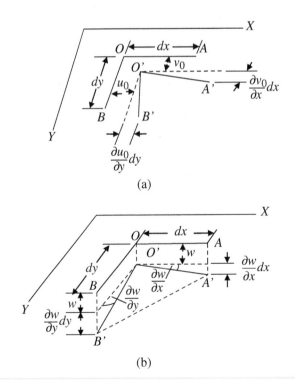

(a)

(b)

Figure 7.36 Shear strain in a plate large deflection analysis: (a) Shear strain due to middle surface displacements u_0 and v_0; (b) Shear strain due to displacement w in the z direction.

The shear strain due to u_0 and v_0 from Figure 7.36a is $\dfrac{\partial u_0}{\partial y} + \dfrac{\partial v_0}{\partial x}$. The shear strain due to w is given by the difference between the angles AOB and $A'O'B'$. The angle $A'O'B'$ is equal to $\dfrac{\pi}{2} - \gamma$ in Figure 7.36b. The length $A'B'$ is expressed as

$$(O'A')^2 = dx^2 + \left(\frac{\partial w}{\partial x} dx \right)^2$$

$$(O'B')^2 = dy^2 + \left(\frac{\partial w}{\partial y} dy \right)^2$$

$$(A'B')^2 = dx^2 + dy^2 + \left(\frac{\partial w}{\partial y} dy - \frac{\partial w}{\partial x} dx \right)^2 \tag{7.34c}$$

$A'B'$ can also be written as

$$(A'B')^2 = (O'A')^2 + (O'B')^2 - 2(O'A')(O'B') \cos \left(\frac{\pi}{2} - \gamma \right)$$

$$O'A' = dx \left[1 + \left(\frac{\partial w}{\partial x} \right)^2 \right]^{\frac{1}{2}} = dx \left[1 + \frac{1}{2} \left(\frac{\partial w}{\partial x} \right)^2 + -- \right] \approx dx$$

Similarly, $O'B' \approx dy$

Hence,

$$(A'B')^2 = dx^2 + \left(\frac{\partial w}{\partial x} dx \right)^2 + dy^2 + \left(\frac{\partial w}{\partial y} dy \right)^2 - 2 dx dy \cos \left(\frac{\pi}{2} - \gamma \right) \tag{7.34d}$$

From Eqs. (7.34c) and (7.34d), we have

$$\cos \left(\frac{\pi}{2} - \gamma \right) = \frac{\partial w}{\partial x} \frac{\partial w}{\partial y}$$

For small shear strain of γ

$$\cos \left(\frac{\pi}{2} - \gamma \right) = \cos \frac{\pi}{2} \cos \gamma + \sin \frac{\pi}{2} \sin \gamma \approx \gamma$$

Therefore, the shear strain due to w is $\gamma = \dfrac{\partial w}{\partial x} \dfrac{\partial w}{\partial y}$, and the total middle surface shear strain is

$$\gamma_{xy0} = \frac{\partial u_0}{\partial y} + \frac{\partial v_0}{\partial x} + \frac{\partial w}{\partial x} \frac{\partial w}{\partial y} \tag{7.34e}$$

The relations between middle surface strains and middle surface forces are

$$\varepsilon_{x0} = \frac{1}{Eh} (N_x - v N_y) \tag{7.34f}$$

$$\varepsilon_{y0} = \frac{1}{Eh} (N_y - v N_x) \tag{7.34g}$$

$$\gamma_{xy0} = \frac{2(1+v)}{Eh} N_{xy} \tag{7.34h}$$

Equations (7.2a), (7.2b), (7.2j), (7.34a), (7.34b), (7.34e)–(7.34h) are a set of nine equations in the nine unknowns of N_x, N_y, N_{xy}, w, u_0, v_0, ε_{x0}, ε_{y0}, and γ_{xy0}. We wish to reduce the number of equations by uncoupling some unknowns. Differentiate Eq. (7.34a) twice with respect to y, Eq. (7.34b) with respect to x, and differentiate Eq. (7.34e) successfully with respect to x and y to get

$$\frac{\partial^2 \varepsilon_{x0}}{\partial y^2} = \frac{\partial^3 u_0}{\partial x \partial y^2} + \left(\frac{\partial^2 w}{\partial x \partial y}\right)^2 + \frac{\partial w}{\partial x}\frac{\partial^3 w}{\partial x \partial y^2}$$

$$\frac{\partial^2 \varepsilon_{y0}}{\partial x^2} = \frac{\partial^3 v_0}{\partial x^2 \partial y} + \left(\frac{\partial^2 w}{\partial x \partial y}\right)^2 + \frac{\partial w}{\partial y}\frac{\partial^3 w}{\partial x^2 \partial y}$$

$$\frac{\partial^2 \gamma_{xy0}}{\partial x \partial y} = \frac{\partial^3 u_0}{\partial x \partial y^2} + \frac{\partial^3 v_0}{\partial x^2 \partial y} + \frac{\partial^3 w}{\partial x^2 \partial y}\frac{\partial w}{\partial y} + \frac{\partial^2 w}{\partial x^2}\frac{\partial^2 w}{\partial y^2} + \left(\frac{\partial^2 w}{\partial x \partial y}\right)^2 + \frac{\partial w}{\partial x}\frac{\partial^3 w}{\partial x \partial y^2}$$

$$\frac{\partial^2 \varepsilon_{x0}}{\partial y^2} + \frac{\partial^2 \varepsilon_{y0}}{\partial x^2} - \frac{\partial^2 \gamma_{xy0}}{\partial x \partial y} = \left(\frac{\partial^2 w}{\partial x \partial y}\right)^2 - \frac{\partial^2 w}{\partial x^2}\frac{\partial^2 w}{\partial y^2} \tag{7.34i}$$

To further reduce the number of equations, a stress function $F(x, y)$ is introduced. Equations (7.2a) and (7.2b) are satisfied if the in-plane forces are defined as

$$N_x = h\frac{\partial^2 F}{\partial y^2} \tag{7.34j}$$

$$N_y = h\frac{\partial^2 F}{\partial x^2} \tag{7.34k}$$

$$N_{xy} = -h\frac{\partial^2 F}{\partial x \partial y} \tag{7.34l}$$

Substitute the in-plane forces in terms of the function $F(x, y)$ given by Eqs. (7.34j)–(7.34l) in Eqs.(7.34f)–(7.34h) to get

$$\varepsilon_{x0} = \frac{1}{E}\left(\frac{\partial^2 F}{\partial y^2} - v\frac{\partial^2 F}{\partial x^2}\right) \tag{7.34m}$$

$$\varepsilon_{y0} = \frac{1}{E}\left(\frac{\partial^2 F}{\partial x^2} - v\frac{\partial^2 F}{\partial y^2}\right) \tag{7.34n}$$

$$\gamma_{xy0} = -\frac{2(1 + v)}{E}\frac{\partial^2 F}{\partial x \partial y} \tag{7.34o}$$

Substitute Eqs. (7.34m)–(7.34o) into Eq. (7.34i) and we obtain

$$\frac{\partial^4 F}{\partial x^4} + 2\frac{\partial^4 F}{\partial x^2 \partial y^2} + \frac{\partial^4 F}{\partial y^4} = E\left[\left(\frac{\partial^2 w}{\partial x \partial y}\right)^2 - \frac{\partial^2 w}{\partial x^2}\frac{\partial^2 w}{\partial y^2}\right] \tag{7.34p}$$

Substitute Eqs. (7.34j)–(7.34l) into Eqs. (7.2j) and we get

$$\frac{\partial^4 w}{\partial x^4} + 2\frac{\partial^4 w}{\partial x^2 \partial y^2} + \frac{\partial^4 w}{\partial y^4} - \frac{h}{D}\left(\frac{\partial^2 F}{\partial y^2}\frac{\partial^2 w}{\partial x^2} + \frac{\partial^2 F}{\partial x^2}\frac{\partial^2 w}{\partial y^2} - 2\frac{\partial^2 F}{\partial x \partial y}\frac{\partial^2 w}{\partial x \partial y}\right) = 0 \tag{7.34q}$$

We have reduced the number of equations to be simultaneously solved to two given by Eqs. (7.34p) and (7.34q). These equations are the governing equations and were first derived by von Kármán and are known as von Kármán large deflection plate equations [14].

7.10.1 Post-buckling Behavior of Plates

The large deflection theory of plates is used to find the post-buckling behavior of plates. It is not possible to obtain the closed form solution of the two governing equations. Hence, approximate or numerical methods are used to find the solution. The numerical solutions involve lengthy computations. Approximate methods that are less accurate than the numerical solutions are sometimes used because of their relative simplicity. This method is illustrated here for a simply supported square plate subjected to a uniaxial compressive force N_x shown in Figure 7.37. The finite deflection involves deflections in the middle surface of the plate as well as transverse bending deformations. Both in-plane and transverse boundary conditions must be satisfied. The solution given here is based on the works of Volmir [15].

Transverse boundary conditions for a simply supported plate are

$$w = 0, \text{ and } \frac{\partial^2 w}{\partial x^2} = 0 \quad \text{at } x = 0 \text{ and } x = a \tag{7.7f}$$

$$w = 0, \text{ and } \frac{\partial^2 w}{\partial y^2} = 0 \quad \text{at } y = 0 \text{ and } y = a \tag{7.7g}$$

The following assumptions are made for the in-plane boundary conditions:

1. All edges remain straight and retain the original 90° angles during bending.
2. The shearing forces N_{xy} and N_{yx} are zero on the four edges of the plate.
3. The edges $y = 0$ and $y = a$ are free to move in the y direction.

It is assumed that the displacement u remains constant along the edges $y = 0$ and $y = a$, whereas distribution of the force N_x is not known in the x direction. N_x is assumed to vary because of large displacement during post-buckling. Hence, use the average value of the compressive stress in the x direction given by

$$\sigma_{x \text{ avg}} = -\frac{1}{ah} \int_0^a N_x dy \tag{7.35a}$$

The negative sign indicates that the stress is compressive. The problem is solved by assuming the lateral deflection as follows:

$$w = f \sin \frac{\pi x}{a} \sin \frac{\pi y}{a} \tag{7.35b}$$

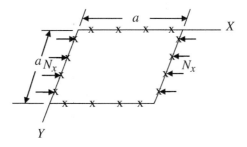

Figure 7.37 Simply supported plate under uniaxial compression.

The assumed deflection expression satisfies the boundary conditions given by Eqs.(7.7f) and (7.7g) and hence is a good approximation of w. Substitute Eqs. (7.35b) in (7.34p) and we get

$$\frac{\partial^4 F}{\partial x^4} + 2\frac{\partial^4 F}{\partial x^2 \partial y^2} + \frac{\partial^4 F}{\partial y^4} = f^2 \frac{E\pi^4}{a^4}\left(\cos^2\frac{\pi x}{a}\cos^2\frac{\pi y}{a} - \sin^2\frac{\pi x}{a}\sin^2\frac{\pi y}{a}\right) \tag{7.35c}$$

Use the following trigonometry relations

$$\cos^2\frac{\pi x}{a} = \frac{1}{2}\left(1 + \cos\frac{2\pi x}{a}\right), \quad \sin^2\frac{\pi x}{a} = \frac{1}{2}\left(1 - \cos\frac{2\pi x}{a}\right)$$

in Eq. (7.35c) to obtain

$$\frac{\partial^4 F}{\partial x^4} + 2\frac{\partial^4 F}{\partial x^2 \partial y^2} + \frac{\partial^4 F}{\partial y^4} = f^2 \frac{E\pi^4}{2a^4}\left(\cos\frac{2\pi x}{a} + \cos\frac{2\pi y}{a}\right) \tag{7.35d}$$

The solution of the Eq. (7.35d) is

$$F = F_c + F_p \tag{7.35e}$$

where F_c and F_p are the complementary and particular solutions of Eq. (7.35d). The complementary solution is obtained by taking the right-hand side of Eq. (7.35c) equal to zero. It implies that the transverse deflection $w = 0$ in the plate. Hence, the complementary solution corresponds to the in-plane stress distribution in the plate before buckling. The in-plane forces consist of constant N_x at any point in the plate, and $N_y = N_{xy} = 0$ before buckling. Now

$$N_x = h\frac{\partial^2 F}{\partial y^2} = \text{Constant}$$

or

$$F_c = Ay^2$$

$$\sigma_{x\,\text{avg}} = -\frac{N_x}{h} = -\frac{\partial^2 F}{\partial y^2} = -2A$$

Hence, the complementary solution of Eq. (7.35d) is

$$F_c = -\frac{\sigma_{x\,\text{avg}}}{2}y^2 \tag{7.35f}$$

The complementary solution gives the in-plane stress distribution before buckling takes place in the plate. Thus, the particular solution gives the changes in the plane stress distribution due to the buckling of the plate. We can assume the particular solution of Eq. (7.35d) by considering the terms on the right-hand side of the equation. Assume

$$F_p = B\cos\frac{2\pi x}{a} + C\cos\frac{2\pi y}{a} \tag{7.35g}$$

Substitute the derivatives of F_p from Eqs. (7.35g) into Eq. (7.35d) to give

$$\frac{16B\pi^4}{a^4}\cos\frac{2\pi x}{a} + \frac{16C\pi^4}{a^4}\cos\frac{2\pi y}{a} = f^2\frac{E\pi^4}{2a^4}\left(\cos\frac{2\pi x}{a} + \cos\frac{2\pi y}{a}\right)$$

Equate the coefficients of like terms to obtain

$$B = C = \frac{Ef^2}{32} \tag{7.35h}$$

Hence, the particular solution of Eq. (7.35d) is

$$F_p = \frac{Ef^2}{32}\left(\cos\frac{2\pi x}{a} + \cos\frac{2\pi y}{a}\right)$$

The complete solution of Eq. (7.35d) is given by

$$F = \frac{Ef^2}{32}\left(\cos\frac{2\pi x}{a} + \cos\frac{2\pi y}{a}\right) - \frac{\sigma_{x\,avg}y^2}{2} \tag{7.35i}$$

To determine the coefficient f, solve Eq. (7.34q) by the Galerkin method shown in Chapter 2, Section 2.11.3. For the plate under consideration, the Galerkin equation is given by

$$\int_0^a \int_0^a Q(\phi)g(x,y)dxdy = 0 \tag{7.35j}$$

where, $Q(\phi)$ is the left-hand side of the differential equation of the plate in the z direction given by Eq. (7.34q), and

$$g(x,y) = \sin\frac{\pi x}{a}\sin\frac{\pi y}{a} \tag{7.35k}$$

Substitute w from Eq. (7.35b) and F from Eq. (7.35i) to obtain

$$Q(\phi) = \left[\frac{4f\pi^4 D}{4} - \frac{Ehf^3\pi^4}{8a^4}\left(\cos\frac{2\pi x}{a} + \cos\frac{2\pi y}{a}\right) - \sigma_{x\,avg}\frac{hf\pi^2}{a^2}\right]\sin\frac{\pi x}{a}\sin\frac{\pi y}{a} \tag{7.35l}$$

The Galerkin Eq. (7.35j) can be written as

$$\int_0^a \int_0^a \left[\left(\frac{4fD\pi^4}{a^4} - \sigma_{x\,avg}\frac{hf\pi^2}{a^2}\right)\left(\sin^2\frac{\pi x}{a}\sin^2\frac{\pi y}{a}\right)\right.$$
$$\left. - \frac{Ehf^3\pi^4}{8a^4}\left(\cos\frac{2\pi x}{a} + \cos\frac{2\pi y}{a}\right)\left(\sin^2\frac{\pi x}{a}\sin^2\frac{\pi y}{b}\right)\right]dxdy = 0 \tag{7.35m}$$

The following definite integrals have the values given by

$$\int_0^a \sin^2\frac{\pi x}{a}dx = \int_0^a \sin^2\frac{\pi y}{a}dy = \frac{a}{2}$$

$$\left(\frac{4fD\pi^4}{a^4} - \sigma_{x\,avg}\frac{hf\pi^2}{a^2}\right)\frac{a^2}{4} - \frac{Ehf^3\pi^4}{16a^3}\left(\int_0^a \cos\frac{2\pi x}{a}\sin^2\frac{\pi x}{a}dx + \int_0^a \cos\frac{2\pi y}{a}\sin^2\frac{\pi y}{a}\right) = 0 \tag{7.35n}$$

$$\cos\frac{2\pi x}{a}\sin^2\frac{\pi x}{a} = \frac{1}{2}\left(\cos\frac{2\pi x}{a} - \cos^2\frac{2\pi x}{a}\right)$$

$$\cos\frac{2\pi y}{a}\sin^2\frac{\pi y}{a} = \frac{1}{2}\left(\cos\frac{2\pi y}{a} - \cos^2\frac{2\pi y}{a}\right)$$

$$\int_0^a \cos^2\frac{2\pi x}{a}dx = \frac{a}{2}, \int_0^a \cos\frac{2\pi x}{a}dx = 0$$

Hence, Eq. (7.35n) can be written as

$$\frac{fD\pi^4}{a^2} - \sigma_{x\,avg}\frac{hf\pi^2}{4} + \frac{Ehf^3\pi^4}{32a^2} = 0 \tag{7.35o}$$

$$\sigma_{x\,avg} = \frac{4D\pi^2}{ha^2} + \frac{E\pi^2f^2}{8a^2}$$

or

$$\sigma_{x\,avg} = \sigma_{cr} + \frac{E\pi^2f^2}{8a^2} \tag{7.35p}$$

where for the plate the critical stress is given by

$$\sigma_{cr} = \frac{4D\pi^2}{ha^2}$$

see Eq. (7.7m), or

$$f^2 = \frac{8a^2}{E\pi^2}(\sigma_{x\,avg} - \sigma_{cr}) \tag{7.35q}$$

It is shown in Eq. (7.35q) that the lateral deflection given by the coefficient f continues to increase after buckling with the increase in the average stress $\sigma_{x\,avg}$. It shows that a plate can resist axial loads more than its critical load and it is known as its post-buckling strength. The post-buckling behavior of plates is different from that of columns. Columns collapse as soon as the critical load is reached whereas plates can take a load in excess of their critical load.

From Eqs. (7.34j) and (7.35i) we have

$$\sigma_x = -\frac{\partial^2 F}{\partial y^2} = \frac{E\pi^2 f^2}{8a^2}\cos\frac{2\pi y}{a} + \sigma_{x\,avg}$$

The negative sign is taken because σ_x is a compressive stress.

From Eq. (7.35p) we have

$$\sigma_x = \sigma_{x\,avg} + (\sigma_{x\,avg} - \sigma_{cr})\cos\frac{2\pi y}{a} \tag{7.35r}$$

Similarly, use Eqs. (7.34k) and (7.35i) to obtain

$$\sigma_y = \frac{-\partial^2 f}{\partial x^2} = \frac{E\pi^2 f^2}{8a^2}\cos\frac{2\pi x}{a}$$

or from Eq. (7.35p) we have

$$\sigma_y = (\sigma_{x\,avg} - \sigma_{cr})\cos\frac{2\pi x}{a} \tag{7.35s}$$

Figure 7.38 shows the stress distribution in the middle surface of the plate given by Eqs. (7.35r) and (7.35s). In the post-buckling range the stress σ_x varies across the width of the plate whereas it was constant in the plate before buckling occurs. The stress σ_y varies from compression at the edges to tension in the middle of the length of the plate in the post-buckling stage, whereas it was not there before buckling. The presence of the tensile stress stiffens the plate and prevents its collapse after buckling.

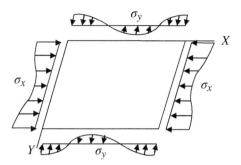

Figure 7.38 Variation of stress in post-buckling range in the middle surface of the plate.

It is seen in Figure 7.38 that the ends of the plate withstand higher stresses than the middle portion of the plate. This indicates that the sides of the plate are stiffer than the middle portion.

7.11 Inelastic Buckling of Plates

So far we have assumed that stress remains within the elastic range when buckling occurs. We showed in Chapter 3 that stresses at buckling may exceed the elastic limit in columns. In plates too, the stresses may exceed the elastic limit at buckling, depending on the thickness of the plate and the load applied. In the case of plates, the applied load may be applied along the longitudinal direction only, then the stress in this direction is likely to increase beyond the elastic limit. In that case, the nonlinear behavior of the plate is governed by either the tangent modulus or reduced modulus theory in the longitudinal direction, whereas in the transverse direction, it may remain within the elastic limit and is governed by the modulus of elasticity. Hence, the plate response is governed by the two moduli. On the other hand if the load acting on the plate is in both directions, the entire plate will show the inelastic behavior. Here the approach given by Bleich and Bleich [16] is used to solve the problem. According to this theory, when the stress in the longitudinal direction exceeds the elastic limit, the tangent modulus, E_t, is used to calculate deformation along this direction. In the transverse direction the modulus of elasticity E is used. If the stress exceeds the elastic limit in both directions, then the tangent modulus governs the deformation in both the X and Y directions.

The expressions for moments M and shears Q in terms of the curvature were derived in Eqs. (7.1r), (7.1s), (7.1t), and (7.1z) assuming the plate remained elastic. When the buckling stress is higher than the elastic limit, the expressions for moment M and shear Q are modified by introducing a quantity $\tau = E_t/E$, where E_t is the tangent modulus and E is the modulus of elasticity. If a plate is subjected to a compressive force of N_x per unit length only ($N_y = N_{xy} = 0$) in the X direction acting on the edges $x = 0$ and $x = a$ in Figure 7.39, then the modulus E_t governs the plate behavior in the X direction and modulus of elasticity E governs the plate behavior in the Y direction. The moment curvature and transverse shear force equations are

$$M_x = -D\left(\tau\frac{\partial^2 w}{\partial x^2} + v\sqrt{\tau}\frac{\partial^2 w}{\partial y^2}\right) \tag{7.36a}$$

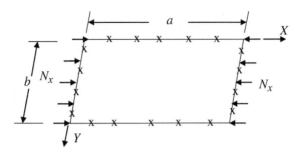

Figure 7.39 Simply supported rectangular plate under axial compression.

$$M_y = -D\left(v\sqrt{\tau}\frac{\partial^2 w}{\partial x^2} + \frac{\partial^2 w}{\partial y^2}\right) \tag{7.36b}$$

$$M_{xy} = -M_{yx} = D\sqrt{\tau}(1-v)\frac{\partial^2 w}{\partial x \partial y} \tag{7.36c}$$

$$Q_x = -D\frac{\partial}{\partial x}\left(\tau\frac{\partial^2 w}{\partial x^2} + \sqrt{\tau}\frac{\partial^2 w}{\partial y^2}\right) \tag{7.36d}$$

$$Q_y = -D\frac{\partial}{\partial y}\left(\sqrt{\tau}\frac{\partial^2 w}{\partial x^2} + \frac{\partial^2 w}{\partial y^2}\right) \tag{7.36e}$$

where $D = \dfrac{Eh^3}{12(1-v^2)}$ is called the flexural rigidity of the plate per unit width. In Eq. (7.36a) the first and second terms within the parentheses are the contribution to M_x from the curvatures along the X axis and Y axis respectively. Hence the first term is multiplied by the ratio $E_t/E = \tau$, the second term is multiplied by the mean value between 1 and τ which is chosen arbitrarily by Bleich and Bleich [16] as $\sqrt{\tau}$. Substitute the expressions from Eqs. (7.36d) and (7.36e) into Eq. (7.2i) to obtain

$$D\left(\tau\frac{\partial^4 w}{\partial x^4} + 2\sqrt{\tau}\frac{\partial^4 w}{\partial x^2 \partial y^2} + \frac{\partial^4 w}{\partial y^4}\right) + N_x\frac{\partial^2 w}{\partial x^2} = 0 \tag{7.36f}$$

If a plate is subjected to a compressive force of N_y per unit length only in the Y direction acting on the edges $y = 0$ and $y = b$, then the moment curvature and transverse shear force equations are

$$M_x = -D\left(\frac{\partial^2 w}{\partial x^2} + v\sqrt{\tau}\frac{\partial^2 w}{\partial y^2}\right) \tag{7.36g}$$

$$M_y = -D\left(v\sqrt{\tau}\frac{\partial^2 w}{\partial x^2} + \tau\frac{\partial^2 w}{\partial y^2}\right) \tag{7.36h}$$

$$M_{xy} = -M_{yx} = D\sqrt{\tau}(1-v)\frac{\partial^2 w}{\partial x \partial y} \tag{7.36i}$$

$$Q_x = -D\frac{\partial}{\partial x}\left(\frac{\partial^2 w}{\partial x^2} + \sqrt{\tau}\frac{\partial^2 w}{\partial y^2}\right) \tag{7.36j}$$

$$Q_y = -D\frac{\partial}{\partial y}\left(\sqrt{\tau}\frac{\partial^2 w}{\partial x^2} + \tau\frac{\partial^2 w}{\partial y^2}\right)$$ (7.36k)

The governing equation is given by

$$D\left(\frac{\partial^4 w}{\partial x^4} + 2\sqrt{\tau}\frac{\partial^4 w}{\partial x^2 \partial y^2} + \tau\frac{\partial^4 w}{\partial y^4}\right) + N_y\frac{\partial^2 w}{\partial y^2} = 0$$ (7.36l)

If a plate is subjected to forces in the X and Y directions both, then Eqs. (7.36g)–(7.36l) reduce to those derived in Section 7.3 with $D\tau$ replacing D. Eqs. (7.36f) and (7.36l) are considered valid because the theoretical values of critical stress agree with the test results.

7.11.1 Rectangular Plates with Simply Supported Edges

Consider a simply supported plate with sides of a and b subjected to a compressive force of N_x per unit length only acting on the edges $x = 0$ and $x = a$ shown in Figure 7.39. Take the tangent modulus, E_t, as the modulus governing the plate behavior in the X direction and the modulus of elasticity, E, the modulus governing the plate behavior in the Y direction when the critical load exceeds the elastic limit. There are no applied loads in the Y direction, $N_y = 0$, and also $N_{xy} = 0$. The governing equation is

$$D\left(\tau\frac{\partial^4 w}{\partial x^4} + 2\sqrt{\tau}\frac{\partial^4 w}{\partial x^2 \partial y^2} + \frac{\partial^4 w}{\partial y^4}\right) + N_x\frac{\partial^2 w}{\partial x^2} = 0$$ (7.36f)

Boundary conditions are

$$w = 0, \quad \frac{\partial^2 w}{\partial x^2} = 0 \quad \text{at } x = 0 \text{ and } x = a$$ (7.37a)

and

$$w = 0, \quad \frac{\partial^2 w}{\partial y^2} = 0 \quad \text{at } y = 0 \text{ and } y = b$$ (7.37b)

The solution is

$$w = \sum_{m=1}^{\infty}\sum_{n=1}^{\infty} A_{mn}\sin\frac{m\pi x}{a}\sin\frac{n\pi y}{b} \quad m = 1, 2, 3\ldots, n = 1, 2, 3, \ldots$$ (7.37c)

where m and n define the number of half-waves that the plate buckles in the x and y directions respectively. A_{mn} gives the amplitudes of the mode shapes or shape functions. The double trigonometric series in Eq. (7.37c) satisfies the boundary conditions given by Eqs. (7.37a) and (7.37b). The substitution of w and its derivatives from Eq. (7.37c) into Eq. (7.36f) gives

$$\sum_{m=1}^{\infty}\sum_{n=1}^{\infty} A_{mn}\left[\tau\frac{m^4\pi^4}{a^4} + 2\sqrt{\tau}\frac{m^2 n^2 \pi^4}{a^2 b^2} + \frac{n^4\pi^4}{b^4} - \frac{N_x}{D}\frac{m^2\pi^2}{a^2}\right]\sin\frac{m\pi x}{a}\sin\frac{n\pi y}{b} = 0$$ (7.37d)

or

$$A_{mn}\left[\pi^4\left(\sqrt{\tau}\frac{m^2}{a^2} + \frac{n^2}{b^2}\right)^2 - \frac{N_x}{D}\frac{m^2\pi^2}{a^2}\right] = 0$$ (7.37e)

For Eq. (7.37e) to be satisfied, either A_{mn} is zero, which leads to a trivial solution, where the plate remains flat for all loads. For nontrivial solution the quantity in the parenthesis should be zero, which gives

$$(N_x)_{cr} = \frac{Da^2\pi^2}{m^2}\left(\sqrt{\tau}\frac{m^2}{a^2} + \frac{n^2}{b^2}\right)^2$$

or

$$(N_x)_{cr} = \frac{D\pi^2}{b^2}\left(\sqrt{\tau}\frac{mb}{a} + \frac{n^2a}{mb}\right)^2 \tag{7.37f}$$

We wish to know the lowest value of N_x called the critical load at which the equilibrium of the plate can change from the plane to a bent shape. The values of m and n, the number of half-waves that will minimize N_x is to be found. In Eq. (7.37f) as n increases N_x increases, therefore $n = 1$ gives the smallest value of N_x. This shows that the plate buckles in a single half sine wave in the Y direction. The number of half sine waves in the X direction corresponding to the minimum value of N_x is found by taking the derivative of the expression in Eq. (7.37f) with respect to m at $n = 1$, and equating it to zero.

$$\frac{dN_x}{dm} = \frac{2D\pi^2}{b^2}\left(\tau\frac{mb}{a} + \frac{a}{mb}\right)\left(\tau\frac{b}{a} - \frac{a}{bm^2}\right) = 0$$

This gives

$$m = \frac{a}{b\tau^{\frac{1}{4}}} \tag{7.37g}$$

Substitute Eqs. (7.37g) into (7.37f) and we obtain

$$(N_x)_{cr} = \frac{4D\pi^2}{b^2}\sqrt{\tau} \tag{7.37h}$$

In Eq. (7.37g) the wave length given by m in the X direction depends on the material properties and is larger than that given by Eq. (7.71) for the elastic case. In general, the load $(N_x)_{cr}$ can be written as

$$(N_x)_{cr} = \frac{k_1 D\pi^2}{b^2} \tag{7.37i}$$

where

$$k_1 = \left(\sqrt{\tau}\frac{mb}{a} + \frac{n^2a}{mb}\right)^2 \tag{7.37j}$$

Find the lowest k_1 for a particular aspect ratio a/b, τ, and $n = 1$, to get the critical load. Hence, find the value of m that gives the lowest value of k_1 as was done in Section 7.5.1. To find the transition point, where the number of half-waves m change to $m+1$ half-waves for the same value of k_1, is given by

$$\sqrt{\tau}\frac{mb}{a} + \frac{a}{mb} = \sqrt{\tau}\frac{(m+1)b}{a} + \frac{a}{(m+1)b} \tag{7.37k}$$

Let $p = a/b$, then

$$\frac{p}{m} - \frac{\sqrt{\tau}}{p} - \frac{p}{m+1} = 0$$

or

$$p = \frac{a}{b} = \tau^{\frac{1}{4}} \sqrt{m(m+1)} \qquad (7.37l)$$

Therefore, the aspect ratio a/b at which the transition takes place also depends on the material property.

7.11.2 Plate with Loading Edges Simply Supported and the Sides $y = 0$ and $y = b$ Are Clamped

Consider the plate shown in Figure 7.40. The boundary conditions are

$$w = 0, \quad \frac{\partial^2 w}{\partial x^2} = 0 \quad \text{at } x = 0 \text{ and } x = a \qquad (7.38a)$$

and

$$w = 0, \quad \frac{\partial w}{\partial y} = 0 \quad \text{at } y = 0 \text{ and } y = b \qquad (7.38b)$$

The governing differential equation is

$$D\left(\tau\frac{\partial^4 w}{\partial x^4} + 2\sqrt{\tau}\frac{\partial^4 w}{\partial x^2 \partial y^2} + \frac{\partial^4 w}{\partial y^4} \right) + N_x\frac{\partial^2 w}{\partial x^2} = 0 \qquad (7.36f)$$

Take the solution of Eq. (7.36f) as

$$w = f(y)\sin\frac{m\pi x}{a} \qquad (7.38c)$$

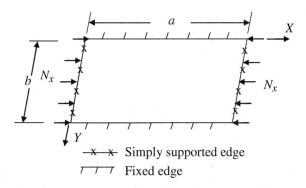

—x—x— Simply supported edge

Г Г Г Fixed edge

Figure 7.40 Plate with loading edges simply supported and others clamped.

It satisfies the boundary conditions given by Eq. (7.38a). Substitute Eq. (7.38c) and its derivatives in Eq. (7.36f) to get

$$D\left[\tau f\frac{m^4\pi^4}{a^4} - 2\sqrt{\tau}\frac{m^2\pi^2}{a^2}\frac{d^2f}{dy^2} + \frac{d^4f}{dy^4}\right]\sin\frac{m\pi x}{a} - N_x f\frac{m^2\pi^2}{a^2}\sin\frac{m\pi x}{a} = 0 \tag{7.38d}$$

Since $\sin\dfrac{m\pi x}{a}$ is not zero, hence

$$\frac{d^4f}{dy^4} - 2\sqrt{\tau}\frac{m^2\pi^2}{a^2}\frac{d^2f}{dy^2} + \left(\tau\frac{m^4\pi^4}{a^4} - \frac{N_x}{D}\frac{m^2\pi^2}{a^2}\right)f = 0 \tag{7.38e}$$

The solution of Eq. (7.38e) is

$$f(y) = A_1 \sinh\alpha y + A_2 \cosh\alpha y + A_3 \sin\beta y + A_4 \cos\beta y \tag{7.38f}$$

where

$$\alpha = \sqrt{\sqrt{\tau}\frac{m^2\pi^2}{a^2} + \sqrt{\frac{N_x}{D}\frac{m^2\pi^2}{a^2}}}, \quad \beta = \sqrt{-\sqrt{\tau}\frac{m^2\pi^2}{a^2} + \sqrt{\frac{N_x}{D}\frac{m^2\pi^2}{a^2}}} \tag{7.38g}$$

The constants A_1, A_2, A_3 and A_4 are to be determined from the boundary conditions given by Eq. (7.38b). The same steps as in Section 7.5.4 can be followed to obtain the transcendental equation

$$2(1 - \cosh\alpha b \cos\beta b) = \left(\frac{\beta}{\alpha} - \frac{\alpha}{\beta}\right)\sinh\alpha b \sin\beta b = 0 \tag{7.38h}$$

where α and β are given by Eq. (7.38g). As before in Section 7.5.1, the critical load is given by

$$(N_x)_{cr} = \frac{kD\pi^2}{b^2} \tag{7.38i}$$

Substitute the critical load N_x in Eq. (7.38g) to get

$$\alpha b = \frac{m\pi}{a/b}\sqrt{\sqrt{\tau} + \frac{a}{mb}\sqrt{k}}, \quad \text{and} \quad \beta b = \frac{m\pi}{a/b}\sqrt{-\sqrt{\tau} + \frac{a}{mb}\sqrt{k}} \tag{7.38j}$$

Find the lowest value of k and the corresponding m that satisfied Eq. (7.38h) for a particular aspect ratio a/b, τ, and $n = 1$. The critical load $(N_x)_{cr}$ is then found from Eq. (7.38i) once the lowest value of k is known, as was done in Section 7.5.4.

7.12 Ultimate Strength of Plates in Compression

It is desirable in plate design to eliminate buckling under service loads. However, it was shown in Section 7.10 that a plate carries after buckling a much larger force before failure. Therefore, in situations where lightweight structures are required, as in aircraft design, it is important to consider the ultimate load a plate can carry in addition to the critical load. A theoretical calculation of ultimate strength will be difficult because it will involve both material and geometric nonlinearities. Thus, the ultimate load at which a plate collapses is usually determined by an approximate method. It can be seen in Figure 7.38 that as the axial compressive force is increased on a plate after initial buckling, the stresses on the edges are much higher than in the central region. It is observed that a plate fails when the stresses in the direction of the applied compressive load reach the yield point strength of the material at the edges of the plate. It is assumed that the load on the plate is carried by two strips of width b_e and the stress is considered uniform over the width of these strips as shown in Figure 7.41. The middle portion of the plate is disregarded and the simply supported rectangular plate is considered of equivalent width $2b_e$. The concept of the effective width was introduced by von Kármán, Sechler, and Donnell [17].

The ultimate load the plate can support is given by

$$P_u = 2b_e h\sigma_y \tag{7.39a}$$

Von Kármán suggested the following formula for the effective width of simply supported plates

$$2b_e = b\sqrt{\frac{\sigma_{cr}}{\sigma y}} \tag{7.39b}$$

For simply supported plates, we have from Eq. (7.38i)

$$(N_x)_{cr} = \frac{kD\pi^2}{b^2}, k = 4$$

or

$$\sigma_{cr} = \frac{4\pi^2}{b^2} \frac{Eh^3}{12(1-v^2)h} = \frac{\pi^2 E}{3(1-v^2)} \frac{1}{(b/h)^2} \tag{7.39c}$$

Substitute Eqs. (7.39c) into Eq. (7.39b) to obtain

$$b_e = \frac{\pi h}{\sqrt{12(1-v^2)}} \sqrt{\frac{E}{\sigma_y}} \tag{7.39d}$$

or

$$P_u = \frac{\pi h}{\sqrt{3(1-v^2)}} \sqrt{\frac{E}{\sigma_y}}(h\sigma_y) = \frac{\pi h^2}{\sqrt{3(1-v^2)}} \sqrt{E\sigma_y}$$

or

$$P_u = 1.90 h^2 \sqrt{E\sigma_y} \text{ for } v = 0.3 \tag{7.39e}$$

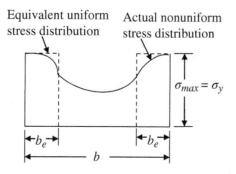

Figure 7.41 Stress variation across the plate width.

An alternate approximate formula was given by G. Winter [18] to find the effective width based on test results as follows:

$$2b_e = b\sqrt{\frac{\sigma_{cr}}{\sigma_y}}\left(1 - 0.25\sqrt{\frac{\sigma_{cr}}{\sigma_y}}\right) \tag{7.39f}$$

From Eq. (7.39c)

$$\sigma_{cr} = \frac{4\pi^2}{b^2}\frac{Eh^3}{12(1-v^2)h} = \frac{\pi^2 E}{3(1-v^2)}\frac{1}{(b/h)^2}$$

$$\sqrt{\frac{\sigma_{cr}}{\sigma_y}} = \frac{\pi h}{\sqrt{3b}}\sqrt{\frac{E}{(1-v^2)\sigma_y}}$$

$$b_e = \frac{\pi h}{\sqrt{12(1-v^2)}}\sqrt{\frac{E}{\sigma_y}}\left[1 - 0.25\frac{\pi h}{b\sqrt{3(1-v^2)}}\sqrt{\frac{E}{\sigma_y}}\right] \tag{7.39g}$$

$$P_u = \frac{\pi h}{\sqrt{3(1-v^2)}}\sqrt{\frac{E}{\sigma_y}}\left[1 - 0.25\frac{\pi h}{b\sqrt{3(1-v^2)}}\sqrt{\frac{E}{\sigma_y}}\right]h\sigma_y$$

or

$$P_u = 1.90h^2\sqrt{E\sigma_y}\left[1 - 0.475\frac{h}{b}\sqrt{\frac{E}{\sigma_y}}\right] \tag{7.39h}$$

Example 7.4 A steel plate panel is axially loaded and is simply supported along the loaded edges. The plate is provided with stiffeners, as shown in Figure 7.42. The panel is considered simply supported along the bolt lines and free at the sides. Each stiffener has the area of cross-section 0.1 in.^{2} (64.52 mm^2). Take the modulus of elasticity of the plate and stiffener material, $E = 29 \times 10^6$ psi (200 000 MPa). Find the compressive force P when the plate (i) first buckles and (ii) when the stress in the stiffeners is $\sigma_u = 36$ ksi (248 MPa).

Panel 1: 10 in. × 5 in. × 1/16 in. (254 mm × 127 mm × 1.59 mm) – simply supported on all four sides.

Panel 2: 10 in. × 2 in. × 1/16 in. (254 mm × 50.8 mm × 1.59 mm) – simply supported on three sides and free on the fourth side.

(1) When the plate buckles first

Panel 1: $a = 10$ in. (254 mm), $b = 5$ in. (127 mm), $\sigma_{cr1} = \dfrac{kD\pi^2}{b^2 h}$

$$\frac{a}{b} = \frac{10}{5} = 2 \left(\frac{254}{127} = 2\right)$$

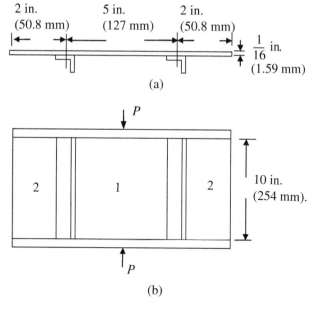

Figure 7.42 Stiffened plate panel: (a) Top view; (b) Front view.

From Figure 7.9 for a plate simply supported on all four sides, $k = 4$.

$$\sigma_{cr1} = \frac{4\pi^2}{b^2} \frac{Eh^3}{12(1-v^2)h} = \frac{\pi^2 E}{3(1-v^2)} \frac{1}{(b/h)^2} \tag{7.39c}$$

$$= \frac{4\pi^2 E}{12(1-0.3^2)}\left(\frac{h}{b}\right)^2 = 3.615E\left(\frac{h}{b}\right)^2$$

$$= 3.615(29x10^6)\left(\frac{1}{16x5}\right)^2 = 16380.47 \, psi$$

$$\left[3.615(200,000)\left(\frac{1.59}{127}\right)^2 = 113.32 \, MPa\right]$$

Panel 2: $a = 10$ in.(254 mm), $b = 2$ in. (50.8 mm),

$$\sigma_{cr2} = \frac{kD\pi^2}{b^2h}$$

$$\frac{a}{b} = \frac{10}{2} = 5\left(\frac{254}{50.8} = 5\right)$$

From Table 7.3 for a plate simply supported on two loaded sides, simply supported on the third side and free on the fourth side, $k = 0.47$.

$$\sigma_{cr2} = \frac{0.47\pi^2}{b^2} \frac{Eh^3}{12(1-v^2)h} = 0.425E\left(\frac{h}{b}\right)^2$$

$$= 0.425(29x10^6)\left(\frac{1}{16x2}\right)^2 = 12036.13psi$$

$$\left[0.425(200,000)\left(\frac{1.59}{50.8}\right)^2 = 83.27MPa\right]$$

This means the plate panels 2 buckle first. Before buckling, the entire width of the plate carries the load. The area of the plate including stiffeners is given by

$$A_{plate} = (5 + 2 + 2)x\frac{1}{16} + 2x0.1 = 0.76 \text{ in.}^2$$
$$\times \left[(127 + 50.8 + 50.8)x1.59 + 2x64.52 = 492.51 \text{ mm}^2\right]$$

$$P_{cr} = 12036.13x0.76 = 9147.46 \text{ lbs.} \quad (492.51x83.27 = 41011.31 \text{ N} = 41 \text{ kN})$$

(2) The post-buckling region when the stress in the stiffeners is 36 ksi (248 MPa)
In the post-buckling region, use the effective width of the plate
Panel 1:

$$2b_{e1} = \frac{\pi h}{\sqrt{3(1 - v^2)}}\sqrt{\frac{E}{\sigma_y}} = 1.90h\sqrt{\frac{E}{\sigma_y}}$$
$$= 1.90\left(\frac{1}{16}\right)\sqrt{\frac{29x10^6}{36000}} = 3.37 \text{ in.} \quad \left[1.90(1.59)\sqrt{\frac{200,000}{248}} = 85.79 \text{ mm}\right]$$

Panel 2:

$$\sigma_{cr2} = 0.425E\left(\frac{h}{b}\right)^2$$

$$2b_{e2} = b\sqrt{\frac{\sigma_{cr}}{\sigma_y}} = b\sqrt{\frac{0.425E}{\sigma_y}}\left(\frac{h}{b}\right) = 0.65h\sqrt{\frac{E}{\sigma_y}}$$
$$= 0.65\left(\frac{1}{16}\right)\sqrt{\frac{29x10^6}{36000}} = 1.15 \text{ in.} \quad \left[0.65(1.59)\sqrt{\frac{200,000}{248}} = 29.35 \text{ mm}\right]$$

The effective area of the plate including stiffeners is given by

$$A_{effectiive} = (3.37 + 2x1.15)\frac{1}{16} + 2x0.1 = 0.55 \text{ in.}^2$$
$$\left[(85.79 + 2x29.35)1.59 + 2x64.52 = 358.78 \text{ mm}^2\right]$$

The ultimate load is

$$P_u = 0.55x36000 = 19,800 \text{ lbs.} \quad (358.78x248 = 88977.44 \text{ N} = 89 \text{ kN})$$

The post-buckling load is about two times the initial buckling load.

7.13 Local Buckling of Compression Elements: Design

In the design of hot rolled structural elements, the width to thickness ratio of flanges and webs is proportioned such that overall failure occurs before local buckling takes place. In the design of steel columns and beams according to the AISC Specifications [19], the local buckling stress for these elements is kept higher than the yield stress for the material. If local buckling is not to occur for stresses smaller than yield stress for the material, then

$$\sigma_{cr} > \sigma_y \tag{7.40a}$$

where σ_{cr} is the critical stress for the plate element depending on how it is supported, and σ_y is the yield strength of the material of the plate element. Use Eq. (7.7s) for the critical stress in Eq. (7.40a) to obtain

$$\frac{k\pi^2 E}{12(1-v^2)}\left(\frac{h}{b}\right)^2 > \sigma_y \tag{7.40b}$$

If

$$v = 0.3$$

we get

$$\frac{b}{h} < 0.904\sqrt{\frac{kE}{\sigma_y}} \tag{7.40c}$$

If the plate element has one edge simply supported and the other edge as free, then $k = 0.47$ for $a/b = 5$ from Section 7.5.5, hence

$$\frac{b}{h} < 0.62\sqrt{\frac{E}{\sigma_y}} \tag{7.40d}$$

The ratio given in Eq. (7.40d) is further reduced by the AISC Specifications [19] to account for initial imperfections and residual stresses. In the Specifications, a member is defined as compact where the overall buckling failure occurs before local buckling. For flanges in rolled I-shaped sections, channels and tees, the thirteenth edition of AISC specifies

$$\frac{b}{h} < 0.38\sqrt{\frac{E}{\sigma_y}} \tag{7.40e}$$

and for the legs of single angles, AISC specifies

$$\frac{b}{h} < 0.54\sqrt{\frac{E}{\sigma_y}} \tag{7.40f}$$

If the plate element is simply supported along both of its unloaded edges, then $k = 4.0$ from Figure 7.9, hence from Eq. (7.40b) we have

$$\frac{b}{h} < 1.81\sqrt{\frac{E}{\sigma_y}} \tag{7.40g}$$

For the flanges of a rectangular box section and hollow structural sections of uniform thickness, the AISC specifies

$$\frac{b}{h} < 1.12\sqrt{\frac{E}{\sigma_y}}$$ (7.40h)

Problems

7.1 Determine the critical load of a simply supported rectangular plate compressed in two perpendicular directions by uniformly distributed forces shown in Figure P7.1. Find the solution by
(a) the differential equation method
(b) the energy method

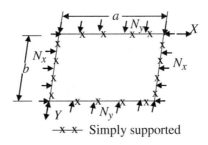

—×—× Simply supported

Figure P7.1

7.2 Find the buckling load of a rectangular plate simply supported on two opposite edges. The uniformly distributed compressive force is acting on the perpendicular faces that are fixed as shown in Figure P7.2 by the energy method.

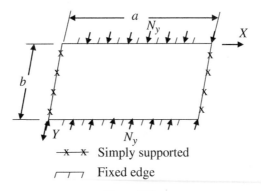

—×—× Simply supported
⌐⌐ Fixed edge

Figure P7.2

7.3 Find the critical load of a simply supported plate subjected to in-plane sinusoidal load N_x per unit length shown in Figure P7.3. Use the energy method.

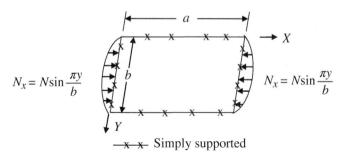

$$N_x = N\sin\frac{\pi y}{b} \qquad\qquad N_x = N\sin\frac{\pi y}{b}$$

—x—x— Simply supported

Figure P7.3

7.4 Obtain the buckling force for a rectangular plate clamped on all sides, which is stiffened by two transverse stiffeners shown in Figure P7.4. The plate and the stiffeners are of the same material. Take I as the second moment of area for each stiffener, E as the modulus of elasticity and v as the Poisson's ratio of the material.

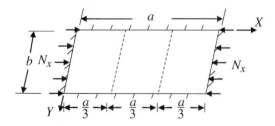

Figure P7.4

7.5 A simply supported plate of 20 in. × 10 in. (508 mm × 254 mm) is subjected to a shear force of 500 lb./in. (87.5 kN/m) shown in Figure P7.5. Find the thickness h required to resist buckling. Modulus of elasticity $E = 29 \times 10^6$ psi (200 000 MPa), Poisson's ratio $v = 0.3$.

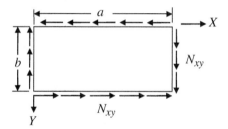

Figure P7.5

7.6 Use the finite difference method to find the critical load for a rectangular plate which is clamped on all four edges. It is subjected to uniformly distributed forces N_x and N_y as shown in Figure P7.6. Divide the plate into eight elements.

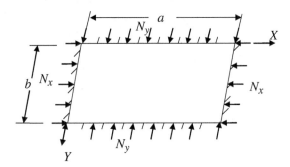

Figure P7.6

7.7 The stresses in the rectangular plate exceed the elastic limit at buckling. The plate is simply supported on the loaded edges by uniformly distributed force N_x as shown in Figure P7.7. The side $y = 0$ is clamped and the side $y = b$ is free. Find the critical stress in the plate by using the modified plate equilibrium equation taking into account inelasticity. Assume $a/b = 1.5$, $h/b = 0.02$, $E_t/E = 0.414$, $h = $ thickness of the plate, $E_t = $ tangent modulus, $E = $ modulus of elasticity.

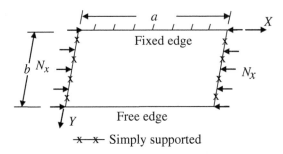

Figure P7.7

References

1 Timoshenko, S. and Woinowsky-Krieger, S. (1959). *Theory of Plates and Shells*, 2e. New York: McGraw Hill.

2 Gerard, G. (1962). *Introduction to Structural Stability Theory*. New York: McGraw Hill.

3 Timoshenko, S.P. and Gere, J.M. (1961). *Theory of Elastic Stability*, 2e, 330. New York: McGraw Hill.

4 Bryan, G.H. (1891). On the stability of a plane plate under thrusts in its own plane, with applications to the bucking of the sides of a ship. *The Proceedings of the London Mathematical Society* 22: 54–67.

5 Levy, S. (1942). Buckling of rectangular plates with built-in edges. *Journal of Applied Mechanics* 9: 171–174.

6 Kumar, A. (1998). *Stability of Structures*. New Delhi, India: Allied Publishers Ltd.

7 Kreyszig, E. (1972). *Advanced Engineering Mathematics*, 3e. New York: Wiley.

8 Zwillinger, D. (2018). *CRC Standard Mathematical Tables and Formulas*, 33e. Boca Raton, FL: CRC Press.

9 Salvadori, M.G. and Baron, M.L. (1961). *Numerical Methods in Engineering*, 2e. Englewood Cliffs, NJ: Prentice Hall, Inc.

10 Ralston, A. (1965). *A First Course in Numerical Analysis*. New York: McGraw Hill.

11 Szilard, R. (1973). *Theory and Analysis of Plates: Classical and Numerical Methods*. Englewood Cliffs, NJ: Prentice Hall, Inc.

12 Zienkiewicz, O.C. (1977). *The Finite Element Method in Engineering Science*, 3e. London, UK: McGraw Hill.

13 Desai, C.S. (1979). *Elementary Finite Element Method*. Englewood Cliffs, NJ: Prentice Hall, Inc.

14 Chajes, A. (1974). *Principles of Structural Stability Theory*. Englewood Cliffs, NJ: Prentice Hall, Inc.

15 Volmir, A.S. (1967). A translation of flexible plates and shells, Air Force Flight Dynamics Laboratory, Technical Report No. 66-216, Wright-Patterson Air Force Base, Ohio.

16 Bleich, F. and Bleich, H.H. (1952). *Buckling Strength of Metal Structures*. New York: McGraw Hill.

17 Von Kármán, T., Sechler, E.E., and Donnell, L.H. (1932). The strength of thin plates in compression. *Transactions of the ASME* 54: 53–57.

18 Winter, G. (1947). Strength of thin steel compression flanges. *Transactions of the ASCE* 112: 527–554.

19 AISC (2017). *Steel Construction Manual.*, 15e. Chicago, IL: AISC.

8

Buckling of Shells

8.1 Introduction

Shells are three-dimensional structures but the stresses can be considered in two planes only if the thickness of the shells is small in comparison to other dimensions. The difference between a shell and a plate is that a shell structure has an initial curvature whereas a plate is considered flat before an external load is applied. Because of their curved shape, shell structures are very efficient in supporting the external force. Shell structures are used in many industries to store and handle different materials. The materials stored by shell structures include solids, as in agricultural industries, and liquids such as gasoline and other petroleum products. Shell structures are used as water storage tanks and cooling towers in nuclear power plants. In recent times the size of the storage facilities has increased along with the use of high strength materials. In many cases, failure of these structures occurs due to instability before the material reaches its strength capacity. Therefore, there is a great deal of interest in the process of shell buckling nowadays. In the case of cylindrical shells, the buckling load reduces considerably when slight imperfections are present, which makes the analysis more important. In this chapter, the applications considered involve examples that can be solved by analytical methods. Practical structures and loads are much more complicated, which can now be solved by numerical methods with the help of computers. The stability of cylindrical shells under static and earthquake loads was considered by Jerath and Lee [1]. Comparison was made between static and earthquake buckling loads for cylindrical shells using the finite element large deflection method.

Shells have two types of forces: primary and secondary forces. Primary forces are in-plane forces that cause a membrane action whether the initial curvature exists or not. The secondary forces are caused by flexural deformations. If the structure is initially flat, the secondary forces do not cause appreciable membrane action if the deflections are small. That is why in small deflection theory the membrane action due to secondary forces is ignored, but in the large deflection theory it is considered for plates. If there is initial curvature, the membrane action caused by the secondary forces is considerable, regardless of the magnitude of the deformation. Hence, the membrane action due to secondary forces is considered in shells for both the small and the large deflection theories.

8.2 The Large Deflection Theory of Cylindrical Shells

The governing equations for shell buckling analysis are derived based on the following assumptions:

1. The shell is thin, i.e. its thickness is small in comparison to other dimensions.
2. The shell is made of homogeneous, isotropic, elastic material that obeys Hooke's law.
3. Lines perpendicular to the middle surface of the shell before bending remain straight and normal during bending.
4. It is a perfect cylinder initially and is loaded concentrically at every cross-section.

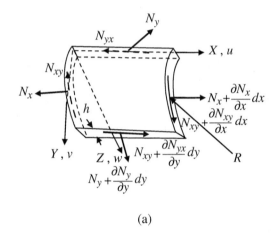

(a)

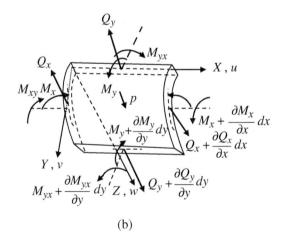

(b)

Figure 8.1 Cylindrical shell displacements and forces: (a) Cylindrical shell displacements and in-plane forces; (b) Cylindrical shell moments, transverse shears and external force.

Consider a small element of a cylindrical shell of thickness h and radius of curvature R in Figure 8.1. The origin of the coordinates is in the middle surface of the shell. The X axis is parallel to the axis of the cylinder, the Y axis is along the tangent to the circular curve, and the Z axis is normal to the middle surface pointing to the center of the curvature. Shell displacements are given by the components u, v, and w in the X, Y, and Z directions shown in Figure 8.1. The in-plane forces are shown in Figure 8.1a and the transverse shears, bending moments, and twisting moments are shown in Figure 8.1b.

The positive directions of the in-plane forces, bending moments, twisting moments, transverse shears, and external force p are shown in Figure 8.1. If the initial curvature of the shell surface and the curvature due to bending in the X and Y directions are neglected, the equations of equilibrium in the X and Y directions are simplified. That means we assume the transverse shear forces have negligible components in the X and Y directions. Also, the components of the in-plane forces in the X and Y directions are equal to the forces themselves. Then the summation of the forces in the X and Y directions leads to the same equations (Eqs. 7.2a and 7.2b) as in the case of plates as follows:

$$\frac{\partial N_x}{\partial x} + \frac{\partial Nyx}{\partial y} = 0 \tag{8.1a}$$

$$\frac{\partial N_y}{\partial y} + \frac{\partial N_{xy}}{\partial x} = 0 \tag{8.1b}$$

It is necessary to take into account the initial curvature of the shell element and the curvature due to bending to write the equation of equilibrium in the Z direction. The force N_y has a component in the Z direction due to the initial curvature of the shell given by $\frac{N_y}{R} dxdy$, as shown in Figure 8.2. None of the other in-plane forces have components in the Z direction due to the initial curvature of the shell, whereas all the in-plane forces have components in the Z direction due to curvature from bending. These components are identical to the ones in a bent plate that are given by Eq. (7.2g). If we add all the components of the in-plane forces in the Z direction including the component of N_y we get

$$\left[N_x \frac{\partial^2 w}{\partial x^2} + N_y \left(\frac{\partial^2 w}{\partial y^2} + \frac{1}{R} \right) + 2N_{xy} \frac{\partial^2 w}{\partial x \partial y} \right] dxdy \tag{8.1c}$$

To the Z direction forces we should add the transverse shear forces and the external force p shown in Figure 8.1b to obtain

$$\left[N_x \frac{\partial^2 w}{\partial x^2} + N_y \left(\frac{\partial^2 w}{\partial y^2} + \frac{1}{R} \right) + 2N_{xy} \frac{\partial^2 w}{\partial x \partial y} + \left(\frac{\partial Q_x}{\partial x} + \frac{\partial Q_y}{\partial y} \right) + p \right] dxdy = 0 \tag{8.1d}$$

The moment equilibrium equations about the X and Y axes are the same as for the plates given by Eqs. (7.1v) and (7.1w) as follows:

$$\frac{\partial M_{xy}}{\partial x} - \frac{\partial M_y}{\partial y} + Q_y = 0 \tag{8.1e}$$

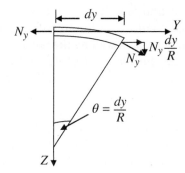

Figure 8.2 Component of N_y in the Z direction due to initial curvature.

or

$$\frac{\partial^2 M_{xy}}{\partial x \partial y} - \frac{\partial^2 M_y}{\partial y^2} + \frac{\partial Q_y}{\partial y} = 0 \tag{8.1f}$$

$$\frac{\partial M_{yx}}{\partial y} + \frac{\partial M_x}{\partial x} - Q_x = 0 \tag{8.1g}$$

or

$$\frac{\partial^2 M_{yx}}{\partial x \partial y} + \frac{\partial^2 M_x}{\partial x^2} - \frac{\partial Q_x}{\partial x} = 0 \tag{8.1h}$$

Substitute $\dfrac{\partial Q_x}{\partial x}$ and $\dfrac{\partial Q_y}{\partial y}$ from Eqs. (8.1f) and (8.1h) into Eq. (8.1d), and $M_{yx} = -M_{xy}$ to get

$$\frac{\partial^2 M_x}{\partial x^2} - \frac{2\partial^2 M_{xy}}{\partial x \partial y} + \frac{\partial^2 M_y}{\partial y^2} + N_x \frac{\partial^2 w}{\partial x^2} + +2N_{xy}\frac{\partial^2 w}{\partial x \partial y} + N_y\left(\frac{\partial^2 w}{\partial y^2} + \frac{1}{R}\right) + p = 0 \tag{8.1i}$$

The moment curvature relations in the shell can be assumed to be the same as given by Eqs. (7.1r)–(7.1t) in plates, thus

$$M_x = -D\left(\frac{\partial^2 w}{\partial x^2} + v\frac{\partial^2 w}{\partial y^2}\right) \tag{7.1r}$$

$$M_y = -D\left(\frac{\partial^2 w}{\partial y^2} + v\frac{\partial^2 w}{\partial x^2}\right) \tag{7.1s}$$

$$M_{xy} = -M_{yx} = D(1-v)\frac{\partial^2 w}{\partial x \partial y} \tag{7.1t}$$

where $D = \dfrac{Eh^3}{12(1-v^2)}$ is called the flexural rigidity of the shell per unit width. Substitute these expressions into Eq. (8.1i) to give

$$D\left(\frac{\partial^4 w}{\partial x^4} + 2\frac{\partial^4 w}{\partial x^2 \partial y^2} + \frac{\partial^4 w}{\partial y^4}\right) - N_x\frac{\partial^2 w}{\partial x^2} - 2N_{xy}\frac{\partial^2 w}{\partial x \partial y} - N_y\left(\frac{\partial^2 w}{\partial y^2} + \frac{1}{R}\right) - p = 0 \tag{8.1j}$$

Equations (8.1a), (8.1b), and (8.1j) are nonlinear equilibrium equations for thin cylindrical shells. There are four unknowns N_x, N_y, N_{xy}, and w in Eqs.(8.1a), (8.1b), and (8.1j). The strain displacement relations for the middle surface of the shell are identical to those given for plates (Eqs. 7.34a, 7.34b, and 7.34e) except for ε_y because in the Y direction the shell has a curvature which was not present in a plate. The strain in the X direction is given by

$$\varepsilon_x = \frac{\partial u}{\partial x} + \frac{1}{2}\left(\frac{\partial w}{\partial x}\right)^2 \tag{8.1k}$$

The additional term in the shell to be added in the middle surface strain ε_y is calculated from Figure 8.3. The element AB moves to $A'B'$ due to the radial deformation w. The strain due to this displacement of the element is given by

$$\frac{A'B' - AB}{AB} = \frac{(R-w)d\theta - Rd\theta}{Rd\theta} = -\frac{w}{R} \tag{8.1l}$$

Adding the expression in Eq. (8.1l) to the middle surface strain in the Y direction given for plates by Eq. (7.34b) we obtain

$$\varepsilon_y = \frac{\partial v}{\partial y} - \frac{w}{R} + \frac{1}{2}\left(\frac{\partial w}{\partial y}\right)^2 \tag{8.1m}$$

Since the initial curvature of the shell has no effect on the middle surface shear strain, it is given by Eq. (7.34e) for the middle surface strain of plates as follows:

$$\gamma_{xy} = \frac{\partial u}{\partial y} + \frac{\partial v}{\partial x} + \frac{\partial w}{\partial x}\frac{\partial w}{\partial y} \tag{8.1n}$$

The constitutive relations for the thin walled elastic isotropic cylindrical shells are the same as for plates (Eqs. 7.34f–7.34h)

$$\varepsilon_x = \frac{1}{Eh}(N_x - \nu N_y) \tag{8.1o}$$

$$\varepsilon_y = \frac{1}{Eh}(N_y - \nu N_x) \tag{8.1p}$$

$$\gamma_{xy} = \frac{2(1+\nu)}{Eh}N_{xy} \tag{8.1q}$$

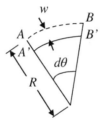

Figure 8.3 Circumferential strain due to radial displacement w

Three equations in three unknowns, u, v, and w can be obtained by using the kinematic and constitutive relations of Eqs. (8.1k), (8.1m), and (8.1n) and (8.1o)–(8.1q) respectively. These equations can be reduced further as follows:

$$\frac{\partial^2 \varepsilon_x}{\partial y^2} = \frac{\partial^3 u}{\partial x \partial y^2} + \left(\frac{\partial^2 w}{\partial x \partial y}\right)^2 + \frac{\partial w}{\partial x}\frac{\partial^3 w}{\partial x \partial y^2} \tag{8.1r}$$

$$\frac{\partial^2 \varepsilon_y}{\partial x^2} = \frac{\partial^3 v}{\partial x^2 \partial y} - \frac{1}{R}\frac{\partial^2 w}{\partial x^2} + \left(\frac{\partial^2 w}{\partial x \partial y}\right)^2 + \frac{\partial w}{\partial y}\frac{\partial^3 w}{\partial x^2 \partial y} \tag{8.1s}$$

$$\frac{\partial^2 \gamma_{xy}}{\partial x \partial y} = \frac{\partial^3 u}{\partial x \partial y^2} + \frac{\partial^3 v}{\partial x^2 \partial y} + \frac{\partial^3 w}{\partial x^2 \partial y}\frac{\partial w}{\partial y} + \frac{\partial^2 w}{\partial x^2}\frac{\partial^2 w}{\partial y^2} + \left(\frac{\partial^2 w}{\partial x \partial y}\right)^2 + \frac{\partial w}{\partial x}\frac{\partial^3 w}{\partial x \partial y^2} \tag{8.1t}$$

Equations (8.1r)–(8.1t) are used to obtain the compatibility relation

$$\frac{\partial^2 \varepsilon_x}{\partial y^2} + \frac{\partial^2 \varepsilon_y}{\partial x^2} - \frac{\partial^2 \gamma_{xy}}{\partial x \partial y} = \left(\frac{\partial^2 w}{\partial x \partial y}\right)^2 - \frac{\partial^2 w}{\partial x^2}\frac{\partial^2 w}{\partial y^2} - \frac{1}{R}\frac{\partial^2 w}{\partial x^2} \tag{8.1u}$$

Now a stress function $F(x, y)$ is introduced that satisfies Eqs. (8.1a) and (8.1b) such that

$$N_x = h\frac{\partial^2 F}{\partial y^2} \tag{8.1v}$$

$$N_y = h\frac{\partial^2 F}{\partial x^2} \tag{8.1w}$$

$$N_{xy} = -h\frac{\partial^2 F}{\partial x \partial y} \tag{8.1x}$$

Substitute the forces from Eqs. (8.1v)–(8.1x) in terms of the function $F(x, y)$ into Eqs. (8.1o)–(8.1q) to give

$$\varepsilon_x = \frac{1}{E}\left(\frac{\partial^2 F}{\partial y^2} - v\frac{\partial^2 F}{\partial x^2}\right)$$

$$\varepsilon_y = \frac{1}{E}\left(\frac{\partial^2 F}{\partial x^2} - v\frac{\partial^2 F}{\partial y^2}\right)$$

$$\gamma_{xy} = -\frac{2(1+v)}{E}\left(\frac{\partial^2 F}{\partial x \partial y}\right)$$

Substitute these relations into Eq. (8.1u) to obtain

$$\frac{\partial^4 F}{\partial x^4} + 2\frac{\partial^4 F}{\partial x^2 \partial y^2} + \frac{\partial^4 F}{\partial y^4} = E\left[\left(\frac{\partial^2 w}{\partial x \partial y}\right)^2 - \frac{\partial^2 w}{\partial x^2}\frac{\partial^2 w}{\partial y^2} - \frac{1}{R}\frac{\partial^2 w}{\partial x^2}\right] \tag{8.1y}$$

Substitute Eqs. (8.1v)–(8.1x) into Eq. (8.1j) to get

$$\frac{\partial^4 w}{\partial x^4} + 2\frac{\partial^4 w}{\partial x^2 \partial y^2} + \frac{\partial^4 w}{\partial y^4} - \frac{h}{D}\left[\frac{\partial^2 F}{\partial y^2}\frac{\partial^2 w}{\partial x^2} - 2\frac{\partial^2 F}{\partial x \partial y}\frac{\partial^2 w}{\partial x \partial y} + \frac{\partial^2 F}{\partial x^2}\left(\frac{1}{R} + \frac{\partial^2 w}{\partial y^2}\right)\right] - \frac{p}{D} = 0 \tag{8.1z}$$

We have reduced the number of equations to be simultaneously solved to two in two variables, F and w, given by Eqs. (8.1y) and (8.1z). These equations were first derived by Donnell in 1934 [2] using the von Kármán large deflection plate theory. Therefore, the equations are called the von Kármán–Donnell large displacement equilibrium equation for thin cylindrical shells.

8.3 The Linear Theory of Cylindrical Shells

Linear equilibrium equations can be obtained by omitting all quadratic and higher-order terms in u, v, and w from the nonlinear Eqs. (8.1a), (8.1b), and (8.1j) as follows:

$$\frac{\partial N_x}{\partial x} + \frac{\partial Nyx}{\partial y} = 0 \tag{8.2a}$$

$$\frac{\partial N_y}{\partial y} + \frac{\partial N_{xy}}{\partial x} = 0 \tag{8.2b}$$

and

$$D\nabla^4 w - \frac{N_y}{R} = p \tag{8.2c}$$

where

$$\nabla^4 w = \frac{\partial^4 w}{\partial x^4} + 2\frac{\partial^4 w}{\partial x^2 \partial y^2} + \frac{\partial^4 w}{\partial y^4}$$

The stress-strain and strain-displacement relations for the small displacements in the shell are

$$N_x = \sigma_x h = \frac{Eh}{1 - v^2}(\varepsilon_x + v\varepsilon_y) \tag{8.2d}$$

$$N_y = \sigma_y h = \frac{Eh}{1 - v^2}(\varepsilon_y + v\varepsilon_x) \tag{8.2e}$$

$$N_{xy} = \tau_{xy} h = \frac{Eh}{2(1 + v)}\gamma_{xy} \tag{8.2f}$$

$$\varepsilon_x = \frac{\partial u}{\partial x}, \quad \varepsilon_y = \frac{\partial v}{\partial y} - \frac{w}{R}, \quad \gamma_{xy} = \frac{\partial u}{\partial y} + \frac{\partial v}{\partial x} \tag{8.2g}$$

Equations (8.2a)–(8.2c) are a set of three coupled equations in four unknowns N_x, N_y, N_{xy}, and w. Substitute the constitutive and kinematic relations (Eqs. 8.2d–8.2g) into Eqs. (8.2a)–(8.2c) to get the following three equations in three unknowns u, v, and w:

$$\frac{\partial^2 u}{\partial x^2} + \frac{1 - v}{2}\frac{\partial^2 u}{\partial y^2} + \frac{1 + v}{2}\frac{\partial^2 v}{\partial x \partial y} - \frac{v}{R}\frac{\partial w}{\partial x} = 0 \tag{8.2h}$$

$$\frac{\partial^2 v}{\partial y^2} + \frac{1 - v}{2}\frac{\partial^2 v}{\partial x^2} + \frac{1 + v}{2}\frac{\partial^2 u}{\partial x \partial y} - \frac{1}{R}\frac{\partial w}{\partial y} = 0 \tag{8.2i}$$

$$D\nabla^4 w - \frac{1}{R}\frac{Eh}{1-v^2}\left(\frac{\partial v}{\partial y} - \frac{w}{R} + v\frac{\partial u}{\partial x}\right) = p \tag{8.2j}$$

Equations (8.2h)–(8.2j) were partially uncoupled by Donnell [3] to get

$$\nabla^4 u = \frac{v}{R}\frac{\partial^3 w}{\partial x^3} - \frac{1}{R}\frac{\partial^3 w}{\partial x \partial y^2} \tag{8.2k}$$

$$\nabla^4 v = \frac{v+2}{R}\frac{\partial^3 w}{\partial x^2 \partial y} + \frac{1}{R}\frac{\partial^3 w}{\partial y^3} \tag{8.2l}$$

Apply the operator ∇^4 to Eq. (8.2j) to get

$$D\nabla^8 w - \frac{1}{R}\frac{Eh}{1-v^2}\left(\nabla^4\frac{\partial v}{\partial y} - \frac{1}{R}\nabla^4 w + v\nabla^4\frac{\partial u}{\partial x}\right) = \nabla^4 p \tag{8.2m}$$

Now operate Eq. (8.2k) by $\dfrac{\partial}{\partial x}$ and Eq. (8.2l) by $\dfrac{\partial}{\partial y}$ to give

$$\nabla^4\frac{\partial u}{\partial x} = \frac{v}{R}\frac{\partial^4 w}{\partial x^4} - \frac{1}{R}\frac{\partial^4 w}{\partial x^2 \partial y^2} \tag{8.2n}$$

and

$$\nabla^4\frac{\partial v}{\partial y} = \frac{v+2}{R}\frac{\partial^4 w}{\partial x^2 \partial y^2} + \frac{1}{R}\frac{\partial^4 w}{\partial y^4} \tag{8.2o}$$

Substitute Eqs. (8.2n) and (8.2o) into Eq. (8.2m) to obtain

$$D\nabla^8 w + \frac{Eh}{R^2}\frac{\partial^4 w}{\partial x^4} = \nabla^4 p \tag{8.2p}$$

Equations (8.2k), (8.2l), and (8.2p) are partially uncoupled linear equilibrium equations.

8.3.1   Linear Membrane Equations for Cylindrical Shells

If we set the bending rigidity of a shell element to be zero, we get from Eqs. (8.2a)–(8.2c) linear membrane equations for a cylindrical shell below

$$\frac{\partial N_x}{\partial x} + \frac{\partial Nxy}{\partial y} = 0 \tag{8.3a}$$

$$\frac{\partial N_y}{\partial y} + \frac{\partial N_{xy}}{\partial x} = 0 \tag{8.3b}$$

$$-\frac{N_y}{R} = p \tag{8.3c}$$

These are three equations in three unknowns N_x, N_y, and N_{xy}, and are statically determinate. From Eq. (8.3c) we get $N_y = -pR$ and hoop stress

$$\sigma_y = -\frac{pR}{t} \tag{8.3d}$$

Where p is the external radial pressure, R is the radius, and t is the thickness of a cylinder.

8.4 Donnell's Linear Equations of Stability of Cylindrical Shells

There are relatively small quantities in the shell equilibrium equations that can be neglected to simplify the solution. The nonlinear equations give the bifurcation point and the corresponding critical load as well as the post-buckling behavior of the shells shown in Figure 8.4. However, the solution of the nonlinear equations becomes difficult, so there has been a tendency to simplify the nonlinear equations. To simplify, investigators have omitted different terms leading to various types of shell equations. The linear equations give the bifurcation point and the corresponding critical load, but do not give the post-buckling behavior of a shell. The curve of the axial compressive strain versus the axial compressive stress is plotted in Figure 8.4 [4].

The most notable characteristic of the curve is that, as the cylinder bends further after reaching the critical load, the axial load drops. This shows that at finite deflection an equilibrium position exists at loads that are considerably below the critical load. The phenomenon of sudden drop in the axial load is responsible for the discrepancy between the critical load calculated by the linear theory and the experimental results. In addition, if there are initial imperfections in the shell, then the critical load drops further, as shown in Figure 8.4 by the dotted line. Hence, it is believed now that the major reason for the difference between the theoretical and experimental critical load values lies in the presence of imperfections.

Here we are using the Donnell equations [3] to develop the linear stability theory for cylindrical shells. All the idealizations made previously for the nonlinear theory are valid except that the transverse deflections are assumed to be small in comparison to the shell thickness. As a consequence, quadratic terms that are functions of the deformations and their derivatives are negligible and are ignored when formulating the linear governing differential equations of stability for cylindrical shells.

The equations for determining the bifurcation point load are derived using the adjacent equilibrium criterion [5]. To the equilibrium configuration at the bifurcation point give small displacement increments that represent the bending of the shell. The total displacements and

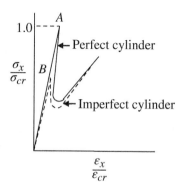

Figure 8.4 Axially compressed cylinder.

strains at a point in the shell are assumed to consist of two parts. These total displacements and strains are

$$u = u_o + u_b$$
$$v = v_o + v_b$$
$$w = w_o + w_b$$
$$\varepsilon_x = \varepsilon_{xo} + \varepsilon_{xb}$$
$$\varepsilon_y = \varepsilon_{yo} + \varepsilon_{yb}$$
$$\gamma_{xy} = \gamma_{xyo} + \gamma_{xyb} \tag{8.4a}$$

The terms with the subscript o are middle surface pre-buckling terms and the terms with the subscript b denote bending or buckling mode terms relative to the middle surface. It is assumed that the displacements (u_o, v_o, w_o) and (u, v, w) are adjacent equilibrium configurations when there has been no increment in the applied pressure p. The linearization is applied to the nonlinear Eqs. (8.1a), (8.1b), and (8.1j). The increment given to the displacements changes the internal forces as follows:

$$N_x = N_{xo} + \Delta N_x$$
$$N_y = N_{yo} + \Delta N_y$$
$$N_{xy} = N_{xyo} + \Delta N_{xy} \tag{8.4b}$$

where the terms with the subscripts o correspond to the displacements u_o, v_o, w_o and the incremental quantities with prefixes Δ correspond to the incremental displacements u_b, v_b, w_b. Let N_{xb}, N_{yb}, and N_{xyb} be parts of $\Delta N_x, \Delta N_y$, and ΔN_{xy} that are linear in u_b, v_b, w_b. The strain displacement relations are given by

$$\varepsilon_x = \frac{\partial u}{\partial x} + \frac{1}{2}\left(\frac{\partial w}{\partial x}\right)^2$$

or $$\varepsilon_x = \left(\frac{\partial u_o}{\partial x} + \frac{\partial u_b}{\partial x}\right) + \frac{1}{2}\left(\frac{\partial w_o}{\partial x}\right)^2 + \left(\frac{\partial w_o}{\partial x}\right)\left(\frac{\partial w_b}{\partial x}\right) + \frac{1}{2}\left(\frac{\partial w_b}{\partial x}\right)^2 \tag{8.4c}$$

$$\varepsilon_y = \frac{\partial v}{\partial y} - \frac{w}{R} + \frac{1}{2}\left(\frac{\partial w}{\partial y}\right)^2$$

$$\varepsilon_y = \left(\frac{\partial v_o}{\partial y} + \frac{\partial v_b}{\partial y}\right) - \left(\frac{w_o}{R} + \frac{w_b}{R}\right) + \frac{1}{2}\left(\frac{\partial w_o}{\partial y}\right)^2 + \left(\frac{\partial w_o}{\partial y}\right)\left(\frac{\partial w_b}{\partial y}\right) + \frac{1}{2}\left(\frac{\partial w_b}{\partial y}\right)^2 \tag{8.4d}$$

$$\gamma_{xy} = \frac{\partial u}{\partial y} + \frac{\partial v}{\partial x} + \frac{\partial w}{\partial x}\frac{\partial w}{\partial y}$$

$$\gamma_{xy} = \frac{\partial u_o}{\partial y} + \frac{\partial u_b}{\partial y} + \frac{\partial v_o}{\partial x} + \frac{\partial v_b}{\partial x} + \frac{\partial w_o}{\partial x}\frac{\partial w_o}{\partial y} + \frac{\partial w_o}{\partial x}\frac{\partial w_b}{\partial y} + \frac{\partial w_b}{\partial x}\frac{\partial w_o}{\partial y} + \frac{\partial w_b}{\partial x}\frac{\partial w_b}{\partial y} \tag{8.4e}$$

The constitutive relations for thin-walled isotropic elastic cylindrical shells are given as before

$$N_x = \sigma_x h = \frac{Eh}{1-v^2}(\varepsilon_x + v\varepsilon_y), \tag{8.2d}$$

$$N_y = \sigma_y h = \frac{Eh}{1-v^2}(\varepsilon_y + v\varepsilon_x) \tag{8.2e}$$

$$N_{xy} = \tau_{xy}h = \frac{Eh}{2(1+v)}\gamma_{xy} \tag{8.2f}$$

Equations (8.4c)–(8.4e) and Eqs. (8.2d)–(8.2f) give

$$N_x = N_{xo} + \Delta N_x$$

$$= \frac{Eh}{1-v^2}\left[\left(\frac{\partial u_o}{\partial x} + \frac{\partial u_b}{\partial x}\right) + \frac{1}{2}\left(\frac{\partial w_o}{\partial x}\right)^2 + \left(\frac{\partial w_o}{\partial x}\right)\left(\frac{\partial w_b}{\partial x}\right) + \frac{1}{2}\left(\frac{\partial w_b}{\partial x}\right)^2\right.$$

$$\left. + v\left\{\left(\frac{\partial v_o}{\partial y} + \frac{\partial v_b}{\partial y}\right) - \left(\frac{w_o}{R} + \frac{w_b}{R}\right) + \frac{1}{2}\left(\frac{\partial w_o}{\partial y}\right)^2 + \left(\frac{\partial w_o}{\partial y}\right)\left(\frac{\partial w_b}{\partial y}\right) + \frac{1}{2}\left(\frac{\partial w_b}{\partial y}\right)^2\right\}\right] \tag{8.4f}$$

$$N_{xo} = \frac{Eh}{1-v^2}\left[\frac{\partial u_o}{\partial x} + \frac{1}{2}\left(\frac{\partial w_o}{\partial x}\right)^2 + v\left\{\frac{\partial v_o}{\partial y} - \frac{w_o}{R} + \frac{1}{2}\left(\frac{\partial w_o}{\partial y}\right)^2\right\}\right]$$

$$\Delta N_x = \frac{Eh}{1-v^2}\left[\frac{\partial u_b}{\partial x} + \left(\frac{\partial w_o}{\partial x}\right)\left(\frac{\partial w_b}{\partial x}\right) + \frac{1}{2}\left(\frac{\partial w_b}{\partial x}\right)^2\right.$$

$$\left. + v\left\{\frac{\partial v_b}{\partial y} - \frac{w_b}{R} + \left(\frac{\partial w_o}{\partial y}\right)\left(\frac{\partial w_b}{\partial y}\right) + \frac{1}{2}\left(\frac{\partial w_b}{\partial y}\right)^2\right\}\right] \tag{8.4g}$$

Hence,

$$N_{xb} == \frac{Eh}{1-v^2}\left[\frac{\partial u_b}{\partial x} + \left(\frac{\partial w_o}{\partial x}\right)\left(\frac{\partial w_b}{\partial x}\right) + v\left\{\frac{\partial v_b}{\partial y} - \frac{w_b}{R} + \left(\frac{\partial w_o}{\partial y}\right)\left(\frac{\partial w_b}{\partial y}\right)\right\}\right] \tag{8.4h}$$

Similarly,

$$N_y = N_{yo} + \Delta N_y$$

$$N_{yo} = \frac{Eh}{1-v^2}\left[\frac{\partial v_o}{\partial y} - \frac{w_o}{R} + \frac{1}{2}\left(\frac{\partial w_o}{\partial y}\right)^2 + v\left\{\frac{\partial u_o}{\partial x} + \frac{1}{2}\left(\frac{\partial w_o}{\partial x}\right)^2\right\}\right] \tag{8.4i}$$

$$\Delta N_y = \frac{Eh}{1-v^2}\left[\frac{\partial v_b}{\partial y} - \frac{w_b}{R} + \left(\frac{\partial w_o}{\partial y}\right)\left(\frac{\partial w_b}{\partial y}\right) + \frac{1}{2}\left(\frac{\partial w_b}{\partial y}\right)^2\right.$$

$$\left. + v\left\{\frac{\partial u_b}{\partial x} + \left(\frac{\partial w_o}{\partial x}\right)\left(\frac{\partial w_b}{\partial x}\right) + \frac{1}{2}\left(\frac{\partial w_b}{\partial x}\right)^2\right\}\right]$$

Hence,

$$N_{yb} == \frac{Eh}{1-v^2}\left[\frac{\partial v_b}{\partial y} - \frac{w_b}{R} + \left(\frac{\partial w_o}{\partial y}\right)\left(\frac{\partial w_b}{\partial y}\right) + v\left\{\frac{\partial u_b}{\partial x} + \left(\frac{\partial w_o}{\partial x}\right)\left(\frac{\partial w_b}{\partial x}\right)\right\}\right] \tag{8.4j}$$

$$N_{xy} = N_{xyo} + \Delta N_{xy}$$

$$N_{xyo} = \frac{Eh}{2(1+v)} \left[\frac{\partial u_o}{\partial y} + \frac{\partial v_o}{\partial x} + \left(\frac{\partial w_o}{\partial x} \right) \left(\frac{\partial w_o}{\partial y} \right) \right] \tag{8.4k}$$

$$\Delta N_{xy} = \left[\frac{\partial u_b}{\partial y} + \frac{\partial v_b}{\partial x} + \left(\frac{\partial w_o}{\partial x} \right) \left(\frac{\partial w_b}{\partial y} \right) + \left(\frac{\partial w_o}{\partial y} \right) \left(\frac{\partial w_b}{\partial x} \right) + \left(\frac{\partial w_b}{\partial x} \right) \left(\frac{\partial w_b}{\partial y} \right) \right]$$

Hence,

$$N_{xyb} = \left[\frac{\partial u_b}{\partial y} + \frac{\partial v_b}{\partial x} + \left(\frac{\partial w_o}{\partial x} \right) \left(\frac{\partial w_b}{\partial y} \right) + \left(\frac{\partial w_o}{\partial y} \right) \left(\frac{\partial w_b}{\partial x} \right) \right] \tag{8.4l}$$

Substitute Eqs. (8.4a) and (8.4b) into Eqs. (8.1a), (8.1b), and (8.1j). All terms containing u_o, v_o, and w_o alone and the corresponding N_{xo}, N_{yo}, and N_{xyo} and the radial pressure term p drop out because u_o, v_o, and w_o is an equilibrium configuration. In addition, the quadratic and higher-order terms in u_b, v_b, w_b or their counterpart in the form of N_{xb}, N_{yb}, and N_{xyb} may be neglected because u_b, v_b, w_b are small. The resulting equations are

$$\frac{\partial N_{xb}}{\partial x} + \frac{\partial N_{xyb}}{\partial y} = 0 \tag{8.4m}$$

$$\frac{\partial N_{yb}}{\partial y} + \frac{\partial N_{xyb}}{\partial x} = 0 \tag{8.4n}$$

$$D\nabla^4 w_b - \left[\left(N_{xo} \frac{\partial^2 w_b}{\partial x^2} + N_{xb} \frac{\partial^2 w_o}{\partial x^2} \right) + \left(N_{yo} \frac{\partial^2 w_b}{\partial y^2} + N_{yb} \frac{\partial^2 w_o}{\partial y^2} \right) + \left(2N_{xyo} \frac{\partial^2 w_b}{\partial x \partial y} + 2N_{xyb} \frac{\partial^2 w_o}{\partial x \partial y} \right) \right]$$

$$- \frac{N_{yb}}{R} = 0 \tag{8.4o}$$

where

$$N_{xo} = \sigma_{xo} h = \frac{Eh}{1-v^2}(\varepsilon_{xo} + v\varepsilon_{yo}) \qquad N_{xb} = \sigma_{xb} h = \frac{Eh}{1-v^2}(\varepsilon_{xb} + v\varepsilon_{yb})$$

$$N_{yo} = \sigma_{yo} h = \frac{Eh}{1-v^2}(\varepsilon_{yo} + v\varepsilon_{xo}) \qquad N_{yb} = \sigma_{yb} h = \frac{Eh}{1-v^2}(\varepsilon_{yb} + v\varepsilon_{xb})$$

$$N_{xyo} = \tau_{xyo} h = \frac{Eh}{2(1+v)}\gamma_{xyo} \qquad N_{xyb} = \tau_{xyb} h = \frac{Eh}{2(1+v)}\gamma_{xyb} \tag{8.4p}$$

and the strain displacement elations are

$$\varepsilon_{xo} = \frac{\partial u_o}{\partial x} + \frac{1}{2}\left(\frac{\partial w_o}{\partial x} \right)^2 \qquad\qquad \varepsilon_{xb} = \frac{\partial u_b}{\partial x} + \left(\frac{\partial w_o}{\partial x} \right)\left(\frac{\partial w_b}{\partial x} \right)$$

$$\varepsilon_{yo} = \frac{\partial v_o}{\partial y} - \frac{w}{R} + \frac{1}{2}\left(\frac{\partial w_o}{\partial y} \right)^2 \qquad \varepsilon_{yb} = \frac{\partial v_b}{\partial y} - \frac{w_b}{R} + \left(\frac{\partial w_o}{\partial y} \right)\left(\frac{\partial w_b}{\partial y} \right) \tag{8.4q}$$

$$\gamma_{xyo} = \frac{\partial u_o}{\partial y} + \frac{\partial v_o}{\partial x} + \left(\frac{\partial w_o}{\partial x} \right)\left(\frac{\partial w_o}{\partial y} \right)$$

$$\gamma_{xyb} = \frac{\partial u_b}{\partial y} + \frac{\partial v_b}{\partial x} + \left(\frac{\partial w_o}{\partial x} \right)\left(\frac{\partial w_b}{\partial y} \right) + \left(\frac{\partial w_o}{\partial y} \right)\left(\frac{\partial w_b}{\partial x} \right)$$

The influence of pre-buckling rotations is very small in many cases, hence we neglect the terms containing pre-buckling rotations $\dfrac{\partial w_o}{\partial x}$ and $\dfrac{\partial w_o}{\partial y}$. The terms containing the rotations, $\dfrac{\partial w_o}{\partial x}, \dfrac{\partial w_o}{\partial y}$ are eliminated from Eqs. (8.4m)–(8.4o) to get the stability equations

$$\frac{\partial N_{xb}}{\partial x} + \frac{\partial N_{xyb}}{\partial y} = 0$$

$$\frac{\partial N_{yb}}{\partial y} + \frac{\partial N_{xyb}}{\partial x} = 0$$

$$D\nabla^4 w_b - \left[N_{xo}\frac{\partial^2 w_b}{\partial x^2} + N_{yo}\frac{\partial^2 w_b}{\partial y^2} + 2N_{xyo}\frac{\partial^2 w_b}{\partial x \partial y} \right] - \frac{N_{yb}}{R} = 0 \tag{8.4r}$$

Similarly, neglecting the pre-buckling rotations $\dfrac{\partial w_o}{\partial x}$ and $\dfrac{\partial w_o}{\partial y}$ in the strain displacement relations (Eq. (8.4q)) gives

$$\varepsilon_{xo} = \frac{\partial u_o}{\partial x} \qquad\qquad \varepsilon_{xb} = \frac{\partial u_b}{\partial x}$$

$$\varepsilon_y = \frac{\partial v_o}{\partial y} - \frac{w}{R} \qquad \varepsilon_{yb} = \frac{\partial v_b}{\partial y} - \frac{w_b}{R}$$

$$\gamma_{xyo} = \frac{\partial u_o}{\partial y} + \frac{\partial v_o}{\partial x} \qquad \gamma_{xyb} = \frac{\partial u_b}{\partial y} + \frac{\partial v_b}{\partial x} \tag{8.4s}$$

Equations (8.4p), (8.4r), and (8.4s) lead to a coupled set of three linear homogeneous equations in the variables u_b, v_b, w_b.

By substituting constitutive and kinematic relations from Eqs. (8.4p) and (8.4s) in Eq.(8.4r) we get

$$\frac{\partial^2 u_b}{\partial x^2} + \frac{1-v}{2}\frac{\partial^2 u_b}{\partial y^2} + \frac{1+v}{2}\frac{\partial^2 v_b}{\partial x \partial y} - \frac{v}{R}\frac{\partial w_b}{\partial x} = 0$$

$$\frac{1+v}{2}\frac{\partial^2 u_b}{\partial x \partial y} + \frac{1-v}{2}\frac{\partial^2 v_b}{\partial x^2} + \frac{\partial^2 v_b}{\partial y^2} - \frac{1}{R}\frac{\partial w_b}{\partial y} = 0 \tag{8.4t}$$

$$D\nabla^4 w_b - \left[N_{xo}\frac{\partial^2 w_b}{\partial x^2} + N_{yo}\frac{\partial^2 w_b}{\partial y^2} + 2N_{xyo}\frac{\partial^2 w_b}{\partial x \partial y} \right]$$

$$- \frac{1}{R}\frac{Eh}{1-v^2}\left(\frac{\partial v_b}{\partial y} - \frac{w_b}{R} + v\frac{\partial u_b}{\partial x} \right) = 0$$

Equation (8.4t) are the Donnell's coupled equations in the variables $u_b, v_b,$ and w_b. Donnell [3] uncoupled these equations partially and wrote them in the following form

$$\nabla^4 u_b = \frac{v}{R}\frac{\partial^3 w_b}{\partial x^3} - \frac{1}{R}\frac{\partial^3 w_b}{\partial x \partial y^2} \tag{8.4u}$$

$$\nabla^4 v_b = \frac{v+2}{R}\frac{\partial^3 w_b}{\partial x^2 \partial y} + \frac{1}{R}\frac{\partial^3 w_b}{\partial y^3} \tag{8.4v}$$

$$D\nabla^8 w_b + \frac{Eh}{R^2}\frac{\partial^4 w_b}{\partial x^4} - \nabla^4\left(N_{x0}\frac{\partial^2 w_b}{\partial x^2} + 2N_{xy0}\frac{\partial^2 w_b}{\partial x \partial y} + N_{y0}\frac{\partial^2 w_b}{\partial y^2}\right) = 0 \tag{8.4w}$$

Equations (8.4u)–(8.4w) are called Donnell's uncoupled stability equations that can be compared with Eqs. (8.2k), (8.2l), and (8.2p). Equation (8.4w) is a homogeneous equation in w_b only, with variable coefficients N_{x0}, N_{y0}, and N_{xy0}. The coefficients N_{x0}, N_{y0}, and N_{xy0} are governed by the linear equilibrium equations given by Eqs. (8.2a)–(8.2c) because the influence of pre-buckling rotations has been neglected. The cylindrical shell stability equations where pre-buckling rotation terms are retained are given by Eqs. (8.4m)–(8.4o). The corresponding equations where pre-buckling terms are omitted are given in equivalent forms by Eqs. (8.4r), (8.4t) and (8.4u)–(8.4w).

8.5 The Energy Method

The nonlinear equilibrium equations for the cylindrical shells are now derived using the principle of minimum potential energy. The total potential of the cylindrical shell is the sum of the strain energy U of the shell and the potential energy V of the applied load and can be written as

$$\Pi = U + V \tag{8.5a}$$

The strain energy of a deformed shell can be expressed as the sum of strain energy due to membrane action U_m and the strain energy due to bending U_b:

$$U = U_m + U_b \tag{8.5b}$$

$$U_m = \frac{1}{2}\int_{-\frac{h}{2}}^{\frac{h}{2}}\iint(\sigma_x\varepsilon_x + \sigma_y\varepsilon_y + \tau_{xy}\gamma_{xy})dxdydz$$

Substitute Eqs. (8.2d)–(8.2f) to get

$$U_m = \frac{1}{2}\frac{Eh}{1-v^2}\iint\left(\varepsilon_x^2 + \varepsilon_y^2 + 2v\varepsilon_x\varepsilon_y + \frac{1-v}{2}\gamma_{xy}^2\right)dxdy \tag{8.5c}$$

The strain energy due to bending is given by Eq. (7.17c) as in the case of plates

$$U_b = \frac{D}{2}\iint\left[\left(\frac{\partial^2 w}{\partial x^2}\right)^2 + \left(\frac{\partial^2 w}{\partial y^2}\right)^2 + 2v\left(\frac{\partial^2 w}{\partial x^2}\right)\left(\frac{\partial^2 w}{\partial y^2}\right) + 2(1-v)\left(\frac{\partial^2 w}{\partial x \partial y}\right)^2\right]dxdy \tag{8.5d}$$

For a cylindrical shell subjected to lateral pressure p, the potential energy of the applied load is

$$V = -\iint pwdxdy \tag{8.5e}$$

The total potential energy of the shell is given by using Eqs. (8.5b)–(8.5e) as

$$\Pi = \int\int \left[\frac{1}{2}\frac{Eh}{1-v^2}\left(\varepsilon_x^2 + \varepsilon_y^2 + 2v\varepsilon_x\varepsilon_y + \frac{1-v}{2}\gamma_{xy}^2\right) + \right.$$
$$\left. \frac{D}{2}\left\{\left(\frac{\partial^2 w}{\partial x^2}\right)^2 + \left(\frac{\partial^2 w}{\partial y^2}\right)^2 + 2v\left(\frac{\partial^2 w}{\partial x^2}\right)\left(\frac{\partial^2 w}{\partial y^2}\right) + 2(1-v)\left(\frac{\partial^2 w}{\partial x\partial y}\right)^2\right\} - pw \right] dxdy$$

(8.5f)

The integrant F is given by

$$F = \frac{1}{2}\frac{Eh}{1-v^2}\left(\varepsilon_x^2 + \varepsilon_y^2 + 2v\varepsilon_x\varepsilon_y + \frac{1-v}{2}\gamma_{xy}^2\right)$$
$$+\frac{D}{2}\left\{\left(\frac{\partial^2 w}{\partial x^2}\right)^2 + \left(\frac{\partial^2 w}{\partial y^2}\right)^2 + 2v\left(\frac{\partial^2 w}{\partial x^2}\right)\left(\frac{\partial^2 w}{\partial y^2}\right) + 2(1-v)\left(\frac{\partial^2 w}{\partial x\partial y}\right)^2\right\} - pw$$

(8.5g)

The Euler equations for the integrant F in the total potential energy expression in Eq. (8.5f) are given by Eq. (D.10) in Appendix D

$$\frac{\partial F}{\partial u} - \frac{\partial}{\partial x}\frac{\partial F}{\partial u_x} - \frac{\partial}{\partial y}\frac{\partial F}{\partial u_y} = 0$$

$$\frac{\partial F}{\partial v} - \frac{\partial}{\partial x}\frac{\partial F}{\partial v_x} - \frac{\partial}{\partial y}\frac{\partial F}{\partial v_y} = 0$$

$$\frac{\partial F}{\partial w} - \frac{\partial}{\partial x}\frac{\partial F}{\partial w_x} - \frac{\partial}{\partial y}\frac{\partial F}{\partial w_y} + \frac{\partial^2}{\partial x^2}\frac{\partial F}{\partial w_{xx}} + \frac{\partial^2}{\partial x\partial y}\frac{\partial F}{\partial w_{xy}} + \frac{\partial^2}{\partial y^2}\frac{\partial F}{\partial w_{yy}} = 0 \qquad \text{(D.10)}$$

$$\varepsilon_x^2 = \left(u_x + \frac{1}{2}w_x^2\right)^2 = u_x^2 + \frac{1}{4}w_x^4 + u_x w_x^2$$

$$\varepsilon_y^2 = \left(v_y - \frac{w}{R} + \frac{1}{2}w_y^2\right)^2$$

$$\gamma_{xy}^2 = (u_y + v_x + w_x w_y)^2 \qquad \text{(8.5h)}$$

Equations (8.5g) and (8.5h) give

$$\frac{\partial F}{\partial u} = 0$$

$$\frac{\partial F}{\partial u_x} = \frac{1}{2}\frac{Eh}{1-v^2}\left\{2\left(u_x + \frac{1}{2}w_x^2\right) + 2v\left(v_y - \frac{w}{R} + \frac{1}{2}w_y^2\right)\right\}$$

$$\frac{\partial}{\partial x}\frac{\partial F}{\partial u_x} = \frac{1}{2}\frac{Eh}{1-v^2}\left\{2u_{xx} + 2w_x w_{xx} + 2v\left(v_{yx} - \frac{1}{R}w_x + w_y w_{yx}\right)\right\}$$

$$\frac{\partial F}{\partial u_y} = \frac{1}{2}\frac{Eh}{1-v^2}\{(1-v)(u_y + v_x + w_x w_y)\}$$

$$\frac{\partial}{\partial y}\frac{\partial F}{\partial u_y} = \frac{1}{2}\frac{Eh}{1-v^2}\{(1-v)(u_{yy} + v_{xy} + w_x w_{yy} + w_y w_{xy})\}$$

Substitute in the first equation of Eq. (D.10) to obtain

$$\frac{1}{2}\frac{Eh}{1-v^2}\left\{2u_{,xx} + 2w_{,x}w_{,xx} + 2v\left(v_{,yx} - \frac{1}{R}w_{,x} + w_{,y}w_{,yx}\right) + \right.$$

$$\left. (1-v)(u_{,yy} + v_{,xy} + w_{,x}w_{,yy} + w_{,y}w_{,xy})\right\} = 0$$

$$\frac{Eh}{1-v^2}\left[u_{,xx} + w_{,x}w_{,xx} + v\left(v_{,xy} - \frac{1}{R}w_{,x} + w_{,y}w_{,yx}\right)\right] + $$

$$\frac{Eh}{2(1+v)}[u_{,yy} + v_{,xy} + w_{,x}w_{,yy} + w_{,y}w_{,xy}] = 0 \qquad (8.5i)$$

Equation (8.5i) is the same as Eq. (8.1a), the nonlinear equilibrium equation for thin cylindrical shells

$$\frac{\partial N_x}{\partial x} + \frac{\partial N_{xy}}{\partial y} = 0 \qquad (8.1a)$$

Similarly, the second and third Euler equations in Eq. (D.10) will lead to the nonlinear equilibrium equations of Eqs. (8.1b) and (8.1j) respectively.

$$\frac{\partial N_y}{\partial y} + \frac{\partial N_{xy}}{\partial x} = 0 \qquad (8.1b)$$

$$D\left(\frac{\partial^4 w}{\partial x^4} + 2\frac{\partial^4 w}{\partial x^2 \partial y^2} + \frac{\partial^4 w}{\partial y^4}\right) - N_x\frac{\partial^2 w}{\partial x^2} - 2N_{xy}\frac{\partial^2 w}{\partial x \partial y} - N_y\left(\frac{\partial^2 w}{\partial y^2} + \frac{1}{R}\right) - p = 0 \qquad (8.1j)$$

8.6 Application of the Linear Stability Equations

Donnell's stability equations are used in this section to obtain critical loads for circular cylinders. Pre-buckling rotations are omitted in these applications. The subscript b, for example in the incremental quantities w_b, N_{xyb}, etc. is omitted for simplicity.

8.6.1 Circular Cylinders Under Axial Compression

Consider a cylindrical shell of length L and radius R that is simply supported at its ends and is subjected to a uniformly distributed axial compressive force P in lbs. (Newtons). The cylinder shortens under the action of the compressive force and increases in diameter except at the ends and the pre-buckling deformation of the shell is axisymmetric as shown in Figure 8.5a. The radial displacement w_0 is a function of x with the localized bending near the cylinder ends but we assume it to be uniform for simplicity, as shown in Figure 8.5b. Then the pre-buckling deformation may be determined by the linear membrane equations with sufficient accuracy. From membrane analysis of the unbuckled cylinder

$$N_{x0} = -\frac{P}{2\pi R} \quad \text{and} \quad N_{xy0} = N_{y0} = 0 \qquad (8.6a)$$

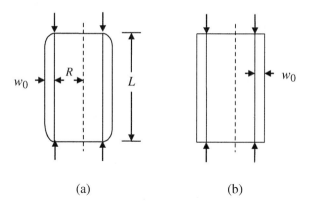

(a)   (b)

Figure 8.5 Cylinder under the action of axial compression: (a) Buckled shape: (b) Buckled shape idealization.

The critical load P_{cr} is the lowest load at which the equilibrium ceases to be stable in the axisymmetric form. The critical load is determined by solving the Donnell equation, Eq. (8.4w), as given by Batdorf [6]. Substitute Eq. (8.6a) into Eq. (8.4w) to get

$$D\nabla^8 w + \frac{Eh}{R^2}\frac{\partial^4 w}{\partial x^4} + \frac{P}{2\pi R}\nabla^4 \frac{\partial^2 w}{\partial x^2} = 0 \tag{8.6b}$$

Equation (8.6b) is a linear partial differential equation with constant coefficients. For simply supported ends, the boundary conditions are

$$w = \frac{\partial^2 w}{\partial x^2} = 0 \text{ at } x = 0 \text{ and } x = L \tag{8.6c}$$

Both the governing differential equation and the boundary conditions are satisfied if the lateral displacement is taken, of the form

$$w = A\sin\frac{m\pi x}{L}\sin\frac{n\pi y}{\pi R} \tag{8.6d}$$

where m and n are the number of half-waves in the longitudinal and the circumferential directions respectively. Let

$$\beta = \frac{nL}{\pi R} \tag{8.6e}$$

then Eq. (8.6d) can be written as

$$w = A\sin\frac{m\pi x}{L}\sin\frac{\beta\pi y}{L} \tag{8.6f}$$

From Eq. (8.6f), we have

$$\nabla^8 w = \nabla^4 \nabla^4 w = \nabla^4\left(\frac{\partial^4 w}{\partial x^4} + 2\frac{\partial^4 w}{\partial x^2 \partial y^2} + \frac{\partial^4 w}{\partial y^4}\right)$$

$$\frac{\partial^2 w}{\partial x^2} = -\frac{m^2\pi^2}{L^2}A\sin\frac{m\pi x}{L}\sin\frac{\beta\pi y}{L}$$

$$\frac{\partial^4 w}{\partial x^4} = \frac{m^4 \pi^4}{L^4} A \sin \frac{m \pi x}{L} \sin \frac{\beta \pi y}{L}$$

$$\frac{\partial^4 w}{\partial y^4} = \frac{\beta^4 \pi^4}{L^4} A \sin \frac{m \pi x}{L} \sin \frac{\beta \pi y}{L}$$

$$2\frac{\partial^4 w}{\partial x^2 \partial y^2} = 2\frac{m^2 \pi^2}{L^2} \frac{\beta^2 \pi^2}{L^2} A \sin \frac{m \pi x}{L} \sin \frac{\beta \pi y}{L}$$

$$\nabla^4 w = \frac{\pi^4}{L^4}(m^2 + \beta^2)^2 A \sin \frac{m \pi x}{L} \sin \frac{\beta \pi y}{L}$$

$$\nabla^8 w = \frac{\pi^8}{L^8}(m^2 + \beta^2)^4 A \sin \frac{m \pi x}{L} \sin \frac{\beta \pi y}{L}$$

$$\nabla^4 \frac{\partial^2 w}{\partial x^2} = -\frac{m^2 \pi^2}{L^2} \left(\frac{\pi}{L}\right)^4 (m^2 + \beta^2)^2 A \sin \frac{m \pi x}{L} \sin \frac{\beta \pi y}{L} \tag{8.6g}$$

Substitute Eq. (8.6g) into Eq. (8.6b) to obtain

$$D\left(\frac{\pi}{L}\right)^8 (m^2 + \beta^2)^4 + \frac{Eh}{R^2} m^4 \left(\frac{\pi}{L}\right)^4 - \frac{P}{2\pi R}\left(\frac{\pi}{L}\right)^6 m^2 (m^2 + \beta^2)^2 = 0 \tag{8.6h}$$

Divide Eq. (8.6h) by $D\left(\frac{\pi}{L}\right)^8$, and substitute, $D = \dfrac{Eh^3}{12(1 - v^2)}$, and we can write

$$(m^2 + \beta^2)^4 + \frac{12Z^2 m^4}{\pi^4} - k_x m^2 (m^2 + \beta^2)^2 = 0 \tag{8.6i}$$

where

$$Z = \frac{L^2}{Rh}(1 - v^2)^{\frac{1}{2}} \tag{8.6j}$$

and

$$k_x = \frac{P}{2\pi R} \frac{L^2}{D\pi^2} \tag{8.6k}$$

Z is a nondimensional Bartdorf parameter useful for differentiating between long and short cylinders. k_x is a buckling stress parameter similar to that appears in plate buckling equation (Eq. (7.7n)).

Solving for k_x in Eq. (8.6i) we have

$$k_x = \frac{(m^2 + \beta^2)^2}{m^2} + \frac{12Z^2 m^2}{\pi^4 (m^2 + \beta^2)^2} \tag{8.6l}$$

Differentiate k_x with respect to $\dfrac{(m^2 + \beta^2)^2}{m^2}$ and set equal to zero to get minimum k_x.

Let $s = \dfrac{(m^2 + \beta^2)^2}{m^2}$, and substitute in Eq. (8.6l) to get

$$k_x = s + \frac{12Z^2}{\pi^4 s}$$

$$\frac{dk_x}{ds} = 1 + \frac{12Z^2}{\pi^4}\left(-\frac{1}{s^2}\right) = 0$$

or

$$s = \left(\frac{12Z^2}{\pi^4} \right)^{\frac{1}{2}}$$

or

$$\frac{(m^2 + \beta^2)^2}{m^2} = \left(\frac{12Z^2}{\pi^4} \right)^{\frac{1}{2}}$$ (8.6m)

Substitute Eq. (8.6m) into Eq. (8.6l) to get

$$k_x = 2 \left(\frac{12Z^2}{\pi^4} \right)^{\frac{1}{2}}$$

or

$$k_x = \frac{4\sqrt{3}}{\pi^2} Z$$ (8.6n)

From Eqs. (8.6k) and (8.6n) we obtain

$$k_x = \frac{P}{2\pi R} \frac{L^2}{D\pi^2} = \frac{4\sqrt{3}}{\pi^2} Z$$ (8.6o)

or

$$\sigma_{cr} = \frac{P}{2\pi R h} = \frac{Eh}{R} \frac{1}{\sqrt{3(1 - v^2)}}$$ (8.6p)

If

$$v = 0.3$$

$$\sigma_{cr} = 0.605 \frac{Eh}{R}$$ (8.6q)

From Eq. (8.6m), β is given by

$$\beta = \left[\frac{(12Z^2)^{\frac{1}{4}}}{\pi} m - m^2 \right]^{\frac{1}{2}}$$ (8.6r)

At

$$Z = 2.85$$

$$\beta = m^{\frac{1}{2}} (1 - m)^{\frac{1}{2}}$$ (8.6s)

which means Z can be smaller than 2.85 if either $m < 1$ or β is imaginary. Neither of these conditions are true, hence Eqs. (8.6o) and (8.6q) are only applicable to cylinders for which $Z > 2.85$. Equation (8.6m) indicates that cylindrical shells subjected to axial compression can have a large number of instability modes corresponding to a single bifurcation point. Since m and n are positive integers, for $Z < 2.85$, Eq. (8.6h) and the trial and error procedure may be used to determine the critical load. The calculations are simplified if for $Z < 2.85$, the critical stress coefficient k_x is determined by setting $m = 1$ and $\beta = 0$ in Eq. (8.6l).

$$k_x = 1 + \frac{12Z^2}{\pi^4} \tag{8.6t}$$

The critical stress for cylinders with $Z < 2.85$ is given by Eq. (8.6k) as

$$\sigma_{cr} = \frac{P}{2\pi Rh} = \frac{k_x D\pi^2}{L^2 h} \tag{8.6u}$$

where k_x is given by Eq. (8.6t). The graph of Z versus k_x found from Eqs. (8.6t) and (8.6o) is plotted in Figure 8.6. As Z approaches zero (cylinder radius approaches infinity), the k_x approaches 1. Thus, Eq. (8.6u) gives

$$\sigma_{cr} = \frac{D\pi^2}{L^2 h} \tag{8.6v}$$

This indicates the critical stress of a short cylinder approaches that of a wide column, i.e. a flat plate which is simply supported on the loaded ends and free on the unloaded edges. In the analysis so far, it is assumed that the failure of a cylinder would occur due to local surface buckling. A very long cylinder can buckle as a column with undeformed cross-section ($m = n = 1$) before local surface buckling occurs. The Donnell equations do not give accurate results for moderately long columns. The critical load is determined by setting, $I = 2\pi Rh(R^2)/2 = \pi R^3 h$, in the appropriate column equations in Chapter 2 to determine the critical load.

In summary, the critical load for an axially loaded cylindrical shell depends on its length to radius ratio. Short length and large radius cylinders behave like plates and buckle into a single half-wave along the length and with no waves along the circumference. On the other hand, very

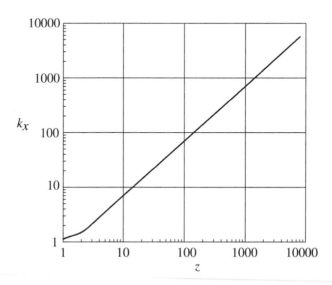

Figure 8.6 Critical axial pressure for cylindrical shells.

long cylinders do not involve surface buckling and behave like Euler columns. The cylinders whose length lies between these two extremes are called intermediate length or moderately long cylinders, the majority of actual cylinders lie in this category. These cylinders buckle by developing surface deformations both in the longitudinal and circumferential directions. The critical stress for the intermediate length cylinders is given by Eq. (8.6p) derived from the Donnell linear theory, but the results are not very accurate. Results obtained from the experiments are usually much smaller than those given by Eq. (8.6q), more accurate results are given in the literature [7]. The buckled shape of the axially compressed cylindrical shell is shown in Figure 8.5a. It shows the applied load is eccentric to the shell wall in the mid-length. The pre-buckling deformations are nonlinear due to this eccentricity and the primary buckling path is curved from the beginning. The inclusion of this eccentricity is taken care of by a non-linear large deflection analysis.

8.6.2 Circular Cylinders Under Uniform Lateral Pressure

Consider a cylindrical shell of length L and radius R that is simply supported at its ends and is subjected to uniform external pressure p in lbs./in.^{2} (Newtons per square mm). The pre-buckling deformation is axisymmetric under such loading, as shown in Figure 8.7a. The radial displacement, w_0, is a function of x with the localized bending near the cylinder ends but we assume it to be uniform for simplicity, as shown in Figure 8.7b. Then the coefficient N_{y0} is governed by the linear membrane equations (Eqs. 8.3a–8.3c), and it is constant. There is no moment at the ends because of simply supported ends, but the boundary condition allows longitudinal and radial translations. If the cylinder is free to expand longitudinally as the lateral pressure is applied, $N_{xo} = 0$. Also, $N_{xyo} = 0$ if no torsional load is applied. The membrane analysis of the unbuckled cylinder gives from Eqs. (8.3a–8.3c).

$$N_{y0} = -pR \tag{8.7a}$$

The critical pressure is defined as the lowest pressure p_{cr} at which the axisymmetric form loses its stability. Substituting Eq. (8.7a) into Eq. (8.4w) gives

$$D\nabla^8 w_b + \frac{Eh}{R^2}\frac{\partial^4 w}{\partial x^4} + pR\nabla^4 \frac{\partial^2 w}{\partial y^2} = 0 \tag{8.7b}$$

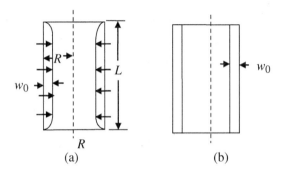

(a)　　　　(b)

Figure 8.7 Cylinder subjected to uniform external pressure: (a) Buckled shape; (b) Buckled shape idealization.

The boundary conditions at the simply supported ends are

$$w = \frac{\partial^2 w}{\partial x^2} = 0 \text{ at } x = 0 \text{ and } x = L \tag{8.7c}$$

Both the governing equation and the boundary conditions are satisfied by the lateral displacement of the form

$$w = A \sin \frac{m\pi x}{L} \sin \frac{n\pi y}{\pi R} \quad m = 1, 2, 3 \text{---} \text{ and } n = 1, 2, 3 \text{---} \tag{8.7d}$$

where m is the number of half-waves in the longitudinal direction and n is the number of half-waves in the circumferential direction. Let us introduce a variable

$$\beta = \frac{nL}{\pi R} \tag{8.7e}$$

Then Eq. (8.7d) can be written as

$$w = A \sin \frac{m\pi x}{L} \sin \frac{\beta \pi y}{L} \tag{8.7f}$$

Similar to the case of axial compression on the cylinders, we can get the partial derivatives of w with respect to x and y.

$$\frac{\partial^2 w}{\partial y^2} = -\frac{\beta^2 \pi^2}{L^2} A \sin \frac{m\pi x}{L} \sin \frac{\beta \pi y}{L}$$

$$\nabla^4 \frac{\partial^2 w}{\partial y^2} = -\frac{\beta^2 \pi^2}{L^2} \left(\frac{\pi}{L}\right)^4 (m^2 + \beta^2)^2 A \sin \frac{m\pi x}{L} \sin \frac{\beta \pi y}{L} \tag{8.7g}$$

Substitute Eq. (8.7g) into Eq. (8.7b) to have

$$D\left(\frac{\pi}{L}\right)^8 (m^2 + \beta^2)^4 + \frac{Eh}{R^2} m^4 \left(\frac{\pi}{L}\right)^4 - pR\left(\frac{\pi}{L}\right)^6 \beta^2 (m^2 + \beta^2)^2 = 0 \tag{8.7h}$$

Divide Eq. (8.7h) by $\left(\frac{\pi}{L}\right)^6$ and substitute $D = \frac{Eh^3}{12(1-v^2)}$ to obtain

$$\frac{pR}{Eh} = \frac{h^2(m^2 + \beta^2)^2}{12(1 - v^2)\beta^2} \left(\frac{\pi}{L}\right)^2 + \frac{m^4}{R^2 \beta^2 (m^2 + \beta^2)^2} \left(\frac{L}{\pi}\right)^2 \tag{8.7i}$$

For particular values of L/R and R/h, the m and n corresponding to the smallest eigenvalue that gives the external pressure p may be calculated by trial and error. It is indicated by Brush and Almroth [5] that for the external pressure p to be minimum, the secondary equilibrium path will have one half sine wave in the axial direction, i.e. $m = 1$, and the Eq. (8.7i) can be written as

$$\frac{pR}{Eh} = \frac{h^2(1 + \beta^2)^2}{12(1 - v^2)\beta^2} \left(\frac{\pi}{L}\right)^2 + \frac{1}{R^2 \beta^2 (1 + \beta^2)^2} \left(\frac{L}{\pi}\right)^2 \tag{8.7j}$$

Equation (8.7j) can be written as

$$\frac{pR}{Eh} = \frac{h^2 \left(\frac{1}{\beta^2} + 1\right)^2 \beta^2}{12(1 - v^2)} \left(\frac{\pi}{L}\right)^2 + \frac{1}{R^2 \beta^6 \left(\frac{1}{\beta^2} + 1\right)^2} \left(\frac{L}{\pi}\right)^2 \tag{8.7k}$$

If L/R approaches infinity, β also approaches infinity and Eq. (8.7k) becomes

$$p = \frac{Eh^3}{12(1-v^2)}\frac{n^2}{R^3} = n^2\frac{D}{R^3} \tag{8.7l}$$

Example 8.1 A circular cylinder of diameter, $D = 48$ in. (1219.2 mm), length, $L = 24$ in. (609.6 mm), and a wall thickness of $h = 0.24$ in. (6.1 mm) is subjected to the uniform external pressure, p. Find the critical pressure p_{cr} if the cylinder is made of steel with $E = 29 \times 10^6$ psi (200 000 MPa), and the Poisson's ratio for the material is $v = 0.3$.

Assume $n = 7$, then $\beta = \dfrac{nL}{\pi R} = \dfrac{7}{\pi}(1) = 2.23$

$L/R = 1$ and $R/h = 24/0.24\,[609.6/6.1] = 100$

From Eq. (8.7j) we have

$$p = \frac{29x10^6}{100}\left\{\frac{(1+2.23^2)^2}{12(1-0.3^2)2.23^2}\frac{\pi^2}{(100)^2} + \frac{1}{2.23^2(1+2.23^2)^2\pi^2}\right\} = 353.66\text{ lb/in.}^2$$

$$\left[\frac{200{,}000}{100}\left\{\frac{(1+2.23^2)^2}{12(1-0.3^2)2.23^2}\frac{\pi^2}{(100)^2} + \frac{1}{2.23^2(1+2.23^2)^2\pi^2}\right\} = 2.44\text{ N/mm}^2\right]$$

Similarly for $n = 8$, $\beta = \dfrac{nL}{\pi R} = \dfrac{8}{\pi}(1) = 2.55$, and $p_{cr} = 307.34$ lb/in.^{2} [2.12 N/mm^2]

For $n = 9$, $\beta = \dfrac{nL}{\pi R} = \dfrac{9}{\pi}(1) = 2.87$, and $p = 312.98$ lb/in.^{2} [2.16 N/mm^2]

For other values of n, p is higher, hence $p_{cr} = 307.34$ lb/in.^{2} [2.12 N/mm^2]

8.6.2.1 Critical Pressures for Cylinders Subjected to External Pressure

Let $\bar{p} = \dfrac{L^2 R}{\pi^2 D}p$, then, from Eq. (8.7j) we have

$$\bar{p} = \frac{(1+\beta^2)^2}{\beta^2} + \frac{1}{\beta^2(1+\beta^2)^2}\frac{12Z^2}{\pi^4} \tag{8.8a}$$

Minimize $\bar{p}$ with respect to β in Eq. (8.8a) to get $\bar{p}$ as a function of single geometric parameter Z instead of the two parameters L/R and R/h.

$$\frac{d\bar{p}}{d\beta} = \frac{\beta^2[2(1+\beta^2)(2\beta)] - (1+\beta^2)^2(2\beta)}{\beta^4} - \frac{2\beta(1+\beta^2)^2 + \beta^2(1+\beta^2)(4\beta)}{[\beta^2(1+\beta^2)^2]^2}\frac{12Z^2}{\pi^4} = 0$$

or

$$\frac{(\beta^2-1)(\beta^2+1)^4}{1+3\beta^2} = \frac{12Z^2}{\pi^4} \tag{8.8b}$$

Let $\beta = 2$, then, from Eq. (8.8b)

$$Z^2 = \frac{3(5)^4}{13}\frac{\pi^4}{12} = 1170.78, \quad \text{and } Z = 34.22$$

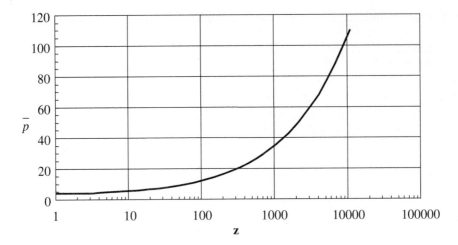

Figure 8.8 Critical external lateral pressure for cylindrical shells.

Use Eq. (8.8a) to get

$$\bar{p} = \frac{(1+4)^2}{4} + \frac{1}{4(1+4)^2}\frac{12}{\pi^4}(1170.78) = 7.69$$

Similarly, we can calculate Z and $\bar{p}$ for other assumed values of β. The graph of Z versus $\bar{p}$ is plotted in Figure 8.8. From the given dimensions of a cylindrical shell and properties of the material, we can find the critical external pressure for the cylinder from Figure 8.8 and the critical stress is given by, $\sigma_{cr} = \frac{P_{cr}R}{h}$. For the long cylinders which will have large values of L and Z, the minimization shown here is not very accurate, because n, the number of half-wave lengths in the circumferential direction, are small and cannot be treated as a continuous variable. As the cylinder radius approaches infinity, the parameter Z approaches zero, and the critical stress σ_{cr} approaches that of long plates in compression given by $\sigma_{cr} = 4\pi^2D/(L^2h)$ from Figure 7.9.

In the literature Jerath, Sadid, and Ghosh have investigated the buckling of isotropic and orthotropic cylinders under non-uniform external pressure due to wind load [8, 9].

8.6.3 Cylinders Subjected to Torsion

Consider a simple supported cylindrical shell at its ends of length L and radius R that is subjected to a twisting moment. For simplicity, assume that the membrane analysis can be used for the pre-buckling deformation. In this case, N_{xy0} is constant and Eq. (8.4w) can be written as

$$DV^8w + \frac{Eh}{R^2}\frac{\partial^4 w}{\partial x^4} - 2N_{xy0}\nabla^4\frac{\partial^2 w}{\partial x\partial y} = 0 \tag{8.9a}$$

There is a difference between Eq. (8.9a) and Eqs. (8.6b) and (8.7b) obtained for axial or lateral compressive forces acting on a cylinder. The difference is that there are odd orders of derivatives with respect to each coordinate in the third term, and even orders of derivatives in the first two terms in Eq. (8.9a). Therefore, we cannot satisfy the equation by using the deflection as a product of sine functions. It means there are no generators that remain straight during buckling due to torsion, instead they are transformed into helical form. The buckling deformation under the torsional moment consists of a number of waves that spiral around the cylinder from one end to the other. These waves can be expressed by the displacement function given by the following expression:

$$w = A \sin \left(\frac{m\pi x}{L} - \frac{\beta \pi y}{L} \right) \qquad (8.9b)$$

where $\beta = \dfrac{nL}{\pi R}$, m and n are the number of half-waves in the longitudinal and circumferential directions respectively. Equation (8.9b) does not satisfy the boundary conditions at the cylinder ends, but satisfies the differential equation and the periodicity in the circumferential direction. Hence, Eq. (8.9b) can be used only for long cylinders in which the constraints at the ends have little influence on the magnitude of the critical stresses. Now substitute Eq. (8.9b) into Eq. (8.9a). We can write the partial derivatives of w from Eq. (8.9b) as follows:

$$\frac{\partial^4 w}{\partial x^4} = \frac{m^4 \pi^4}{L^4} A \sin \left(\frac{m\pi x}{L} - \frac{\beta \pi y}{L} \right)$$

$$\frac{\partial^4 w}{\partial y^4} = \frac{\beta^4 \pi^4}{L^4} A \sin \left(\frac{m\pi x}{L} - \frac{\beta \pi y}{L} \right)$$

$$2 \frac{\partial^4 w}{\partial x^2 \partial y^2} = 2 \frac{m^2 \beta^2 \pi^4}{L^4} A \sin \left(\frac{m\pi x}{L} - \frac{\beta \pi y}{L} \right) \qquad (8.9c)$$

$$\nabla^4 w = \frac{\pi^4}{L^4} (m^2 + \beta^2)^2 A \sin \left(\frac{m\pi x}{L} - \frac{\beta \pi y}{L} \right)$$

$$\nabla^8 w = \frac{\pi^8}{L^8} (m^2 + \beta^2)^4 A \sin \left(\frac{m\pi x}{L} - \frac{\beta \pi y}{L} \right)$$

$$\nabla^4 \frac{\partial^2 w}{\partial x \partial y} = m\beta \frac{\pi^6}{L^6} (m^2 + \beta^2)^2 A \sin \left(\frac{m\pi x}{L} - \frac{\beta \pi y}{L} \right)$$

Substitute Eq. (8.9c) into Eq. (8.9a)

$$\left[D \frac{\pi^8}{L^8} (m^2 + \beta^2)^4 + \frac{Eh}{R^2} \frac{m^4 \pi^4}{L^4} - 2N_{xy0} \frac{m\beta \pi^6}{L^6} (m^2 + \beta^2)^2 \right] A \sin \left(\frac{m\pi x}{L} - \frac{\beta \pi y}{L} \right) = 0 \quad (8.9d)$$

or

$$N_{xy0} = \frac{D}{2m\beta} \frac{\pi^2}{L^2} (m^2 + \beta^2)^2 + \frac{Eh}{2R^2} \frac{m^3 L^2}{\pi^2 \beta} \frac{1}{(m^2 + \beta^2)^2} \qquad (8.9e)$$

For long cylinders the shell buckles in two circumferential waves, i.e. the smallest value of N_{xy0} corresponds to $n = 2$ [3]. Substitution of $n = 2$ gives

$$N_{xy0} = \frac{DL}{4m\pi R^3} \left(\frac{m^2 \pi^2 R^2}{L^2} + 4 \right)^2 + \frac{Ehm^3 \pi^3 R^3}{4L^3} \frac{1}{\left(\dfrac{m^2 \pi^2 R^2}{L^2} + 4 \right)^2}$$

Also for long cylindrical shells, $\left(\dfrac{m\pi R}{L}\right)^2 \ll 4$, hence the N_{xy0} can be expressed approximately as

$$N_{xy0} = \frac{4DL}{\pi R^3}\frac{1}{m} + \frac{Eh}{64}\frac{\pi^3 R^3}{L^3}m^3 \tag{8.9f}$$

The value of m to make N_{xy0} minimum can be found from

$$\frac{dN_{xy0}}{dm} = \frac{4DL}{\pi R^3}\left(-\frac{1}{m^2}\right) + \frac{Eh}{64}\frac{\pi^3 R^3}{L^3}(3m^2) = 0$$

or

$$m^4 = \frac{256}{3}\frac{1}{Eh}\frac{DL^4}{\pi^4 R^6} \tag{8.9g}$$

Substitute $D = \dfrac{Eh^3}{12(1-v^2)}$ to get

$$m^4 = \frac{64}{9(1-v^2)}\left(\frac{h}{R}\right)^2\left(\frac{L}{\pi R}\right)^4 \tag{8.9h}$$

Equation (8.9f) can be written as

$$N_{xy0} = \frac{Eh}{m}\left[\frac{h^2}{3(1-v^2)}\frac{L}{\pi R^3} + \frac{\pi^3 R^3 m^4}{64L^3}\right]$$

Substitute the value of m from Eq. (8.9h) to get

$$N_{xy0} = \frac{0.272E}{(1-v^2)^{\frac{3}{4}}}\left(\frac{h}{R}\right)^{\frac{3}{2}}h \tag{8.9i}$$

or

$$\frac{N_{xy0}}{h} = \tau_{cr} = \frac{0.272E}{(1-v^2)^{\frac{3}{4}}}\left(\frac{h}{R}\right)^{\frac{3}{2}} \tag{8.9j}$$

τ_{cr} in Eq. (8.9j) is the approximate value given by Donnell [3] which is 15.25% higher than a more accurate value given by Eq. (8.9k) in Timoshenko and Gere [7].

$$\tau_{cr} = \frac{0.236E}{(1-v^2)^{\frac{3}{4}}}\left(\frac{h}{R}\right)^{\frac{3}{2}} \tag{8.9k}$$

The formula in Eq. (8.9k) gives correct critical stress for long cylinders under torsion.

For shorter cylinders the boundary conditions cannot be ignored and finding critical stresses is more complicated. The solution for shorter cylinders can be found by using the deflection function composed of a finite sum of terms in Eq. (8.9b). This procedure was followed by Donnell [3] to give results for simply supported and clamped cylinders at their ends. The analysis is simplified by omitting small terms from the equations [7]. Equations (8.9l) and (8.9m) were obtained for the short and moderately long cylinders based on these simplifications, where the parameter, $Z < 10\left(\dfrac{R}{h}\right)^2$. For simply supported ends:

$$(1 - v^2)\frac{\tau_{cr}}{E}\frac{L^2}{h^2} = 2.8 + \sqrt{2.6 + 1.40\left(\sqrt{1 - v^2}\frac{L^2}{2hR}\right)^{\frac{3}{2}}} \tag{8.9l}$$

and for clamped ends:

$$(1 - v^2)\frac{\tau_{cr}}{E}\frac{L^2}{h^2} = 4.6 + \sqrt{7.8 + 1.67\left(\sqrt{1 - v^2}\frac{L^2}{2hR}\right)^{\frac{3}{2}}} \tag{8.9m}$$

Define a parameter, $k = \dfrac{L^2 h \tau_{cr}}{\pi^2 D} = (1 - v^2)\dfrac{\tau_{cr}}{E}\dfrac{L^2}{h^2}\dfrac{12}{\pi^2}$ and use $Z = \dfrac{L^2}{Rh}(1 - v^2)^{\frac{1}{2}}$ to get, for simply supported ends:

$$k = \left(2.8 + \sqrt{2.6 + 1.40\left(\frac{Z}{2}\right)^{\frac{3}{2}}}\right)\frac{12}{\pi^2} \tag{8.9n}$$

and for clamped ends:

$$k = \left(4.6 + \sqrt{7.8 + 1.67\left(\frac{Z}{2}\right)^{\frac{3}{2}}}\right)\frac{12}{\pi^2} \tag{8.9o}$$

The critical shear stress τ_{cr} is plotted in terms of a non-dimensional coefficient k versus Z in Figure 8.9 for short and moderately long cylinders defined by $Z < 10\left(\dfrac{R}{h}\right)^2$. As the cylinder radius approaches infinity, the parameter Z approaches zero. In this case, the critical coefficient k approaches the values of 5.35 and 8.98 obtained for infinitely long flat plates [5] whose loaded edges are simply supported, whereas the other edges are simply supported and clamped respectively.

8.6.4 Cylinders Subjected to Combined Axial Compression and Uniform External Lateral Pressure

Consider a cylindrical shell with simply supported ends subjected to a combined loading consisting of an axial compressive load P in lbs. (Newtons), and a uniform external pressure p in lbs./in.^{2} (N/mm^2). Under the action of these forces, the cylindrical shell may retain its cylindrical form, but at a certain critical combination of forces the cylindrical form of equilibrium may become unstable and the cylinder may buckle. If the linear membrane analysis is assumed to be satisfactory for the axisymmetric pre-buckling deformation, then Eq. (8.4w) for this type of loading can be written as

$$D\nabla^8 w + \frac{Eh}{R^2}\frac{\partial^4 w}{\partial x^4} + \nabla^4\left(\frac{P}{2\pi R}\frac{\partial^2 w}{\partial x^2} + pR\frac{\partial^2 w}{\partial y^2}\right) = 0 \tag{8.10a}$$

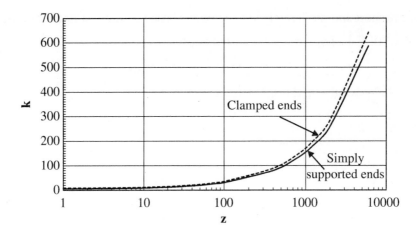

Figure 8.9 Critical shear stress for cylindrical shells subjected to torsion.

where the axial compression is given by $N_{x0} = -\dfrac{P}{2\pi R}$, and the uniform external pressure is given by $N_{y0} = -pR$. Assume

$$\frac{P}{2\pi R} = qpR \tag{8.10b}$$

where q is a non-dimensional constant giving the value of load ratio. Substitute Eq. (8.10b) into Eq. (8.10a) to obtain

$$D\nabla^8 w + \frac{Eh}{R^2}\frac{\partial^4 w}{\partial x^4} + pR\nabla^4\left(q\frac{\partial^2 w}{\partial x^2} + \frac{\partial^2 w}{\partial y^2}\right) = 0 \tag{8.10c}$$

Substitute the deflection "w" given by Eq. (8.7f)

$$w = A \sin\frac{m\pi x}{L}\sin\frac{\beta\pi y}{L} \tag{8.7f}$$

in Eq. (8.10c) to get

$$D\left(\frac{\pi}{L}\right)^8 (m^2 + \beta^2)^4 + \frac{Eh}{R^2} m^4 \left(\frac{\pi}{L}\right)^4 - pR(qm^2 + \beta^2)\frac{\pi^6}{L^6}(m^2 + \beta^2)^2 = 0 \tag{8.10d}$$

or

$$pR = \frac{D(m^2 + \beta^2)^2}{(qm^2 + \beta^2)}\left(\frac{\pi}{L}\right)^2 + Eh\left(\frac{L}{\pi R}\right)^2 \frac{m^4}{(m^2 + \beta^2)^2(qm^2 + \beta^2)} \tag{8.10e}$$

For any given load ratio "q," a distinct eigenvalue corresponds to each pair of m and n values. The minimum eigenvalue may be determined by trial. When the ratio $q = 1/2$, then $P = p\pi R^2$, and the cylinder is acted on by the same uniform pressure p on both of its lateral surface and the ends. This is called hydrostatic pressure loading. Interaction curves for different combinations of loads are given in [10].

8.6.5 Cylindrical Shells with Different End Conditions

The assumed solution of

$$w = C \sin \frac{m\pi x}{L} \sin \frac{\beta \pi y}{L} \qquad (8.11a)$$

satisfies the differential Eq. (8.4w), and is satisfactory for the simply supported cylinders with end conditions of $w = 0$, and $\frac{\partial^2 w}{\partial x^2} = 0$. Substitute Eq. (8.11a) into the uncoupled stability equations given by Eqs. (8.4u) and (8.4v), and we will see that the displacements u and v must be expressed as

$$u = A \cos \frac{m\pi x}{L} \sin \frac{\beta \pi y}{L} \qquad (8.11b)$$

$$v = B \sin \frac{m\pi x}{L} \cos \frac{\beta \pi y}{L} \qquad (8.11c)$$

Substitute the displacements u, v, and w into the uncoupled stability equations, and we will see that the equations are reduced to a set of three homogeneous ordinary differential equations with constant coefficients. These displacements satisfy only the end conditions $\frac{\partial u}{\partial x} = v = 0$ at $x = 0$, L. The boundary conditions $u = v = w = \frac{\partial^2 w}{\partial x^2} = 0$ are not satisfied. Hence uncoupled form of Donnell stability Eqs. (8.4u)–(8.4w) are not suitable for solving cylindrical shell problems with different end conditions. For general end conditions in cylinders, the coupled Donnell equations (Eq. 8.4t) may be used. These coupled equations are second order in u and v and fourth order in w. Therefore, the assumed displacement functions have to satisfy periodicity requirement along the circumference of the cylinder, and four boundary conditions at each end of the cylindrical shell with a total of eight end conditions. The boundary conditions can be different at each end including displacements and rotations that have elastic restraints. Consider a cylinder that is not simply supported at the ends and is subjected to external hydrostatic pressure p. The membrane analysis of the pre-buckled shell under hydrostatic pressure gives, from Eqs. (8.3a)–(8.3c)

$$N_{x0} = -\frac{1}{2}pR \left(\text{see Eq.8.10b}, q = \frac{1}{2} \right), \quad N_{y0} = -pR, \quad N_{xy0} = 0 \qquad (8.11d)$$

Substitute these values in Eqs. (8.4t) to have

$$\frac{\partial^2 u}{\partial x^2} + \frac{1-v}{2}\frac{\partial^2 u}{\partial y^2} + \frac{1+v}{2}\frac{\partial^2 v}{\partial x \partial y} - \frac{v}{R}\frac{\partial w}{\partial x} = 0 \qquad (8.11e)$$

$$\frac{1+v}{2}\frac{\partial^2 u}{\partial x \partial y} + \frac{1-v}{2}\frac{\partial^2 v}{\partial x^2} + \frac{\partial^2 v}{\partial y^2} - \frac{1}{R}\frac{\partial w}{\partial y} = 0 \qquad (8.11f)$$

$$D\nabla^4 w - \frac{1}{R}\frac{Eh}{1-v^2}\left(\frac{\partial v}{\partial y} - \frac{w}{R} + v\frac{\partial u}{\partial x} \right) + pR\left(\frac{1}{2}\frac{\partial^2 w}{\partial x^2} + \frac{\partial^2 w}{\partial y^2} \right) = 0 \qquad (8.11g)$$

Assume the displacement functions as

$$u = u_n(x) \cos \frac{\beta \pi y}{L} \qquad (8.11h)$$

$$v = v_n(x) \sin \frac{\beta \pi y}{L} \qquad (8.11i)$$

$$w = w_n(x) \cos \frac{\beta \pi y}{L} \qquad (8.11j)$$

Substituting Eqs. (8.11h)–(8.11j) into Eqs. (8.11e)–(8.11g) gives the following coupled ordinary differential equations

$$\frac{d^2 u_n}{dx^2} - \frac{1-v}{2}\frac{\beta^2 \pi^2}{L^2}u_n + \frac{1+v}{2}\frac{\beta \pi}{L}\frac{dv_n}{dx} - \frac{v}{R}\frac{dw_n}{dx} = 0 \qquad (8.11k)$$

$$-\frac{1+v}{2}\frac{\beta \pi}{L}\frac{du_n}{dx} + \frac{1-v}{2}\frac{d^2 v_n}{dx^2} - \frac{\beta^2 \pi^2}{L^2}v_n + \frac{1}{R}\frac{\beta \pi}{L}w_n = 0 \qquad (8.11l)$$

$$D\left(\frac{d^4 w_n}{dx^4} - 2\frac{\beta^2 \pi^2}{L^2}\frac{d^2 w_n}{dx^2} + \frac{\beta^4 \pi^4}{L^4}w_n\right) - \frac{Eh}{R(1-v^2)}\left(v\frac{du_n}{dx} + \frac{\beta \pi}{L}v_n - \frac{1}{R}w_n\right)$$
$$+ pR\left(-\frac{\beta^2 \pi^2}{L^2}w_n + \frac{1}{2}\frac{d^2 w_n}{dx^2}\right) = 0 \qquad (8.11m)$$

Sobel [11] obtained the solution of the ordinary differential equations for different sets of boundary conditions. This procedure is applicable for stability equations with constant coefficients and is not applicable for the stability equations with variable coefficients. For such problems, numerical methods of analysis are used. Even for the stability equations with constant coefficients, the amount of work involved is so great that the numerical methods are used in applications.

8.7 Failure and Post-buckling Behavior of Cylindrical Shells

The equilibrium paths of initially perfect and imperfect cylindrical shells subjected to axial compression are shown in Figure 8.4. It is evident from Figure 8.4 that the buckling load represents the ultimate strength of the cylinder. It also shows that the buckling load of an imperfect cylindrical shell is substantially lower than that of the perfect shell. In Figure 8.4 the plot is for axial compression load but the drop in the buckling load in the presence of imperfections occurs for other loading conditions also. In the case of cylinders subjected to external pressure or torsion, the effect of imperfections is small in comparison to the axially loaded case. The classical solution by the small deflection linear theory was first obtained by Lorenz in 1908 [12] and was later independently solved by Timoshenko in 1910 [13], Southwell in 1914 [14] and Flügge in 1932 [15]. The buckling loads obtained from experiments were as low as 30% of the loads given by the classical theories. Donnell in 1934 [2] proposed a non-linear large deflection theory to understand the discrepancy between the theoretical and the experimental results.

Donnell added the same nonlinear terms that von Kármán used in solving nonlinear plate equations and derived the von Kármán–Donnell large displacement cylindrical shell equations. However, this analysis did not give satisfactory buckling loads. In 1942, von Kármán and Tsien [16] used the same nonlinear equations that were used by Donnell but approximated the lateral deflection that represented more adequately the deformed shape of the buckled shell. Their analysis gave the stress versus strain solid curve shown in Figure 8.4. It shows that the actual load a cylinder in practice could support would be less than the critical load in the post-buckling stage. Several other researchers: Leggett and Jones [17], Michielsen [18], Kempner [19], Hoff, Madsen, and Mayers [20], have improved on the von Kármán–Tsien work by including more terms in the nonlinear equations and the assumed deflection function.

Donnell and Wan [4] in 1950 took into account the initial imperfections in the analysis of cylindrical shells. They showed that the main reason for the difference in the theoretical and experimental buckling loads is due to the presence of initial imperfections as shown by the dotted curve in Figure 8.4. It is seen from the two curves that the maximum load reached by an imperfect cylinder represented by point B is much smaller than the critical load represented by the point A for the perfect cylinder. Donnell and Wan's analysis is based on the nonlinear differential equilibrium equations and assumes a deflected shape of the shell that is slightly different from the cylindrical shape. Koiter [21] in 1945 confirmed the influence of initial imperfections and his analysis focused on the initial post-buckling behavior. Koiter's analysis showed that a limited amount of information can be obtained about the secondary path by examining the state of equilibrium at the bifurcation point. When the initial slope of the secondary path at the bifurcation point is negative, the post-buckling path is unstable and is more sensitive to the imperfections. In the case of cylinders, the slope of the initial post-buckling path at the bifurcation point is negative, hence the cylinders are very sensitive to the presence of imperfections. This point was emphasized in Chapter 1 also.

8.7.1 Post-Buckling Behavior of Cylindrical Shells

The von Kármán–Donnell large displacement Eqs. (8.1y) and (8.1z) were derived for initially perfect cylinders. These equations are modified to include the initial imperfections in the shells. Assume there is an initial distortion w_0 in addition to the deformation w produced by the applied forces in the lateral direction. The initial distortion will change Eq. (8.1z), which is an equation of equilibrium in the radial direction of the cylinder. The first three terms in Eq. (8.1z) are related to the lateral shear forces that depend only on the curvature caused by bending due to external forces, the initial distortion does not affect them. The remaining terms are obtained by multiplying the components of the middle surface forces by the surface curvature of the shell. Since the lateral surface curvature is to be used, the total lateral deformation $w + w_0$ is substituted for w in Eq. (8.1z) to get [22]

$$\frac{\partial^4 w}{\partial x^4} + 2\frac{\partial^4 w}{\partial x^2 \partial y^2} + \frac{\partial^4 w}{\partial y^4} - \frac{h}{D}\left[\frac{\partial^2 F}{\partial y^2}\left(\frac{\partial^2 w}{\partial x^2} + \frac{\partial^2 w_0}{\partial x^2}\right) - 2\frac{\partial^2 F}{\partial x \partial y}\left(\frac{\partial^2 w}{\partial x \partial y} + \frac{\partial^2 w_0}{\partial x \partial y}\right)\right.$$
$$\left. + \frac{\partial^2 F}{\partial x^2}\left(\frac{1}{R} + \frac{\partial^2 w}{\partial y^2} + \frac{\partial^2 w_0}{\partial y^2}\right)\right] = \frac{p}{D} \qquad (8.12a)$$

The strain–displacement relations (Eqs. 8.1k, 8.1m, 8.1n) are rewritten by replacing w by $w + w_0$ as follows:

$$\varepsilon_x = \frac{\partial u}{\partial x} + \frac{1}{2}\left(\frac{\partial(w + w_0)}{\partial x}\right)^2 = \frac{\partial u}{\partial x} + \frac{1}{2}\left(\frac{\partial w}{\partial x} + \frac{\partial w_0}{\partial x}\right)^2$$

or

$$\varepsilon_x = \frac{\partial u}{\partial x} + \frac{1}{2}\left(\frac{\partial w}{\partial x}\right)^2 + \frac{1}{2}\left(\frac{\partial w_0}{\partial x}\right)^2 + \frac{\partial w}{\partial x}\frac{\partial w_0}{\partial x}$$

Neglecting the small term we get

$$\varepsilon_x = \frac{\partial u}{\partial x} + \frac{1}{2}\left(\frac{\partial w}{\partial x}\right)^2 + \frac{\partial w}{\partial x}\frac{\partial w_0}{\partial x} \tag{8.12b}$$

Similarly, we can write

$$\varepsilon_y = \frac{\partial v}{\partial y} - \frac{w}{R} + \frac{1}{2}\left(\frac{\partial w}{\partial y}\right)^2 + \frac{\partial w}{\partial y}\frac{\partial w_0}{\partial y} \tag{8.12c}$$

$$\gamma_{xy} = \frac{\partial u}{\partial y} + \frac{\partial v}{\partial x} + \frac{\partial w}{\partial x}\frac{\partial w}{\partial y} + \frac{\partial w}{\partial x}\frac{\partial w_0}{\partial y} + \frac{\partial w_0}{\partial x}\frac{\partial w}{\partial y} \tag{8.12d}$$

$$\frac{\partial^2 \varepsilon_x}{\partial y^2} = \frac{\partial^3 u}{\partial x \partial y^2} + \frac{\partial w}{\partial x}\frac{\partial^3 w}{\partial x \partial y^2} + \left(\frac{\partial^2 w}{\partial x \partial y}\right)^2 + 2\frac{\partial^2 w_0}{\partial x \partial y}\frac{\partial^2 w}{\partial x \partial y} + \frac{\partial w}{\partial x}\frac{\partial^3 w_0}{\partial x \partial y^2} + \frac{\partial w_0}{\partial x}\frac{\partial^3 w}{\partial x \partial y^2} \tag{8.12e}$$

$$\frac{\partial^2 \varepsilon_y}{\partial x^2} = \frac{\partial^3 v}{\partial x^2 \partial y} - \frac{1}{R}\frac{\partial^2 w}{\partial x^2} + \frac{\partial w}{\partial y}\frac{\partial^3 w}{\partial y \partial x^2} + \left(\frac{\partial^2 w}{\partial x \partial y}\right)^2 + 2\frac{\partial^2 w_0}{\partial x \partial y}\frac{\partial^2 w}{\partial x \partial y} + \frac{\partial^3 w}{\partial y \partial x^2}\frac{\partial w_0}{\partial y} + \frac{\partial w}{\partial y}\frac{\partial^3 w_0}{\partial y \partial x^2} \tag{8.12f}$$

$$\frac{\partial^2 \gamma_{xy}}{\partial x \partial y} = \frac{\partial^3 u}{\partial x \partial y^2} + \frac{\partial^3 v}{\partial x^2 \partial y} + \frac{\partial w}{\partial y}\frac{\partial^3 w}{\partial y \partial x^2} + \frac{\partial^2 w}{\partial x^2}\frac{\partial^2 w}{\partial y^2} + \left(\frac{\partial^2 w}{\partial x \partial y}\right)^2 + \frac{\partial w}{\partial x}\frac{\partial^3 w}{\partial x \partial y^2} + \frac{\partial^3 w}{\partial x^2 \partial y}\frac{\partial w_0}{\partial y}$$
$$+ \frac{\partial^2 w}{\partial x^2}\frac{\partial^2 w_0}{\partial y^2} + 2\frac{\partial^2 w}{\partial x \partial y}\frac{\partial^2 w_0}{\partial x \partial y} + \frac{\partial w}{\partial x}\frac{\partial^3 w_0}{\partial x \partial y^2} + \frac{\partial w}{\partial y}\frac{\partial^3 w_0}{\partial x^2 \partial y} + \frac{\partial^2 w_0}{\partial x^2}\frac{\partial^2 w}{\partial y^2} + \frac{\partial w_0}{\partial x}\frac{\partial^3 w}{\partial y^2 \partial x} \tag{8.12g}$$

Equations (8.12e)–(8.12g) are used to obtain the compatibility relation

$$\frac{\partial^2 \varepsilon_x}{\partial y^2} + \frac{\partial^2 \varepsilon_y}{\partial x^2} - \frac{\partial^2 \gamma_{xy}}{\partial x \partial y} = \left(\frac{\partial^2 w}{\partial x \partial y}\right)^2 + 2\frac{\partial^2 w_0}{\partial x \partial y}\frac{\partial^2 w}{\partial x \partial y} - \frac{\partial^2 w}{\partial x^2}\frac{\partial^2 w}{\partial y^2}$$
$$- \frac{\partial^2 w}{\partial x^2}\frac{\partial^2 w_0}{\partial y^2} - \frac{\partial^2 w_0}{\partial x^2}\frac{\partial^2 w}{\partial y^2} - \frac{1}{R}\frac{\partial^2 w}{\partial x^2} \tag{8.12h}$$

Use the constitutive relations (Eqs. 8.1o–8.1q) and substitute the stress function from Eqs. (8.1v)–(8.1x) into Eq. (8.12h) to have

$$\frac{\partial^4 F}{\partial x^4} + 2\frac{\partial^4 F}{\partial x^2 \partial y^2} + \frac{\partial^4 F}{\partial y^4} = E\left[\left(\frac{\partial^2 w}{\partial x \partial y} \right)^2 + 2\frac{\partial^2 w_0}{\partial x \partial y}\frac{\partial^2 w}{\partial x \partial y} - \frac{\partial^2 w}{\partial x^2}\frac{\partial^2 w}{\partial y^2} \right.$$
$$\left. - \frac{\partial^2 w}{\partial x^2}\frac{\partial^2 w_0}{\partial y^2} - \frac{\partial^2 w_0}{\partial x^2}\frac{\partial^2 w}{\partial y^2} - \frac{1}{R}\frac{\partial^2 w}{\partial x^2} \right] \tag{8.12i}$$

Equations (8.12a) and (8.12i) are the governing differential equations for an initially imperfect shell.

8.7.2 Post-buckling Behavior of Cylindrical Panels

The post-buckling behavior of a cylindrical shell is shown here by depicting the post-buckling behavior of a rectangular cylindrical panel because its behavior is similar to that of a complete cylinder. It is easier to study the behavior of the panel than that of a complete cylinder and avoid complex calculations. The rectangular circular panel analysis given here follows that given by Volmir [23]. A rectangular panel is basically a portion of the entire cylinder bounded by two generators and two circular arcs. Consider a panel acted upon by a uniform axial compression p_x. Let R be the radius of curvature of the panel. Its thickness is h, and the lengths along the generator (x direction), and along the circumference (y direction) are a. The x and y axes are taken along the generator and the circumference of the cylindrical panel as shown in Figure 8.10.

It is assumed the panel is simply supported on its edges, the shear force N_{xy} is absent on each edge, the edges $y = 0, a$ are free to move in the y direction, and the panel remains rectangular during buckling. The displacement shape that satisfies these conditions is

$$w = g \sin\frac{\pi x}{a} \sin\frac{\pi y}{a} \tag{8.13a}$$

and assumes the initial distortion in the panel as

$$w_0 = g_0 \sin\frac{\pi x}{a} \sin\frac{\pi y}{a} \tag{8.13b}$$

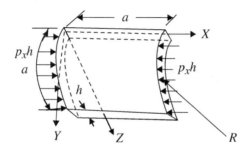

Figure 8.10 Cylindrical panel under axial compressive load.

Subsitute Eqs. (8.13a) and (8.13b) in (8.12i) to get

$$\frac{\partial^4 F}{\partial x^4} + 2\frac{\partial^4 F}{\partial x^2 \partial y^2} + \frac{\partial^4 F}{\partial y^4} = E\left[(g^2 + 2gg_0)\frac{\pi^4}{2a^4}\left(\cos\frac{2\pi x}{a} + \frac{\cos 2\pi y}{a}\right) + \frac{g}{R}\frac{\pi^2}{a^2}\sin\frac{\pi x}{a}\sin\frac{\pi y}{a}\right]$$

(8.13c)

Use the method of undetermined coefficients to get the particular solution F_p of Eq. (8.13c). Assume

$$F_p = A\left(\cos\frac{2\pi x}{a} + \cos\frac{2\pi y}{a}\right) + B\sin\frac{\pi x}{a}\sin\frac{\pi y}{a}$$

(8.13d)

Obtain the necessary derivatives of F_p from Eq. (8.13d) and substitute in Eq. (8.13c) on the left side. Now, equate the relevant terms on the left and the right sides of (Eq. 8.13c) to obtain

$$A = E\frac{g^2 + 2gg_0}{32} \quad \text{and} \quad B = E\frac{g}{R}\frac{a^2}{4\pi^2}$$

(8.13e)

Hence,

$$F_p = E\frac{g^2 + 2gg_0}{32}\left(\cos\frac{2\pi x}{a} + \cos\frac{2\pi y}{a}\right) + \frac{Eg}{4R}\frac{a^2}{\pi^2}\sin\frac{\pi x}{a}\sin\frac{\pi y}{a}$$

(8.13f)

The homogeneous solution of Eq. (8.13c) should satisfy $\nabla^4 F = 0$ and the boundary conditions for F. The homogeneous solution can be obtained by considering the primary middle surface stresses that exist prior to buckling. Prior to buckling $N_x = -p_x h$, $N_y = N_{xy} = 0$. Eq. (8.1v) gives $N_x = h\frac{\partial^2 F}{\partial y^2}$. Hence, the homogeneous solution F_h is given by

$$h\frac{\partial^2 F}{\partial y^2} = -p_x h, \text{ or } \quad F_h = -\frac{p_x y^2}{2}$$

(8.13g)

Therefore, the total solution of Eq. (8.13c) is given by

$$F = -\frac{p_x y^2}{2} + E\frac{g^2 + 2gg_0}{32}\left(\cos\frac{2\pi x}{a} + \cos\frac{2\pi y}{a}\right) + \frac{Eg}{4R}\frac{a^2}{\pi^2}\sin\frac{\pi x}{a}\sin\frac{\pi y}{a}$$

(8.13h)

The quantities g, g_0, and p_x are obtained using the Galerkin method from Eq. (8.12a). Here, the Galerkin equation is

$$\int_0^a \int_0^a Q(g)f(x,y)dxdy = 0$$

(8.13i)

$Q(g)$ is the left-hand side of Eq. (8.12a) and $f(x,y) = \sin\frac{\pi x}{a}\sin\frac{\pi y}{a}$.

$$Q(g) = \frac{\partial^4 w}{\partial x^4} + 2\frac{\partial^4 w}{\partial x^2 \partial y^2} + \frac{\partial^4 w}{\partial y^4} - \frac{h}{D}\left[\frac{\partial^2 F}{\partial y^2}\left(\frac{\partial^2 w}{\partial x^2} + \frac{\partial^2 w_0}{\partial x^2}\right) - 2\frac{\partial^2 F}{\partial x \partial y}\left(\frac{\partial^2 w}{\partial x \partial y} + \frac{\partial^2 w_0}{\partial x \partial y}\right)\right.$$

$$\left. + \frac{\partial^2 F}{\partial x^2}\left(\frac{1}{R} + \frac{\partial^2 w}{\partial y^2} + \frac{\partial^2 w_0}{\partial y^2}\right)\right]$$

(8.13j)

Note that p is zero because there is no radial force in this problem. Use Eqs. (8.13a), (8.13b), (8.13h) for w, w_0, and F to obtain

$$\frac{\partial^4 w}{\partial x^4} = g\frac{\pi^4}{a^4}\sin\frac{\pi x}{a}\sin\frac{\pi y}{a}, \quad \frac{\partial^4 w}{\partial y^4} = g\frac{\pi^4}{a^4}\sin\frac{\pi x}{a}\sin\frac{\pi y}{a}, \quad \frac{\partial^4 w}{\partial x^2 \partial y^2} = g\frac{\pi^4}{a^4}\sin\frac{\pi x}{a}\sin\frac{\pi y}{a}$$

$$\frac{\partial^2 w}{\partial x^2} = -g\frac{\pi^2}{a^2}\sin\frac{\pi x}{a}\sin\frac{\pi y}{a}, \quad \frac{\partial^2 w}{\partial y^2} = -g\frac{\pi^2}{a^2}\sin\frac{\pi x}{a}\sin\frac{\pi y}{a}, \quad \frac{\partial^2 w}{\partial x \partial y} = g\frac{\pi^2}{a^2}\cos\frac{\pi x}{a}\cos\frac{\pi y}{a}$$

$$\frac{\partial^2 w_0}{\partial x^2} = -g_0\frac{\pi^2}{a^2}\sin\frac{\pi x}{a}\sin\frac{\pi y}{a}, \quad \frac{\partial^2 w_0}{\partial y^2} = -g_0\frac{\pi^2}{a^2}\sin\frac{\pi x}{a}\sin\frac{\pi y}{a}, \quad \frac{\partial^2 w_0}{\partial x \partial y} = g_0\frac{\pi^2}{a^2}\cos\frac{\pi x}{a}\cos\frac{\pi y}{a}$$

(8.13k)

$$\frac{\partial^2 F}{\partial x^2} = -E\frac{g^2 + 2gg_0}{8}\frac{\pi^2}{a^2}\cos\frac{2\pi x}{a} - \frac{Eg}{4R}\sin\frac{\pi x}{a}\sin\frac{\pi y}{a}$$

$$\frac{\partial^2 F}{\partial y^2} = -P_x - E\frac{g^2 + 2gg_0}{8}\frac{\pi^2}{a^2}\cos\frac{2\pi y}{a} - \frac{Eg}{4R}\sin\frac{\pi x}{a}\sin\frac{\pi y}{a}$$

$$\frac{\partial^2 F}{\partial x \partial y} = \frac{Eg}{4R}\cos\frac{\pi x}{a}\cos\frac{\pi y}{a}$$

(8.13l)

Substitute Eqs. (8.13k) and (8.13l) into Eq. (8.13j) to have

$$Q(g) = 4g\frac{\pi^4}{a^4}\sin\frac{\pi x}{a}\sin\frac{\pi y}{a} - \frac{h}{D}\left\{\left[E\frac{g^2 + 2gg_0}{8}\frac{\pi^2}{a^2}\cos\frac{2\pi y}{a} + \frac{Eg}{4R}\sin\frac{\pi x}{a}\sin\frac{\pi y}{a} + P_x\right]\right.$$

$$\times\left[(g + g_0)\frac{\pi^2}{a^2}\sin\frac{\pi x}{a}\sin\frac{\pi y}{a}\right] - \frac{Eg}{2R}\cos^2\frac{\pi x}{a}\cos^2\frac{\pi y}{a}(g + g_0)\frac{\pi^2}{a^2}$$

$$\left. + \left[E\frac{(g^2 + 2gg_0)}{8}\frac{\pi^2}{a^2}\cos\frac{2\pi x}{a} + \frac{Eg}{4R}\sin\frac{\pi x}{a}\sin\frac{\pi y}{a}\right]\left[-\frac{1}{R} + (g + g_0)\frac{\pi^2}{a^2}\sin\frac{\pi x}{a}\sin\frac{\pi y}{a}\right]\right\}$$

(8.13m)

Substitute Eq. (8.13m) into Eq. (8.13i) to have

$$\int_0^a \int_0^a \left\{\left[4Dg\frac{\pi^4}{a^4} + \frac{Ehg}{4R^2} - \frac{P_x h\pi^2}{a^2}(g + g_0)\right]\sin^2\frac{\pi x}{a}\sin^2\frac{\pi y}{a}\right.$$

$$- \frac{Eh\pi^4}{8a^4}(g^2 + 2gg_0)(g + g_0)\left(\cos\frac{2\pi y}{a}\sin^2\frac{\pi x}{a}\sin^2\frac{\pi y}{a} + \cos\frac{2\pi x}{a}\sin^2\frac{\pi x}{a}\sin^2\frac{\pi y}{a}\right)$$

$$- \frac{Ehg}{2R}(g + g_0)\frac{\pi^2}{a^2}\sin^3\frac{\pi x}{a}\sin^3\frac{\pi y}{a} + \frac{Eh(g^2 + 2gg_0)}{8R}\frac{\pi^2}{a^2}\cos\frac{2\pi x}{a}\sin\frac{\pi x}{a}\sin\frac{\pi y}{a}$$

$$\left. + \frac{Egh}{2R}(g + g_0)\frac{\pi^2}{a^2}\left(\cos^2\frac{\pi x}{a}\cos^2\frac{\pi y}{a}\sin\frac{\pi x}{a}\sin\frac{\pi y}{a}\right)\right\}dxdy = 0$$

(8.13n)

Use the following trigonometric quantities and integrals

$$\int_0^a \int_0^a \sin^2\frac{\pi x}{a}\sin^2\frac{\pi y}{a}dxdy = \frac{a^2}{4}$$

$$\int_0^a \int_0^a \cos\frac{2\pi y}{a}\sin^2\frac{\pi y}{a}\sin^2\frac{\pi x}{a}dxdy = \int_0^a \int_0^a \cos\frac{2\pi x}{a}\sin^2\frac{\pi x}{a}\sin^2\frac{\pi y}{a}dxdy = -\frac{a^2}{8}$$

$$\int_0^a \int_0^a \sin^3 \frac{\pi x}{a} \sin^3 \frac{\pi y}{a} dxdy = \frac{16a^2}{9\pi^2}$$

$$\int_0^a \int_0^a \cos \frac{2\pi x}{a} \sin \frac{\pi x}{a} \sin \frac{\pi y}{a} dxdy = -\frac{4a^2}{3\pi^2}$$

$$\int_0^a \int_0^a \cos \frac{2\pi x}{a} \cos \frac{2\pi y}{a} \sin \frac{\pi x}{a} \sin \frac{\pi y}{a} dxdy = \frac{4a^2}{9\pi^2}$$

Substituting the integrals into Eq. (8.13n) we get

$$\frac{Dg\pi^4}{a^2} - P_x \frac{h\pi^2}{4}(g + g_0) + \frac{Ehga^2}{16R^2} - \frac{Eh}{R}\left(\frac{5}{6}g^2 + gg_0\right) + \frac{Eh\pi^4}{32a^2}(g^3 + 3g^2 g_0 + 2g_0^2 g) = 0$$

$$(8.13o)$$

or

$$P_x = \left[\frac{4D\pi^2}{ha^2} + \frac{Ea^2}{4R^2\pi^2} - \frac{4E}{\pi^2 R}\left(\frac{5}{6}g + g_0\right) + \frac{E\pi^2}{8a^2}(g^2 + 3gg_0 + 2g_0^2)\right]\frac{g}{g + g_0} \qquad (8.13p)$$

Assume the following non-dimensional parameters

$$\bar{p}_x = \frac{p_x a^2}{Eh^2}, \quad k = \frac{a^2}{Rh}, \quad \delta = \frac{g}{h}, \quad \delta_0 = \frac{g_0}{h} \qquad (8.13q)$$

These parameters measure the loading, the curvature and the deflection. Substitute the parameters from Eq. (8.13q) into Eq. (8.13p) to give

$$\bar{p}_x = \left[\frac{\pi^2}{3(1 - v^2)} + \frac{k^2}{4\pi^2} - \frac{4k}{\pi^2}\left(\frac{5}{6}\delta + \delta_0\right) + \frac{\pi^2}{8}(\delta^2 + 3\delta\delta_0 + 2\delta_0^2)\right]\frac{\delta}{\delta + \delta_0} \qquad (8.13r)$$

$$\bar{p}_x = \left[3.615 + \frac{k^2}{39.478} + 1.234(\delta^2 + 3\delta\delta_0 + 2\delta_0^2) - 0.405k(0.833\delta + \delta_0)\right]\frac{\delta}{\delta + \delta_0}, \text{If } v = 0.3$$

$$(8.13s)$$

The relation between the load parameter $\bar{p}_x$ and the total deflection parameter $\delta + \delta_0$ given by Eq. (8.13s) is plotted in Figure 8.11 for $k = 0$ $(R = \infty)$, which is a flat plate. The relationship curves are plotted for initial imperfections of $\delta_0 = 0$, and 0.1–0.25 for plates. The curves in Figure 8.11 show that the imperfect plates with an initial deformation start bending immediately as soon as the external load is applied. The deflections increase slowly in the beginning as the load is applied and then increase more rapidly as the critical load is reached. In the case of perfect and imperfect plates, the maximum load-carrying capacity is much higher than the critical load. In the post-buckling stage, as the deflection increases, the load continues to increase. The curves of the imperfect plates approach that of the perfect plates as the deflections increase further. Thus, one can infer that minor imperfections do not affect the buckling behavior very seriously in the case of plates. Hence, plates exhibit a stable equilibrium at buckling and the results obtained from the perfect plates can be used for plates containing minor imperfections. Koiter [21] called this an imperfection insensitive phenomenon.

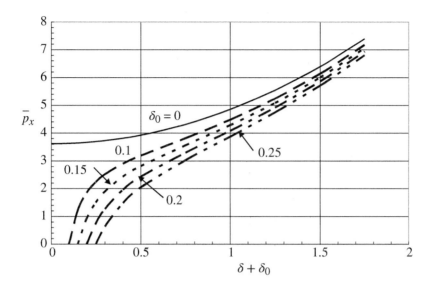

Figure 8.11 Post-buckling behavior of flat plates, $k = 0$.

The behavior of curved panels is described in Figure 8.12. The relation between the load parameter $\bar{p}_x$ and the total deflection parameter $\delta + \delta_0$ given by Eq. (8.13s) is plotted in Figure 8.12 for curved panels where $k = 30$. The relationship curves are plotted for initial imperfections of $\delta_0 = 0$, and 0.1–0.25. The curves in Figure 8.12 show that the imperfect cylindrical panels with an initial deformation start bending immediately as soon as the external load is applied. The deflections increase slowly in the beginning as the load is applied and then the load drops as the critical load is reached. In the post-buckling stage, as the deflection increases, the load continues to decrease. The curves of the imperfect curved panels approach that of the perfect panels as the deflections increase further. The most important conclusion that can be drawn is that the maximum load that an imperfect curved panel can support is much less than the critical load obtained from the classical shell theory for initially perfect panels. For $k = 30$, the maximum load for initial deformations of $\delta_0 = 0.1$ and 0.25 are 66.29 and 51.14% respectively of the maximum load for a perfect panel. The critical loads for the curved panels represent the maximum carrying capacity for the panels, which is in contrast to the plates where the maximum load-carrying capacity is higher than the critical loads. The initial imperfections have a significant effect on the load-carrying capacity of the curved panels. The curves obtained here for the curved panels compare well with the curves obtained by Donnell and Wan [4] and thus can be used to draw conclusions about cylindrical shells. This case is imperfection sensitive because at the critical load, the cylindrical shell has unstable equilibrium. Koiter [24] studied the effect of initial imperfections on the post-buckling behavior of cylindrical shells and his work is considered an important contribution in this area. It is interesting to note that the cylinders subjected to the external pressure or torsion are not very imperfection sensitive. In these cases, the failure stress is fairly close to the critical stress calculated by the linear theory.

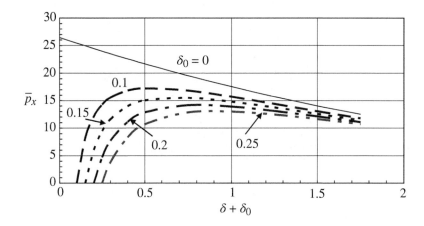

Figure 8.12   Post-buckling behavior of cylindrical shells, $k = 30$.

8.8   General Shells

The theory of thin elastic shells of arbitrary shape is considered here, it means the distance from any point inside the shell thickness to some reference surface, usually the middle surface, is small in comparison to other dimensions of the shell, such as the radius of curvature of a shell. The thin shell theory reduces the three-dimensional problem to a two-dimensional problem. The displacement of any point in the shell is expressed in terms of the displacement of a corresponding point on the middle surface. The basic equations for the behavior of a thin elastic shell were first derived by Love [25] in 1888. Love's theory is based on the following assumptions:

1. The shell is thin.
2. The deflections of the shell are small.
3. Transverse normal stress is small so it can be neglected.
4. Normals to the reference surface remain straight and normal during the deformation.

The entire theory is based on the thickness of the shell being small, so the ratio of the thickness to the radius of curvature of the reference surface is neglected in comparison to unity. Usually the shells are considered thin, if the ratio of the thickness to the radius of curvature of the shell reference surface is less than one-tenth. The assumption of small deflections allows all derivations with reference to the original configuration of the shell. The assumption that normal remain normal to the deformed surface implies that the resistance to the deformation under transverse shear is infinite. This assumption is an extension of the Bernoulii–Euler beam theory which states that plane sections remain plane after deformation.

8.8.1   Nonlinear Equations of Equilibrium

The nonlinear equilibrium equations and linear stability equations discussed here are based on the analyses done by Brush and Almroth [5], who followed the approach given by Koiter [26]

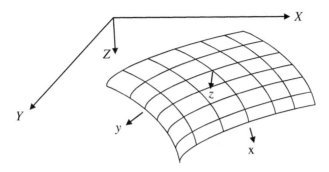

Figure 8.13   General shell coordinates.

and Sanders [27]. The derivations that follow are valid for small strains. The curvilinear coordinates x, y, and z are used in the shell theory as shown in Figure 8.13. If X, Y, and Z are the rectangular Cartesian coordinates of a point on the middle surface, then $X = X\,(x, y)$, $Y = Y\,(x, y)$, $Z = Z\,(x, y)$ define the middle surface in Figure 8.13. The middle surface is taken as a reference surface. The coordinates of a point in the shell are given by x, y, and z, where z is the normal distance from the middle surface in the thickness of the shell. The positive distance z is in the direction of positive normal n to the middle surface. The shell coordinates x, y, and z are orthogonal and are in the direction of the lines of principal curvature. The R_x and R_y are the principal radii of curvature. The distances ds_x and ds_y along the coordinate lines are given by Eq. (E.19) in Appendix E as

$$ds_x = A\,dx \quad \text{and} \quad ds_y = B\,dy \tag{8.14a}$$

where A and B are called Lamé coefficients for the coordinate system. A and B are given by Eqs. (E.21) and (E.22) in Appendix E as

$$A = \sqrt{\left(\frac{\partial X}{\partial x}\right)^2 + \left(\frac{\partial Y}{\partial x}\right)^2 + \left(\frac{\partial Z}{\partial x}\right)^2} \tag{E.21}$$

$$B = \sqrt{\left(\frac{\partial X}{\partial y}\right)^2 + \left(\frac{\partial Y}{\partial y}\right)^2 + \left(\frac{\partial Z}{\partial y}\right)^2} \tag{E.22}$$

The area of the surface is given by

$$\text{area} = \iint ds_x ds_y = \iint AB\,dx\,dy \tag{8.14b}$$

The strain energy of thin elastic shells is given by

$$U = U_m + U_b \tag{8.14c}$$

$$U_m = \frac{C}{2} \iint \left(\varepsilon_x^2 + \varepsilon_y^2 + 2\nu\varepsilon_x\varepsilon_y + \frac{1-\nu}{2}\gamma_{xy}^2 \right) AB\,dx\,dy \tag{8.14d}$$

$$U_b = \frac{D}{2} \iint \left(\kappa_x^2 + \kappa_y^2 + 2\nu\kappa_x\kappa_y + 2(1-\nu)\kappa_{xy}^2 \right) AB\,dx\,dy \tag{8.14e}$$

where ε_x, ε_y, and γ_{xy} are middle surface normal and shear strains, and κ_x, κ_y, and κ_{xy} are middle surface curvature changes and twist. $C = \dfrac{Eh}{1-v^2}$, and $D = \dfrac{Eh^3}{12(1-v^2)}$, where h is the thickness of the shell. U_m and U_b are the membrane and bending strain energies of the shell. The total potential energy of the shell Π is given by

$$\Pi = U + V \tag{8.14f}$$

where V is the potential energy of the external forces. Let p_x, p_y, and p_z be the x, y, and z components, respectively, of the load distributed over the surface of the shell element, and let u, v, and w be the corresponding x, y, and z components of the displacement of a point on the middle surface. The potential energy of applied load is

$$V = -\iint (p_x u + p_y v + p_z w) AB dx dy \tag{8.14g}$$

The nonlinear middle surface kinematic relations used here were derived by Sanders [27]

$$\varepsilon_x = e_{xx} + \frac{1}{2}\beta_x^2, \quad \varepsilon_y = e_{yy} + \frac{1}{2}\beta_y^2, \quad \gamma_{xy} = e_{xy} + \beta_x \beta_y$$

$$\kappa_x = \chi_{xx}, \quad \kappa_y = \chi_{yy}, \quad \kappa_{xy} = \chi_{xy} \tag{8.14h}$$

Sanders [27] kinematic relations are given by

$$e_{xx} = \frac{u_{,x}}{A} + \frac{A_{,y}v}{AB} - \frac{w}{R_x}$$

$$e_{yy} = \frac{v_{,y}}{B} + \frac{B_{,x}u}{AB} - \frac{w}{R_y}$$

$$e_{xy} = \frac{v_{,x}}{A} + \frac{u_{,y}}{B} - \frac{B_{,x}v + A_{,y}u}{AB} \tag{8.14i}$$

$$\beta_x = -\frac{w_{,x}}{A} - \frac{u}{R_x}, \quad \beta_y = -\frac{w_{,y}}{B} - \frac{v}{R_y} \tag{8.14j}$$

$$\chi_{xx} = \frac{\beta_{x,x}}{A} + \frac{A_{,y}\beta_y}{AB}$$

$$\chi_{yy} = \frac{\beta_{y,y}}{B} + \frac{B_{,x}\beta_x}{AB}$$

$$2\chi_{xy} = \frac{\beta_{y,x}}{A} + \frac{\beta_{x,y}}{B} - \frac{A_{,y}\beta_x + B_{,x}\beta_y}{AB} \tag{8.14k}$$

It has been seen in numerical examples that the u and v terms in Eq. (8.14j) have negligible influence in shell segments that are almost flat and for shells whose displacements are rapidly varying functions of shell coordinates, such as small buckles where the bases are significantly smaller than the radius. Such shells are called "quasi-shallow." By discarding u and v terms in Eq. (8.14j) we obtain

$$\beta_x = -\frac{w_{,x}}{A}, \quad \beta_y = -\frac{w_{,y}}{B} \tag{8.14l}$$

Use Eq. (8.14l) to obtain

$$\beta_{x,x} = -\frac{w_{,xx}}{A} + \frac{w_{,x}A_{,x}}{A^2}, \quad \beta_{y,y} = -\frac{w_{,yy}}{B} + \frac{w_{,y}B_{,y}}{B^2}$$

$$\beta_{x,y} = -\frac{Aw_{,xy} - w_{,x}A_{,y}}{A^2} \quad \beta_{y,x} = -\frac{Bw_{,yx} - w_{,y}B_{,x}}{B^2} \tag{8.14m}$$

Substitute Eqs. (8.14i), and (8.14l) into Eq. (8.14h) to have

$$\varepsilon_x = \frac{u_{,x}}{A} + \frac{A_{,y}v}{AB} - \frac{w}{R_x} + \frac{1}{2}\frac{w_{,x}^2}{A^2}$$

$$\varepsilon_y = \frac{v_{,y}}{B} + \frac{B_{,x}u}{AB} - \frac{w}{R_y} + \frac{1}{2}\frac{w_{,y}^2}{B^2}$$

$$\gamma_{xy} = \frac{v_{,x}}{A} + \frac{u_{,y}}{B} - \frac{B_{,x}v + A_{,y}u}{AB} + \frac{w_{,x}w_{,y}}{AB} \tag{8.14n}$$

Substitute Eq. (8.14m) into Eq. (8.14k) and the expressions for the curvature changes and the twist for the quasi-shallow shells are given by

$$\kappa_x = \chi_{xx} = -\frac{w_{,xx}}{A^2} + \frac{w_{,x}A_{,x}}{A^3} - \frac{A_{,y}w_{,y}}{AB^2}$$

$$\kappa_y = \chi_{yy} - \frac{w_{,yy}}{B^2} + \frac{w_{,y}B_{,y}}{B^3} - \frac{B_{,x}w_{,x}}{A^2B}$$

$$\kappa_{xy} = \chi_{xy} = -\frac{w_{,xy}}{AB} + \frac{w_{,x}A_{,y}}{A^2B} + \frac{B_{,x}w_{,y}}{AB^2} \tag{8.14o}$$

Substitute Eqs. (8.14n) and (8.14o) into Eqs. (8.14d) and (8.14e) to get the total strain energy U, and add to it the potential energy of the external forces V to get the total potential energy Π as follows:

$$\Pi = \frac{C}{2}\int\int\left[\left(\frac{u_{,x}}{A} + \frac{A_{,y}v}{AB} - \frac{w}{R_x} + \frac{1}{2}\frac{w_{,x}^2}{A^2}\right)^2 + \left(\frac{v_{,y}}{B} + \frac{B_{,x}u}{AB} - \frac{w}{R_y} + \frac{1}{2}\frac{w_{,y}^2}{B^2}\right)^2 \right.$$

$$+2\nu\left(\frac{u_{,x}}{A} + \frac{A_{,y}v}{AB} - \frac{w}{R_x} + \frac{1}{2}\frac{w_{,x}^2}{A^2}\right)\left(\frac{v_{,y}}{B} + \frac{B_{,x}u}{AB} - \frac{w}{R_y} + \frac{1}{2}\frac{w_{,y}^2}{B^2}\right)$$

$$\left. +\frac{1-\nu}{2}\left(\frac{v_{,x}}{A} + \frac{u_{,y}}{B} - \frac{B_{,x}v + A_{,y}u}{AB} + \frac{w_{,x}w_{,y}}{AB}\right)^2\right]ABdxdy$$

$$+\frac{D}{2}\int\int\left[\left(-\frac{w_{,xx}}{A^2} + \frac{w_{,x}A_{,x}}{A^3} - \frac{A_{,y}w_{,y}}{AB^2}\right)^2 + \left(-\frac{w_{,yy}}{B^2} + \frac{w_{,y}B_{,y}}{B^3} - \frac{B_{,x}w_{,x}}{A^2B}\right)^2\right.$$

$$+2\nu\left(-\frac{w_{,xx}}{A^2} + \frac{w_{,x}A_{,x}}{A^3} - \frac{A_{,y}w_{,y}}{AB^2}\right)\left(-\frac{w_{,yy}}{B^2} + \frac{w_{,y}B_{,y}}{B^3} - \frac{B_{,x}w_{,x}}{A^2B}\right) +$$

$$\left. + 2(1-\nu)\left(-\frac{w_{,xy}}{AB} + \frac{w_{,x}A_{,y}}{A^2B} + \frac{B_{,x}w_{,y}}{AB^2}\right)^2\right]ABdxdy - \int\int(p_xu + p_yv + p_zw)ABdxdy$$

$$\tag{8.14p}$$

Equation (8.14p) is the total potential energy expression for the Donnell-Mushtari-Vlasov (DMV) form of the general shell equations. The nonlinear equilibrium equations can be derived from Eq. (8.14p) by applying the principle of stationary potential energy. The Euler equations derived by the calculus of variations in Appendix D are given by Eq. (D.10)

$$\frac{\partial F}{\partial u} - \frac{\partial}{\partial x}\frac{\partial F}{\partial u_x} - \frac{\partial}{\partial y}\frac{\partial F}{\partial u_y} = 0$$

$$\frac{\partial F}{\partial v} - \frac{\partial}{\partial x}\frac{\partial F}{\partial v_x} - \frac{\partial}{\partial y}\frac{\partial F}{\partial v_y} = 0$$

$$\frac{\partial F}{\partial w} - \frac{\partial}{\partial x}\frac{\partial F}{\partial w_x} - \frac{\partial}{\partial y}\frac{\partial F}{\partial w_y} + \frac{\partial^2}{\partial x^2}\frac{\partial F}{\partial w_{xx}} + \frac{\partial^2}{\partial x\partial y}\frac{\partial F}{\partial w_{xy}} + \frac{\partial^2}{\partial y^2}\frac{\partial F}{\partial w_{yy}} = 0 \qquad \text{(D.10)}$$

where F is the integrand in Eq. (8.14p).

$$\frac{\partial F}{\partial u} = B_{,x}C(\varepsilon_y + v\varepsilon_x) - A_{,y}\frac{C}{2}(1-v)\gamma_{xy} - p_x AB$$

or

$$\frac{\partial F}{\partial u} = B_{,x}N_y - A_{,y}N_{xy} - p_x AB$$

$$\frac{\partial F}{\partial u_x} = CB(\varepsilon_x + v\varepsilon_y) = BN_x$$

or

$$\frac{\partial}{\partial x}\left(\frac{\partial F}{\partial u_x}\right) = B_{,x}N_x + BN_{x,x}$$

$$\frac{\partial F}{\partial u_{,y}} = A\frac{C}{2}(1-v)\gamma_{xy} = AN_{xy}$$

or

$$\frac{\partial}{\partial y}\left(\frac{\partial F}{\partial u_{,y}}\right) = A_{,y}N_{xy} + AN_{xy,y}$$

Substitute the partial derivatives of F in the first equation of Eq. (D.10) to get

$$B_{,x}N_x + BN_{x,x} + 2A_{,y}N_{xy} + AN_{xy,y} - B_{,x}N_y = -ABp_x$$

or

$$(BN_x)_{,x} + (AN_{xy})_{,y} - B_{,x}N_y + A_{,y}N_{xy} = -ABp_x \qquad \text{(8.14q)}$$

Similarly, substitute the partial derivatives of F in the second and third equations of Eq. (D.10) to obtain

$$(AN_y)_{,y} + (BN_{xy})_{,x} - A_{,y}N_x + B_{,x}N_{xy} = -ABp_y \qquad \text{(8.14r)}$$

$$\left[\frac{1}{A}(BM_x)_{,x}\right]_{,x} - \left(\frac{A_{,y}}{B}M_x\right)_{,y} + \left[\frac{1}{B}(AM_y)_{,y}\right]_{,y} - \left(\frac{B_{,x}}{A}M_y\right)_{,x}$$

$$+2\left[M_{xy,xy} + \left(\frac{A_{,y}}{A}M_{xy}\right)_{,x} + \left(\frac{B_{,x}}{B}M_{xy}\right)_{,y}\right] + AB\left(\frac{N_x}{R_x} + \frac{N_y}{R_y}\right)$$

$$-[(BN_x\beta_x + BN_{xy}\beta_y)_{,x} + (AN_y\beta_y + AN_{xy}\beta_x)_{,y}] = -ABp_z \tag{8.14s}$$

where

$$
\begin{aligned}
N_x &= C(\varepsilon_x + v\varepsilon_y) & M_x &= D(\kappa_x + v\kappa_y)\\
N_y &= C(\varepsilon_y + v\varepsilon_x) & M_y &= D(\kappa_y + v\kappa_x)\\
N_{xy} &= C\frac{1-v}{2}\gamma_{xy} & M_{xy} &= D(1-v)\kappa_{xy}
\end{aligned}
\tag{8.14t}
$$

Equations (8.14q)–(8.14s) are the nonlinear equations of equilibrium for thin shells of general shape. These equations can be converted to the rectangular flat plate equations by assuming $A = B = 1$ and $1/R_x = 1/R_y = 0$. If we substitute the new assumed values in Eqs. (8.14q)–(8.14s) we have

$$
\begin{aligned}
&N_{x,x} + N_{xy,y} = 0\\
&N_{y,y} + N_{xy,x} = 0\\
&D\nabla^4 w - N_x w_{,xx} - N_y w_{,yy} - 2N_{xy}w_{,xy} = p
\end{aligned}
\tag{8.14u}
$$

Equation (8.14u) are the flat plate equations given by Eqs. (7.2a), (7.2b), and (7.2j) subjected to a normal external force of intensity p. We can use Eqs. (8.14q)–(8.14s) to derive equations of cylindrical shells by assuming $A = B = 1$ and $1/R_x = 0$, $R_y = R$ to obtain

$$
\begin{aligned}
&N_{x,x} + N_{xy,y} = 0\\
&N_{y,y} + N_{xy,x} = 0\\
&D\nabla^4 w - N_x w_{,xx} - N_y\left(w_{,yy} + \frac{1}{R}\right) - 2N_{xy}w_{,xy} = p
\end{aligned}
\tag{8.14v}
$$

Equations (8.14v) are the circular cylindrical shell equations given by Eqs. (8.1a), (8.1b), and (8.1j) subjected to a radial pressure p. Similarly, we can derive column equations by setting $A = B = 1$, $1/R_x = 1/R_y = 0$, $N_y = N_{xy} = M_y = M_{xy} = 0$, and $D = EI$. Hence,

$$EI\frac{d^4 w}{dx^4} + P\frac{d^2 w}{dx^2} = 0 \tag{8.14w}$$

which is the same as Eq. (2.11e) for columns.

8.8.2 Linear Equations of Stability (the Donnell Type)

Linear equilibrium equations can be obtained from the nonlinear equations of Eqs. (8.14q)–(8.14s) as outlined by Brush and Almroth [5] from the adjacent equilibrium configurations.

Consider the equilibrium at the bifurcation point given by the displacements u_0, v_0, and w_0. This equilibrium position is perturbed by the small arbitrary incremental displacements u_1, v_1, w_1, and we get an adjacent equilibrium configuration represented by the displacements u, v, and w at the same applied load such that

$$u \rightarrow u_0 + u_1$$
$$v \rightarrow v_0 + v_1$$
$$w \rightarrow w_0 + w_1 \tag{8.15a}$$

The increments in the displacement components change the internal forces and rotations such as

$$N_x \rightarrow N_{x0} + \Delta N_x$$
$$N_y \rightarrow N_{y0} + \Delta N_y$$
$$N_{xy} \rightarrow N_{xy0} + \Delta N_{xy}$$
$$\beta_x \rightarrow \beta_{x0} + \beta_{x1}$$
$$\beta_y \rightarrow \beta_{y0} + \beta_{y1} \tag{8.15b}$$

The terms with 0 subscript correspond to u_0, v_0, w_0 displacements, whereas ΔN_x, ΔN_y, ΔN_{xy}, β_{x1}, and β_{y1} are the increments corresponding to the displacements u_1, v_1, and w_1. Substitution of Eqs. (8.15a) into Eqs. (8.14q)–(8.14s) give equations that contain terms that are linear, quadratic, and cubic in u_0, v_0, w_0 and u_1, v_1, w_1 displacement components. In these equations, the terms containing u_0, v_0, w_0 equate to zero because they constitute an equilibrium position. The quadratic and cubic terms of u_1, v_1, w_1 can be neglected because the arbitrary displacement increments are small. Therefore, the resulting equations are homogeneous and linear in u_1, v_1, w_1. Let N_{x1}, N_{y1}, N_{xy1} be the portions of ΔN_x, ΔN_y, ΔN_{xy} respectively that are linear in u_1, v_1, w_1. Substitute Eqs. (8.14h), (8.14i), (8.14k), and (8.14l) into Eqs. (8.14t) to obtain

$$N_x = C\left\{\frac{1}{A}u_{,x} + \frac{A_{,y}}{AB}v - \frac{w}{R_x} + \frac{1}{2A^2}(w_{,x})^2 + v\left[\frac{v_{,y}}{B} + \frac{B_{,x}}{AB}u - \frac{w}{R_y} + \frac{1}{2B^2}(w_{,y})^2\right]\right\} \tag{8.15c}$$

$$N_{x0} + \Delta N_x = C\left\{\frac{1}{A}(u_{0,x} + u_{1,x}) + \frac{A_{,y}}{AB}(v_0 + v_1) - \frac{1}{R_x}(w_0 + w_1)\right.$$
$$+ \frac{1}{2A^2}(w_{0,x})^2 + \frac{1}{A^2}(w_{0,x})(w_{1,x}) + \frac{1}{2A^2}(w_{1,x})^2$$
$$+ v\left[\frac{1}{B}(v_{0,y} + v_{1,y}) + \frac{B_{,x}}{AB}(u_0 + u_1) - \frac{1}{R_y}(w_0 + w_1)\right.$$
$$\left.\left. + \frac{1}{2B^2}(w_{0,y})^2 + \frac{1}{B^2}(w_{0,y})(w_{1,y}) + \frac{1}{2B^2}(w_{1,y})^2\right]\right\}$$

$$N_{x0} = C \left\{ \frac{1}{A} u_{0,x} + \frac{A_{,y}}{AB} v_0 - \frac{1}{R_x} w_0 + \frac{1}{2A^2} (w_{0,x})^2 + v \left[\frac{1}{B} v_{0,y} + \frac{B_{,x}}{AB} u_0 - \frac{w_0}{R_y} + \frac{1}{2B^2} (w_{0,y})^2 \right] \right\}$$

$$\Delta N_x = C \left\{ \frac{1}{A} u_{1,x} + \frac{A_{,y}}{AB} v_1 - \frac{1}{R_x} w_1 + \frac{1}{A^2} (w_{0,x})(w_{1,x}) + \frac{1}{2A^2} (w_{1,x})^2 \right.$$
$$\left. + v \left[\frac{1}{B} v_{1,y} + \frac{B_{,x}}{AB} u_1 - \frac{1}{R_y} w_1 + \frac{1}{B^2} (w_{0,y})(w_{1,y}) + \frac{1}{2B^2} (w_{1,y})^2 \right] \right\}$$

$$N_{x1} = C \left\{ \frac{1}{A} u_{1,x} + \frac{A_{,y}}{AB} v_1 - \frac{1}{R_x} w_1 + \frac{1}{A^2} (w_{0,x})(w_{1,x}) \right.$$
$$\left. + v \left[\frac{1}{B} v_{1,y} + \frac{B_{,x}}{AB} u_1 - \frac{1}{R_y} w_1 + \frac{1}{B^2} (w_{0,y})(w_{1,y}) \right] \right\}$$

$$(8.15d)$$

Similarly,

$$N_y = C \left\{ \frac{1}{B} v_{,y} + \frac{B_{,x}}{AB} u - \frac{w}{R_y} + \frac{1}{2B^2} (w_{,y})^2 + v \left[\frac{u_{,x}}{A} + \frac{A_{,y}}{AB} v - \frac{w}{R_x} + \frac{1}{2A^2} (w_{,x})^2 \right] \right\} \quad (8.15e)$$

$$N_{y0} + \Delta N_y = C \left\{ \frac{1}{B} (v_{0,y} + v_{1,y}) + \frac{B_{,x}}{AB} (u_0 + u_1) - \frac{1}{R_y} (w_0 + w_1) \right.$$
$$+ \frac{1}{2B^2} (w_{0,y})^2 + \frac{1}{B^2} (w_{0,y})(w_{1,y}) + \frac{1}{2B^2} (w_{1,y})^2$$
$$+ v \left[\frac{1}{A} (u_{0,x} + u_{1,x}) + \frac{A_{,y}}{AB} (v_0 + v_1) - \frac{1}{R_x} (w_0 + w_1) \right.$$
$$\left. \left. + \frac{1}{2A^2} (w_{0,x})^2 + \frac{1}{A^2} (w_{0,x})(w_{1,x}) + \frac{1}{2A^2} (w_{1,x})^2 \right] \right\}$$

$$N_{y0} = C \left\{ \frac{1}{B} v_{0,y} + \frac{B_{,x}}{AB} u_0 - \frac{1}{R_y} w_0 + \frac{1}{2B^2} (w_{0,y})^2 + v \left[\frac{1}{A} u_{0,x} + \frac{A_{,y}}{AB} v_0 - \frac{w_0}{R_x} + \frac{1}{2A^2} (w_{0,x})^2 \right] \right\}$$

$$\Delta N_y = C \left\{ \frac{1}{B} v_{1,y} + \frac{B_{,x}}{AB} u_1 - \frac{1}{R_y} w_1 + \frac{1}{B^2} (w_{0,y})(w_{1,y}) + \frac{1}{2B^2} (w_{1,y})^2 \right.$$
$$\left. + v \left[\frac{1}{A} u_{1,x} + \frac{A_{,y}}{AB} v_1 - \frac{1}{R_x} w_1 + \frac{1}{A^2} (w_{0,x})(w_{1,x}) + \frac{1}{2A^2} (w_{1,x})^2 \right] \right\}$$

$$N_{y1} = C \left\{ \frac{1}{B} v_{1,y} + \frac{B_{,x}}{AB} u_1 - \frac{1}{R_y} w_1 + \frac{1}{B^2} (w_{0,y})(w_{1,y}) \right.$$
$$\left. + v \left[\frac{1}{A} u_{1,x} + \frac{A_{,y}}{AB} v_1 - \frac{1}{R_x} w_1 + \frac{1}{A^2} (w_{0,x})(w_{1,x}) \right] \right\}$$

$$(8.15f)$$

$$N_{xy} = C \frac{1-v}{2} \left(\frac{1}{A} v_{,x} + \frac{1}{B} u_{,y} - \frac{B_{,x} v + A_{,y} u}{AB} + \frac{1}{AB} w_{,x} w_{,y} \right) \quad (8.15g)$$

$$N_{xy0} + \Delta N_{xy} = C\frac{1-\nu}{2}\left[\frac{1}{A}(v_{0,x} + v_{1,x}) + \frac{1}{B}(u_{0,y} + u_{1,y}) - \frac{B_{,x}(v_0 + v_1) + A_{,y}(u_0 + u_1)}{AB}\right.$$

$$\left. + \frac{1}{AB}(w_0 + w_1)_{,x}(w_0 + w_1)_{,y}\right]$$

$$N_{xy0} = C\frac{1-\nu}{2}\left[\frac{1}{A}(v_{0,x}) + \frac{1}{B}(u_{0,y}) - \frac{B_{,x}(v_0) + A_{,y}(u_0)}{AB} + \frac{1}{AB}(w_0)_{,x}(w_0)_{,y}\right]$$

$$\Delta N_{xy} = C\frac{1-\nu}{2}\left[\frac{1}{A}(v_{1,x}) + \frac{1}{B}(u_{1,y}) - \frac{B_{,x}(v_1) + A_{,y}(u_1)}{AB}\right.$$

$$\left. + \frac{1}{AB}(w_{0,x}w_{1,y} + w_{1,x}w_{0,y} + w_{1,x}w_{1,y})\right]$$

$$N_{xy1} = C\frac{1-\nu}{2}\left[\frac{1}{A}(v_{1,x}) + \frac{1}{B}(u_{1,y}) - \frac{B_{,x}(v_1) + A_{,y}(u_1)}{AB} + \frac{1}{AB}(w_{0,x}w_{1,y} + w_{1,x}w_{0,y})\right]$$

(8.15h)

For moments,

$$M_x = D\left[-\frac{w_{,xx}}{A^2} + \frac{w_{,x}A_{,x}}{A^3} - \frac{A_{,y}w_{,y}}{AB^2} + \nu\left(-\frac{w_{,yy}}{B^2} + \frac{w_{,y}B_{,y}}{B^3} - \frac{B_{,x}w_{,x}}{A^2B}\right)\right] \qquad (8.15i)$$

$$M_{x0} + \Delta M_x = D\left[-\frac{1}{A^2}(w_{0,xx} + w_{1,xx}) + \frac{A_{,x}}{A^3}(w_{0,x} + w_{1,x}) - \frac{A_{,y}}{AB^2}(w_{0,y} + w_{1,y})\right.$$

$$\left. + \nu\left\{-\frac{1}{B^2}(w_{0,yy} + w_{1,yy}) + \frac{B_{,y}}{B^3}(w_{0,y} + w_{1,y}) - \frac{B_{,x}}{A^2B}(w_{0,x} + w_{1,x})\right\}\right]$$

$$M_{x0} = D\left[-\frac{w_{0,xx}}{A^2} + \frac{A_{,x}}{A^3}w_{0,x} - \frac{A_{,y}}{AB^2}w_{0,y} + \nu\left(-\frac{1}{B^2}w_{0,yy} + \frac{B_{,y}}{B^3}w_{0,y} - \frac{B_{,x}}{A^2B}w_{0,x}\right)\right]$$

$$\Delta M_x = M_{x1} = D\left[-\frac{w_{1,xx}}{A^2} + \frac{A_{,x}}{A^3}w_{1,x} - \frac{A_{,y}}{AB^2}w_{1,y} + \nu\left(-\frac{w_{1,yy}}{B^2} + \frac{B_{,y}}{B^3}w_{1,y} - \frac{B_{,x}}{A^2B}w_{1,x}\right)\right]$$

$$M_{x1} = D\left[\left(\frac{\beta_{x1,x}}{A} + \frac{A_{,y}}{AB}\beta_{y1}\right) + \nu\left(\frac{\beta_{y1,y}}{B} + \frac{B_{,x}\beta_{x1}}{AB}\right)\right] \qquad (8.15j)$$

Similarly,

$$M_{y1} = D\left[\left(\frac{\beta_{y1,y}}{B} + \frac{B_{,x}}{AB}\beta_{x1}\right) + \nu\left(\frac{\beta_{x1,x}}{A} + \frac{A_{,y}\beta_{y1}}{AB}\right)\right] \qquad (8.15k)$$

$$M_{xy1} = D\frac{1-\nu}{2}\left(\frac{\beta_{y1,x}}{A} + \frac{\beta_{x1,y}}{B} - \frac{A_{,y}\beta_{x1} + B_{,x}\beta_{y1}}{AB}\right) \qquad (8.15l)$$

Substitution of Eqs. (8.15d), (8.15f), (8.15h), (8.15j)–(8.15l) into Eqs. (8.14q)–(8.14s) gives

$$(BN_{x1})_{,x} + (AN_{xy1})_{,y} - B_{,x}N_{y1} + A_{,y}N_{xy1} = 0 \qquad (8.15m)$$

$$(AN_{y1})_{,y} + (BN_{xy1})_{,x} - A_{,y}N_{x1} + B_{,x}N_{xy1} = 0 \qquad (8.15n)$$

$$\left[\frac{1}{A}(BM_{x1})_{,x}\right]_{,x} - \left(\frac{A_{,y}}{B}M_{x1}\right)_{,y} + \left[\frac{1}{B}(AM_{y1})_{,y}\right]_{,y} - \left(\frac{B_{,x}}{A}M_{y1}\right)_{,x}$$

$$+2\left[M_{xy1,xy} + \left(\frac{A_{,y}}{A}M_{xy1}\right)_{,x} + \left(\frac{B_{,x}}{B}M_{xy1}\right)_{,y}\right] + AB\left(\frac{N_{x1}}{R_x} + \frac{N_{y1}}{R_y}\right)$$

$$-[(BN_{x0}\beta_{x1} + BN_{xy0}\beta_{y1})_{,x} + (B\beta_{x0}N_{x1} + B\beta_{y0}N_{xy1})_{,x}$$

$$+(AN_{y0}\beta_{y1} + AN_{xy0}\beta_{x1})_{,y} + (A\beta_{y0}N_{y1} + A\beta_{x0}N_{xy1})_{,y}] = 0 \qquad (8.15o)$$

where

$$e_{xx1} = \frac{u_{1,x}}{A} + \frac{A_{,y}v_1}{AB} - \frac{w_1}{R_x}$$

$$e_{yy1} = \frac{v_{1,y}}{B} + \frac{B_{,x}u_1}{AB} - \frac{w_1}{R_y}$$

$$e_{xy1} = \frac{v_{1,x}}{A} + \frac{u_{1,y}}{B} - \frac{B_{,x}v_1 + A_{,y}u_1}{AB} \qquad (8.15p)$$

$$\beta_{x0} = -\frac{w_{0,x}}{A} \qquad \beta_{y0} = -\frac{w_{0,y}}{B}$$

$$\beta_{x1} = -\frac{w_{1,x}}{A} \qquad \beta_{y1} = -\frac{w_{1,y}}{B} \qquad (8.15q)$$

$$N_{x0} = C(\varepsilon_{x0} + v\varepsilon_{y0})$$

$$N_{y0} = C(\varepsilon_{y0} + v\varepsilon_{x0})$$

$$N_{xy0} = C\frac{1-v}{2}\gamma_{xy0} \qquad (8.15r)$$

$$N_{x1} = C[(e_{xx1} + \beta_{x0}\beta_{x1}) + v(e_{yy1} + \beta_{y0}\beta_{y1})]$$

$$N_{y1} = C[(e_{yy1} + \beta_{y0}\beta_{y1}) + v(e_{xx1} + \beta_{x0}\beta_{x1})]$$

$$N_{xy1} = C\frac{1-v}{2}(e_{xy1} + \beta_{x0}\beta_{y1} + \beta_{y0}\beta_{x1}) \qquad (8.15s)$$

$$\kappa_{x1} = \chi_{xx1} = \frac{\beta_{x1,x}}{A} + \frac{A_{,y}\beta_{y1}}{AB} = -\frac{w_{1,xx}}{A^2} + \frac{A_{,x}}{A^3}w_{1,x} - \frac{A_{,y}}{AB^2}w_{1,y}$$

$$\kappa_{y1} = \chi_{yy1} = \frac{\beta_{y1,y}}{B} + \frac{B_{,x}\beta_{x1}}{AB} = -\frac{w_{1,yy}}{B^2} + \frac{B_{,y}}{B^3}w_{1,y} - \frac{B_{,x}}{A^2B}w_{1,x}$$

$$2\kappa_{xy} = 2\chi_{xy1} = \frac{\beta_{y1,x}}{A} + \frac{\beta_{x1,y}}{B} - \frac{A_{,y}\beta_{x1} + B_{,x}\beta_{y1}}{AB} = 2\left(-\frac{w_{1,xy}}{AB} + \frac{A_{,y}w_{1,x}}{A^2B} + \frac{B_{,x}w_{1,y}}{AB^2}\right) \qquad (8.15t)$$

Equations (8.15m)–(8.15o) are the linear stability equations for shells of general shape under the DMV approximations.

8.9   Shells of Revolution

Shells of revolution are common types of structural shells. The middle surface of a shell of revolution is formed by the rotation of a plane curve about an axis that lies in the plane of the curve. The surface curves formed by the intersection of planes containing the axis of rotation with the surface are called the lines of principal curvature or meridians. The surface curves formed by the intersection of planes perpendicular to the axis of rotation with the surface are called parallels. Points on the middle surface are expressed by the coordinates ϕ and θ, where ϕ is the angle between the axis of rotation and a normal to the surface, and θ is a circumferential coordinate shown in Figure 8.14a. The principal radii of curvature of the surface of revolution in the ϕ and θ direction are denoted by R_ϕ and R_θ, respectively.

An additional variable r is defined in Figure 8.14b as

$$r = R_\theta \sin \phi \tag{8.16a}$$

The distances ds_ϕ and ds_θ along the coordinate lines are given by

$$ds_\phi = R_\phi d\phi, \qquad ds_\theta = rd\theta \tag{8.16b}$$

If ϕ and θ correspond to the x and y curvilinear coordinates respectively, then the Lamé coefficients are

$$A = R_\phi \quad B = r \tag{8.16c}$$

and

$$R_x = R_\phi, \quad R_y = R_\theta \tag{8.16d}$$

From Eq. (E.87) we have

$$\frac{\partial}{\partial x}\left(\frac{B}{R_y}\right) = \frac{1}{R_x}\frac{\partial B}{\partial x} \tag{E.87}$$

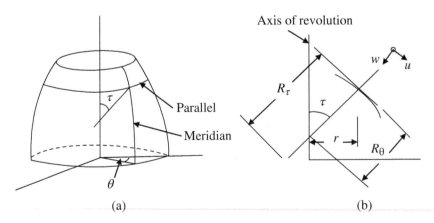

(a)   (b)

Figure 8.14   Shell of revolution: (a) Coordinates of shell of revolution; (b) Meridian of a shell of revolution.

or

$$\frac{d}{d\phi}\left(\frac{R_\theta \sin \phi}{R_\theta}\right) = \frac{1}{R_\phi}\frac{dr}{d\phi}$$

or

$$\frac{dr}{d\phi} = R_\phi \cos \phi$$

or

$$dr = ds_\phi \cos \phi \tag{8.16e}$$

The variables R_ϕ, R_θ, and r are the functions of ϕ only, and define the shape of the middle surface of the undeformed shell. Let u, v, w be the middle surface displacement components in the ϕ, θ, and the normal directions, respectively. The displacement components are functions of both ϕ and θ.

8.9.1 Stability Equations Where Pre-buckling Rotations Are Retained

The linear equations of stability for the shells of revolution are obtained by substituting Eqs. (8.16a)–(8.16e) into Eqs. (8.15m)–(8.15o) as follows [5]:

$$(rN_{\phi 1})_{,\phi} + (R_\phi N_{\phi\theta 1})_{,\theta} - r_{,\phi}N_{\theta 1} + R_{\phi,\theta}N_{\phi\theta 1} = 0$$

or

$$(rN_{\phi 1})_{,\phi} + R_{\phi,\theta}N_{\phi\theta 1} + R_\phi N_{\phi\theta 1,\theta} - r_{,\phi}N_{\theta 1} + R_{\phi,\theta}N_{\phi\theta 1} = 0$$

$$r_{,\phi} = \frac{dr}{d\phi} = R_\phi Cos\phi, \text{ and } R_{\phi,\theta} = 0$$

or

$$(rN_{\phi 1})_{,\phi} + R_\phi N_{\phi\theta 1,\theta} - R_\phi N_{\theta 1} \cos \phi = 0 \tag{8.17a}$$

Similarly, Eqs. (8.15n) and (8.15o) give

$$(rN_{\phi\theta 1})_{,\phi} + R_\phi N_{\theta 1,\theta} + R_\phi N_{\phi\theta 1} \cos \phi = 0 \tag{8.17b}$$

$$\left[\frac{1}{R_\phi}(rM_{\phi 1})_{,\phi}\right]_{,\phi} + 2\left(M_{\phi\theta 1,\phi\theta} + \frac{R_\phi}{r}M_{\phi\theta 1,\theta} \cos \phi\right)$$
$$+ \left[\frac{R_\phi}{r}(M_{\theta 1,\theta\theta}) - (M_{\theta 1} \cos \phi)_{,\phi}\right] + (rN_{\phi 1} + R_\phi N_{\theta 1} \sin \phi)$$
$$- [(rN_{\phi 0}\beta_{\phi 1} + rN_{\phi\theta 0}\beta_{\theta 1})_{,\phi} + (r\beta_{\phi 0}N_{\phi 1} + r\beta_{\theta 0}N_{\phi\theta 1})_{,\phi}$$
$$+ R_\phi(N_{\theta 0}\beta_{\theta 1} + N_{\phi\theta 0}\beta_{\phi 1})_{,\theta} + R_\phi(\beta_{\theta 0}N_{\theta 1} + \beta_{\phi 0}N_{\phi\theta 1})_{,\theta}] = 0 \tag{8.17c}$$

where

$$N_{\phi 1} = C[(e_{\phi\phi 1} + \beta_{\phi 0}\beta_{\phi 1}) + v(e_{\theta\theta 1} + \beta_{\theta 0}\beta_{\theta 1})]$$
$$N_{\theta 1} = C[(e_{\theta\theta 1} + \beta_{\theta 0}\beta_{\theta 1}) + v(e_{\phi\phi 1} + \beta_{\phi 0}\beta_{\phi 1})]$$
$$N_{\phi\theta 1} = C\frac{1-v}{2}(e_{\phi\theta 1} + \beta_{\phi 0}\beta_{\theta 1} + \beta_{\theta 0}\beta_{\phi 1}) \tag{8.17d}$$

$$M_{\phi 1} = D\left[\frac{\beta_{\phi 1,\phi}}{R_\phi} + \frac{v}{r}(\beta_{\theta 1,\theta} + \beta_{\phi 1}\cos\phi)\right]$$

$$M_{\theta 1} = D\left[\frac{1}{r}(\beta_{\theta 1,\theta} + \beta_{\phi 1}\cos\phi) + \frac{v}{R_\phi}(\beta_{\phi 1,\phi})\right]$$

$$M_{\phi\theta 1} = D\frac{1-v}{2}\left[\frac{r}{R_\phi}\left(\frac{\beta_{\theta 1}}{r}\right)_{,\phi} + \frac{\beta_{\phi 1,\theta}}{r}\right] \tag{8.17e}$$

$$e_{\phi\phi 1} = \frac{1}{R_\phi}(u_{1,\phi} - w_1)$$

$$e_{\theta\theta 1} = \frac{1}{r}(v_{1,\theta} + u_1\cos\phi - w_1\sin\phi)$$

$$e_{\phi\theta 1} = \frac{v_{1,\phi}}{R_\phi} + \frac{u_{1,\theta}}{r} - \frac{r_{,\phi}v_1}{R_\phi r} = \frac{r}{R_\phi}\left[\frac{rv_{1,\phi} - v_1 r_{,\phi}}{r^2}\right] + \frac{u_{1,\theta}}{r} \tag{8.17f}$$

or

$$e_{\phi\theta 1} = \frac{r}{R_\phi}\left(\frac{v_1}{r}\right)_{,\phi} + \frac{u_{1,\theta}}{r}$$

$$\beta_{\phi 1} = -\frac{w_{1,\phi}}{R_\phi} \qquad \beta_{\theta 1} = -\frac{w_{1,\theta}}{r} \tag{8.17g}$$

Equations (8.14q)–(8.14s) are specialized for a shell of revolution, and are used to obtain the coefficients $N_{\phi 0}, N_{\theta 0}, N_{\phi\theta 0}, \beta_{\phi 0},$ and $\beta_{\theta 0}$ in Eqs. (8.17a)–(8.17c). If the applied load is axisymmetric, the deformation prior to loss of stability is also axisymmetric. Then, $\beta_{\theta 0} = 0$, and $N_{\phi 0}, N_{\theta 0}, N_{\phi\theta 0},$ and $\beta_{\phi 0}$ are functions of ϕ only. For axisymmetric deformation of a shell of revolution, specialization of the nonlinear equilibrium equations in (8.14q)–(8.14s) reduces to

$$\frac{d}{d\phi}(rN_\phi) - R_\phi N_\theta\cos\phi = -rR_\phi p_\phi \tag{8.17h}$$

$$\frac{d}{d\phi}(rN_{\phi\theta}) + R_\phi N_{\phi\theta}\cos\phi = -rR_\phi p_\theta \tag{8.17i}$$

$$\frac{d}{d\phi}\left[\frac{1}{R_\phi}\frac{d}{d\phi}(rM_\phi)\right] - \frac{d}{d\phi}(M_\theta\cos\phi) + (rN_\phi + R_\phi N_\theta\sin\phi) - \frac{d}{d\phi}(rN_\phi\beta_\phi) = -rR_\phi p \tag{8.17j}$$

where p_ϕ, p_θ, and p are surface load components in the ϕ, θ, and normal directions, respectively. The constitutive relations are specialized for the shell of revolutions from Eq. (8.14t) as

$$N_\phi = C(\varepsilon_\phi + v\varepsilon_\theta), \qquad M_\phi = D(\kappa_\phi + v\kappa_\theta)$$

$$N_\theta = C(\varepsilon_\theta + v\varepsilon_\phi), \qquad M_\theta = D(\kappa_\theta + v\kappa_\phi) \qquad (8.17k)$$

$$N_{\phi\theta} = C\frac{1-v}{2}\gamma_{\phi\theta}$$

The kinematic relations are obtained from Eqs. (8.14h), (8.14i), (8.14k), and (8.14l) by specializing these equations for the shells of revolution as

$$\varepsilon_\phi = e_{\phi\phi} + \frac{1}{2}\beta_\phi^2 \qquad \varepsilon_\theta = e_{\theta\theta} \qquad \gamma_{\phi\theta} = e_{\phi\theta} \qquad (8.17l)$$

$$e_{\phi\phi} = \frac{1}{R_\phi}\left(\frac{du}{d\phi} - w\right) \qquad e_{\theta\theta} = \frac{1}{r}(u\cos\phi - w\sin\phi) \qquad e_{\phi\theta} = \frac{r}{R_\phi}\frac{d}{d\phi}\left(\frac{v}{r}\right) \qquad (8.17m)$$

$$\kappa_\phi = \chi_{\phi\phi} = \frac{1}{R_\phi}\frac{d\beta_\phi}{d\phi}, \qquad \kappa_\theta = \chi_{\theta\theta} = \frac{1}{r}\beta_\phi\cos\phi \qquad (8.17n)$$

$$\beta_\phi = -\frac{1}{R_\phi}\frac{dw}{d\phi} \qquad (8.17o)$$

If the shell is not subjected to torsion, the coefficient $N_{\phi\theta0} = 0$ in Eq. (8.17c), and (8.17i) is omitted.

8.9.2 Stability Equations with Pre-buckling Rotations Neglected

When the pre-buckling rotations are small, the terms containing $\beta_{\phi0}$ and $\beta_{\theta0}$ are neglected in Eqs. (8.17a)–(8.17c) to obtain the stability equations as follows:

$$(rN_{\phi1})_{,\phi} + R_\phi N_{\phi\theta1,\theta} - R_\phi N_{\theta1}\cos\phi = 0 \qquad (8.18a)$$

$$(rN_{\phi\theta1})_{,\phi} + R_\phi N_{\theta1,\theta} + R_\phi N_{\phi\theta1}\cos\phi = 0 \qquad (8.18b)$$

$$\left[\frac{1}{R_\phi}(rM_{\phi1})_{,\phi}\right]_{,\phi} + 2\left(M_{\phi\theta1,\phi\theta} + \frac{R_\phi}{r}M_{\phi\theta1,\theta}\cos\phi\right)$$

$$+ \left[\frac{R_\phi}{r}(M_{\theta1,\theta\theta}) - (M_{\theta1}\cos\phi)_{,\phi}\right] + (rN_{\phi1} + R_\phi N_{\theta1}\sin\phi)$$

$$- [(rN_{\phi0}\beta_{\phi1} + rN_{\phi\theta0}\beta_{\theta1})_{,\phi} + (R_\phi N_{\theta0}\beta_{\theta1} + R_\phi N_{\phi\theta0}\beta_{\phi1})_{,\theta}] = 0 \qquad (8.18c)$$

where

$$N_{\phi1} = C(e_{\phi\phi1} + ve_{\theta\theta1})$$

or

$$N_{\phi1} = C\left[\frac{1}{R_\phi}(u_{1,\phi} - w_1) + \frac{v}{r}(v_{1,\theta} + u_1\cos\phi - w_1\sin\phi)\right]$$

$$N_{\theta 1} = C(e_{\theta\theta 1} + v e_{\phi\phi 1})$$

or

$$N_{\theta 1} = C\left[\frac{1}{r}(v_{1,\theta} + u_1 \cos\phi - w_1 \sin\phi) + \frac{v}{R_\phi}(u_{1,\phi} - w_1)\right] \tag{8.18d}$$

$$N_{\phi\theta 1} = C\frac{1-v}{2}e_{\phi\theta 1} = C\frac{1-v}{2}\left[\frac{r}{R_\phi}\left(\frac{v_1}{r}\right)_{,\phi} + \frac{u_{1,\theta}}{r}\right]$$

or

$$N_{\phi\theta 1} = C\frac{1-v}{2}\left[\frac{r}{R_\phi}\left(\frac{rv_{1,\phi} - v_1 r_{,\phi}}{r^2}\right) + \frac{u_{1,\theta}}{r}\right] = C\frac{1-v}{2}\left[\frac{v_{1,\phi}}{R_\phi} - \frac{v_1}{r}\cos\phi + \frac{u_{1,\theta}}{r}\right]$$

$$M_{\phi 1} = D\left[\frac{\beta_{\phi 1,\phi}}{R_\phi} + \frac{v}{r}(\beta_{\theta 1,\theta} + \beta_{\phi 1}\cos\phi)\right]$$

$$M_{\theta 1} = D\left[\frac{1}{r}(\beta_{\theta 1,\theta} + \beta_{\phi 1}\cos\phi) + \frac{v}{R_\phi}(\beta_{\phi 1,\phi})\right]$$

$$M_{\phi\theta 1} = D\frac{1-v}{2}\left[\frac{r}{R_\phi}\left(\frac{\beta_{\theta 1}}{r}\right)_{,\phi} + \frac{\beta_{\phi 1,\theta}}{r}\right] \tag{8.18e}$$

$$\beta_{\phi 1} = -\frac{w_{1,\phi}}{R_\phi} \qquad \beta_{\theta 1} = -\frac{w_{1,\theta}}{r} \tag{8.18f}$$

The coefficients, $N_{\phi 0}$, $N_{\theta 0}$, $N_{\phi\theta 0}$, in Eq. (8.18a)–(8.18c) are determined by the linear equilibrium equations obtained by specializing Eqs. (8.14q)–(8.14s). For axisymmetric loads, omit the nonlinear terms from Eqs. (8.17h)–(8.17j) to obtain

$$\frac{d}{d\phi}(rN_\phi) - R_\phi N_\theta \cos\phi = -rR_\phi p_\phi \tag{8.18g}$$

$$\frac{d}{d\phi}(rN_{\phi\theta}) + R_\phi N_{\phi\theta} \cos\phi = -rR_\phi p_\theta \tag{8.18h}$$

$$\frac{d}{d\phi}\left[\frac{1}{R_\phi}\frac{d}{d\phi}(rM_\phi)\right] - \frac{d}{d\phi}(M_\theta \cos\phi) + (rN_\phi + R_\phi N_\theta \sin\phi) = -rR_\phi p \tag{8.18i}$$

where the constitutive and kinematic relations are given by (8.17k)–(8.17o), except that

$$\varepsilon_\phi = e_{\phi\phi} \tag{8.18j}$$

To simplify the determination of the coefficients in the stability equations, the bending terms in Eq. (8.18i) are considered small in comparison to other terms and are neglected to get the linear membrane equations

$$\frac{d}{d\phi}(rN_\phi) - R_\phi N_\theta \cos\phi = -rR_\phi p_\phi \tag{8.18k}$$

$$\frac{d}{d\phi}(rN_{\phi\theta}) + R_{\phi}N_{\phi\theta}\cos\phi = -rR_{\phi}p_{\theta} \tag{8.18l}$$

$$rN_{\phi} + R_{\phi}N_{\theta}\sin\phi = -rR_{\phi}p \tag{8.18m}$$

Equations (8.18k)–(8.18m) are statically determinate, so the solutions can be obtained without the use of constitutive and kinematics equations.

If the shell is not subjected to torsion, the coefficient $N_{\phi\theta 0} = 0$ in Eqs. (8.18a)–(8.18c), and (8.18h) and (8.18l) is discarded. In these cases, if we substitute Eqs. (8.18d)–(8.18f) into Eqs. (8.18a)–(8.18c), the stability equations are reduced to ordinary differential equations by selecting the solution of the form

$$u_1 = u_n(\phi)\cos n\theta$$
$$v_1 = v_n(\phi)\sin n\theta$$
$$w_1 = w_n(\phi)\cos n\theta \tag{8.18n}$$

To summarize, the stability equations for the shells of revolution are given by Eqs. (8.17a)–(8.17c) if pre-buckling rotations are retained, and are given by Eqs. (8.18a)–(8.18c) if pre-buckling rotations are neglected. Nonlinear equations of equilibrium for symmetrically loaded shells of revolutions are given by Eqs. (8.17h)–(8.17j), the corresponding linear bending equations are given by Eqs. (8.18g)–(8.18i), and the corresponding linear membrane equations are given by Eqs. (8.18k)–(8.18m).

Now we apply the equations derived for the shells of revolution to some common structural forms such as circular plates, shallow spherical caps, conical shells, and toroidal shells.

8.9.3 Circular Flat Plates

The middle plane of a circular flat plate is defined by polar coordinates r and θ shown in Figure 8.15. To apply the stability equations for the shells of revolution to the circular plates, we assign the following values to different parameters:

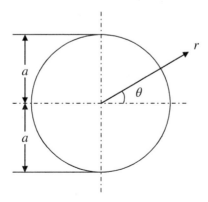

Figure 8.15 Circular flat plate.

$R_\phi \to \infty$, $R_\theta \to \infty$, $\phi \to 0$, and the limit $(R_\phi \, d\phi) = dr$

$$R_\phi \to \infty$$

where the arrow is read as "go to." Hence, $\sin \phi = 0$ and $\cos \phi = 1$. The subscript 1 is omitted for the incremental quantities for notational simplicity. Substitute these values into Eqs. (8.18a)–(8.18c) to get

$$(rN\phi)_{,\phi} + R_\phi N_{\phi\theta,\theta} - R_\phi N_\theta \cos \phi = 0$$

or

$$\frac{\partial}{R_\phi \partial \phi}(rN_\phi) + N_{\phi\theta,\theta} - N_\theta = 0$$

Substitute

$$R_\phi d\phi = dr$$

or

$$\frac{\partial}{\partial r}(rN_\phi) + N_{\phi\theta,\theta} - N_\theta = 0$$

Replace the subscript ϕ by r, then

$$(rN_r)_{,r} + N_{r\theta,\theta} - N_\theta = 0 \tag{8.19a}$$

Similarly, Eqs. (8.18b) and (8.18c) can be written as

$$(rN_{r\theta})_{,r} + N_{\theta,\theta} + N_{r\theta} = 0 \tag{8.19b}$$

$$(rM_r)_{,rr} + 2\left(M_{r\theta,r\theta} + \frac{1}{r}M_{r\theta,\theta}\right) + \left(\frac{1}{r}M_{\theta,\theta\theta} - M_{\theta,r}\right)$$
$$- [(rN_{r0}\beta_r + rN_{r\theta0}\beta_\theta)_{,r} + (N_{r\theta0}\beta_r + N_{\theta0}\beta_\theta)_{,\theta}] = 0 \tag{8.19c}$$

The corresponding constitutive and kinematic relations are from Eqs. (8.18d)–(8.18f) and given by

$$N_r = C\left[u_{,r} + \frac{v}{r}(v_{,\theta} + u)\right]$$

$$N_\theta = C\left[\frac{1}{r}(v_{,\theta} + u) + vu_{,r}\right]$$

$$N_{r\theta} = C\frac{1-v}{2}\left[\left(v_{,r} - \frac{v}{r}\right) + \frac{u_{,\theta}}{r}\right] \tag{8.19d}$$

$$M_r = D\left[\beta_{r,r} + \frac{v}{r}(\beta_{\theta,\theta} + \beta_r)\right]$$

$$M_\theta = D\left[\frac{1}{r}(\beta_{\theta,\theta} + \beta_r) + v(\beta_{r,r})\right]$$

$$M_{\phi\theta} = D\frac{1-v}{2}\left[r\left(\frac{\beta_\theta}{r}\right)_{,r} + \frac{\beta_{r,\theta}}{r}\right] \tag{8.19e}$$

$$\beta_r = -w_{,r} \qquad \beta_\theta = -\frac{w_{,\theta}}{r} \qquad\qquad (8.19f)$$

If we substitute Eqs. (8.19d)–(8.19f) into Eqs. (8.19a)–(8.19c), we get three homogeneous equations in u, v, and w, in which the third equation is uncoupled from the first two as it was for rectangular plates.

Example 8.2 Find the critical load of a circular plate subjected to a uniform compressive force around the circumference of $N_{r0} = -N$ lb./in. (N/mm).

$\beta_\theta = N_{r\theta 0} = 0$. Let us assume, $\beta_r \equiv \beta$ to simplify the notation, then Eq. (8.19c) can be written as

$$\frac{d^2}{dr^2}(rM_r) - \frac{d}{dr}(M_\theta) + \frac{d}{dr}(rN\beta) = 0 \qquad\qquad (8.20a)$$

where we have from Eq. (8.19e)

$$M_r = D\left(\frac{d\beta}{dr} + \frac{v}{r}\beta\right) \qquad M_\theta = D\left(\frac{\beta}{r} + v\frac{d\beta}{dr}\right) \qquad\qquad (8.20b)$$

Integrate Eq. (8.20a) to get

$$\frac{d}{dr}(rM_r) - M_\theta + rN\beta = C_1 \qquad\qquad (8.20c)$$

where C_1 is a constant. Now, $M_r = M_\theta = 0$, when $N = 0$. Therefore, $C_1 = 0$. Substitute Eq. (8.20b) into Eq. (8.20c) to get

$$\frac{d}{dr}\left[D\left(r\frac{d\beta}{dr} + v\beta\right)\right] - D\left(\frac{1}{r}\beta + v\frac{d\beta}{dr}\right) + rN\beta = 0$$

or

$$D\left[\frac{d\beta}{dr} + r\frac{d^2\beta}{dr^2} + v\frac{d\beta}{dr}\right] - D\left(\frac{\beta}{r} + v\frac{d\beta}{dr}\right) + rN\beta = 0$$

or

$$r^2\frac{d^2\beta}{dr^2} + r\frac{d\beta}{dr} - \left(1 - \frac{r^2N}{D}\right)\beta = 0 \qquad\qquad (8.20d)$$

This is the same equation as Eq. (7.28l) in Chapter 7 for a circular plate subjected to a uniform compressive force N_r distributed around the edge of the plate shown in Figure 7.28. The solution of Eq. (8.20d) is repeated here for convenience.

Let

$$\lambda^2 = \frac{N}{D}, \text{ and } u = \lambda r \qquad\qquad (8.20e)$$

Thus,

$$\frac{du}{dr} = \lambda,$$

hence,

$$\frac{d\beta}{dr} = \frac{d\beta}{du}\frac{du}{dr} = \lambda\frac{d\beta}{du}, \text{and } r\frac{d\beta}{dr} = u\frac{d\beta}{du}$$

Similarly,

$$r^2\frac{d^2\beta}{dr^2} = u^2\frac{d^2\beta}{du^2}$$

Now, we can write Eq. (8.20d) in terms of the new variable, u, and its derivatives

$$u^2\frac{d^2\beta}{du^2} + u\frac{d\beta}{du} + (u^2 - 1)\beta = 0 \tag{8.20f}$$

It is a Bessel differential equation of order 1 for real variable u. If instead of $(u^2 - 1)$, the term was $(u^2 - n)$, the Bessel differential equation would be classified of order n. The general solution of Eq. (8.20f) is given by Timoshenko and Gere [7]

$$\beta = C_1 J_1(u) + C_2 Y_1(u) \tag{8.20g}$$

where $J_1(u)$ and $Y_1(u)$ are Bessel functions of first order of the first and second kinds, respectively. The coefficients C_1 and C_2 are obtained from the boundary conditions.

8.9.3.1 Clamped Plate

For a circular plate that is fixed around its edge, the boundary conditions are:

1. At the center of the plate at $r = u = 0$, the angle $\beta = 0$ because of the symmetry of deflected plate. From the table of Bessel function [28], $Y_1(0) \to \infty$, therefore, $C_2 = 0$. Hence

$$\beta = C_1 J_1(u) \tag{8.21a}$$

2. The slope at the edge of the plate is zero, i.e. $\beta|_{r=a} = 0$. Hence from Eq. (8.21a) we have

$$C_1 J_1(\lambda a) = 0$$

or

$$J_1(\lambda a) = 0 \tag{8.21b}$$

From the table of function $J_1(u)$ [28], the smallest root of Eq. (8.21b) is

$$\lambda a = 3.832$$

Substitute for λ in Eq. (8.20e) to get

$$(N_r)_{cr} = \frac{14.684D}{a^2} \tag{8.21c}$$

8.9.3.2 Simply Supported Plate

For a circular plate that is simply supported around its edge, the boundary conditions are

1. From the symmetry of deflection at the center of the plate we have

$$C_2 = 0$$

or

$$\beta = C_1 J_1(u) \tag{8.22a}$$

2. The bending moment along the edge is zero, hence

$$M_r = \left(\frac{d\beta}{dr} + v\frac{\beta}{r}\right) = 0$$

$$u = \lambda r, \frac{du}{dr} = \lambda$$

$$\frac{d\beta}{dr} = \frac{d\beta}{du}\frac{du}{dr} = \lambda\frac{d\beta}{du} = C_1\lambda\frac{dJ_1(u)}{du}$$

or

$$M_r = C_1\lambda\left[\frac{dJ_1(u)}{du} + v\frac{J_1(u)}{u}\right]_{r=a} = 0 \tag{8.22b}$$

The derivative of $J_1(u)$ is written as [29]

$$\frac{d}{du}[uJ_1(u)] = uJ_0(u)$$

or

$$\frac{dJ_1(u)}{du} = J_0(u) - \frac{J_1(u)}{u} \tag{8.22c}$$

From Eqs. (8.22b) and (8.22c) we obtain

$$C_1\lambda\left[J_0(u) - \frac{J_1(u)}{u} + v\frac{J_1(u)}{u}\right]_{r=a} = 0 \tag{8.22d}$$

where J_0 is the Bessel function of the zero order.

$$\lambda J_0(u) - \frac{J_1(u)}{a} + v\frac{J_1(u)}{a} = 0$$

Assume the Poisson's ratio $v = 0.3$ and we get

$$\lambda a J_0(\lambda a) - 0.7 J_1(\lambda a) = 0 \tag{8.22e}$$

The smallest root of Eq. (8.22e) by using tables of functions J_o and J_1 is found to be, $\lambda a = 2.05$. $[J_0(2.05) = 0.1953$ and $J_1(2.05) = 0.5725]$. Substitute for λ in $\lambda^2 = \frac{N}{D}$ to get

$$(N_r)_{cr} = \frac{4.2025D}{a^2} \tag{8.22f}$$

The critical force for the clamped circular plate in Eq. (8.21c) is 3.494 times the critical force for the simply supported circular plate given by Eq. (8.22f).

8.9.4 Shallow Spherical Caps

Shallow spherical caps are segments of spherical shells shown in Figure 8.16. The points on the middle surface are described by curvilinear coordinates, r and θ. The rise of the shell, H, is much smaller than the base radius "a."

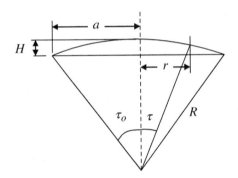

Figure 8.16 Shallow spherical cap.

The parameter, $R_\phi = R$ is constant, $\sin\phi = r/R$, $\cos\phi \approx 1$, and $R_\phi d\phi = dr$. Substitute these values into Eq. (8.18a) (the subscript 1 is omitted from the incremental quantities for notational simplicity) to get the following:

$$(rN_\phi)_{,\phi} + R_\phi N_{\phi\theta,\theta} - R_\phi N_\theta \cos\phi = 0$$

$$\frac{\partial r}{R_\phi \partial\phi} N_\phi + r\frac{\partial N_\phi}{R_\phi \partial\phi} + N_{\phi\theta,\theta} - N_\theta = 0$$

Substitute $R_\phi d\phi = dr$ and replace the subscript ϕ by r as before to get

$$(rN_r)_{,r} + N_{r\theta,\theta} - N_\theta = 0 \tag{8.23a}$$

Similarly, we have from Eqs. (8.18b) and (8.18c)

$$(rN_{r\theta})_{,r} + N_{\theta,\theta} + N_{r\theta} = 0 \tag{8.23b}$$

$$(rM_r)_{,rr} + 2\left(M_{r\theta,r\theta} + \frac{1}{r}M_{r\theta,\theta}\right) + \left(\frac{1}{r}M_{\theta,\theta\theta} - M_{\theta,r}\right) + \frac{r}{R}(N_r + N_\theta)$$
$$-[(rN_{r0}\beta_r + rN_{r\theta0}\beta_\theta)_{,r} + (N_{\theta0}\beta_\theta + N_{r\theta0}\beta_r)_{,\theta}] = 0 \tag{8.23c}$$

We have from Eqs. (8.18d) and (8.18e)

$$N_r = C(e_{rr} + ve_{\theta\theta}) = C\left[\left(u_{,r} - \frac{w}{R}\right) + v\left(\frac{v_{,\theta} + u}{r} - \frac{w}{R}\right)\right]$$

$$N_\theta = C(e_{\theta\theta} + ve_{rr}) = C\left[\left(\frac{v_{,\theta} + u}{r} - \frac{w}{R}\right) + v\left(u_{,r} - \frac{w}{R}\right)\right]$$

$$N_{r\theta} = C\frac{1-v}{2}e_{r\theta} = C\frac{1-v}{2}\left[v_{,r} - \frac{v}{r} + \frac{u_{,\theta}}{r}\right] \tag{8.23d}$$

$$M_r = D\left[\beta_{r,r} + \frac{v}{r}(\beta_{\theta,\theta} + \beta_r)\right] = -D\left[w_{,rr} + \frac{v}{r}\left(\frac{w_{,\theta\theta}}{r} + w_{,r}\right)\right]$$

$$M_\theta = D\left[\frac{1}{r}(\beta_{\theta,\theta} + \beta_r) + v\beta_{r,r}\right] = -D\left[\frac{1}{r}\left(\frac{w_{,\theta\theta}}{r} + w_{,r}\right) + vw_{,rr}\right]$$

$$M_{r\theta} = D\frac{1-v}{2}\left[r\left(\frac{\beta_\theta}{r}\right)_{,r} + \frac{\beta_{r,\theta}}{r}\right]$$

or

$$M_{r\theta} = D\frac{1-v}{2}\left[\beta_{\theta,r} - \frac{\beta_\theta}{r} + \frac{\beta_{r,\theta}}{r}\right] = D\frac{1-v}{2}\left[\left(-\frac{w_{,\theta}}{r}\right)_{,r} + \left(\frac{w_{,\theta}}{r^2}\right) - \frac{w_{,r\theta}}{r}\right]$$

or

$$M_{r\theta} = -D(1-v)\left[\frac{w_{,\theta r}}{r} - \frac{w_{,\theta}}{r^2}\right] \qquad (8.23e)$$

If $R \to \infty$, Eqs. (8.23a)–(8.23e) reduce to that of the circular plate (Eqs. (8.19a)–(8.19e)). Substitution of M_r, M_θ, and $M_{r\theta}$ from Eq. (8.23e) into Eq. (8.23c) leads to

$$D\nabla^4 w - \frac{N_r + N_\theta}{R} - \frac{1}{r}\left[(rN_{r0}w_{,r} + N_{r\theta 0}w_{,\theta})_{,r} + \left(N_{r\theta 0}w_{,r} + N_{\theta 0}\frac{w_{,\theta}}{r}\right)_{,\theta}\right] = 0 \qquad (8.23f)$$

where

$$\nabla^2() = \left[()_{,rr} + \frac{1}{r}()_{,r} + \frac{1}{r^2}()_{,\theta\theta}\right]$$

and

$$\nabla^4() = \nabla^2\nabla^2()$$

Substitute Eq. (8.23d) into Eqs. (8.23a) and (8.23b), the resulting equations along with Eq. (8.23f) form a set of three homogeneous equations in u, v, and w. Assume that the spherical cap is subjected to a uniform external pressure p_e normal to the middle surface and that the pre-buckling state can be analyzed by membrane analysis. Then, $N_{r0} = N_{\theta 0} = -p_e R/2$, and $N_{r\theta 0} = 0$. Substituting these values into Eq. (8.23f) gives

$$D\nabla^4 w - \frac{N_r + N_\theta}{R} - \frac{1}{r}\left[N_{r0}w_{,r} + rN_{r0}w_{,rr} + N_{r\theta 0}w_{,\theta r} + N_{r\theta 0}w_{,r\theta} + N_{\theta 0}\frac{w_{,\theta\theta}}{r}\right] = 0$$

or

$$D\nabla^4 w - \frac{N_r + N_\theta}{R} + \frac{p_e R}{2}\left[\frac{1}{r}w_{,r} + w_{,rr} + \frac{w_{,\theta\theta}}{r^2}\right] = 0$$

or

$$D\nabla^4 w - \frac{N_r + N_\theta}{R} + \frac{p_e R}{2}\nabla^2 w = 0 \qquad (8.23g)$$

A stress function f was introduced by Vlasov [30] such that

$$N_r = \frac{f_{,r}}{r} + \frac{1}{r^2}f_{,\theta\theta} \qquad N_\theta = f_{,rr} \qquad N_{r\theta} = -\left(\frac{f_{,\theta}}{r}\right)_{,r} \qquad (8.23h)$$

Substitute the values of N_r and N_θ from Eq. (8.23h) into Eq. (8.23g) to have

$$D\nabla^4 w - \frac{1}{R}\left(\frac{f_{,r}}{r} + \frac{1}{r^2}f_{,\theta\theta} + f_{,rr}\right) + \frac{p_e R}{2}\nabla^2 w = 0$$

or

$$DV^4w - \frac{1}{R}\nabla^2 f + \frac{P_e R}{2}\nabla^2 w = 0 \tag{8.23i}$$

From Eq. (8.23d) we have

$$e_{rr} = u_{,r} - \frac{w}{R}$$

$$e_{rr,r} = u_{,rr} - \frac{1}{R}w_{,r}$$

$$e_{rr,\theta\theta} = u_{,r\theta\theta} - \frac{1}{R}w_{,\theta\theta}$$

$$e_{\theta\theta} = \frac{v_{,\theta} + u}{r} - \frac{w}{R}, \qquad e_{\theta\theta,r} = \frac{1}{r}(v_{,\theta r} + u_{,r}) - \frac{1}{r^2}(v_{,\theta} + u) - \frac{1}{R}w_{,r}$$

$$r^2 e_{\theta\theta,r} = r(v_{,\theta r} + u_{,r}) - (v_{,\theta} + u) - \frac{1}{R}r^2 w_{,r}$$

$$(r^2 e_{\theta\theta,r})_{,r} = r(v_{,\theta rr} + u_{,rr}) - \frac{1}{R}(2rw_{,r} + r^2 w_{,rr})$$

$$\frac{1}{r^2}(r^2 e_{\theta\theta,r}) = \frac{v_{,\theta rr} + u_{,rr}}{r} - \frac{1}{R}\left(\frac{2w_{,r}}{r} + w_{,rr}\right)$$

$$e_{r\theta} = r\left(\frac{v}{r}\right)_{,r} + \frac{u_{,\theta}}{r} = v_{,r} - \frac{v}{r} + \frac{u_{,\theta}}{r}$$

$$(re_{r\theta})_{,r} = v_{,r} + rv_{,rr} - v_{,r} + u_{,\theta r}$$

$$\frac{1}{r^2}(re_{r\theta})_{,r\theta} = \frac{v_{,rr\theta}}{r} + \frac{u_{,\theta\theta r}}{r^2}$$

Using the relations derived we can write

$$\frac{1}{r^2}e_{rr,\theta\theta} - \frac{1}{r}e_{rr,r} + \frac{1}{r^2}(r^2 e_{\theta\theta,r})_{,r} - \frac{1}{r^2}(re_{r\theta})_{,r\theta} = -\frac{1}{R}\left(w_{,rr} + \frac{w_{,r}}{r} + \frac{w_{,\theta\theta}}{r^2}\right)$$

or

$$\frac{1}{r^2}e_{rr,\theta\theta} - \frac{1}{r}e_{rr,r} + \frac{1}{r^2}(r^2 e_{\theta\theta,r})_{,r} - \frac{1}{r^2}(re_{r\theta})_{,r\theta} = -\frac{1}{R}\nabla^2 w \tag{8.23j}$$

$$e_{rr} = \frac{1}{Eh}(N_r - vN_\theta) \qquad e_{\theta\theta} = \frac{1}{Eh}(N_\theta - vN_r) \qquad e_{r\theta} = \frac{2(1+v)}{Eh}N_{r\theta}$$

$$\frac{1}{r^2}e_{rr,\theta\theta} = \frac{1}{r^2 Eh}(N_{r,\theta\theta} - vN_{\theta,\theta\theta})$$

$$\frac{1}{r}e_{rr,r} = \frac{1}{rEh}(N_{r,r} - vN_{\theta,r})$$

$$\frac{1}{r^2}(r^2 e_{\theta\theta,r})_{,r} = \frac{1}{r^2}(2re_{\theta\theta,r} + r^2 e_{\theta\theta,rr}) = \frac{2}{r}e_{\theta\theta,r} + e_{\theta\theta,rr}$$

or

$$\frac{1}{r^2}(r^2 e_{\theta\theta,r})_{,r} = \frac{2}{rEh}(N_{\theta,r} - vN_{r,r}) + \frac{1}{Eh}(N_{\theta,rr} - vN_{r,rr})$$

$$\frac{1}{r^2}(re_{r\theta})_{,r} = \frac{1}{r^2}(e_{r\theta} + re_{r\theta,r})$$

$$\frac{1}{r^2}(re_{r\theta})_{,r\theta} = \frac{1}{r^2}(e_{r\theta,\theta} + r_{,\theta}e_{r\theta,r} + re_{r\theta,r\theta}) = \frac{1}{r^2}e_{r\theta,\theta} + \frac{1}{r}e_{r\theta,r\theta}$$

or

$$\frac{1}{r^2}(re_{r\theta})_{,r\theta} = \frac{2(1+v)}{r^2 Eh}N_{r\theta,\theta} + \frac{2(1+v)}{rEh}N_{r\theta,r\theta}$$

Equation (8.23j) can be written as

$$\frac{1}{r^2 Eh}(N_{r,\theta\theta} - vN_{\theta,\theta\theta}) - \frac{1}{rEh}(N_{r,r} - vN_{\theta,r}) + \frac{2}{rEh}(N_{\theta,r} - vN_{r,r}) + \frac{1}{Eh}(N_{\theta,rr} - vN_{r,rr})$$

$$-\frac{2(1+v)}{r^2 Eh}N_{r\theta,\theta} - \frac{2(1+v)}{rEh}N_{r\theta,r\theta} = -\frac{\nabla^2 w}{R} \qquad (8.23k)$$

Substitute the in-plane normal and shearing force intensities of N_r, N_θ, and $N_{r\theta}$ in Eq. (8.23k) in terms of the stress function f, as given in Eq. (8.23h), and we get

$$\nabla^4 f = -\frac{Eh}{R}\nabla^2 w \qquad (8.23l)$$

Now the problem has been reduced to the solution of two homogeneous differential equations in w and f given by Eqs. (8.23i) and (8.23l). Hutchinson [31] gave a solution by first transforming the coordinates into Cartesian coordinates as follows:

Let

$$x = r\cos\theta$$

and

$$y = r\sin\theta$$

then

$$\nabla^2() = \frac{\partial}{\partial r^2}() + \frac{1}{r}\frac{\partial}{\partial r}() + \frac{1}{r^2}\frac{\partial^2}{\partial\theta^2}()$$

For the case of Cartesian coordinates it reduces to

$$\nabla^2() = \frac{\partial}{\partial x^2}() + \frac{\partial}{\partial y^2}()$$

Equations (8.23i) and (8.23l) are satisfied by the sinusoidal functions of the form

$$w = \cos\left(k_x\frac{x}{R}\right)\cos\left(k_y\frac{y}{R}\right) \qquad (8.23m)$$

$$f = A_1\cos\left(k_x\frac{x}{R}\right)\cos\left(k_y\frac{y}{R}\right) \qquad (8.23n)$$

where k_x and k_y are mode shape parameters and A_1 is a constant. In Cartesian coordinates

$$\nabla^4() = \frac{\partial^4}{\partial x^4}() + 2\frac{\partial^4}{\partial x^2\partial y^2}() + \frac{\partial^4}{\partial y^4}()$$

Substitute Eqs. (8.23m) and (8.23n) into Eq. (8.23l) to obtain

$$A_1 \left(\frac{k_x^4}{R^4} + \frac{2k_x^2 k_y^2}{R^4} + \frac{k_y^4}{R^4} \right) \cos k_x \frac{x}{R} \cos k_y \frac{y}{R} = \frac{Eh}{R} \left(\frac{k_x^2}{R^2} + \frac{k_y^2}{R^2} \right) \cos k_x \frac{x}{R} \cos k_y \frac{y}{R}$$

or

$$A_1 = EhR(k_x^2 + k_y^2)^{-1} \tag{8.23o}$$

Substitute Eq. (8.23o) and the expression $D = Eh^3/12(1 - v^2)$ into Eq. (8.23i) to obtain

$$D \left(\frac{k_x^4}{R^4} + \frac{2k_x^2 k_y^2}{R^4} + \frac{k_y^4}{R^4} \right) \cos k_x \frac{x}{R} \cos k_y \frac{y}{R} + \frac{A_1}{R} \left(\frac{k_x^2}{R^2} + \frac{k_y^2}{R^2} \right) \cos k_x \frac{x}{R} \cos k_y \frac{y}{R}$$
$$- \frac{p_e R}{2} \left(\frac{k_x^2}{R^2} + \frac{k_y^2}{R^2} \right) \cos k_x \frac{x}{R} \cos k_y \frac{y}{R} = 0$$

or

$$p_e = \frac{2Eh}{R} \left[(k_x^2 + k_y^2)^{-1} + \frac{\left(\frac{h}{R} \right)^2}{12(1 - v^2)} (k_x^2 + k_y^2) \right] \tag{8.23p}$$

The critical pressure p_{cr} is obtained by finding the minimum of p_e with respect to $k_x^2 + k_y^2$. For simplicity of calculations, assume $k = k_x^2 + k_y^2$, then

$$p_e = \frac{2Eh}{R} \left[\frac{1}{k} + \frac{\left(\frac{h}{R} \right)^2}{12(1 - v^2)} k \right] \tag{8.23q}$$

or

$$\frac{dp_e}{dk} = \frac{2Eh}{R} \left[-\frac{1}{k^2} + \frac{\left(\frac{h}{R} \right)^2}{12(1 - v^2)} \right] = 0$$

or

$$k = 2\sqrt{3(1 - v^2)} \frac{R}{h}$$

or

$$k_x^2 + k_y^2 = k = 2\sqrt{3(1 - v^2)} \frac{R}{h} \tag{8.23r}$$

Substitute this $k_x^2 + k_y^2$ value into Eq. (8.23p) to get

$$P_{cr} = \frac{2Eh}{R}\left[\frac{h}{R}\frac{1}{2\sqrt{3(1-v^2)}} + \frac{\left(\dfrac{h}{R}\right)^2}{12(1-v^2)}2\sqrt{3(1-v^2)}\frac{R}{h}\right]$$

or

$$P_{cr} = \frac{2E}{\sqrt{3(1-v^2)}}\left(\frac{h}{R}\right)^2 \tag{8.23s}$$

Equation (8.23s) is the same as given in [7] for a complete spherical shell. The solution functions in Eqs. (8.23m) and (8.23n) do not satisfy the boundary conditions at the edges of the spherical cap. Hence, the present buckling analysis is applicable to buckling mode shape wavelengths that are small in comparison to the radius of the shell. Even there, it is found to be not in good agreement with the experimental results. The error is because of neglecting the nonlinear terms in the buckling analysis and due to initial imperfections. That is why in the design procedures a large safety factor of five or more is used.

8.9.5 Conical Shells

In Section 8.9 a shell of revolution was formed by the rotation of a plane curve about an axis in the plane of the curve. If the plane curve is an inclined straight line, the shell of revolution is called a conical shell. A truncated conical shell having a vertex angle of 2α is shown in Figure 8.17. Points on the middle surface of the shell are defined by the longitudinal coordinate s along the incline of the cone, and a circumferential coordinate θ. The coordinates s and θ are the orthogonal curvilinear coordinates for conical shells.

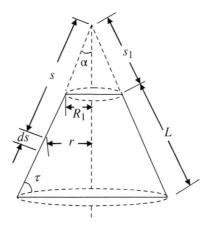

Figure 8.17 Conical shell.

The equations of shells of revolution when applied to conical shells have the following values for different parameters:

$$R_\phi \to \infty, \quad \text{and limit } (R_\phi \, d\phi) = ds, \quad r = s \sin \alpha, \quad \phi = (\pi/2)-\alpha \text{ (constant)},$$

Then, $\sin \phi = \cos \alpha$ and $\cos \phi = \sin \alpha$. Substitute these values in Eqs. (8.18a)–(8.18c) to obtain

$$\frac{1}{R_\phi}(rN_\phi)_{,\phi} + N_{\phi\theta,\theta} - N_\theta \cos \phi = 0 \tag{8.18a}$$

or

$$\sin \alpha (sN_s)_{,s} + N_{s\theta,\theta} - N_\theta \sin \alpha = 0$$

or

$$(sN_s)_{,s} + \frac{1}{\sin \alpha} N_{s\theta,\theta} - N_\theta = 0 \tag{8.24a}$$

$$\frac{1}{R_\phi}(rN_{\phi\theta})_{,\phi} + N_{\theta,\theta} + N_{\phi\theta} \cos \phi = 0 \tag{8.18b}$$

or

$$\sin \alpha (sN_{s\theta})_{,s} + N_{\theta,\theta} + N_{s\theta} \sin \alpha = 0$$

or

$$\frac{1}{\sin \alpha} N_{\theta,\theta} + \frac{1}{s}(s^2 N_{s\theta})_{,s} = 0 \tag{8.24b}$$

$$\frac{1}{R_\phi}\left[\frac{1}{R_\phi}(rM_\phi)_{,\phi}\right]_{,\phi} + 2\left(\frac{1}{R_\phi}M_{\phi\theta,\phi\theta} + \frac{1}{r}M_{\phi\theta,\theta}\cos\phi\right)$$
$$+ \left[\frac{1}{r}(M_{\theta,\theta\theta}) - \frac{1}{R_\phi}(M_\theta \cos\phi)_{,\phi}\right] + \left(\frac{1}{R_\phi}rN_\phi + N_\theta \sin\phi\right)$$
$$- \left[\frac{1}{R_\phi}(rN_{\phi0}\beta_\phi + rN_{\phi\theta0}\beta_\theta)_{,\phi} + (N_{\theta0}\beta_\theta + N_{\phi\theta0}\beta_\phi)_{,\theta}\right] = 0 \tag{8.18c}$$

or

$$\frac{1}{\partial s}\left[\frac{1}{\partial s}(s\sin\alpha M_s)\right] + 2\left(M_{s\theta,s\theta} + \frac{1}{s}M_{s\theta,\theta}\right) + \left[\frac{1}{s\sin\alpha}(M_{\theta,\theta\theta}) - (M_\theta \sin\alpha)_{,s}\right]$$
$$+ N_\theta \cos\alpha - [(s\sin\alpha N_{s0}\beta_s + s\sin\alpha N_{s\theta0}\beta_\theta)_{,s} + (N_{\theta0}\beta_\theta + N_{s\theta0}\beta_s)_{,\theta}] = 0$$

or

$$(sM_s)_{,ss} + \frac{2}{\sin\alpha}\left(M_{s\theta,s\theta} + \frac{1}{s}M_{s\theta,\theta}\right) + \frac{1}{s\sin^2\alpha}(M_{\theta,\theta\theta}) - M_{\theta,s}$$
$$+ N_\theta \cot\alpha - \left[(sN_{s0}\beta_s + sN_{s\theta0}\beta_\theta)_{,s} + \frac{1}{\sin\alpha}(N_{\theta0}\beta_\theta + N_{s\theta0}\beta_s)_{,\theta}\right] = 0 \tag{8.24c}$$

The subscript ϕ has been replaced by s in Eqs. (8.24a)–(8.24c). From Eqs. (8.18d)–(8.18f), the constitutive relations for the conical shell are

$$N_s = C\left[u_{,s} + \frac{v}{s}\left(\frac{v_{,\theta}}{\sin\alpha} + u - w\cot\alpha\right)\right]$$

$$N_\theta = C\left[\frac{1}{s}\left(\frac{v_{,\theta}}{\sin\alpha} + u - w\cot\alpha\right) + vu_{,s}\right]$$

$$N_{s\theta} = C\frac{1-v}{2}\left[v_{,s} - \frac{v}{s} + \frac{u_{,\theta}}{s\sin\alpha}\right] \tag{8.24d}$$

$$M_s = D\left[\beta_{s,s} + \frac{v}{s}\left(\frac{\beta_{\theta,\theta}}{\sin\alpha} + \beta_s\right)\right]$$

$$M_\theta = D\left[\frac{1}{s}\left(\frac{\beta_{\theta,\theta}}{\sin\alpha} + \beta_s\right) + v\beta_{s,s}\right]$$

$$M_{s\theta} = D\frac{1-v}{2}\left[s\left(\frac{\beta_\theta}{s}\right)_{,s} + \frac{\beta_{s,\theta}}{s\sin\alpha}\right]$$

$$M_{s\theta} = D\frac{1-v}{2}\left[\beta_{\theta,s} - \frac{\beta_\theta}{s} + \frac{\beta_{s,\theta}}{s\sin\alpha}\right] \tag{8.24e}$$

$$\beta_\theta = -w_{,s}, \qquad \beta_s = -\frac{w_{,\theta}}{s\sin\alpha} \tag{8.24f}$$

If $\alpha = \pi/2$, then $s \to r$. Equations (8.24a)–(8.24c) can be reduced to the corresponding expressions for flat plates as follows:

$$(sN_s)_{,s} + \frac{1}{\sin\alpha}N_{s\theta,\theta} - N_\theta = 0 \tag{8.24a}$$

$$(rN_r)_{,r} + N_{r\theta,\theta} - N_\theta = 0 \tag{8.24g}$$

$$\frac{1}{\sin\alpha}N_{\theta,\theta} + \frac{1}{s}(s^2N_{s\theta})_{,s} = 0 \tag{8.24b}$$

or

$$N_{\theta,\theta} + N_{s\theta} + (sN_{s\theta})_{,s} = 0$$

or

$$N_{\theta,\theta} + N_{r\theta} + (rN_{r\theta})_{,r} = 0 \tag{8.24h}$$

$$(sM_s)_{,ss} + \frac{2}{\sin\alpha}\left(M_{s\theta,s\theta} + \frac{1}{s}M_{s\theta,\theta}\right) + \frac{1}{s\sin^2\alpha}(M_{\theta,\theta\theta}) - M_{\theta,s}$$

$$+N_\theta\cot\alpha - \left[(sN_{s0}\beta_s + sN_{s\theta 0}\beta_\theta)_{,s} + \frac{1}{\sin\alpha}(N_{\theta 0}\beta_\theta + N_{s\theta 0}\beta_s)_{,\theta}\right] = 0 \tag{8.24c}$$

or

$$(rM_r)_{,rr} + 2\left(M_{r\theta,r\theta} + \frac{1}{r}M_{s\theta,\theta}\right) + \frac{1}{r}(M_{\theta,\theta\theta}) - M_{\theta,r}$$

$$-[(rN_{r0}\beta_r + rN_{r\theta 0}\beta_\theta)_{,r} + (N_{\theta 0}\beta_\theta + N_{r\theta 0}\beta_r)_{,\theta}] = 0 \tag{8.24i}$$

Equations (8.24g)–(8.24i) are the same as Eqs. (8.19a)–(8.19c) derived for the circular flat plates. For $\alpha = 0$, and replacing $s \sin \alpha$ by the radius R of a cylinder, Eqs. (8.24a)–(8.24c) give the Donnell equations for the cylindrical shells.

Substitute Eqs. (8.24d)–(8.24f) in (8.24a)–(8.24c) to obtain

$$(sN_s)_{,s} + \frac{1}{\sin \alpha} N_{s\theta,\theta} - N_\theta = 0$$

$$(sN_s)_{,s} = N_s + sN_{s,s} \tag{8.24a}$$

$$N_s = C \left[u_{,s} + \frac{v}{s} \left(\frac{v_{,\theta}}{\sin \alpha} + u - w \cot \alpha \right) \right]$$

$$N_\theta = C \left[\frac{1}{s} \left(\frac{v_{,\theta}}{\sin \alpha} + u - w \cot \alpha \right) + v u_{,s} \right]$$

$$N_{s\theta} = C \frac{1-v}{2} \left[v_{,s} - \frac{v}{s} + \frac{u_{,\theta}}{s \sin \alpha} \right] \tag{8.24d}$$

$$(sN_s)_{,s} = C \left[u_{,s} + \frac{v}{s} \left(\frac{v_{,\theta}}{\sin \alpha} + u - w \cot \alpha \right) \right]$$

$$+ Cs \left[u_{,ss} + \frac{v}{s} \left(\frac{v_{,\theta s}}{\sin \alpha} - \frac{v_{,\theta}}{s \sin \alpha} + u_{,s} - \frac{u}{s} - w_{,s} \cot \alpha + \frac{w}{s} \cot \alpha \right) \right]$$

$$\frac{1}{\sin \alpha} N_{s\theta,\theta} = C \frac{1-v}{2 \sin \alpha} \left[v_{,s\theta} - \frac{v_{,\theta}}{s} + \frac{u_{,\theta\theta}}{s \sin \alpha} \right]$$

Therefore, Eq. (8.24a) can be written as

$$su_{,ss} + u_{,s} - \frac{u}{s} + \frac{1-v}{2} \frac{u_{,\theta\theta}}{s \sin^2 \alpha} + \frac{1+v}{2} \frac{v_{,s\theta}}{\sin \alpha} - \frac{3-v}{2} \frac{v_{,\theta}}{s \sin \alpha} - \left(vw_{,s} - \frac{w}{s} \right) \cot \alpha = 0 \tag{8.24j}$$

Similarly, Eqs. (8.24b) and (8.24c) can be written as

$$\frac{1+v}{2} \frac{u_{,s\theta}}{\sin \alpha} + \frac{3-v}{2} \frac{u_{,\theta}}{s \sin \alpha} + \frac{1-v}{2} sv_{,ss} + \frac{1-v}{2} \left(v_{,s} - \frac{v}{s} \right) + \frac{v_{,\theta\theta}}{s \sin^2 \alpha} - \frac{w_{,\theta} \cot \alpha}{s \sin \alpha} = 0 \tag{8.24k}$$

and

$$Ds \left(w_{,ssss} + \frac{2w_{,sss}}{s} - \frac{w_{,ss}}{s^2} + \frac{w_{,s}}{s^3} - 2 \frac{w_{,s\theta\theta}}{s^3 \sin^2 \alpha} + 2 \frac{w_{,ss\theta\theta}}{s^2 \sin^2 \alpha} + 4 \frac{w_{,\theta\theta}}{s^4 \sin^2 \alpha} + \frac{w_{,\theta\theta\theta\theta}}{s^4 \sin^4 \alpha} \right)$$

$$- C \left(\frac{v_{,\theta}}{s \sin \alpha} + \frac{u}{s} - \frac{w \cot \alpha}{s} + v u_{,s} \right) \cot \alpha$$

$$- \left[\left(N_{s0} sw_{,s} + N_{s\theta 0} \frac{w_{,\theta}}{\sin \alpha} \right)_{,s} + \frac{1}{\sin \alpha} \left(N_{s\theta 0} w_{,s} + N_{\theta 0} \frac{w_{,\theta}}{s \sin \alpha} \right)_{,\theta} \right] = 0 \tag{8.24l}$$

Equations (8.24j)–(8.24l) are the coupled three homogeneous equations in u, v, and w.

Example 8.3 Consider a conical shell subjected to uniform external hydrostatic pressure of p_h in lbs./in.^{2} (MPa), and the self weight of the shell p_w in lbs./in.^{2} (MPa). If we assume that the membrane analysis is accurate enough for the pre-buckling analysis, then the coefficients N_{s0}, $N_{\theta 0}$, and $N_{s\theta 0}$ can be determined by Eqs. (8.18k)–(8.18m). For a cone $R_\phi \to \infty$, limit $(R_\phi\, d\phi) = ds$, $r = s \sin \alpha$, $\phi = (\pi/2) - \alpha$ (constant), $\sin \phi = \cos \alpha$, and $\cos \phi = \sin \alpha$. Substitute these values into Eqs. (8.18k)–(8.18m) to obtain

$$(sN_{s0})_{,s} - N_{\theta 0} = -sp_s \tag{8.25a}$$

$$(sN_{s\theta 0})_{,s} + N_{s\theta 0} = -sp_\theta \tag{8.25b}$$

$$N_{\theta 0} = -sp \tan \alpha \tag{8.25c}$$

where p_s, p_θ, and p are the surface load components in the s, θ, and normal directions, respectively. The subscript ϕ has been replaced by s in Eqs. (8.25a)–(8.25c).The self-weight of the shell p_w has the components

$$p_s = p_w \cos \alpha, \qquad p_1 = p_w \sin \alpha$$

Therefore, the total surface load normal to the conical surface direction is

$$p = p_1 + p_h = p_w \sin \alpha + p_h$$

From Eq. (8.25c) we get

$$N_{\theta 0} = -s(p_w \sin \alpha + p_h) \tan \alpha \tag{8.25d}$$

The load is axisymmetric,

$$N_{s\theta 0} = 0 \tag{8.25e}$$

From Eq. (8.25a) we get

$$\frac{d}{ds}(sN_{s0}) = -s(p_w \sin \alpha + p_h) \tan \alpha - sp_w \cos \alpha$$

or

$$sN_{s0} = -\int [s(p_w \sin \alpha + p_h) \tan \alpha + sp_w \cos \alpha]ds + C$$

or

$$sN_{s0} = -\frac{s^2}{2}[(p_w \sin \alpha + p_h) \tan \alpha + p_w \cos \alpha] + C$$

At $s = 0$, $N_s = 0$, therefore, $C = 0$

$$N_{s0} = -\frac{s}{2}\left(\frac{p_w}{\cos \alpha} + p_h \tan \alpha\right) \tag{8.25f}$$

Substitute the values from Eqs. (8.25d)–(8.25f) into the stability Eqs. (8.24j)–(8.24l), but these are variable coefficient equations. Hence, numerical techniques have to be used to obtain the critical load. Baruch, Harari, and Singer [32] analyzed the problem by using the Galerkin procedure for the hydrostatic pressure p_h ($p_w = 0$) and reported the results for a wide range of parameters.

8.9.6 Toroidal Shells

The shells of revolution are formed by rotating a plane curve about an axis that lies in the plane of the curve as described in Section 8.9 and is shown in Figure 8.14. If the plane curve is a circular arc of radius b shown in Figure 8.18, the surface of revolution obtained is a segment of a torous or toroidal shell. The various parameters of the middle surface of the toroidal shell as a shell of revolution in Figure 8.18 are:

$$R_\phi = b, r = a - b(1 - \sin \phi), \text{and} \frac{\partial r}{\partial \phi} = b \cos \phi \tag{8.26a}$$

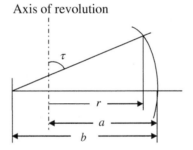

Figure 8.18 Toroidal shell meridian.

The analysis of toroidal shells was provided by Stein and McElman [33] and by Hutchinson [34] and is taken here from Brush and Almroth [5]. For a small segment near the central horizontal axis of the toroidal shell the angle ϕ is approximately equal to $\pi/2$. Then,

$$\cos \phi = 0, \sin \phi = 1, \frac{\partial r}{\partial \phi} = 0, \text{and } r = a \tag{8.26b}$$

Also, $dx = b\, d\phi$ and $dy = a\, d\theta$, where x and y are the axial and circumferential coordinates respectively. Substitute these values in Eqs. (8.18a)–(8.18c) and we get the stability equations for the torus segment. Eqs. (8.18a)–(8.18c) are repeated here for convenience, the subscript 1 is omitted for the incremental quantities for notational simplicity as before.

$$(rN_\phi)_{,\phi} + R_\phi N_{\phi\theta,\theta} - R_\phi N_\theta \cos \phi = 0 \tag{8.18a}$$

or

$$\frac{r}{R_\phi} N_{\phi,\phi} + N_{\phi\theta,\theta} = 0$$

or

$$\frac{1}{b}\frac{\partial N_\phi}{\partial \phi} + \frac{1}{a}\frac{\partial N_{\phi\theta}}{\partial \theta} = 0$$

or

$$N_{x,x} + N_{xy,y} = 0 \tag{8.26c}$$

$$(rN_{\phi\theta})_{,\phi} + R_\phi N_{\theta,\theta} + R_\phi N_{\phi\theta} \cos\phi = 0 \tag{8.18b}$$

Similarly Eq. (8.18b) can be expressed as

$$N_{xy,x} + N_{y,y} = 0 \tag{8.26d}$$

$$\left[\frac{1}{R_\phi}(rM_\phi)_{,\phi}\right]_{,\phi} + 2\left(M_{\phi\theta,\phi\theta} + \frac{R_\phi}{r}M_{\phi\theta,\theta}\cos\phi\right)$$
$$+ \left[\frac{R_\phi}{r}(M_{\theta,\theta\theta}) - (M_\theta\cos\phi)_{,\phi}\right] + (rN_\phi + R_\phi N_\theta\sin\phi)$$
$$- [(rN_{\phi 0}\beta_\phi + rN_{\phi\theta 0}\beta_\theta)_{,\phi} + (R_\phi N_{\theta 0}\beta_\theta + R_\phi N_{\phi\theta 0}\beta_\phi)_{,\theta}] = 0 \tag{8.18c}$$

or

$$\frac{1}{R_\phi}\left[\frac{1}{R_\phi}(rM_\phi)_{,\phi}\right]_{,\phi} + 2\left(\frac{1}{R_\phi}M_{\phi\theta,\phi\theta} + \frac{1}{r}M_{\phi\theta,\theta}\cos\phi\right)$$
$$+ \left[\frac{1}{r}M_{\theta,\theta\theta} - \frac{1}{R_\phi}(M_\theta\cos\phi)_{,\phi}\right] + \left(\frac{r}{R_\phi}N_\phi + N_\theta\sin\phi\right)$$
$$- \left[\frac{1}{R_\phi}(rN_{\phi 0}\beta_\phi + rN_{\phi\theta 0}\beta_\theta)_{,\phi} + (N_{\theta 0}\beta_\theta + N_{\phi\theta 0}\beta_\phi)_{,\theta}\right] = 0$$

or

$$(aM_{x,xx}) + 2a(M_{xy,xy}) + (aM_{y,yy}) + \left(\frac{a}{b}N_x + N_y\right)$$
$$- [(aN_{x0}\beta_x + aN_{xy0}\beta_y)_{,x} + (aN_{y0}\beta_y + aN_{xy0}\beta_x)_{,y}] = 0 \tag{8.26e}$$

Equations (8.18d)–(8.18f) giving the constitutive and kinematic relations can be expressed as

$$N_x = C\left[\left(u_x - \frac{w}{b}\right) + v\left(v_y - \frac{w}{a}\right)\right]$$
$$N_y = C\left[\left(v_y - \frac{w}{a}\right) + v\left(u_x - \frac{w}{b}\right)\right]$$
$$N_{xy} = C\frac{1-v}{2}[v_x + u_y] \tag{8.26f}$$
$$M_x = D[\beta_{x,x} + v\beta_{y,y}] = -D[w_{xx} + vw_{yy}]$$
$$M_y = D[\beta_{y,y} + v\beta_{x,x}] = -D[w_{yy} + vw_{xx}]$$
$$M_{xy} = D\frac{1-v}{2}[\beta_{y,x} + \beta_{x,y}] = -D\frac{1-v}{2}[w_{yx} + w_{xy}] = -D(1-v)w_{xy} \tag{8.26g}$$
$$\beta_x = -w_x, \quad \beta_y = -w_y \tag{8.26h}$$

Substitute the constitutive and the kinematic relations in Eqs. (8.26f)–(8.26h) into Eq. (8.26e) to give

$$D\nabla^4 w - \frac{N_x}{b} - \frac{N_y}{a} - (N_{x0}w_{,xx} + 2N_{xy0}w_{,xy} + N_{y0}w_{,yy}) = 0 \tag{8.26i}$$

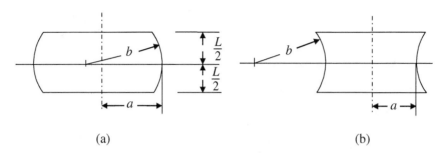

Figure 8.19 Segments of toroidal shell: (a) Bowed-out segment; (b) Bowed-in segment.

Equations (8.26c), (8.26d), and (8.26i) are the stability equations of the toroidal shell. Stein and McElman [33] and Hutchinson [34] provided the analyses for both the bowed-out and bowed-in segments shown in Figure 8.19. For the bowed-out case in Figure 8.19a, the radius of curvature b of the shell is positive and for the bowed-in case in Figure 8.19b, the radius of curvature of the shell is negative. When $b \to \infty$, the stability equation of the torus reduces to the Donnell equations of cylindrical shells given by Eq. (8.4r).

Example 8.4 Consider a toroidal shell segment that is subjected to uniform external lateral pressure of p_e in lbs./in.^{2} (MPa). The shell is simply supported at its ends at $x = 0$ and $x = L$. If we assume that the membrane analysis is accurate enough for pre-buckling analysis, then the coefficients N_{x0}, N_{y0}, and N_{xy0} can be determined from Eqs. (8.18k)–(8.18m) to have

$$\frac{d}{d\phi}(rN_\phi) - R_\phi N_\theta \cos \phi = -rR_\phi p_\phi \tag{8.18k}$$

$$\cos \phi = 0$$

or

$$\frac{1}{R_\phi}\frac{d}{d\phi}(rN_\phi) + rp_\phi = 0$$

or

$$(rN_{x0})_{,x} + rp_x = 0 \tag{8.27a}$$

Since

$$p_x = 0$$

$$N_{x0} = 0$$

$$\frac{d}{d\phi}(rN_{\phi\theta}) + R_\phi N_{\phi\theta} \cos \phi = -rR_\phi p_\theta \tag{8.18l}$$

Similarly, Eq. (8.18l) can be written as

$$(rN_{xy0})_{,x} + rp_y = 0 \tag{8.27b}$$

Since

$$p_y = 0$$
$$N_{xy0} = 0$$
$$rN_\phi + R_\phi N_\theta \sin \phi = -rR_\phi p \tag{8.18m}$$

or

$$\frac{a}{b} N_{x0} + N_{y0} + p_e a = 0 \tag{8.27c}$$

Since

$$N_{x0} = 0$$
$$N_{y0} = -p_e a$$

Substitute the values of N_{xo}, N_{yo}, and N_{xyo} into Eq. (8.26i) to obtain

$$D\nabla^4 w - \frac{N_x}{b} - \frac{N_y}{a} + p_e a w_{,yy} = 0 \tag{8.27d}$$

At the simply supported ends at $x = 0$ and $x = L$, the boundary conditions are

$$w = w_{,xx} = N_x = v = 0$$

The solution of this problem was obtained by Hutchinson [34] and is presented here as given by Bush and Almroth [5]. Assume a stress function f, such that

$$N_x = f_{,yy} \qquad N_{xy} = -f_{,xy} \qquad N_y = f_{,xx} \tag{8.27e}$$

Substitute these values into Eq. (8.27d) and we get

$$D\nabla^4 w - \frac{f_{,yy}}{b} - \frac{f_{,xx}}{a} + p_e a w_{,yy} = 0 \tag{8.27f}$$

Now

$$\nabla^4 f = \frac{\partial^4 f}{\partial x^4} + 2\frac{\partial^4 f}{\partial x^2 \partial y^2} + \frac{\partial^4 f}{\partial y^4} \tag{8.27g}$$

Substitute values from Eqs. (8.26f) and (8.27e) into Eq. (8.27g) to obtain

$$\frac{\nabla^4 f}{Eh} + \frac{w_{,xx}}{a} + \frac{w_{,yy}}{b} = 0 \tag{8.27h}$$

The problem is now reduced to the solution of two homogeneous differential equations in w and f given by Eqs. (8.27f) and (8.27h). The boundary conditions can be written as

$$w = w_{,xx} = f_{,xx} = f = 0 \tag{8.27i}$$

Assume the displacement function

$$w = C_1 \sin \frac{m\pi x}{L} \sin \frac{ny}{a} \tag{8.27j}$$

This displacement function satisfies the differential equations and the boundary conditions, where C_1 is a constant, and m and n are integers. From Eqs. (8.27h) and (8.27j), we get

$$f = \frac{EhL^2}{\pi^2 a} \frac{m^2 + \beta^2 a/b}{(m^2 + \beta^2)^2} C_1 \sin \frac{m\pi x}{L} \sin \frac{ny}{a} \tag{8.27k}$$

where $\beta = nL/\pi a$. Introduction into Eq. (8.27f) gives the eigenvalues as

$$\bar{p} = \frac{(m^2 + \beta^2)^2}{\beta^2} + \frac{(m^2 + \beta^2 a/b)^2}{\beta^2(m^2 + \beta^2)^2} \frac{12}{\pi^4} Z^2 \tag{8.27l}$$

where

$$\bar{p} = \frac{L^2 a}{\pi^2 D} p_e \quad \text{and} \quad Z = \frac{L^2}{ah}(1 - v^2)^{\frac{1}{2}}$$

When $m = 1$, the smallest value of eigenvalue is obtained from Eq. (8.27l) as

$$\bar{p} = \frac{(1 + \beta^2)^2}{\beta^2} + \frac{(1 + \beta^2 a/b)^2}{\beta^2(1 + \beta^2)^2} \frac{12}{\pi^4} Z^2 \tag{8.27m}$$

When $b \to \infty$, Eq. (8.27m) reduces to Eq. (8.8a) of a cylindrical shell. Minimize $\bar{p}$ with respect to β in Eq. (8.27m) to get $\bar{p}_{cr}$ as a function of single geometric parameter Z instead of the two parameters L/R and R/h. Hence,

$$\frac{d\bar{p}}{d\beta} = \frac{\beta^2[2(1 + \beta^2)(2\beta)] - 2\beta(1 + \beta^2)^2}{\beta^4}$$

$$+ \frac{\beta^2(1 + \beta^2)^2 2\left(1 + \beta^2 \frac{a}{b}\right)\left(2\beta \frac{a}{b}\right) - \left(1 + \beta^2 \frac{a}{b}\right)^2 [2\beta(1 + \beta^2)^2 + \beta^2(1 + \beta^2)(2)(2\beta)]}{[\beta^2(1 + \beta^2)^2]^2} \frac{12Z^2}{\pi^4} = 0$$

or

$$4\beta^3 - 2\beta(1 + \beta^2) + \frac{4\beta^3 \left(\frac{a}{b}\right)(1 + \beta^2)\left(1 + \beta^2 \frac{a}{b}\right) - \left(1 + \beta^2 \frac{a}{b}\right)^2 [2\beta(1 + \beta^2) + 4\beta^3]}{(1 + \beta^2)^4} \frac{12Z^2}{\pi^4} = 0$$

or

$$\frac{(\beta^2 - 1)(1 + \beta^2)^4}{\left(1 + \beta^2 \frac{a}{b}\right)\left[\beta^2 \frac{a}{b}(\beta^2 - 1) + (1 + 3\beta^2)\right]} = \frac{12Z^2}{\pi^4} \tag{8.27n}$$

When $a/b = 0$, Eq. (8.27n) reduces to that of Eq. (8.8b) for the cylindrical shells. In these equations, $\beta = \frac{nL}{\pi a}$, and $Z = \frac{L^2}{ah}(1 - v^2)^{\frac{1}{2}}$. For particular parameters of shell geometry, a, b, and Z, Eq. (8.27n) gives a relation between β (the integer n) and the Z value corresponding to the critical pressure. Substitute the β and Z values so obtained to get the critical $\bar{p}$ from Eq. (8.27m). The graph of Z versus $\bar{p}$ is plotted in Figure 8.20 for various values of a/b.

Once the critical $\bar{p}$ is known, the critical external pressure p_e is given by

$$p_e = \frac{\pi^2 D \bar{p}}{L^2 a} \tag{8.27o}$$

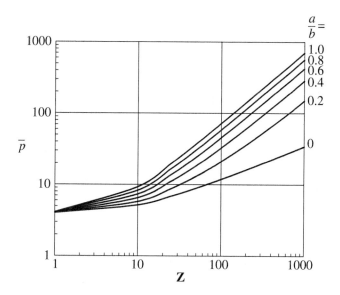

Figure 8.20 Critical external lateral pressure for toroidal shells.

For $a/b = 0$ the curve in Figure 8.20 is the same as for cylindrical shells in Figure 8.8.

Problems

8.1 A steel cylinder simply supported at the ends has a diameter of 48 in. (1219.2 mm), length of 24 in. (609.6 mm), and a wall thickness of 0.1 in. (2.54 mm). Find:
(a) the critical external lateral pressure; (b) the critical external hydrostatic pressure; and
(c) compare the values of external lateral and external hydrostatic pressures. The modulus of elasticity $E = 29 \times 10^6$ psi (200 000 MPa), Poisson's ratio $v = 0.3$.

8.2 An aluminum cylinder has a diameter of 30 in. (762 mm), length of 15 in. (381 mm), and a wall thickness of 0.03 in. (0.762 mm). It is subjected to torsion, find the critical shearing stress when the ends of the cylinder are: (a) simply supported and (b) clamped. The modulus of elasticity $E = 10 \times 10^6$ psi (68 950 MPa), Poisson's ratio $v = 0.3$.

8.3 A simply supported cylinder with a radius to thickness (R/h) ratio of 60 and length to radius ratio (L/R) of 0.2 is subjected to an axially compressive force. Find the critical pressure and mode if the modulus of elasticity is $E = 10 \times 10^6$ psi (68 950 MPa), and the Poisson's ratio $v = 0.3$.

8.4 A simply supported steel cylinder has a diameter of 48 in. (1219.2 mm), length of 24 in. (609.6 mm), and wall thickness of 0.1 in. (2.54 mm). It is subjected to external lateral pressure p and axial tension force P, and $P/2\pi R = -pR/2$. R is the radius of the cylinder. Find the critical pressure p and the axial tension force P. The modulus of elasticity $E = 29 \times 10^6$ psi (200, 000 MPa), and Poisson's ratio $v = 0.3$.

8.5 Use the potential energy Eq. (8.14p) to derive the nonlinear equilibrium Eqs. (8.14q)–(8.14s) for the general shells by using the principle of stationary potential energy.

8.6 (a) Derive nonlinear equilibrium equations of (8.1a), (8.1b), and (8.1j) in the text for thin cylindrical shells from the nonlinear equilibrium equations of (8.14q)–(8.14s) for thin shells of general shape.
(b) Show similarly the derivations of nonlinear rectangular flat plate equations, also called von Kármán plate equations, from the general shell equations.

8.7 Find the critical pressure of a simply supported aluminum toroid shell segment in Figure P8.7 when it is acted on by uniform pressure p_e. The dimensions of the shell are: $a = 12$ in. (305 mm), $L = 3$ in. (76.2 mm), thickness $h = 0.025$ in.(0.635 mm), the modulus of elasticity $E = 10 \times 10^6$ psi (68 950 MPa), Poisson's ratio $v = 0.3$. (a) $a/b = 0.6$ (b) $a/b = -0.6$.

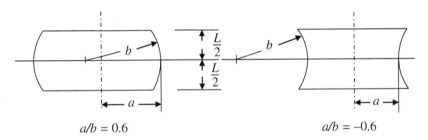

$a/b = 0.6$ $a/b = -0.6$

Figure P8.7

References

1 Jerath, S. and Lee, M. (2015). Stability of cylindrical tanks under static and earthquake loading. *Journal of Civil Engineering and Architecture* 9 (1): 72–79.
2 Donnell, L.H. (1934). A new theory for buckling of thin cylinders under axial compression and bending. *Transactions of the ASME* 56: 795–806.
3 Donnell, L.H. (1933) Stability of thin walled tubes under torsion, NACA, Technical Report, No. 479, Washington, D.C.
4 Donnell, L.H. and Wan, C.C. (1950). Effect of imperfections on buckling of thin cylinders and columns under axial compression. *Journal of Applied Mechanics ASME* 17 (1): 73–79.
5 Brush, D.O. and Almroth, B.O. (1975). *Buckling of Bars, Plates, and Shells*. New York: McGraw Hill.
6 Batdorf, S.B. (1947) A simplified method of elastic-stability analysis for thin cylindrical shells, Technical report, No. 874. NACA, Washington, D.C.

7 Timoshenko, S.P. and Gere, J.M. (1961). *Theory of Elastic Stability*, 2e. New York: McGraw Hill.

8 Jerath, S. and Sadid, H. (1985). Buckling of orthotropic cylinders due to wind load. *Journal of Engineering Mechanics, ASCE* 111 (5): 610–622.

9 Jerath, S. and Ghosh, A.K. (1987). Buckling of cylindrical shells under the action of non-uniform external pressure. *Journal of Computers and Structures* 25 (4): 607–614.

10 Gerard, G. and Becker, H. (1957) Buckling of curved plates and shells. In: *Handbook of Structural Stability*, TN 3783. NACA, Washington, DC.

11 Sobel, L.H. (1964). Effects of boundary conditions on the stability of cylinders subject to lateral and axial pressures. *AIAA Journal* 2: 1437–1440.

12 Lorenz, R. (1908). Achsensymmetrische Verzerrungen in dünnwandigen Hohlzylindern. *Zeitschrift des Vereines Deutscher Ingenieure* 52 (43).

13 Timoshenko, S.P. (1910). Einige Stabilitätsprobleme der Elastizitäts – Theorie. *Zeitschrift für Mathematik und Physik* 58 (4).

14 Southwell, R.V. (1914). On the general theory of elastic stability. *Philosophical Transactions of the Royal Society of London*, Series A 213.

15 Flügge, W. (1932). Die Stabilität der Kreiszylinderschale. *Ingenieurarchiv* 3.

16 von Kármán, T. and Tsien, H.S. (1941). The buckling of thin cylindrical shells under axial compression. *Journal of the Aeronautical Sciences* 8 (8): 303–312.

17 Leggett, D.M.A. and Jones, R.P.N. (1942). The behavior of a cylindrical shell under axial compression when the buckling load has been exceeded. *Aeronautical Research Council Memorandum* 2190.

18 Michielsen, H.F. (1948). The behavior of thin cylindrical shells after buckling under axial compression. *Journal of the Aeronautical Sciences* 15: 738–744.

19 Kempner, J. (1954). Post-buckling behavior of axially compressed circular cylindrical sheets. *Journal of the Aeronautical Sciences* 21 (5): 329–335.

20 Hoff, N.J., Madsen, W.A., and Mayers, J. (1965) The post buckling equilibrium of axially compressed circular cylindrical shells, Report SUDAER No. 221. Department of Aeronautics and Astronautics, Stanford University, Palo Alto, CA.

21 Koiter, W.T. (1945) On the stability of elastic equilibrium (in Dutch), thesis, Delft, H.J. Paris, Amsterdam. English translation, Air Force Flight Dynamics Laboratory, Technical Report AFFDL-TR-70-25, 1970.

22 Yoo, C.H. and Lee, S.C. (2011). *Stability of Structures Principles and Applications*. Burlington, MA: Elsevier.

23 Volmir, A.S. (1967) A translation of flexible plates and shells, air force flight dynamics laboratory, Technical Report No. 66-216, Wright-Patterson Air Force Base, Ohio.

24 Koiter, W.T. (1963) Elastic stability and post-buckling behavior, Proceedings, Symposium Non-linear Problems, University of Wisconsin Press, Madison, WI, pp. 257–275.

25 Love, A.E.H. (1888). On the small free vibrations and deformations of thin elastic shells. *Philosophical Transactions of the Royal Society of London* 17A: 491–546.

26 Koiter, W.T. (1967) General equations of elastic stability for thin shells, Proceedings Symposium Theory of Shells to Honor Lloyd Hamilton Donnell, University of Houston, Houston, TX, pp. 187–223.

27 Sanders, J.L. (1963). Non-linear theories of thin shells. *Quarterly of Applied Mathematics.* 21 (1): 21–36.

28 CRC (2018). *CRC Standard Mathematical Tables and Formulas*, 33e. Boca Raton, FL: CRC Press.

29 Kreyszig, E. (1972). *Advanced Engineering Mathematics*, 3e. New York: Wiley.

30 Vlasov, V.Z. (1944) General theory of shells and its application in engineering, NASA TTF-99, Washington, D.C.

31 Hutchinson, J.W. (1967). Imperfection sensitivity of externally pressurized spherical shells. *Journal of Applied Mechanics* 34: 49–55.

32 Baruch, M., Harari, O., and Singer, J. (1967). Influence of in-plane boundary conditions on the stability of conical shells under hydrostatic pressure. *Israel Journal of Technology* 5 (1–2): 12–24.

33 Stein, M. and McElman, J.A. (1965). Buckling of segments of toroidal shells. *AIAA Journal* 3: 1704–1709.

34 Hutchinson, J.W. (1967). Initial post-buckling behavior of toroidal shell segments. *International Journal of Solids Structures* 3: 97–115.

Answers to the Problems

Chapter 1

1.1 $P_{cr} = \dfrac{c}{L}$

1.2 $P_{cr} = \dfrac{c}{L}$

1.3 (a) $P_{cr} = \dfrac{kL}{2}$

(b) $P_{cr} = \dfrac{k_1 k_2}{k_1 + k_2} L$

1.4 (a) $P_{cr} = \dfrac{kL}{2}$

(b) $P_{cr} = \dfrac{k_1 k_2}{k_1 + k_2} L$

1.5 $P_{cr} = \dfrac{c_1 + 4c_2}{2L}$

1.6 $\dfrac{P}{P_{cr}} = \dfrac{2}{\sin \theta} \left[\dfrac{1}{2} \cos 2\theta - \cos(45^\circ - \alpha) \sin(45^\circ - \theta) \right]$

Chapter 2

2.1 $P_{cr} = \dfrac{\pi^2 EI}{L^2}$

Mode 1: $y = \dfrac{\Delta}{2}\left(1 - \cos\dfrac{\pi x}{L}\right)$

Mode 2: $y = \dfrac{\Delta}{2}\left(1 - \cos\dfrac{3\pi x}{L}\right)$

2.2 $\tan kL_c = \dfrac{4kL_c I_b}{4I_b + k^2 L_c L_b I_c}$

$L_b = 2L_c, I_b = I_c$

$P_{cr} = \dfrac{1.14\pi^2 EI_c}{L_c^2}$

2.3 $P_{cr} = 32.69$ kips (145.26 kN)

2.4 $P_{cr} = \dfrac{\pi^2 EI_1}{4L^2}\dfrac{1}{\dfrac{L_1}{L} + \dfrac{I_2}{I_1}\dfrac{L_2}{L} + \dfrac{1}{\pi}\sin\dfrac{\pi L_1}{L}\left(1 - \dfrac{I_2}{I_1}\right)}$

2.5 $P_{cr} = \dfrac{\pi^2 EI_1}{4L^2}\dfrac{-\dfrac{L_1}{2} - \dfrac{L_2}{2}\dfrac{I_2}{I_1} + \dfrac{2L}{\pi}\sin\dfrac{\pi L_1}{2L}\left(1 - \dfrac{I_2}{I_1}\right) + \dfrac{L}{2\pi}\sin\dfrac{\pi L_1}{L}\left(\dfrac{I_2}{I_1} - 1\right) + \dfrac{I_2}{I_1}\dfrac{2L}{\pi}}{-\dfrac{L_1}{2} - \dfrac{L_2}{2} + \dfrac{2L}{\pi}}$

2.6 $P_{cr} = \dfrac{\pi^2 EI}{(0.7L)^2}$

2.7 $P_{cr} = \dfrac{\pi^2 EI}{(1.548L)^2}$

2.8 $P_{cr} = EI_0\dfrac{3\pi^2}{2L^2}$

Chapter 3

3.3 Allowable strength, $P_a = 254.74$ kips (1143.23 kN)
Design strength, $P_u = 382.87$ kips (1718.28 kN)

Chapter 4

4.1 $y|_{x=\frac{L}{2}} = \dfrac{5w_oL^4}{768EI} \dfrac{12(2\sec u - 2 - u^2)}{5u^4}$

$u = \dfrac{\pi}{2}\sqrt{\dfrac{P}{P_e}}, P_e = \dfrac{\pi^2EI}{L^2}$

$M|_{x=\frac{L}{2}} = \dfrac{w_oL^2}{16}\left(\dfrac{2\sec u - 2}{u^2}\right)$

4.2 $y|_{x=\frac{L}{2}} = \dfrac{w}{2EIk^4\cos u}[\cos a + \cos k(L-b) - 2\cos u] - \dfrac{w}{2EIk^2}\left(Lb - \dfrac{a^2}{2} - \dfrac{b^2}{2} - \dfrac{L^2}{4}\right)$

$+ \dfrac{QL^3}{48EI}\left[\dfrac{3(\tan u - u)}{u^3}\right], \qquad \dfrac{kL}{2} = u, k^2 = \dfrac{P}{EI}$

4.3   (a) $P/P_{ek} = 0.6$, $\dfrac{Mmax_{exact}}{Mmax_{formula}} = \dfrac{1.949}{1.900}$

      (b) $P/P_{ek} = 0.6$, $\dfrac{Mmax_{exact}}{Mmax_{formula}} = \dfrac{1.959}{1.900}$

4.4 $P_{cr} = \dfrac{\pi^2EI}{1.676L^2}$

$L_2 = L, L_1 = 1.5\,L, I_1 = I_2 = I$

4.6   (a) $M_{max} = M_B$ at the end of the member
      (b) $M_{max} = M_B$ at $x = 0.975\,L$
      (c) $M_{max} = M_B$ at the end of the member

4.7   LRFD – Section satisfactory
      ASD – Section satisfactory

Chapter 5

5.1 $P_{cr} = 1.31\dfrac{\pi^2EI}{L^2}$

$L_b = L_c = L, I_b = I_c = I$

5.2 $P_{cr} = \dfrac{\pi^2EI}{(2.33L)^2}$

$L_b = L_c = L, I_b = I_c = I$

5.3 $P_{cr} = 2.52 \dfrac{\pi^2 EI}{L^2}$

5.4 $P_{cr} = \dfrac{\pi^2 EI}{L^2}$

5.5 $P_{cr} = \dfrac{16.8 EI}{L^2}$

5.6 (a) $G_{modified} = \dfrac{\sum \dfrac{EI_c}{L_c}}{\dfrac{3}{2} \sum \dfrac{EI_b}{L_b}}$ Sway inhibited

(b) $G_{modified} = \dfrac{\sum \dfrac{EI_c}{L_c}}{2 \sum \dfrac{EI_b}{L_b}}$ Sway inhibited

(a) $G_{modified} = \dfrac{\sum \dfrac{EI_c}{L_c}}{\dfrac{1}{2} \sum \dfrac{EI_b}{L_b}}$ Sway not inhibited

(b) $G_{modified} = \dfrac{\sum \dfrac{EI_c}{L_c}}{\dfrac{2}{3} \sum \dfrac{EI_b}{L_b}}$ Sway not inhibited

Chapter 6

6.1 $P_{cr} = P_{cry} = 555.72$ kips (2468.36 kN)

6.2 $P_{cr} = 23.86$ kips (107.23 kN)

6.3 (a) $P_{cr} = 70.87$ kips (312.74 kN)
(b) $P_{cr} = 84.92$ kips (375.40 kN)
(c) $P_{cr} = 101.76$ kips (450.71 kN)

6.4 (a) $C_b = 1.33$
(b) $C_b = 1.17$

6.5 $\quad GJ\dfrac{d^2\beta}{dz^2} - EC_w\dfrac{d^4\beta}{dz^4} + wz\dfrac{du}{dz} + \dfrac{\beta}{EI_y}\left[\dfrac{w}{2}(L-z)^2\right]^2 = 0$

6.6 $\quad GJ\dfrac{d^2\beta}{dz^2} - EC_w\dfrac{d^4\beta}{dz^4} - w\left(\dfrac{L}{2}-z\right)\dfrac{du}{dz} + \left[\dfrac{w}{2}\left(\dfrac{L^2}{4}-z^2\right)\right]^2\dfrac{\beta}{EI_y} = 0$

6.7 (a) Design moment strength, $M_u = 498.75$ kip. ft (676.89 kN.m)
 Allowable moment, $M_a = 331.84$ kip. ft (450.36 kN.m)
 (b) Same as in part (a)
 (c) Design moment strength, $M_u = 271.41$ kip. ft (366.11 kN.m)
 Allowable moment, $M_a = 180.58$ kip. ft (243.59 kN.m)

Chapter 7

7.1 (a) $\sigma_{cr} = \dfrac{\pi^2 D}{a^2 h}\left(1 + \dfrac{a^2}{b^2}\right), D = \dfrac{Eh^3}{12(1-v^2)}$
 (b) Same as in (a).

7.2 $(N_y)_{cr} = \dfrac{D\pi^2}{4a^2}\left(\dfrac{3}{\beta^2} + 16\beta^2 + 8 - 16v\right)$

 $\beta = \dfrac{a}{b}$

7.3 $N_{cr} = \dfrac{3}{8}\dfrac{\pi^3 D(a^2+b^2)^2}{a^2 b^4}$

7.4 $(N_x)_{cr} = \dfrac{4D\pi^2}{3\beta^2 b^2}(3 + 2\beta^2 + 3\beta^4 + 9\gamma\beta^3)$

 $\beta = \dfrac{a}{b}, \gamma = \dfrac{EI}{bD}$

7.5 $h = 0.066$ in. (1.68 mm)

7.6 $N = 3.45\dfrac{\pi^2 D}{a^2}$

 $\dfrac{a}{b} = 1, h = \dfrac{a}{4}, k = \dfrac{b}{2}, N_x = N_y = N$

7.7 $\sigma_{cr} = 8796.24$ psi (60.66 MPa)

 $\dfrac{a}{b} = 1.5, \dfrac{h}{b} = 0.02, \dfrac{E_t}{E} = 0.414, h = $ Thickness of the plate,

 $E_t = $ Tangent modulus, $E = $ Modulus of elasticity

Chapter 8

8.1 (a) 32.94 psi (0.227 MPa)

(b) 31.39 psi (0.22 MPa)

(c) $\dfrac{\text{External lateral pressure}}{\text{Hydrostatic pressure}} = 1.05$

8.2 (a) 3280 psi (22.61 MPa)

(b) 3651.63 psi (25.18 MPa)

8.3 103.3 ksi (712.33 MPa)

It is short cylinder with $Z < 2.85$. Its mode in buckling is single half wave along the length and no waves along the circumference.

8.4 $p = 34.65$ psi (0.24 Mpa)

$P = 62.7$ kips (280.2 kN)

8.7 (a) 19.92 psi (0.137 MPa)

(b) 5.54 psi (0.0382 MPa)

Appendix A

Slope Deflection Coefficients for Beam Column Buckling

kL	k_{ii}	k_{ij}
0.0000	4.0000	2.0000
0.1000	3.9987	2.0003
0.2000	3.9947	2.0013
0.3000	3.9880	2.0030
0.4000	3.9786	2.0054
0.5000	3.9666	2.0084
0.6000	3.9518	2.0121
0.7000	3.9342	2.0166
0.8000	3.9139	2.0218
0.9000	3.8908	2.0277
1.0000	3.8649	2.0344
1.1000	3.8360	2.0419
1.2000	3.8043	2.0502
1.3000	3.7695	2.0594
1.4000	3.7317	2.0695
1.5000	3.6907	2.0806
1.6000	3.6466	2.0926
1.7000	3.5991	2.1057
1.8000	3.5483	2.1199
1.9000	3.4940	2.1353

kL	k_{ii}	k_{ij}
2.0000	3.4361	2.1519
2.1000	3.3745	2.1699
2.2000	3.3090	2.1893
2.3000	3.2395	2.2102
2.4000	3.1659	2.2328
2.5000	3.0878	2.2572
2.6000	3.0052	2.2834
2.7000	2.9178	2.3118
2.8000	2.8254	2.3425
2.9000	2.7276	2.3756
3.0000	2.6242	2.4115
3.1000	2.5148	2.4503
3.2000	2.3990	2.4924
3.3000	2.2763	2.5382
3.4000	2.1463	2.5880
3.5000	2.0083	2.6424
3.6000	1.8618	2.7017
3.7000	1.7060	2.7668
3.8000	1.5400	2.8382
3.9000	1.3627	2.9168
4.0000	1.1731	3.0037
4.1000	0.9698	3.1001
4.2000	0.7510	3.2074
4.3000	0.5149	3.3273
4.4000	0.2592	3.4619
4.5000	−0.0191	3.6140
4.6000	−0.3234	3.7866
4.7000	−0.6582	3.9839
4.8000	−1.0289	4.2112
4.9000	−1.4427	4.4751
5.0000	−1.9087	4.7845
5.1000	−2.4394	5.1514
5.2000	−3.0516	5.5921
5.3000	−3.7688	6.1296
5.4000	−4.6253	6.7977
5.5000	−5.6726	7.6472

kL	k_{ii}	k_{ij}
5.6000	−6.9922	8.7589
5.7000	−8.7214	10.2692
5.8000	−11.1106	12.4278
5.9000	−14.6715	15.7453
6.0000	−20.6375	21.4540
6.1000	−32.9345	33.4784
6.2000	−74.3621	74.6167
6.3000	374.6369	−374.6900
6.4000	54.5347	−54.9159
6.5000	29.4960	−30.2280
6.6000	20.1038	−21.2118
6.7000	15.0847	−16.5971
6.8000	11.8889	−13.8378
6.9000	9.6187	−12.0405
7.0000	7.8755	−10.8118

Appendix B

Torsion Properties of Thin-Walled Open Cross-Sections

Cross-section	Shear center O	Torsion constant J	Warping constant C_w
		$J = \dfrac{2bt_f^{\,3} + ht_w^3}{3}$ If $t = t_f = t_w$ $J = \dfrac{t^3}{3}(2b + h)$	$C_w = \dfrac{t_f h^2 b^3}{24}$
	$e = h\dfrac{b_1^3}{b_1^3 + b_2^3}$	$J = \dfrac{(b_1 + b_2)t_f^3 + ht_w^3}{3}$ If $t = t_f = t_w$ $J = \dfrac{t^3}{3}(b_1 + b_2 + h)$	$C_w = \dfrac{t_f h^2}{12}\dfrac{b_1^3 b_2^3}{b_1^3 + b_2^3}$

Cross-section	Shear center O	Torsion constant J	Warping constant C_w
	$e = \dfrac{3b^2 t_f}{6bt_f + ht_w}$	$J = \dfrac{2bt_f^3 + ht_w^3}{3}$	$C_w = \dfrac{t_f b^3 h^2}{12}\dfrac{3bt_f + 2ht_w}{6bt_f + ht_w}$
	If $t = t_f = t_w$	If $t = t_f = t_w$	If $t = t_f = t_w$
	$e = \dfrac{3b^2}{6b + h}$	$J = \dfrac{t^3}{3}(2b + h)$	$C_w = \dfrac{tb^3 h^2}{12}\dfrac{3b + 2h}{6b + h}$
		$J = \dfrac{2bt_f^3 + ht_w^3}{3}$	$C_w = \dfrac{b^3 h^2}{12(2b + h)^2}$ $\times [2t_f(b^2 + bh + h^2) + 3t_w bh]$
		If $t = t_f = t_w$	If $t = t_f = t_w$
		$J = \dfrac{t^3}{3}(2b + h)$	$C_w = \dfrac{tb^3 h^2}{12}\dfrac{b + 2h}{2b + h}$
	$e = 2a\dfrac{\sin \alpha - \alpha \cos \alpha}{\alpha - \sin \alpha \cos \alpha}$	$J = \dfrac{2a\alpha t^3}{3}$	$C_w = \dfrac{2ta^5}{3}\left[\alpha^3 - \dfrac{6(\sin \alpha - \alpha \cos \alpha)^2}{\alpha - \sin \alpha \cos \alpha}\right]$
	If $2\alpha = 2\pi$	If $2\alpha = 2\pi$	If $2\alpha = 2\pi$
	$e = \dfrac{4a}{\pi}$	$J = \dfrac{\pi a t^3}{3}$	$C_w = \dfrac{2ta^5}{3}\left(\dfrac{\pi^3}{8} - \dfrac{12}{\pi}\right)$ $= 0.0374ta^5$

Source: Timoshenko.

Appendix C

Calculus of Variations

C.1 Calculus of Variations

The calculus of variation is a generalization of the minimum and maximum problem of the ordinary calculus. It seeks to determine a function $y = f(x)$ that minimizes/maximizes a definite integral called functional (function of functions) and whose integrant contains y and its derivatives and the independent variable x given by Eq. (C.1).

$$I = \int_{x_1}^{x_2} F(x, y, y', y'', \dots \dots \dots, y^n)\, dx \tag{C.1}$$

In ordinary calculus, one obtains the actual value of a variable at which a given function has a stationary value. In the calculus of variations, one does not get a function that extremizes a given integral. Here, one only gets the differential equation that the function must satisfy so that the function has a stationary value. Thus, the calculus of variations is used to obtain the governing differential equation of a stationary value problem. It is not a computational tool to solve the problem.

In structural mechanics, the method is used to find the deformed shape of a system at which the system has a stationary potential energy or in other words finding the deformation corresponding to the equilibrium state of the system. To illustrate the calculus of variation consider a pinned-pinned column in Figure C.1 and find the conditions under which it will be in equilibrium under a deformed shape. The strain energy for bending for the column is

$$U = \int_0^L \frac{M_x^2 dx}{2EI} = \frac{1}{2} \int_0^L EI \left(\frac{d^2 y}{dx^2} \right)^2 \tag{C.2}$$

$$V = -P\Delta \tag{C.3}$$

From Figure C.1c

$$ds^2 = dx^2 + dy^2$$

or

$$ds^2 = \left[1 + \left(\frac{dy}{dx} \right)^2 \right] dx^2$$

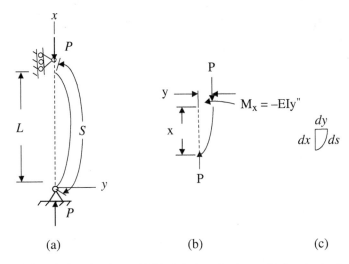

Figure C.1 Buckling of pinned-pinned column: (a) Pinned-pinned column; (b) Free body diagram; (c) Differential element.

or

$$ds = \left[1 + \left(\frac{dy}{dx} \right)^2 \right]^{\frac{1}{2}} dx$$

or

$$ds = \left[1 + \frac{1}{2} \left(\frac{dy}{dx} \right)^2 \right] dx$$

Hence,

$$\int_0^S ds = \int_0^L \left[1 + \frac{1}{2} \left(\frac{dy}{dx} \right)^2 \right] dx \qquad \text{(C.4)}$$

$$\Delta = S - L = \frac{1}{2} \int_0^L \left(\frac{dy}{dx} \right)^2 dx \qquad \text{(C.5)}$$

$$V = -\frac{P}{2} \int_0^L \left(\frac{dy}{dx} \right)^2 dx \qquad \text{(C.6)}$$

$$\Pi = \frac{EI}{2} \int_0^L \left(\frac{d^2y}{dx^2} \right)^2 dx - \frac{P}{2} \int_0^L \left(\frac{dy}{dx} \right)^2 dx \qquad \text{(C.7)}$$

It is intended to find $y(x)$ which will make the total potential energy of the system stationary, that is

$$\delta(U + V) = 0 \qquad \text{(C.8)}$$

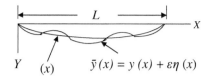

Figure C.2 Deflected shape.

$y(x)$ must be continuous and it must satisfy the boundary conditions $y(0) = y(L) = 0$. Assume

$$\bar{y}(x) = y(x) + \varepsilon \ \eta(x) \tag{C.9}$$

which satisfies only the geometric boundary conditions. $\eta(x)$ is an arbitrary function satisfying boundary conditions and is twice differentiable, and ε is a small parameter.

$$\eta(0) = \eta(L) = 0 \tag{C.10}$$

The function $\bar{y}(x)$ is drawn graphically in Figure C.2.

The total potential energy in terms of the displacement $\bar{y}(x)$ is

$$\Pi = U + V = \int_0^L \left[\frac{EI}{2}(y'' + +\varepsilon\eta'')^2 - \frac{P}{2}(y' + \varepsilon\eta')^2 \right] dx \tag{C.11}$$

Π is a function of ε for a given $\eta(x)$. If $\varepsilon = 0$, then $\bar{y}(x) = y(x)$, which is the curve that provides a stationary value to Π. Hence,

$$\left| \frac{d(U+V)}{d\varepsilon} \right|_{\varepsilon=0} = 0 \tag{C.12}$$

$$\frac{d(U+V)}{d\varepsilon} = \int_0^L [EI(y'' + \varepsilon\eta'')\eta'' - P(y' + \varepsilon\eta')\eta'] \, dx \tag{C.13}$$

Equation (C.13) is zero at $\varepsilon = 0$, hence,

$$\int_0^L [EIy''\eta'' - Py'\eta'] \, dx = 0 \tag{C.14}$$

Integrate Eq. (C.14) by parts using $\int u\,dv = uv - \int v\,du$

$$\int_0^L \eta'y' \, dx = y'\eta|_0^L - \int_0^L \eta y'' \, dx$$

Use Eq. (C.10) to get

$$\int_0^L \eta'y' \, dx = - \int_0^L \eta y'' \, dx$$

$$\int_0^L y''\eta'' \, dx = y''\eta'|_0^L - \int_0^L \eta'y''' \, dx = y''\eta'|_0^L - y'''\eta|_0^L + \int_0^L \eta y^{IV} \, dx$$

Thus, Eq. (C.14) becomes

$$\int_0^L (EIy^{IV} + Py'') \eta\,dx + (EIy''\eta')_0^L = 0 \tag{C.15}$$

Each of the two parts of Eq. (C.15) is separately equal to zero because η is arbitrary. Hence

$$\int_0^L (EIy^{IV} + Py'') \, \eta dx = 0 \tag{C.16}$$

$$(EIy''\eta')_0^L = 0 \tag{C.17}$$

Since $\eta'(0), \eta'(L)$ are not zero, $\eta(x)$ is arbitrary, also $\eta'(0) \neq \eta'(L)$, therefore, $y(x)$ must satisfy

$$EIy^{IV} + Py'' = 0 \tag{C.18}$$

$$EIy''|_{x=0} = 0 \tag{C.19}$$

$$EIy''|_{x=L} = 0 \tag{C.20}$$

Equation (C.18) is the Eulerian differential equation of an axially loaded column as was found in Eq. (2.11e) by considering the moment equilibrium of the deformed column. Eqs. (C.19) and (C.20) give the natural boundary conditions and indicate that the bending moments at the ends of a simply supported column are zero. For simple systems, such as simply supported columns, the governing differential equation can be obtained by considering the equilibrium of the deformed shape. For complex systems, such as plate and shell buckling, the Stationary Potential Energy method is simpler to obtain the governing differential equation. The geometric or kinematic boundary conditions involve displacements (deflection and slope), where natural boundary conditions give force conditions (bending moment or shear force) at the boundary.

Appendix D

Euler Equations

Let us consider a structure for which the integrand F is a given function of one independent variable x, and one dependent variable w and its derivatives w' and w'', then say the potential energy V is

$$V = \int_{x_0}^{x_1} F(x, w, w', w'') \, dx \tag{D.1}$$

To study the behavior of V for a particular configuration $w = w_0$, let $w \to w_0 + w_1$ where $w_1(x) = \varepsilon \zeta(x)$, ε is an arbitrary small constant, and $\zeta(x)$ is any arbitrary function that satisfies the necessary continuity and forced boundary conditions at x_0 and x_1. Substitute into Eq. (D.1) to give

$$\Delta V = \int_{x_0}^{x_1} [F(x, w_0 + \varepsilon\zeta, w_0' + \varepsilon\zeta', w_0'' + \varepsilon\zeta'') - F(x, w_0, w', w_0'')] dx \tag{D.2}$$

The Taylor series of a function of n variables is given by

$$\Delta V = \delta V + \delta^2 V + \delta^3 V + - - -$$

where

$$\delta V = \sum_{i=1}^{n} \frac{\partial V}{\partial q_i} \delta q_i, \quad \delta^2 V = \frac{1}{2!} \sum_{i=1}^{n} \sum_{j=1}^{n} \frac{\partial^2 V}{\partial q_i \partial q_j} \delta q_i \delta q_j, \quad \delta^3 V = \frac{1}{3!} \sum_{i=1}^{n} \sum_{j=1}^{n} \sum_{k=1}^{n} \frac{\partial^3 V}{\partial q_i \partial q_j \partial q_k} \delta q_i \delta q_j \delta q_k$$

The expansion of the integrand in Eq. (D.2) in the Taylor series gives

$$\delta V = \varepsilon \int_{x_0}^{x_1} \left(\frac{\partial F}{\partial w_0} \zeta + \frac{\partial F}{\partial w_0'} \zeta' + \frac{\partial F}{\partial w_0''} \zeta'' \right) dx \tag{D.3}$$

where $\dfrac{\partial F}{\partial w_0}$ is the value of $\dfrac{\partial F}{\partial w}$ at $w = w_0$, etc. For equilibrium, V should be a relative minimum which gives, $\delta V = 0$. Since ε is arbitrary, we have

$$\int_{x_0}^{x_1} \left(\frac{\partial F}{\partial w_0} \zeta + \frac{\partial F}{\partial w_0'} \zeta' + \frac{\partial F}{\partial w_0''} \zeta'' \right) dx = 0 \tag{D.4}$$

Equation (D.4) is repeatedly integrated by parts as follows:

$$\int u\,dv = uv - \int v\,du$$

$$\int_{x_0}^{x_1} \frac{\partial F}{\partial w_0'}\zeta'\,dx = \frac{\partial F}{\partial w_0'}\zeta - \int_{x_0}^{x_1}\zeta\frac{d}{dx}\left(\frac{\partial F}{\partial w_0'}\right)dx$$

$$\int_{x_0}^{x_1}\frac{\partial F}{\partial w_0''}\zeta''\,dx = \frac{\partial F}{\partial w_0''}\zeta' - \int_{x_0}^{x_1}\zeta'\frac{d}{dx}\left(\frac{\partial F}{\partial w_0''}\right)dx = \frac{\partial F}{\partial w_0''}\zeta'$$

$$- \left[\frac{d}{dx}\left(\frac{\partial F}{\partial w_0''}\right)\zeta - \int_{x_0}^{x_1}\zeta\frac{d^2}{dx^2}\left(\frac{\partial F}{\partial w_0''}\right)dx\right]$$

Now Eq. (D.4) can be written as

$$\zeta\frac{\partial F}{\partial w_0'} + \zeta'\frac{\partial F}{\partial w_0''} - \zeta\frac{d}{dx}\left(\frac{\partial F}{\partial w_0''}\right) + \int_{x_0}^{x_1}\zeta\left[\frac{\partial F}{\partial w_0} - \frac{d}{dx}\left(\frac{\partial F}{\partial w_0'}\right) + \frac{d^2}{dx^2}\left(\frac{\partial F}{\partial w_0''}\right)\right]dx = 0$$

$$(D.5)$$

For Eq. (D.5) to be zero, each of the three terms and the integral must be separately zero for equilibrium. That means the multiplier of ζ in the integrant must be zero for all values of x, hence

$$\frac{\partial F}{\partial w_0} - \frac{d}{dx}\left(\frac{\partial F}{\partial w_0'}\right) + \frac{d^2}{dx^2}\left(\frac{\partial F}{\partial w_0''}\right) = 0 \text{ for } x_0 \le x \le x_1 \tag{D.6}$$

The subscript 0 can be omitted in the Eq. (D.6), so that $w(x)$ represents the configuration at which the potential energy V is stationary. Hence,

$$\frac{\partial F}{\partial w} - \frac{d}{dx}\left(\frac{\partial F}{\partial w'}\right) + \frac{d^2}{dx^2}\left(\frac{\partial F}{\partial w''}\right) = 0 \tag{D.7}$$

Equation (D.7) is known as the Euler equation of the calculus of variations. The Euler equation for the integrant in the total potential energy expression gives the criterion for equilibrium of the continuous systems. For example, in the case of an axially loaded simply supported column in Figure 2.26 in Chapter 2, the total potential energy, Π, in the deformed configuration is given by

$$\Pi = \frac{EI}{2}\int_0^L (w'')^2\,dx - \frac{P}{2}\int_0^L (w')^2\,dx \tag{2.28g}$$

The integrand, F, for the example of the column is

$$F = \frac{EI}{2}(w'')^2 - \frac{P}{2}(w')^2$$

Therefore, $\dfrac{\partial F}{\partial w} = 0, \dfrac{\partial F}{\partial w'} = -Pw'$, and $\dfrac{\partial F}{\partial w''} = EIw''$. Substitute in Eq. (D.7) to get

$$EIw^{IV} + Pw'' = 0 \tag{D.8}$$

Equation (D.8) is the same differential equation of equilibrium of the column in its deformed position as was given by Eq. (2.28r). Eq. (D.7) is the Euler equation for a functional of one dependent variable given by Eq. (D.1). When the functional contains two dependent variables, $u(x)$ and $w(x)$, where the highest derivatives in u and w are of the first and second order respectively, the Euler equations are given by [1].

$$\frac{\partial F}{\partial u} - \frac{d}{dx}\frac{\partial F}{\partial u'} = 0$$

$$\frac{\partial F}{\partial w} - \frac{d}{dx}\frac{\partial F}{\partial w'} + \frac{d^2}{dx^2}\frac{\partial F}{\partial w''} = 0 \tag{D.9}$$

When there are three dependent variables u, v, w, and two independent variables x, y, and the highest derivatives are of first order in u and v, and second order in w, then the Euler equations are given by [1].

$$\frac{\partial F}{\partial u} - \frac{\partial}{\partial x}\frac{\partial F}{\partial u_x} - \frac{\partial}{\partial y}\frac{\partial F}{\partial u_y} = 0$$

$$\frac{\partial F}{\partial v} - \frac{\partial}{\partial x}\frac{\partial F}{\partial v_x} - \frac{\partial}{\partial y}\frac{\partial F}{\partial v_y} = 0$$

$$\frac{\partial F}{\partial w} - \frac{\partial}{\partial x}\frac{\partial F}{\partial w_x} - \frac{\partial}{\partial y}\frac{\partial F}{\partial w_y} + \frac{\partial^2}{\partial x^2}\frac{\partial F}{\partial w_{xx}} + \frac{\partial^2}{\partial x\partial y}\frac{\partial F}{\partial w_{xy}} + \frac{\partial^2}{\partial y^2}\frac{\partial F}{\partial w_{yy}} = 0 \tag{D.10}$$

Reference

1 Brush, D.O. and Almroth, B.O. (1975). Buckling of Bars. *Plates, and Shells*, McGraw Hill, New York, p 365.

Appendix E

Differential Geometry in Curvilinear Coordinates

A three-dimensional curve in a rectangular coordinate system (x, y, z) can be represented by the locus of the end point of the position vector r in Figure E.1.

$$r = x\,i + y\,j + z\,k \tag{E.1}$$

Let s be the arc length along the space curve, then

$$\frac{d}{ds}(r) = \frac{dx}{ds}i + \frac{dy}{ds}j + \frac{dz}{ds}k \tag{E.2}$$

From the dot product of the foregoing derivative with itself we get

$$\frac{d}{ds}(r)\cdot\frac{d}{ds}(r) = \left(\frac{dx}{ds}\right)^2 + \left(\frac{dy}{ds}\right)^2 + \left(\frac{dz}{ds}\right)^2 \tag{E.3}$$

$$(ds)^2 = (dx)^2 + (dy)^2 + (dz)^2 \tag{E.4}$$

Hence,

$$\frac{d}{ds}(r)\cdot\frac{d}{ds}(\mathbf{r}) = 1 \tag{E.5}$$

This shows that dr/ds is a unit vector. The vector $\Delta r/\Delta s$ in Figure E.1 becomes the vector tangent to the curve at the point P as Δs approaches zero. Therefore, $t = dr/ds$ is a unit tangent vector. The vector

$$\frac{d}{dt}(r) = \frac{d}{ds}(r)\frac{ds}{dt} \tag{E.6}$$

is also a tangent vector in the direction dr/ds but is not necessarily a unit vector.

E.1 Curvature

By Eq. (E.5)

$$\frac{d}{ds}(r).\frac{d}{ds}(\mathbf{r}) = t\,.\,t = 1$$

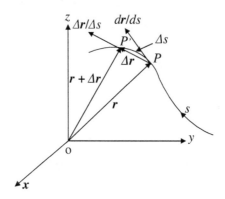

Figure E.1 Position vectors on a curve.

$$\frac{d}{ds}(\boldsymbol{t}.\boldsymbol{t}) = 2\,\boldsymbol{t}.\boldsymbol{t}' = 0 \tag{E.7}$$

where prime denotes differentiation with respect to s. It shows that $\boldsymbol{t}$ and $\boldsymbol{t}'$ are perpendicular vectors. From the definition of $\boldsymbol{t}$, we have

$$\boldsymbol{t} = \frac{d}{ds}(\boldsymbol{r}) = \frac{d}{dt}(\boldsymbol{r})\frac{dt}{ds} = \dot{\boldsymbol{r}}\,t' \tag{E.8}$$

$$\frac{d}{ds}(\boldsymbol{t}) = \boldsymbol{t}' = \dot{\boldsymbol{r}}\,t'' + \frac{d^2}{dt^2}(\boldsymbol{r})\,(t')^2 \tag{E.9}$$

Equation (E.9) indicates that the vector $\boldsymbol{t}'$ lies in the plane of the vectors $\dot{\boldsymbol{r}}$ and $\dfrac{d^2}{dt^2}\boldsymbol{r}$, i.e. in the osculating plane. Since $\boldsymbol{t}'$ is perpendicular to tangent $\boldsymbol{t}$, it is concluded that $\boldsymbol{t}'$ is parallel to the principal normal and can be written as

$$\boldsymbol{t}' = \boldsymbol{K} = KN \tag{E.10}$$

where N is a unit normal vector in the direction of the principal normal to the curve at a point. The vector $\boldsymbol{K}$ is called the curvature vector and gives the rate of change of the tangent vector as a point moves along the curve. The factor K is called the curvature, and its reciprocal is the radius of curvature, i.e. $R = \dfrac{1}{K}$. It is assumed that the normal vector N points away from the center of curvature. From Eq. (E.10) when the sense of $\boldsymbol{K}$ and N are the same, $K > 0$, and when the sense of N is opposite to that of $\boldsymbol{K}$, we have $K < 0$.

E.2 Surfaces

A thin shell is bounded by two closely spaced curved surfaces. The middle surface of a shell is defined as $X = X(x, y)$, $Y = Y(x, y)$, and $Z = Z(x, y)$, where X, Y, Z is a rectangular coordinate system and x, y, z are surface coordinates in Figure E.2. We can write

$$\boldsymbol{r} = X(x,y)\,\boldsymbol{i} + Y(x,y)\boldsymbol{j} + Z(x,y)\,\boldsymbol{k} \tag{E.11}$$

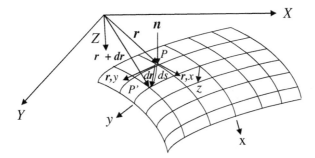

Figure E.2 Shell coordinate system.

From differential calculus

$$dr = r_x dx + r_y dy \tag{E.12}$$

where

$$r_x = \frac{\partial}{\partial x}(r), \quad r_y = \frac{\partial}{\partial y}(r) \tag{E.13}$$

$$dr \cdot dr = (ds)^2 = E(dx)^2 + 2F\,dx\,dy + G\,(dy)^2 \tag{E.14}$$

where

$$E = r_x \cdot r_x, \quad F = r_x \cdot r_y, \quad G = r_y \cdot r_y \tag{E.15}$$

The differential length of the arc along the surface coordinates is expressed as

$$ds_x = \sqrt{E}\,dx \text{ along the } x \text{ curvilinear coordinate} \tag{E.16}$$

$$ds_y = \sqrt{G}\,dy \text{ along the } y \text{ curvilinear coordinate}$$

The quantities r_x and r_y are tangents to the curvilinear coordinates x and y at P respectively. If these coordinates are orthogonal, then F will be zero, and it is usually written as

$$ds^2 = A^2(dx)^2 + B^2(dy)^2 \tag{E.17}$$

or

$$ds^2 = ds_x^2 + ds_y^2 \tag{E.18}$$

where

$$A = \sqrt{E}, B = \sqrt{G}, \text{ and } F = 0$$

$$ds_x = A\,dx \quad \text{and} \quad ds_y = B\,dy \tag{E.19}$$

From Eq. (E.11) we have

$$r_x = \frac{\partial}{\partial x}(r) = \frac{\partial X}{\partial x}i + \frac{\partial Y}{\partial x}j + \frac{\partial Z}{\partial x}k$$

$$E = \frac{\partial}{\partial x}(r) \cdot \frac{\partial}{\partial x}(r) = \left(\frac{\partial X}{\partial x}\right)^2 + \left(\frac{\partial Y}{\partial y}\right)^2 + \left(\frac{\partial Z}{\partial x}\right)^2 \tag{E.20}$$

or

$$A = \sqrt{E} = \sqrt{\left(\frac{\partial X}{\partial x}\right)^2 + \left(\frac{\partial Y}{\partial x}\right)^2 + \left(\frac{\partial Z}{\partial x}\right)^2} \tag{E.21}$$

Similarly,

$$B = \sqrt{G} = \sqrt{\left(\frac{\partial X}{\partial y}\right)^2 + \left(\frac{\partial Y}{\partial y}\right)^2 + \left(\frac{\partial Z}{\partial y}\right)^2} \tag{E.22}$$

For orthogonal surface coordinates, the magnitudes of vectors r_x and r_y are given by A and B respectively. A unit normal to the surface at a point P is perpendicular to the plane that contains vectors r_x and r_y, the tangent plane at P. The unit normal is thus parallel to the cross-product of r_x and r_y and is given by

$$n(x,y) = (r_x \times r_y)/|\, r_x \times r_y\,| \tag{E.23}$$

From vector algebra

$$|\, r_x \times r_y\,| = |\, r_x\,|\,|r_y|\, \sin\theta = AB \sin\theta \tag{E.24}$$

and

$$|\, r_x \cdot r_y\,| = |\, r_x\,|\,|r_y|\, \cos\theta = AB \cos\theta \tag{E.25}$$

where θ is the angle between the vectors r_x and r_y. From Eq. (E.15) we have

$$F = AB \cos\theta$$

or

$$\cos\theta = \frac{F}{AB} = \frac{F}{\sqrt{EG}} \tag{E.26}$$

$$\sin\theta = \sqrt{1 - \cos^2\theta} = \sqrt{\frac{EG - F^2}{EG}} \tag{E.27}$$

The unit normal vector $n\,(x,y)$ is thus given by

$$n(x,y) = (r_x \times r_y)/AB \sin\theta = \frac{r_x \times r_y}{\sqrt{EG - F^2}} = \frac{r_x \times r_y}{H} \tag{E.28}$$

where $H = \sqrt{EG - F^2}$, provided H does not vanish.

It should be noted that the principal normal N of a curve on a surface need not be normal to the surface, i.e. in general $N.\,n \neq 1$. Like the principal normal of a curve, the sense of a normal to a surface is arbitrary. Therefore, we adopt the convention that the parametric curves should always be arranged in such a manner that the normal n points from the concave side to the convex side of the surface.

From Eq. (E.10)

$$K = \frac{d}{ds}(t) = K_n + K_t \tag{E.29}$$

where K_n and K_t are the normal curvature vector and tangential vectors respectively. The normal curvature vector can be written as

$$K_n = -K_n \, n \tag{E.30}$$

where K_n is called the normal curvature. The minus sign is there because the sense of the curvature vector K is opposite to that of the unit normal vector n. Since n is perpendicular to t

$$n \cdot t = 0$$

Differentiate the scalar product to get

$$\frac{d}{ds}(n) \cdot t + n \cdot \frac{d}{ds}(t) = 0$$

or

$$\frac{d}{ds}(n) \cdot t = -n \cdot \frac{d}{ds}(t) \tag{E.31}$$

From Eq. (E.30) the dot product with n gives

$$-K_n \cdot n = K_n \tag{E.32}$$

From Eq. (E.29) the scalar product with n gives

$$\frac{d}{ds}(t) \cdot n = K_n \cdot n \tag{E.33}$$

because $K_t \cdot n = 0$ (Perpendicular to each other).
From Eqs. (E.31), (E.32), and (E.33) we have

$$\frac{d}{ds}(n) \cdot t = K_n$$

or

$$\frac{d}{ds}(n) \cdot \frac{d}{ds}(r) = K_n$$

or

$$K_n = (dn \cdot dr)/(ds^2)$$

or

$$K_n = (dn \cdot dr)/(dr \cdot dr) \tag{E.34}$$

where

$$(ds)^2 = dr \cdot dr$$

From Eq. (E.12)

$$dr = r_x dx + r_y \, dy \tag{E.12}$$

Similarly,

$$dn = n_x \, dx + n_y \, dy \tag{E.35}$$

Substitute Eqs. (E.12) and (E.35) into Eq. (E.34) to get

$$K_n = (\boldsymbol{n}_x\, dx + \boldsymbol{n}_y\, dy) \cdot (\boldsymbol{r}_x dx + \boldsymbol{r}_y\, dy)/(\boldsymbol{r}_x dx + \boldsymbol{r}_y dy) \cdot (\boldsymbol{r}_x dx + \boldsymbol{r}_y\, dy)$$

or

$$K_n = \frac{L(dx)^2 + 2Mdxdy + N(dy)^2}{E(dx)^2 + 2Fdxdy + G(dy)^2} \qquad (E.36)$$

where, L, M, and N are defined as

$$L = \boldsymbol{n}_x \cdot \boldsymbol{r}_x, \quad 2M = (\boldsymbol{n}_x \cdot \boldsymbol{r}_y + \boldsymbol{n}_y \cdot \boldsymbol{r}_x), \quad N = \boldsymbol{n}_y \cdot \boldsymbol{r}_y$$

In Eq. (E.36) E, F, G, L, M, and N are all functions of x and y and are constant at a given point, therefore, the normal curvature K_n in Eq. (E.36) depends only on the direction dx / dy.

E.3 Principal Curvatures

It is intended to seek those directions dy / dx for which the normal curvature K_n is either maximum or minimum. Dropping the subscript n, the normal curvature can be written as

$$K(\lambda) = \frac{L + 2M\lambda + N\lambda^2}{E + 2F\lambda + G\lambda^2} \qquad (E.37)$$

where

$$, \lambda = \frac{dy}{dx}$$

To find the direction dy / dx for the normal curvature to be maximum or minimum, $\dfrac{dK}{d\lambda} = 0$. Hence from Eq. (E.37) we have

$$(E + 2F\lambda + G\lambda^2)(M + N\lambda) - (L + 2M\lambda + N\lambda^2)(F + G\lambda) = 0 \qquad (E.38)$$

We can write

$$E + 2F\lambda + G\lambda^2 = (E + F\lambda) + \lambda(F + G\lambda)$$
$$L + 2M\lambda + N\lambda^2 = (L + M\lambda) + \lambda(M + N\lambda) \qquad (E.39)$$

Using Eq. (E.39) we can write Eq. (E.38) as

$$(E + F\lambda)(M + N\lambda) = (L + M\lambda)(F + G\lambda) \qquad (E.40)$$

Substitute Eq. (E.38) into Eq. (E.37) to obtain

$$K = \frac{M + N\lambda}{F + G\lambda}$$

Now make use of Eq. (E.40) to get

$$K = \frac{M + N\lambda}{F + G\lambda} = \frac{L + M\lambda}{E + F\lambda} \qquad (E.41)$$

From Eq. (E.40) we have

$$(MG - NF)\lambda^2 + (LG - NE)\lambda + (LF - ME) = 0 \tag{E.42}$$

Equation (E.42) is a quadratic equation in λ giving two roots λ_1 and λ_2, that give two directions $(dy/dx)_1$ and $(dy/dx)_2$, corresponding to the maximum and minimum values of the normal curvature, K_1 and K_2 respectively. These are called principal curvatures, and $R_1 = \frac{1}{K_1}$ and $R_2 = \frac{1}{K_2}$ are called the principal radii of curvature. We can prove that the directions of the principal curvatures are orthogonal. Consider the angle θ between two directions tangent to a surface given by dy/dx and $\delta y/\delta x$. Along these directions differential change in the position vector $\boldsymbol{r}$ is given by

$$dr = r_x dx + r_y dy$$
$$\delta r = r_x \delta x + r_y \delta y \tag{E.43}$$

The cosine of the angle between the dr and δr can be found from the dot product of the two vectors as

$$\cos\theta = (dr \cdot \delta r)/|dr| \, |\delta r|$$

or

$$\cos\theta = [(r_x \, dx + r_y \, dy) \cdot (r_x \, \delta x + r_y \, \delta y)]/|\,dr\,|\,|\,\delta r\,| \tag{E.44}$$

or

$$\cos\theta = \frac{E \, dx \, \delta x + F(dx \, \delta y + dy \, \delta x) + G \, dy \, \delta y}{\sqrt{dx^2 + dy^2 + dz^2}\sqrt{\delta x^2 + \delta y^2 + \delta z^2}} \tag{E.45}$$

Substitute $ds = \sqrt{dx^2 + dy^2 + dz^2}$ and $\delta s = \sqrt{\delta x^2 + \delta y^2 + \delta z^2}$
Hence,

$$\cos\theta = E\frac{dx}{ds}\frac{\delta x}{\delta s} + F\left(\frac{dx}{ds}\frac{\delta y}{\delta s} + \frac{dy}{ds}\frac{\delta x}{\delta s}\right) + G\left(\frac{dy}{ds}\frac{\delta y}{\delta s}\right) \tag{E.46}$$

If $\theta = \frac{\pi}{2}$, we get the orthogonality condition for two directions on a surface as

$$E \, dx \, \delta x + F(dx \, \delta y + dy \, \delta x) + G \, dy \, \delta y = 0 \tag{E.47}$$

Divide by $dx\delta x$ and let $\lambda_1 = \frac{dy}{dx}$ and $\lambda_2 = \frac{\delta y}{\delta x}$ to obtain

$$E + F(\lambda_1 + \lambda_2) + G\lambda_1\lambda_2 = 0 \tag{E.48}$$

From Eq. (E.42) the two roots are given by

$$\lambda_{1,2} = \frac{-(LG - NE) \pm \sqrt{(LG - NE)^2 - 4(MG - NF)(LF - ME)}}{2(MG - NF)} \tag{E.49}$$

From Eq. (E.49) we can obtain

$$\lambda_1 + \lambda_2 = -\frac{(LG - NE)}{(MG - NF)}$$

$$\lambda_1\lambda_2 = \frac{(LF - ME)}{(MG - NF)} \tag{E.50}$$

Substituting $(\lambda_1 + \lambda_2)$ and $\lambda_1\lambda_2$ into Eq. (E.48) we see that the orthogonality condition is satisfied. Therefore, the directions of the principal curvatures are orthogonal.

If the lines of curvature are taken as parametric lines (curves) of the surface, then the Eq. (E.42) must be satisfied by both $dx/dy = 0$ and $dy/dx = 0$. This is possible if

$$LF\text{--}ME = 0 \quad \text{and} \quad MG\text{--}NF = 0 \tag{E.51}$$

We have postulated that the parametric lines are to be the lines of curvature and the latter are orthogonal, therefore, $F = 0$. It can be shown that $EG - F^2 > 0$, so that for $F = 0$, neither E nor G can be zero. Thus, from Eq. (E.51), $M = 0$. Hence, the conditions under which the parametric lines are also lines of curvature are

$$F = M = 0. \tag{E.52}$$

When parametric lines are also lines of curvature, then we can find their curvature by setting $F = M = 0$ in Eq. (E.36), then letting $dx = 0$ and $dy = 0$, in turn to get

$$K_x = \frac{1}{R_x} = \frac{L}{E}, \quad \text{and} \quad K_y = \frac{1}{R_y} = \frac{N}{G} \tag{E.53}$$

In the theory of thin elastic shells, the lines of curvature of the reference surface are also used as parametric lines. Hence we assume that Eq. (E.52) is satisfied.

E.4 Derivatives of Unit Vectors along Parametric Lines

Consider mutually orthogonal unit vectors t_x, t_y, and n at a point on a surface so that these are tangent to the x and y directions and normal to the surface, respectively. As these vectors are moved over the surface, the magnitudes of these unit vectors remain constant and they remain mutually perpendicular, but their directions change. A unit vector can be obtained by dividing the vector by its magnitude, hence

$$t_x = r_{,x}/|r_{,x}| = r_{,x}/A \tag{E.54}$$

$$t_y = r_{,y}/|r_{,y}| = r_{,y}/B \tag{E.55}$$

$$n = (t_x \times t_y) = (r_{,x} \times r_{,y})/AB \tag{E.56}$$

The vector derivatives $n_{,x}$ and $n_{,y}$ are perpendicular to n, thus they lie in the plane formed by t_x and t_y and each vector derivative can be decomposed into its components along t_x and t_y. Hence,

$$n_{,x} = a\,t_x + b\,t_y \tag{E.57}$$

Where a and b represent the projections of $\boldsymbol{n}_{,x}$ on $\boldsymbol{t}_x$ and $\boldsymbol{t}_y$ respectively and are unknown. a and b are determined by forming the following scalar products

$$\boldsymbol{t}_x \cdot \boldsymbol{n}_{,x} = \boldsymbol{r}_{,x} \cdot \boldsymbol{n}_{,x}/A = L/A \tag{E.58}$$

$$\boldsymbol{t}_x \cdot \boldsymbol{n}_{,x} = \boldsymbol{t}_x \cdot (a\,\boldsymbol{t}_x + b\boldsymbol{t}_y) = a(\boldsymbol{t}_x \cdot \boldsymbol{t}_x) + b(\boldsymbol{t}_x \cdot \boldsymbol{t}_y) \tag{E.59}$$

$$\boldsymbol{t}_y \cdot \boldsymbol{n}_{,x} = \boldsymbol{r}_{,y} \cdot \boldsymbol{n}_{,x}/B = M/B \tag{E.60}$$

$$\boldsymbol{t}_y \cdot \boldsymbol{n}_{,x} = \boldsymbol{t}_y \cdot (a\boldsymbol{t}_x + b\boldsymbol{t}_y) = a\,(\boldsymbol{t}_y \cdot \boldsymbol{t}_x) + b\,(\boldsymbol{t}_y \cdot \boldsymbol{t}_y) \tag{E.61}$$

From Eqs. (E.58) and (E.59) we have

$$a = \frac{L}{A} \tag{E.62}$$

For orthogonal systems, $M = 0$, hence from Eqs. (E.60) and (E.61) we get

$$b = 0 \tag{E.63}$$

Therefore, from Eq. (E.57) we obtain

$$\boldsymbol{n}_{,x} = \frac{L}{A}\,\boldsymbol{t}_x \tag{E.64}$$

From Eq. (E.53)

$$K_x = \frac{1}{R_x} = \frac{L}{A^2} \tag{E.65}$$

or

$$\boldsymbol{n}_{,x} = \frac{A}{R_x}\,\boldsymbol{t}_x \tag{E.66}$$

Similarly,

$$\boldsymbol{n}_{,y} = \frac{B}{R_y}\,\boldsymbol{t}_y \tag{E.67}$$

Let us now find the derivatives of the unit vectors $\boldsymbol{t}_x$ and $\boldsymbol{t}_y$ by noting first that $\boldsymbol{r}_{,xy} = \boldsymbol{r}_{,yx}$. From Eqs. (E.54) and (E.55) we have

$$\boldsymbol{r}_{,x} = A\,\boldsymbol{t}_x, \quad \text{and} \quad \boldsymbol{r}_{,y} = B\,\boldsymbol{t}_y \tag{E.68}$$

or

$$(A\,\boldsymbol{t}_x)_{,y} = (B\,\boldsymbol{t}_y)_{,x}$$

or

$$A\,\boldsymbol{t}_{xy} + \boldsymbol{t}_x A_{,y} = B\,\boldsymbol{t}_{y,x} + \boldsymbol{t}_y B_{,x}$$

or

$$\boldsymbol{t}_{yx} = \frac{1}{B}(A\,\boldsymbol{t}_{x,y} + \boldsymbol{t}_x A_{,y} - \boldsymbol{t}_y B_{,x}) \tag{E.69}$$

$t_{x,x}$ is perpendicular to t_x and will lie in the plane formed by t_y and n. Thus, we can express $t_{x,x}$ in terms of t_y and n as

$$t_{,xx} = c\,n + d\,t_y \tag{E.70}$$

Where c and d are projections of $t_{,xx}$ on n and t_y and are unknown. c and d are determined by forming the following scalar products

$$n \cdot t_{xx} = c\,(n \cdot n) + d(t_y \cdot n) = c \tag{E.71}$$

Now,

$$t_x \cdot n = 0$$

or

$$(t_x \cdot n)_{,x} = t_x \cdot n_{,x} + n \cdot t_{x,x} = 0 \tag{E.72}$$

From Eq. (E.71) and Eq. (E.72) we have

$$c = n \cdot t_{x,x} = -t_x \cdot n_{,x} = -t_x \cdot \frac{A}{R_x} t_x = -\frac{A}{R_x} \tag{E.73}$$

$$t_y \cdot t_{x,x} = t_y \cdot (cn + dt_y) = c(t_y \cdot n) + d(t_y \cdot t_y) = d \tag{E.74}$$

$$(t_y \cdot t_x)_{,x} = t_{y,x} \cdot t_x + t_y \cdot t_{xx} = 0 \tag{E.75}$$

From Eq.(E.74) and Eq. (E.75) we have

$$d = t_y \cdot t_{xx} = -t_x \cdot t_{y,x} \tag{E.76}$$

Using Eq. (E.68) and $r_{,xy} = r_{,yx}$ we get

$$(A\,t_x)_{,y} = (Bt_y)_{,x}$$

or

$$A_{,y}\,t_x + A\,t_{x,y} = B_{,x}\,t_y + Bt_{y,x}$$

or

$$t_{y,x} = (A_{,y}t_x + A\,t_{x,y} - B_{,x}t_y)/B \tag{E.77}$$

Using Eq. (E.76) we obtain

$$d = -t_x \cdot (A_{,y}\,t_x + A\,t_{x,y} - B_{,x}t_y)/B \tag{E.78}$$

$t_{x,y}$ is perpendicular to t_x, hence

$$d = -\frac{A_{,y}}{B} \tag{E.79}$$

Substitute c and d from Eq. (E.73) and Eq. (E.79) in Eq. (E.70) to get

$$t_{x,x} = -\frac{A}{R_x}n - \frac{1}{B}\frac{\partial A}{\partial y}t_y \tag{E.80}$$

Similarly

$$t_{xy} = \frac{1}{A}\frac{\partial B}{\partial x}t_y \tag{E.81}$$

$$t_{yx} = \frac{1}{B}\frac{\partial A}{\partial y}t_x \tag{E.82}$$

$$t_{yy} = -\frac{B}{R_y}n - \frac{1}{A}\frac{\partial B}{\partial x}t_x \tag{E.83}$$

We can now drive equations that relate the quantities A, B, R_x, and R_y of a given surface. From the equality of mixed second derivatives of the unit vectors we have

$$n_{,xy} = n_{,yx} \tag{E.84}$$

From Eqs. (E.66) and (E.67) we have

$$\left(\frac{A}{R_x}t_x\right)_{,y} - \left(\frac{B}{R_y}t_y\right)_{,x} = 0 \tag{E.85}$$

or

$$\frac{A}{R_x}t_{x,y} + \left(\frac{A}{R_x}\right)_y t_x - \frac{B}{R_y}t_{y,x} - \left(\frac{B}{R_y}\right)_x t_y = 0$$

Substitute for $t_{x,y}$ and $t_{y,x}$ from Eqs. (E.81) and (E.82) to get

$$\frac{A}{R_x}\frac{1}{A}B_{,x}t_y + \left(\frac{A}{R_x}\right)_y t_x - \frac{B}{R_y}\frac{1}{B}A_{,y}t_x - \left(\frac{B}{R_y}\right)_x t_y = 0$$

or

$$t_x\left[\left(\frac{A}{R_x}\right)_y - \left(\frac{A_{,y}}{R_y}\right)\right] + t_y\left[\left(\frac{B_{,x}}{R_x}\right) - \left(\frac{B}{R_y}\right)_x\right] = 0 \tag{E.86}$$

The above vector equation is true only if the quantities in the square brackets are zero. Hence,

$$\left(\frac{A}{R_x}\right)_y = \frac{A_{,y}}{R_y}, \quad \left(\frac{B}{R_y}\right)_x = \frac{B_{,x}}{R_x} \tag{E.87}$$

Eq. (E.87) is known as the Codazzi condition.

In a similar manner, if we start with the equation

$$t_{x,xy} = t_{x,yx}$$

we will obtain

$$\left(\frac{1}{A}B_{,x}\right)_x + \left(\frac{1}{B}A_{,y}\right)_y = -\frac{AB}{R_x R_y} \tag{E.88}$$

Equation (E.88) is known as the Gauss condition.

Index